Heat Transfer and Fluid Flow in Minichannels and Microchannels

T0383266

Elsevier Internet Homepage

http://www.elsevier.com

Consult the Elsevier homepage for full catalogue information on all books, journals and electronic products and services.

Elsevier Titles of Related Interest

Cho & Greene, Advances in Heat Transfer, 2011, ISBN 9780123815293

Serth, Process Heat Transfer: Principles, Applications and Rules of Thumb, 2007, 978-0-12-373588-1

Upp & LaNasa, Fluid Flow Measurement, 2nd Edition, 2002, ISBN 9780884157588

Related Journals

The following titles can all be found at: http://www.sciencedirect.com

Applied Thermal Engineering
Experimental Thermal and Fluid Science
Fluid Abstracts: Process Engineering
Fluid Dynamics Research
International Communications in Heat and Mass Transfer
International Journal of Heat and Fluid Flow
International Journal of Heat and Mass Transfer
International Journal of Multiphase Flow
International Journal of Refrigeration
International Journal of Thermal Sciences

To Contact the Publisher

Elsevier welcomes enquiries concerning publishing proposals: books, journal special issues, conference, proceedings, etc. All formats and media can be considered. Should you have a publishing proposal you wish to discuss, please contact, without obligation, the editor responsible for Elsevier's Mechanical Engineering publishing programme:

Hayley Gray
Senior Acquisitions Editor
S&T Books
Elsevier LTD
The Boulevard
Langford Lane
Kidlington
Oxford
OX5 1GB, UK
Tel: +44 1865 844731
E-mail: h.gray@elsevier.com

General enquiries, including placing orders, should be directed to Elsevier's Regional Sales Offices – please access the Elsevier Internet homepage for full contact details.

Heat Transfer and Fluid Flow in Minichannels and Microchannels

Contributing Authors

Satish G. Kandlikar
Editor and Contributing Author
Mechanical Engineering Department
Rochester Institute of Technology, NY, USA

Srinivas Garimella
George W. Woodruff School of Mechanical Engineering
Georgia Institute of Technology, Atlanta, USA

Dongqing Li
Department of Mechanical and Industrial Engineering
University of Toronto, Ontario, Canada

Stéphane Colin
Department of General Mechanic National Institute of Applied
Sciences of Toulouse Toulouse cedex, France

Michael R. King
Departments of Biomedical Engineering, Chemical Engineering
and Surgery University of Rochester, NY, USA

AMSTERDAM • BOSTON • HEIDELBERG • LONDON
NEW YORK • OXFORD • PARIS • SAN DIEGO
SAN FRANCISCO • SINGAPORE • SYDNEY • TOKYO
Butterworth-Heinemann is an imprint of Elsevier

Butterworth-Heinemann is an imprint of Elsevier
The Boulevard, Langford Lane, Kidlington, Oxford OX5 1GB, UK
225 Wyman Street, Waltham, MA 02451, USA

First edition 2006

Copyright © 2014 Elsevier Ltd. All rights reserved.

No part of this publication may be reproduced, stored in a retrieval system or transmitted in any form or by any means electronic, mechanical, photocopying, recording or otherwise without the prior written permission of the publisher.

Permissions may be sought directly from Elsevier's Science & Technology Rights Department in Oxford, UK: phone (+ 44) (0) 1865 843830; fax (+ 44)(0)1865 853333; email: permissions@elsevier.com. Alternatively you can submit your request online by visiting the Elsevier web site at http://elsevier.com/locate/permissions, and selecting *Obtaining permission to use Elsevier material.*

Notice
No responsibility is assumed by the publisher for any injury and/or damage to persons or property as a matter of products liability, negligence or otherwise, or from any use or operation of any methods, products, instructions or ideas contained in the material herein. Because of rapid advances in the medical sciences, in particular, independent verification of diagnoses and drug dosages should be made.

Library of Congress Cataloging-in-Publication Data
A catalogue record for this book is available from the Library of Congress

British Library Cataloguing-in-Publication Data
A catalogue record for this book is available from the British Library

ISBN: 978-0-08-101326-7

For information on all Butterworth-Heinemann publications
visit our website at books.elsevier.com

Printed and bound in the United States

14 15 16 17 18 10 9 8 7 6 5 4 3 2 1

Contents

About the Authors

Satish Kandlikar Dr. Satish Kandlikar is the Gleason Professor of Mechanical Engineering at Rochester Institute of Technology. He obtained his B.E. degree from Marathawada University and M.Tech. and Ph.D. degrees from I.I.T. Bombay. His research focuses on flow boiling and single-phase heat transfer and fluid flow in microchannels, high flux cooling, and fundamentals of interfacial phenomena. He has published over 130 conference and journal papers, presented over 25 invited and keynote papers, has written contributed chapters in several handbooks, and has been editor-in-chief of a handbook on boiling and condensation. He is the recipient of the IBM Faculty award for the past three consecutive years. He received the Eisenhart Outstanding Teaching Award at RIT in 1997. He is an Associate Editor of several journals, including the *Journal of Heat Transfer*, *Heat Transfer Engineering*, *Journal of Microfluidics and Nanofluidics*, *International Journal of Heat and Technology*, and *Microscale Thermophysical Engineering*. He is a Fellow member of ASME.

Srinivas Garimella Dr. Srinivas Garimella is the Hightower Chair in Engineering and Director of the Sustainable Thermal Systems Laboratory at Georgia Institute of Technology. He received M. S. and Ph.D. degrees from The Ohio State University, and a Bachelor's degree from the Indian Institute of Technology, Kanpur. He has held prior positions as Research Scientist at Battelle Memorial Institute, Senior Engineer at General Motors Corp., and Associate Professor at Western Michigan University and Iowa State University. He is a Fellow of the American Society of Mechanical Engineers, past Associate Editor of the ASME Journal of Heat Transfer, ASME Journal of Energy Resources Technology, and the ASHRAE HVAC&R Research Journal. He is currently the Editor of the International Journal of Air-conditioning and Refrigeration. He was Chair of the Advanced Energy Systems Division of ASME, and the ASHRAE Research Administration Committee. He conducts research in the areas of vapor-compression and absorption heat pumps, phase-change in microchannels, heat and mass transfer in binary fluids, waste heat recovery, and sustainable energy systems. He has mentored over 60 graduate students, with his research resulting in over 175 archival journal and conference publications, and he has been awarded five patents. He is the recipient of the NSF CAREER Award, the ASHRAE New Investigator Award, the SAE Ralph E. Teetor Educational Award for Engineering Educators, and was the Iowa State University Miller Faculty Fellow and Woodruff Faculty Fellow at Georgia Tech. He received the ASME Award for Outstanding Research Contributions in the Field of Two-Phase Flow and Condensation in Microchannels, 2012. He also received the Thomas French Distinguished Educator Achievement Award from The Ohio State University, and the Zeigler Outstanding Educator Award (2012) at Georgia Tech.

Dongqing Li Dr. Dongqing Li obtained his B.A. and M.Sc. degrees in Thermophysics Engineering in China, and his Ph.D. degree in Thermodynamics from the University of Toronto, Canada, in 1991. He was a professor at the University of Alberta and later at the University of Toronto from 1993 to 2005. Currently, Dr. Li is the H. Fort Flowers Professor of Mechanical Engineering, Vanderbilt University. His research is in the areas of microfluidics and lab-on-a-chip. Dr. Li has published one book, 11 book chapters, and over 160 journal papers. He is the Editor-in-Chief of the international journal *Microfluidics and Nanofluidics.*

Stéphane Colin Dr. Stéphane Colin is a Professor of Mechanical Engineering at the National Institute of Applied Sciences (INSA) of Toulouse, France. He obtained his Engineering degree in 1987 and received his Ph.D. degree in Fluid Mechanics from the Polytechnic National Institute of Toulouse in 1992. In 1999, he created the Microfluidics Group of the Hydrotechnic Society of France, and he currently leads this group. He is the Assistant Director of the Mechanical Engineering Laboratory of Toulouse. His research is in the area of microfluidics. Dr. Colin is editor of the book *Microfluidique*, published by Hermes Science Publications.

Michael King Dr. Michael King is an Assistant Professor of Biomedical Engineering and Chemical Engineering at the University of Rochester. He received a B.S. degree from the University of Rochester and a Ph.D. from the University of Notre Dame, both in chemical engineering. At the University of Pennsylvania, Dr. King received an Individual National Research Service Award from the NIH. He is a Whitaker Investigator, a James D. Watson Investigator of New York State, and is a recipient of the NSF CAREER Award. He is editor of the book *Principles of Cellular Engineering: Understanding the Biomolecular Interface*, published by Academic Press. His research interests include biofluid mechanics and cell adhesion.

Preface

We are pleased to bring this second edition with an updated coverage of the recent developments in this field. After the initial questions regarding the applicability of macroscopic formulation were addressed, the research has focused on topics that utilize the microscale behavior for understanding the transport processes. New advances are reported on the fundamental understanding and experimental analysis of the gas flow and heat transfer at microscale, transport modeling and roughness effects. Sections on enhancement of single-phase liquid heat transfer, axial conduction effects, entrance region effects, nanofluids, and roughness effects have been updated to include some of the recent work. The chapter on electrokinetic flows has been updated to include a new section on electroosmotic flows considering the Joule heating effects. Significant new developments made in flow boiling in microchannels, including scale effects, nanofluids, enhancements, stability considerations, CHF modeling, flow boiling enhancement techniques, and a new configuration providing a new level of enhancement exceeding 500 W/cm2 limit have been included. The chapter on condensation has been streamlined to focus more closely on mini and microchannel geometries, and expanded to include some of the latest developments in experimentation and modeling efforts. The biomedical applications in Chapter 7 have been expanded significantly to present the advancements in fluid mechanics modeling in this field.

The authors are thankful to many researchers who have provided feedback on the second edition. We believe that the readers will find this edition as a timely and updated resource that is useful to researchers as well as designers of microfluidic transport devices.

Satish G. Kandlikar
Srinivas Garimella
Dongqing Li
Stéphane Colin
Michael King

Nomenclature

$\boldsymbol{A}$	Section area, m^2 (Chapters 2 and 6).
$\boldsymbol{A, B, C, D, F}$	Equation coefficients and exponents (Chapters 3 and 7).
$\boldsymbol{A_1}$	First-order slip coefficient, dimensionless (Chapter 2).
$\boldsymbol{A_2}$	Second-order slip coefficient, dimensionless (Chapter 2).
$\boldsymbol{A_3}$	High-order slip coefficient, dimensionless (Chapter 2).
$\boldsymbol{A_c}$	Cross-sectional area, m^2 (Chapter 3).
$\boldsymbol{A_p}$	Total plenum cross-sectional area, m^2 (Chapter 3).
$\boldsymbol{A_T}$	Total heat transfer surface area (Chapter 5).
$\boldsymbol{a}$	Speed of sound, m/s (Chapter 2); channel width, m (Chapter 3); equation constant in Eqs. (6.71), (6.83), and (6.87) (Chapter 6); coefficient in entrance length equations, dimensionless (Chapter 7).
$\boldsymbol{a^*}$	Aspect ratio of rectangular sections, dimensionless, $a^* = h/b$ (Chapter 2).
$\boldsymbol{a_1, a_2, a_3}$	Coefficients for the mass flow rate in a rectangular microchannel, dimensionless (Chapter 2).
$\boldsymbol{a_1 \ldots a_5}$	Coefficients in Eq. (6.107) (Chapter 6).
$\boldsymbol{B}$	Parameter used in Eqs. (6.9) and (6.41) (Chapter 6).
$\boldsymbol{B_B}$	Parameter used in Eq. (6.21) (Chapter 6).
$\boldsymbol{B_n}$	Coefficient in cell surface oxygen concentration equation (Chapter 7).
$\boldsymbol{b}$	Half-channel width, m (Chapter 2); channel height, m (Chapter 3); constant in Eqs. (6.71), (6.83), and (6.87) (Chapter 6).
Bo	Boiling number, dimensionless, $\mathrm{Bo} = q''/(Gh_{LV})$ (Chapter 5).
Bo	Bond number, dimensionless, $Bo = (\rho_L - \rho_V)gD_h^2/\sigma$ (Chapter 5); $Bo = g(\rho_L - \rho_G)((d/2)^2/\sigma)$ (Chapter 6).
C, $\boldsymbol{C}$	Constant, dimensionless (Chapter 1); coefficient in a Nusselt number correlation (Chapter 3); concentration, mol/m^3 (Chapters 4 and 7); Chisholm's parameter, dimensionless (Chapter 5); constant used in Eqs. (6.39), (6.40), and (6.157) (Chapter 6).
$\boldsymbol{C}$	Reference concentration, mol/m^3 (Chapter 4).
$\boldsymbol{C^*}$	Ratio of experimental and theoretical apparent friction factors, dimensionless, $C^* = f_{app,ex}/f_{app,th}$, (Chapter 3); nondimensionalized concentration, $C^* = C(x, y)/C_{in}$ (Chapter 7).
$\boldsymbol{C_0}$	Oxygen concentration at the lower channel wall, mol/m^3 (Chapter 7).
$\boldsymbol{C_1, C_2}$	Empirically derived constants in Eq. (6.35) (Chapter 6); parameter, used in Eq. (6.54) (Chapter 6).
$\boldsymbol{C_C}$	Coefficient of contraction, dimensionless (Chapter 6).
$\boldsymbol{C_{in}}$	Nondimensionalized gas phase oxygen concentration, $C_{in} = \tilde{C}_g/C_g$ (Chapter 7).

C_f	Friction factor, dimensionless (Chapter 2).
Co	Convection number, dimensionless, $C_o = [(1-x)/x]^{0.9}\ [\rho_V/\rho_L]^{0.5}$ (Chapter 5).
Co	Confinement number, dimensionless, $C_o = \sqrt{\sigma/g(\rho_l - \rho_v)}/D_h$ (Chapter 6).
C_o	Contraction coefficient, dimensionless (Chapter 5); distribution parameter in drift flux model, dimensionless, $C_o = <\alpha j>/(<\alpha><j>)$ (Chapter 6).
C_p	Specific heat capacity at a constant pressure, J/kg K (Chapter 6).
C_S	Saturation oxygen concentration, mol/m^3 (Chapter 7).
c	Mean square molecular speed, m/s (Chapter 2); constant in the thermal entry length equation, dimensionless (Chapter 3); constant in Eq. (6.57) (Chapter 6).
$\mathbf{c}'$	Molecular thermal velocity vector m/s (Chapter 2).
$\overline{c'}$	Mean thermal velocity, m/s (Chapter 2).
c_1, c_2	Coefficients used in Eq. (6.133) (Chapter 6).
c_p	Specific heat at a constant pressure, J/kg K (Chapters 2, 3, and 5).
c_v	Specific heat at a constant volume, J/kg K (Chapter 2).
Ca	Capillary number, dimensionless, $Ca = \mu V/\sigma$ (Chapter 5).
CHF	Critical heat flux, W/m^2 (Chapter 5).
D	Diameter, m (Chapters 1, 3, 5 and 6).
D, D^+, D^-	Diffusion coefficient, diffusivity, m^2/s (Chapters 2 and 4).
D_{cf}	Diameter constricted by channel roughness, m, $D_{cf} = D - 2\varepsilon$ (Chapter 3).
D_h, D_H	Hydraulic diameter, m (Chapters 1–5).
D_{le}	Laminar equivalent diameter, m (Chapter 3).
d	Mean molecular diameter, m (Chapter 2); diameter, m (Chapter 6).
d_B	Departure bubble diameter, m (Chapter 5).
E	Applied electrical field strength, V/m (Chapter 4); total energy per unit volume, J/m^3 (Chapter 2); diode efficiency, dimensionless (Chapter 2); parameter used in Eq. (6.141) (Chapter 6).
E_1, E_2	Parameter used in Eq. (6.22) (Chapter 6).
E_x	Electric field strength, V/m (Chapter 4).
e	Internal specific energy, J/kg (Chapter 2); charge of a proton, $e = 1.602 \times 10^{-19}$ C (Chapter 4).
$Eo, E\ddot{o}$	Eotvos number, dimensionless, $Eo = g(\rho_L - \rho_V)L^2/\sigma$ in case of liquid gas contact (Chapters 5 and 6).
F, F	Nondimensional constant accounting for an electrokinetic body force (Chapter 4); general periodic function of unit magnitude (Chapter 4); force, N (Chapters 5 and 6); modified Froude number, dimensionless, $F = \sqrt{\left(\frac{\rho_g}{\rho_l - \rho_g}\right)\left(\frac{U_{GS}}{\sqrt{D_g}}\right)}$ (Chapter 6); stress ratio, dimensionless, $F = \tau_w/(\rho_L g \delta)$ (Chapter 6); parameter used in

	Eq. (6.141) (Chapter 6); external force acting on a spherical cell, N (Chapter 7).
F	External force per unit mass vector, N/kg (Chapter 2).
F'_M	Interfacial force created by evaporation momentum, N (Chapter 5).
F'_S	Interfacial force created by surface tension, N (Chapter 5).
F_{F1}	Fluid-surface parameter accounting for the nucleation characteristics of different fluid surface combinations, dimensionless (Chapter 5).
F_g	Function of the liquid volume fraction and the vapor Reynolds number, used in Eq. (6.128) (Chapter 6).
F_T	Dimensionless parameter of Eq. (6.112) (Chapter 6).
F_x	Electrical force per unit volume of the liquid, N/m^3 (Chapter 4).
f	Volume force vector, N/m^3 (Chapter 2).
f	Fanning friction factor, dimensionless (Chapters 1, 3, and 5); single-phase friction factor, dimensionless (Chapter 6).
f_{app}	Apparent friction factor accounting for developing flows, dimensionless (Chapter 3).
f	Frequency, Hz (Chapter 2); velocity distribution function (Chapter 2).
f_{ls}	Superficial liquid phase friction factor, dimensionless (Chapter 6).
F_p	Floor distance to mean line in roughness elements, m (Chapter 3).
Fr_l	Liquid Froude number, dimensionless, $Fr_l^2 = \overline{V}_l^2/g\delta$ (Chapter 6).
Fr_m	Modified Froude number, dimensionless (Chapter 6).
Fr_{so}	Soliman modified Froude number, dimensionless (Chapter 6).
F_t	Froude rate, dimensionless $F_t = [G^2x^3/(1-x)\rho_g^2 gD]^{0.5}$ (Chapter 6).
G	Mass flux, kg/m^2 s (Chapters 1, 5, and 6).
G_{eq}	Equivalent mass flux, kg/m^{2}s, $G_{eq} = G_l + G'_l$ (Chapter 6).
G'_l	Mass flux that produces the same interfacial shear stress as a vapor core, kg/m^2 s, $G'_l = G_v\sqrt{\rho_l/\rho_v}\sqrt{f_v/f_l}$ (Chapter 6).
G_t	Total mass flux, kg/m^2 s (Chapter 6).
g	Acceleration due to gravity, m/s^2 (Chapters 5 and 6).
Ga_l	Liquid Galileo number, dimensionless, $Ga_l = gD^3/\nu_l^2$ (Chapter 6).
H	Maximum height, m (Chapter 4); distance between parallel plates or height, m (Chapter 7); parameter used in Eq. (6.141) (Chapter 6).
h	Heat transfer coefficient, W/m^2 K (Chapters 1, 3, 5, and 6); channel half-depth, m (Chapter 2); specific enthalpy, J/kg (Chapters 2 and 5); wave height, m (Chapter 7).
$\overline{h}$	Average heat transfer coefficient, W/m^2 K (Chapters 3 and 6).
h_c	Film heat transfer coefficient, W/m^2 K (Chapter 6).
h_{fg}	Latent heat of vaporization, J/kg (Chapter 6).
h_G	Gas-phase height in channel, m, $h_G \leq \pi/4\sqrt{\sigma/\rho g(1-\pi/4)}$ (Chapter 6).

h_{LV}	Latent heat of vaporization at p_L, J/kg (Chapter 5).
h_{lv}	Specific enthalpy of vaporization, J/kg (Chapter 6).
h'_{lv}	Modified specific enthalpy of vaporization, J/kg (Chapter 6).
I	Unit tensor, dimensionless (Chapter 2).
I	Current, A (Chapter 3).
i	Enthalpy, J/kg (Chapter 5).
J	Mass flux vector, kg/m^2 s (Chapter 2).
J	Electrical current, A (Chapter 4).
j	Superficial velocity, m/s (Chapters 5 and 6).
j^*_g, j^*_G	Wallis dimensionless gas velocity, $j^*_g = G_t x / \sqrt{D g \rho_v (\rho_l - \rho_v)}$ (Chapter 6).
Ja	Jakob number, dimensionless, $Ja = (\rho_L/\rho_v)\ (c_{p,L} \Delta T / h_{LV})$ (Chapter 5).
Ja_l	Liquid Jakob number, dimensionless, $Ja_l = c_{pL}(T_{sat} - T_s)/h_{lv}$ (Chapter 6).
K	Nondimensional double layer thickness, $K = D_h \kappa$ (Chapter 4); constant in Eqs. (6.56) and (6.95) (Chapter 6).
$K(x)$	Incremental pressure defect, dimensionless (Chapter 3).
$K(\infty)$	Hagenbach's factor, dimensionless, $K(x)$ when $x > L_h$ (Chapter 3).
$\mathbf{K}_1$	Ratio of evaporation momentum to inertia forces at the liquid–vapor interface, dimensionless, $K_1 = (q''/G h_{LV})^2 \rho_L / \rho_V$ (Chapter 5).
$\mathbf{K}_2$	Ratio of evaporation momentum to surface tension forces at the liquid–vapor interface, dimensionless, $\mathbf{K}_2 = (q''/h_{LV})^2\ D/\rho_V \sigma$ (Chapter 5).
K_{90}	Loss coefficient at a 90° bend, dimensionless (Chapter 3).
K_c	Contraction loss coefficient due to an area change, dimensionless (Chapter 3).
K_e	Expansion loss coefficient due to an area change, dimensionless (Chapter 3).
K_m	Michaelis constant, mol/m^3 (Chapter 7).
k	Thermal conductivity, W/mK (Chapters 1–3, 5, and 6); constant, dimensionless (Chapter 7).
k_1	Coefficient in the collision rate expression, dimensionless (Chapter 2).
k_2	Coefficient in the mean free path expression, dimensionless (Chapter 2).
k_B, κ_b	Boltzmann constant, $k_B = 1.38065$ J/K (Chapters 2 and 4).
Kn	Knudsen number, $Kn = \lambda/L$, dimensionless (Chapters 1 and 2).
Kn'	Minimal representative length Knudsen number, $Kn' = \lambda/L_{min}$ (Chapter 2).
Ku	Kutateladze number, dimensionless $Ku = C_p\ \Delta\ T/h_{fg}$ (Chapter 6).
L	Length or characteristic length in a given system, m (Chapters 1–3 and 5–7); Laplace constant, m, $L = \sqrt{\sigma / g(\rho_l - \rho_v)}$ (Chapter 6).

L_G, L_L	Gas and liquid slug lengths in the slug flow regime, m (Chapter 6).
L_{ent}, L_h, L_{hd}	Hydrodynamically developing entrance length, m, $L_{ent} = aHRe$ (Chapters 2, 3, and 7).
L_t	Thermally developing entrance length, m, $L_t = cRePrD_h$ (Chapter 3).
L_{eq}	Total pipette length, m (Chapter 7).
l	Microchannel length, m (Chapter 2).
l_{SV}	Characteristic length of a sampling volume, m (Chapter 2).
$l_{x,y,z}$	Channel half height, m (Chapter 4).
LHS	Left-hand side (Chapter 7).
M	Molecular weight, kg/mol (Chapter 2).
M	Ratio of the electrical force to frictional force per unit volume, dimensionless, $M = 2n_\infty ze\xi D_h^2/\mu\, UL$ (Chapter 4).
M, N	Equation exponents, dimensionless (Chapter 3).
MW	Molecular weight, g/mol (Chapter 7).
m	Molecular mass, kg (Chapter 2); liquid volume fraction, dimensionless (Chapter 6); dimensionless constant in Eq. (6.57) (Chapter 6).
$\dot{m}$	Mass flow rate, kg/s (Chapters 2, 3, and 5).
$\dot{m}^*$	Mass flow rate, $\dot{m}/\dot{m}_{ns}$, dimensionless (Chapter 2).
$\dot{m}_{ns}$	Mass flow rate for a no-slip flow, kg/s (Chapter 2).
Ma	Mach number, dimensionless, $Ma = u/a$ (Chapter 2).
N	Avogadro's number, $6.022137 \cdot 10^{23}$ mol^{-1} (Chapter 2).
$\dot{N}$	Molecular flux, s^{-1} (Chapter 2).
N^+	Nondimensional positive species concentration (Chapter 4).
N^-	Nondimensional negative species concentration (Chapter 4).
N_{conf}	Confinement number, dimensionless, $N_{conf} = \sqrt{\sigma/(g(\rho_l - \rho_v))}/D_h$ (Chapter 6).
N_0	Cellular uptake rate, mol/m^2 s (Chapter 7).
n	Number density, m^{-3} (Chapter 2); number or number of channels, dimensionless (Chapter 3); number of channels (Chapter 5); constant in Eqs. (6.41) and (6.57) (Chapter 6); number (Chapter 7).
n_1, n_2, n_3	Constant in Eq. (6.21) (Chapter 6).
n_i	Number concentration of type-i ion (Chapter 4).
n_{io}	Bulk ionic concentration of type-i ions (Chapter 4).
$\mathbf{n}_x$	Normal vector in the x direction (Chapter 4).
Nu	Nusselt number, dimensionless, $Nu = hD_h/k$ (Chapters 1–3, 5, and 6).
Nu_H	Nusselt number under a constant heat flux boundary condition, dimensionless (Chapter 3).
Nu_i	Nusselt number for high interfacial shear condensation, dimensionless (Chapter 6).
$\overline{Nu}_L$	Average Nusselt number along a plate of length L, dimensionless (Chapter 6).

$\boldsymbol{Nu_o}$	Nusselt number for quiescent vapor condensation, dimensionless, $Nu_o = [(Nu_L^{n_1}) + (Nu_T^{n_1})]^{1/n_1}$ (Chapter 6).
$\boldsymbol{Nu_T}$	Nusselt number for a turbulent film, dimensionless (Chapter 6).
$\boldsymbol{Nu_T}$	Nusselt number under a constant wall temperature boundary condition, dimensionless (Chapter 3).
$\boldsymbol{Nu_x}$	Combined Nusselt number, dimensionless, $Nu_x = [(Nu_o^{n_2}) + (Nu_i^{n_2})]^{1/n_2}$ (Chapter 6).
ONB	Onset of nucleate boiling (Chapter 5).
$\boldsymbol{P}$	Wetted perimeter, m (Chapter 2); dimensionless pressure (Chapter 4); heated perimeter, m (Chapter 5); pressure, Pa (Chapter 6).
$\boldsymbol{P_w}$	Wetted perimeter, m (Chapter 3).
$\boldsymbol{p}$	Pressure, Pa (Chapters 1–3 and 5–7).
$\boldsymbol{p_R}$	Reduced pressure, dimensionless (Chapter 6).
$\boldsymbol{Pe}$	Péclet number, dimensionless, Pe = UH/D (Chapter 7).
$\boldsymbol{Pe_F}$	Péclet number of fluid, dimensionless (Chapter 4).
$\boldsymbol{Po}$	Poiseuille number, dimensionless, $Po = f\,Re$ (Chapters 2 and 3).
$\boldsymbol{Pr}$, $\boldsymbol{Pr}$	Prandtl number, dimensionless, $Pr = \mu c_p/k$ (Chapters 2, 3, 5, and 6).
$\boldsymbol{Q}$	Heat load, W (Chapter 3).
$\boldsymbol{Q}$	Volumetric flow rate, m^3/s (Chapters 2, 3 and 7).
q	Heat flux vector, W/m^2 (Chapter 2).
$\boldsymbol{q}$	Heat flux, W/m^2 (Chapter 2); dissipated power, W (Chapter 3); constant in Eq. (6.60) (Chapter 6).
$\boldsymbol{q}$	Volumetric flow rate per unit width, m^2/s (Chapter 7).
$\boldsymbol{q}$	Oxygen uptake rate on a per-cell basis, mol/s (Chapter 7).
$\boldsymbol{q''}$	Heat flux, W/m^2 (Chapters 5 and 6).
$\boldsymbol{q''_{CHF}}$	Critical heat flux, W/m^2 (Chapter 5).
$\boldsymbol{R}$	Gas constant (Chapter 1); upstream to downstream flow resistance, dimensionless (Chapter 5).
$\boldsymbol{R}$	Specific gas constant, J/kg K, $R = c_p - c_v$ (Chapter 2); radius, m (Chapter 6).
R	Universal gas constant, 8.314511 J/mol K (Chapter 2).
$\boldsymbol{R^+}$	Dimensionless pipe radius (Chapter 6).
$\boldsymbol{R_1}$, $\boldsymbol{R_2}$	Radii of curvature of fluid–liquid interface, m.
$\boldsymbol{R_p}$	Mean profile peak height (Chapter 3); pipette radius, m (Chapter 7).
$\boldsymbol{R_{p,i}}$	Maximum profile peak height of individual roughness elements, m (Chapter 3).
$\boldsymbol{R_{pm}}$	Average maximum profile peak height of roughness elements, m (Chapter 3).
$\boldsymbol{r}$	Distance between two molecular centers, m (Chapter 2); radial coordinate, radius, radius of cavity, m (Chapters 2 and 4–7); constant in Eq. (6.60) (Chapter 6).
$\boldsymbol{r_b}$	Bubble radius, m (Chapter 5).

r_c	Cavity radius, m (Chapter 5).
r_1	Inner radius of an annular microtube, m (Chapter 2).
r_2	Outer radius of annular microtube or a circular microtube radius, m (Chapter 2).
R_a	Average surface roughness, m (Chapter 3).
Re, Re	Reynolds number, dimensionless, $Re = GD/\mu$ (Chapters 1–5 and 7).
Re^*	Laminar equivalent Reynolds number, dimensionless, $Re^* = \rho u_m D_{le}/\mu$ (Chapter 3).
Re^+	Friction Reynolds number, dimensionless (Chapter 6).
Re_{Dh}	Reynolds number based on hydraulic diameter, dimensionless (Chapter 6).
$Re_{g,\ si}$	Reynolds number based on superficial gas velocity at the inlet, dimensionless (Chapter 6).
$Re_{l,\ si}$	Reynolds number based on superficial liquid velocity at the inlet, dimensionless (Chapter 6).
Re_l	Liquid film Reynolds number, dimensionless, $Re_l = G(1-x)D/\mu_l$ (Chapter 6).
Re_m	Mixture Reynolds number, dimensionless, $Re_m = GD/\mu_m$ (Chapter 6).
Re_t	Transitional Reynolds number, dimensionless (Chapter 3).
RS_m	Mean spacing of profile irregularities in roughness elements, m (Chapter 3).
S	Slip ratio, dimensionless, $S = U_G/U_L$ (Chapter 6).
s	Fin width or distance between channels, m (Chapter 3); constant in Eq. (6.60) (Chapter 6).
Sc	Schmidt number, dimensionless, $Sc = \mu/(\rho D)$ (Chapter 2).
Sh	Sherwood number, dimensionless, $Sh = \alpha H/D$ (Chapter 7).
Sm	Distance between two roughness element peaks, m (Chapter 3).
St	Stanton number, dimensionless, $St = h/c_p G$ (Chapter 3).
T	Temperature, K or °C (Chapters 1–6).
T_s	Liquid surface temperature, K or °C (Chapters 3 and 5); surface temperature of tube wall, K or °C (Chapter 6).
T_{sat}	Saturation temperature, K or °C (Chapters 5 and 6).
ΔT_{Sat}	Wall superheat, K, $\Delta T_{Sat} = T_W - T_{Sat}$ (Chapter 5).
ΔT_{Sub}	Liquid subcooling, K, $\Delta T_{Sub} = T_{Sat} - T_B$ (Chapter 5).
T_δ^+	Dimensionless temperature in condensate film (Chapter 6).
t	Time, s (Chapters 2 and 7).
U	Uncertainty (Chapter 3); reference velocity, m/s (Chapter 4); potential, such as gravity (Chapter 7); average velocity, m/s (Chapter 7).
U_{SL}, $V_{L,\ S}$	Superficial liquid velocity, m/s (Chapter 6).
U_{Gj}, V_{Gj}	Drift velocity in drift flux model, m/s, $\nu_G = j_G/\alpha = C_{oj} + V_{Gj}$ (Chapter 6).

U_{GS}, $V_{G,S}$	Superficial gas velocity, m/s (Chapter 6).
u	Velocity, m/s (Chapters 2–4, 6, and 7).
u	Velocity vector, m/s (Chapter 2).
$\overline{u}_{ave}$	Average electroosmotic velocity, m/s (Chapter 4).
$\overline{u}_z$	Mean axial velocity, m/s (Chapter 2).
$\overline{u_z^*}$	Mean axial velocity, $\overline{u_z^*} = \overline{u_z}/u_{z0}$, dimensionless (Chapter 2).
u^*	Friction velocity, m/s, $u^* = \sqrt{\tau_i/\rho_l}$ (Chapter 6).
u_m	Mean flow velocity, m/s (Chapters 3 and 5).
u_r	Relative velocity between a large gas bubble and liquid in the slug flow regime, m/s, $u_r = u_S - (j_G + j_L)$ (Chapter 6).
u_S	Velocity of large gas bubble in slug flow regime, m/s (Chapter 6).
u_{z0}	Maximum axial velocity with no-slip conditions, m/s (Chapter 2).
U_A	Overall heat transfer conductance, W/K (Chapter 6).
V	Voltage, V (Chapter 3); velocity, m/s (Chapters 5 and 6).
V	Nondimensional velocity, $V = \nu/\nu_0$ (Chapter 4).
$\overline{V}_l$	Average velocity of a liquid film, m/s (Chapter 6).
V_m	Zeroth-order uptake of oxygen by the hepatocytes (Chapter 7).
v	Velocity, m/s (Chapter 4).
ν	Specific volume, $\nu = 1/\rho$, m^3/kg (Chapters 2 and 5); velocity, m/s (Chapters 6 and 7).
ν_0	Reference velocity, m/s (Chapter 4).
ν_{LV}	Difference between the specific volumes of the vapor and liquid phases, m^3/kg, $\nu_{LG} = \nu_G - \nu_L$ (Chapter 5).
W	Maximum width, m (Chapters 3 and 4).
w	Velocity, m/s (Chapter 7).
We	Weber number, dimensionless, $We = LG^2/\rho\sigma$ (Chapter 5).
We	Weber number, dimensionless, $We = \rho V_S^2 D/\sigma$ (Chapter 6).
X	Cell density (Chapter 7).
X	Martinelli parameter, dimensionless, $X = \{(dp/d_Z)_L/(d_p/d_Z)_G\}^{1/2}$ (Chapters 5 and 6).
X, Y, Z	Coordinate axes (Chapter 4).
X_{tt}	Martinelli parameter for turbulent flow in the gas and liquid phases, dimensionless (Chapter 6).
x	Mass quality, dimensionless (Chapters 5 and 6).
x	Position vector, m (Chapter 2).
x, y, z	Coordinate axes (Chapters 2–7); length (Chapter 6).
x^*, y^*	Cross-sectional coordinates, dimensionless (Chapter 2).
x^*	Dimensionless version of x, $x^* = x/RePrD_h$ (Chapter 3).
x^+	Dimensionless version of x, $x^+ = x/D_h/Re$ (Chapter 3).
Y	Chisholm parameter, dimensionless, $y = [dP_F/dz]_{GO}/(dP_F/dz)_{LO}]$ (Chapter 6).
y	Dimensionless parameter in Eq. (6.22) (Chapter 6).
y_b	Bubble height, m (Chapter 5).

y_s — Distance to bubble stagnation point from heated wall, m (Chapter 5).
$\mathbf{Z}$ — Ohnesorge number, dimensionless, $Z = \mu/(\rho L\sigma)^{1/2}$ (Chapter 5).
z — Heated length from the channel entrance, m (Chapter 5).
z^* — Axial coordinate, dimensionless (Chapter 2).
z_i — Valence of type-i ions (Chapter 4).

Greek Symbols

α — Convection heat transfer coefficient, W/m^2 K (Chapter 2); coefficient in the VSS molecular model, dimensionless (Chapter 2); aspect ratio, dimensionless (Chapter 6); void fraction, dimensionless (Chapter 6).
$\alpha_1, \alpha_2, \alpha_3$ — Coefficients for the pressure distribution along a plane microchannel, dimensionless (Chapter 2).
α_c — Channel aspect ratio, dimensionless, $\alpha_c = a/b$ (Chapter 3).
α_i — Eigenvalues for the velocity distribution in a rectangular microchannel, dimensionless (Chapter 2).
α_r — Radial void fraction, dimensionless, $a_r = 0.8372 + [1 - (r/r_w)^{7.316}]$ (Chapter 6).
β — Coefficient in the VSS molecular model, dimensionless (Chapter 2); fin spacing ratio, dimensionless, $\beta = s/a$ (Chapter 3); angle with horizontal (Chapter 5); homogeneous void fraction, dimensionless (Chapter 6); velocity ratio, dimensionless (Chapter 6); multiplier to transition line, dimensionless, $\beta(F, X) =$ constant, used by Sardesai et al. (1981) (Chapter 6).
$\beta_1, \beta_2, \beta_3$ — Coefficients for the pressure distribution along a circular microtube, dimensionless (Chapter 2).
β_A, β_B — Empirically derived transition points for the Kariyasaki et al. void fraction correlation (Chapter 6).
Γ — Euler or gamma function (Chapter 2).
γ — Area ratio, dimensionless (Chapter 6); dimensionless length ratio, $\gamma = L/H$ (Chapter 7); Specific heat ratio, dimensionless, $\gamma = c_p/c_\nu$ (Chapter 2).
ΔP — Pressure drop, Pa (Chapter 6).
$\Delta P_2/\Delta P_1$ — Ratio of differential pressure between two system conditions (Chapter 6).
Δp — Pressure drop, pressure difference, Pa (Chapters 1–3, 5, and 7).
ΔT — Temperature difference, K (Chapter 6).
ΔT_{Sat} — Wall superheat, K, $\Delta T_{Sub} = T_{Sat} - T_B$ (Chapter 5).
ΔT_{Sub} — Liquid subcooling, K, $\Delta T_{Sat} = T_W - T_{Sat}$ (Chapter 5).
Δt — Elapsed time, s (Chapter 4).
Δx — Quality change, dimensionless (Chapter 6).

δ	Mean molecular spacing, m (Chapter 2); film thickness, m (Chapter 6).
δ^+	Nondimensional film thickness (Chapter 6).
δ_t	Thermal boundary layer thickness, m (Chapter 5).
ε	Average roughness, m (Chapter 3).
ε	Dielectric constant of a solution (Chapter 4).
$\overline{\varepsilon}$	Dimensionless gap spacing (Chapter 7).
ε_h	Turbulent thermal diffusivity, m^2/s (Chapter 6).
ε_m	Momentum eddy diffusivity, m^2/s (Chapter 6).
ε_w	Electrical permittivity of the solution (Chapter 4).
ζ	Zeta potential (Chapter 4).
ζ	Dimensionless zeta potential, $\zeta = ze\zeta/k_bT$ (Chapter 4).
ζ	Second coefficient of viscosity or Lamé coefficient, kg/m s, (Chapter 2).
ς	Temperature jump distance, m (Chapter 2).
ς^*	Temperature jump distance, dimensionless (Chapter 2).
η	Exponent in the inverse power law model, dimensionless (Chapter 2).
η'	Exponent in the Lennard–Jones potential, dimensionless (Chapter 2).
η_f	Fin efficiency, dimensionless (Chapter 3).
Θ	Dimensionless surface charge density (Chapter 4); angle, degrees (Chapter 6).
θ	Dimensionless time (Chapter 4).
θ_r	Receding contact angle, degrees (Chapter 5).
κ	Dimensionless Michaelis constant, $\kappa = K_m/C^*$ (Chapter 7).
κ	Debye–Hückel parameter, m^{-1}, $\kappa = (2n_\infty z^2e^2/\varepsilon\varepsilon_0k_bT)^{1/2}$ (Chapter 4).
κ	Constant in the inverse power law model, N m^η (Chapter 2).
κ'	Constant in the Lennard–Jones potential, dimensionless (Chapter 2).
λ	Wavelength, m (Chapter 7); mean free path, m (Chapters 1 and 2); dimensionless parameter, $\lambda = {\mu_L}^2/(\rho_L\sigma D_h)$ (Chapter 6).
λ_b	Bulk conductivity (Chapter 4).
λ_n	Roots of the transcendental equation tan $(\lambda_n) = Sh/\lambda_n$ (Chapter 7).
λ_n	Eigenvalues (Chapter 4).
μ	Dynamic viscosity, kg/ms (Chapters 1–7).
μ	Mobility (Chapter 4).
μ_n	Eigenvalues (Chapter 4).
μ_H	Homogeneous dynamic viscosity, kg/ms, $\mu_H = \beta\mu_G + (1-\beta)\mu_L$ (Chapter 6).
μ_m	Mixture dynamic viscosity, kg/ms, $1/\mu_m = x/\mu_G + (1-x)/\mu_L$ (Chapter 6).
ν	Collision rate, s (Chapter 2).

ξ Coefficient of slip, m (Chapter 2).
ξ^* Coefficient of slip, dimensionless (Chapter 2).
Π Inlet over outlet pressures ratio, dimensionless (Chapter 2).
ρ Density, kg/m^3 (Chapters 1–7).
ρ_e Local net charge density per unit volume (Chapter 4).
ρ_m Mixture density, kg/m^3, $1/\rho_m = (1/\rho_l(1-x) + (1/\rho_v)x)$ (Chapter 6).
ρ_{TP} Two-phase mixture density, kg/m^3, $\rho TP = [x/\rho_G + 1 - x/\rho_L]^{-1}$ (Chapter 6).
σ Viscous stress tensor, Pa (Chapter 2); Area ratio, dimensionless (Chapter 3); surface charge density (Chapter 4); surface tension, N/m (Chapter 5); fractional saturation, $s = C/C^*$ (Chapter 7); cell membrane permeability to oxygen (Chapter 7).
σ Tangential momentum accommodation coefficient, dimensionless (Chapter 2); surface tension, N/m (Chapter 6).
σ' Stress tensor, Pa (Chapter 2).
S_c Contraction area ratio (header to channel, >1), dimensionless (Chapter 5).
S_e Expansion area ratio (channel to header, <1), dimensionless (Chapter 5).
σT Thermal accommodation coefficient, dimensionless (Chapter 2).
σ_t Total collision cross-section, m^2 (Chapter 2).
τ Dimensionless time (Chapter 4); shear stress, Pa (Chapters 6 and 7); time scale, s (Chapter 7).
τ Characteristic time for QGD and QHD equations, s (Chapter 2).
τ_i Shear stress at vapor–liquid interface, Pa (Chapter 6).
τ_i^* Dimensionless shear stress, Pa (Chapter 6).
τ_m Shear stress due to momentum change, Pa (Chapter 6).
τ_W Frictional wall shear stress, Pa (Chapters 2, 3, and 6).
$\overline{\tau_w}$ Average wall shear, Pa (Chapter 2).
ϕ Intermolecular potential, J (Chapter 2); ratio of the characteristic diffusion time to the characteristic cellular oxygen uptake time, dimensionless (Chapter 7); velocity potential, m (Chapter 7).
ϕ_m Angle, degrees (Chapter 6).
Φ Dimensionless electric field strength (Chapter 4).
φ_i Eigenvalues for the velocity distribution in a rectangular microchannel, dimensionless (Chapter 2).
ϕ_L Two-phase friction multiplier, dimensionless, ${\phi_L}^2 = \Delta p_{f,\,TP}/\Delta p_{f,\,L}$, ratio of two-phase frictional pressure drop against frictional pressure drop of liquid flow (Chapters 5 and 6).
ψ Electrical potential (Chapter 4); dimensionless parameter, $\psi = (\sigma_w/\sigma)[\mu_L/\mu_W(\rho_W/\rho_L)^2]^{1/3}$ (Chapter 6).
ψ_h Two-phase homogeneous flow multiplier, dimensionless (Chapter 5).

ψ_j	Eigenvalues for the velocity distribution in a rectangular microchannel, dimensionless (Chapter 2).
ψ_s, ψ_S	Two-phase separated flow multiplier, dimensionless (Chapters 5 and 6).
ψ	Dimensionless double layer potential (Chapter 4).
Ω	Dimensionless frequency (Chapter 4); correction parameter used in Eqs. (6.52) and (6.53) (Chapter 6).
ω	Angular speed or vorticity, rad/s (Chapter 7); frequency, Hz (Chapter 4); temperature exponent of the coefficient of viscosity, dimensionless (Chapter 2).

Subscripts

0	Lowest boundary condition (Chapters 4 and 7).
1-ph	Single phase (Chapter 5).
a	Air, acceleration, ambient (Chapter 5); air (Chapter 6).
AB	Augmented Burnett equations (Chapter 2).
an	Annular (Chapter 6).
annu	Flow in an annular microduct (Chapter 2).
app	Apparent (Chapter 3).
av	Average (Chapter 4).
avg	Average (Chapters 5 and 7).
B	Bulk (Chapter 5); gas bubble (Chapter 6).
B	Burnett equations, (Chapter 2).
b	Bubble (Chapter 5); bulk (Chapter 3).
BGKB	Bhatnagar–Gross–Krook–Burnett equations (Chapter 2).
c, cr, crit	Critical condition (Chapters 3, 5, and 6).
c	Channel or in a single channel (Chapter 3); cavity mouth (Chapter 5); entrance contraction (Chapter 5).
cf	Calculated based on a flow diameter constricted by roughness elements (D_{cf}) (Chapter 3).
CBD	Convective boiling dominant (Chapter 5).
CHF	Critical heat flux (Chapter 5).
circ	Flow in a circular microtube (Chapter 2).
const	Constant (Chapter 4).
cp	Constant property (Chapter 3).
crit	Critical (Chapter 5).
cst	Constant (Chapter 7).
E	Euler equations (Chapter 2).
e	Outlet expansion, exit (Chapter 5).
eo	Electroosmostic (Chapter 4).
ep	Electrophoretic (Chapter 4).
EQ	Set of equations (Chapter 2).

eq	Equivalent (Chapter 6).
ex	Experimental (Chapter 3).
F	Frictional (Chapter 5).
f	Fluid (Chapter 3).
f	Fluid (Chapters 1, 3, and 4); frictional (Chapters 5 and 6); flooded (Chapter 6).
F/B	Film-bubble region (Chapter 6).
f/d	Film-bubble region (Chapter 6).
fd	Fully developed (Chapter 3).
G	Gas (Chapter 6).
g	Gas (Chapters 6 and 7).
g	Gravitational (Chapters 5 and 6).
GHS	Generalized hard sphere (Chapter 2).
Gn	Refers to Gnielinski's correlation (Chapter 3).
H	Homogeneous (Chapter 6).
h	Hydraulic (Chapter 6).
H1	Boundary condition with constant circumferential wall temperature and axial heat flux (Chapter 3).
H2	Boundary condition with constant wall heat flux, both circumferentially and axially (Chapter 3).
hetero	Heterogeneous solution (Chapter 4).
homo	Homogeneous solution (Chapter 4).
HS	Hard spheres (Chapter 2).
i	Species number (Chapters 4 and 7); vapor–liquid interface (Chapter 6).
in, i	Inlet (Chapters 2, 3, 5, and 7).
L	Liquid (Chapters 5 and 6).
l	Liquid (Chapter 6).
LG	Gas-superficial (Chapter 6).
LS	Liquid-superficial (Chapter 6).
LO	Entire flow as liquid (Chapters 5 and 6).
lv	Liquid–vapor (Chapter 6).
M	Momentum (Chapter 5).
M	Maxwell model (Chapter 2).
m	Mean (Chapter 3).
max	Maximum (Chapters 4 and 5).
min	Minimum (Chapter 5).
MM	Maxwell molecules (Chapter 2).
n	Normal direction (Chapter 2).
NBD	Nucleate boiling dominant (Chapter 5).
NS	Navier–Stokes equations (Chapter 2).
ns	No-slip (Chapter 2).
o	Out, outlet (Chapters 2 and 3).

ONB	Onset of nucleate boiling (Chapter 5).
plan	Flow between plane parallel plates (Chapter 2).
QGD	Quasi-gasodynamic equations (Chapter 2).
QHD	Quasi-hydrodynamic equations (Chapter 2).
r	Radial coordinate, radius, m (Chapter 4).
rect	Flow in a rectangular microchannel (Chapter 2).
S	Surface tension (Chapter 5).
S	Stagnation (Chapter 5); superficial (Chapter 6); liquid slug (Chapter 6).
s	Surface (Chapter 3); spherical (Chapter 7).
s	Tangential direction (Chapter 2).
Sat, sat	Saturation at system or local pressure (Chapters 5 and 6).
sh	Shear (Chapter 6).
st	Surface tension (Chapter 6).
str	Stratified (Chapter 6).
Sub	Subcooled, subcooling (Chapter 5).
SV	Sampling volume (Chapter 2).
T	Two-phase mixture (Chapter 6).
t	Total (Chapter 3); turbulent (Chapter 6).
th	Theoretical (Chapter 3).
TP	Two-phase (Chapter 5).
tp	Two-phase (Chapter 5).
tr	Transition regime (Chapter 6).
u	Unflooded (Chapter 6).
UC	Unit cell (Chapter 6).
V	Vapor (Chapter 5).
v	Vapor phase (Chapter 6).
VHS	Variable hard spheres (Chapter 2).
VSS	Variable soft spheres (Chapter 2).
W, w	Wall, heated surface (Chapter 5).
w	Fluid at the wall (Chapter 2); at the wall (Chapter 3); wall (Chapter 6).
wall	Wall (Chapter 2).
x, y, z	Local value at a location or as a function of the coordinates (Chapters 3, 5, and 7).
x, y	Cross-sectional coordinates (Chapter 2).
z	Axial coordinate (Chapter 2).
0	Standard conditions (Chapter 2); reference value (Chapter 2).
1	First-order boundary conditions (Chapter 2).
2	Second-order boundary conditions (Chapter 2).
∞	Infinity (Chapter 4).

Superscripts

$+, -$	Charge designation (Chapter 4).
$+$	Dimensionless parameters (Chapter 6).
$*$	Dimensionless parameters (Chapters 2–4, 6, and 7).

Operators

∇	Nabla function (Chapter 2).
$\tilde{\nabla}$	Dimensionless gradient operator (Chapter 4).
$\langle\,\rangle$	Averaged quantities (Chapter 7).

CHAPTER

Introduction

1

Satish G. Kandlikar[a] and Michael R. King[b]

[a]*Mechanical Engineering Department, Rochester Institute of Technology, Rochester, NY, USA;*
[b]*Department of Biomedical Engineering, Cornell University, Ithaca, NY, USA*

1.1 Need for smaller flow passages

Fluid flow inside channels is at the heart of many natural and man-made systems. Heat and mass transfer is accomplished across channel walls in biological systems, such as the brain, lungs, kidneys, intestines, and blood vessels, as well as in many man-made systems, such as heat exchangers, nuclear reactors, desalination units, and air separation units. In general, transport processes occur across the channel walls, whereas bulk flow takes place through the cross-sectional area of the channel. The channel cross-section thus serves as a conduit to transport fluid to and away from the channel walls.

A channel serves to accomplish two objectives: (i) to bring a fluid into intimate contact with the channel walls, and (ii) to bring fresh fluid to the walls and remove fluid away from the walls as the transport process is accomplished. The rate of the transport process depends on the surface area, which varies with the diameter D for a circular tube, whereas the flow rate depends on the cross-sectional area, which varies linearly with D^2. Thus, the tube surface area to volume ratio varies as $1/D$. Clearly, as the diameter decreases, the surface area to volume ratio increases. In the human body, two of the most efficient heat and mass transfer processes occur inside the lungs and the kidneys, with the flow channels approaching capillary dimensions of around 4 μm.

Figure 1.1 shows the ranges of channel dimensions employed in various systems. Interestingly, biological systems with mass transport processes employ much smaller dimensions, whereas larger channels are used for fluid transportation. From an engineering standpoint, there has been a steady shift from larger diameters, on the order of 10–20 mm, to smaller-diameter channels. Since the dimensions of interest are in the range of a few tens or hundreds of micrometers, use of the term "microscale" has become an accepted classifier for science and engineering associated with processes at this scale.

As the channel size becomes smaller, some of the conventional theories for (bulk) fluid, energy, and mass transport need to be revisited for validation. There

© 2014 Elsevier Ltd. All rights reserved.

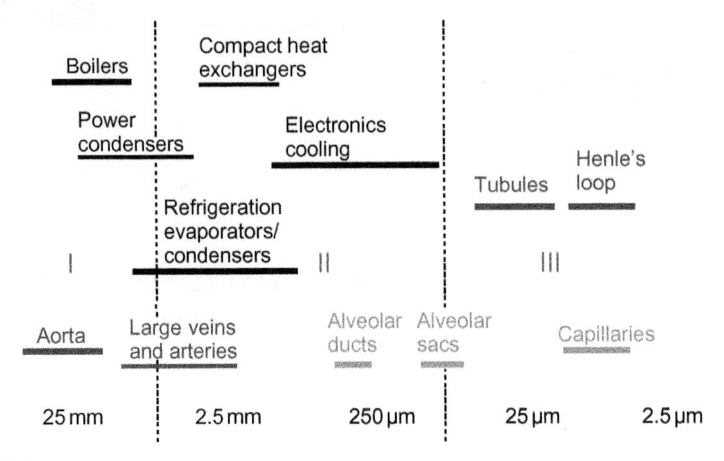

FIGURE 1.1

Ranges of channel diameters employed in various applications (Kandlikar and Steinke, 2003).

are two fundamental elements responsible for departure from "conventional" theories at the microscale. For example, differences in modeling fluid flow in small-diameter channels may arise as a result of

1. a change in the fundamental process, such as a deviation from the continuum assumption for gas flow, or an increased influence of some additional forces, such as electrokinetic forces;
2. uncertainty regarding the applicability of empirical factors derived from experiments conducted at larger scales, such as entrance and exit loss coefficients for fluid flow in pipes; or
3. uncertainty in measurements at microscale, including geometrical dimensions and operating parameters.

In this book, the potential changes in fundamental processes are discussed in detail, and the needs for experimental validation of empirical constants and correlations are identified if they are not available for small-diameter channels.

1.2 Flow channel classification

Channel classification based on hydraulic diameter is intended to serve as a simple guide for conveying the dimensional range under consideration. Channel size reduction has different effects on different processes. Deriving specific criteria based on the process parameters may seem to be an attractive option, but considering the number of processes and parameters that govern transitions from regular to microscale phenomena (if present), a simple dimensional classification is generally adopted in

Table 1.1 Channel Dimensions for Different Types of Flow for Gases at 1 atm (Kandlikar and Grande, 2003)

	Channel Dimensions (μm)			
Gas	**Continuum Flow**	**Slip Flow**	**Transition Flow**	**Free Molecular Flow**
Air	>67	0.67–67	0.0067–0.67	<0.0067
Helium	>194	1.94–194	0.0194–1.94	<0.0194
Hydrogen	>123	1.23–123	0.0123–1.23	<0.0123

Table 1.2 Channel Classification Scheme (Kandlikar and Grande, 2003)

Conventional channels	>3 mm
Minichannels	3 mm $\geq D >$ 200 μm
Microchannels	200 μm $\geq D >$ 10 μm
Transitional microchannels	10 μm $\geq D >$ 1 μm
Transitional nanochannels	1 μm $\geq D >$ 0.1 μm
Nanochannels	0.1 μm $\geq D$

D: smallest channel dimension.

the literature. The classification proposed by Mehendale et al. (2000) described the range from 1 to 100 μm as microchannels, 100 μm to 1 mm as mesochannels, 1–6 mm as compact passages, and greater than 6 mm as conventional passages.

Kandlikar and Grande (2003) considered the rarefaction effect of common gases at atmospheric pressure. Table 1.1 shows the ranges of channel dimensions that would fall under different flow types.

In biological systems, the flow in capillaries occurs at very low Reynolds numbers. A different modeling approach is needed in such cases. Also, the influence of electrokinetic forces begins to play an important role. Two-phase flow in channels below 10 μm remains unexplored. In a slight modification of the earlier channel classification scheme of Kandlikar and Grande (2003), a more general scheme based on the smallest channel dimension is presented in Table 1.2. Although one may be able to classify the channels depending on the relevant physical phenomena, such as single-phase flow, boiling, condensation, or cell transport, the scheme given in Table 1.2 is recommended for wider application.

In Table 1.2, D is the channel diameter. In the case of noncircular channels, it is recommended that the minimum channel dimension (e.g., the short side of a rectangular cross-section) should be used in place of the diameter D. We will use the above classification scheme for defining minichannels and microchannels. This classification scheme is employed for simplicity in terminology; the applicability of continuum theory or slip flow conditions for gas flow needs to be checked for the actual operating conditions in any channel.

1.3 Basic heat transfer and pressure drop considerations

The effect of hydraulic diameter on heat transfer and pressure drop is illustrated in Figures 1.2 and 1.3 for water and air flowing in a square channel under constant heat flux and fully developed laminar flow conditions. The heat transfer coefficient h is unaffected by the flow Reynolds number (Re) in the fully developed laminar region. It is given by:

$$h = Nu \frac{k}{D_h} \tag{1.1}$$

where k is the thermal conductivity of the fluid and D_h is the hydraulic diameter of the channel. The Nusselt number (Nu) for fully developed laminar flow in a square channel under constant heat flux conditions is 3.61. Figure 1.2 shows the variation of h for flow of water and air with channel hydraulic diameter under these conditions. The dramatic enhancement in h with a reduction in channel size is clearly demonstrated.

On the other hand, the friction factor f varies inversely with Re, since the product $f \cdot Re$ remains constant during fully developed laminar flow. The frictional pressure drop per unit length for the flow of an incompressible fluid is given by:

$$\frac{\Delta p_f}{L} = \frac{2fG^2}{\rho D} \tag{1.2}$$

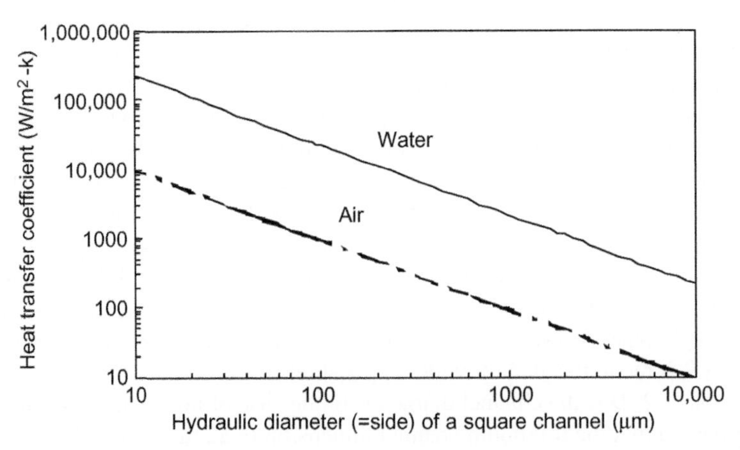

FIGURE 1.2

Variation of the heat transfer coefficient with channel size for fully developed laminar flow of air and water.

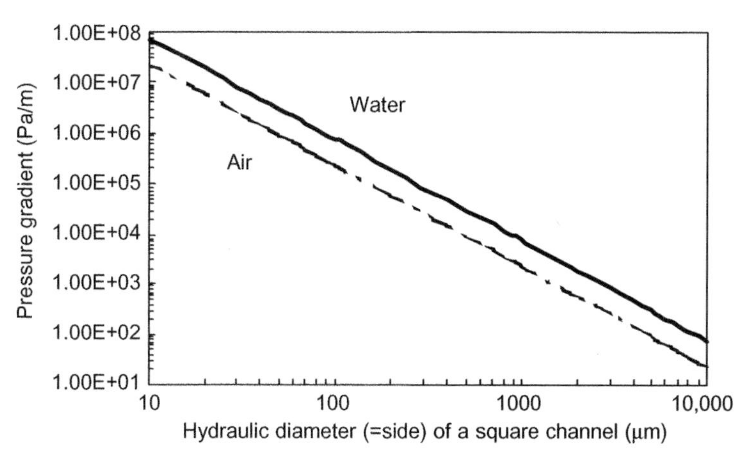

FIGURE 1.3

Variation of pressure gradient with channel size for fully developed laminar flow of air and water.

where $\Delta p_f/L$ is the frictional pressure gradient, f is the Fanning friction factor, G is the mass flux, and ρ is the fluid density. For fully developed laminar flow, we can write:

$$f \cdot Re = C \tag{1.3}$$

where Re is the Reynolds number, $Re = GD_h/\mu$, and C is a constant, $C = 14.23$ for a square channel.

Figure 1.3 shows the variation of pressure gradient with the channel size for a square channel with $G = 200$ kg/m^2 s, and for air and water assuming incompressible flow conditions. These plots are for illustrative purposes only, as the above assumptions may not be valid for the flow of air, especially in smaller-diameter channels. It can be seen from Figure 1.3 that the pressure gradient increases dramatically with a reduction in the channel size.

The balance between the heat transfer rate and pressure drop becomes an important issue in designing the coolant flow passages for high-flux heat removal in microprocessor chip cooling. These issues are addressed in detail in Chapter 3 under single-phase liquid cooling.

1.4 The potential and special demands of fluidic biological applications

The potential for minichannels, microchannels, and nanochannels to contribute to our understanding of biological systems and impact human health lies in the ability to manipulate individual cells, molecules, or multicellular aggregates, and in the ability to use engineering methods to recreate the dazzling degree of complexity in living beings. The study and manipulation of living cells goes beyond fluid mechanics, and is a multidisciplinary field that borrows from biology, engineering, applied mathematics, physics, and chemistry. Some of the more common applications include improved quantification and more minute measurement to enable basic research, biosensors for pathogen detection or to mimic physiological sensing of the environment, and hydrogel constructs for drug delivery and tissue engineering. Cellular engineering research and development is also inherently *multiscale* (like this book!) because to gain a mechanistic rather than "black box" understanding of cell behavior it is necessary to concern oneself with the phenomena that occur one length below the scale you most care about: How are proteins employed by the cell to carry out chemical and mechanical processes? How do cells communicate with each other and organize into functional tissues? How does a drug's chemical properties affect how it is distributed to the different organs of the body? Or between a pregnant mother and child? These are all questions that minichannels, microchannels, and nanochannels can help to address.

The use of minichannels, microchannels, and microfluidics in general is becoming increasingly important to the biomedical community. However, the transport and manipulation of living cells and biological macromolecules place increasingly critical demands on maintaining system conditions within acceptable ranges. For instance, human cells require an environment of 37°C and a pH = 7.4 to ensure their continued viability. If these parameter values stray more than 10%, then cell death will result. All protein molecules have their own preferred pH environments, and large variations from this can cause (sometimes irreversible) denaturation and loss of biological activity due to unfolding.

High temperatures can also cause irreversible protein denaturation. However, in a polymerase chain reaction, or PCR, the rapid cycling (~1 min/cycle) of temperature between 94°C and 54°C for 30–40 cycles is necessary to induce repeated denaturation and annealing of DNA chains. Here, the microchannel geometry can be exploited due to the ease of changing the temperature of small liquid volumes, and in fact thermally driven natural convection can be used to achieve such temperature cycles (Krishnan et al., 2002).

Additionally, the concentrations of solutes, such as dissolved gases, nutrients, and metabolic products, must be maintained within specified tolerances to ensure cell proliferation in microchannel bioreactors. Finally, local shear stresses can be critical in suspensions of biological particles. For instance, many cell types such as blood platelets become activated to a highly adhesive state upon exposure to elevated shear stresses above 10 dynes/cm^2. Endothelial cells, which line the cardiovascular system, require a certain level of laminar shear stress on their luminal surface ($\geq$0.5 dynes/cm^2) or else they will not align themselves properly and can express surface receptor molecules which induce chronic inflammation.

At much higher shear stresses such as 1500 dynes/cm^2, as can be produced around sharp corners for very high liquid flow rates, red blood cells are known to "lyse" or rupture. This is especially critical when designing artificial blood pumps or implantable replacement valves. All of these concerns must be separately addressed when developing new microscale-flow applications for the transport and manipulation of biological materials.

One example of a successful microchannel application in biomedicine that has exploded in interest over the past 5 years is the isolation of circulating tumor cells (CTCs) from blood collected from cancer patients. Metastasis, or the spread of cancer to remote organs, is responsible for 90% of cancer deaths in the United States. Many forms of cancer spread by disseminating individual cancer cells or small aggregates through the bloodstream, and these rare cells can be detected in blood. The ability to detect and count CTCs in blood is a minimally invasive way to track disease progression; furthermore, if cells can be obtained in a live, intact form, then crucial new information can be gained on an individual patient-by-patient basis. Mehmet Toner and colleagues first employed a microfluidic chip featuring an array of microfabricated posts to isolate CTCs from blood samples (Nagrath et al., 2007). The posts are coated with an antibody that binds to an epithelial cell adhesion molecule (EpCAM) that is present on the surface of most CTCs, and the large surface area per volume of the device is well suited to this type of immunocapture. Variations of the CTC microchip have since emerged from other researchers, the most noteworthy advance being a "herringbone chip" design by Toner and colleagues that promotes better mixing of the fluid flowing through the chip (Stott et al., 2010). A recent microscale-flow device for CTC capture developed by the King Lab utilizes common polymeric tubing coated with halloysite nanotubes and functionalized with E-selectin adhesion protein, which mimics the natural process of cell rolling in blood vessels to isolate these cells to enable clinical research and diagnostics (Hughes et al., 2012a,b). This field has seen a wave of collaborative studies involving technologists and clinical researchers to obtain meaningful new insights into human disease.

1.5 Summary

Microchannels are found in many biological systems where extremely efficient heat and mass transfer processes occur, such as lungs and kidneys. Channel size classification is based on observations of many different processes, but its use is mainly in arriving at a common terminology. The heat and mass transfer rates in small-diameter channels are also associated with a high pressure drop penalty for fluid flow. The fundamentals of microchannels and minichannels in various applications are presented in this book.

1.6 Practice problems

Problem 1.1

Calculate the heat and mass transfer coefficients for airflow in a human lung at various branches. Assume fully developed laminar flow conditions. Comment on the differences between this idealized case and realistic flow conditions.

> *Hint:* Consult a basic anatomy book for the dimensions of the airflow passages and flow rates, and a heat and mass transfer book for laminar fully developed Nusselt number and Sherwood number in a circular tube.

Problem 1.2

Calculate the Knudsen number for flow of helium at a pressure of 1 mTorr (1 Torr = 1 mmHg) in 0.1-, 1-, 10-, and 100-mm diameter tubes. What type of flow model is applicable for each case?

> *Hint:* Knudsen number is given by $Kn = \lambda/D_h$, and the mean free path λ is given by $\lambda = (\mu\sqrt{\pi}/\rho\sqrt{2RT})$, where μ is the fluid viscosity, ρ is the fluid density, T is the absolute temperature, and R is the gas constant.

Problem 1.3

Calculate the pressure drop for flow of water in a 15-mm-long 100-μm circular microchannel flowing at a temperature of 300 K and with a flow Reynolds number of (a) 10, (b) 100, and (c) 1000. Also calculate the corresponding water flow rates in kilogram per second and milliliter per minute.

References

Hughes, A.D., Mattison, J., Powderly, J.D., Greene, B.T., King, M.R., 2012a. Rapid isolation of viable circulating tumor cells from patient blood samples. J. Vis. Exp. 64, e4248. Available from: http://dx.doi.org/10.3791/4248.

Hughes, A.D., Mattison, J., Western, L.T., Powderly, J.D., Greene, B.T., King, M.R., 2012b. Microtube device for selectin-mediated capture of viable circulating tumor cells from blood. Clin. Chem. 58 (5), 846–853.

Kandlikar, S.G., Grande, W.J., 2003. Evolution of microchannel flow passages—thermohydraulic performance and fabrication technology. Heat Trans. Eng. 24 (1), 3–17.

Kandlikar, S.G., Steinke, M.E., 2003. Examples of microchannel mass transfer processes in biological systems. Proceedings of First International Conference on Minichannels and Microchannels. ASME, Rochester, NY, April 24–25, Paper ICMM2003-1124 pp. 933–943.

Krishnan, M., Ugaz, V.M., Burns, M.A., 2002. PCR in a Rayleigh–Benard convection cell. Science 298, 793.

Mehendale, S.S., Jacobi, A.M., Shah, R.K., 2000. Fluid flow and heat transfer at micro- and meso-scales with applications to heat exchanger design. Appl. Mech. Rev. 53, 175–193.

Nagrath, S., Sequist, L.V., Maheswaran, S., Bell, D.W., Irimia, D., Ulkus, L., et al., 2007. Isolation of rare circulating tumour cells in cancer patients by microchip technology. Nature 450, 1235–1239.

Stott, S.L., Hsu, C.H., Tsukrov, D.I., Yu, M., Miyamoto, D.T., Waltman, B.A., et al., 2010. Isolation of circulating tumor cells using a microvortex-generating herringbone-chip. Proc. Natl. Acad. Sci. USA 107, 18392–18397.

CHAPTER

2 Single-Phase Gas Flow in Microchannels

Stéphane Colin

Department of Mechanical Engineering, National Institute of Applied Sciences of Toulouse, University of Toulouse, Toulouse, France

Microfluidics is a rather young research field, born in the early 1980s. Its older relative *fluidics* was in fashion in the 1960s and 1970s. Fluidics seems to have started in the USSR in 1958, then developed in the United States and Europe, first for military purposes, with civil applications appearing later. At that time, fluidics was mainly concerned with inner gas flows in devices involving millimetric or sub-millimetric sizes. These devices were designed to perform the same actions (amplification, logic operations, diode effects, etc.) as their electric counterparts. The idea was to design pneumatically, in place of electrically, supplied computers. The main applications were concerned with the spatial domain, for which electric power overload was indeed an issue due to the electric components of the time, which generated excessive magnetic fields and dissipated too much thermal energy to be safe in a confined space. Most of the fluidic devices were etched in a substrate by means of conventional machining techniques, or by insolation techniques applied on specific resins where masks protected the parts to be preserved. The rapid development of microelectronics put a sudden end to pneumatic computers, but these two decades were particularly useful in enhancing our knowledge about gas flows in minichannels or mini-pneumatic devices.

As microfluidics concerns smaller sizes—the inner sizes of microelectromechanical systems (MEMS)—new issues must be considered in order to accurately model gas microflows. These issues are mainly due to rarefaction effects, which typically must be taken into account when characteristic lengths are of the order of 1 μm, under usual temperature and pressure conditions.

In this chapter, the role of rarefaction is explained and its consequences for the behavior of gas flows in microchannels are detailed. The main theoretical and experimental results from the literature about pressure-driven, steady, or pulsed gas microflows are summarized. Heat transfer in microchannels and thermally driven gas microflows are also described. They are particularly interesting for vacuum generation, using microsystems without moving parts.

Heat Transfer and Fluid Flow in Minichannels and Microchannels.
© 2014 Elsevier Ltd. All rights reserved.

2.1 Rarefaction and wall effects in microflows

In microfluidics, theoretical knowledge of gas flows is currently more advanced than that for liquid flows (Colin, 2010). Concerning gases, the issues are actually more clearly identified: the main micro-effect that results from shrinking the size of devices is rarefaction. This allows us to exploit the strong, although not complete, analogy between microflows and low-pressure flows that has been extensively studied for more than 50 years, particularly for aerospace applications.

2.1.1 Gas at the molecular level

2.1.1.1 Microscopic length scales

Modeling gas microflows requires us to take into account several characteristic length scales. At the molecular level, we may consider the mean molecular diameter d, the mean molecular spacing δ, and the mean free path λ (Figure 2.1).

The mean free path is the average distance traveled by a molecule between two consecutive collisions, in a frame moving with the stream speed of the gas. If we consider a *simple gas*, that is, a gas consisting of a single chemical species with molecules having the same structure, then the mean free path depends on their mean diameter d and on the number density:

$$n = \delta^{-3} \tag{2.1}$$

and the inverse $1/n$ represents the mean volume available for one molecule.

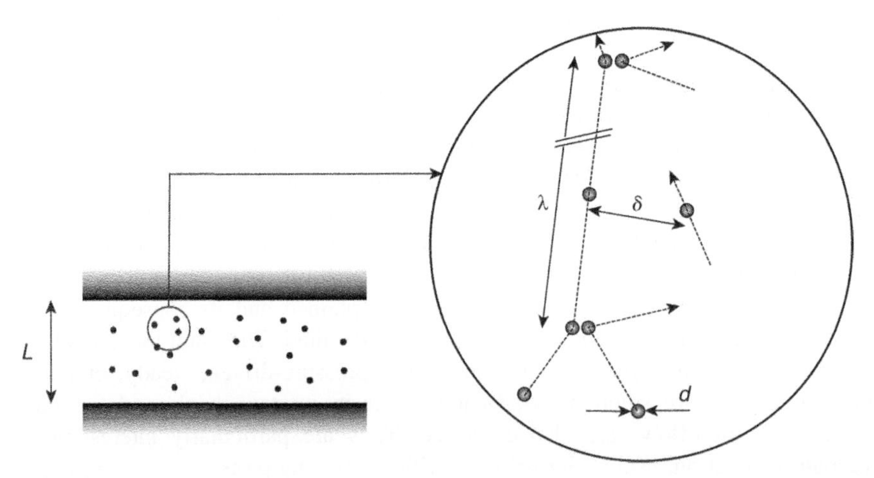

FIGURE 2.1

Mean characteristic length scales to take into account at the molecular level.

For example, a gas under standard conditions (i.e., for a temperature $T_0 = 273.15$ K and a pressure $p_0 = 1.013 \times 10^5$ Pa) has about 27 million molecules in a cube of 1 μm in width ($n_0 = 2.687 \times 10^{25}/\text{m}^3$ and $\delta_0 = 3.34$ nm); in the case of air, for which $d_0 \approx 0.42$ nm, the mean free path is $\lambda_0 \approx 49$ nm. When compared to the case of liquid water, the mean molecular diameters of each are nearly equal, but the mean molecular spacing is about 10 times smaller and the mean free path is 10^5 times smaller than air!

2.1.1.2 Binary intermolecular collisions in dilute simple gases

Gases that satisfy

$$\frac{d}{\delta} \ll 1 \tag{2.2}$$

are said to be *dilute gases*. In that case, most of the intermolecular interactions are binary collisions. Conversely, if Eq. (2.2) is not satisfied, the gas is said to be a *dense gas*. The dilute gas approximation, along with the equipartition of energy principle, leads to the classic kinetic theory and the Boltzmann transport equation.

With this approximation, the mean free path of the molecules may be expressed as the ratio of the mean thermal velocity magnitude $\overline{c'}$ to the collision rate ν:

$$\lambda = \frac{\overline{c'}}{\nu} = \frac{\sqrt{8RT/\pi}}{\nu} \tag{2.3}$$

The thermal velocity c' of a molecule is the difference between its total velocity c and the local macroscopic velocity $\mathbf{u}$ of the flow (cf. Section 2.1.2), and its mean value $\overline{c'} = \sqrt{8RT/\pi}$, which depends on the temperature T and on the specific gas constant R, is calculated from the Boltzmann equation. The estimation of the collision rate and consequently of the mean free path depends on the model chosen for describing the elastic binary collision between two molecules. Such a model also allows estimation of the total collision cross-section

$$\sigma_t = \pi d^2 \tag{2.4}$$

in the collision plane and the mean molecular diameter d (Figure 2.2), as well as the dynamic viscosity μ as a function of the temperature.

A collision model generally requires the definition of the force $\mathbf{F}$ exerted between the two molecules being considered. This force is actually repulsive at short distances and weakly attractive at large distances. Different approximated models are proposed to describe this force. The classic one is the so-called *inverse power law* (IPL) *model* or *point center of repulsion model*

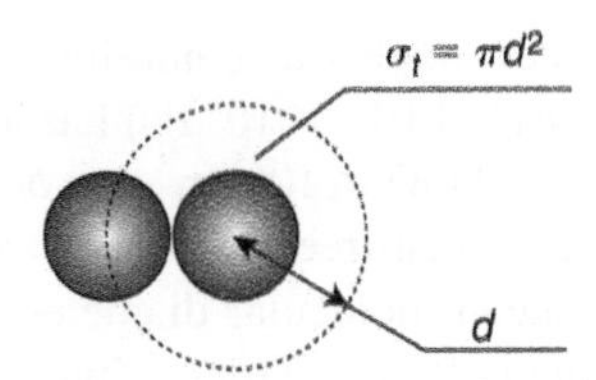

FIGURE 2.2

Total collision cross-section in the collision plane.

(Bird, 1998). It only takes into account the repulsive part of the force and assumes that

$$F = |\mathbf{F}| = \frac{\kappa}{r^{\eta}} \tag{2.5}$$

This derives from an intermolecular potential

$$\varphi = \frac{\kappa}{(\eta - 1)r^{\eta - 1}} \tag{2.6}$$

where κ and the exponent η have a constant value and r denotes the distance between the two molecules' centers. Using the Chapman–Enskog theory, this model leads to a law of viscosity in the form

$$\mu = \mu_0 \left(\frac{T}{T_0}\right)^{\omega} \tag{2.7}$$

with

$$\omega = \frac{\eta + 3}{2(\eta - 1)} \tag{2.8}$$

The collision rate may be deduced as (Lengrand and Elizarova, 2010)

$$v = 4\left(\frac{5}{2} - \omega\right)^{\omega - \frac{1}{2}} \Gamma\left(\frac{5}{2} - \omega\right) n\,\sigma_{t0} \left(\frac{T}{T_0}\right)^{\frac{1}{2} - \omega} \sqrt{\frac{RT}{\pi}} \tag{2.9}$$

where $\Gamma(j) = \int_0^{\infty} x^{j-1} \exp(-x)\mathrm{d}x$ is the Euler, or gamma, function and T_0 is a reference temperature for which the total collision cross-section σ_{t0} is calculated. From Eq. (2.3), we can see that the mean free path

$$\lambda = \frac{1}{\sqrt{2}((5/2) - \omega)^{\omega - (1/2)} \Gamma((5/2) - \omega) n \sigma_{t0}} \left(\frac{T}{T_0}\right)^{\omega - \frac{1}{2}} \tag{2.10}$$

is proportional to $T^{\omega - (1/2)}$ and inversely proportional to the number density n.

The simplest collision model is the *hard sphere* (HS) model, which assumes that the total collision cross-section σ_t is constant and for which the viscosity is proportional to the square root of the temperature. Actually, the HS model may be described from the IPL with the exponent $\eta \to \infty$ and $\omega = 1/2$.

Bird, who proposed the variable hard sphere (VHS) model for applications to the direct simulation Monte Carlo method (DSMC; cf. Section 2.2.4.2), has improved the HS model. The VHS model may be considered as a HS model with a diameter d that is a function of the relative velocity between the two colliding molecules. The Chapman–Enskog method leads to a viscosity

$$\mu_{\mathrm{VHS}} = \frac{15m\sqrt{\pi RT}}{8((5/2)-\omega)^{\omega-(1/2)}\Gamma((9/2)-\omega)\sigma_{t0}}\left(\frac{T}{T_0}\right)^{\omega-\frac{1}{2}} \tag{2.11}$$

which satisfies Eq. (2.7) and where m is the molecular mass. By eliminating σ_{t0} between Eqs. (2.9) or (2.10) and (2.11) and noting that $\Gamma(j+1) = j\,\Gamma(j)$, the mean free path and the collision rate for a VHS model may be expressed as functions of the viscosity and the temperature:

$$\nu_{\mathrm{VHS}} = \frac{\rho RT}{\mu}\frac{30}{(7-2\omega)(5-2\omega)} \tag{2.12}$$

and

$$\lambda_{\mathrm{VHS}} = \frac{\mu}{\rho\sqrt{2\pi RT}}\frac{2(7-2\omega)(5-2\omega)}{15} \tag{2.13}$$

with $\rho = mn$ as the local density.

Another classic model is the Maxwell molecules (MM) model. This is a special case of the IPL model with $\eta = 5$ and $\omega = 1$. Actually, the HS and the MM models may be considered as the limits of the more realistic VHS model, since real molecules generally exhibit a behavior that corresponds to an intermediate value $1/2 \le \omega \le 1$. Another expression is frequently encountered in the literature; for example, Karniadakis and Beskok (2002) or Nguyen and Wereley (2002) use the formula $\lambda_{\mathrm{M}} = \sqrt{\pi/2}\mu/(\rho\sqrt{RT})$ proposed by Maxwell in 1879 (Eq. (2.57)).

Other binary collision models based on the IPL assumption are described in the literature. Koura and Matsumoto (1991, 1992) introduced the variable soft sphere (VSS) model, which differs from the VHS model by a different expression of the deflection angle taken by the molecule after a collision. The VSS model leads to a correction of the mean free path and the collision rate values:

$$\lambda_{\mathrm{VSS}} = \beta\lambda_{\mathrm{VHS}}; \quad \nu_{\mathrm{VSS}} = \frac{\nu_{\mathrm{VHS}}}{\beta} \tag{2.14}$$

where $\beta = 6\alpha/[(\alpha+1)(\alpha+2)]$ and the value of α is generally between 1 and 2 (Bird, 1998). Thus, the correction introduced by the VSS model in the mean free path and in the collision rate remains limited, less than 3%. For $\alpha = 1$, the VSS model reduces to the VHS model.

Finally, some models can take into account the long-range attractive part of the force between two molecules by adding a uniform attractive potential (square-well model), or an IPL attractive component (Sutherland, Hassé, and Cook or Lennard–Jones potential models) to the HS model. For example, the Lennard–Jones potential is

$$\phi = \frac{\kappa}{(\eta-1)r^{\eta-1}} - \frac{\kappa'}{(\eta'-1)r^{\eta'-1}} \tag{2.15}$$

and the widely used Lennard–Jones 12–6 model corresponds to $\eta = 13$ and $\eta' = 7$. The generalized hard sphere (GHS) model (Hassan and Hash, 1993) is a generalized model that combines the computational simplicity of the VHS model and the accuracy of complicated attractive–repulsive interaction potentials. More recently, the variable sphere (VS) molecular model proposed by Matsumoto (2002) provides consistency for diffusion and viscosity coefficients with those of any realistic intermolecular potential.

Table 2.1 resumes the relationships between the collision rate, the mean free path, the viscosity, the density, and the temperature for classic IPL collision

Table 2.1 Mean Free Path, Dynamic Viscosity, and Collision Rate for Classic IPL Collision Models

	$F = \frac{\kappa}{r^\eta}$	$\mu \propto T^\omega$	$\nu = k_1 \frac{\rho RT}{\mu}$	$\lambda = k_2 \frac{\mu}{\rho\sqrt{RT}}$
Model	η	ω	k_1	k_2
HS	∞	$\frac{1}{2}$	$\frac{5}{4}$	$\frac{16}{5\sqrt{2\pi}} \approx 1.277$
VHS	η	$\frac{\eta+3}{2(\eta-1)}$	$\frac{30}{(7-2\omega)(5-2\omega)}$	$\frac{2(7-2\omega)(5-2\omega)}{15\sqrt{2\pi}}$
MM	5	1	2	$\sqrt{\frac{2}{\pi}} \approx 0.798$
VSS	η	$\frac{\eta+3}{2(\eta-1)}$	$\frac{5(\alpha+1)(\alpha+2)}{\alpha(7-2\omega)(5-2\omega)}$	$\frac{4\alpha(7-2\omega)(5-2\omega)}{5(\alpha+1)(\alpha+2)\sqrt{2\pi}}$

models. As the mean free path λ is an important parameter for the simulation of gas microflows (cf. Section 2.1.2), we should be careful when comparing some theoretical results from the literature and verify from which model λ has been calculated. Note that for the formula of Maxwell, $k_{2,M} = \sqrt{2/\pi}$ is 2% lower than the value $k_{2,HS} = 16/(5\sqrt{2\pi})$ obtained from a HS model.

Equation (2.11) is also interesting because it allows estimation of the mean molecular diameter d from viscosity data and a VHS model. Table 2.2 gives the value of d for different gases under standard conditions, obtained by Bird (1998), from a VHS or a VSS hypothesis. The ratio δ_0/d is calculated for the mean value d of d_{VHS} and d_{VSS}. For any gas, we can see that condition (2.2) is roughly satisfied and the different gases can reasonably be considered as dilute gases under standard conditions. However, the dilute gas approximation will be better for He than for SO_2.

Table 2.2 Molecular Weight, Dynamic Viscosity, and mean Molecular Diameters Under Standard Conditions ($p_0 = 1.013 \times 10^5$ Pa and $T_0 = 273.15$ K) Estimated from a VHS or a VSS Model

Gas	$M \times 10^3$ (kg/mol)	$\mu_0 \times 10^7$ (N s/m^2)	ω	d_{VHS} (pm)	α	d_{VSS} (pm)	δ_0/d
Sea level air	28.97	171.9	0.77	419	–	–	7.97
Ar	39.948	211.7	0.81	417	1.40	411	8.06
CH_4	16.043	102.4	0.84	483	1.60	478	6.95
Cl_2	70.905	123.3	1.01	698	–	–	4.78
CO	28.010	163.5	0.73	419	1.49	412	8.04
CO_2	44.010	138.0	0.93	562	1.61	554	5.98
H_2	2.0159	84.5	0.67	292	1.35	288	11.51
HCl	36.461	132.8	1.00	576	1.59	559	5.88
He	4.0026	186.5	0.66	233	1.26	230	14.42
Kr	83.80	232.8	0.80	476	1.32	470	7.06
N_2	28.013	165.6	0.74	417	1.36	411	8.06
N_2O	44.013	135.1	0.94	571	–	–	5.85
Ne	20.180	297.5	0.66	277	1.31	272	12.16
NH_3	17.031	92.3	1.10	594	–	–	5.62
NO	30.006	177.4	0.79	420	–	–	7.95
O_2	31.999	191.9	0.77	407	1.40	401	8.26
SO_2	64.065	116.4	1.05	716	–	–	4.66
Xe	131.29	210.7	0.85	574	1.44	565	5.86

Source: *Data from Bird (1998) with corresponding values of the ratio* δ_0/d.

2.1.2 Continuum assumption and thermodynamic equilibrium

When applicable, the continuum assumption is very convenient since it erases the molecular discontinuities by averaging the microscopic quantities on a small sampling volume. All macroscopic quantities of interest in classic fluid mechanics (density ρ, velocity $\mathbf{u}$, pressure p, temperature T, etc.) are assumed to vary continuously from point to point within the flow.

For example, if we consider an air flow in a duct, for which the macroscopic velocity varies from 0 to 1 m/s and is parallel to the axis of the duct, the velocity of a molecule is on the order of 1 km/s and may take any direction. Similar considerations also concern the other mechanical and thermodynamic quantities.

In order to respect the continuum assumption, the microscopic fluctuations should not generate significant fluctuations of the averaged quantities. Consequently, the size of a representative sampling volume must be large enough to erase the microscopic fluctuations, but it must also be small enough to point out the macroscopic variations, such as velocity or pressure gradients of interest in the control volume (Figure 2.3). If the shaded area in Figure 2.3 does not exist, the sampling volume is not representative and the continuum assumption is not valid.

It may be considered that sampling a volume containing 10,000 molecules leads to 1% statistical fluctuations in the macroscopic quantities (Karniadakis and Beskok, 2002). Such a fluctuation level needs a sampling volume, which characteristic length l_{SV} verifies $l_{SV}/\delta = 10^{4/3} \approx 22$. Consequently, the control volume must have a much higher characteristic length L, that is

$$\frac{L}{\delta} \gg 10^{4/3} \tag{2.16}$$

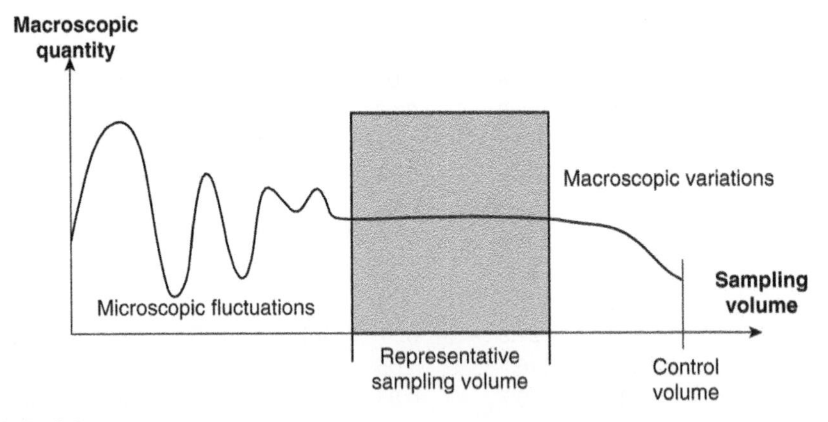

FIGURE 2.3

The existence of a representative sampling volume (shaded area) is necessary for the continuum assumption to be valid.

so that the *statistical fluctuations* can be *neglected.* For example, for air under standard conditions, the value of l_{SV} corresponding to 1% statistical fluctuations is 72 nm. This is comparable to the value of the mean free path, $\lambda_0 = 49$ nm.

Moreover, the continuum approach requires that the sampling volume be in *thermodynamic equilibrium.* For the thermodynamic equilibrium to be respected, the frequency of the intermolecular collisions inside the sampling volume must be high enough. This implies that the mean free path $\lambda = \overline{c'}/\nu$ must be small compared with the characteristic length l_{SV} of the sampling volume, itself being small compared with the characteristic length L of the control volume. As a consequence, the thermodynamic equilibrium requires that

$$\frac{\lambda}{L} \ll 1 \tag{2.17}$$

The ratio

$$Kn = \frac{\lambda}{L} \tag{2.18}$$

is called the Knudsen number; it plays a very important role in gaseous microflows (see Section 2.1.3). If λ is obtained from an IPL collision model,

$$\lambda = k_2 \mu / \rho \sqrt{RT} \tag{2.19}$$

with k_2 given by Table 2.1, and the Knudsen number can be related to the Reynolds number

$$Re = \frac{\rho u L}{\mu} \tag{2.20}$$

and the Mach number

$$Ma = \frac{u}{a} \tag{2.21}$$

by the relationship

$$Kn = k_2 \sqrt{\gamma} \frac{Ma}{Re} \tag{2.22}$$

Here, γ is the ratio of the specific heats of the gas and a, the local speed of sound, with $a = \sqrt{\gamma RT}$ for an ideal gas (Anderson, 1990), which is verified for a dilute gas (cf. Section 2.2.1). Equation (2.22) shows the link between rarefaction (characterized by Kn) and compressibility (characterized by Ma) effects, the latter having to be taken into account if $Ma > 0.3$.

The limits that correspond to Eqs. (2.2), (2.16), and (2.17), with the indicative values ($\delta/d = 7$, $L/\delta = 100$, and $\lambda/L = 0.1$) proposed by Bird (1998), are shown in Figure 2.4. In this figure, the mean free path has been estimated with a HS model, for which Eq. (2.10) can be written using Eq. (2.4) and $\omega = 1/2$, in the form

$$\lambda_{HS} = \frac{\delta^3}{\sqrt{2}\pi d^2} \tag{2.23}$$

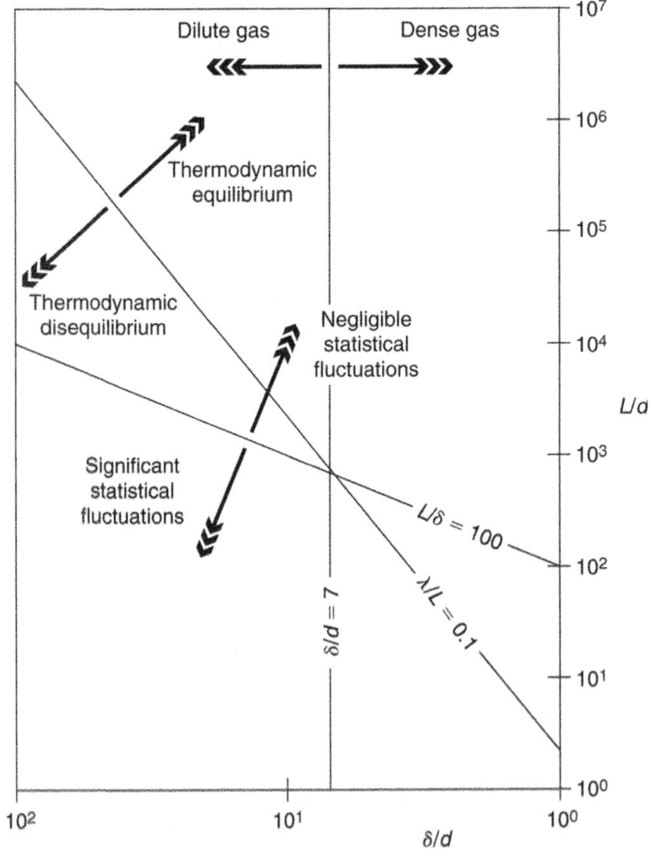

FIGURE 2.4

Limits of the mean approximations for the modeling of gas microflows.

Source: *From Bird (1998).*

Thus, $\lambda/L = (\delta/d)^3\ (d/L)(1/\sqrt{2}\pi)$, and the thermodynamic equilibrium limit $\lambda/L = 0.1$ is a straight line with a slope equal to -3 in the log–log plot of Figure 2.4.

2.1.3 Rarefaction and Knudsen analogy

It is now clear that the similitude between low pressure and confined flows is not complete, since the Knudsen number is not the only parameter to take into account (cf. Figure 2.4.).

However, it is convenient to differentiate the flow regimes as a function of Kn, and the following classification, although tinged with empiricism, is usually accepted:

- For $Kn < 10^{-3}$, the flow is a *continuum flow* (**C**), and it is accurately modeled by the compressible Navier–Stokes equations with classical no-slip boundary conditions.
- For $10^{-3} < Kn < 10^{-1}$, the flow is a *slip flow* (**S**), and the Navier–Stokes equations remain applicable, provided a velocity slip and a temperature jump are taken into account at the walls. These new boundary conditions point out that rarefaction effects first become sensitive at the wall.
- For $10^{-1} < Kn < 10$, the flow is a *transition flow* (**T**), and the continuum approach of the Navier–Stokes equations is no longer valid. However, the intermolecular collisions are not yet negligible and should be taken into account.
- For $Kn > 10$, the flow is a *free molecular flow* (**M**), and the occurrence of intermolecular collisions is negligible compared with collisions between the gas molecules and the walls.

These different regimes will be detailed in the following sections. Their limits are only indicative and could vary from one case to another, partly because the choice of the characteristic length L is rarely unique. For flows in channels, L is generally the hydraulic diameter or the depth of the channel. In complex geometrical configurations, it is generally preferable to define L from local gradients (e.g., of the density ρ: $L = 1/|\nabla\rho/\rho|$) rather than from simple geometrical considerations (Gad-el-Hak, 1999); the Knudsen number based on this characteristic length is the so-called local Knudsen number (Lengrand and Elizarova, 2010). Figure 2.5 locates these different regimes for air under standard conditions. For comparison, the cases of helium and sulfur dioxide are also represented. For SO_2, the dilute gas assumption is less valid than for air and He, and the binary intermolecular collision models described in Section 2.1.1.2 could be inaccurate. The shaded area roughly represents the domain of validity for classic gas flow models (**C**).

The relationship with the characteristic length L expressed in micrometers is illustrated in Figure 2.6, which shows the typical ranges covered by fluidic

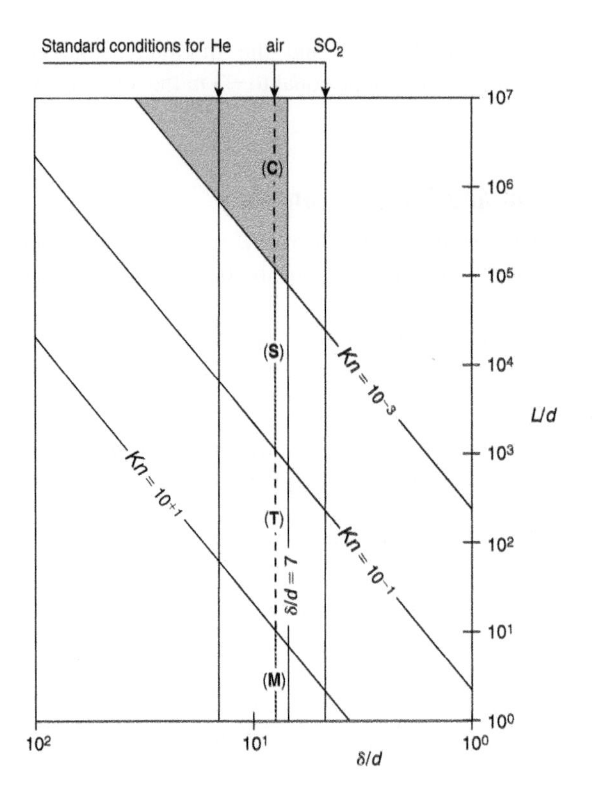

FIGURE 2.5

Gas flow regimes as a function of the Knudsen number.

microsystems described in the literature. Typically, most of the microsystems that use gases work in the slip flow regime, or in the early transition regime. In simple configurations, such flows can be analytically or semi-analytically modeled. The core of the transition regime relates to more specific flows involving lengths less than 100 nm, as is the case for a Couette flow between a hard disk and a read–write head. In that regime, the only theoretical models are molecular models that require numerical simulations. Finally, under the effect of both low pressures and small dimensions, more rarefied regimes can occur, notably inside microsystems dedicated to vacuum generation.

2.1.4 Wall effects

We have seen that initial deviations from the classic continuum models appear when we shrink the characteristic length L of the flow with respect to the boundary conditions. It is logical that the first sources of thermodynamic disequilibria appear

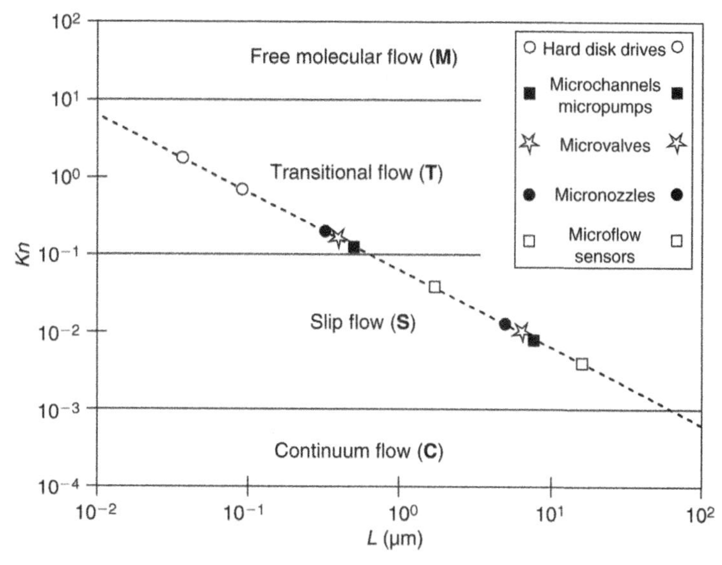

FIGURE 2.6

Gas flow regimes for usual microsystems.

Source: *From Karniadakis and Beskok (2002).*

on the boundary because at the wall there are fewer interactions between the gas molecules than in the core of the flow.

In fact, the walls play a crucial role in gas microflows. First, shrinking system sizes results in an increase of surface over volume effects. It is interesting for heat transfer enhancement, but it also requires thorough knowledge of the velocity, as well as the temperature and boundary conditions. The velocity slip and the temperature jump observed at the wall in the slip flow regime strongly depend on the nature and on the state of the wall, in relation to the nature of the gas. Thus, the roughness of the wall and the chemical affinity between the wall and the gas, although often not properly known, could strongly affect the fluid flow and the heat transfer.

2.2 Gas flow regimes in microchannels

Usually, the *regime* of gas flow in a macrochannel refers to *viscous* and *compressibility* effects, respectively quantified by the Reynolds number Re and the Mach number Ma. For low Reynolds numbers (typically for $Re < 2000$), the flow is *laminar*; it becomes *turbulent* for higher values of Re. For $Ma < 1$, the flow is

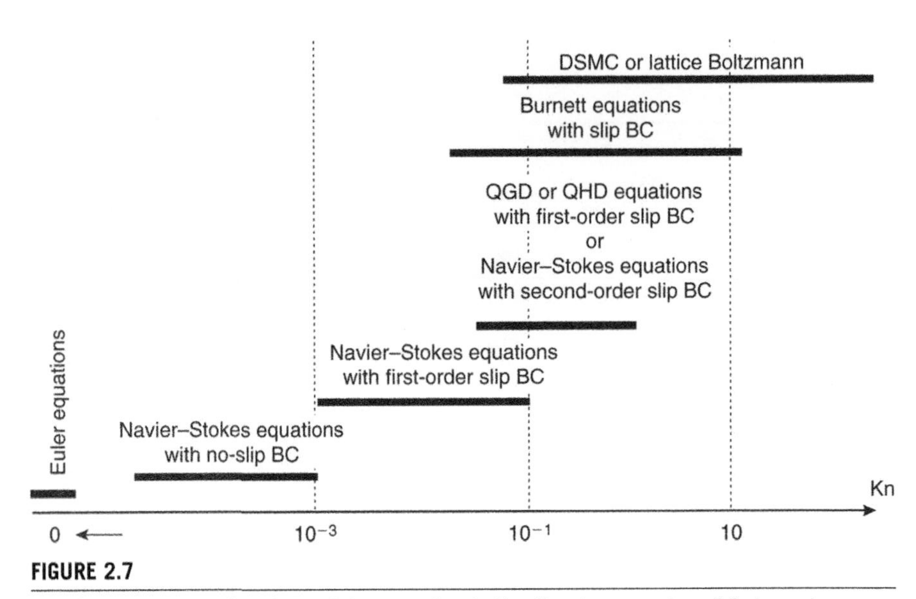

FIGURE 2.7

Gas flow regimes and main models according to the Knudsen number. BC: boundary condition.

subsonic and for $Ma > 1$, it is *supersonic*. Typically, if $Ma > 0.3$, compressibility effects should be taken into account.

In microchannels, on the other hand, the main effects are *rarefaction* effects, essentially quantified by the Knudsen number Kn. Consequently, the *regime* usually refers to these effects, according to the classification given in Section 2.1.3, and flows are considered microflows for $Kn > 10^{-3}$. Actually, Eq. (2.22) shows that for a dilute gas, $Re = k_2\sqrt{\gamma}Ma/Kn$, where $k_2\sqrt{\gamma}$ is of the order of unity, which implies that for subsonic microflows, viscous effects are predominant and the flows are laminar.

The models used to describe gas flows in microchannels depend on the rarefaction regime. The main models are presented in Figure 2.7, with the rough values of Kn for which each of them is applicable, that is, valid and interesting to use. The limits are indicative and may vary according to the configurations under consideration. The Boltzmann equation (see Section 2.2.4) is valid on the whole range of Kn.

These models are described in the following sections. Among the three main rarefied regimes (slip flow, transition flow, and molecular flow), the slip flow regime is more specially detailed, since it is the most frequently encountered (cf. Figure 2.6) and it allows analytic or semi-analytic developments. Moreover, the gas is considered a dilute gas, which makes valid the ideal gas assumption.

2.2.1 Ideal gas model

The classic equation of state for an ideal gas,

$$p = \rho RT = nk_B T \tag{2.24}$$

is also called the Boyle–Mariotte's law. Here, p is the pressure and $\rho = mn$ is the density, $k_B = R_u/N = (M/N)(R_u/M) = mR = 1.381 \times 10^{-23}$ J/K is the Boltzmann constant, $R_u = 8.315$ J/mol/K being the universal gas constant, $N = 6.022 \times 10^{23}$ mol^{-1} the Avogadro's number, and M the molecular weight. Equation (2.24) may be obtained from a molecular analysis, assuming that the gas is a dilute gas in thermodynamic equilibrium.

If the thermodynamic equilibrium is not verified, Eq. (2.24) remains valid, providing the thermodynamic temperature T is replaced with the translational kinetic temperature T_{tr}, which measures the mean translational kinetic energy of the molecules. The specific ideal gas constant

$$R = c_p - c_v \tag{2.25}$$

is the difference between the specific heats

$$c_p = \left.\frac{\partial h}{\partial T}\right|_p \tag{2.26}$$

and

$$c_v = \left.\frac{\partial e}{\partial T}\right|_v \tag{2.27}$$

where e is the internal specific energy, $v = 1/\rho$ is the specific volume, and h is the specific enthalpy. For an ideal gas, the specific heats are constant and their ratio is the index

$$\gamma = \frac{c_p}{c_v} \tag{2.28}$$

Therefore, for an ideal gas,

$$c_v = \frac{1}{\gamma - 1} R; \quad c_p = \frac{\gamma}{\gamma - 1} R \tag{2.29}$$

The specific heat ratio is constant for a monatomic gas ($\gamma = 5/3$) but it is dependent on the temperature for a diatomic gas, with a value close to 7/5 near atmospheric conditions.

Finally, we can note that for an ideal gas and an IPL collision model, the Knudsen number can be written as

$$Kn = \frac{k_2 \mu(T)}{\rho \sqrt{RT} L} = \frac{k_2 \mu(T) \sqrt{RT}}{pL} \tag{2.30}$$

with the value of k_2 given in Table 2.1.

2.2.2 Continuum flow regime

The classic continuum flow regime (which corresponds to the shaded area in Figure 2.5) may be accurately modeled by the *compressible Navier–Stokes equations*, the ideal gas equation of state, and classic boundary conditions that express the continuity of temperature and velocity between the fluid and the wall.

2.2.2.1 Compressible Navier–Stokes equations

The compressible Navier–Stokes equations are the governing conservation laws for mass, momentum, and energy. These laws are written assuming that the fluid is Newtonian, so that the stress tensor

$$\boldsymbol{\sigma}' = -p\mathbf{I} + \boldsymbol{\sigma} \tag{2.31}$$

is a linear function of the velocity gradients. Thus, the viscous stress tensor has the form $\boldsymbol{\sigma} = \mu[\nabla \otimes \mathbf{u} + (\nabla \otimes \mathbf{u})^T] + \zeta(\nabla \cdot \mathbf{u})\mathbf{I}$, where $\mathbf{u}$ is the velocity vector, $\mathbf{I}$ is the unit tensor, μ is the dynamic viscosity, and ζ is the second coefficient of viscosity, or Lamé coefficient. With Stokes' hypothesis, $2\mu + 3\zeta = 0$, which expresses that the changes of volume do not involve viscosity, the viscous stress tensor may be written as $\boldsymbol{\sigma} = \mu[\nabla \otimes \mathbf{u} + (\nabla \otimes \mathbf{u})^T - (2/3)(\nabla \cdot \mathbf{u})\mathbf{I}]$. The validity of this assumption is discussed in Gad-el-Hak (1995).

The Navier–Stokes equations also assume that the fluid follows the Fourier law of diffusion. Thus, the heat flux vector $\mathbf{q}$ is related to the temperature gradient by $\mathbf{q} = -k\nabla T$, where the thermal conductivity k is a function of the temperature.

With the above relations, the compressible Navier–Stokes equations may be written as

$$\frac{\partial \rho}{\partial t} + \nabla \cdot (\rho \mathbf{u}) = 0 \tag{2.32}$$

$$\frac{\partial (\rho \mathbf{u})}{\partial t} + \nabla \cdot \left(\rho \mathbf{u} \otimes \mathbf{u} - \mu \left[\nabla \otimes \mathbf{u} + (\nabla \otimes \mathbf{u})^T - \frac{2}{3}(\nabla \cdot \mathbf{u})\mathbf{I} \right] \right) + \nabla p = \mathbf{f} \tag{2.33}$$

$$\frac{\partial E}{\partial t} + \nabla \cdot \left((E + p)\mathbf{u} - \mu \left[\nabla \otimes \mathbf{u} + (\nabla \otimes \mathbf{u})^T - \frac{2}{3}(\nabla \cdot \mathbf{u})\mathbf{I} \right] \cdot \mathbf{u} - k\nabla T \right) = \mathbf{f} \cdot \mathbf{u} \tag{2.34}$$

These conservation equations are respectively the continuity (2.32), the momentum (2.33), and the energy (2.34) equations. The total energy per unit volume

$$E = \rho\left(\frac{1}{2}\mathbf{u}\cdot\mathbf{u} + e\right) \tag{2.35}$$

is a function of the internal specific energy e that may be written for an ideal gas, using Eqs. (2.24)–(2.26) and (2.28):

$$e = \frac{p}{\rho(\gamma - 1)} \tag{2.36}$$

The external volume forces $\mathbf{f}$ (gravitational forces, magnetic forces, etc.) are generally negligible in gas flow. It is even more the case in gas microflows, since the volume over surface ratio decreases with the characteristic length L and consequently, in microscale geometries, volume effects may be neglected when compared to surface effects.

2.2.2.2 Classic boundary conditions

The previous set of Eqs. (2.24) and (2.32)–(2.34) must be completed with appropriate boundary conditions. In the continuum flow regime, they express the continuity of the velocity,

$$\mathbf{u}|_{\mathrm{w}} = \mathbf{u}_{\mathrm{wall}} \tag{2.37}$$

and of the temperature,

$$T|_{\mathrm{w}} = T_{\mathrm{wall}} \tag{2.38}$$

at the wall, the subscripts "wall" relating to the wall itself, and "w" to the conditions in the fluid at the wall.

2.2.3 Slip flow regime

Actually, whatever the Knudsen number (even for very low values of Kn), there is in the neighborhood of the wall a domain in which the gas is out of equilibrium. This domain is called the *Knudsen layer* and has a thickness in the order of the mean free path. For very low Knudsen numbers (in the continuum flow regime), the effect of the Knudsen layer is negligible. In the slip flow regime, which is roughly in the range $10^{-3} < Kn < 10^{-1}$, the Knudsen layer must be taken into account. Actually, the flow in the Knudsen layer cannot be captured by continuum models based on the Navier–Stokes equations, for example (Dongari et al., 2011; Zhang et al., 2006). But for $Kn < 10^{-1}$, its thickness is small enough and the Knudsen layer can be neglected, providing the boundary conditions (2.37)–(2.38) are modified and express a velocity slip, as well as a temperature jump at the wall (Section 2.2.3.2).

Thus, the Navier–Stokes (NS) equations remain applicable, but it could also be convenient to replace them by another set of conservation equations, such as the quasi-hydrodynamic (QHD) equations or the quasi-gasdynamic (QGD) equations (see Section 2.2.3.1).

2.2.3.1 Continuum NS–QGD–QHD equations

In generic form, the conservation equations may be written in the following way:

$$\frac{\partial \rho}{\partial t} + \nabla \cdot \mathbf{J}_{\mathrm{EQ}} = \mathbf{0} \tag{2.39}$$

for the continuity equation,

$$\frac{\partial (\rho \mathbf{u})}{\partial t} + \nabla \cdot (\mathbf{J}_{\mathrm{EQ}} \otimes \mathbf{u} - \boldsymbol{\sigma}_{\mathrm{EQ}}) + \nabla p = 0 \tag{2.40}$$

for the momentum equation, and

$$\frac{\partial E}{\partial t} + \nabla \cdot \left(\frac{E+p}{\rho} \mathbf{J}_{\mathrm{EQ}} - \boldsymbol{\sigma}_{\mathrm{EQ}} \cdot \mathbf{u} + \mathbf{q}_{\mathrm{EQ}} \right) = 0 \tag{2.41}$$

for the energy equation. In these equations, $\mathbf{J}_{\mathrm{EQ}}$ is the mass flux vector, $\boldsymbol{\sigma}_{\mathrm{EQ}}$ the viscous stress tensor, and $\mathbf{q}_{\mathrm{EQ}}$, the heat flux vector. The external volume forces $\mathbf{f}$ have been neglected.

In the case of the Navier–Stokes equations (EQ≡NS), these variables have the form:

$$\mathbf{J}_{\mathrm{NS}} = \rho \mathbf{u} \tag{2.42}$$

$$\boldsymbol{\sigma}_{\mathrm{NS}} = \mu\left[\nabla \otimes \mathbf{u} + (\nabla \otimes \mathbf{u})^T\right] + \zeta(\nabla \cdot \mathbf{u})\mathbf{I} = \mu\left[\nabla \otimes \mathbf{u} + (\nabla \otimes \mathbf{u})^T - \frac{2}{3}(\nabla \cdot \mathbf{u})\mathbf{I}\right] \tag{2.43}$$

$$\mathbf{q}_{\mathrm{NS}} = -k\nabla T \tag{2.44}$$

These lead to the previous Eqs. (2.32)–(2.34). In the Navier–Stokes equations, the density ρ, the pressure p, and the velocity $\mathbf{u}$ correspond to space averaged instantaneous quantities.

From spatiotemporal considerations, the QGD and the QHD equations may be obtained (Lengrand and Elizarova, 2010). They are based on the same governing conservation laws (2.39)–(2.41), but these laws are closed with a different definition of $\mathbf{J}_{\mathrm{EQ}}$, $\boldsymbol{\sigma}_{\mathrm{EQ}}$, and $\mathbf{q}_{\mathrm{EQ}}$. Dissipative terms are added, which involve a parameter τ. This parameter

$$\tau = \frac{\gamma \mu}{Sc\, p a^2} \tag{2.45}$$

is a time characteristic of the temporal averaging. Here, $Sc = \mu/(\rho D)$ is the Schmidt number, which represents the ratio of the diffusive mass over the diffusive momentum fluxes. In the expression of Sc, D is the diffusion coefficient, which can be expressed as a function of α and ω for a VSS molecule. Thus, for a VSS molecule,

$$Sc = \frac{5(2+\alpha)}{3(7-2\omega)\alpha} \tag{2.46}$$

which reduces to 5/6 for a HS model. For an ideal gas, $a^2 = \gamma RT$, and τ may be written as

$$\tau = \frac{1}{Sc}\frac{\mu}{p} \tag{2.47}$$

The QGD equations are appropriate for dilute gases. For these equations (EQ≡QGD),

$$\mathbf{J}_{\mathrm{QGD}} = \mathbf{J}_{\mathrm{NS}} - \tau[\nabla \cdot (\rho\mathbf{u} \otimes \mathbf{u}) + \nabla p] \tag{2.48}$$

$$\boldsymbol{\sigma}_{\mathrm{QGD}} = \boldsymbol{\sigma}_{\mathrm{NS}} + \tau[\mathbf{u} \otimes (\rho\mathbf{u} \cdot (\nabla \otimes \mathbf{u}) + \nabla p) + \mathbf{I}(\mathbf{u} \cdot \nabla p + \gamma p \nabla \cdot \mathbf{u})] \tag{2.49}$$

$$\mathbf{q}_{\mathrm{QGD}} = \mathbf{q}_{\mathrm{NS}} - \tau\rho\mathbf{u}[\mathbf{u} \cdot \nabla e + p\mathbf{u} \cdot \nabla(1/\rho)] \tag{2.50}$$

From Eqs. (2.18), (2.19), (2.24), and (2.47), we can deduce that for a dilute gas and an IPL collision model,

$$\tau = \frac{Kn}{Sc}\frac{1}{k_2}\frac{L}{\sqrt{RT}} \tag{2.51}$$

Therefore, $[\tau\nabla]_{(Kn\to 0)} \to 0$, which shows that the above QGD equations degenerate into the classic NS equations when rarefaction is negligible.

Another set of equations, the QHD equations, are appropriate to dense gases. In that case (EQ≡QHD),

$$\mathbf{J}_{\mathrm{QHD}} = \mathbf{J}_{\mathrm{NS}} - \tau[\rho\mathbf{u} \cdot (\nabla \otimes \mathbf{u}) + \nabla p] \tag{2.52}$$

$$\boldsymbol{\sigma}_{\mathrm{QHD}} = \boldsymbol{\sigma}_{\mathrm{NS}} + \tau\mathbf{u} \otimes [\rho\mathbf{u} \cdot (\nabla \otimes \mathbf{u}) + \nabla p] \tag{2.53}$$

$$\mathbf{q}_{\mathrm{QHD}} = \mathbf{q}_{\mathrm{NS}} \tag{2.54}$$

They also bring additional terms to the NS equations, which are proportional to the parameter τ.

2.2.3.2 First-order slip boundary conditions

From a series of experiments at low pressures performed in 1875, Kundt and Warburg were probably the first who pointed out that for a flow of rarefied gas, slipping occurs at the walls (Kennard, 1938). For a gas flowing in the direction s parallel to the wall, the slip velocity may be written as

$$u_{\text{slip}} = u_{\text{s}} - u_{\text{wall}} = \xi \frac{\partial u_{\text{s}}}{\partial n}\bigg|_{\text{w}} \tag{2.55}$$

where ξ is a length commonly called the *coefficient of slip* and n is the normal direction, exiting the wall. With appropriate assumptions, it is possible to calculate the magnitude of the coefficient of slip from the kinetic theory of gases. The slip velocity u_{slip} is not the actual relative velocity ($u_{\text{gas}} - u_{\text{wall}}$) at the wall, but the fictitious velocity required to correctly predict the velocity profile out of the Knudsen layer, as illustrated by Figure 2.8.

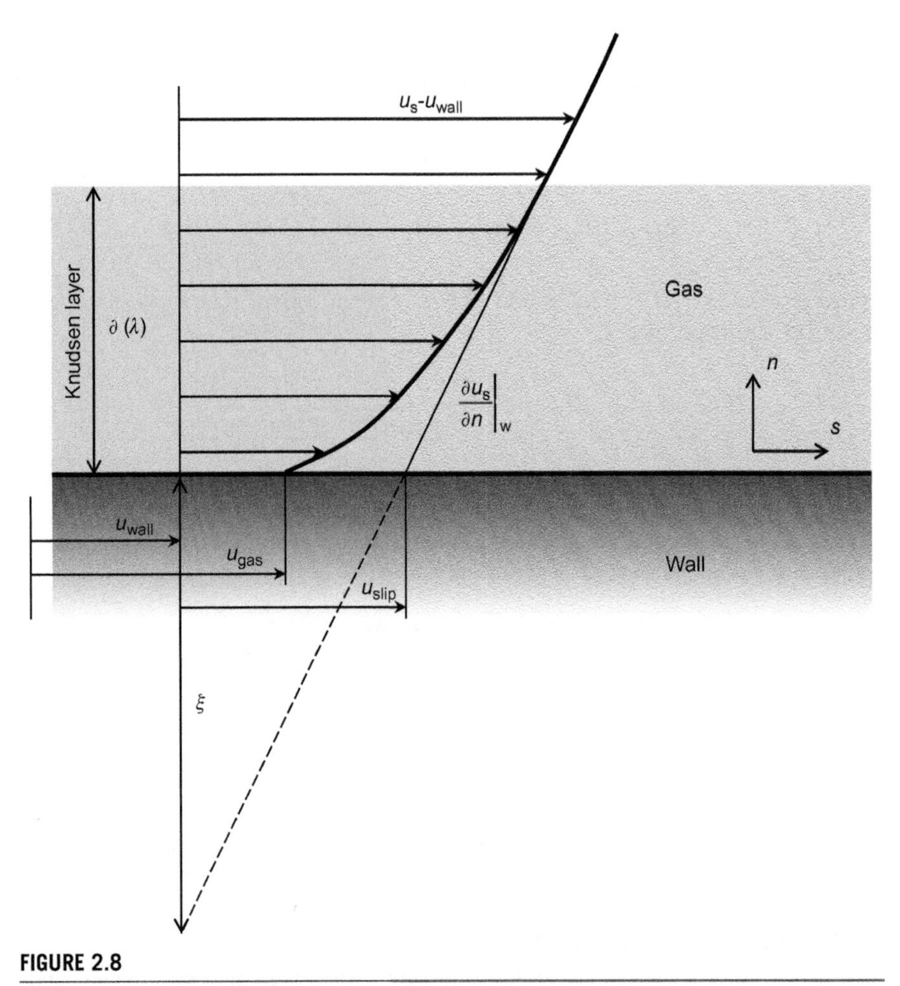

FIGURE 2.8

Structure of the Knudsen layer.

Following a paper initially dealing with stresses in rarefied gases due to temperature gradients (Maxwell, 1879), Maxwell added an appendix on the advice of a referee, with the aim of expressing the conditions which must be satisfied by a gas in contact with a solid body. Maxwell treated the surface as something intermediate between a perfectly reflecting and a perfectly absorbing surface. Therefore, he assumed that for every unit of area, a fraction σ of the molecules is absorbed by the surface (due to the roughness of the wall, or to a condensation–evaporation process; Kennard, 1938) and is afterwards reemitted with velocities corresponding to those in still gas at the temperature of the wall. The other fraction, $1-\sigma$, of the molecules is perfectly reflected by the wall (Figure 2.9). The dimensionless coefficient σ is called the *tangential momentum accommodation coefficient*. When $\sigma = 0$, the tangential momentum of the incident molecules equals that of the reflected molecules and no momentum is transmitted to the wall, as if the flow was inviscid. This kind of reflection is called *specular reflection*. Conversely, when $\sigma = 1$, the gas molecules transmit all their tangential momentum to the wall and the reflection is a *diffuse reflection*.

From a momentum balance at the wall, Maxwell finally demonstrated that there is a slip at the wall, which takes the form:

$$u_{\text{slip}} = u_{\text{s}} - u_{\text{wall}} = \frac{2-\sigma}{\sigma}\lambda_{\text{M}}\left[\frac{\partial u_{\text{s}}}{\partial n} - \frac{3}{2}\frac{\mu}{\rho T}\frac{\partial^2 T}{\partial s\,\partial n}\right]_{\text{w}} + \frac{3}{4}\left[\frac{\mu}{\rho T}\frac{\partial T}{\partial s}\right]_{\text{w}} \tag{2.56}$$

with

$$\lambda_{\text{M}} = \sqrt{\frac{\pi}{2}}\frac{\mu}{\sqrt{p\rho}} = \sqrt{\frac{\pi}{2}}\frac{\mu}{\rho\sqrt{RT}} \tag{2.57}$$

which is very close to the value of the mean free path λ_{HS} deduced from a HS model.

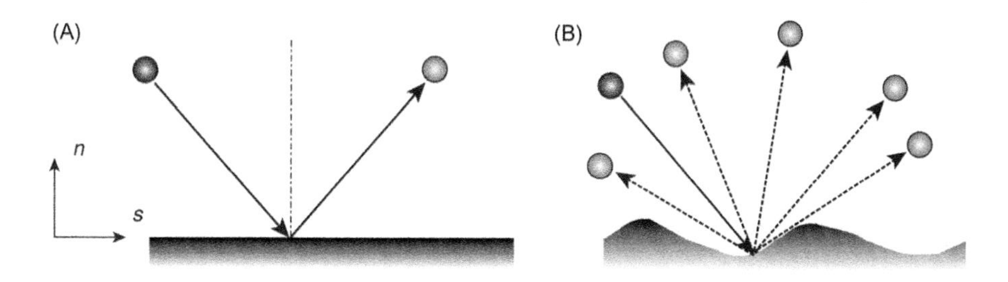

FIGURE 2.9

Maxwell hypothesis: (A) perfect (specular) reflection of a fraction $(1-\sigma)$ of the molecules and (B) absorption and diffuse reemission of a fraction σ of the molecules.

The equation of Maxwell is generally cited in the literature with the form

$$u_{\text{slip}} = u_{\text{s}} - u_{\text{wall}} = \frac{2-\sigma}{\sigma}\lambda\left.\frac{\partial u_{\text{s}}}{\partial n}\right|_{\text{w}} + \frac{3}{4}\frac{\mu}{\rho T}\left.\frac{\partial T}{\partial s}\right|_{\text{w}}$$
$$= \frac{2-\sigma}{\sigma}\lambda\left.\frac{\partial u_{\text{s}}}{\partial n}\right|_{\text{w}} + \frac{3}{4}\frac{\lambda}{k_2}\left.\sqrt{\frac{R}{T}}\frac{\partial T}{\partial s}\right|_{\text{w}} \tag{2.58}$$

which omits the second term on the right-hand side of Eq. (2.56).

After nondimensionalization with the characteristic length L, a reference velocity u_0, and a reference temperature T_0, Eq. (2.58) is written as follows:

$$u_{\text{s}}^* - u_{\text{wall}}^* = \frac{2-\sigma}{\sigma}Kn\left.\frac{\partial u_{\text{s}}^*}{\partial n^*}\right|_{\text{w}} + \frac{3}{4\gamma k_2^2}\frac{Kn^2 Re_0}{Ma_0^2}\left.\frac{\partial T^*}{\partial s^*}\right|_{\text{w}} \tag{2.59}$$

where $u_{\text{s}}^* = u_{\text{s}}/u_0$, $T^* = T/T_0$, $n^* = n/L$, and $s^* = s/L$. Note that this nondimensionalization is valid only if the reference lengths are the same in the s and n directions, which is generally not the case for a flow in a microchannel. In that case, it is preferable to keep the dimensional form of Eq. (2.58). The second term on the right-hand side, which is proportional to the square of the Knudsen number $Kn = \lambda/L$, involves the Reynolds number $Re_0 = \rho u_0 L/\mu$ and the Mach number $Ma_0 = u_0/\sqrt{\gamma R T_0}$ defined at the reference conditions. Coefficient k_2 depends on the intermolecular collision model; its value is given in Table 2.1. This second term is associated with the thermal creep or transpiration phenomenon. It shows that a flow can be caused in the sole presence of a tangential temperature gradient, without any pressure gradient. In that case, the gas experiences a slip velocity and moves from the colder toward the warmer region. This can also result in pressure variation in a microchannel only submitted to tangential temperature gradients.

The nondimensional form of Eq. (2.59) clearly shows that slip is negligible for very low Knudsen numbers, but should be taken into account when Kn is no longer small, typically as soon as it becomes higher than 10^{-3}.

By analogy with the slip phenomenon, Poisson suggested that there might also be a temperature jump at the wall, which could be described by an equation equivalent to

$$T - T_{\text{wall}} = \varsigma\left.\frac{\partial T}{\partial n}\right|_{\text{w}} \tag{2.60}$$

where ς is called the *temperature jump distance*. Smoluchowski experimentally confirmed this hypothesis (Smoluchowski, 1898) and showed that ς was proportional to the mean free path λ. As in the case of slip, it can be assumed that a fraction σ_T of the molecules have a long contact with the wall and that the wall adjusts their mean thermal energy. Therefore, these molecules are reemitted as if they were issuing from a gas at the temperature of the wall. The other fraction, $1-\sigma_T$, is reflected keeping its incident thermal energy. The dimensionless coefficient σ_T is called the *energy accommodation coefficient*. From an energy balance at the wall, it can be shown that (Kennard, 1938)

$$T - T_{\text{wall}} = \frac{2-\sigma_T}{\sigma_T} \frac{2\gamma}{\gamma+1} \frac{k}{\mu c_{\text{p}}} \lambda \frac{\partial T}{\partial n}\bigg|_{\text{w}} \tag{2.61}$$

After nondimensionalization with the characteristic length L and the reference temperature T_0, Eq. (2.61) is written as follows:

$$T^* - T^*_{\text{wall}} = \frac{2-\sigma_T}{\sigma_T} \frac{2\gamma}{\gamma+1} \frac{Kn}{Pr} \frac{\partial T}{\partial n^*}\bigg|_{\text{w}} \tag{2.62}$$

where

$$Pr = \frac{\nu \rho c_{\text{p}}}{k} = \frac{\mu c_{\text{p}}}{k} \tag{2.63}$$

is the Prandtl number.

In the absence of tangential temperature gradients ($\partial T^*/\partial s^* = 0$), the boundary conditions (2.59) and (2.62) are called first-order (i.e., $\vartheta(Kn)$) boundary conditions. The velocity slip (respectively the temperature jump) is then proportional to the normal velocity (respectively temperature) gradient and to the Knudsen number Kn. These boundary conditions are classically associated with the Navier–Stokes equations, but they can also lead to interesting results when used with the QGD or QHD equations.

Equations (2.58) and (2.61) are extensively used in the literature. In case of wall curvature, however, Eq. (2.58) should be extended as

$$u_{\text{s}} - u_{\text{wall}} = \beta_{u1} \frac{2-\sigma}{\sigma} \lambda \left(\frac{\partial u_{\text{s}}}{\partial n} + \frac{\partial u_{\text{n}}}{\partial s} \right)\bigg|_{\text{w}} + \frac{3}{4} \frac{\mu}{\rho T} \frac{\partial T}{\partial s}\bigg|_{\text{w}} \tag{2.64}$$

since the first term of the RHS derives from the tangential shear stress at the wall, which also depends on the derivative $\partial u_n/\partial s$ of the normal velocity (Barber et al., 2004; Lockerby et al., 2004). This derivative should also be taken into account in microchannels with significant roughness inducing two components of the velocity in the vicinity of the wall. In addition, numerical simulations (Cercignani et al., 1994) have shown that a corrective coefficient β_{u1} should be added for better prediction of the flow out of the Knudsen layer. This coefficient is slightly different from unity, unlike initially proposed by Maxwell. Its value depends on the expression of λ, which can be calculated with various assumptions (see Table 2.1), as well as on the numerical model used for the simulation. A review of the different values of β_{u1} proposed in the literature can be found in Zhang et al. (2012).

Similarly, the accuracy of Eqs. (2.61) and (2.62) can be improved, adding a corrective coefficient β_{T1}.

2.2.3.3 Higher-order slip boundary conditions

From a theoretical point of view, the slip flow regime is particularly interesting because it generally leads to analytical or semi-analytical models (cf. Section 2.3). For example, these analytical models allow us to calculate velocities and flow rates for isothermal and locally fully developed microflows between plane plates or in cylindrical ducts with simple sections: circular (Kennard, 1938), annular (Ebert and Sparrow, 1965), and rectangular (Ebert and Sparrow, 1965; Morini and Spiga, 1998). These models proved to be quite precise for moderate Knudsen numbers, typically up to about 0.1 (Arkilic et al., 2001; Harley et al., 1995; Liu et al., 1995; Shih et al., 1996).

For $Kn > 0.1$, experimental studies (Sreekanth, 1969) or numerical studies (Piekos and Breuer, 1996) using the DSMC method show significant deviations with models based on first-order boundary conditions. Therefore, since 1947, several authors have proposed second-order boundary conditions, hoping to extend the validity of the slip flow regime to higher Knudsen numbers. Second-order boundary conditions take more or less complicated forms, which are difficult to group together in a single equation. Actually, according to the assumptions, the second-order terms ($\vartheta(Kn^2)$) may be dependent on σ (Chapman and Cowling, 1952; Karniadakis and Beskok, 2002) and involve tangential second derivatives $\partial^2 u_s^* / \partial s^{*2}$ (Deissler, 1964). In the simple case of a developed flow between plane plates, the tangential second derivatives are zero and one may compare most of the second-order models that take the generic form

$$u_s^* - u_{\text{wall}}^* = A_1 Kn \frac{\partial u_s^*}{\partial n^*} + A_2 Kn^2 \frac{\partial^2 u_s^*}{\partial n^{*2}} \tag{2.65}$$

In the particular case of a fully diffuse reflection ($\sigma = 1$), some examples of coefficients A_1 and A_2 proposed in the literature are compared in Table 2.3. In this table, the Knudsen number is based on the expression λ_M of the mean free path provided by Maxwell and given by Eq. (2.57). We note significant differences, essentially for the second-order term, depending on the assumptions. Moreover, some models based on a simple mathematical extension of Maxwell's condition predict a decrease of the slip compared to the first-order model, while other models predict an increase in the slip. This point is discussed in Section 2.3. Recent papers (Cercignani and Lorenzani, 2010; Lorenzani, 2011) also provide accurate values of A_1 and A_2 according to the actual value of the accommodation coefficient σ.

Implementation of the boundary conditions (2.65) with the Navier–Stokes equations is not always easy and can lead to computational difficulties. The obvious interest in using higher-order boundary conditions has led some authors to propose new conditions, which may be more appropriate for a treatment with

Table 2.3 Coefficients of the Main Models of Second-Order Boundary Conditions Proposed in the Literature, for $\sigma = 1$

Author, Year (reference)	A_1	A_2
Maxwell (Maxwell, 1879)	1	0
Schamberg, 1947 (Karniadakis and Beskok, 2002)	1	$-5\pi/12$
Chapman and Cowling (Chapman and Cowling, 1952)	κ_0 (≈ 1)	$\kappa_0^2/2$($\approx 1/2$)
Deissler (Deissler, 1964)	1	−9/8
Cercignani, 1964 (Sreekanth, 1969)	1.1466	−0.9756
Hsia and Domoto (Hsia and Domoto, 1983)	1	−1/2
Mitsuya (Mitsuya, 1993)	1	−2/9
Karniadakis and Beskok (Karniadakis and Beskok, 2002)	1	1/2
Hadjiconstantinou (Hadjiconstantinou, 2003)	1.11	−0.61
Pitakarnnop et al. (Pitakarnnop et al., 2010)	1.1466	−0.647
Cercignani and Lorenzani, (Cercignani and Lorenzani, 2010)	1.1209	−0.2347
Lorenzani (Lorenzani, 2011)	1.1366	−0.69261

the Navier–Stokes equations. Thus, Beskok and Karniadakis (1999) suggested a high-order form

$$u_s^* - u_{wall}^* = \frac{2-\sigma}{\sigma}\,\frac{Kn}{1 - A_3 Kn}\,\frac{\partial u_s^*}{\partial n^*} \tag{2.66}$$

that involves the sole first derivative of the velocity and an empirical parameter $A_3(Kn)$. As for Xue and Fan (2000), they proposed

$$u_s^* - u_{wall}^* = \frac{2-\sigma}{\sigma}\tanh(Kn)\frac{\partial u_s^*}{\partial n^*} \tag{2.67}$$

which leads to results close to that calculated by the DSMC method, up to high Knudsen numbers, of the order of 3. Other hybrid boundary conditions, such as

$$u_s^* - u_{wall}^* = \frac{2-\sigma}{\sigma}\left[Kn\frac{\partial u_s^*}{\partial n^*} + \frac{Kn}{2}\left(Re\frac{\partial p^*}{\partial s^*}\right)\right] \tag{2.68}$$

(Jie et al., 2000), were also proposed in order to allow a more stable numerical solution, while giving results comparable to that from Eq. (2.66) for the cases that were tested.

The above first-order and higher-order boundary conditions only give a partial idea of the numerous formulations currently published. A review of the different slip boundary conditions proposed in the literature is available in Zhang et al. (2012). More in-depth discussion can be found in Sharipov (2011), which includes analysis of temperature jump boundary conditions.

2.2.3.4 Accommodation coefficients

The accommodation coefficients depend on various parameters that affect surface interaction, such as the magnitude and the direction of the velocity. Fortunately, it seems that these coefficients are reasonably constant for a given gas and surface combination (Schaaf, 1963). Many measurements of σ_T and σ have been made (Agrawal and Prabhu, 2008), most of them from indirect macroscopic measurements, though a few of them are from direct measurements using molecular beam techniques. Schaaf (1963) reported values of σ ranging from 0.60 to 1.00, most of them being between 0.85 and 1.00. Recently, good correlations for microchannel flows were found in the slip flow regime, with values of σ between 0.80 and 1.00 (Section 2.3.5.2). Schaaf also reported values of σ_T in a wide range, from 0.0109 to 0.990.

2.2.4 Transition flow and free molecular flow

Typically, for Knudsen numbers higher than unity, the continuum Navier–Stokes or QGD/QHD equations become invalid. However, in the early transition regime (for $Kn \sim 1$), a slip flow model may still be used provided the continuum model is corrected by replacing the Navier–Stokes equations with the Burnett equations. For higher Knudsen numbers, in the full transition and in the free molecular regimes, the Boltzmann equation must be directly treated by appropriate numerical techniques such as the DSMC or the lattice Boltzmann methods (LBM).

2.2.4.1 Burnett equations

Higher-order fluid dynamic models, also called extended hydrodynamic equations (EHE), may be obtained from the conservation Eqs. (2.39)–(2.41), deriving the form of the viscous stress tensor $\boldsymbol{\sigma}_{\mathrm{EQ}}$ and of the heat flux vector $\mathbf{q}_{\mathrm{EQ}}$ from a molecular approach and the Boltzmann equation.

In a molecular description with the assumption of molecular chaos, a flow is entirely described if, for each time t and each position $\mathbf{x}$, we know the number density n, the velocity, and the internal energy distribution functions (Lengrand and Elizarova, 2010). For example, if we consider a monatomic gas, the velocity distribution function $f(t, \mathbf{x}, \mathbf{u})$ represents the number of particles in the six-dimensional phase space $d\mathbf{x}\, d\mathbf{u}$ at time t. It verifies the Boltzmann equation

$$\frac{\partial f}{\partial t} + \mathbf{u} \cdot \frac{\partial f}{\partial \mathbf{x}} + \mathbf{F} \cdot \frac{\partial f}{\partial \mathbf{u}} = Q(f, f^*) \tag{2.69}$$

which is valid in the entire Knudsen regime, that is, for $0 \leq Kn < \infty$, since it does not require a local thermodynamic equilibrium hypothesis. In Eq. (2.69), $\mathbf{F}$ is an external body force per unit mass and $Q(f, f^*)$ represents the intermolecular collisions. From a Chapman–Enskog expansion of the Boltzmann equation with the Knudsen number as a small parameter (i.e., $f = f_0 + Kn\, f_1 + Kn^2 f_2 + \ldots$), the form of the viscous stress tensor and of the heat flux vector may be obtained as follows:

$$\boldsymbol{\sigma}_{\mathrm{EQ}} = \boldsymbol{\sigma}_{\mathrm{EQ}}^{(0)} + \boldsymbol{\sigma}_{\mathrm{EQ}}^{(1)} + \boldsymbol{\sigma}_{\mathrm{EQ}}^{(2)} + \boldsymbol{\sigma}_{\mathrm{EQ}}^{(3)} + \cdots + \boldsymbol{\sigma}_{\mathrm{EQ}}^{(i)} + \vartheta(Kn^{i+1}) \tag{2.70}$$

$$\mathbf{q}_{\mathrm{EQ}} = \mathbf{q}_{\mathrm{EQ}}^{(0)} + \mathbf{q}_{\mathrm{EQ}}^{(1)} + \mathbf{q}_{\mathrm{EQ}}^{(2)} + \mathbf{q}_{\mathrm{EQ}}^{(3)} + \cdots + \mathbf{q}_{\mathrm{EQ}}^{(i)} + \vartheta(Kn^{i+1}) \tag{2.71}$$

In the EHE, the mass flux vector $\mathbf{J}_{\mathrm{EQ}} = \mathbf{J}_{\mathrm{NS}} = \rho\mathbf{u}$ is the same as for the Navier–Stokes equations, but according to the number of terms kept in Eqs. (2.70) and (2.71), we obtain the Euler, Navier–Stokes, Burnett, or super-Burnett equations.

Keeping only the first term in Eqs. (2.70) and (2.71), we obtain the zeroth-order approximation ($i = 0$) which results in the Euler (E) equations, with

$$\boldsymbol{\sigma}_{\mathrm{E}} = \boldsymbol{\sigma}_{\mathrm{EQ}}^{(0)} = \mathbf{0} \tag{2.72}$$

$$\mathbf{q}_{\mathrm{E}} = \mathbf{q}_{\mathrm{EQ}}^{(0)} = \mathbf{0} \tag{2.73}$$

The Euler equations correspond to an inviscid flow ($\mu = 0$) for which heat losses by thermal diffusion are neglected ($k = 0$). This system of equations is a hyperbolic system that can describe subsonic or supersonic flows, with the transition for $Ma = 1$ being modeled by a shock wave, which represents a discontinuity of the flow variables. The Euler equations are widely used for high-speed flows in macrosystems such as nozzles and ejectors (Anderson, 1990).

With the second terms in Eqs. (2.70) and (2.71), that is, for a first-order approximation, we obtain the compressible NS equations, with

$$\boldsymbol{\sigma}_{\mathrm{NS}} = \boldsymbol{\sigma}_{\mathrm{EQ}}^{(0)} + \boldsymbol{\sigma}_{\mathrm{EQ}}^{(1)} = \mathbf{0} + \mu\left[\nabla\otimes\mathbf{u} + (\nabla\otimes\mathbf{u})^{T} - \frac{2}{3}(\nabla\cdot\mathbf{u})\mathbf{I}\right] \tag{2.74}$$

$$\mathbf{q}_{\mathrm{NS}} = \mathbf{q}_{\mathrm{EQ}}^{(0)} + \mathbf{q}_{\mathrm{EQ}}^{(1)} = \mathbf{0} - k\nabla T \tag{2.75}$$

These classic equations are detailed in Section 2.2.2.1. According to the associated boundary conditions, they can be valid up to Knudsen numbers on the order of 10^{-1} or 1 (cf. Figure 2.7). In the case of a plane flow in the (x, y) plane, Eq. (2.74) yields

$$\begin{aligned} \sigma_{xx}^{(1)} &= \mu\left(\delta_1\frac{\partial u_x}{\partial x} + \delta_2\frac{\partial u_y}{\partial y}\right) \\ \sigma_{xy}^{(1)} &= \sigma_{yx}^{(1)} = \mu\left(\frac{\partial u_x}{\partial y} + \frac{\partial u_y}{\partial x}\right) \\ \sigma_{yy}^{(1)} &= \mu\left(\delta_1\frac{\partial u_y}{\partial y} + \delta_2\frac{\partial u_x}{\partial x}\right) \end{aligned} \tag{2.76}$$

with $\delta_1 = 4/3$ and $\delta_2 = -2/3$.

As the Knudsen number becomes higher, additional higher-order terms from Eqs. (2.70) and (2.71) are required. The Burnett (B) equations correspond to a second-order approximation, with

$$\boldsymbol{\sigma}_B = \boldsymbol{\sigma}_{\mathrm{EQ}}^{(0)} + \boldsymbol{\sigma}_{\mathrm{EQ}}^{(1)} + \boldsymbol{\sigma}_{\mathrm{EQ}}^{(2)} \tag{2.77}$$

$$\mathbf{q}_B = \mathbf{q}_{\mathrm{EQ}}^{(0)} + \mathbf{q}_{\mathrm{EQ}}^{(1)} + \mathbf{q}_{\mathrm{EQ}}^{(2)} \tag{2.78}$$

These new expressions for stress and heat-flux terms dramatically complicate the EHE systems. Yun et al. (1998) provide the expressions of these terms in the case of a plane flow:

$$\begin{aligned}
\sigma_{xx}^{(2)} = \frac{u^2}{p}\Bigg[& \alpha_1\left(\frac{\partial u_x}{\partial x}\right)^2 + \alpha_2\frac{\partial u_x}{\partial x}\frac{\partial u_y}{\partial y} + \alpha_3\left(\frac{\partial u_y}{\partial y}\right)^2 + \alpha_4\frac{\partial u_x}{\partial y}\frac{\partial u_y}{\partial x} + \alpha_5\left(\frac{\partial u_x}{\partial y}\right)^2 \\
& + \alpha_6\left(\frac{\partial u_y}{\partial x}\right)^2 + \alpha_7 R\frac{\partial^2 T}{\partial x^2} + \alpha_8 R\frac{\partial^2 T}{\partial y^2} + \alpha_9\frac{RT}{\rho}\frac{\partial^2 \rho}{\partial x^2} + \alpha_{10}\frac{RT}{\rho}\frac{\partial^2 \rho}{\partial y^2} \\
& + \alpha_{11}\frac{RT}{\rho^2}\left(\frac{\partial \rho}{\partial x}\right)^2 + \alpha_{12}\frac{R}{\rho}\frac{\partial T}{\partial x}\frac{\partial \rho}{\partial x} + \alpha_{13}\frac{R}{T}\left(\frac{\partial T}{\partial x}\right)^2 + \alpha_{14}\frac{RT}{\rho^2}\left(\frac{\partial \rho}{\partial y}\right)^2 \\
& + \alpha_{15}\frac{R}{\rho}\frac{\partial T}{\partial y}\frac{\partial \rho}{\partial y} + \alpha_{16}\frac{R}{T}\left(\frac{\partial T}{\partial y}\right)^2 \Bigg]
\end{aligned}$$

$$\begin{aligned}
\sigma_{yy}^{(2)} = \frac{u^2}{p}\Bigg[& \alpha_1\left(\frac{\partial u_y}{\partial y}\right)^2 + \alpha_2\frac{\partial u_x}{\partial x}\frac{\partial u_y}{\partial y} + \alpha_3\left(\frac{\partial u_x}{\partial x}\right)^2 + \alpha_4\frac{\partial u_x}{\partial y}\frac{\partial u_y}{\partial x} + \alpha_5\left(\frac{\partial u_y}{\partial x}\right)^2 \\
& + \alpha_6\left(\frac{\partial u_x}{\partial y}\right)^2 + \alpha_7 R\frac{\partial^2 T}{\partial y^2} + \alpha_8 R\frac{\partial^2 T}{\partial x^2} + \alpha_9\frac{RT}{\rho}\frac{\partial^2 \rho}{\partial y^2} + \alpha_{10}\frac{RT}{\rho}\frac{\partial^2 \rho}{\partial x^2} \\
& + \alpha_{11}\frac{RT}{\rho^2}\left(\frac{\partial \rho}{\partial y}\right)^2 + \alpha_{12}\frac{R}{\rho}\frac{\partial T}{\partial y}\frac{\partial \rho}{\partial y} + \alpha_{13}\frac{R}{T}\left(\frac{\partial T}{\partial y}\right)^2 + \alpha_{14}\frac{RT}{\rho^2}\left(\frac{\partial \rho}{\partial x}\right)^2 \\
& + \alpha_{15}\frac{R}{\rho}\frac{\partial T}{\partial x}\frac{\partial \rho}{\partial x} + \alpha_{16}\frac{R}{T}\left(\frac{\partial T}{\partial x}\right)^2 \Bigg]
\end{aligned} \tag{2.79}$$

$$\begin{aligned}
\sigma_{xy}^{(2)} = \sigma_{yx}^{(2)} = \frac{\mu^2}{p}\Bigg[& \beta_1\left(\frac{\partial u_x}{\partial x}\frac{\partial u_x}{\partial y} + \frac{\partial u_y}{\partial x}\frac{\partial u_y}{\partial y}\right) + \beta_2\left(\frac{\partial u_x}{\partial y}\frac{\partial u_y}{\partial y} + \frac{\partial u_x}{\partial x}\frac{\partial u_y}{\partial x}\right) \\
& + \beta_3 R\frac{\partial^2 T}{\partial x \partial y} + \beta_4\frac{RT}{\rho}\frac{\partial^2 \rho}{\partial x \partial y} + \beta_5\frac{R}{T}\frac{\partial T}{\partial x}\frac{\partial T}{\partial y} + \beta_6\frac{RT}{\rho^2}\frac{\partial \rho}{\partial x}\frac{\partial \rho}{\partial y} \\
& + \beta_7\frac{R}{\rho}\left(\frac{\partial T}{\partial x}\frac{\partial \rho}{\partial y} + \frac{\partial T}{\partial y}\frac{\partial \rho}{\partial x}\right) \Bigg]
\end{aligned}$$

$$q_x^{(2)} = \frac{\mu^2}{\rho}\left[\gamma_1 \frac{1}{T}\frac{\partial T}{\partial x}\frac{\partial u_x}{\partial x} + \gamma_2 \frac{1}{T}\frac{\partial T}{\partial x}\frac{\partial u_y}{\partial y} + \gamma_3 \frac{\partial^2 u_x}{\partial x^2} + \gamma_4 \frac{\partial^2 u_x}{\partial y^2} + \gamma_5 \frac{\partial^2 u_y}{\partial x \partial y} + \gamma_6 \frac{1}{T}\frac{\partial T}{\partial y}\frac{\partial u_y}{\partial x}\right.$$
$$\left. + \gamma_7 \frac{1}{T}\frac{\partial T}{\partial y}\frac{\partial u_x}{\partial y} + \gamma_8 \frac{1}{\rho}\frac{\partial \rho}{\partial x}\frac{\partial u_x}{\partial x} + \gamma_9 \frac{1}{\rho}\frac{\partial \rho}{\partial x}\frac{\partial u_y}{\partial y} + \gamma_{10} \frac{1}{\rho}\frac{\partial \rho}{\partial y}\frac{\partial u_x}{\partial y} + \gamma_{11} \frac{1}{\rho}\frac{\partial \rho}{\partial y}\frac{\partial u_y}{\partial x}\right]$$

$$q_y^{(2)} = \frac{\mu^2}{\rho}\left[\gamma_1 \frac{1}{T}\frac{\partial T}{\partial y}\frac{\partial u_y}{\partial y} + \gamma_2 \frac{1}{T}\frac{\partial T}{\partial y}\frac{\partial u_x}{\partial x} + \gamma_3 \frac{\partial^2 u_y}{\partial y^2} + \gamma_4 \frac{\partial^2 u_y}{\partial x^2} + \gamma_5 \frac{\partial^2 u_x}{\partial x \partial y} + \gamma_6 \frac{1}{T}\frac{\partial T}{\partial x}\frac{\partial u_x}{\partial y}\right.$$
$$\left. + \gamma_7 \frac{1}{T}\frac{\partial T}{\partial x}\frac{\partial u_y}{\partial x} + \gamma_8 \frac{1}{\rho}\frac{\partial \rho}{\partial y}\frac{\partial u_y}{\partial y} + \gamma_9 \frac{1}{\rho}\frac{\partial \rho}{\partial y}\frac{\partial u_x}{\partial x} + \gamma_{10} \frac{1}{\rho}\frac{\partial \rho}{\partial x}\frac{\partial u_y}{\partial x} + \gamma_{11} \frac{1}{\rho}\frac{\partial \rho}{\partial x}\frac{\partial u_x}{\partial y}\right] \tag{2.80}$$

Unfortunately, these Burnett equations experience stability problems for very fine computational grids. Zhong proposed in 1991 to add linear third-order terms from the super-Burnett equations (which corresponds to $i = 3$) in order to stabilize the conventional Burnett equations and maintain second-order accuracy for the stress and heat flux terms (Agarwal et al., 2001). This set of equations is called the augmented Burnett (AB) equations. Moreover, Welder et al. (1993) have shown that a linear stability analysis is not sufficient to explain the instability of the Burnett equations. This instability could be due to the fact that the conventional Burnett equations can violate the second law of thermodynamics at high Knudsen numbers (Comeaux et al., 1995). To overcome this difficulty, Balakrishnan and Agarwal have proposed another set of equations termed the Bhatnagar–Gross–Krook–Burnett or BGK–Burnett (BGKB) equations, which are consistent with the second principle and unconditionally stable for both monatomic and polyatomic gases. The BGKB equations also use third-order terms but the second-order terms $\{\alpha_1, \ldots, \alpha_{16}, \beta_1, \ldots, \beta_7, \gamma_1, \ldots, \gamma_{11}\}$ are different from those of the conventional Burnett or AB equations, as well as the first-order terms ($\delta_1 = -1.6$ and $\delta_2 = -0.4$ for $\gamma = 1.4$). The additional terms for the AB equations are

$$\sigma_{xx}^{(a)} = \frac{\mu^3}{p^2} RT \left[\alpha_{17}\left(\frac{\partial^3 u_x}{\partial x^3} + \frac{\partial^3 u_x}{\partial x \partial y^2}\right) + \alpha_{18}\left(\frac{\partial^3 u_y}{\partial y^3} + \frac{\partial^3 u_y}{\partial y \partial x^2}\right)\right]$$
$$\sigma_{yy}^{(a)} = \frac{\mu^3}{p^2} RT \left[\alpha_{17}\left(\frac{\partial^3 u_y}{\partial y^3} + \frac{\partial^3 u_y}{\partial y \partial x^2}\right) + \alpha_{18}\left(\frac{\partial^3 u_x}{\partial x^3} + \frac{\partial^3 u_x}{\partial x \partial y^2}\right)\right] \tag{2.81}$$
$$\sigma_{xy}^{(a)} = \sigma_{yx}^{(a)} = \frac{\mu^3}{p^2} RT \beta_8 \left(\frac{\partial^3 u_x}{\partial y^3} + \frac{\partial^3 u_y}{\partial x^3} + \frac{\partial^3 u_x}{\partial y \partial x^2} + \frac{\partial^3 u_y}{\partial x \partial y^2}\right)$$

$$
\begin{aligned}
q_x^{(a)} &= \frac{\mu^3}{p\rho} R \left[\gamma_{12} \left(\frac{\partial^3 T}{\partial x^3} + \frac{\partial^3 T}{\partial x \partial y^2} \right) + \gamma_{13} \frac{T}{\rho} \left(\frac{\partial^3 \rho}{\partial x^3} + \frac{\partial^3 \rho}{\partial x \partial y^2} \right) \right] \\
q_y^{(a)} &= \frac{\mu^3}{p\rho} R \left[\gamma_{12} \left(\frac{\partial^3 T}{\partial y^3} + \frac{\partial^3 T}{\partial y \partial x^2} \right) + \gamma_{13} \frac{T}{\rho} \left(\frac{\partial^3 \rho}{\partial y^3} + \frac{\partial^3 \rho}{\partial y \partial x^2} \right) \right]
\end{aligned}
\tag{2.82}
$$

and for the BGK–Burnett equations:

$$
\begin{aligned}
\sigma_{xx}^{(B)} = {} & \frac{\mu^3}{p^2} RT \left[\theta_1 \frac{\partial^3 u_x}{\partial x^3} + \theta_2 \frac{\partial^3 u_x}{\partial x \partial y^2} + \theta_3 \frac{\partial^3 u_y}{\partial x^2 \partial y} + \theta_4 \frac{\partial^3 u_y}{\partial y^3} \right] \\
& - \frac{\mu^3 RT}{p^2 \rho} \left[\frac{\partial \rho}{\partial x} \left(\theta_1 \frac{\partial^2 u_x}{\partial x^2} + \theta_5 \frac{\partial^2 u_y}{\partial x \partial y} + \theta_6 \frac{\partial^2 u_x}{\partial y^2} \right) + \frac{\partial \rho}{\partial y} \left(\theta_7 \frac{\partial^2 u_y}{\partial x^2} + \theta_8 \frac{\partial^2 u_x}{\partial x \partial y} + \theta_4 \frac{\partial^2 u_y}{\partial y^2} \right) \right] \\
& + \frac{\mu^3}{p^2} \left[\theta_9 \left(\frac{\partial u_x}{\partial x} \right)^3 + \theta_{10} \left\{ 3 \left(\frac{\partial u_x}{\partial x} \right)^2 \frac{\partial u_y}{\partial y} + \left(\frac{\partial u_y}{\partial y} \right)^3 \right\} + \theta_{11} \frac{\partial u_x}{\partial x} \left(\frac{\partial u_y}{\partial y} \right)^2 \right. \\
& \left. - \left(\theta_4 \frac{\partial u_x}{\partial x} + \theta_{12} \frac{\partial u_y}{\partial y} \right) \left\{ \left(\frac{\partial u_x}{\partial y} \right)^2 + 2 \frac{\partial u_x}{\partial y} \frac{\partial u_y}{\partial x} + \left(\frac{\partial u_y}{\partial x} \right)^2 \right\} \right] \\
& + \frac{\mu^3}{p^2} R \left[\left(\theta_{13} \frac{\partial u_x}{\partial x} + \theta_{14} \frac{\partial u_y}{\partial y} \right) \left(\frac{\partial^2 T}{\partial x^2} + \frac{\partial^2 T}{\partial y^2} \right) \right] \\
\sigma_{yy}^{(B)} = {} & \frac{\mu^3}{p^2} RT \left[\theta_1 \frac{\partial^3 u_y}{\partial y^3} + \theta_2 \frac{\partial^3 u_y}{\partial y \partial x^2} + \theta_3 \frac{\partial^3 u_x}{\partial y^2 \partial x} + \theta_4 \frac{\partial^3 u_x}{\partial x^3} \right] \\
& - \frac{\mu^3 RT}{p^2 \rho} \left[\frac{\partial \rho}{\partial y} \left(\theta_1 \frac{\partial^2 u_y}{\partial y^2} + \theta_5 \frac{\partial^2 u_x}{\partial x \partial y} + \theta_6 \frac{\partial^2 u_y}{\partial x^2} \right) + \frac{\partial \rho}{\partial x} \left(\theta_7 \frac{\partial^2 u_x}{\partial y^2} + \theta_8 \frac{\partial^2 u_y}{\partial x \partial y} + \theta_4 \frac{\partial^2 u_x}{\partial x^2} \right) \right] \\
& + \frac{\mu^3}{p^2} \left[\theta_9 \left(\frac{\partial u_y}{\partial y} \right)^3 + \theta_{10} \left\{ 3 \left(\frac{\partial u_y}{\partial y} \right)^2 \frac{\partial u_x}{\partial x} + \left(\frac{\partial u_x}{\partial x} \right)^3 \right\} + \theta_{11} \frac{\partial u_y}{\partial y} \left(\frac{\partial u_x}{\partial x} \right)^2 \right. \\
& \left. - \left(\theta_4 \frac{\partial u_y}{\partial y} + \theta_{12} \frac{\partial u_x}{\partial x} \right) \left\{ \left(\frac{\partial u_y}{\partial x} \right)^2 + 2 \frac{\partial u_x}{\partial y} \frac{\partial u_y}{\partial x} + \left(\frac{\partial u_x}{\partial y} \right)^2 \right\} \right] \\
& + \frac{\mu^3}{p^2} R \left[\left(\theta_{13} \frac{\partial u_y}{\partial y} + \theta_{14} \frac{\partial u_x}{\partial x} \right) \left(\frac{\partial^2 T}{\partial x^2} + \frac{\partial^2 T}{\partial y^2} \right) \right]
\end{aligned}
\tag{2.83}
$$

$$\sigma_{xy}^{(B)} = \frac{\mu^3}{p^2} RT \left[\theta_{15} \left(\frac{\partial^3 u_x}{\partial y \partial x^2} + \frac{\partial^3 u_y}{\partial y^2 \partial x} \right) + \frac{\partial^3 u_x}{\partial y^3} + \frac{\partial^3 u_y}{\partial x^3} \right]$$

$$- \frac{\mu^3 RT}{p^2 \rho} \left[\frac{\partial \rho}{\partial y} \left(\theta_6 \frac{\partial^2 u_x}{\partial x^2} + \theta_{16} \frac{\partial^2 u_y}{\partial x \partial y} + \frac{\partial^2 u_x}{\partial y^2} \right) + \frac{\partial \rho}{\partial x} \left(\frac{\partial^2 u_y}{\partial x^2} + \theta_{16} \frac{\partial^2 u_x}{\partial x \partial y} + \theta_6 \frac{\partial^2 u_y}{\partial y^2} \right) \right]$$

$$- \frac{\mu^3}{p^2} \left(\frac{\partial u_x}{\partial y} + \frac{\partial u_y}{\partial x} \right) \left[\theta_4 \left\{ \left(\frac{\partial u_x}{\partial x} \right)^2 + \left(\frac{\partial u_y}{\partial y} \right)^2 \right\} + 2\theta_{12} \frac{\partial u_x}{\partial x} \frac{\partial u_y}{\partial y} + \theta_7 \left\{ \left(\frac{\partial u_y}{\partial x} \right)^2 \right. \right.$$

$$\left. \left. + 2 \frac{\partial u_x}{\partial y} \frac{\partial u_y}{\partial x} + \left(\frac{\partial u_x}{\partial y} \right)^2 \right\} \right] + \frac{\mu^3}{p^2} R \left[\theta_{17} \left(\frac{\partial u_x}{\partial y} + \frac{\partial u_y}{\partial x} \right) \left(\frac{\partial^2 T}{\partial x^2} + \frac{\partial^2 T}{\partial y^2} \right) \right]$$

$$q_x^{(B)} = \frac{\mu^3}{p\rho} R \theta_{18} \left[\frac{\partial^3 T}{\partial x^3} + \frac{\partial^3 T}{\partial y^2 \partial x} - \frac{1}{\rho} \frac{\partial \rho}{\partial x} \left(\frac{\partial^2 T}{\partial x^2} + \frac{\partial^2 T}{\partial y^2} \right) \right]$$

$$+ \frac{\mu^3}{p\rho} \left[\frac{\partial u_x}{\partial x} \left(\theta_{19} \frac{\partial^2 u_x}{\partial x^2} + \theta_{20} \frac{\partial^2 u_y}{\partial x \partial y} + \theta_6 \frac{\partial^2 u_x}{\partial y^2} \right) \right.$$

$$+ \frac{\partial u_y}{\partial y} \left(\theta_{21} \frac{\partial^2 u_x}{\partial x^2} + \theta_{22} \frac{\partial^2 u_y}{\partial x \partial y} + \theta_7 \frac{\partial^2 u_x}{\partial y^2} \right)$$

$$\left. + \left(\frac{\partial u_x}{\partial y} + \frac{\partial u_y}{\partial x} \right) \left(\theta_{23} \frac{\partial^2 u_y}{\partial x^2} + \theta_{24} \frac{\partial^2 u_x}{\partial x \partial y} + \theta_6 \frac{\partial^2 u_y}{\partial y^2} \right) \right] \tag{2.84}$$

$$- \frac{\mu^3}{p\rho} \left(\frac{1}{\rho} \frac{\partial \rho}{\partial x} + \frac{1}{T} \frac{\partial T}{\partial x} \right) \left[\theta_{13} \left\{ \left(\frac{\partial u_x}{\partial x} \right)^2 + \left(\frac{\partial u_y}{\partial y} \right)^2 \right\} \right.$$

$$\left. + 2\theta_{14} \frac{\partial u_x}{\partial x} \frac{\partial u}{\partial y} + \theta_{17} \left\{ 2 \frac{\partial u_x}{\partial y} \frac{\partial u_y}{\partial x} + \left(\frac{\partial u_x}{\partial y} \right)^2 + \left(\frac{\partial u_y}{\partial x} \right)^2 \right\} \right]$$

$$+ \frac{\mu^3}{p\rho} \frac{R}{T} \frac{\partial T}{\partial x} \theta_{18} \left(\frac{\partial^2 T}{\partial x^2} + \frac{\partial^2 T}{\partial y^2} \right)$$

$$
\begin{aligned}
q_y^{(B)} = &\frac{\mu^3}{p\rho} R\theta_{18}\left[\frac{\partial^3 T}{\partial y^3}\frac{\partial^3 T}{\partial x^2 \partial y} - \frac{1}{\rho}\frac{\partial \rho}{\partial y}\left(\frac{\partial^2 T}{\partial x^2} + \frac{\partial^2 T}{\partial y^2}\right)\right] \\
&+ \frac{\mu^3}{p\rho}\left[\frac{\partial u_y}{\partial y}\left(\theta_{19}\frac{\partial^2 u_y}{\partial y^2} + \theta_{20}\frac{\partial^2 u_x}{\partial x \partial y} + \theta_6 \frac{\partial^2 u_y}{\partial x^2}\right)\right. \\
&\quad + \frac{\partial u_x}{\partial x}\left(\theta_{21}\frac{\partial^2 u_y}{\partial y^2} + \theta_{22}\frac{\partial^2 u_x}{\partial x \partial y} + \theta_7 \frac{\partial^2 u_y}{\partial x^2}\right) \\
&\quad \left. + \left(\frac{\partial u_x}{\partial y} + \frac{\partial u_y}{\partial x}\right)\left(\theta_{23}\frac{\partial^2 u_x}{\partial y^2} + \theta_{24}\frac{\partial^2 u_y}{\partial x \partial y} + \theta_6 \frac{\partial^2 u_x}{\partial x^2}\right)\right] \\
&- \frac{\mu^3}{p\rho}\left(\frac{1}{\rho}\frac{\partial \rho}{\partial y} + \frac{1}{T}\frac{\partial T}{\partial y}\right)\left[\theta_{13}\left\{\left(\frac{\partial u_x}{\partial x}\right)^2 + \left(\frac{\partial u_y}{\partial y}\right)^2\right\}\right. \\
&\quad \left. + 2\theta_{14}\frac{\partial u_x}{\partial x}\frac{\partial u_y}{\partial y} + \theta_{17}\left\{2\frac{\partial u_x}{\partial y}\frac{\partial u_y}{\partial x} + \left(\frac{\partial u_x}{\partial y}\right)^2 + \left(\frac{\partial u_y}{\partial x}\right)^2\right\}\right] \\
&+ \frac{\mu^3}{p\rho}\frac{R}{T}\frac{\partial T}{\partial y}\theta_{18}\left(\frac{\partial^2 T}{\partial x^2} + \frac{\partial^2 T}{\partial y^2}\right)
\end{aligned}
$$

The different coefficients for the above equations are given in Table 2.4.

2.2.4.2 DSMC method

For high Knudsen numbers, the continuum approach (NS, QGD, QHD, or B equations) is no longer valid, even with slip boundary conditions. A molecular approach is then required and the flow may be described from the Boltzmann equation (2.69).

However, the integral formulation of $Q(f,f^*)$, which represents the binary intermolecular collisions, is complex in the general case due to nonlinearities and a great number of independent variables (Karniadakis and Beskok, 2002). Therefore, the numerical resolution of the Boltzmann equation is reserved for problems with a simple geometry or problems for which the rarefaction level allows simplification. Thus, $Q(f,f^*)$ is zero in the free molecular regime when $Kn \rightarrow \infty$. Also, when $Kn \rightarrow 0$, a semi-analytical resolution of the Boltzmann equation can be obtained using the method of moments of Grad, for which the distribution function is represented by a series of orthogonal Hermite polynomials (Grad, 1949) or the Chapman–Enskog method (Chapman and Cowling, 1952), for which the distribution function is

Table 2.4 Coefficients for AB and BGKB Equations; the coefficients for AB equations correspond to a HS collision model, and the coefficients for BGKB equations correspond to $\gamma = 1.4$

	α_1	α_2	α_3	α_4	α_5	α_6	α_7	α_8	α_9	α_{10}
AB	1.199	0.153	−0.600	−0.115	1.295	−0.733	0.260	−0.130	−1.352	0.676
BGKB	−2.24	−0.48	0.56	−1.2	0.0	0.0	−19.6	−5.6	−1.6	0.4
	α_{11}	α_{12}	α_{13}	α_{14}	α_{15}	α_{16}	α_{17}	α_{18}	β_1	β_2
AB	1.352	−0.898	0.600	−0.676	0.449	−0.300	0.2222	−0.1111	−0.115	1.913
BGKB	1.6	−19.6	−18.0	−0.4	−5.6	−6.9	—	—	1.4	−1.4
	β_3	β_4	β_5	β_6	β_7	β_8	γ_1	γ_2	γ_3	γ_4
AB	0.390	−2.028	0.900	2.028	−0.676	0.1667	10.830	0.407	−2.269	1.209
BGKB	0.0	−2.0	2.0	2.0	0.0	—	−25.241	−0.2	−1.071	−2.0
	γ_5	γ_6	γ_7	γ_8	γ_9	γ_{10}	γ_{11}	γ_{12}	γ_{13}	
AB	−3.478	−0.611	11.033	−2.060	1.030	−1.545	−1.545	0.6875	−0.625	
BGKB	−2.8	−7.5	−11.0	−1.271	1.0	−3.0	−3.0	—	—	
	θ_1	θ_2	θ_3	θ_4	θ_5	θ_6	θ_7	θ_8	θ_9	θ_{10}
BGKB	2.56	1.36	0.56	−0.64	0.96	1.6	−0.4	−0.24	1.024	−0.256
	θ_{11}	θ_{12}	θ_{13}	θ_{14}	θ_{15}	θ_{16}	θ_{17}	θ_{18}	θ_{19}	θ_{20}
BGKB	1.152	0.16	2.24	−0.56	3.6	0.6	1.4	4.9	7.04	−0.16
	θ_{21}	θ_{22}	θ_{23}	θ_{24}	δ_1	δ_2				
BGKB	−1.76	4.24	3.8	3.4	−1.6	−0.4				

Source: *From Yun et al. (1998)*

expanded into a perturbation series with the Knudsen number being the small parameter (cf. Section 2.2.4.1). In the transition regime, on the other hand, the numerical resolution of the Boltzmann equation requires approximate methods based on the simplification of $Q(f,f^*)$. Sharipov and Seleznev (1998) describe the different available methods—BGK equation (Bhatnagar et al., 1954), linearized Boltzmann equation (Cercignani et al., 1994)—with their conditions of validity.

The DSMC method is actually better suited to the transition regime. This method was developed by Bird (1978, 1998). Initially widely used for the simulation of low-pressure rarefied flows (Cheng and Emmanuel, 1995; Muntz, 1989), it is now often used for microfluidic applications (Chen et al., 1998; Hudson and Bartel, 1999; Mavriplis et al., 1997; Oran et al., 1998; Pan et al., 1999, 2000, 2001; Piekos and Breuer, 1996; Stefanov and Cercignani, 1994; Wang and Li, 2004; Wu and Tseng, 2001), and it is compatible with the simulation of multicomponent gas microflows (Qazi Zade et al., 2012).

As in classical kinetic theory, the DSMC method assumes a molecular chaos and a dilute gas. The technique is equivalent to solving the Boltzmann equation for a monatomic gas with binary collisions. The DSMC technique consists of splitting off the simulations of the intermolecular collisions and of the molecular motion, with a time step smaller than the mean collision time $1/\nu$. The simulation is performed using a limited number of molecules, with each simulated molecule representing a great number W of real molecules. The control volume is divided into cells, whose size is of the order of $\lambda/3$, in order to correctly treat large gradient regions. The simulation involves four main steps (Oran et al., 1998):

1. Moving simulated molecules and modeling molecule–surface interactions, applying conservation laws.
2. Indexing and tracking molecules.
3. Simulating collisions with a probabilistic process, generally with a VHS model.
4. Sampling the macroscopic flow properties at the geometric center of the cells.

The statistical error of the DSMC solution is inversely proportional to the square root of the total number of simulated molecules. The technique being explicit and time-marching, the simulation is unsteady. The solution of an unsteady problem is obtained from averaging an ensemble of many independent computations to reduce statistical errors. For a steady problem, each independent computation is run until a steady flow is established (Figure 2.10).

2.2.4.3 Lattice Boltzmann method

The LBM is appropriate for complex geometries and seems particularly effective for the simulation of flows in microsystems, where both mesoscopic dynamics and microscopic statistics are important (Karniadakis and Beskok, 2002). The method, whose detailed description is proposed by Chen and Doolen (1998), solves a simplified Boltzmann equation on a discrete lattice. The use of LBM for the simulation of gas microflows is recent, but LBM is destined to be more widely used for such applications in the future. For example, Nie et al. (2002) demonstrated that the

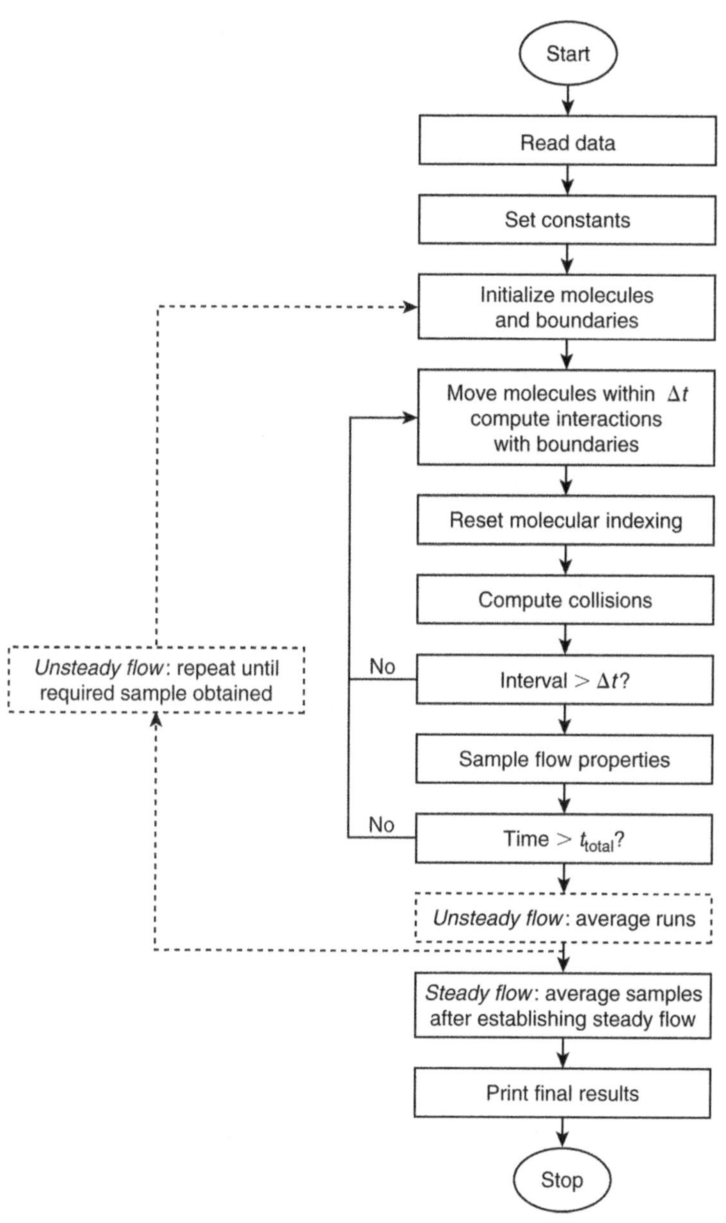

FIGURE 2.10

DSMC flowchart.

Source: *From Oran et al. (1998).*

LBM can capture the fundamental behaviors in microchannel flows, including velocity slip, nonlinear pressure drop along the channel, and mass flow rate variation with Knudsen number. The numerical results found by Lim et al. (2002) are also in good agreement with the data obtained analytically and experimentally for a two-dimensional isothermal pressure-driven microchannel flow. In the transition regime, the simulations of a microchannel flow by Li et al. (2011) are in good agreement with the solution of the linearized Boltzmann equation, DSMC, and experimental data over a broad range of Knudsen numbers.

2.3 Pressure-driven steady slip flows in microchannels

In this section, we consider the flow of an ideal gas through different microchannels in the slip flow regime. The flow is steady, isothermal, and the volume forces are neglected. The flow is also assumed to be locally fully developed: the velocity profile in a section is the same as the one obtained for a fully developed incompressible flow, but the density is recalculated in each cross-section and depends on the pressure and the temperature via the equation of state (2.24). This assumption is no longer valid if the Mach number is greater than 0.3 (Guo and Wu, 1998; Harley et al., 1995). The main sections of interest are the rectangular and the circular cross-sections. Microchannels with rectangular cross-sections are easily etched in silicon wafers, for example by reactive ion etching (RIE) or by deep reactive ion etching (DRIE). If the aspect ratio of the microchannel is small enough, the flow may be considered as a plane flow between parallel plates and the modeling is much simpler. Gas flows in circular microtubes are also frequently encountered, for example, in chromatography applications. Fused silica microtubes are proposed by different suppliers with a variety of internal diameters, from one micrometer to hundreds of micrometers. Due to their low cost, they are an interesting solution as connecting lines for a number of microfluidics devices or prototypes.

The axial coordinate is z and the cross-section is in the (x, y) plane. The different sections considered in this section are represented in Figure 2.11.

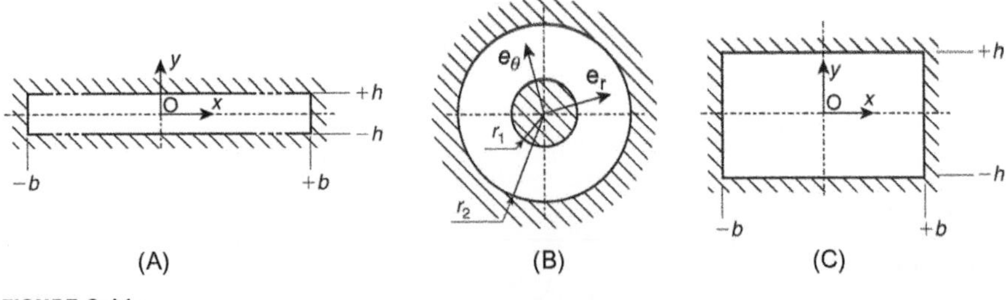

FIGURE 2.11

Different microchannel sections: (A) plane channel limited by parallel plates ($b \gg h$); (B) circular or annular duct; and (C) rectangular channel.

Two different Knudsen numbers are usually defined for microchannel flows.

$$Kn = \frac{\lambda}{D_{\mathrm{H}}} = \frac{\lambda}{4A/P} \tag{2.85}$$

is defined from the hydraulic diameter as a reference length. The section area is noted as A and the wetted perimeter is noted as P.

$$Kn' = \frac{\lambda}{L_{\min}} \tag{2.86}$$

is defined from the minimal representative length of the section.

2.3.1 Plane flow between parallel plates

Here, $h/b \to 0$ and the velocity depends only on the transverse coordinate y. With the previous assumptions, the momentum Eq. (2.33) reduces to

$$\frac{\mathrm{d}^2 u_z}{\mathrm{d}y^2} = \frac{1}{\mu}\frac{\mathrm{d}p}{\mathrm{d}z} \tag{2.87}$$

2.3.1.1 First-order solution

The first-order boundary condition (2.64) for a fixed wall and an isothermal flow is written for a flow between parallel plates:

$$u_z|_{y=h} = -\beta_{u1}\frac{2-\sigma}{\sigma}\lambda\left.\frac{\mathrm{d}u_z}{\mathrm{d}y}\right|_{y=h} \tag{2.88}$$

to be associated with the condition of symmetry (cf. Figure 2.11)

$$\left.\frac{\mathrm{d}u_z}{\mathrm{d}y}\right|_{y=0} = 0 \tag{2.89}$$

The general solution of Eq. (2.87) is

$$u_z = \frac{1}{\mu}\frac{\mathrm{d}p}{\mathrm{d}z}\left(\frac{y^2}{2} + a_1 y + a_2\right) \tag{2.90}$$

The condition of symmetry (2.89) yields $a_1 = 0$ and the boundary condition (2.88) leads to $a_2 = -\beta_{u1}\lambda\, h(2-\sigma)/\sigma - h^2/2$. Therefore, in a dimensionless form, the velocity distribution can be written as

$$u_z^* = 1 - y^{*2} + 8\beta_{u1}\frac{2-\sigma}{\sigma}Kn = 1 - y^{*2} + 4\beta_{u1}\frac{2-\sigma}{\sigma}Kn' \tag{2.91}$$

where $y^* = y/h$, $u_z^* = u_z/u_{z0}$ and $u_{z0} = u_{z(y=0,\ Kn=0)}$ represents the velocity at the center of the microchannel when rarefaction is not taken into account:

$$u_{z0} = -\frac{h^2}{2\mu}\frac{dp}{dz} \tag{2.92}$$

The Knudsen number $Kn = \lambda/L = \lambda/D_H$ in Eq. (2.91) is defined from the hydraulic diameter $D_H = 4A/P = 4h$, noting $A = 4hb$ as the area and $P = 4(b+h)$ as the wetted perimeter of the channel cross-section, with $b \gg h$ (Figure 2.11(a)). $Kn' = \lambda/2h$ is defined from the depth of the microchannel. Equation (2.91) shows that, in this simple case of plane flow, the usual Poiseuille parabolic profile ($u_z^* = 1 - y^{*2}$) is globally translated with a quantity $2(2-\sigma)\beta_{u1}\lambda/(\sigma h)$, due to the slip at the wall. The integration of the velocity distribution given by Eq. (2.91) over the cross-section leads to the mean velocity

$$\overline{u_z^*} = \frac{1}{2}\int_{-1}^{1} u_z^* dy^* \tag{2.93}$$

that is, with Eq. (2.91),

$$\overline{u_z^*} = \frac{\overline{u_z}}{u_{z0}} = \frac{2}{3} + 8\beta_{u1}\frac{2-\sigma}{\sigma}Kn = \frac{2}{3} + 4\beta_{u1}\frac{2-\sigma}{\sigma}Kn' \tag{2.94}$$

We can now calculate the Poiseuille number Po, defined as

$$Po = C_f Re \tag{2.95}$$

where

$$C_f = \frac{\overline{\tau_w}}{(1/2)\rho\overline{u_z}^2} \tag{2.96}$$

is the friction factor that represents a dimensionless value of the average wall shear $\overline{\tau_w}$ and

$$Re = \frac{\rho\overline{u_z}D_H}{\mu} \tag{2.97}$$

is the Reynolds number. The average wall shear is obtained from the force balance

$$\overline{\tau_w}P\,dz = -A\,dp \tag{2.98}$$

on a control volume that spans the channel and has an axial extension dz. Therefore,

$$\overline{\tau_{\mathrm{w}}} = -\frac{D_{\mathrm{H}}}{4}\frac{\mathrm{d}p}{\mathrm{d}z} \tag{2.99}$$

and

$$Po = -\frac{D_{\mathrm{H}}^2}{2\mu}\frac{\mathrm{d}p}{\mathrm{d}z}\frac{1}{\overline{u_z}} \tag{2.100}$$

which leads, with Eqs. (2.92) and (2.94), to:

$$Po_{\mathrm{NS1,plan}} = \frac{24}{1 + 12\beta_{u1}\frac{2-\sigma}{\sigma}Kn} = \frac{24}{1 + 6\beta_{u1}\frac{2-\sigma}{\sigma}Kn'} \tag{2.101}$$

Equation (2.101) shows that the Poiseuille number is less than its usual value of 24 for a Poiseuille flow between parallel plates (Shah, 1975), as soon as the Knudsen number is no longer negligible. This result shows that slip at the wall logically reduces friction and consequently, for a given pressure gradient, the flow rate is increased.

Considering a long microchannel of length l with a pressure p_{i} at the inlet and a pressure p_{o} at the outlet and neglecting the entrance effects, we can deduce from Eq. (2.94) the mass flow rate $\dot{m} = \rho\overline{u_z}A = p\overline{u_z}A/RT$, which is independent of z. According to Eq. (2.30), the quantity ($Kn\ p$) is constant for an isothermal flow; it can be written as a function of the outlet conditions: $Kn\ p = Kn_{\mathrm{o}}p_{\mathrm{o}}$. Therefore, the integration of Eq. (2.94) yields

$$\int_{p_{\mathrm{i}}}^{p_{\mathrm{o}}} \left(\frac{2}{3}p + 8\beta_{u1}\frac{2-\sigma}{\sigma}Kn_{\mathrm{o}}p_{\mathrm{o}}\right)\mathrm{d}p = -\frac{2\mu\dot{m}RTl}{h^2A} \tag{2.102}$$

which leads to

$$\dot{m}_{\mathrm{NS1,plan}} = \frac{2bh^3p_{\mathrm{o}}^2}{\mu RTl}\left[\frac{\Pi^2-1}{3} + 8\beta_{u1}\frac{2-\sigma}{\sigma}Kn_{\mathrm{o}}(\Pi-1)\right] \tag{2.103}$$

where $\Pi = p_{\mathrm{i}}/p_{\mathrm{o}}$ is the ratio of the inlet over outlet pressures and Kn_{o} is the outlet Knudsen number. The subscript NS1 refers to the Navier–Stokes equations with first-order slip flow boundary conditions. In a dimensionless form that compares the actual mass flow rate $\dot{m}$ with the mass flow rate $\dot{m}_{\mathrm{ns}}$ obtained from a no-slip hypothesis, Eq. (2.103) becomes:

$$\dot{m}^*_{\mathrm{NS1,plan}} = \frac{\dot{m}_{\mathrm{NS1,plan}}}{\dot{m}_{\mathrm{ns,plan}}} = 1 + 24\beta_{u1}\frac{2-\sigma}{\sigma}\frac{Kn_{\mathrm{o}}}{\Pi+1} = 1 + 12\beta_{u1}\frac{2-\sigma}{\sigma}\frac{Kn'_{\mathrm{o}}}{\Pi+1} \tag{2.104}$$

The pressure distribution along the z-axis can be found by expressing the conservation of mass flow rate through the channel:

$$\dot{m}_{\mathrm{NS1,plan}} = \frac{2bh^3p_o^2}{\mu RTl}\left[\frac{\Pi^2-1}{3}+8\beta_{u1}\frac{2-\sigma}{\sigma}Kn_o(\Pi-1)\right]$$
$$= \frac{2bh^3p_o^2}{\mu RTl(l-z)}\left[\frac{(p/p_o)^2-1}{3}+8\beta_{u1}\frac{2-\sigma}{\sigma}Kn_o(p/p_o-1)\right] \tag{2.105}$$

Equation (2.105) shows that the pressure $p^* = p/p_o$ is the solution of the polynomial

$$p^{*2}+\alpha_1 p^* + \alpha_2 + \alpha_3 z^* = 0 \tag{2.106}$$

where $z^* = z/l$ and

$$\alpha_1 = 24\beta_{u1}\frac{2-\sigma}{\sigma}Kn_o;\quad \alpha_2 = -\Pi(\Pi+\alpha_1);\quad \alpha_3 = (\Pi-1)(\Pi+1+\alpha_1) \tag{2.107}$$

Equation (2.106) is plotted in Figure 2.12 for an inlet over outlet pressure ratio $\Pi = 2$. This figure shows that rarefaction and compressibility have opposite

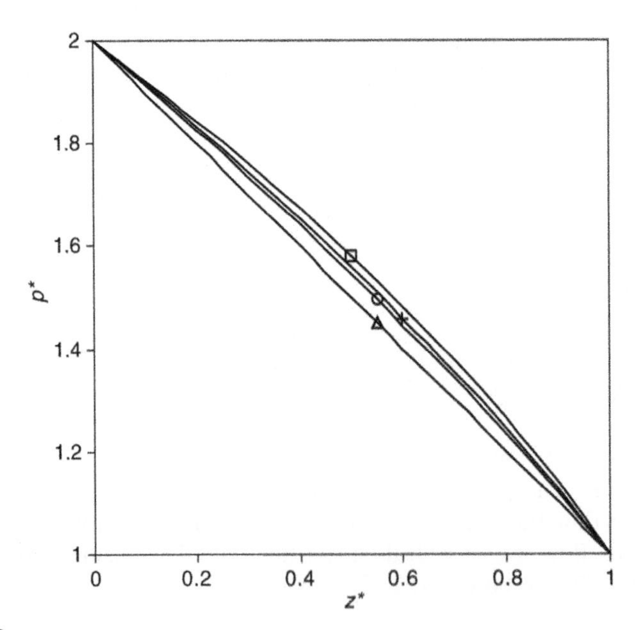

FIGURE 2.12

Pressure distribution along a plane microchannel. Δ: incompressible flow; □: $Kn_o = 0$; +: $Kn_o = 0.05$; o: $Kn_o = 0.1$.

effects on pressure distribution. Rarefaction effects (characterized by Kn) reduce the curvature in pressure distribution due to compressibility effects (characterized by Ma).

The first-order mass flow rate relation (2.103) accurately predicts the experimental data available in the literature for moderate outlet Knudsen numbers and typically for $Kn_o < 0.03$. For higher values of Kn_o, some deviations appear (cf. Section 2.3.5).

2.3.1.2 Second-order solutions

Thus, the previous model can be improved to remain accurate for higher Knudsen numbers. For this purpose, second-order—$\vartheta(Kn^2)$—models, based on second-order boundary conditions or the QHD equations, may be used.

The use of second-order boundary conditions

$$u_z|_{y=h} = -A_1\lambda\frac{du_z}{dy}\bigg|_{y=h} + A_2\lambda^2\frac{d^2u_z}{dy^2}\bigg|_{y=h} = -\beta_{u1}\frac{2-\sigma}{\sigma}\lambda\frac{du_z}{dy}\bigg|_{y=h} + A_2\lambda^2\frac{d^2u_z}{dy^2}\bigg|_{y=h} \tag{2.108}$$

leads to a velocity distribution

$$u_z^* = 1 - y^{*2} + 8\beta_{u1}\frac{2-\sigma}{\sigma}Kn - 32A_2Kn^2 = 1 - y^{*2} + 4\beta_{u1}\frac{2-\sigma}{\sigma}Kn' - 8A_2Kn'^2 \tag{2.109}$$

and to a Poiseuille number

$$Po_{NS2,plan} = \frac{24}{1 + 12\beta_{u1}\frac{2-\sigma}{\sigma}Kn - 48A_2Kn^2} = \frac{24}{1 + 6\beta_{u1}\frac{2-\sigma}{\sigma}Kn' - 12A_2Kn'^2} \tag{2.110}$$

It is easy to verify that the mass flow rate is consequently corrected as

$$\begin{aligned} \dot{m}^*_{NS2,plan} = \frac{\dot{m}_{NS2,plan}}{\dot{m}_{ns,plan}} &= 1 + 24\beta_{u1}\frac{2-\sigma}{\sigma}\frac{Kn_o}{\Pi+1} - 96A_2Kn_o^2\frac{\ln\Pi}{\Pi^2-1} \\ &= 1 + 12\beta_{u1}\frac{2-\sigma}{\sigma}\frac{Kn'_o}{\Pi+1} - 24A_2Kn_o'^2\frac{\ln\Pi}{\Pi^2-1} \end{aligned} \tag{2.111}$$

It is interesting to note that a close result can also be found keeping the first-order boundary conditions, but using the QHD equations in place of the NS equations (Elizarova and Sheretov, 2003). The solution of the momentum Eq. (2.40), with the expression (2.53) of the viscous stress tensor and the boundary condition (2.88), yields

$$\dot{m}^*_{\mathrm{QHD1,plan}} = \frac{\dot{m}_{\mathrm{QHD1,plan}}}{\dot{m}_{\mathrm{ns,plan}}} = 1 + 24\beta_{u1}\frac{2-\sigma}{\sigma}\frac{Kn_o}{\Pi+1} + \frac{96}{k_2^2 Sc}Kn_o^2\frac{\ln\Pi}{\Pi^2-1}$$
$$= 1 + 12\beta_{u1}\frac{2-\sigma}{\sigma}\frac{Kn'_o}{\Pi+1} + \frac{24}{k_2^2 Sc}Kn_o'^2\frac{\ln\Pi}{\Pi^2-1} \tag{2.112}$$

Equations (2.111) and (2.112) differ only in their last term. For example, if we consider the second-order boundary conditions of Deissler, $A_2 = -9/8$ and $-96\ A_2 = 108$. This value must be compared with $96/(k_2^2\ Sc)$. For a HS model, $k_2 = 16/(5\sqrt{2\pi})$ and $Sc = 5/6$. Therefore, $96/(k_2^2 Sc) = 70.7$. The deviation on the second-order term between the two models is then around 42%. But, if we consider the more accurate VSS model, with $\omega = 0.74$ for nitrogen under standard conditions (cf. Table 2.2), $k_2 = 1.064$, $Sc = 0.746$, and $96/(k_2^2 Sc) = 114$, which corresponds to a deviation of about 5% between the second-order terms of NS2 and QDH1 models. Of course, this deviation is a function of the gas, the temperature, and the choice of the second-order boundary condition and of the IPL model.

2.3.2 Gas flow in circular microtubes

With the general assumptions of this section, the momentum Eq. (2.33) in polar coordinates reduces to

$$\frac{1}{r}\frac{\mathrm{d}}{\mathrm{d}r}\left(r\frac{\mathrm{d}u_z}{\mathrm{d}r}\right) = \frac{1}{\mu}\frac{\mathrm{d}p}{\mathrm{d}z} \tag{2.113}$$

2.3.2.1 First-order solution

The first-order boundary condition (2.64) for a non-moving wall and an isothermal flow is now:

$$u_z|_{r=r_2} = -\beta_{u1}\frac{2-\sigma}{\sigma}\lambda\frac{\mathrm{d}u_z}{\mathrm{d}r}\bigg|_{r=r_2} \tag{2.114}$$

The general solution of Eq. (2.113) is

$$u_z = \frac{1}{\mu}\frac{\mathrm{d}p}{\mathrm{d}z}\left(\frac{r^2}{4} + c_1\ln r + c_2\right) \tag{2.115}$$

where $c_1 = 0$ since the velocity has a finite value in $r = 0$. Using the boundary condition (2.114), we find

$$c_2 = -\beta_{u1}\frac{2-\sigma}{\sigma}\lambda\frac{r_2}{2} - \frac{r_2^2}{4} \tag{2.116}$$

which leads to the dimensionless velocity profile

$$u_z^* = 1 - r^{*2} + 4\beta_{u1}\frac{2-\sigma}{\sigma}Kn \tag{2.117}$$

where the Knudsen number $Kn = \lambda/D_H = \lambda/(2r_2) = Kn' = \lambda/L_{min}$ is defined from a characteristic length $L = 2r_2$. Here, $r^* = r/r_2$ and $u_z^* = u_z/u_{z0}$, where

$$u_{z0} = \frac{r_2^2}{4\mu}\left(-\frac{\mathrm{d}p}{\mathrm{d}z}\right) \tag{2.118}$$

is the velocity at the center of the microchannel when rarefaction is not taken into account. Equation (2.117) shows that the usual Hagen–Poiseuille velocity profile ($u_z^* = 1 - r^{*2}$) is globally translated with a quantity $2\beta_{u1}(2-\sigma)\lambda/(\sigma r_2)$ due to the slip at the wall. The integration of the velocity distribution given by Eq. (2.117) over the cross-section leads to the mean velocity

$$\overline{u_z^*} = \frac{1}{\pi}\int_0^1 u_z^* 2\pi r^* \mathrm{d}r^* \tag{2.119}$$

that is, with Eq. (2.117),

$$\overline{u_z^*} = \frac{1}{2} + 4\beta_{u1}\frac{2-\sigma}{\sigma}Kn \tag{2.120}$$

The Poiseuille number Po is still defined by Eq. (2.100), which leads with Eqs. (2.118) and (2.120) to:

$$Po_{\mathrm{NS1,circ}} = \frac{16}{1 + 8\beta_{u1}((2-\sigma)/\sigma)Kn} \tag{2.121}$$

If Kn is not negligible, the Poiseuille number is less than its usual value of 16 for a circular cross-section (Shah, 1975). Following the same method as for a plane flow, the integration of Eq. (2.119) with $\dot{m} = p\overline{u_z}A/(\mathrm{RT})$ yields

$$\dot{m}_{\mathrm{NS1,circ}} = \frac{\pi r_2^4 p_o^2}{4\mu RTl}\left[\frac{\Pi^2 - 1}{4} + 4\beta_{u1}\frac{2-\sigma}{\sigma}Kn_o(\Pi - 1)\right] \tag{2.122}$$

for a long microtube, neglecting entrance effects, and

$$\dot{m}^*_{\text{NS1,circ}} = \frac{\dot{m}_{\text{NS1,circ}}}{\dot{m}_{\text{ns,circ}}} = 1 + 16\beta_{u1}\frac{2-\sigma}{\sigma}\frac{Kn_o}{\Pi+1} \tag{2.123}$$

The pressure distribution $p^* = p/p_o$ along the microtube is the solution of the polynomial

$$p^{*2} + \beta_1 p^* + \beta_2 + \beta_3 z^* = 0 \tag{2.124}$$

where $z^* = z/l$ and

$$\beta_1 = 16\beta_{u1}\frac{2-\sigma}{\sigma}Kn_o;\ \ \beta_3 = (\Pi-1)(\Pi+1+\beta_1);\ \ \beta_2 = -1-\beta_1-\beta_3 \tag{2.125}$$

which qualitatively leads to the same conclusions as for a plane flow.

2.3.2.2 Second-order solution

The first-order solution can easily be extended to second-order boundary conditions, following the same reasoning as in Section 2.3.1.2. We find

$$\overline{u_z^*} = \frac{1}{2} + 4\beta_{u1}\frac{2-\sigma}{\sigma}Kn - 8A_2 Kn^2 \tag{2.126}$$

$$Po_{\text{NS2,circ}} = \frac{16}{1 + 8\beta_{u1}((2-\sigma)/\sigma)Kn - 16A_2 Kn^2} \tag{2.127}$$

and

$$\dot{m}^*_{\text{NS2,circ}} = \frac{\dot{m}_{\text{NS2,circ}}}{\dot{m}_{\text{ns,circ}}} = 1 + 16\beta_{u1}\frac{2-\sigma}{\sigma}\frac{Kn_o}{\Pi+1} - 32A_2 Kn_o^2\frac{\ln\Pi}{\Pi^2-1} \tag{2.128}$$

The problem can also be solved from the QHD equations with first-order boundary conditions. Lengrand et al. (2004) showed that

$$\dot{m}^*_{\text{QHD1,circ}} = \frac{\dot{m}_{\text{QHD1,circ}}}{\dot{m}_{\text{ns,circ}}} = 1 + 16\beta_{u1}\frac{2-\sigma}{\sigma}\frac{Kn_o}{\Pi+1} + \frac{32}{k_2^2 Sc}Kn_o^2\frac{\ln\Pi}{\Pi^2-1} \tag{2.129}$$

The deviation between the QHD1 and the NS2 models with respect to the second-order terms is exactly the same as in the case of a plane flow (Section 2.3.1.2).

2.3.3 Gas flow in annular ducts

The solution for the slip flow of a gas in a circular microduct is easily extensible to the case of an annular duct (Ebert and Sparrow, 1965). The inner radius r_1 and the outer radius r_2 are not necessarily small, but the distance $r_2 - r_1 = D_H/2$ must be small enough so that the flow between the two cylinders is rarefied. The first-order boundary conditions are

$$u_z = \beta_{u1} \frac{2-\sigma}{\sigma} \lambda \frac{du_z}{dr}\bigg|_{r=r_1} \quad ; \quad u_z = -\beta_{u1} \frac{2-\sigma}{\sigma} \lambda \frac{du_z}{dr}\bigg|_{r=r_2} \tag{2.130}$$

The integration of Eq. (2.113) leads to the velocity distribution

$$u_z^* = 1 - r^{*2} + 4(1 - r_1^*)\beta_{u1} \frac{2-\sigma}{\sigma} Kn + \frac{r_1^*(1 - r_1^{*2})(1 + 4\beta_{u1}((2-\sigma)/\sigma)Kn)[2(1 - r_1^*)\beta_{u1}((2-\sigma)/\sigma)Kn - \ln r_1^*]}{r_1^* \ln r_1^* - 2(1 - r_1^{*2})\beta_{u1}((2-\sigma)/\sigma)Kn} \tag{2.131}$$

with $Kn = Kn'/2 = \lambda/(2(r_2 - r_1))$ and to its mean value

$$\overline{u_z^*} = \frac{1}{2}\left[1 + r_1^{*2} + 8(1 - r_1^* + r_1^{*2})\beta_{u1} \frac{2-\sigma}{\sigma} Kn + \frac{r_1^*(1 - r_1^{*2})(1 + 4\beta_{u1}((2-\sigma)/\sigma)Kn)^2}{r_1^* \ln r_1^* - 2(1 - r_1^{*2})\beta_{u1}((2-\sigma)/\sigma)Kn}\right] \tag{2.132}$$

Thus, the Poiseuille number takes the form

$$Po_{NS1,annu} = \frac{16(1 - r_1^*)^2}{1 + r_1^{*2} + 8(1 - r_1^* + r_1^{*2})\beta_{u1}\left(\frac{2-\sigma}{\sigma}\right)Kn + \left(\frac{r_1^*(1 - r_1^{*2})\left(1 + 4\beta_{u1}\left(\frac{2-\sigma}{\sigma}\right)Kn\right)^2}{r_1^* \ln r_1^* - 2(1 - r_1^{*2})\beta_{u1}\left(\frac{2-\sigma}{\sigma}\right)Kn}\right)} \tag{2.133}$$

which generalizes the expression (2.121) obtained for a circular microtube.

2.3.4 Gas flow in rectangular microchannels

With the rectangular cross-section defined in Figure 2.11(c) and the general assumptions of Section 2.3, the momentum equation takes the form

$$\frac{\partial^2 u_z}{\partial x^2} + \frac{\partial^2 u_z}{\partial y^2} = \frac{1}{\mu} \frac{dp}{dz} \tag{2.134}$$

In a dimensionless form, the momentum equation can be written

$$a^{*2}\frac{\partial^2 u_z^*}{\partial x^{*2}} + \frac{\partial^2 u_z^*}{\partial y^{*2}} = -1 \tag{2.135}$$

with

$$x^* = \frac{x}{b};\quad y^* = \frac{y}{h};\quad u_z^* = \frac{u_z}{-(h^2/\mu)(\mathrm{d}p/\mathrm{d}z)} \tag{2.136}$$

and $a^* = h/b$ is the aspect ratio of the section with $0 < a^* \leq 1$, if we assume $h \leq b$.

2.3.4.1 First-order solution

The first-order boundary conditions are

$$u_z^*\Big|_{y^*=1} = -\beta_{u1}\frac{2-\sigma}{\sigma}2Kn'\frac{\partial u_z^*}{\partial y^*}\Big|_{y^*=1} \tag{2.137}$$

$$u_z^*\Big|_{x^*=1} = -\beta_{u1}\frac{2-\sigma}{\sigma}2Kn'\frac{\partial u_z^*}{\partial x^*}\Big|_{x^*=1} \tag{2.138}$$

with $Kn' = \lambda/(2h)$ and the conditions of symmetry

$$\frac{\partial u_z^*}{\partial y^*}\Big|_{y^*=0} = 0 \tag{2.139}$$

$$\frac{\partial u_z^*}{\partial x^*}\Big|_{x^*=0} = 0 \tag{2.140}$$

Ebert and Sparrow (1965) proposed a solution for the velocity distribution, compatible with the condition of symmetry (2.139):

$$u_z^* = -\sum_{i=1}^{\infty}\Psi_i(x^*)\cos(\alpha_i y^*) \tag{2.141}$$

The eigenvalues α_i must satisfy

$$\alpha_i \tan\alpha_i = \beta_{u1}\frac{\sigma}{2-\sigma}\frac{1}{2Kn'} \tag{2.142}$$

in order to be compatible with the boundary condition (2.137) without leading to a trivial velocity solution. The x^*-dependent Ψ_i functions are found by expanding the right-hand side of Eq. (2.135), that is unity, in terms of the orthogonal functions $\cos(\alpha_i y^*)$:

$$-1 = -\sum_{i=1}^{\infty} \Omega_i \cos(\alpha_i y^*) = -\sum_{i=1}^{\infty} \frac{2 \sin \alpha_i}{\alpha_i + \sin \alpha_i \cos \alpha_i} \cos(\alpha_i y^*) \tag{2.143}$$

Therefore, to verify the momentum Eq. (2.135), the functions Ψ_i must obey

$$\frac{d^2\Psi_i}{dx^{*2}} - \left(\frac{\alpha_i}{a^*}\right)^2 \Psi_i - \frac{\Omega_i}{a^{*2}} = 0 \tag{2.144}$$

The solution of Eq. (2.144), with the conditions (2.138) and (2.140) and using Eq. (2.142), is

$$\Psi_i = -\frac{\Omega_i}{\alpha_i^2}\left(1 - \frac{\cosh(\alpha_i x^*/a^*)}{\cosh(\alpha_i/a^*) + 2\beta_{u1}((2-\sigma)/\sigma)Kn'\alpha_i \sinh(\alpha_i/a^*)}\right) \tag{2.145}$$

Consequently, the velocity distribution is given by Eq. (2.141). This solution is given as a function of a Knudsen number $Kn' = \lambda/(2h)$ based on the channel depth, but it can also be expressed as a function of the previous Knudsen number $Kn = \lambda/D_H$ based on the hydraulic diameter by using the relation

$$2Kn = Kn'(1 + a^*) \tag{2.146}$$

In the special case of a square section, for which $a^* = 1$, both Knudsen numbers Kn and Kn' are identical.

2.3.4.2 Second-order solution

The above solution from Ebert and Sparrow was improved by Aubert and Colin (2001) in order to be more accurate for higher Knudsen numbers. This model is based on the second-order boundary conditions proposed by Deissler (1964):

$$u_z|_{y=h} = -\frac{2-\sigma}{\sigma}\lambda \left.\frac{\partial u_z}{\partial y}\right|_{y=h} - \frac{9}{16}\lambda^2\left(2\left.\frac{\partial^2 u_z}{\partial y^2}\right|_{y=h} + \left.\frac{\partial^2 u_z}{\partial x^2}\right|_{y=h} + \left.\frac{\partial^2 u_z}{\partial z^2}\right|_{y=h}\right) \tag{2.147}$$

and

$$u_z|_{x=b} = -\frac{2-\sigma}{\sigma}\lambda \left.\frac{\partial u_z}{\partial x}\right|_{x=b} - \frac{9}{16}\lambda^2\left(2\left.\frac{\partial^2 u_z}{\partial x^2}\right|_{x=b} + \left.\frac{\partial^2 u_z}{\partial y^2}\right|_{x=b} + \left.\frac{\partial^2 u_z}{\partial z^2}\right|_{x=b}\right) \tag{2.148}$$

This choice of Deissler boundary conditions, rather than other second-order formulations such as the ones proposed in Section 2.2.3.3, is due to the physical approach proposed by Deissler. The above equations are obtained from a three-dimensional local momentum balance at the wall, and the shear stress satisfies the definition of a Newtonian fluid used in the Navier–Stokes equations.

In Eqs. (2.147) and (2.148), the term $\partial^2 u_z/\partial z^2$ vanishes according to the continuity equation, which reduces to $\partial u_z/\partial z = 0$ for a locally fully developed flow. Finally, using the momentum Eq. (2.135), these equations may be written in a nondimensional form as follows:

$$u_z^*\big|_{y^*=1} = -\frac{2-\sigma}{\sigma}2Kn'\frac{\partial u_z^*}{\partial y^*}\bigg|_{y^*=1} - \frac{9}{4}Kn'^2\left(\frac{\partial^2 u_z^*}{\partial y^{*2}}\bigg|_{y^*=1} - 1\right) \tag{2.149}$$

and

$$u_z^*\big|_{x^*=1} = -\frac{2-\sigma}{\sigma}2a^*Kn'\frac{\partial u_z^*}{\partial x^*}\bigg|_{x^*=1} - \frac{9}{4}Kn'^2\left(a^{*2}\frac{\partial^2 u_z^*}{\partial x^{*2}}\bigg|_{x^*=1} - 1\right) \tag{2.150}$$

The velocity distribution (2.141) proposed by Ebert and Sparrow, in the form of a single Fourier series, does not converge with second-order boundary conditions. Actually, the eigenvalues α_i obtained from the boundary condition (2.149) are such that the functions cos $(\alpha_i y^*)$ are no longer orthogonal on the interval $[-1, 1]$. Therefore, a new form

$$u_z^*(x^*, y^*) = \sum_{i,j=1}^{\infty} A_{ij}N_{ij}\cos\left(\varphi_i\frac{x^*}{a^*}\right)\cos(\psi_j y^*) + \frac{9}{4}Kn'^2 \tag{2.151}$$

is required, based on a double Fourier series. The functions $\omega_{ij}(x^*, y^*) = \cos(\varphi_i x^*/a^*)$ $\cos(\psi_j y^*)$ are orthogonal over the domain $[-1, 1]\times[-1, 1]$. They are solutions to the eigenvalues problem stemming from Eqs. (2.135), (2.149), and (2.150). The terms N_{ij} are such that the functions $N_{ij}\cos(\varphi_i x^*/a^*)\cos(\psi_j y^*)$ are normed over the domain $[-1, 1]\times[-1, 1]$.

The conditions (2.149) and (2.150) require ψ_j and φ_i to be solutions of (Aubert, 1999)

$$1 - 2Kn'\frac{(2-\sigma)}{\sigma}\psi_j\tan\psi_j - \frac{9}{4}Kn'^2\psi_j^2 = 0 \tag{2.152}$$

and

$$1 - 2Kn'\frac{(2-\sigma)}{\sigma}\varphi_i\tan\frac{\varphi_i}{a^*} - \frac{9}{4}Kn'^2\varphi_i^2 = 0 \tag{2.153}$$

In order to determine the terms A_{ij}, the solution (2.151) is injected into the momentum Eq. (2.135). Furthermore, the right-hand side of Eq. (2.135) is expanded in a double Fourier series. Finally, it is found that

$$
\begin{aligned}
&A_{ij}N_{ij} \\
&= \frac{1}{\varphi_i^2+\psi_j^2}\left\{\frac{4\sin(\varphi_i/a^*)\sin\psi_j}{\varphi_i\psi_j}+\frac{9}{2}\frac{\sigma}{2-\sigma}Kn'\left(\frac{\sin(\varphi_i/a^*)\cos\psi_j}{\varphi_i}+\frac{\cos(\varphi_i/a^*)\sin\psi_j}{\psi_j}\right)\right\} \\
&\times\left\{\left(\frac{1}{a^*}+\frac{\cos(\varphi_i/a^*)\sin(\varphi_i/a^*)}{\varphi_i}\right)\left(1+\frac{\sin\psi_j\cos\psi_j}{\psi_j}\right)\right. \\
&\quad+\frac{9}{4}\frac{\sigma}{2-\sigma}Kn'\left(\cos^2\psi_j\left[\frac{1}{a^*}+\frac{\cos(\varphi_i/a^*)\sin(\varphi_i/a^*)}{\varphi_i}\right]\right. \\
&\quad\left.\left.+\cos^2(\varphi_i/a^*)\left[1+\frac{\sin\psi_j\cos\psi_j}{\psi_j}\right]\right)\right\}^{-1}
\end{aligned}
\tag{2.154}
$$

Whatever the aspect ratio a^*, the convergence of the series (2.151) was verified. In every case, the asymptotic value of the sum is reached with good precision from a number of roots ψ_j (Kn') and φ_i (a^*, Kn') from Eqs. (2.152) and (2.153) lower than 50. In Figure 2.13, the normalized velocity distribution u_z^* is plotted on the

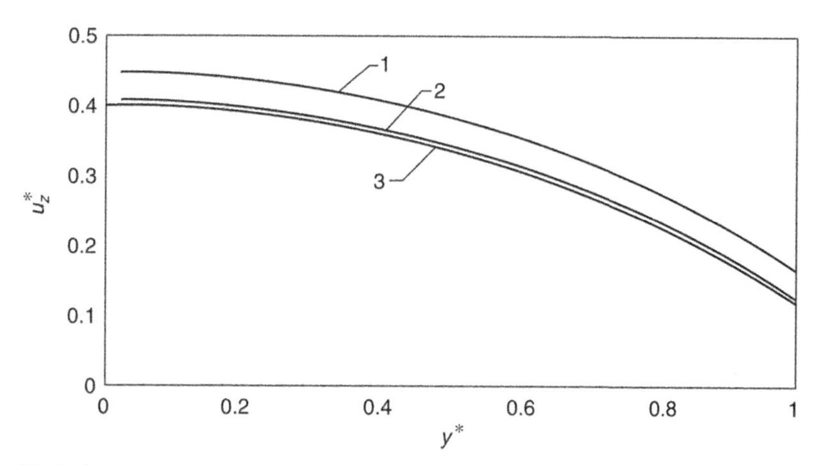

FIGURE 2.13

Normalized velocity distribution in a square cross-section with diffuse reflection at the wall for $Kn'=0.1$. 1: second-order model; 2: first-order model (Ebert and Sparrow, 1965); 3: first-order model (Morini and Spiga, 1998).

axis $x^* = 0$. The solutions proposed by Morini and Spiga (1998) and by Ebert and Sparrow (1965) and the second-order solution given by Eq. (2.151) are compared for $\sigma = 1$, $Kn' = 0.1$, and for a square section ($a^* = 1$). The slight deviation observed between the velocity distributions found by Morini and Spiga and by Ebert and Sparrow results from their two different ways of solving the same first-order problem. It is pointed out that the flow rate is underestimated (about 10% in the case of Figure 2.13) when the second-order terms are not taken into account. Moreover, it is verified that the velocity distribution obtained with a second-order model cannot be deduced from a simple translation of the first-order velocity distribution.

The mass flow rate through the cross-section (Σ) is $\dot{m} = \iint_{(\Sigma)} \rho u_z \, dx \, dy$, that is, with Eqs. (2.136) and (2.151),

$$\dot{m}_{\mathrm{NS2,rect}} = -\frac{\rho h^2}{\mu}\frac{\mathrm{d}p}{\mathrm{d}z}\frac{4h^2}{a^*}\left[\sum_{i,j=1}^{\infty} A_{ij}N_{ij}\frac{a^*\sin(\varphi_i/a^*)\sin\psi_j}{\varphi_i\psi_j} + \frac{9}{4}Kn'^2\right] \tag{2.155}$$

In order to obtain $\dot{m}$ as a function of the inlet pressure p_{i} and the outlet pressure p_{o} of a long rectangular microchannel of length l, Eq. (2.155) must be integrated along the microchannel. For an isothermal flow and known geometric dimensions, the Knudsen number only depends on the pressure $p(z)$. Therefore, the solutions φ_i and ψ_j of Eqs. (2.152) and (2.153) depend on the position along the z-axis. The bracketed term of Eq. (2.155) can be precisely fitted by a polynomial function $a_1 + a_2Kn' + a_3Kn'^2$, where a_1, a_2, and a_3 are coefficients depending on the aspect ratio a^* of the cross-section and on the momentum accommodation coefficient σ for a_2 and a_3. Thus,

$$\dot{m}_{\mathrm{NS2,rect}} = -\frac{4ph^4}{a^*RT\mu}\frac{\mathrm{d}p}{\mathrm{d}z}\left[a_1 + a_2Kn' + a_3Kn'^2\right] \tag{2.156}$$

Some numerical values of the coefficients a_1, a_2, and a_3, obtained with a least squares fitting method from values calculated using Eq. (2.155) for $Kn' \in [0; 1]$, are given in Table 2.5. Note that the numerical values of the coefficient a_1 correspond closely to the analytical expression

$$a_1 = 2\sum_{i=1}^{\infty}\frac{\sin^2\alpha_{i,\mathrm{ns}}}{\alpha_{i,\mathrm{ns}}^4(\alpha_{i,\mathrm{ns}} + \cos\alpha_{i,\mathrm{ns}}\sin\alpha_{i,\mathrm{ns}})}\left(\alpha_{i,\mathrm{ns}} - a^*\tanh\frac{\alpha_{i,\mathrm{ns}}}{a^*}\right) \tag{2.157}$$

which can be deduced from the no-slip problem. In Eq. (2.157), $\alpha_{i,\mathrm{ns}} = (2i-1)\pi/2$.

Since for an isothermal flow, the quantity $Kn'p$ is constant, the integration of Eq. (2.156) along the microchannel yields

$$\dot{m}_{\mathrm{NS2.rect}} = \frac{4h^4p_{\mathrm{o}}^2}{a^*\mu RTl}\left[\frac{a_1}{2}(\Pi^2 - 1) + a_2Kn'_{\mathrm{o}}(\Pi - 1) + a_3Kn'^2_{\mathrm{o}}\ln\Pi\right] \tag{2.158}$$

Table 2.5 Values of Coefficients a_1, a_2, and a_3 for different values of σ and a^*, for Deissler Second-Order Boundary Conditions

a^* σ	0.5	0.6	0.7	0.8	0.9	1
0	$a_2 = 6.0000$	$a_2 = 4.6667$	$a_2 = 3.7143$	$a_2 = 3.0000$	$a_2 = 2.4444$	$a_2 = 2.0000$
$a_1 = 0.33333$	$a_3 = 4.5000$	$a_3 = 4.5000$	$a_3 = 4.5000$	$a_3 = 4.5000$	$a_3 = 4.5000$	$a_3 = 4.5000$
0.01	$a_2 = 5.9694$	$a_2 = 4.6459$	$a_2 = 3.7007$	$a_2 = 2.9917$	$a_2 = 2.4405$	$a_2 = 1.9993$
$a_1 = 0.33123$	$a_3 = 4.4528$	$a_3 = 4.4542$	$a_3 = 4.4548$	$a_3 = 4.4550$	$a_3 = 4.4549$	$a_3 = 4.4549$
0.025	$a_2 = 5.9228$	$a_2 = 4.6141$	$a_2 = 3.6797$	$a_2 = 2.9788$	$a_2 = 2.4337$	$a_2 = 1.9977$
$a_1 = 0.32808$	$a_3 = 4.3840$	$a_3 = 4.3875$	$a_3 = 4.3887$	$a_3 = 4.3895$	$a_3 = 4.3896$	$a_3 = 4.3892$
0.05	$a_2 = 5.8420$	$a_2 = 4.5591$	$a_2 = 3.6423$	$a_2 = 2.9547$	$a_2 = 2.4201$	$a_2 = 1.9925$
$a_1 = 0.32283$	$a_3 = 4.2761$	$a_3 = 4.2813$	$a_3 = 4.2843$	$a_3 = 4.2858$	$a_3 = 4.2858$	$a_3 = 4.2853$
0.1	$a_2 = 5.6774$	$a_2 = 4.4431$	$a_2 = 3.5613$	$a_2 = 2.9000$	$a_2 = 2.3860$	$a_2 = 1.9751$
$a_1 = 0.31233$	$a_3 = 4.0697$	$a_3 = 4.0832$	$a_3 = 4.0906$	$a_3 = 4.0943$	$a_3 = 4.0952$	$a_3 = 4.0944$
0.25	$a_2 = 5.1332$	$a_2 = 4.0464$	$a_2 = 3.2675$	$a_2 = 2.6823$	$a_2 = 2.2267$	$a_2 = 1.8626$
$a_1 = 0.28081$	$a_3 = 3.6073$	$a_3 = 3.6287$	$a_3 = 3.6449$	$a_3 = 3.6560$	$a_3 = 3.6626$	$a_3 = 3.6647$
0.5	$a_2 = 4.2443$	$a_2 = 3.3519$	$a_2 = 2.7120$	$a_2 = 2.2289$	$a_2 = 1.8487$	$a_2 = 1.5395$
$a_1 = 0.22868$	$a_3 = 3.3489$	$a_3 = 3.3588$	$a_3 = 3.3708$	$a_3 = 3.3856$	$a_3 = 3.4048$	$a_3 = 3.4292$
1	$a_2 = 3.1207$	$a_2 = 2.4552$	$a_2 = 1.9782$	$a_2 = 1.6187$	$a_2 = 1.3364$	$a_2 = 1.1077$
$a_1 = 0.14058$	$a_3 = 3.3104$	$a_3 = 3.3176$	$a_3 = 3.3287$	$a_3 = 3.3438$	$a_3 = 3.3644$	$a_3 = 3.3908$

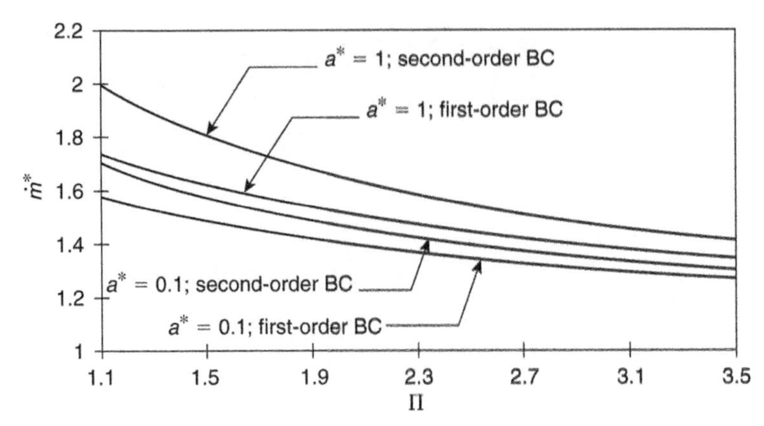

FIGURE 2.14

Influence of the aspect ratio and of the slip flow model on the normalized mass flow rate in a rectangular microchannel, for $Kn'_o = 0.1$ and $\sigma = 1$. BC: boundary condition.

and consequently

$$\dot{m}^*_{\mathrm{NS2,rect}} = \frac{\dot{m}_{\mathrm{NS2,rect}}}{\dot{m}_{\mathrm{ns,rect}}} = \left[1 + 2\frac{a_2}{a_1}Kn'_o\frac{1}{\Pi+1} + 2\frac{a_3}{a_1}Kn'^2_o\frac{\ln\Pi}{\Pi^2-1}\right] \tag{2.159}$$

The mass flow rate is strongly dependent on the aspect ratio (cf. Figure 2.14). If a^* is less than 0.01, the simpler plane flow model of Section 2.3.1 is accurate enough, but for higher values of a^*, it is no longer the case and the model presented in this section should be used. As an example, let us consider a pressure-driven flow in a long rectangular microchannel with an inlet over outlet pressure ratio $\Pi = 2$, an outlet Knudsen number $Kn'_o = 0.1$, and an accommodation coefficient $\sigma = 1$. The flow rate deduced from Eq. (2.111),

$$\dot{m}_{\mathrm{NS2,plan}} = [(2bh^3p_o^2)/(\mu RTl)][(\Pi^2-1)/3 + 4((2-\sigma)/\sigma)Kn'_o(\Pi-1) + 9Kn'^2_o\ln\Pi]$$

which was calculated from a plane model with Deissler boundary conditions (i.e., with $\beta_{u1} = 1$ and $A_2 = -9/8$), overestimates the flow rate $\dot{m}_{\mathrm{NS2,rect}}$ given by Eq. (2.158) and calculated from a rectangular model. The overestimation $((\dot{m}_{\mathrm{plan}} - \dot{m}_{\mathrm{rect}})/\dot{m}_{\mathrm{rect}})_{\mathrm{NS2}}$ is about 0.5% for an aspect ratio $a^* = 0.01$, rising to 5.3% for an aspect ratio $a^* = 0.1$, and reaching 112% for a square cross-section with $a^* = 1$!

The pressure distribution $p(z)$ can also be deduced from the mass conservation. It is the solution of the equation

$$\frac{(a_1/2)[(p/p_o)^2-1] + a_2Kn'_o[(p/p_o)-1] + a_3Kn'^2_o\ln(p/p_o)}{(a_1/2)(\Pi^2-1) + a_2Kn'_o(\Pi-1) + a_3Kn'^2_o\ln\Pi} = 1 - z^* \tag{2.160}$$

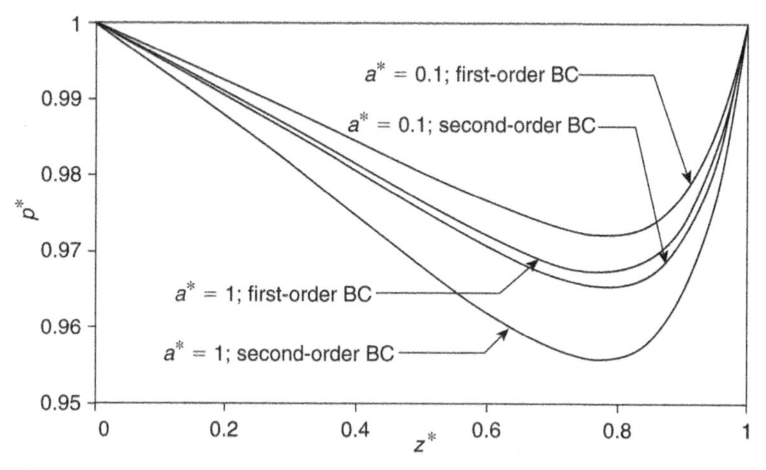

FIGURE 2.15

Influence of the aspect ratio and of the slip flow model on the normalized pressure distribution along a rectangular microchannel for $Kn'_o = 0.1$ and $\sigma = 1$. BC: boundary condition.

which can be numerically solved. An example of the nondimensional pressure distribution $p^* = p/p_{ns}$ as a function of $z^* = z/l$ is given in Figure 2.15 for an outlet Knudsen number $Kn'_o = 0.1$, an accommodation coefficient $\sigma = 1$, and for $a^* = 1$ (square duct) or $a^* = 0.1$. The maximum influence of the slip at the wall corresponds to $z^* = 0.77$. A no-slip model overestimates the pressure by a few percent, as is the case for a plane microchannel (Figure 2.12). Moreover, the greater the aspect ratio, the larger the pressure overestimation is.

The solution provided in Section 2.3.4.2 is based on Deissler boundary conditions, which assume that $\beta_{u1} = 1$ and $A_2 = -9/8$. This solution can be improved using more accurate values of β_{u1} and A_2, following the correction proposed by Hadjiconstantinou (2003). Pitakarnnop et al. (2010) obtained a more accurate solution using an improved form of the boundary conditions (2.147) and (2.148), respectively, as

$$u_z|_{y=h} = -1.1466\frac{2-\sigma}{\sigma}\lambda\left.\frac{\partial u_z}{\partial y}\right|_{y=h} - 0.647\lambda^2\left(\left.\frac{\partial^2 u_z}{\partial y^2}\right|_{y=h} + \frac{1}{2}\left.\frac{\partial^2 u_z}{\partial x^2}\right|_{y=h} + \frac{1}{2}\left.\frac{\partial^2 u_z}{\partial z^2}\right|_{y=h}\right) \tag{2.161}$$

and

$$u_z|_{x=b} = -1.1466\frac{2-\sigma}{\sigma}\lambda\left.\frac{\partial u_z}{\partial x}\right|_{x=b} - 0.647\lambda^2\left(\left.\frac{\partial^2 u_z}{\partial x^2}\right|_{x=b} + \frac{1}{2}\left.\frac{\partial^2 u_z}{\partial y^2}\right|_{x=b} + \frac{1}{2}\left.\frac{\partial^2 u_z}{\partial z^2}\right|_{x=b}\right) \tag{2.162}$$

The solution keeps the same form but with values of the coefficients (a_1, a_2, a_3) different from those given in Table 2.5. Recently, Meolans et al. (2012) proposed another approach appropriate for the treatment of second-order boundary conditions in rectangular microchannels. The solution is fully analytical but requires a second-order term in the slip boundary condition in the form of a Laplacian, which is not the case for the Deissler-based boundary conditions.

2.3.5 Experimental data

A discussion of the accuracy of the different models described in the previous section requires fine experimental data, notably to know which second-order boundary conditions are more appropriate.

2.3.5.1 Experimental setups for flow rate measurements

A few experimental setups are described in the literature. The first techniques used to provide experimental data on gas flow rates in a microchannel were proposed by Arkilic et al. (1994) or by Pong et al. (1994). Both studies used accumulation techniques: the gas flowing through the microchannel accumulates in a reservoir, resulting in either a change in pressure (constant-volume technique) or a change in volume (constant-pressure technique). Actually, this method is limited because the tiny mass flows are very difficult to measure accurately using a single tank. In particular, very small changes in temperature can overwhelm the mass-flow measurement due to thermal expansion of the gas. To overcome this difficulty, a differential technique was used afterwards (Arkilic et al., 1998). The system is schematically represented in Figure 2.16. Gas flows through the test microchannel and into the accumulation tank A. The mass flow through the system can be inferred from a measurement of the differential pressure between the accumulation tank A and a reference tank B. This arrangement is very insensitive to temperature fluctuations and allows accurate measurements of mass flow rates, with a sensitivity announced to be as low as 7×10^{-15} kg/s.

Another experimental setup was designed by Lalonde (2001) for the measurement of gaseous microflow rates under controlled temperature and pressure conditions, in the range 10^{-7} to 10^{-13} m^3/s. Its principle is based on the tracking of a drop meniscus in a calibrated pipette connected in series with the microchannel. Its main advantage, compared to other similar setups presented in the literature (Maurer et al., 2003; Shih et al., 1996; Zohar et al., 2002), is the simultaneous measurement of the flow rate both upstream and downstream of the microchannel (Figure 2.17). The inlet and the outlet pressures can both be independently tuned. The volume flow rate is measured by means of opto-electronic sensors that detect the passage of the two menisci of a liquid drop injected into the calibrated pipettes connected upstream and downstream of the microchannel.

A series of individual measurements of the volumetric flow rate can therefore be obtained, which allows the determination of the mean volumetric flow rates upstream and downstream of the microchannel and, consequently, the mass flow

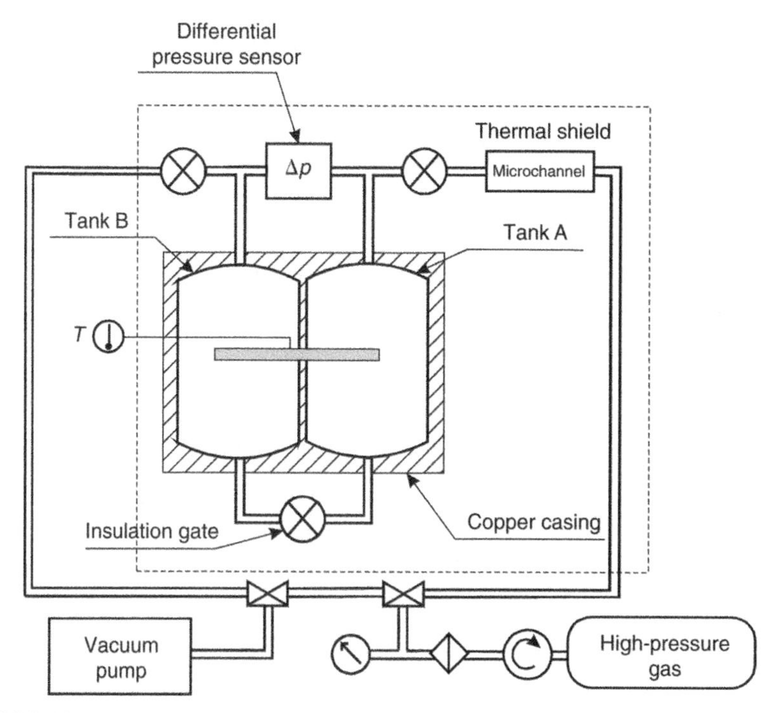

FIGURE 2.16

Schematic of an accumulation mass flow measurement system.

Source: *Adapted from Arkilic et al. (1998).*

rate through the microchannel. The comparison of the individual data given by each pipette, as well as the comparison of the two mean mass flow rates deduced from the two mean volumetric flow rates, must be consistent with the experimental uncertainties to allow the validation of the acquisition. A typical example is shown in Figure 2.18. The upper data (circles) correspond to the downstream pipette and the lower ones (triangles) to the upstream pipette, through which the volumetric flow rate is lower, since the pressure is higher. The white symbols are relative to the first, the dark symbols to the second, of the two menisci of the drop in the pipette. In total, 44 individual data values are obtained for each test. Experimental uncertainties are reported by vertical bars.

Other similar setups are described by Zohar et al. (2002) or by Maurer et al. (2003). They both provide a single but accurate measurement downstream of the microchannel. In the first case (Zohar et al., 2002), the outlet pressure is always the atmospheric pressure, whereas in the second case (Maurer et al., 2003), the outlet can be tuned as well as the upstream pressure.

The more recent and accurate setups are generally based on the constant volume technique (Ewart et al., 2007; Pitakarnnop et al., 2010; Yamaguchi et al., 2011). Some of them (Pitakarnnop et al., 2010) are able to couple the drop tracking method

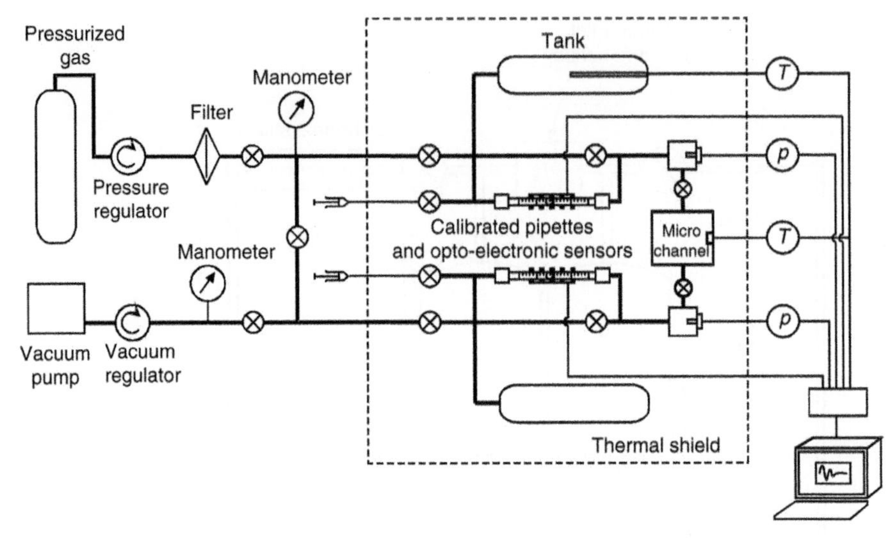

FIGURE 2.17

Schematic of a droplet tracking system for the measurement of gas microflow rates.

Source: *From Lalonde (2001).*

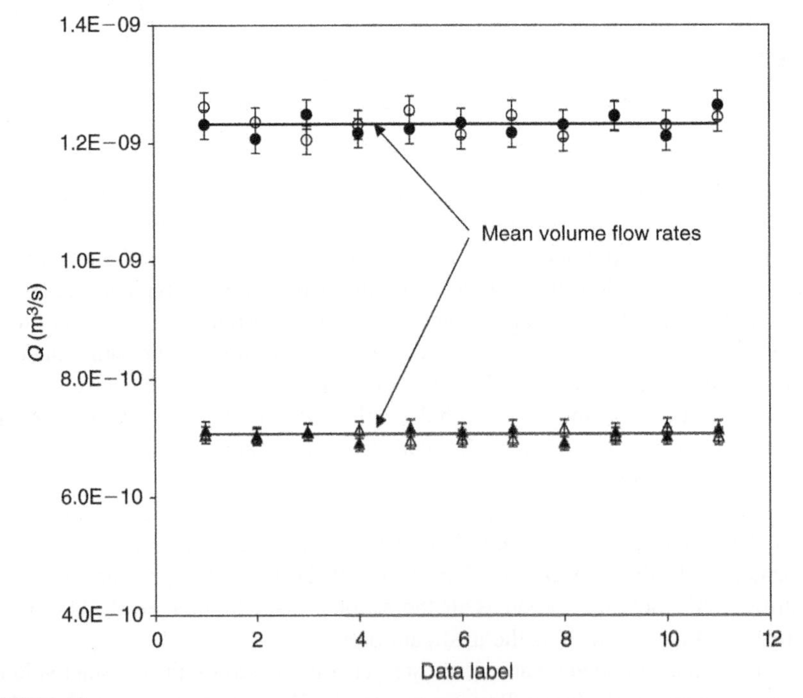

FIGURE 2.18

Typical example of volumetric flow rate (Q) measurement with the setup.

Source: *Lalonde (2001).*

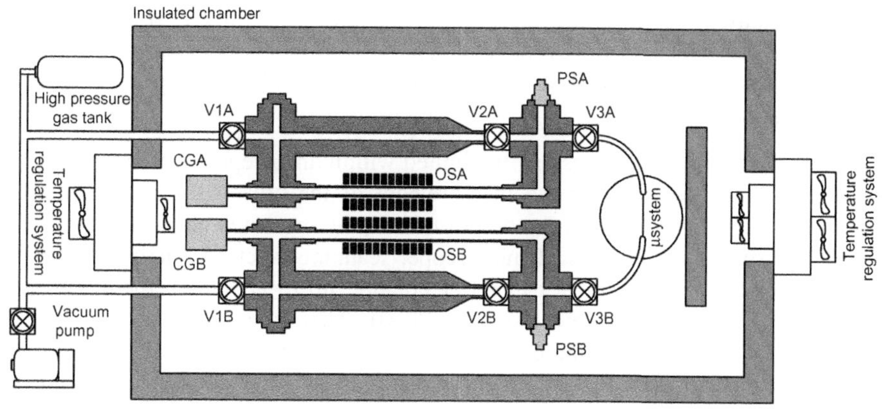

FIGURE 2.19

Schematic of a coupled constant-volume and droplet tracking system for the measurement of gas microflow rates.

Source: *Pitakamnop et al. (2010).*

and the constant volume method in an insulated and thermally regulated chamber (Figure 2.19).

2.3.5.2 Flow rate data

Comparing the different models available to describe the slip flow regime in microchannels, we note significant differences. For example, in a plane flow between parallel plates, the mass flow rate predicted by Eq. (2.111) depends significantly on the choice of the first-order correcting coefficient β_{u1}, as well as on the value of the coefficient A_2 (see Table 2.3). Even from a qualitative point of view, the role of the second-order term is not totally clarified. Some models generally based on a simple mathematical extension of Maxwell's condition (2.58), and for which $A_2 > 0$, predict a decrease of the slip compared to the first-order model; other models, for which $A_2 < 0$, predict an increase of the slip. It would also be interesting to test the accuracy of a QHD model that does not involve second-order boundary conditions, but is dependent on the Knudsen number at the second order (cf. Eq. (2.112)).

The increase of the slip due to second-order terms was first confirmed by Lalonde (2001), who measured the flow rate of helium and nitrogen in rectangular microchannels and showed that for outlet Knudsen numbers $Kn'_{\mathrm{o,M}}$ higher than 0.05, a first-order model underestimates the slip. For comparison of the experimental flow rate data with the predictions from the different models presented above, the sources of error or uncertainty must be considered. They can globally be divided into three classes:

1. Uncertainties inherent in the fluid properties and in their dependence on the experimental parameters (temperature, pressure). One can also note that some properties such as the momentum accommodation coefficient, σ, depend on both the fluid and the wall.

2. Uncertainties relative to the microchannel geometrical characteristics (dimensions of the section, roughness, etc.) and to their dispersion.
3. Uncertainties due to the flow-rate measurement (metrology, leakage, operating conditions, etc.).

Most of these uncertainties can be limited within an acceptable precision. This is generally not the case for the measurement of the cross-section dimensions. When these dimensions are of the order of 1 μm, the measures obtained by different means (profilometer, optical microscope, scanning electron microscopy) that require preliminary calibration can differ notably: up to about 5% for the width and about 10% for the depth (Anduze, 2000). Therefore, the data are hardly exploitable, since the hydraulic diameter plays a part to the power of 3 or 4 in the calculation of the flow rate. Also, an inaccuracy of only 10 nm when measuring a depth of 1 μm leads to an error of 4% in the estimation of the flow rate. The second parameter that can pose a problem is the accommodation coefficient σ, the value of which is not usually known.

In order to avoid interpretation errors due to these two main sources of uncertainty, the idea is to compare among themselves data measured in a same microchannel, in a Knudsen number range that covers at least two of the following regimes:

1. In the *little rarefied* regime, differences between the flow rates predicted by the no-slip (ns), the NS1, and the NS2 models are negligible, on the order of the experimental uncertainties. The value of σ consequently does not play any significant role.
2. In the *lightly rarefied* regime, the difference between NS1 and NS2 models remains negligible, but the difference with the ns model is now significant. The influence of the accommodation coefficient σ may also become significant for the higher values of Kn'_{o}.
3. In the *moderately rarefied* regime, the ns, NS1, and NS2 models predict significantly different flow rates, and σ is now a very sensitive parameter.

Figure 2.20 represents these regimes in a series of microchannels experimentally studied by Lalonde (Colin et al., 2004).

Microchannel no. 1, with a measured depth $2h = 4.48$ μm, allows a check of the transition from ns to NS1 models. First, the flow rate is measured for low values of the outlet Knudsen number Kn'_{o}, that is, for high values of p_{o}. For $Kn'_{o,VSS} = 0.007$, the differences between the flow rates predicted from the NS1 and NS2 models are negligible, whatever the value of the accommodation coefficient σ, and the deviation with the ns model remains on the order of the experimental uncertainties. The depth $2h$ kept for the simulation is adjusted to 4.48 μm—exactly the measured value in that case—for a good correlation between the experimental data and the NS1 or NS2 model. Then rarefaction is increased by decreasing the outlet pressure; the difference between ns and NS1 or NS2 models becomes significant and the experimental data are in very good agreement with the slip flow models. The influence of σ is still negligible, as is the deviation between NS1 and NS2 models.

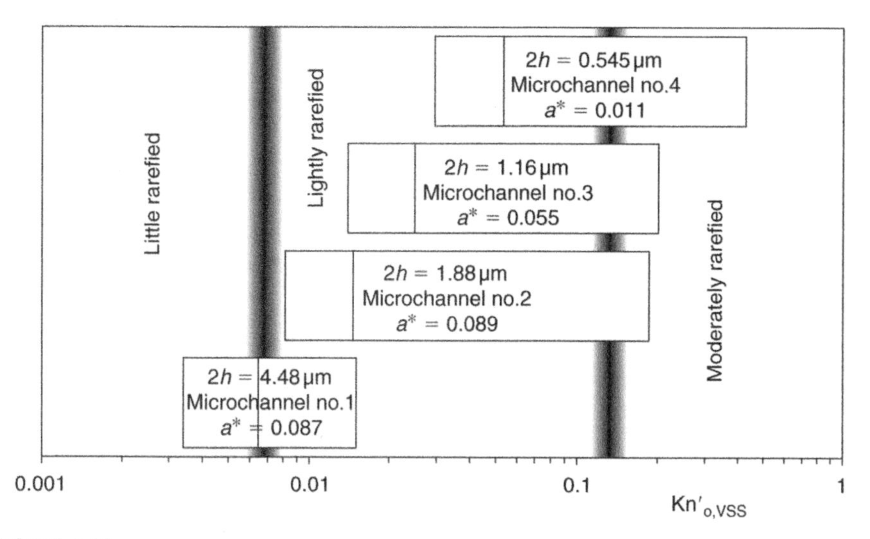

FIGURE 2.20

Knudsen number ranges covered by a series of rectangular microchannels tested by Lalonde (2001); the right part of the rectangles corresponds to the outlet Knudsen number values.

Microchannels nos. 3 and 4 allow a check of the transition from NS1 to NS2 models. As an illustration, Figure 2.21 shows some flow rate data relative to microchannel no. 3, whose measured depth is $2h = 1.15$ μm. The depth for the simulation is adjusted to 1.16 μm, that is, 0.01 μm more than the measured value, from data at low values of Kn'_o, for which both NS1 and NS2 models accurately predict the experimental flow rates. An increase of the rarefaction leads to a deviation between these two models. The agreement between the experimental data and the NS2 model is good, whatever the gas—nitrogen (Figure 2.21(b)) or helium (Figure 2.21(c))—with $\sigma = 0.93$. The rarefaction is increased even more, and the difference between NS1 and NS2 models becomes very significant (Figure 2.21(d)). The experiment shows the validity of the second-order slip flow model with Knudsen numbers higher than 0.2. This observation is confirmed with microchannel no. 4, up to Knudsen numbers around 0.25.

For nitrogen and helium flows, an attempt to summarize the results obtained with microchannels nos. 2, 3, and 4 is shown in Figure 2.22. The dimensionless flow rate $1/\dot{m}^*$, little sensitive to the aspect ratio a^*, is plotted in function of $K\acute{n}_{o,M}$ for an inlet over outlet pressure ratio $\Pi = 1.8$. Up to $Kn'_{o,M} = 0.05$, the first-order (NS1) and the second-order (NS2) slip flow models predict the same flow rate. For higher values of Kn'_o, the deviation is significant and the experimental data are in very good agreement with the NS2 model, up to $Kn'_{o,M} = 0.25$. Beyond that, the agreement is not so fine, although the experimental data are closer to

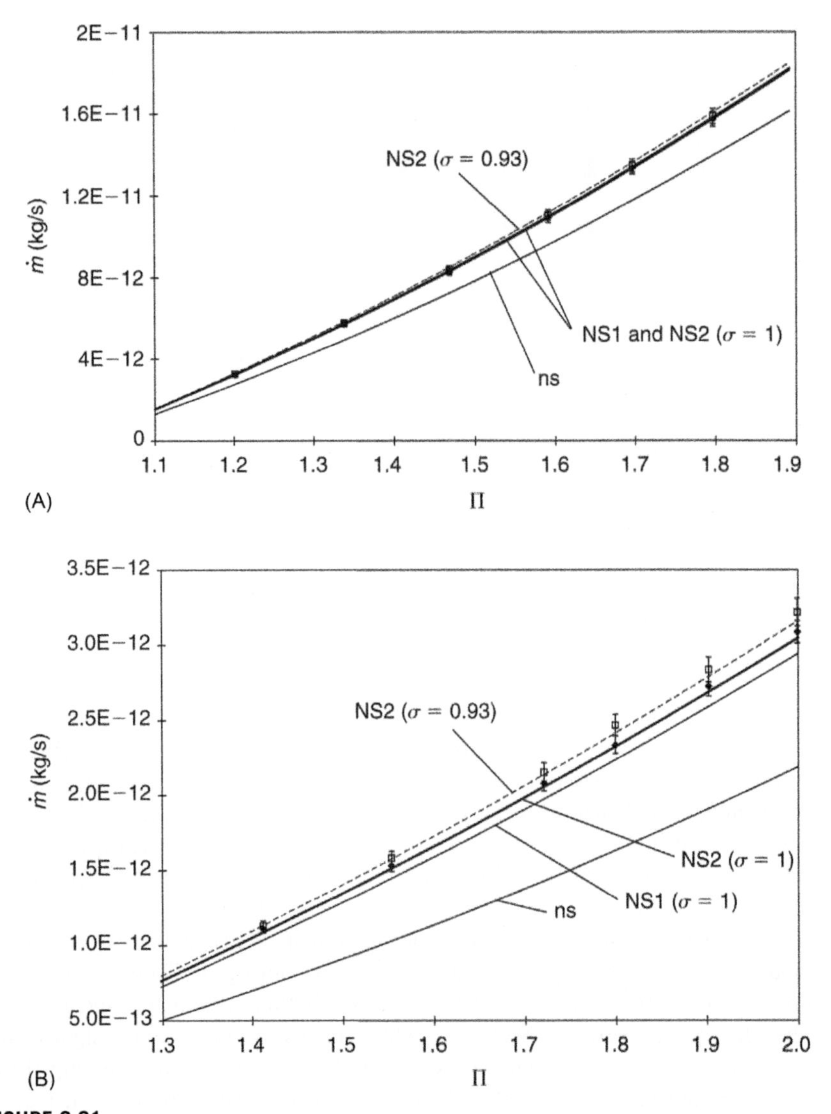

FIGURE 2.21

Theoretical and experimental flow rates for a rectangular microchannel. $2h = 1.16\ \mu m$, $2b = 21\ \mu m$, $l = 5000\ \mu m$, $T = 294.2$ K. (A) Gas: N_2, $p_o = 1.9 \times 10^5$ Pa, $Kn'_{o,VSS} = 2.5 \times 10^{-2}$; (B) gas: N_2, $p_o = 0.65 \times 10^5$ Pa, $Kn'_{o,VSS} = 7.3 \times 10^{-2}$; (C) gas: He, $p_o = 1.9 \times 10^5$ Pa, $Kn'_{o,VSS} = 7.9 \times 10^{-2}$; (D) gas: He, $p_o = 0.75 \times 10^5$ Pa, $Kn'_{o,VSS} = 2.0 \times 10^{-1}$.

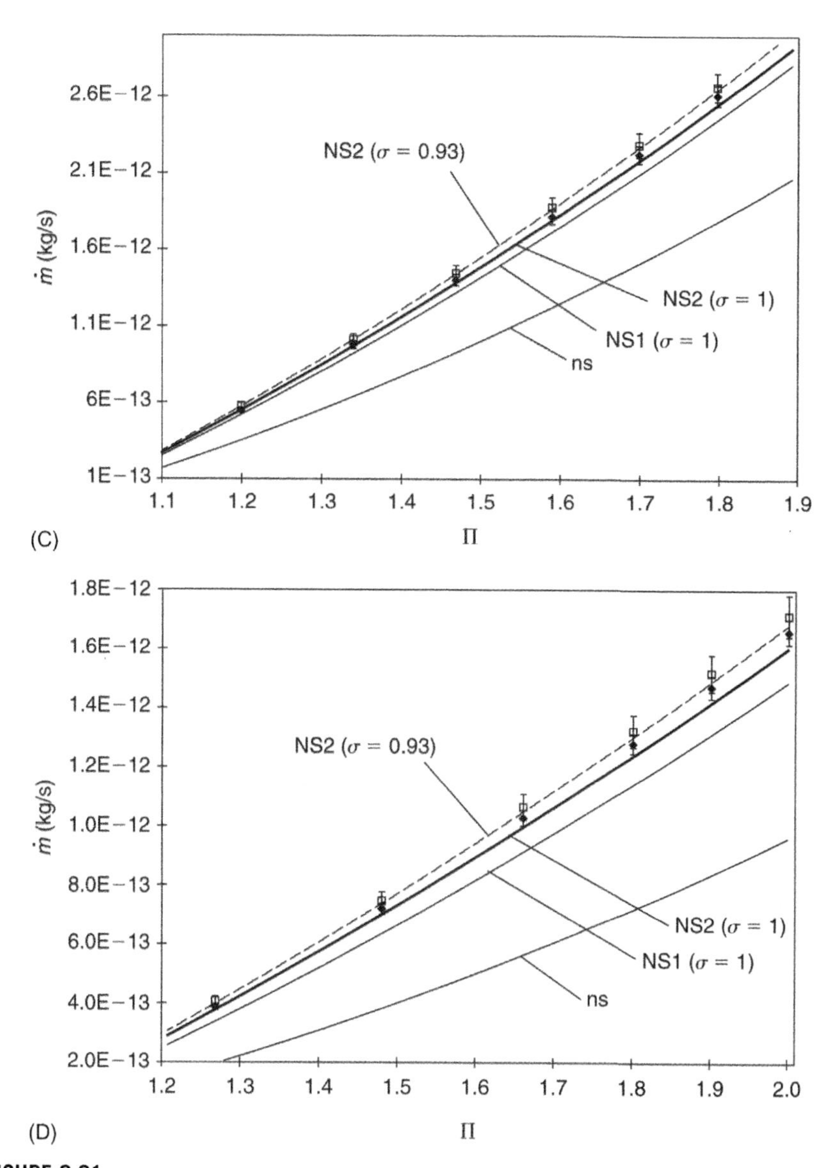

FIGURE 2.21

(Continued)

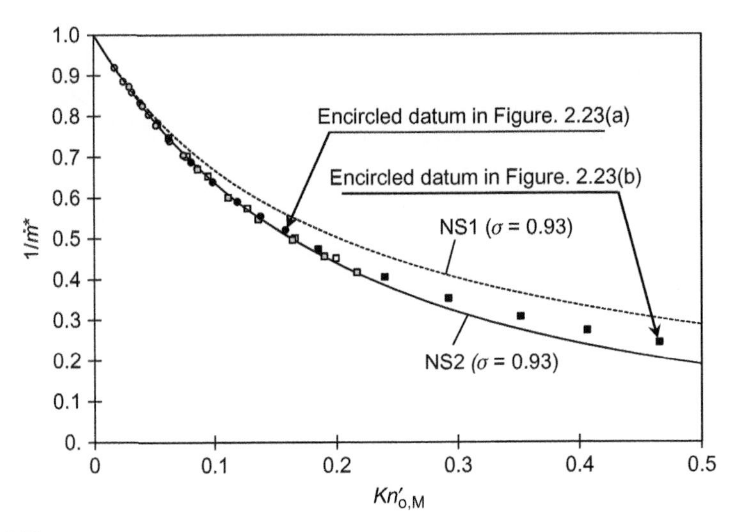

FIGURE 2.22

Inverse reduced flow rate ($1/\dot{m}^*$) in rectangular microchannels: comparison of experimental data with NS1 and NS2 slip flow models; $\Pi = 1.8$; $T = 294.2$ K; microchannels no. 2 (white), no. 3 (gray), no. 4 (black); gas: N_2 (circle) and He (square).

NS2 than to NS1 predictions. In the paper (Colin et al., 2004), the authors used the NS2 model based on Deissler boundary conditions (2.147)–(2.148); they found good agreement for an accommodation coefficient $\sigma = 0.93$. Using more accurate second-order boundary conditions, such as (2.161)–(2.162), the same conclusions hold, but with an accommodation coefficient close to unity (Pitakarnnop et al., 2010).

The experiments of Maurer et al. (2003) follow the same trend. The latest precise experimental data provided in the literature (Table 2.6) concern different gases and are relative to microchannels with cross-sections of comparable lengths l and depths $2h$, but manufactured with different processes. Arkilic et al. (2001) used two silicon wafers facing each other, Colin et al. (2004) and Pitakarnnop et al. (2010) studied microchannels etched by DRIE in silicon and covered with a glass sealed by anodic bonding, and Maurer et al. (2003) used microchannels etched in glass and covered with silicon. The lower and upper walls of the microchannels fabricated by Zohar et al. (2002) were obtained by deposition of low-stress silicon nitride layers on a silicon wafer.

A smart comparison of the data provided by these authors is also tricky, since the aspect ratios a^* of the sections are different from one study to another. The more recent studies (Colin et al., 2004; Ewart et al., 2007; Maurer et al., 2003; Pitakarnnop et al., 2010) confirm that an adequate slip flow model based on second-order boundary conditions can be accurate for high Knudsen numbers that usually apply to the transition regime.

Table 2.6 Recent Experimental Data of Gas Flows in Rectangular Microchannels

	Microchannel Dimensions			Experimental Conditions		Theoretical Comparison	
Reference	**l (μm)**	**$2h$ (μm)**	**a^* (%)**	**Gas**	**Kn'_o**	**σ**	**Model**
Shih et al. (1996)	4000	1.2	3.0	He	0.16	1.16	Plane, NS1,
				N_2	0.055	0.99	$\beta_{u1} = 1$
Arkilic et al. (2001)	7490	1.33	2.5	Ar	0.05–0.41	0.80	Plane, NS1,
				N_2	0.05–0.34	0.80	$\beta_{u1} = 1$
				CO_2	0.03–0.44	0.80	
Lalonde (2001); Colin et al. (2004)	5000	0.54–4.48	1.1–8.7	N_2	0.004–0.16	0.93; 1.00	Rectangular,
				He	0.029–0.47	0.93; 1.00	NS2, $k_2 = k_{2,M}$, $A_2 = -9/8$, $\beta_{u1} = 1$; Plane, QHD1, $k_2 = k_{2,VSS}$
Zohar et al. (2002)	4000	0.53; 0.97	1.3; 2.4	He	0.21; 0.38	1.00	Plane, NS1,
				Ar	0.11; 0.20	1.00	$k_2 = k_{2,M}$,
				N_2	0.065; 0.12	1.00	$\beta_{u1} = 1$,
Maurer et al. (2003)	10000	1.14	0.6	N_2	0.054–1.1	0.87	Plane, NS2
				He	0.17–1.46	0.91	$\beta_{u1} = 1$ $A_2 = -0.23_{(N2)}$, $A_2 = -0.26_{(He)}$
Ewart et al. (2007)	9390	9.38	1.9	He	0.064–171	0.91	Plane, NS2
Pitakarnnop et al. (2010)	5000	1.88	8.9	Ar	0.079–1.98	1.00	Rectangular
				He	0.203–5.38	1.00	$\beta_{u1} = 1.1466$,
				Ar–He (70%–30%)	0.088	1.00	$A_2 = -0.647$

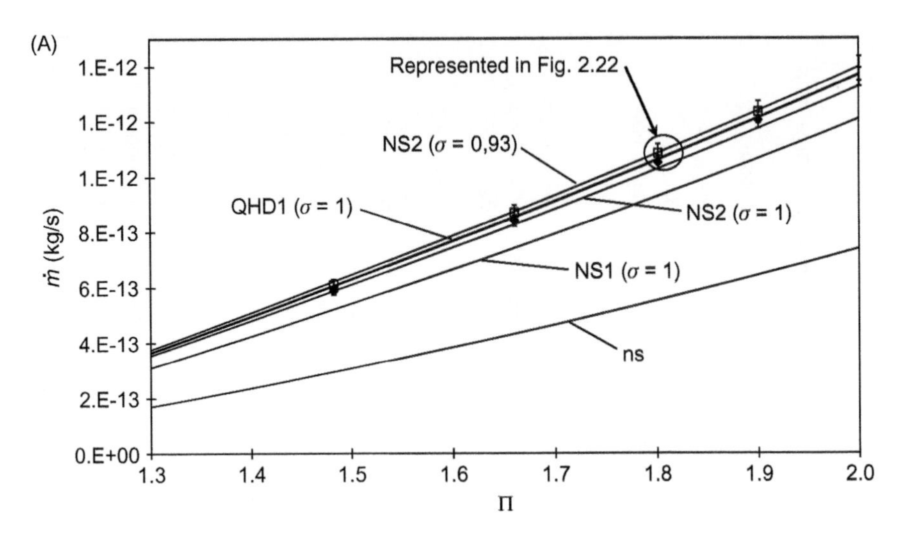

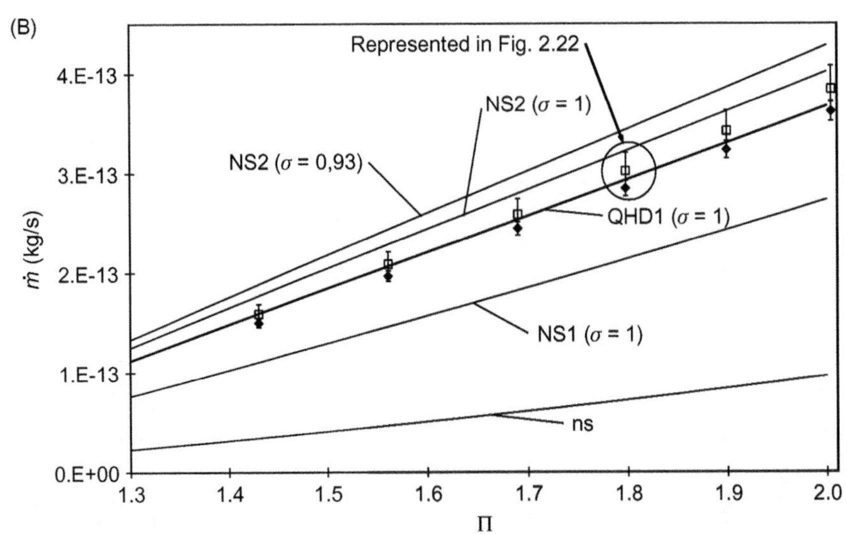

FIGURE 2.23

Experimental data from Lalonde (outlet flow rate □ and inlet flow rate ♦), compared to NS2 and QHD1 models. Microchannel no. 4. $p_o = 75$ kPa, $T = 294.2$ K. (A) Gas: N_2; $Sc = 0.74$; the encircled data correspond to $Kn'_{o,M} = 0.15$. (B) Gas: He; $Sc = 0.77$; the encircled data correspond to $Kn'_{o,M} = 0.47$.

The experimental data from Lalonde have also been compared (Colin et al., 2003) to the mass flow predicted by a QHD model with first-order boundary conditions (QHD1). For microchannel no. 4, with a low aspect ratio $a^* = 0.011$, the plane assumption is acceptable and Eq. (2.112) is valid. Figure 2.23 shows that the QHD1 model could be accurate for Knudsen numbers higher than the NS2 model, with an accommodation coefficient $\sigma = 1$ both for nitrogen and helium. Actually, the analysis of this result should be qualified. For this comparison, the mean free path was calculated using the formula given by Maxwell ($k_{2,M} = \sqrt{\pi/2}$) for the NS2 as well as for the QHD1 simulations, but the Schmidt value (0.74 for nitrogen and 0.77 for helium) used for the QHD1 simulation is very close to the one obtained from a VSS model, given by Eq. (2.46). Another comparison of the QHD1 model with the data of Lalonde is proposed in Elizarova and Sheretov (2003), using $k_{2,VSS}$ calculated from the VSS model (cf. Table 2.1). In this case, the better fit is found for $\sigma = 1$ for helium and $\sigma = 0.93$ for nitrogen.

Other data show the same trend for flows in circular microtubes (Perrier et al. 2011; Yamaguchi et al., 2011).

To conclude, it is clear that the NS2 or QHD1 models improve the accuracy of the NS1 model, but further studies that include smart experimental data are necessary to dissociate the roles of the accommodation coefficient, the collision model, and the slip flow model.

2.3.5.3 Pressure data

Pressure measurements along a microchannel have been successfully performed by different groups (Lee et al., 2002; Liu et al., 1995; Jang and Wereley, 2004; Zohar et al., 2002) using integrated pressure microsensors connected to the microchannels via capillary connections (Figure 2.24). These experiments confirm the theoretical pressure distribution shown in Figure 2.12. Techniques have been developed recently to measure the pressure field along the wall. They are based on the luminescent properties of pressure-sensitive molecules applied in molecular films at the wall (Matsuda et al. 2007; Matsuda et al., 2011).

2.3.5.4 Flow visualization

Although micro-particle image velocimetry (micro-PIV) has been widely used to visualize liquid microflows for several years (Wereley et al., 1998), this technique is hardly applicable to gaseous microflows due to seeding and inertia issues (Sinton, 2004; Wereley et al., 2002). It is possible to use very small particles, but they are subject to the Brownian motion of the molecules, particularly in the more rarefied regimes and the recognition of a pattern of particles becomes a real issue. The current available data using micro-PIV for gas flows at small scale concern millimetric channels in the continuum regime (Yoon et al., 2006). The molecular tagging velocimetry (MTV) could be an alternative and interesting technique for gas microflow visualization. This method was used to obtain the velocity profile of water in microtubes 705 μm in diameter, with a spatial resolution lower than 4% (Webb and Maynes, 1999). The technique was adapted to the analysis of gaseous subsonic or supersonic microjets generated by a 1-mm diameter nozzle (Lempert et al., 2003). The gas was nitrogen seeded with acetone, and the spatial resolution was around 10 μm, with a maximal uncertainty of 5%

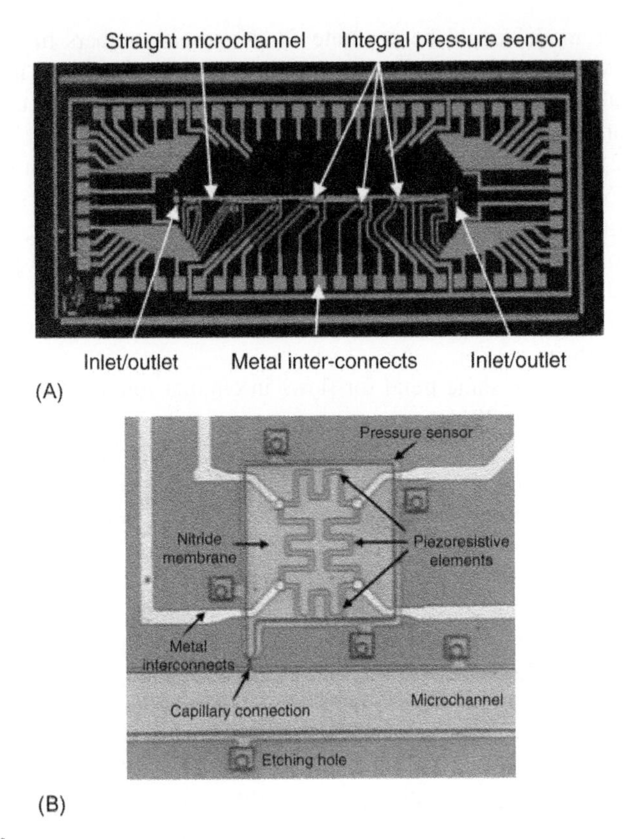

FIGURE 2.24

Microchannel with integrated pressure sensors. (A) Microchannel with inlet and outlet; (B) close-up of the pressure sensor and its capillary connection with the microchannel.

Source: *From Zohar et al. (2002).*

for a Knudsen number between 0.002 and 0.035, that is, in the slip flow regime. Recently, first velocity fields have been successfully measured in millimetric channels in nonrarefied regimes (Samouda et al., 2012).

2.3.6 Entrance effects

In the previous sections, microchannels have been considered as long microchannels and the entrance effects have been neglected. If the microchannel is short, such that the hydrodynamic development length L_{hd} is no longer negligible compared to the microchannel length, entrance effects have to be taken into account by a numerical approach. Rarefaction increases L_{hd} in the slip flow regime, due to slip at the walls. Barber and Emerson (2002) used a two-dimensional finite-volume Navier–Stokes solver to simulate the flow of a gas entering a plane microchannel in the slip-flow regime, with first-order boundary conditions. Their simulated data, for $Kn \in [0; 0.1]$ and $Re \in [0; 400]$, were fitted by a least-squares technique and led to the following expression for the hydrodynamic development length:

$$\frac{L_{\mathrm{hd}}}{D_{\mathrm{H}}} = \frac{0.332}{0.0271Re + 1} + 0.011Re\frac{1 + 14.78Kn}{1 + 9.78Kn} \tag{2.163}$$

which generalizes the expression $L_{\mathrm{hd}}/D_{\mathrm{H}} = 0.315/(0.0175Re + 1) + 0.011Re$ found by Chen (1973) for a no-slip flow with a slight modification of the nonrarefied part of the equation. In Eq. (2.163), the Knudsen number $Kn = \lambda/D_{\mathrm{H}}$ and the Reynolds number $Re = \rho\bar{u}_z D_{\mathrm{H}}/\mu$ are based on the hydraulic diameter $D_{\mathrm{H}} = 4h$ of the microchannel, $2h$ in depth. The hydrodynamic development length is arbitrarily defined as the longitudinal distance required for the centerline velocity to reach 99% of its fully developed value. Equation (2.163) shows that for $Kn = 0.1$ and $Re = 400$, slip at the wall increases the entrance length of about 25%.

2.4 Pulsed gas flows in microchannels

Rarefaction affects steady microflows and modifies the behavior of unsteady microflows. The case of pulsed flows is of particular interest because such flows are encountered in many applications. For example, micropump actuators often generate sinusoidal pressure fluctuations inside a chamber in order to induce a flow in microchannels through microvalves or microdiodes.

Detailed models for pulsed gaseous flows in the slip flow regime can be found in Caen et al. (1996) for circular microtubes or in Colin et al. (1998a) for rectangular microchannels. They are based on the Navier–Stokes equations with first-order slip and temperature jump conditions at the walls. If we consider a microchannel of length l, whose inlet is submitted to a sinusoidal pressure fluctuation with a small amplitude, the gain Δp^* of the microchannel, that is, the ratio of the outlet Δp_{o} over inlet Δp_{i} fluctuating pressures amplitudes, can be calculated. For example, in the simple case of a microchannel closed at its outlet, this gain takes the form

$$\Delta p^* = \left|\frac{\Delta p_{\mathrm{o}}}{\Delta p_{\mathrm{i}}}\right| = \left|\frac{1}{\cosh(\beta l)}\right| \tag{2.164}$$

where $\beta(f, Kn) \approx \sqrt{f}\Psi(Kn)$ depends on the square root of the frequency f and on the Knudsen number. It can be shown that the band pass of the microchannel is underestimated when slip at the walls is not taken into account (Figure 2.25).

Obtaining experimental data to discuss these theoretical results is very hard, due to the very small size of the pressure sensors required for these experiments. However, experimental data were obtained for microtubes down to 50 μm in diameter, using a commercially available pressure microsensor placed in a minichannel connected in series to one or several parallel microtubes (Colin et al., 1998b).

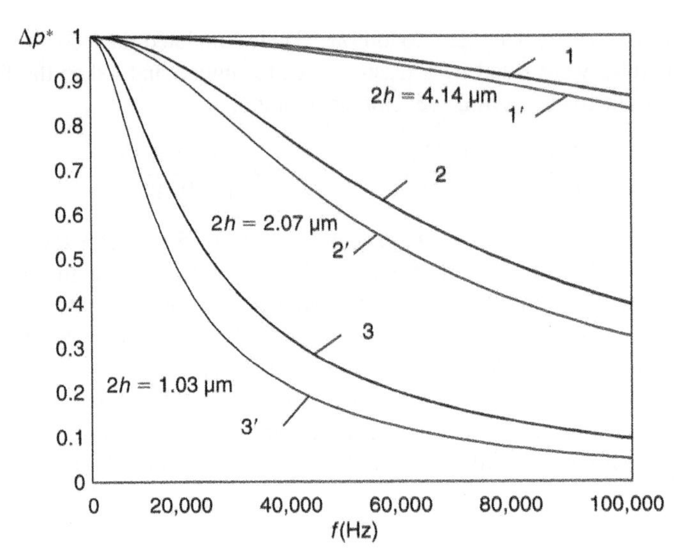

FIGURE 2.25

Influence of microchannel depth $2h$ on the gain Δp^* for a square microchannel closed at its outlet. $l = 100\ \mu\text{m}$, $\overline{T} = 293$ K, $\overline{p} = 0.11$ MPa. 1, 2, 3: slip flow; 1′, 2′, 3′: no-slip flow.

The behavior of the micro- and minichannels' association was simulated and compared to these data. The agreement was good for both the gain and the phase of the association.

The model for pulsed flows in constant-section microchannels can be easily extended to microchannels with slowly varying cross-sections (Aubert et al., 1998). This allows an understanding of the behavior of microdiffusers subjected to sinusoidal pressure fluctuations and to test the diode effect of a microdiffuser/nozzle placed in a microchannel that is subjected to sinusoidal pressure fluctuations. In Figure 2.26, two layouts (A) and (B) are considered.

In layout (A), the tested element is a microdiffuser, with an increasing section from inlet to outlet. In layout (B), the same element is used as a nozzle with a decreasing section from inlet to outlet. To obtain an exploitable comparison between the two layouts, the widths ($2b_1$, $2b_2$, $2b_3$) and the lengths (l_1, l_2, l_3) are the same in layouts (A) and (B). The angle 2α is also the same for the diffuser or the nozzle.

An analysis of the frequency behavior for each layout shows significant differences, which is characteristic of a dynamic diode effect. Therefore, in order to characterize the dissymmetry of the transmission of the pressure fluctuations, an efficiency E of the diode is introduced. This efficiency

$$E = \frac{\Delta p^*_{(a)}}{\Delta p^*_{(b)}} = \frac{|\Delta p_o/\Delta p_i|_{(a)}}{|\Delta p_o/\Delta p_i|_{(b)}} \tag{2.165}$$

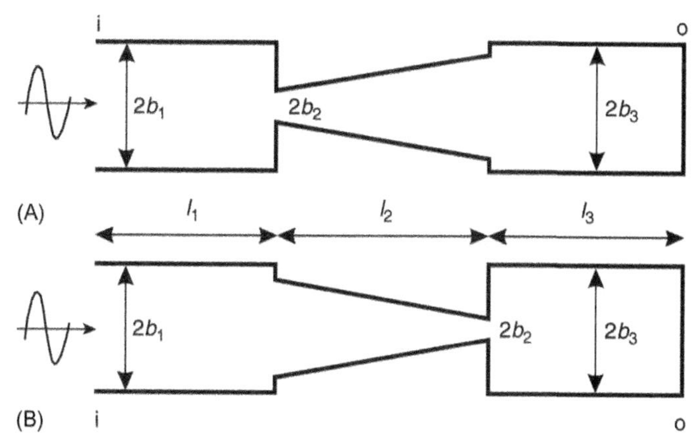

FIGURE 2.26

Two layouts of a microdiode placed in a microchannel.

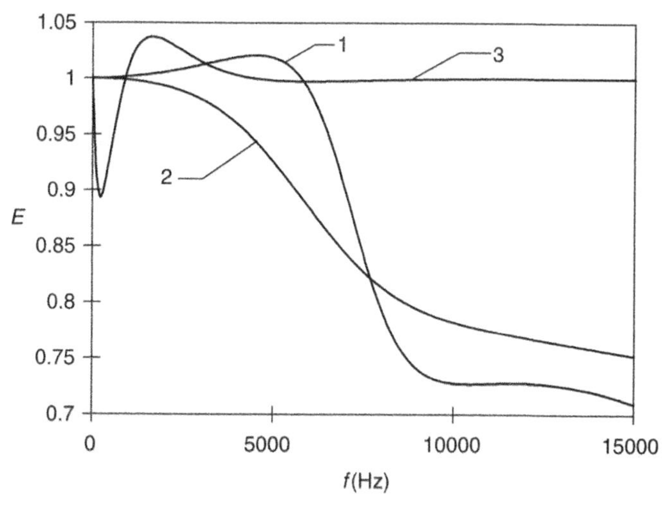

FIGURE 2.27

Diode efficiency for different values of the depth. 1: $2h = 100\ \mu m$; 2: $2h = 62\ \mu m$, 3: $2h = 6\ \mu m$.

defined as the ratio of the fluctuating pressure gain in layout (A) over the fluctuating pressure gain in layout (B), can be studied as a function of the frequency f. Some results are shown in Figure 2.27 for different values of the depth $2h$ with $l_1 = l_2 = l_3 = 3$ mm, $2b_1 = 2b_3 = 467\ \mu m$, $2b_2 = 100\ \mu m$, and $\alpha = 3.5°$.

Figure 2.27 points out that with a microdiffuser ($2h = 6\ \mu m$), E appears to be less than unity *below* a critical frequency. This denotes a reversed diode effect

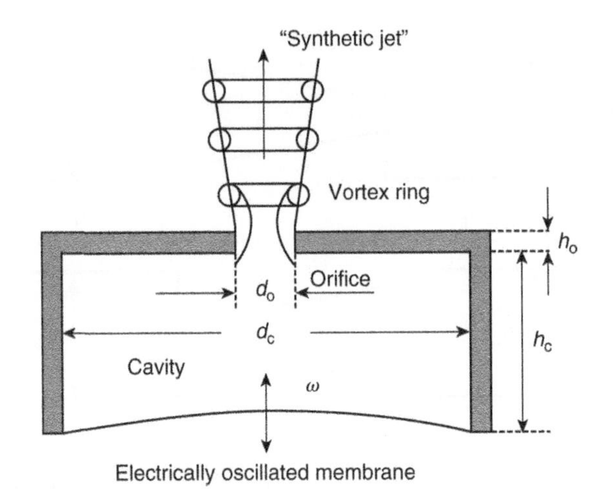

FIGURE 2.28

Schematic of a microsynthetic jet actuator.

Source: *From Mallinson et al. (2003).*

compared to the case of a diffuser with submillimetric dimensions ($2h = 100$ μm), for which E is less than unity *beyond* a critical frequency. Note also that the influence of slip (taken into account in Figure 2.27) is not negligible when $2h = 6$ μm. However, slip has little influence on the values of characteristic frequencies (i.e., the frequencies for which E is an extremum or equal to unity).

Pulsed gaseous flows are also of great interest in short microchannels, or micro-orifices, for example, in synthetic microjets. The synthetic microjet is a low-power, highly compact microfluidic device that has potential for application in boundary layer control (Lee et al., 2003). The oscillation of an actuated membrane at the bottom of a cavity generates a pulsed flow through a micro-slit or orifice (Figure 2.28). It results in a zero net-mass flux, but in a nonzero mean momentum flux. A series of such microjets can be used to control separation or transition in the boundary layer of the flow on a wing in order to reduce drag, to increase lift, or to limit noise generated by the air flow.

2.5 Thermally driven gas microflows and vacuum generation

Generating vacuum by means of microsystems concerns various applications such as the taking of biological or chemical samples, or the control of the vacuum level in the neighborhood of some specific microsystems during working.

The properties of rarefied flows (due to both low pressure and small dimensions) allow unusual pumping techniques. For high Knudsen numbers, the flow may be generated without any moving mechanical component by only using thermal actuation, which is not possible with classical macropumps.

2.5.1 Transpiration pumping

Currently, the most studied technique is based on thermal transpiration. The basic principle requires two chambers filled with gas and linked with an orifice whose hydraulic diameter is small compared with the mean free path of the molecules. Chamber 1 is heated, for example, with an element, so that $T_1 > T_2$. By analyzing the probability that some molecules move from one chamber across the orifice, it can be shown that if the pressure is uniform ($p_1 = p_2 = p$), a molecular flux

$$\dot{N}_{2\to 1} = \frac{Ap}{\sqrt{2\pi m k_B}} \frac{\sqrt{T_1} - \sqrt{T_2}}{\sqrt{T_1 T_2}} \tag{2.166}$$

from chamber 2 toward chamber 1—from cold to hot temperatures—appears (Lengrand and Elizarova, 2010; Muntz and Vargo, 2002). If the net molecular flux is constant, the pressures necessarily satisfy

$$\frac{p_1}{p_2} = \sqrt{\frac{T_1}{T_2}} \tag{2.167}$$

In Eq. (2.166), m is the mass of a molecule and A is the area of the orifice. If $p_2 < p_1 < p_2\sqrt{T_1/T_2}$, there is a net flow from 2 toward 1, which results in a decrease of p_2 and/or an increase of p_1. This indicates that a basic microscale pump is working.

Thus, a Knudsen compressor (Figure 2.29) can be designed by connecting a series of chambers with very small orifices that have a cold region (temperature T_2) on one side and a hot region (temperature T_1) on the other side by means of an adequate local heater placed just downstream of the orifices. This multistage layout leads to important cumulated pressure drops, while keeping a satisfactory flow rate. In practice, the orifices must be replaced by microchannels, and this complicates the modeling (Vargo and Muntz, 1997). Moreover, it is generally difficult to maintain simultaneously a free molecular regime in the microchannels and a continuum regime in the chambers, which are the required conditions for the optimal efficiency corresponding to Eqs. (2.166) and (2.167). More elaborate models, based on the linearization of the Boltzmann equation (Loyalka and Hamoodi, 1990), are able to take into account the transition regime ($0.05 < Kn < 10$), both in the microchannels and in the chambers (Vargo and Muntz, 1999). Several design studies are proposed in the literature that show the theoretical feasibility of microscale thermal transpiration pumps. Original designs combining straight and curved microchannels have also been numerically simulated (Aoki et al., 2009).

The first mesoscale prototypes, heated with an element at the wall (Vargo and Muntz, 1997, 1999; Vargo et al., 1999), were tested. An alternative solution

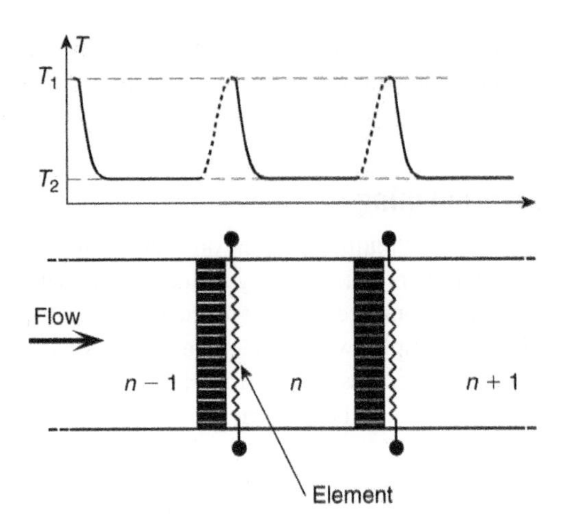

FIGURE 2.29

Multistage Knudsen compressor.

suggesting heating within the gas was also proposed (Young, 1999), but experimental data are not available for this layout.

Actually, there is yet much to do for the theoretical optimization and for the design of thermal transpiration micropumps using microscale manufacturing techniques. The performances of these micropumps remain limited for technological reasons. A high vacuum level may become incompatible with the typical internal sizes of microsystems (Muntz and Vargo, 2002) because the regime in the chambers must be close to a continuum regime, which requires sizes too big for the lower pressures. A 48-stage Knudsen micropump has recently been fabricated with monolithically integrated Pirani gauges (Gupta et al., 2012) and shows promising pumping capabilities. For an input power of 1.35 W, it is able to pump down air from 0.1 Mpa to 7 kPa and from 33 kPa to 0.6 kPa.

2.5.2 Accommodation pumping

Accommodation pumping is another pumping technique. It exploits the property of gas molecules whose reflection on specular walls depends on their temperature. If the wall is warmer than the gas, the mean reflection angle is greater than the incident angle. Conversely, if the wall is colder than the molecules, they have a more tangential reflection. Consequently, in a microchannel with perfectly specular walls connecting two chambers at different temperatures, a flow takes place from the hot toward the cold chambers (Hobson, 1970). If the walls of the microchannel give a diffuse reflection, this effect disappears. This property was confirmed with numerical simulations by the DSMC method (Hudson and Bartel, 1999). By linking two warm chambers (1 and 3) to a cold chamber (2) on one side with a smooth

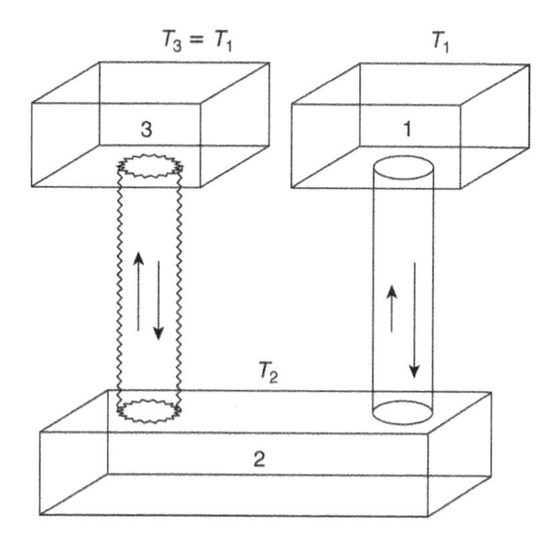

FIGURE 2.30

One stage of an accommodation micropump.

microchannel (1–2) and on the other side with a rough microchannel (3–2), a difference of pressure ($p_1 < p_3$) appears between the two chambers at the same temperature (Figure 2.30).

Connecting in series several stages of that type, the pressure drops relative to each stage can be cumulated, allowing the setup to reach high vacuum levels. The advantage of the accommodation pump is that it is operational without theoretical limitations concerning low pressures and, contrary to the thermal transpiration pumping, the accommodation pumping does not require chambers with high dimensions. To compensate, the accommodation pump requires more stages to reach the same final pressure ratio. A pressure ratio of $\Pi = 100$ requires 125 stages, whereas only 10 stages are enough for a transpiration pump with comparable temperatures. Finally, no operational prototype has been described in the literature so far, although the concept seems attractive. The highest-performance design was proposed by Hobson (1971, 1972); it requires a cold temperature $T_2 = 77$ K for an atmospheric temperature $T_1 = T_3 = 290$ K.

2.6 Heat transfer in microchannels

In the slip flow regime, heat transfer can be treated as well as mass transfer, providing the energy Eq. (2.41) and the temperature boundary conditions (2.61) are solved with the momentum equation and the slip-flow boundary conditions. Such heat transfer problems have an analytical solution only in the simplest cases. For example, if the compressibility of the flow can be neglected, the energy and momentum equations are no longer coupled, and if the flow is developed, longitudinal temperature gradients are constant and longitudinal velocity gradients are zero (see example in Section 2.6.1.1). In the general case, analytical or semi-analytical solutions are not possible and numerical simulations are required. A detailed review of convective heat transfer in the slip flow regime can be found in (Colin, 2012).

Depending on the accommodation coefficients, heat transfer can be found to increase, decrease, or remain unchanged, compared to non-slip-flow conditions.

2.6.1 Heat transfer in a plane microchannel

2.6.1.1 Heat transfer for a fully developed incompressible flow

Let us consider the case of a plane microchannel (Figure 2.11(a)), with a lower adiabatic wall ($q_w|_{y=-h} = 0$) and an upper wall with constant heat flux ($q_w|_{y=h} = q_h$). The flow is assumed to be hydrodynamically and thermally developed and incompressible. Fluid properties are assumed uniform. Noting $y^* = y/h$, the momentum equation reduces to

$$\frac{d^2 u_z}{dy^{*2}} = \frac{h^2}{\mu}\frac{dp}{dz} \tag{2.168}$$

and the energy equation to

$$\frac{k}{h^2}\frac{\partial^2 T}{\partial y^{*2}} = \rho u_z c_p \frac{\partial T}{\partial z} \tag{2.169}$$

neglecting viscous dissipation. We introduce

$$\theta^* = \frac{T - T_{wall}}{q_h h/k} = \frac{4}{Nu}\frac{T - T_{wall}}{\overline{T} - T_{wall}} \tag{2.170}$$

with $q_h = \alpha(\overline{T} - T_{wall})$ and $Nu = \alpha D_H/k = 4\alpha h/k$, noting α as the convection heat transfer coefficient. The flow being thermally developed, $\partial\theta^*/\partial z = 0$, $dNu/dz = 0$, and $dq_h/dz = 0$, which implies $\partial T/\partial z = d\overline{T}/dz = dT_{wall}/dz$. Consequently, the energy equation can be written as

$$\frac{q_h}{h}\frac{d^2\theta^*}{dy^{*2}} = \rho u_z c_p \frac{d\overline{T}}{dz} \tag{2.171}$$

Moreover, the conservation of energy on a control volume of length dz and height $2h$ yields $2h\rho c_p \overline{u_z} d\overline{T} = -q_h\, dz$ and Eq. (2.171) reduces to

$$\frac{d^2\theta^*}{dy^{*2}} = -\frac{u_z}{2\overline{u_z}} \tag{2.172}$$

Neglecting thermal creep, which is a second-order effect, the first-order boundary conditions are

$$\left.\frac{du_z}{dy^*}\right|_{y^*=0} = 0; \quad u_z|_{y^*=1} = -\xi_h^* \left.\frac{du_z}{dy^*}\right|_{y^*=1} \tag{2.173}$$

$$\left.\frac{d\theta^*}{dy^*}\right|_{y^*=-1} = 0; \quad \theta^*|_{y^*=1} = -\varsigma_h^* \left.\frac{d\theta^*}{dy^*}\right|_{y^*=1} \tag{2.174}$$

with a dimensionless coefficient of slip $\xi_h^* = \xi/h = (\lambda/h)\ \beta_{u1}(2-\sigma)/\sigma = 4\ \beta_{u1} Kn\ (2-\sigma)/\sigma$ and a dimensionless temperature jump distance $\varsigma_h^* = \varsigma/h = (\lambda/h)\ \beta_{T1}[(2-\sigma_T)/\sigma_T][2\gamma/(\gamma+1)]/Pr = 8\ \beta_{T1}\ [(2-\sigma_T)/\sigma_T][\gamma/(\gamma+1)][Kn/Pr]$. The resolution of this set of equations gives the velocity distribution

$$u_z = -\frac{h^2}{2\mu}\frac{dp}{dz}(1 - y^{*2} + 2\xi_h^*) \tag{2.175}$$

and, with $\overline{u_z} = \frac{1}{2}\int_{-1}^{1} u_z\, dy^*$,

$$\frac{u_z}{\overline{u_z}} = \frac{3(-y^{*2} + 1 + 2\xi_h^*)}{2(1 + 3\xi_h^*)} \tag{2.176}$$

The integration of Eq. (2.172) with the boundary conditions (2.174) yields

$$\theta^* = \varsigma_h^* + \frac{1}{1+3\xi_h^*}\left[\frac{1}{16}y^{*4} - \frac{3}{8}y^{*2}(1+2\xi_h^*) - \frac{1}{2}y^*(1+3\xi_h^*) + \left(\frac{13}{16} + \frac{9}{4}\xi_h^*\right)\right] \tag{2.177}$$

T and $\overline{T} = (1/2)\int_{-1}^{1}(u_z/\overline{u_z})T\, dy^*$ are calculated and the Nusselt number is deduced as

$$Nu = \left(\frac{\varsigma_h^*}{4} + \frac{26 + 147\xi_h^* + 210\xi_h^{*2}}{140(1+3\xi_h^*)^2}\right)^{-1} \tag{2.178}$$

Equation (2.178) shows that heat transfer depends both on the dimensionless coefficient of slip ξ_h^* and on the dimensionless temperature jump distance ς_h^*; that is, it depends on the two accommodation coefficients σ and σ_T as well as on the Knudsen number Kn. For classical values of the accommodation coefficients, the effect of the temperature jump is greater than the effect of velocity slip, and the heat transfer decreases when rarefaction increases.

The Nusselt number reduces to the classic value 70/13 (Rohsenow et al., 1998) when $Kn = 0$.

For a symmetrically heated microchannel, the solution is

$$Nu = \left(\frac{\varsigma_h^*}{4} + \frac{17 + 84\xi_h^* + 105\xi_h^{*2}}{140(1 + 3\xi_h^*)^2} \right)^{-1} \quad (2.179)$$

with the no-slip limit solution 140/17.

Zhu et al. (2000) extended these solutions to the problem of asymmetrically heated walls ($0 \neq q_0 \neq q_h$) by combining the solutions of two subproblems similar to the above problem.

The problem with constant wall temperature was treated by Hadjiconstantinou and Simek (2002), who extended the study to the transition regime using DSMC. They showed that the slip flow prediction (with first-order boundary conditions) is in good agreement with the DSMC results for $Kn \leq 0.2$ and remains a good approximation beyond its expected range of applicability.

2.6.1.2 Heat transfer for a developing compressible flow

This problem was studied by Kavehpour et al. (1997) using the same first-order slip flow and temperature jump boundary conditions as in Section 2.6.1.1. The flow was developing both hydrodynamically and thermally, and two cases were considered: uniform wall temperature and uniform wall heat flux. Viscous dissipation was neglected, but compressibility was taken into account. The numerical methodology was based on the control volume finite difference scheme. It was found that the Nusselt number was substantially reduced for slip flow compared with continuum flow. Additional effects have been considered by several authors: Yu and Ameel (2000) and Mikhailov and Cotta (2005) took into account axial conduction in the constant wall temperature case, whereas Jeong and Jeong (2006) also took into account viscous dissipation for both uniform wall temperature and uniform wall heat flux cases.

2.6.2 Heat transfer in a circular microtube

Heat transfer in circular microtubes was simulated first by Sparrow and Lin (1962), who considered a hydrodynamically and thermally fully developed incompressible flow, with constant heat flux at the wall. Viscous dissipation and thermal creep were neglected and first-order velocity slip and temperature jump conditions were used. With these assumptions, they obtained the following exact solution:

$$Nu = \left(\frac{\varsigma_r^*}{4} + \frac{11 + 64\xi_r^* + 96\xi_r^{*2}}{48(1 + 4\xi_r^*)^2} \right)^{-1} \quad (2.180)$$

with a dimensionless coefficient of slip $\xi_r^* = \xi/r_2 = (\lambda/r_2)\ \beta_{u1}(2-\sigma)/\sigma = 2Kn$ $\beta_{u1}\ (2-\sigma)/\sigma$ and a dimensionless temperature jump distance $\varsigma_r^* = \varsigma/r_2 = (\lambda/r_2)$ $\beta_{T1}[(2-\sigma_T)/\sigma_T][2\gamma/(\gamma+1)]/Pr = 4\ \beta_{T1}\ [(2-\sigma_T)/\sigma_T][\gamma/(\gamma+1)][Kn/Pr]$. Equation (2.180) reduces to the no-slip value $Nu = 48/11$ when $Kn = 0$. A semi-analytical solution is also given in Ameel et al. (1997) for the entrance region. For a Knudsen number $Kn = 0.04$, it is found that the Nusselt number for the fully developed flow is reduced 14% over that obtained with a no-slip boundary condition, with $\gamma = 1.4$ and $Pr = 0.7$. The effect of viscous dissipation is analyzed in (Aydın and Avcı, 2006).

Sparrow and Lin (1962) also treated the case of constant temperature at the wall. Their solution is tabulated as a function of ξ_r^* and ς_r^* in (Colin, 2012).

2.6.3 Heat transfer in a rectangular microchannel

Yu and Ameel (2001) studied the heat transfer problem for a thermally developing flow in a rectangular microchannel. The flow was assumed to be incompressible and hydrodynamically fully developed, and a constant temperature was imposed at the walls. The velocity distribution was developed in series, as discussed in Section 2.3.4.1, and the energy equation was solved by a modified generalized integral transform method. A competition between ξ and ς was pointed out, with a transition value β_c of $\beta = \xi/\varsigma$ that depends on the aspect ratio a^* (Table 2.7).

For $\beta < \beta_c$, heat transfer is enhanced in comparison with the no-slip flow and if $\beta > \beta_c$, heat transfer is reduced. Moreover, for $\beta < \beta_c$, heat transfer increases with increasing ξ^* and it decreases if $\beta > \beta_c$. According to the values available in the literature, it should be noted that β may vary over a wide range, from unity to 100 or more, for actual wall surface conditions. However, the value $\beta = 1.67$, which corresponds to a diffuse reflection ($\sigma = 1$) and a total thermal accommodation ($\sigma_T = 1$) for air ($\gamma = 1.4$; $Pr = 0.7$), is a representative value for many engineering applications. In this typical case, rarefaction effects result in a decrease of heat transfer, which can be as much as 40%.

Other kinds of thermal boundary conditions have been treated in the literature, including the eight different versions of constant wall heat flux. A review of these studies can be found in (Colin, 2012).

Table 2.7 Transition Values of $\beta = \xi/\varsigma$ as a Function of the Aspect Ratio

a^{*-1}	1	2	3	4	5	6	8	10	∞
β_c	0.67	0.50	0.38	0.32	0.29	0.28	0.27	0.25	0.20

Source: *From Yu and Ameel (2001)*

2.7 Future research needs

Theoretical knowledge is currently well advanced for gas flows in microchannels. However, there is still a need for accurate experimental data for both steady and unsteady gas microflows, with or without heat transfer. For example, in order to definitely validate the choice of the best boundary conditions in the slip flow regime, we need to isolate the influence of the accommodation coefficients, as it remains an open issue. Relationships between their values, the nature of the substrate, and the microfabrication processes involved are currently not available. Theoretical investigations relative to unsteady or thermally driven microflows would also need to be supported by smart experiments.

2.8 Solved examples

Example 2.1

A microchannel with constant rectangular cross-section is submitted to a pressure-driven flow of argon. The inlet and the outlet pressures are $p_i = 0.2$ MPa and $p_o = 25$ kPa, respectively. The geometry is given in Figure 2.11(c), with a width $2b = 20$ μm, a depth $2h = 1$ μm, and a length $l = 5$ mm.

Assume a uniform temperature $T = 350$ K.

- **i.** Calculate the Knudsen numbers Kn and Kn' at the inlet and at the outlet of the microchannel, assuming that the intermolecular collisions are accurately described by a VSS model.
- **ii.** Compare with the values obtained from a HS model.
- **iii.** Assuming a diffuse reflection at the wall, calculate the mass flow rate through the microchannel.
- **iv.** Calculate the mass flow rate increase (in %) due to slip at the wall.
- **v.** What would be the underestimation (in %) of the mass flow rate given by a no-slip model, in the same conditions but with an inlet pressure $p_i = 0.1$ MPa?

Solution

(i) Inlet and outlet Knudsen numbers with a VSS model (Answer: $Kn'_i = 3.48 \times 10^{-2}$; $Kn_i = 1.83 \times 10^{-2}$; $Kn'_o = 2.78 \times 10^{-1}$; $Kn_o = 1.46 \times 10^{-1}$).
From Table 2.2, the temperature exponent of the coefficient of viscosity for argon (Ar) is $\omega = 0.81$, the exponent for the VSS model is $\alpha = 1.40$, and the dynamic viscosity under standard conditions is $\mu_0 = 211.7 \times 10^{-7}$ Pa s.

From Eq. (2.7), the viscosity at $T = 350$ K is

$$\mu(T) = \mu_0 (T/T_0)^{\omega} = 2.117 \times 10^{-5} (350/273.15)^{0.81} = 2.588 \times 10^{-5} \text{ Pa s}$$

The molecular weight of argon is given in Table 2.2: $M = 39.95 \times 10^{-3}$ kg/mol, and its gas constant is $R = R_u/M = 8.315/39.95 \times 10^{-3} = 2.08 \times 10^2$ J/kg/K.

The mean free path is given by Table 2.1: $\lambda = k_2\mu(T)\sqrt{RT}/p$, with

$$k_{2,\text{VSS}} = \frac{4\alpha(7-2\omega)(5-2\omega)}{5(\alpha+1)(\alpha+2)\sqrt{2\pi}} = 0.996$$

At the inlet,

$$\begin{aligned}\lambda_{\text{i,VSS}} &= k_2\mu(T)\sqrt{RT}/p_\text{i} = 0.996 \times 2.588 \times 10^{-5} \times \sqrt{2.08 \times 10^2 \times 350}/(2 \times 10^5) \\ &= 34.8 \text{ nm}\end{aligned}$$

and at the outlet,

$$\begin{aligned}\lambda_{\text{o,VSS}} &= k_2\mu(T)\sqrt{RT}/p_\text{o} = 0.996 \times 2.588 \times 10^{-5} \times \sqrt{2.08 \times 10^2 \times 350}/(2.5 \times 10^4) \\ &= 278 \text{ nm}\end{aligned}$$

The Knudsen number based on the hydraulic diameter

$$D_\text{H} = 4bh/(b+h) = 4 \times 10 \times 10^{-6} \times 0.5 \times 10^{-6}/((10+0.5) \times 10^{-6}) = 1.905\mu m$$

is $Kn = \lambda/D_\text{H}$, that is,

$$Kn_\text{i} = \lambda_i/D_\text{H} = 3.48 \times 10^{-8}/(1.905 \times 10^{-6}) = 1.83 \times 10^{-2} \text{ at the inlet and}$$

$$Kn_o = \lambda_o/D_H = 2.78 \times 10^{-7}/(1.905 \times 10^{-6}) = 1.46 \times 10^{-1} \text{ at the outlet}$$

The Knudsen number based on the microchannel depth is $Kn' = \lambda/(2h)$, that is,

$$Kn'_\text{i} = \lambda_\text{i}/(2h) = 3.48 \times 10^{-8}/(1 \times 10^{-6}) = 3.48 \times 10^{-2} \text{ at the inlet and}$$

$$Kn'_\text{o} = \lambda_\text{o}/(2h) = 2.78 \times 10^{-7}/(1 \times 10^{-6}) = 2.78 \times 10^{-1} \text{ at the outlet}$$

(ii) Inlet and outlet Knudsen numbers with a HS model (Answer: $Kn'_{\text{i,HS}} = 4.13 \times 10^{-2}$; $Kn_{\text{i,HS}} = 2.17 \times 10^{-2}$; $Kn'_{\text{o,HS}} = 3.30 \times 10^{-1}$; $Kn_{\text{o,HS}} = 1.73 \times 10^{-1}$).

With a HS model, $\omega = 0.5$ and the viscosity

$$\begin{aligned}\mu(T) &= \mu_0(T/T_0)^{0.5} = 2.117 \times 10^{-5}(350/273.15)^{0.5} \\ &= 2.396 \times 10^{-5} \text{ Pa s is underestimated at 7.4\%.}\end{aligned}$$

The coefficient $k_{2,\text{HS}} = 16/(5\sqrt{2\pi}) = 1.277$ (cf. Table 2.1). Consequently,

$$\begin{aligned}\lambda_{\text{i,HS}} &= k_{2,\text{HS}}\mu(T)\sqrt{RT}/p_\text{i} = 1.277 \times 2.588 \times 10^{-5} \times \sqrt{2.08 \times 10^2 \times 350}/(2 \times 10^5) \\ &= 41.3 \text{ nm}\end{aligned}$$

and

$$\lambda_{o,HS} = k_{2,HS}\mu(T)\sqrt{RT}/p_o = 1.277 \times 2.588 \times 10^{-5} \times \sqrt{2.08 \times 10^2 \times 350}/(2.5 \times 10^4)$$
$$= 330\,\text{nm}$$

which leads to:

$$Kn_{i,HS} = \lambda_{i,HS}/D_H = 4.13 \times 10^{-8}/(1.905 \times 10^{-6}) = 2.17 \times 10^{-2}$$
$$Kn_{o,HS} = \lambda_{o,HS}/D_H = 3.30 \times 10^{-7}/(1.905 \times 10^{-6}) = 1.73 \times 10^{-1}$$
$$Kn'_{i,HS} = \lambda_{i,HS}/(2h) = 4.13 \times 10^{-8}/(1 \times 10^{-6}) = 4.13 \times 10^{-2}$$
$$Kn'_{o,HS} = \lambda_{o,HS}/(2h) = 3.30 \times 10^{-7}/(1 \times 10^{-6}) = 3.30 \times 10^{-1}$$

The overestimation of the Knudsen number calculated from a HS model, compared with the more accurate solution obtained with a VSS model, is 28.2%, which corresponds to the ratio $(k_{2,HS} - k_{2,VSS})/k_{2,VSS} = 0.282$.

(iii) Mass flow rate (Answer: $\dot{m} = 4.89 \times 10^{-12}$ kg/s).
The rectangular cross-section has a non-negligible aspect ratio $a^* = h/b = 1/20 = 0.05 > 0.01$. Therefore, a rectangular model should be used. Due to the range covered by the Knudsen numbers, a second-order model (NS2) is required (cf. Figure 2.7).

The mass flow rate is obtained from Eq. (2.157):

$$\dot{m}_{NS2,rect} = \frac{4h^4 p_o^2}{a^* \mu R T l}\left[\frac{a_1}{2}(\Pi^2 - 1) + a_2 Kn'_o(\Pi - 1) + a_3 Kn'^2_o \ln \Pi\right]$$

with $a_1 = 0.32283$, $a_2 = 1.9925$, and $a_3 = 4.2853$ given by Table 2.5 for a diffuse reflection ($\sigma = 1$). $\Pi = p_i/p_o = 200/25 = 8$, and

$$\dot{m}_{NS2,rect} = 4(10^{-6})^4(2.5 \times 10^{-4})^2/(0.05 \times 2.588 \times 10^{-5} \times 2.08 \times 10^2 \times 350$$
$$\times 5 \times 10^{-3}) \times [0.32283(8^2 - 1)/2 + 1.9925 \times 2.78 \times 10^{-1}(8 - 1)$$
$$+ 4.2853 \times (2.78 \times 10^{-1})^2 \ln 8] = 4.89 \times 10^{-12}\ \text{kg/s}$$

(iv) Mass flow rate increase (Answer: 44.9%).
From Eq. (2.159),

$$\dot{m}^*_{NS2,rect} = \frac{\dot{m}_{NS2,rect}}{\dot{m}_{ns,rect}} = \left[1 + 2\frac{a_2}{a_1}Kn'_o\frac{1}{\Pi + 1} + 2\frac{a_3}{a_1}Kn'^2_o\frac{\ln \Pi}{\Pi^2 - 1}\right]$$

$$= 1 + 2\frac{1.9925}{0.32283}2.78 \times 10^{-1}\frac{1}{8 + 1} + 2\frac{4.2853}{0.32283}(2.78 \times 10^{-1})^2\frac{\ln 8}{8^2 - 1} = 1.449$$

which means that slip at the wall increases the flow rate of 44.9%.

(v) Underestimation of the mass flow rate (Answer: 46.7%).
If the inlet pressure is decreased to $p_i = 0.1$ MPa, the pressure ratio is $\Pi = 4$ and

$$\dot{m}^*_{NS2,rect} = 1 + 2\frac{1.9925}{0.32283} 2.78 \times 10^{-1} \frac{1}{4+1} + 2\frac{4.2853}{0.32283} (2.78 \times 10^{-1})^2 \frac{\ln 4}{4^2 - 1} = 1.877$$

Therefore, although the outlet Knudsen number is unchanged, the inlet pressure number is increased and rarefaction effects are enhanced. Slip at the wall now increases the flow rate of 87.7%. In others words, the underestimation of the mass flow rate with a no-slip model is 46.7%, since $(\dot{m}_{NS2} - \dot{m}_{ns})/\dot{m}_{NS2} = 1 - 1/\dot{m}^*_{NS2} = 1 - 1/1.877 = 0.467$.

Example 2.2

In a hard disk drive, the flow of gas between the hard disk and the read–write head can locally be comparable to a fully developed micro-Couette flow between parallel plane plates. The shield of the hard disk drive is sealed and the gas inside is pure nitrogen. The distance between the two plates is $2h = 0.2$ μm, the read–write head is at $r = 3$ cm from the center of the disk, whose speed of rotation is $\dot{\omega} = 7200$ rpm. One of the plates ($y = -h$) is fixed and the other ($y = h$) moves with a constant velocity U in the z-direction.

Assume a uniform temperature $T = 300$ K, a uniform pressure $p = 90$ kPa, and an accommodation coefficient $\sigma = 0.9$ associated with Maxwell first-order slip boundary condition.

i. Write the velocity distribution as a function of the Knudsen number.
ii. Calculate the mass flow rate per unit width and compare it to the case of a no-slip Couette flow.
iii. Calculate the friction factor and compare its value to the one obtained with a no-slip assumption.

Solution

(i) Velocity distribution (Answer: $u_z = (U/2)(y^*/4(Kn(2 - \sigma)/\sigma + 1) + 1)$.
The momentum equation reduces to $d^2u_z/dy^2 = 0$, and the boundary conditions are $u_z|_{y=-h} = ((2 - \sigma)/\sigma)\lambda \ du_z/dy|_{y=-h}$ and $u_z|_{y=+h} = U - ((2-\sigma)/\sigma)\lambda du_z/dy|_{y=+h}$. Thus, the velocity profile is linear: $u_z = a_1 y + a_2$. The boundary conditions yield $-a_1 h + a_2 = ((2 - \sigma)/\sigma)\lambda \, a_1$ and $a_1 h + a_2 = U - ((2 - \sigma)/\sigma)\lambda \, a_1$, which leads to $2a_1 = U/(h + \lambda(2-\sigma)/\sigma)$ and $2a_2 = U$. Then $u_z = (U/2)(y/(\lambda(2-\sigma)/\sigma + h) + 1)$.

With $y^* = y/h$ and $Kn = y/D_H = y/(4h)$, the velocity distribution is written as $u_z = (U/2)(y^*/(4Kn(2 - \sigma)/\sigma + 1) + 1)$. The centerline velocity is independent of the slip at the walls and the velocity transverse gradient is reduced as rarefaction is increased.

(ii) Mass flow rate (Answer: $\dot{m}' = (pM/RT)(\dot{\omega}\ r\ 2\pi/60)\ 2h = 4.57 \times 10^{-6}$ kg/s/m).

The mass flow rate is $\dot{m} = \rho \overline{u_z} A = (p/RT)\overline{u_z} A$ with $\overline{u_z} = 1/2 \int_{-1}^{1} u_z\, dy^* = U/2$. Therefore, the mass flow rate per unit width is

$$\dot{m}' = \frac{p}{RT}\frac{U}{2}4h = \frac{pM}{R_u T}\frac{\dot{\omega} r 2\pi}{60}2h$$

$$= \frac{9 \times 10^4 \times 28.013 \times 10^{-3} \times 7200 \times 3 \times 10^{-2} \times 2\pi \times 2 \times 10^{-7}}{8.315 \times 300 \times 60}$$

$$= 4.57 \times 10^{-6}\ \mathrm{kg/s/m}$$

The mass flow rate is the same as the one for a Couette flow without slip at the wall.

(iii) Friction factor (Answer: $C_f^* = C_f/C_{f,\mathrm{ns}} = 0.57$).
From Eq. (2.96), $C_f = 2\overline{\tau_w}/(\rho \overline{u_z}^2)$ and $\overline{\tau_w} = \tau_s = \mu\, du_z/dy$ is uniform since the velocity profile is linear. With $(du_z/dy) = (U/2h)(1/(1 + 4Kn(2-\sigma)/\sigma))$ we obtain:

$$C_f = \frac{4\mu RT}{hpU(1 + 4Kn(2-\sigma)/\sigma)}$$

From Table 2.2, the temperature exponent of the coefficient of viscosity for nitrogen (N_2) is $\omega = 0.74$, the exponent for the VSS model is $\alpha = 1.36$, and the dynamic viscosity under standard conditions is $\mu_0 = 1.656 \times 10^{-5}$ Pa s.

From Eq. (2.7), the viscosity at $T = 300$ K is

$$\mu(T) = \mu_0(T/T_0)^{\omega} = 1.656 \times 10^{-5}(300/273.15)^{0.74} = 1.775 \times 10^{-5}\mathrm{Pa\ s}$$

The molecular weight of nitrogen is given in Table 2.2: $M = 28.013 \times 10^{-3}$ kg/mol, and its gas constant is $R = R_u/M = 8.315/28.013 \times 10^{-3} = 2.97 \times 10^2$ J/kg/K.

The mean free path is given by Table 2.1: $\lambda = k_2\mu(T)\ \sqrt{RT}/p$, with $k_2 = [4\alpha(7-2\omega)(5-2\omega)]/[5(\alpha+1)(\alpha+2)\sqrt{2\pi}] = 1.06$. Therefore, $\lambda = 1.06 \times 1.775 \times 10^{-5} \times \sqrt{2.97 \times 10^2 \times 300}/(9 \times 10^4) = 62.6$ nm, and the Knudsen number based on the hydraulic diameter is

$$Kn = \lambda/(4h) = 6.26 \times 10^{-8}/(4 \times 10^{-7}) = 1.56 \times 10^{-1}$$

The influence of rarefaction on the friction is then given by:

$$C_f^* = C_f/C_{f,\mathrm{ns}} = C_f/C_{f(Kn=0)} = \frac{1}{1 + 4Kn(2-\sigma)/\sigma}$$

$$= \frac{1}{1 + 4 \times 0.156(2-0.9)/0.9} = 0.57$$

The friction factor is reduced of 43% due to slip at the walls.

2.9 Practice problems

Problem 2.1

A flow of helium is generated in a circular microtube by axial pressure and temperature gradients. The microtube has a length $l = 5$ mm and a uniform cross-section with a diameter $2r_2 = 10$ μm. In a section where the flow may be considered as locally fully developed, the gradients are $\mathrm{d}p/\mathrm{d}z = -5$ Pa/μm and $\mathrm{d}T/\mathrm{d}z = 0.5$ K/μm, for a pressure $p = 100$ kPa and a temperature $T = 400$ K.

Assume a tangential momentum accommodation coefficient $\sigma = 0.9$.

- **i.** Write the velocity distribution as a function of the Knudsen number in a dimensionless form: $u_z^*(r^*, Kn)$.
- **ii.** Calculate the part of the velocity increase due to slip at the wall.
- **iii.** Calculate the part of the velocity increase due to thermal creep.

Problem 2.2

A flow of air in a rectangular microchannel is generated by a pressure gradient. The inlet pressure is $p_i = 3.5$ bar and the outlet pressure is $p_o = 150$ mbar. The temperature $T = 350$ K is uniform and the tangential momentum accommodation coefficient is $\sigma = 0.8$. The microchannel length is $l = 10$ mm and its uniform section has a depth $2h = 1.2$ μm and a width $2b = 4.8$ μm.

- **i.** Calculate the mass flow rate with a plane flow assumption and a no-slip model.
- **ii.** Calculate the mass flow rate with a plane flow assumption and a Navier–Stokes model with first-order slip boundary conditions.
- **iii.** Calculate the mass flow rate with a plane flow assumption and a Navier–Stokes model with Deissler second-order slip boundary conditions.
- **iv.** Calculate the mass flow rate with a plane flow assumption and a QHD model with first-order slip boundary conditions.
- **v.** Calculate the mass flow rate with a no-slip model in a rectangular section.
- **vi.** Calculate the mass flow rate with a Navier–Stokes model in a rectangular section with Deissler second-order slip boundary conditions.
- **vii.** Calculate the deviations of the above different models, compared to the most accurate among them.

Problem 2.3

Consider a microchannel with a uniform cross-section whose width is very large compared with its depth $2h = 5$ μm. A pressure-driven flow of helium is generated, the lower wall ($y = -h$) is adiabatic, and the upper wall ($y = h$) has a constant heat flux $q_h = -200$ W/cm^2. Consider a section far from the inlet, such that the flow in this section is thermally and hydrodynamically developed. The flow conditions in this section are such that the Knudsen number based on the hydraulic diameter is $Kn = 2 \times 10^{-2}$ and the accommodation coefficients are $\sigma = 0.9$ and $\sigma_T = 0.6$. The wall temperature $T_{wall} = 400$ K is measured with a microsensor.

i. Calculate the mean temperature $\overline{T}$ of the gas in the section.
ii. Calculate the temperature T_0 at the center of the section ($y = 0$).
iii. Compare with the case of a similar non rarefied flow.

Problem 2.4

Consider a microchannel with a length $l = 5$ mm and whose uniform rectangular cross-section has a low aspect ratio: its width is $2b = 200$ μm and its depth is $2h = 1.5$ μm. A flow of nitrogen is generated with a pressure gradient. The inlet pressure is $p_i = 3 \times 10^5$ Pa and the outlet pressure is $p_o = 1 \times 10^5$ Pa. The temperature $T = 300$ K is uniform and the tangential momentum accommodation coefficient is $\sigma = 0.95$.

i. Calculate the mean velocity in a cross-section of the microchannel as a function of the local pressure gradient and Knudsen number.
ii. Calculate the mass flow rate.
iii. Calculate the pressure value at the middle of the microchannel (at $z = l/2$) and compare it to the no-slip value.
iv. Calculate the maximal velocity reached in the microchannel outside the entrance region.
v. Calculate the maximal Mach number and justify the locally fully developed assumption.

Problem 2.5

A multistage Knudsen compressor is designed to locally tune the pressure level or air inside a microsystem. The geometry of this Knudsen compressor is given in Figure 2.29. Ten chambers are connected via nine series of 100 short microchannels. These microchannels have rectangular cross-sections with a depth $2h = 0.5$ μm and a width $2b = 20$ μm. Each chamber is assumed large enough for the flow inside to be considered as a continuum flow. The temperature in these chambers is $T_2 = 300$ K, and each element is able to raise the temperature at the outlet of the microchannels to the value $T_1 = 600$ K. Chamber 10 is connected to a region whose pressure $p_{10} = 10$ mbar is maintained constant by means of a primary pump. The initial temperature $T_{t=0} = T_2 = 300$ K is uniform before the elements heating. The initial pressure in each chamber is also uniform and is equal to the pressure in chamber 10: $p_{t=0} = p_{10} = 10$ mbar. For $t > 0$, heating of the elements can be regulated in order to tune the pressure in chamber 1 in the range 1 to 10 mbar.

i. Calculate the possible range covered by the Knudsen number in the microchannels during the pumping process and check that the free molecular regime assumption in the microchannel is valid.

ii. Calculate the maximal value of the mass flow rate between two adjacent chambers at the beginning of the pumping progress.
iii. Give the different reasons why the real initial mass flow rate is lower than the value calculated in (ii).
iv. Calculate the minimal pressure value that can be reached in chamber 1 at the end of the pumping process.

Problem 2.6

Consider a pressure-driven flow of different gases (He, O_2, or SO_2) through a microtube with rough walls, such that the accommodation coefficient can reasonably be assumed equal to unity. The microtube length is $l = 3$ mm and its inner diameter is $2r_2 = 1.2$ μm. The inlet pressure is $p_i = 2$ bar and the outlet pressure is $p_o = 1$ bar. Assume that the intermolecular collisions are accurately described by a VSS model.

i. Calculate the mass flow rate through this microtube for the three different gases using a QHD1 model.
ii. Calculate the mass flow rate through this microtube for the different gases using a NS2 model with Deissler boundary conditions.
iii. Calculate the mass flow rate through this microtube for the different gases using a NS2 model with Chapman and Cowling boundary conditions.
iv. Calculate the mass flow rate through this microtube for the different gases using a NS2 model with Mitsuya boundary conditions.
v. Compare and comment on these different results.
vi. Propose an experimental protocol in order to check the more appropriate among the above models.

References

Agarwal, R.K., Yun, K.-Y., Balakrishnan, R., 2001. Beyond Navier–Stokes: Burnett equations for flows in the continuum–transition regime. Phys. Fluids 13 (10), 3061–3085.

Agrawal, A., Prabhu, S.V., 2008. Survey on measurement of tangential momentum accommodation coefficient. J. Vac. Sci. Technol. A 26 (4), 634–645.

Ameel, T.A., Wang, X., Barron, R.F., Warrington, R.O.J., 1997. Laminar forced convection in a circular tube with constant heat flux and slip flow. Microscale Therm. Eng. 1 (4), 303–320.

Anderson, J.D., 1990. Modern Compressible Flow, second edn McGraw-Hill International Editions, New York, NY.

Anduze, M., 2000. Etude expérimentale et numérique de microécoulements liquides dans les microsystèmes fluidiques (PhD thesis). Institut National des Sciences Appliquées de Toulouse, Toulouse.

Aoki, K., Degond, P., Mieussens, L., 2009. Numerical simulations of rarefied gases in curved channels: thermal creep, circulating flow, and pumping effect. Commun. Comput. Phys. 6 (5), 919–954.

Arkilic, E.B., Breuer, K.S., Schmidt, M.A., 1994. Gaseous flow in microchannels, Application of Microfabrication to Fluid Mechanics, Vol. FED-197. ASME, New York, NY, pp. 57–66.

Arkilic, E.B., Schmidt, M.A., Breuer, K.S., 1998. Sub-nanomol per second flow measurement near atmospheric pressure. Exp. Fluid 25 (1), 37–41.

Arkilic, E.B., Breuer, K.S., Schmidt, M.A., 2001. Mass flow and tangential momentum accommodation in silicon micromachined channels. J. Fluid Mech. 437, 29–43.

Aubert, C., 1999. Ecoulements compressibles de gaz dans les microcanaux: effets de raréfaction, effets instationnaires (Ph.D. thesis). Université Paul Sabatier, Toulouse.

Aubert, C., Colin, S., 2001. High-order boundary conditions for gaseous flows in rectangular microchannels. Microscale Therm. Eng. 5 (1), 41–54.

Aubert, C., Colin, S., Caen, R., 1998. Unsteady gaseous flows in tapered microchannels, First International Conference on Modeling and Simulation of Microsystems, Semiconductors, Sensors and Actuators (MSM'98), vol. 1. Computational Publications, Santa Clara Marriot.

Aydin, O., Avcı, M., 2006. Heat and fluid flow characteristics of gases in micropipes. Int. J. Heat Mass Transfer 49 (9-10), 1723–1730.

Barber, R.W., Emerson, D.R., 2002. The influence of Knudsen number on the hydrodynamic development length within parallel plate micro-channels. In: Rahman, M., Verhoeven, R., Brebbia, C.A. (Eds.), Advances in Fluid Mechanics, vol. IV. WIT Press, Southampton.

Barber, R.W., Sun, Y., Gu, X.-J., Emerson, D.R., 2004. Isothermal slip flow over curved surfaces. Vacuum 76, 73–81.

Beskok, A., Karniadakis, G.E., 1999. A model for flows in channels, pipes, and ducts at micro and nano scales. Microscale Therm. Eng. 3 (1), 43–77.

Bhatnagar, P., Gross, E., Krook, K., 1954. A model for collision processes in gasses. Phys. Rev. 94, 511–524.

Bird, G., 1978. Monte Carlo simulation of gas flows. Annu. Rev. Fluid Mech. 10, 11–31.

Bird, G.A., 1998. Molecular Gas Dynamics and the Direct Simulation of Gas Flows. Clarendon Press, Oxford.

Caen, R., Mas, I., Colin, S., 1996. Ecoulements non permanents dans les microcanaux: réponse fréquentielle des microtubes pneumatiques. C. R. Acad. Sci. Sér. II b 323 (12), 805–812.

Cercignani, C., Lorenzani, S., 2010. Variational derivation of second-order slip coefficients on the basis of the Boltzmann equation for hard-sphere molecules. Phys. Fluids 22 (6), 062004.

Cercignani, C., Illner, R., Pulvirenti, M., 1994. The Mathematical Theory of Dilute Gases, vol. 106. Springer-Verlag, New York, NY.

Chapman, S., Cowling, T.G., 1952. The Mathematical Theory of Non-uniform Gases. Cambridge University Press, Cambridge.

Chen, C.S., Lee, S.M., Sheu, J.D., 1998. Numerical analysis of gas flow in microchannels. Numer. Heat Transfer A 33, 749–762.

Chen, R.-Y., 1973. Flow in the entrance region at low Reynolds numbers. J. Fluids Eng. 95, 153–158.

Chen, S., Doolen, G., 1998. Lattice Boltzmann method for fluid flows. Annu. Rev. Fluid Mech. 30, 329–364.

Cheng, H., Emmanuel, G., 1995. Perspectives on hypersonic nonequilibrium flow. AIAA J. 33, 385–400.

Colin, S. (Ed.), 2010. Microfluidics. Wiley, London.

Colin, S., 2012. Gas microflows in the slip flow regime: a critical review on convective heat transfer. J. Heat Transfer 134 (2), 020908.

Colin, S., Aubert, C., Caen, R., 1998a. Unsteady gaseous flows in rectangular microchannels: frequency response of one or two pneumatic lines connected in series. Eur. J. Mech. B Fluid 17 (1), 79–104.

Colin, S., Anduze, M., Caen, R., 1998b. A pneumatic frequency generator for experimental analysis of unsteady microflows. In: Liwei Lin, F.K.F., Aluru, N.R., Zhang, X. (Eds.), Micro-Electro-Mechanical Systems (MEMS)—1998, vol. DSC-66. ASME, New York, NY.

Colin, S., Elizarova, T.G., Sheretov, Y.V., Lengrand, J.-C., Camon, H., 2003. Micro-écoulements gazeux: validation expérimentale de modèles QHD et de Navier–Stokes avec conditions aux limites de glissement. In: 16ème Congrès Français de Mécanique, Proceedings on CDROM, Nice.

Colin, S., Lalonde, P., Caen, R., 2004. Validation of a second-order slip flow model in rectangular microchannels. Heat Transfer Eng. 25 (3), 23–30.

Comeaux, K.A., Chapman, D.R., MacCormack, R.W., 1995. An analysis of the Burnett equations based in the second law of thermodynamics, AIAA Paper, no. 95–0415.

Deissler, R.G., 1964. An analysis of second-order slip flow and temperature-jump boundary conditions for rarefied gases. Int. J. Heat Mass Transfer 7, 681–694.

Dongari, N., Zhang, Y., Reese, J.M., 2011. Modeling of Knudsen layer effects in micro/nanoscale gas flows. J. Fluids Eng. 133 (7), 071101.

Ebert, W.A., Sparrow, E.M., 1965. Slip flow in rectangular and annular ducts. J. Basic Eng. 87, 1018–1024.

Elizarova, T.G., Sheretov, Y.V., 2003. Analyse du problème de l'écoulement gazeux dans les microcanaux par les équations quasi hydrodynamiques. La Houille Blanche 5, 66–72.

Ewart, T., Perrier, P., Graur, I.A., Méolans, J.G., 2007. Mass flow rate measurements in a microchannel, from hydrodynamic to near free molecular regimes. J. Fluid Mech. 584, 337–356.

Gad-el-Hak, M., 1995. Stokes' hypothesis for a newtonian, isotropic fluid. J. Fluids Eng. 117, 3–5.

Gad-el-Hak, M., 1999. The fluid mechanics of microdevices—The Freeman Scholar Lecture. J. Fluids Eng. 121, 5–33.

Grad, H., 1949. On the kinetic theory of rarefied gases. Comm. Pure Appl. Math. 2, 331–407.

Guo, Z.Y., Wu, X.B., 1998. Further study on compressibility effects on the gas flow and heat transfer in a microtube. Microscale Therm. Eng. 2 (2), 111–120.

Gupta, N.K., An, S., Gianchandani, Y.B., 2012. A Si-micromachined 48-stage Knudsen pump for on-chip vacuum. J. Micromech. Microeng. 22 (10), 105026.

Hadjiconstantinou, N.G., 2003. Comment on Cercignani's second-order slip coefficient. Phys. Fluids 15 (8), 2352–2354.

Hadjiconstantinou, N.G., Simek, O., 2002. Constant-wall-temperature Nusselt number in micro and nano-channels. J. Heat Transfer 124, 356–364.

Harley, J.C., Huang, Y., Bau, H.H., Zemel, J.N., 1995. Gas flow in micro-channels. J. Fluid Mech. 284, 257–274.

Hassan, H.A., Hash, D.B., 1993. A generalized hard-sphere model for Monte Carlo simulation. Phys. Fluids A 5 (3), 738–744.

Hobson, J.P., 1970. Accommodation pumping—a new principle for low pressure. J. Vac. Sci. Technol. 7 (2), 301–357.

Hobson, J.P., 1971. Analysis of accommodation pumps. J. Vac. Sci. Technol. 8 (1), 290–293.

Hobson, J.P., 1972. Physical factors influencing accommodation pumps. J. Vac. Sci. Technol. 9 (1), 252–256.

Hsia, Y.-T., Domoto, G.A., 1983. An experimental investigation of molecular rarefaction effects in gas lubricated bearings at ultra-low clearances. J. Lubr. Technol. 105, 120–130.

Hudson, M.L., Bartel, T.J., 1999. DSMC simulation of thermal transpiration and accommodation pumps. In: Brun, R., Campargue, R., Gatignol, R., Lengrand, J.-C. (Eds.), Rarefied Gas Dynamics, vol. 1. Cépaduès Editions, Toulouse, France.

Jang, J., Wereley, S.T., 2004. Pressure distributions of gaseous slip flow in straight and uniform rectangular microchannels. Microfluid. Nanofluid. 1 (1), 41–51.

Jeong, H.-E., Jeong, J.-T., 2006. Extended Graetz problem including streamwise conduction and viscous dissipation in microchannel. Int. J. Heat Mass Transfer 49 (13–14), 2151–2157.

Jie, D., Diao, X., Cheong, K.B., Yong, L.K., 2000. Navier–Stokes simulations of gas flow in micro devices. J. Micromech. Microeng. 10 (3), 372–379.

Karniadakis, G.E., Beskok, A., 2002. Microflows: Fundamentals and Simulation. Springer-Verlag, New York, NY.

Kavehpour, H.P., Faghri, M., Asako, Y., 1997. Effects of compressibility and rarefaction on gaseous flows in microchannels. Numer. Heat Transfer A 32, 677–696.

Kennard, E.H., 1938. Kinetic Theory of Gases, first ed. McGraw-Hill Book Company, New York, NY.

Koura, K., Matsumoto, H., 1991. Variable soft sphere molecular model for inverse-power-law or Lennard–Jones potential. Phys. Fluids A 3 (10), 2459–2465.

Koura, K., Matsumoto, H., 1992. Variable soft sphere molecular model for air species. Phys. Fluids A 4 (5), 1083–1085.

Lalonde, P., 2001. Etude expérimentale d'écoulements gazeux dans les microsystèmes à fluides (Ph.D. thesis). Institut National des Sciences Appliquées de Toulouse, Toulouse.

Lee, C., Hong, G., Ha, Q.P., Mallinson, S.G., 2003. A piezoelectrically actuated micro synthetic jet for active flow control. Sens. Actuators A Phys. 108 (1–3), 168–174.

Lee, W.Y., Wong, M., Zohar, Y., 2002. Microchannels in series connected via a contraction/expansion section. J. Fluid Mech. 459, 187–206.

Lempert, W.R., Boehm, M., Jiang, N., Gimelshein, S., Levin, D., 2003. Comparison of molecular tagging velocimetry data and direct simulation Monte Carlo simulations in supersonic micro jet flows. Exp. Fluids 34 (3), 403–411.

Lengrand, J.-C., Elizarova, T.G., 2010. Gaseous microflows, Chapter 2. In: Colin, S. (Ed.), Microfluidics. Wiley, London

Lengrand, J.-C., Elizarova, T.G., Shirokov, I.A., 2004. Calcul de l'écoulement visqueux compressible d'un gaz dans un microcanal. In: Actes sur CDROM du 2ème Congrès Français de Microfluidique (μFlu'04). SHF, Toulouse.

Li, Q., He, Y., Tang, G., Tao, W., 2011. Lattice Boltzmann modeling of microchannel flows in the transition flow regime. Microfluid. Nanofluid. 10 (3), 607–618.

Lim, C.Y., Shu, C., Niu, X.D., Chew, Y.T., 2002. Application of lattice Boltzmann method to simulate microchannel flows. Phys. Fluids 14 (7), 2299–2308.

Liu, J., Tai, Y.-C., Ho, C.-M., 1995. MEMS for pressure distribution studies of gaseous flows in microchannels. In: 8th Annual International Workshop on Micro-Electro-Mechanical Systems (MEMS'95), An Investigation of Micro Structures, Sensors, Actuators, Machines, and Systems. IEEE, Amsterdam, pp. 209–215.

Lockerby, D.A., Reese, J.M., Emerson, D.R., Barber, R.W., 2004. Velocity boundary condition at solid walls in rarefied gas calculations. Phys. Rev. E 70, 017303.

Lorenzani, S., 2011. Higher order slip according to the linearized Boltzmann equation with general boundary conditions. Philos. Trans. R. Soc. A-Math. Phys. Eng. Sci. 369 (1944), 2228–2236.

Loyalka, S.K., Hamoodi, S.A., 1990. Poiseuille flow of a rarefied gas in a cylindrical tube: solution of linearized Boltzmann equation. Phys. Fluids A 2 (11), 2061–2065.

Mallinson, S.G., Kwok, C.Y., Reizes, J.A., 2003. Numerical simulation of micro-fabricated zero mass-flux jet actuators. Sens. Actuator A Phys. 105 (3), 229–236.

Matsuda, Y., Mori, H., Niimi, T., Uenishi, H., Hirako, M., 2007. Development of pressure sensitive molecular film applicable to pressure measurement for high Knudsen number flows. Exp. Fluids 42 (4), 543–550.

Matsuda, Y., Uchida, T., Suzuki, S., Misaki, R., Yamaguchi, H., Niimi, T., 2011. Pressure-sensitive molecular film for investigation of micro gas flows. Microfluid. Nanofluid. 10 (1), 165–171.

Matsumoto, H., 2002. Variable sphere molecular model for inverse power law and Lennard–Jones potentials in Monte Carlo simulations. Phys. Fluids 14 (12), 4256–4265.

Maurer, J., Tabeling, P., Joseph, P., Willaime, H., 2003. Second-order slip laws in microchannels for helium and nitrogen. Phys. Fluids 15 (9), 2613–2621.

Mavriplis, C., Ahn, J.C., Goulard, R., 1997. Heat transfer and flowfields in short microchannels using direct simulation Monte Carlo. J. Thermophys. Heat Transfer 11 (4), 489–496.

Maxwell, J.C., 1879. On stresses in rarefied gases arising from inequalities of temperature. Philos. Trans. R. Soc. 170, 231–256.

Meolans, J.G., Nacer, M.H., Rojas, M., Perrier, P., Graur, I., 2012. Effects of two transversal finite dimensions in long microchannel: analytical approach in slip regime. Phys. Fluids 24 (11), 112005.

Mikhailov, M.D., Cotta, R.M., 2005. Mixed symbolic-numerical computation of convective heat transfer with slip flow in microchannels. Int. Comm. Heat Mass Transfer 32 (3–4), 341–348.

Mitsuya, Y., 1993. Modified Reynolds equation for ultra-thin film gas lubrication using 1,5-order slip-flow model and considering surface accommodation coefficient. J. Tribol. 115, 289–294.

Morini, G.L., Spiga, M., 1998. Slip flow in rectangular microtubes. Microscale Therm. Eng. 2 (4), 273–282.

Muntz, E.P., 1989. Rarefied gas dynamics. Annu. Rev. Fluid Mech. 21, 387–417.

Muntz, E.P., Vargo, S.E., 2002. Microscale vacuum pumps. In: Gad-el-Hak, M. (Ed.), The MEMS Handbook. CRC Press, New York, NY, pp. 29.1–29.28.

Nguyen, N.-T., Wereley, S.T., 2002. Fundamentals and Applications of Microfluidics. Artech House, Boston, MA.

Nie, X., Doolen, G.D., Chen, S., 2002. Lattice Boltzmann simulations of fluid flows in MEMS. J. Stat. Phys. 107 (1–2), 279–289.

Oran, E.S., Oh, C.K., Cybyk, B.Z., 1998. Direct simulation Monte Carlo: recent advances and applications. Annu. Rev. Fluid Mech. 30, 403–441.

Pan, L.S., Liu, G.R., Lam, K.Y., 1999. Determination of slip coefficient for rarefied gas flows using direct simulation Monte Carlo. J. Micromech. Microeng. 9 (1), 89–96.

Pan, L.S., Liu, G.R., Khoo, B.C., Song, B., 2000. A modified direct simulation Monte Carlo method for low-speed microflows. J. Micromech. Microeng. 10 (1), 21–27.

Pan, L.S., Ng, T.Y., Xu, D., Lam, K.Y., 2001. Molecular block model direct simulation Monte Carlo method for low velocity microgas flows. J. Micromech. Microeng. 11 (3), 181–188.

Perrier, P., Graur, I.A., Ewart, T., Méolans, J.G., 2011. Mass flow rate measurements in microtubes: from hydrodynamic to near free molecular regime. Phys. Fluids 23 (4), 042004.

Piekos, E.S., Breuer, K.S., 1996. Numerical modeling of micromechanical devices using the direct simulation Monte Carlo method. J. Fluid Eng. 118, 464–469.

Pitakarnnop, J., Varoutis, S., Valougeorgis, D., Geoffroy, S., Baldas, L., Colin, S., 2010. A novel experimental setup for gas microflows. Microfluid. Nanofluid. 8 (1), 57–72.

Pong, K.-C., Ho, C.-M., Liu, J., Tai, Y.-C., 1994. Non-linear pressure distribution in uniform microchannels, ASME Winter Annual Meeting, Chicago. In: Bandyopadhyay, P.R., Breuer, K.S., Blechinger, C.J. (Eds.), Application of Microfabrication to Fluid Mechanics, Vol. FED-197. ASME, New York, NY

Qazi Zade, A., Ahmadzadegan, A., Renksizbulut, M., 2012. A detailed comparison between Navier-Stokes and DSMC simulations of multicomponent gaseous flow in microchannels. Int. J. Heat Mass Transfer 55 (17–18), 4673–4681.

Rohsenow, W.M., Hartnett, J.P., Cho, Y.I., 1998. Handbook of Heat Transfer, third edn McGraw-Hill, New York.

Samouda, F., Brandner, J.J., Barrot, C., Colin, S., 2012. Velocity field measurements in gas phase internal flows by molecular tagging velocimetry. J. Phys. Conf. Ser. 362 (1), 012026.

Schaaf, S.A., 1963. Mechanics of rarefied gases. In: Encyclopedia of Physics, Fluid Dynamics II, VII/2, Berlin, pp. 591–624.

Shah, R.K., 1975. Laminar flow friction and forced convection heat transfer in ducts of arbitrary geometry. Int. J. Heat Mass Transfer 18, 849–862.

Sharipov, F., 2011. Data on the velocity slip and temperature jump on a gas–solid interface. J. Phys. Chem. Ref. Data 40 (2), 023101.

Sharipov, F., Seleznev, V., 1998. Data on internal rarefied gas flows. J. Phys. Chem. Ref. Data 27 (3), 657–706.

Shih, J.C., Ho, C.-M., Liu, J., Tai, Y.-C., 1996. Monatomic and polyatomic gas flow through uniform microchannels, Vol. DSC-59. ASME, New York, NY.

Sinton, D., 2004. Microscale flow visualization. Microfluid. Nanofluid. 1 (1), 2–21.

Smoluchowski, M., 1898. Ueber wärmeleitung in verdünnten gasen. Annalen der Physik und Chemie 64, 101–130.

Sparrow, E.M., Lin, S.H., 1962. Laminar heat transfer in tubes under slip-flow conditions. J. Heat Transfer Trans. ASME 84, 363–369.

Sreekanth, A.K., 1969. Slip flow through long circular tubes. In: Trilling, L., Wachman, H.Y. (Eds.), 6th International Symposium on Rarefied Gas Dynamics. Academic Press, New York, NY, pp. 667–680.

Stefanov, S., Cercignani, C., 1994. Monte Carlo simulation of a channel flow of a rarefied gas. Eur. J. Mech. B Fluids 13 (1), 93–114.

Vargo, S.E., Muntz, E.P., 1997. An evaluation of a multiple-stage micromechanical Knudsen compressor and vacuum pump. In: Shen, C. (Ed.), Proceedings of the 20th International Symposium on Rarefied Gas Dynamics. Peking University Press, Beijing, pp. 995–1000.

Vargo, S.E., Muntz, E.P., 1999. Comparison of experiment and prediction for transitional flow in a single-stage micromechanical Knudsen compressor. In: Brun, R., Campargue, R., Gatignol, R., Lengrand, J.-C. (Eds.), Rarefied Gas Dynamics, vol. 1. Cépaduès Editions, Toulouse, France, pp. 711–718.

Vargo, S.E., Muntz, E.P., Shiflett, G.R., Tang, W.C., 1999. Knudsen compressor as a micro- and macroscale vacuum pump without moving parts or fluids. J. Vac. Sci. Technol. A 17 (4), 2308–2313.

Wang, M., Li, Z., 2004. Simulations for gas flows in microgeometries using the direct simulation Monte Carlo method. Int. J. Heat Fluid Flow 25 (6), 975–985.

Webb, A.R., Maynes, D., 1999. Velocity profile measurements in microtubes. In: 30th AIAA Fluid Dynamics Conference, Norfolk, VA: AIAA, no. 99-3803/1-10.

Welder, W.T., Chapman, D.R., MacCormack, R.W., 1993. Evaluation of various forms of the Burnett equations. AIAA Paper, 93–3094.

Wereley, S.T., Meinhart, C.D., Santiago, J.G., Adrian, R.J., 1998. Velocimetry for MEMS applications. In: ASME Annual Meeting, Vol. DSC 66. Anaheim, pp. 453–459.

Wereley, S.T., Gui, L., Meinhart, C.D., 2002. Advanced algorithms for microscale particle image velocimetry. AIAA J. 40 (6), 1047–1055.

Wu, J.-S., Tseng, K.-C., 2001. Analysis of micro-scale gas flows with pressure boundaries using direct simulation Monte Carlo method. Comput. Fluid 30 (6), 711–735.

Xue, H., Fan, Q., 2000. A new analytic solution of the Navier–Stokes equations for micro-channel flow. Microscale Therm. Eng. 4 (2), 125–143.

Yamaguchi, H., Hanawa, T., Yamamoto, O., Matsuda, Y., Egami, Y., Niimi, T., 2011. Experimental measurement on tangential momentum accommodation coefficient in a single microtube. Microfluid. Nanofluid. 11, 57–64.

Yoon, S.Y.R., Ross, J.W., Mench, M.M., Sharp, K.V., 2006. Gas-phase particle image velocimetry (PIV) for application to the design of fuel cell reactant flow channels. J. Power Soc. 160, 1017–1025.

Young, R.M., 1999. Analysis of a micromachine based vacuum pump on a chip actuated by the thermal transpiration effect. J. Vac. Sci. Technol. B 17 (2), 280–287.

Yu, S., Ameel, T.A., 2000. Slip-flow Peclet number thermal entry problem within a flat microchannel subject to constant wall temperature. Proceedings of the International Conference on Heat Transfer and Transport Phenomena in Microscale, Banff, Canada, pp. 101–107.

Yu, S., Ameel, T.A., 2001. Slip-flow heat transfer in rectangular microchannels. Int. J. Heat Mass Transfer 44 (22), 4225–4234.

Yun, K.Y., Agarwal, R.K., Balakrishnan, R., 1998. Augmented Burnett and Bhatnagar–Gross–Krook–Burnett equations for hypersonic flow. J. Therm. Heat Transfer 12 (3), 328–335.

Zhang, Y.H., Gu, X.J., Barber, R.W., Emerson, D.R., 2006. Capturing Knudsen layer phenomena using a lattice Boltzmann method. Phys. Rev. E 74, 046704.

Zhang, W.-M., Meng, G., Wei, X., 2012. A review on slip models for gas microflows. Microfluid. Nanofluid. 13 (6), 845–882.

Zhu, X., Xin, M.D., Liao, Q., 2000. An analysis for heat transfer between two unsymmetrically heated parallel plates with micro spacing in slip flow regime. In: Celata, G.P. (Ed.), Proceedings of the International Conference on Heat Transfer and Transport Phenomena in Microscale, Banff, Canada, pp. 114–120.

Zohar, Y., Lee, S.Y.K., Lee, Y.L., Jiang, L., Tong, P., 2002. Subsonic gas flow in a straight and uniform microchannel. J. Fluid Mech. 472, 125–151.

CHAPTER

Single-Phase Liquid Flow in Minichannels and Microchannels

3

Satish G. Kandlikar

Mechanical Engineering Department, Rochester Institute of Technology, Rochester, NY, USA

3.1 Introduction

3.1.1 Fundamental issues in liquid flow at microscale

Microchannels are used in a variety of devices incorporating single-phase liquid flow. The early applications involved micromachined devices such as micropumps, microvalves, and microsensors. This was followed by a thrust in the biological and life sciences with a need for analyzing biological materials, such as proteins, DNA, cells, embryos, and chemical reagents. The field of micromixers further received attention with developments in microreactors, where two chemical species are mixed prior to introducing them into a reaction chamber. The high flux heat dissipation from high-speed microprocessors provided the impetus for studies on heat transfer in microchannels.

The developments in the microelectromechanical devices naturally require heat removal systems that are equally small. The cooling of mirrors employed in high-power laser systems involves cooling systems that cover very small footprints. Advances in biomedical and genetic engineering require controlled fluid transport and its precise thermal control in passages with dimensions of several micrometers. A proper understanding of fluid flow and heat transfer in these microscale systems is therefore essential for their design and operation.

In dealing with liquid flows in minichannels and microchannels in the absence of any wall surface effects, such as the electrokinetic or electroosmotic forces (covered in Chapter 4), the flow is not expected to experience any fundamental changes from the continuum approximation employed in macrofluidic applications. Gad-el-Hak (1999) argued that liquids such as water should be treated as continuous media, with the results obtained from classical theory being applicable in channels larger than 1 μm. However, there remain a number of unresolved

© 2014 Elsevier Ltd. All rights reserved.

issues that require further study. The main areas of current research are summarized below:

1. Experimental validation of the laminar and turbulent flow transport equations: the laminar flow friction factor and heat transfer equations derived from theoretical considerations are expected to hold in microchannel applications in the absence of any changes in the transport processes or any new physical phenomena. Although explicit equations and experiments for mass transfer are not covered, the conclusions reached for the momentum and heat transfer are expected to be applicable to mass transport processes as well.
2. Verification of the laminar-to-turbulent flow transition at microscale: experimental evidence in this regard needs to be critically evaluated.
3. The effect of large relative roughness values on the flow: large values of relative roughness are more commonly encountered in microchannels. Their effect on the laminar-to-turbulent transition, friction factors, and heat transfer needs to be investigated.
4. Verification of empirical constants derived from macroscale experiments: a number of constants (such as for losses due to flow area changes, bends, etc.) whose values are derived from macroscale fluid flow experiments need to be verified for microscale applications.

This chapter will be devoted to answering the above questions on the basis of the evidence available in the literature.

3.1.2 Need for smaller flow passages

The flow passage dimensions in convective heat transfer applications have been shifting toward smaller dimensions for the following three main reasons:

1. Heat transfer enhancement.
2. Increased heat flux dissipation in microelectronic devices.
3. Emergence of microscale devices that require cooling.

Employing smaller channel dimensions results in higher heat transfer performance, although this is accompanied by a higher pressure drop per unit length. The higher volumetric heat transfer densities require advanced manufacturing techniques and lead to more complex manifold designs. An optimum balance for each application leads to different channel dimensions. For example, in the refrigeration industry, the use of microfin tubes of 6–8 mm diameter has replaced the plain tubes of larger diameters. In automotive applications, the passage dimensions for radiators and evaporators have approached a 1 mm threshold as a balance between the pumping power, heat transfer, and cleanliness constraints imposed by the overall system.

Microelectronic devices, including a variety of applications such as PCs, servers, laser diodes, and RF devices, are constantly pushing the heat flux density requirements to higher levels. What seemed to be an impossibly high limit of 200 W/cm^2 of heat dissipation in 1993 now seems to be a feasible target. The challenge for the coming decade is on the order of 600–1000 W/cm^2. The available temperature differences are becoming smaller, in some cases as low as only a few degrees Celsius with external copper heat sinks. These high levels of heat dissipation require a dramatic reduction in the channel dimensions, matched with suitable coolant loop systems to facilitate the fluid movement away from the heat source.

A cooling system for a microscale device might require cooling channels of a few tens of micrometers as compared to more conventional-size channels with 1–3 mm flow passage dimensions. In addition, several such units may be clustered together and a secondary cooling loop may be employed to remove the heat with a conventional cooling system. Figure 3.1 shows a schematic of a microchannel cooling system configuration for cooling a server application. The combination of (i) the microchannel heat exchangers, mounted directly on the chip or in the heat sink that is bonded to the chip; (ii) the water-cooled cold plates with minichannel or microchannel flow passages; and (iii) the auxiliary localized cooling systems will be able to address the cooling needs of the high-end servers. The cooling system is integrated with the building HVAC system, as described by Kandlikar (2005).

A schematic of direct liquid cooling of a multichip module or a heat sink is shown in Figure 3.2. The liquid flows through the cold plates, which are attached to a substrate cap. In advanced designs, direct cooling of chips is accomplished by circulating water, a water–antifreeze mixture, oil, or a dielectric fluid such as

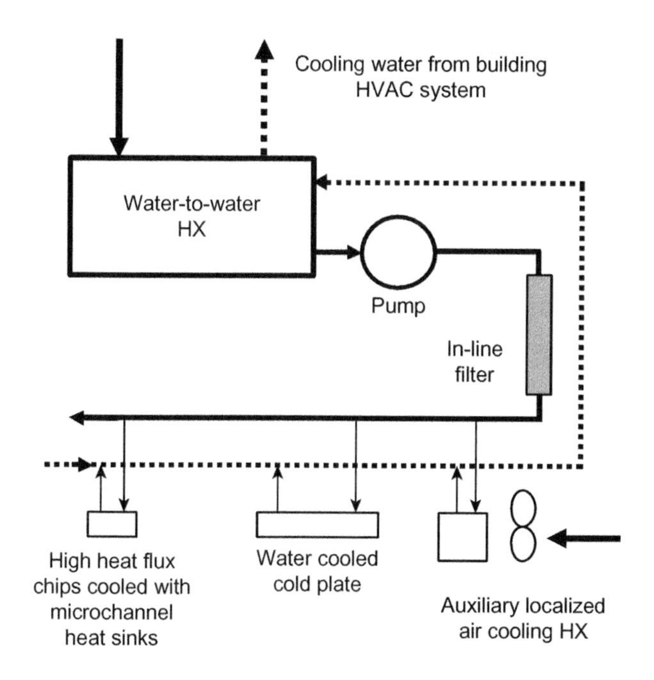

FIGURE 3.1

Schematic of a cluster of servers with high heat flux chips cooled with microchannel heat sinks, cold plates, and localized air cooling integrated with a secondary chilled water loop from the building HVAC system.

Source: *From Kandlikar (2005).*

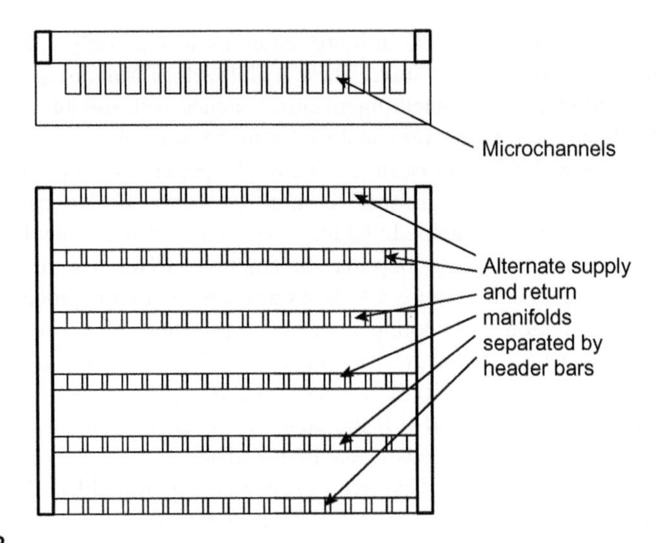

FIGURE 3.2

Schematic of a microchannel cooling arrangement on a chip or a heat sink with alternate supply and return manifolds created by header bars, a design originally proposed by Tuckerman (1984).

FC-72, FC-77, or FC-87 through microchannels that are fabricated on the chip surface. Copper heat sinks with integrated microchannels and minichannels are expected to dominate heat sink applications.

3.2 Pressure drop in single-phase liquid flow

3.2.1 Basic pressure drop relations

One-dimensional flow of an incompressible fluid in a smooth circular pipe forms the basis for the pressure drop analysis in internal flows. The following equations are readily derived based on the continuum assumption for Newtonian liquid flows in minichannels and microchannels.

Considering the equilibrium of a fluid element of length $\mathrm{d}x$ in a pipe of diameter D, the force due to pressure difference $\mathrm{d}p$ is balanced by the frictional force due to shear stress τ_w at the wall.

$$\left(\frac{\pi}{4}D^2\right)\mathrm{d}p = (\pi D\,\mathrm{d}x)\tau_\mathrm{w} \tag{3.1}$$

The pressure gradient and the wall shear stress are thus related by the following equation:

$$\frac{\mathrm{d}p}{\mathrm{d}x} = \frac{4\tau_\mathrm{w}}{D} \tag{3.2}$$

For Newtonian fluids, the wall shear stress τ_w is expressed in terms of the velocity gradient at the wall:

$$\tau_w = \mu \frac{du}{dy}\bigg|_w \tag{3.3}$$

where μ is the dynamic viscosity. The Fanning friction factor f is used in heat transfer literature because of its ability to represent the momentum transfer process of fluid flow in a manner consistent with the heat and mass transfer process representations:

$$f = \frac{\tau_w}{(1/2)\rho u_m^2} \tag{3.4}$$

where u_m is the mean flow velocity in the channel.

The frictional pressure drop Δp over a length L is obtained from Eqs. (3.2) and (3.4), respectively:

$$\Delta p = \frac{2f \rho u_m^2 L}{D} \tag{3.5}$$

The Fanning friction factor f in Eq. (3.5) depends on the flow conditions, the channel wall geometry, and surface conditions:

1. laminar or turbulent flow,
2. flow-channel geometry,
3. fully developed or developing flow,
4. smooth or rough walls.

For noncircular flow channels, the D in Eq. (3.5) is replaced by the hydraulic diameter D_h, represented by the following equation.

$$D_h = \frac{4A_c}{P_w} \tag{3.6}$$

where A_c is the flow-channel cross-sectional area and P_w is the wetted perimeter. For a rectangular channel of sides a and b, D_h is given by

$$D_h = \frac{4ab}{2(a+b)} = \frac{2ab}{(a+b)} \tag{3.7}$$

3.2.2 Fully developed laminar flow

The velocity gradient at the channel wall can be readily calculated from the well-known *Hagen–Poiseuille* parabolic velocity profile for the fully developed laminar flow in a circular pipe. Using this velocity profile, τ_w and f are obtained from Eqs. (3.3) and (3.4). The result for friction factor f is presented in the following form:

$$f = \frac{Po}{Re} \tag{3.8}$$

where Po is the Poiseuille number, ($Po = f\ Re$), which depends on the flow-channel geometry. Table 3.1 gives the $f\ Re$ product and the constant Nusselt number (Nu) in the fully developed laminar flow region for different duct shapes, as derived from Kakac et al. (1987).

It can be seen that for a circular pipe,

$$Po = fRe = 16 \tag{3.9}$$

Shah and London (1978) provided the following equation for a rectangular channel with short side a and long side b, and a channel aspect ratio defined as $\alpha_c = a/b$.

$$fRe = 24(1 - 1.3553\alpha_c + 1.9467\alpha_c^2 - 1.7012\alpha_c^3 + 0.9564\alpha_c^4 - 0.2537\alpha_c^5) \tag{3.10}$$

Table 3.1 Fanning Friction Factor and Nusselt Number for Fully Developed Laminar Flow in Ducts, Derived from Kakac et al. (1987)

Duct Shape			Nu_H	Nu_T	$Po = f\,Re$
	Circular		4.36	3.66	16
b	Flat channel		8.24	7.54	24
a, b	Rectangular, aspect ratio, b/a =	1	3.61	2.98	14.23
		2	4.13	3.39	15.55
		3	4.79	3.96	17.09
		4	5.33	4.44	18.23
		6	6.05	5.14	19.70
		8	6.49	5.60	20.58
		∞	8.24	7.54	24.00
	Hexagon		4.00	3.34	15.05
υ	Isosceles triangle, apex angle θ =	10°	2.45	1.61	12.47
		30°	2.91	2.26	13.07
		60°	3.11	2.47	13.33
		90°	2.98	2.34	13.15
		120°	2.68	2.00	12.74
a, b	Ellipse, major/minor axis a/b =	1	4.36	3.66	16.00
		2	4.56	3.74	16.82
		4	4.88	3.79	18.24
		8	5.09	3.72	19.15
		16	5.18	3.65	19.54

$Nu = hD_h/k$, $Re = \rho u_m D_h/\mu$
Nu_H—*Nu under a constant heat flux boundary condition, constant axial heat flux, and uniform circumferential temperature.*
Nu_T—Nu *under a constant wall temperature boundary condition.*
f—friction factor.

3.2.3 Developing laminar flow

As flow enters a duct, the velocity profile begins to develop along its length, ultimately reaching the fully developed Hagen–Poiseuille velocity profile. Almost all the analyses available in the literature consider a uniform velocity condition at the inlet. The length of the hydrodynamic developing region L_h is given by the following well-accepted equation:

$$\frac{L_h}{D_h} = 0.05Re \tag{3.11}$$

Since the pressure gradients found in small-diameter channels are quite high, the flow lengths are generally kept low. In many applications, the length of channel in the developing region therefore forms a major portion of the flow length through a microchannel. To account for the developing region, the pressure drop equations are presented in terms of an apparent friction factor.

Apparent friction factor f_{app} accounts for the pressure drop due to friction and the developing region effects. It represents an average value of the friction factor over the flow length between the entrance section and the location under consideration. Thus the pressure drop in a channel of hydraulic diameter D_h over a length x from the entrance is expressed as:

$$\Delta p = \frac{2 f_{app} \rho u_m^2 x}{D_h} \tag{3.12}$$

The difference between the apparent friction factor over a length x and the fully developed friction factor f is expressed in terms of an incremental pressure defect $K(x)$:

$$K(x) = (f_{app} - f)\frac{4x}{D_h} \tag{3.13}$$

For $x > L_h$ the incremental pressure defect attains a constant value $K(\infty)$, known as *Hagenbach's factor.*

Combining Eqs. (3.12) and (3.13), the pressure drop can be expressed in terms of the incremental pressure drop:

$$\Delta p = \frac{2(f_{app}Re)\mu u_m x}{D_h^2} = \frac{2(fRe)\mu u_m x}{D_h^2} + K(x)\frac{\rho u_m^2}{2} \tag{3.14}$$

For a circular tube, Hornbeck (1964) obtained the axial velocity distribution and pressure drop in a nondimensional form. He estimated the fully developed region to begin at $x^+ = 0.0565$, with a value of $K(\infty) = 1.28$ for a circular duct. Chen (1972) proposed the following equation for $K(\infty)$ for the circular geometry:

$$K(\infty) = 1.20 + \frac{38}{Re} \tag{3.15}$$

The nondimensionalized length x^+ is given by:

$$x^+ = \frac{x/D_h}{Re} \tag{3.16}$$

Shah and London (1978) showed that the frictional pressure drop in the developing region of a circular duct obtained by Hornbeck (1964) can be accurately described by the following equation:

$$\frac{\Delta p}{(1/2)\rho u_m^2} = 13.74(x^+)^{1/2} + \frac{1.25 + 64x^+ - 13.74(x^+)^{1/2}}{1 + 0.00021(x^+)^{-2}} \tag{3.17}$$

Rectangular geometries are of particular interest in microfluidics applications. Shah and London (1978) and Kakac et al. (1987) presented comprehensive summaries of the available literature. Phillips (1987) reviewed the available information, including that from Curr et al. (1972), and compiled the results for the apparent friction factor in a rectangular duct as shown in Figure 3.3. It can be seen that fully developed flow is attained at different x^+ values, with low-aspect-ratio ducts

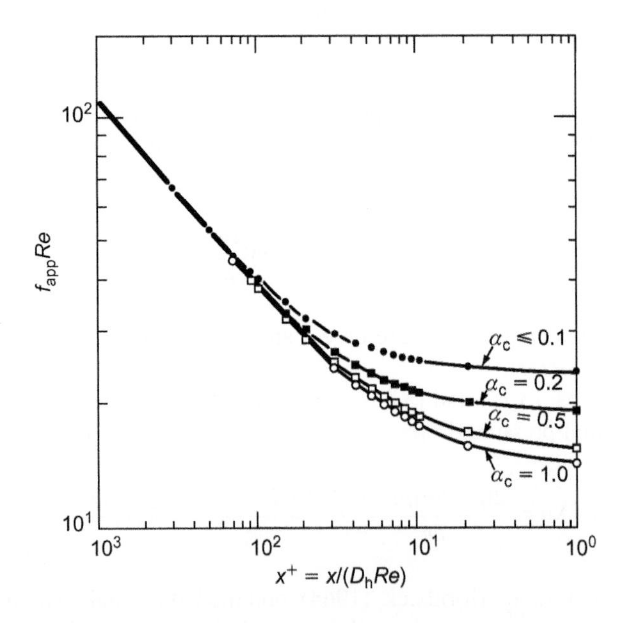

FIGURE 3.3

Apparent friction factors for rectangular ducts in the developing region for different aspect ratios ($\alpha = 1/\alpha_c$).

Source: *From Phillips (1987).*

reaching it earlier. The constant 0.05 in Eq. (3.11) is modified to around 1 ($x^+ = 10^0$) for fully developed value in rectangular ducts, as seen in Figure 3.3. Table 3.2, derived from Phillips (1987), gives the values of the apparent friction factor in tabular form.

For other channel aspect ratios, a linear interpolation is suggested. Alternatively, use the curve-fit equations provided in Appendix A.

The results for the fully developed friction factors and Hagenbach's factors in trapezoidal channels are reported by Kakac et al. (1987). By considering the rectangular channels as a subset of the trapezoidal geometry, Steinke and Kandlikar (2005a) obtained the following curve-fit equation for the Hagenbach's factor for rectangular channels:

$$K(\infty) = 0.6796 + 1.2197\alpha_c + 3.3089\alpha_c^2 - 9.5921\alpha_c^3 + 8.9089\alpha_c^4 - 2.9959\alpha_c^5 \quad (3.18)$$

Table 3.2 Laminar Flow Friction Factor in the Entrance Region of Rectangular Ducts (Phillips, 1987)

	$f_{app}Re$			
$x^+ = (x/D_h)/Re$	$\alpha_c = 1.0$	$\alpha_c = 0.5$	$\alpha_c = 0.2$	$\alpha_c \le 0.1$ $\alpha_c \ge 10$
0	142.0	142.0	142.0	287.0
0.001	111.0	111.0	111.0	112.0
0.003	66.0	66.0	66.1	67.5
0.005	51.8	51.8	52.5	53.0
0.007	44.6	44.6	45.3	46.2
0.009	39.9	40.0	40.6	42.1
0.01	38.0	38.2	38.9	40.4
0.015	32.1	32.5	33.3	35.6
0.02	28.6	29.1	30.2	32.4
0.03	24.6	25.3	26.7	29.7
0.04	22.4	23.2	24.9	28.2
0.05	21.0	21.8	23.7	27.4
0.06	20.0	20.8	22.9	26.8
0.07	19.3	20.1	22.4	26.4
0.08	18.7	19.6	22.0	26.1
0.09	18.2	19.1	21.7	25.8
0.10	17.8	18.8	21.4	25.6
0.20	15.8	17.0	20.1	24.7
>1.0	14.2	15.5	19.1	24.0

For intermediate values use the curve-fit equations provided in Appendix A at the end of the chapter.

The analysis presented in this section assumes a uniform velocity profile at the entrance of the channel. In many microfluidic applications, the channels have the manifold surfaces flush with the two opposing channel surfaces. The effect of such an arrangement was investigated numerically by Gamrat et al. (2004). They showed that the resulting apparent friction factors in the entrance region could be up to 50% lower than the theoretical predictions.

3.2.4 Fully developed and developing turbulent flow

A number of correlations with comparable accuracies are available in literature for fully developed turbulent flow in smooth channels. The following equation by Blasius is used extensively:

$$f = 0.0791 Re^{-0.025} \tag{3.19}$$

A more accurate equation was presented by Phillips (1987) to cover both the developing and fully developed flow regions. He presented the Fanning friction factor for a circular tube in terms of the following equation:

$$f_{\mathrm{app}} = A Re^{B} \tag{3.20}$$

where

$$A = 0.09290 + \frac{1.01612}{x/D_{\mathrm{h}}} \tag{3.21}$$

$$B = 0.26800 \frac{0.32930}{x/D_{\mathrm{h}}} \tag{3.22}$$

For rectangular channel geometries, Re is replaced with the laminar-equivalent Reynolds number (Jones, 1976) given by:

$$Re^{*} = \frac{\rho u_{\mathrm{m}} D_{\mathrm{le}}}{\mu} = \frac{\rho u_{\mathrm{m}}[(2/3) + (11/24)(1/\alpha_{\mathrm{c}})(2 - 1/\alpha_{\mathrm{c}})]D_{\mathrm{h}}}{\mu} \tag{3.23}$$

where D_{le} is the laminar-equivalent diameter given by the term in the brackets in Eq. (3.23).

3.3 Total pressure drop in a microchannel heat exchanger

3.3.1 Entrance and exit loss coefficients

An earnest interest in microchannel flows began with the pioneering work on direct chip cooling with water by Tuckerman and Pease (1981). Recently, a number of investigators including Li et al. (2000), Celata et al. (2002), and Steinke and Kandlikar (2005a) critically evaluated the available literature and presented explanations for the large deviations from classical theory reported by some of the researchers.

A schematic representation of the pressure drop experiments conducted by the researchers is shown in Figure 3.4. Since measuring the local pressure along the flow is difficult in microchannels, researchers have generally measured the pressure drop across the inlet and outlet manifolds. The resulting pressure drop measurement represents the combined effect of the losses in the bends, entrance and exit losses, developing region effects, and core frictional losses. Thus, the measured pressure drop is the sum of these components (Phillips, 1987):

$$\Delta p = \frac{\rho u_{\mathrm{m}}^2}{2}\left[(A_{\mathrm{c}}/A_{\mathrm{p}})^2(2\,K_{90}) + (K_{\mathrm{c}} + K_{\mathrm{e}}) + \frac{4 f_{\mathrm{app}} L}{D_{\mathrm{h}}}\right] \tag{3.24}$$

where A_c and A_p are the total channel area and the total plenum cross-sectional area, K_{90}, is the loss coefficient at the 90-degree bends, K_c and K_e represent the contraction and expansion loss coefficients due to area changes, and f_{app} includes the combined effects of frictional losses and the additional losses in the developing flow region.

Costaschuk et al. (2007) measured pressure drop along a 91.3 μm-deep and 1161.5 μm-wide aluminum microchannel with a hydraulic diameter of 169 μm. They used 27 μm-diameter pressure ports along the flow length. The ports were

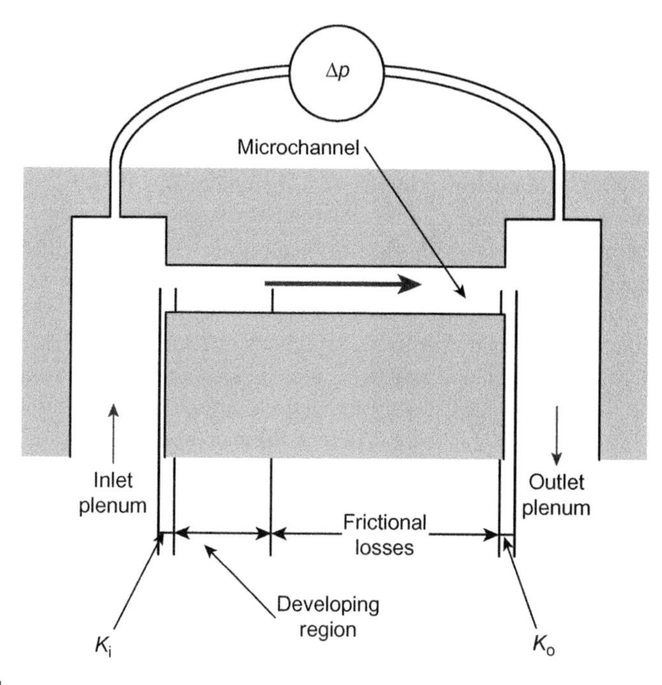

FIGURE 3.4

Schematic representation of the experiments employed by researchers for pressure drop measurements in microchannels.

Source: *From Steinke and Kandlikar (2005a).*

closer together near the entrance to capture the entrance region effect. They also measured the inlet and outlet pressure in the manifold. They noted that the fully developed laminar friction factor for the tested rectangular microchannel with an aspect ratio of 0.0786 was 86.85, while the theoretical value is 86.39. The Haaland equation was found to predict their turbulent data well. The entrance and exit loss coefficients of 0.42 and 1.46, respectively, in the turbulent flow were in agreement with the macroscale values. However, the equivalent lengths for inlet and exit regions in microscale were considerably lower, as given by the following equations in the laminar region:

$$\frac{L_{\mathrm{eq,inlet}}}{D_{\mathrm{h}}} = 0.026Re - 10.7 \tag{3.25}$$

$$\frac{L_{\mathrm{eq,outlet}}}{D_{\mathrm{h}}} = 0.021Re + 8.11 \tag{3.26}$$

In turbulent flow, the equivalent $L_{\mathrm{eq}}/D_{\mathrm{h}}$ ratios for inlet and outlet sections were 14.7 and 38, respectively. The entrance region lengths in the laminar flow were found to be about half of the macroscale values.

Szewczyk (2008) conducted experiments with circular capillary tubes of 327.8 and 280.6 μm and obtained the sum of the entrance and exit loss coefficients. For a sharp-edged orifice, the total loss coefficient was 2.25 in the laminar flow and 1.45 in the turbulent flow. The rounding of the inlet and outlet edges of the capillaries resulted in an increase in the critical Reynolds number and a decrease in the minor losses in the turbulent flow. The minor losses were not affected in the laminar region.

Phillips (1987) studied these losses and recommended that K_{90} be approximately 1.2. Xiong and Chung (2007) obtained the loss coefficients during miter bend. They found it to be 1.1 in the laminar region. They recommended additional experiments to determine the Reynolds number and channel size effects.

The above results from Costaschuk et al. (2007) and Szewczyk (2008) clearly indicate that the entrance and exit loss coefficients and the entrance region lengths are affected by the microscale. The scale effect should not come as a surprise since values for the macroscale have been obtained empirically from measurements. A detailed experimental study covering different entrance and exit conditions for rectangular microchannels of different aspect ratios and other geometries is warranted. Until such experiments are available, the empirical values given by Kays and London (1984) for macroscale application are recommended as described below.

The contraction and exit losses can be read from Figure 3.5, derived from Kays and London (1984) and Phillips (1990). Figure 3.5A is applicable to the low-aspect-ratio channels, $\alpha_c < 0.1$, and Figure 3.5B is for $0.1 \le \alpha_c \le 1.0$.

Equation (3.24) can also be written in terms of the fully developed friction factor f and the pressure drop defect $K(x)$:

$$\Delta p = \frac{\rho u_{\mathrm{m}}^2}{2}\left[(A_{\mathrm{c}}/A_{\mathrm{p}})^2(2K_{90}) + (K_{\mathrm{c}} + K_{\mathrm{e}}) + \frac{4fL}{D_{\mathrm{h}}} + K(x)\right] \tag{3.27}$$

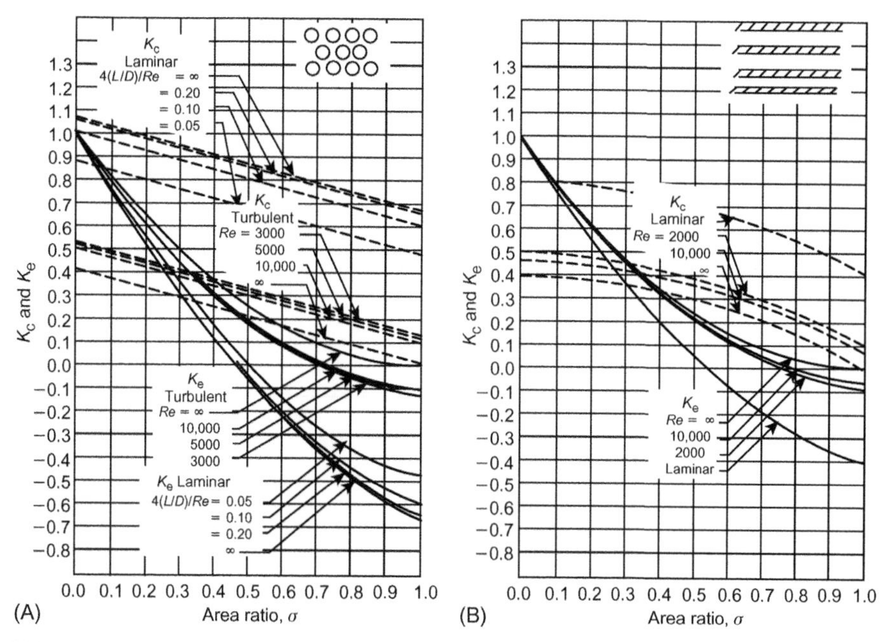

FIGURE 3.5

Contraction and expansion loss coefficients for flow between inlet and outlet manifolds and the microchannels: (A) $\alpha_c < 0.1$; and (B) $0.1 \leq \alpha_c \leq 1.0$.

Source: *Adapted from Kays and London (1984).*

For $L > L_h$, $K(x)$ is replaced by the Hagenbach's factor $K(\infty)$, as discussed in Section 3.2.3. Steinke and Kandlikar (2005a) conducted experiments on square silicon microchannels with 200 μm sides and 10 mm in length, and analyzed the data in detail. A parameter C^* is introduced to represent the ratio of the experimental and theoretical apparent friction factors. Figure 3.6 shows the data before and after applying the corrections in $C^* = f_{app,\ ex}/f_{app,th}$, where the subscripts ex and th refer to experimental and theoretical values, respectively. The uncorrected data converge to within 30% of the theoretical predictions after all corrections are applied. It is suspected that the high errors are due to slight variations in the channel cross-sectional area over its length, different entrance conditions, and errors associated with interpolations during data reduction.

Figure 3.7 shows the excellent agreement between the laminar flow theory and experimental results obtained by Judy et al. (2002) in 15–150 μm round and square microchannels made of fused silica and stainless steel with distilled water, methanol, and isopropanol. Similar agreement has been reported by Bucci et al. (2004) with 172, 290, and 520 μm diameter stainless steel circular tubes with water in the fully developed laminar flow region.

Niklas and Favre-Marinet (2003) analyzed the flow of water in a network of triangular microchannels with $D_h = 110$ μm. The contributions due to various

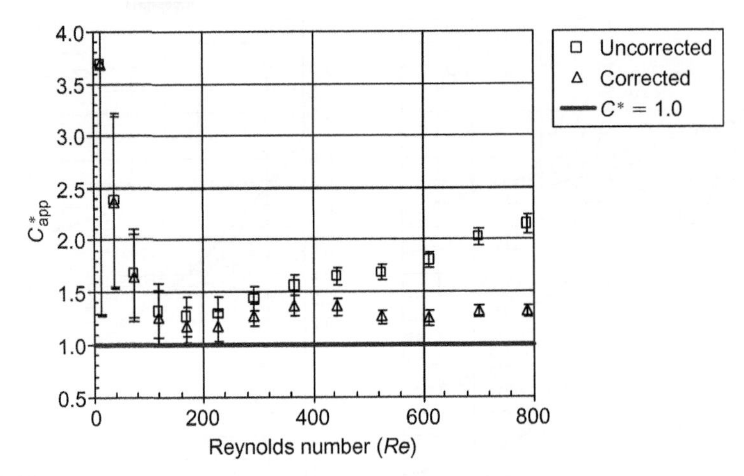

FIGURE 3.6

Variation of $C_{app}^* = f_{app,ex}/f_{app,th}$ for a 200-μm square microchannel array after correcting for entrance and exit losses.

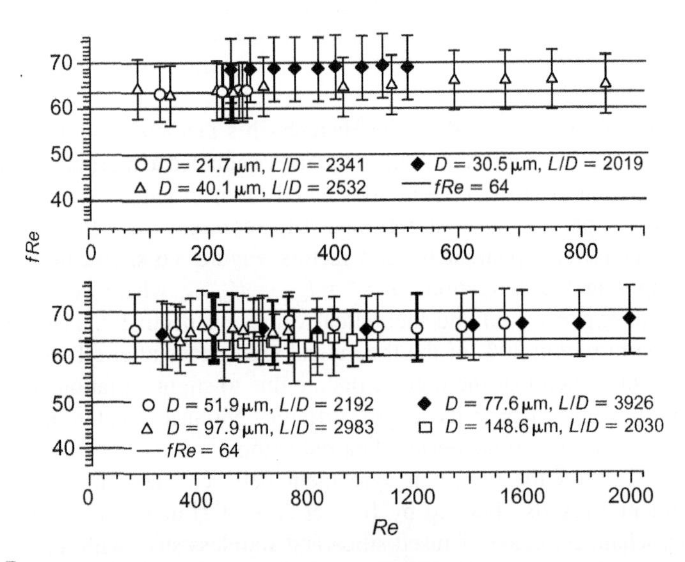

FIGURE 3.7

Comparison between theory and experimental data of Judy et al. (2002) in 50–150 μm circular and square microchannels in the fully developed laminar region. (*Note:* f in this figure represents the Darcy friction factor, which is four times the Fanning friction factor used elsewhere in this text.)

losses were carefully analyzed in both their numerical and experimental work. They concluded that the classical theory is applicable to modeling the flow through the entire system.

Another point that needs to be emphasized is the large errors that are associated with microscale experiments, as pointed out by Judy et al. (2002). A detailed analysis was conducted by Steinke and Kandlikar (2005a) for rectangular microchannels using actual measured parameters. Their final expression for uncertainty in estimating the product $f\,Re$ for rectangular channels is given by:

$$\frac{U_{fRe}}{fRe} = \left[2\left(\frac{U_\rho}{\rho}\right)^2 + \left(\frac{U_\mu}{\mu}\right)^2 + \left(\frac{U_{\Delta p}}{\Delta p}\right)^2 + \left(\frac{U_L}{L}\right)^2 + 3\left(\frac{U_Q}{Q}\right)^2 + \left(\frac{U_a}{a}\right)^2 + 5\left(\frac{U_b}{b}\right)^2 + 2\left(\frac{U_a}{a+b}\right)^2 + 2\left(\frac{U_b}{a+b}\right)^2\right]^{1/2} \tag{3.28}$$

The details of the derivation are given by Steinke (2005). Note that the uncertainties in the measurements of height a and width b have a major influence on the overall uncertainty, followed by the uncertainties in the density and volumetric flow rate measurements. The low-flow Reynolds number data in Figure 3.6 exhibit extremely large uncertainties due to the errors associated with the flow rate measurement. The large discrepancy reported in the literature by some of the earlier investigators is the result of the uncertainties in channel dimensions and flow rate measurements, entrance and exit losses, and the developing region effects.

Figure 3.8 shows a picture of a microchannel cross-section obtained with a scanning electron microscope by Steinke and Kandlikar (2005a). Note that the actual channel profile deviates significantly from the intended rectangular profile. The side walls are undercut by about 20 μm and the corners are rounded. This illustrates the need to accurately measure the flow area for estimating the pressure drop characteristics of microchannels.

Similar conclusions were reached by other investigators, for example, Li et al. (2000), Celata et al. (2002), Judy et al. (2002), Baviere et al. (2004), and Tu and Hrnjak (2003). The increasing deviations from the theoretical values at higher Reynolds numbers in many data sets reported in the literature is believed to be due to the increased length of the developing region. Baviere and Ayela (2004) measured the local pressures along the flow in a microchannel of height 7.5 μm. The range of Reynolds numbers considered was between 0.1 and 15. Their results were found to agree with classical theory. They also identified the errors associated with the channel height measurement as being the largest source of uncertainty.

Baviere et al. (2004, 2006) conducted experiments with bronze-, altuglas-, and silicon-coated microchannels with channel heights between 7.1 and 300 μm and channel widths between 1 and 25 mm. Their results were also in excellent agreement with classical theory.

Other effects that may influence the flow in microchannels are (i) high viscous dissipation causing a change in the fluid viscosity at the wall, and

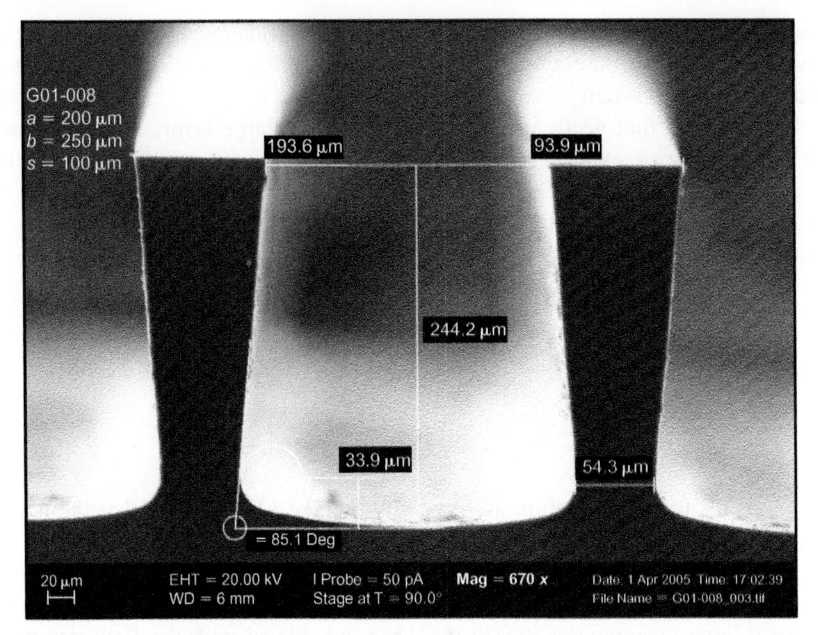

FIGURE 3.8

Cross-sectional view of a rectangular microchannel etched in silicon showing distortion in channel geometry from ideal rectangular profile.

Source: *From Steinke and Kandlikar (2005a).*

(ii) electrokinetic effects, which are covered in Chapter 4. The effect of viscous dissipation is reviewed by Shen et al. (2004). Xu et al. (2000) modeled this effect and found that the velocity profile was modified due to the viscous dissipation. The resulting friction factors were predicted to be lower due to a reduction in the viscosity at higher liquid temperatures. The viscous effect was studied experimentally by Judy et al. (2002). Their pressure drop data correlated well with the theory when they used the average of the inlet and outlet fluid temperatures from their stainless steel microtubes, which were 15–100 μm in diameter. Koo and Kleinstreuer (2004a,b) concluded that viscous dissipation effects increase rapidly with a decrease in the channel dimensions and hence should be considered along with the imposed heat sources at the channel walls.

Costaschuk et al. (2007) also confirmed that the fully developed friction factors were in agreement with conventional theory in both laminar and turbulent flow regions. Ghajar et al. (2010) reviewed the existing literature and reached the same conclusion. They identified relative roughness as an important parameter responsible for deviations from laminar flow theory.

Figure 3.9 shows a plot of the friction factor versus Reynolds number using the experimental data points reported in the literature, as compiled by Steinke and Kandlikar (2005a). The lines represent the theoretical values for different channel

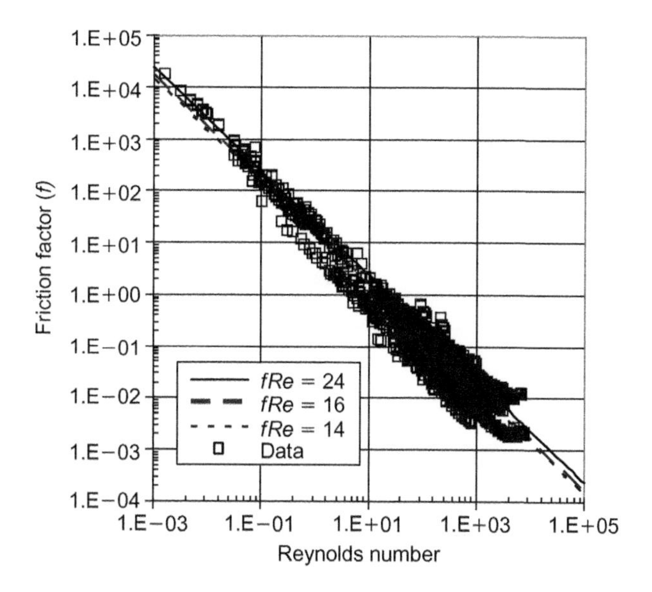

FIGURE 3.9

Comparison of friction factors reported in the literature, with the theoretical values for different microchannel geometries. Entrance and exit losses and the developing region at the entrance were not accounted for in a number of data sets reviewed by Steinke and Kandlikar (2005a).

aspect ratios. These data points were screened to eliminate the data that were obtained with large uncertainties and those that did not account for entrance and exit losses. It can be seen that the general trends are followed, though the errors are still quite large, mainly due to the uncertainties associated with channel size measurement.

3.3.2 Laminar-to-turbulent transition

The laminar-to-turbulent flow transition is another topic that has been analyzed by a number of investigators. The laminar-to-turbulent transition in abrupt entrance rectangular ducts was found to occur at a transition Reynolds number of $Re_t = 2200$ for $\alpha_c = 1$ and at $Re_t = 2500$ (Hartnett et al., 1962) for parallel plates with $\alpha_c = 0$. For other aspect ratios, a linear interpolation between these two values is recommended.

Some of the initial studies indicated an early transition to turbulent flow in microchannels. However, a number of recent studies have shown that the laminar-to-turbulent transition remains unchanged. Figure 3.10, from Bucci et al. (2004), shows that the transition occurred around $Re_t = 2000$ for circular microtubes 171–520 μm in diameter. The results of Baviere et al. (2004, 2006) also indicate that the laminar-to-turbulent transition in smooth microchannels is not influenced by the channel dimensions and occurs around 2300. Similar results were reported by a number of investigators, including Bucci et al. (2004), Schmitt

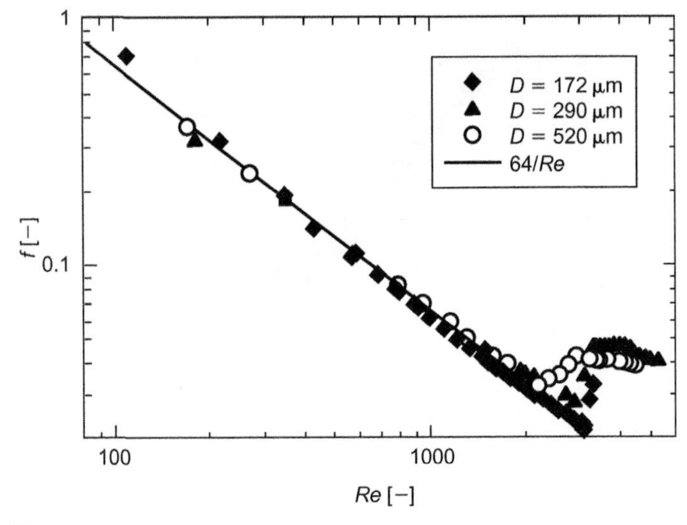

FIGURE 3.10

Comparison between theory and experimental data of Bucci et al. (2004) in 171–520 μm diameter circular tubes in the fully developed laminar region. (*Note: f* in this figure represents the Darcy friction factor, which is four times the Fanning friction factor used elsewhere in this text.)

and Kandlikar (2005), and Kandlikar et al. (2005) for minichannels with $D_h < 1$ mm, and by Li et al. (2000) for 80 μm $\leq D_h \leq$ 166.3 μm.

The transition from the laminar-to-turbulent region is influenced by channel surface roughness. Further details of this effect are discussed in the next section.

3.4 Roughness effects

3.4.1 Roughness representation

Darcy (1857) investigated the effects of surface roughness on the turbulent flow of water in rough pipes made of cast iron, lead, wrought iron, and asphalt-covered cast iron. The pipes were 12–500 mm in diameter and 100 m long. Fanning (1886) proposed a correlation for the pressure drop as a function of roughness. Mises (1914) is credited with introducing the term relative roughness, which was originally defined as the ratio of absolute roughness to the pipe radius (we now use pipe diameter to normalize roughness parameters, such as average roughness R_a). Nikuradse (1937) presented a comprehensive review of the literature covering uniform roughness and roughness structures such as corrugations, and some of the available correlations relating friction factor to relative roughness. He also conducted a systematic study on friction factor by applying uniform-diameter sand grain particles with Japanese lacquer to the inner pipe surface and measuring the pressure drop. He identified three regions which form the basis of the current Moody diagram that is used for friction factor estimation.

The term relative roughness is used for the ratio ε/D_h, where ε is the average roughness. Nikuradse used the diameter of the uniform sand grain particles to represent the roughness ε. The range of relative roughness in Nikuradse's experiments on circular pipes was $0.001 \leq \varepsilon/D \leq 0.033$. Furthermore, Nikuradse identified three ranges to describe his laws of resistances. In Range I, for low Reynolds numbers corresponding to laminar flow, the friction factor was independent of surface roughness and was given by the now-established classical equation $f = 16/Re$ for circular pipes. In Range II, he described the roughness effects on the friction factor in turbulent flow by $f = 0.079/Re^{1/4}$; finally, in Range III, the friction factor is independent of the Reynolds number but depends on the relative roughness, $f = 0.25/[1.74 + 2\log(2\varepsilon/D)]^2$. Note that the original equations by Nikuradse have been modified to express them in terms of the Fanning friction factor (which is one-fourth of Darcy friction factor, $f = 0.25\,f_{\mathrm{Darcy}}$).

Moody (1944) used the available data on the friction factor and developed the well-known Moody diagram covering the relative roughness range of $0 \leq \varepsilon/D \leq 0.05$. Following the earlier investigators, he used the tube root diameter of a circular pipe, similar to Nikuradse's work, in expressing the relative roughness ε/D.

A number of different parameters were studied for their suitability in representing the roughness effects. Kandlikar et al. (2005) investigated parameters based on various roughness characterization schemes and proposed a set of three parameters: mean profile peak height (R_p), mean spacing of profile irregularities (RS_m), and floor distance to mean line (F_p), as shown in Figure 3.11. Two of these parameters (R_p and RS_m) are defined in the ASME B46.1-2002, and the other (F_p) is proposed by Kandlikar et al. (2005).

- *Average maximum profile peak height* (R_{pm}): The distance between the average of the individual highest points of the profile ($R_{p,i}$) and the mean line within the evaluation length. The mean line represents the conventional average roughness value (R_a).

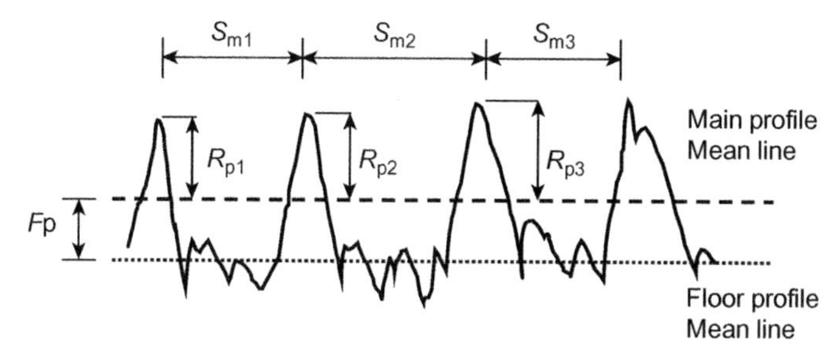

FIGURE 3.11

Maximum profile peak height (R_p), mean spacing of profile irregularities (RS_m), and floor distance to mean line (F_p).

- *Mean spacing of profile irregularities* (RS_m): The mean value of the spacing between profile irregularities within the evaluation length. The irregularities of interest are the peaks, so this is equivalent to the *pitch*.

$$RS_m = \frac{1}{n}\sum_{i=1}^{n} S_{m_i} \tag{3.29}$$

- *Floor distance to mean line* (F_p): The distance between the main profile mean line (determined by R_a) and the floor profile mean line. The floor profile is the portion of the main profile that lies below the main profile mean line.

The three parameters described above allow the characterization of the peak height, peak spacing, and the distance from the floor to the mean line. These parameters will define the characteristics of the surface roughness that influence the location and shape of the fluid flow streamlines (as described by Webb et al., 1971 and Kandlikar et al., 2005) and, consequently, the size of the recirculation flow zones between roughness elements. From the above parameters, the equivalent roughness ε can be estimated by the following relationship:

$$\varepsilon = R_{pm} + F_p \tag{3.30}$$

For the sand grain roughness employed by Nikuradse, the new definition of roughness ε yields the same value, and hence no correction is needed for any of the friction factor correlations or charts. The use of mean spacing between profile irregularities is expected to represent the structured roughness surfaces. Such surfaces may be designed in the future to obtain specific pressure drop and heat transfer performance characteristics.

The roughness parameters for a variety of copper, aluminum, stainless steel, nickel, and silicon wafer surfaces produced using different techniques are listed in Table 3.3. The flycut machining was performed with a single-point, carbide-tipped tool bit with a 1-mm radius. The milling was performed with a 12.7-mm-diameter two-flute high-speed steel end mill. The ground samples were prepared using a 60-grit grinding wheel. The silicon wafers were prepared as with KOH etch, XeF_2 etch, and the DRIE process. The nickel comparator surfaces were prepared with milling, grinding, Blanchard grinding, or lapping processes. The values of various roughness parameters used in deriving the roughness parameter εF_p are tabulated. The roughness varies from 0.147 to 6.847 μm. For microchannel passages, these values may be used as roughness values for the specific process used in manufacturing the surface.

3.4.2 Roughness effect on friction factor

Constricted flow model: For microchannels, the relative roughness values are expected to be higher than the limit of 0.05 used in the Moody diagram. Kandlikar et al. (2005) considered the effect of cross-sectional area reduction due to protruding roughness elements and recommended using the constricted flow

Table 3.3 Summary of Roughness Parameters for All Sample Surface (Parameters in μm)

Roughness Sample	R_a	R_p	R_v	F_p	εF_p
Copper, flycut	0.312	0.817	−0.863	0.225	1.042
Copper, ground	0.290	1.030	−1.818	0.312	1.343
Copper, milled	0.745	0.665	−0.788	0.232	0.896
Aluminum, flycut	0.221	0.792	−0.838	0.228	0.998
Aluminum, ground	0.254	0.849	−1.482	0.269	1.118
Aluminum, milled	0.643	1.908	−1.660	0.213	2.121
Stainless steel, flycut	0.295	0.774	−0.837	0.290	1.065
Stainless steel, ground	0.370	0.999	−1.546	0.328	1.327
Stainless steel, milled	1.195	3.210	−3.098	1.059	4.269
Silicon wafer with KOH etch 0.128	0.337	−0.767	0.149	0.337	
Silicon wafer with XeF_2 etch	–	–	–	–	–
Silicon wafer, DRIE	–	–	–	–	–
Nickel comparator, 2L	0.044	0.118	−0.117	0.028	0.147
Nickel comparator, 4L	0.075	0.182	−0.220	0.056	0.243
Nickel comparator, 8G	0.127	0.449	−0.544	0.174	0.577
Nickel comparator, 8L	0.196	0.548	−0.886	0.218	0.766
Nickel comparator, 16BL	0.452	1.159	−2.410	0.607	1.766
Nickel comparator, 16G	0.389	1.211	−1.441	0.424	1.635
Nickel comparator, 32BL	0.653	1.697	−3.616	0.786	2.484
Nickel comparator, 32G	1.008	2.577	−3.550	1.057	3.634
Nickel comparator, 32ST	0.639	1.770	−2.386	0.649	2.419
Nickel comparator, 63G	1.438	5.293	−8.431	1.555	6.847
Nickel comparator, 63M	1.215	3.668	−4.519	1.380	5.048

area in calculating the friction factor. Using a constricted diameter $D_{cf} = D - 2\varepsilon$, a modified Moody diagram was presented as shown in Figure 3.12. In the turbulent region, it was found that such a representation yielded a constant value of friction factor above $\varepsilon/D_{cf} > 0.03$.

In the turbulent fully rough region, $0.03 < \varepsilon/D_{cf} \leq 0.05$, the friction factor based on the constricted flow diameter is given by:

$$f_{\mathrm{Darcy,cf}} = 0.042 \tag{3.31a}$$

In terms of the Fanning friction factor, we get:

$$f_{\mathrm{cf}} = f_{\mathrm{Darcy,cf}}/4 = 0.042/4 = 0.0105 \tag{3.31b}$$

Since experimental data is not available beyond $\varepsilon/D_{cf} > 0.05$, using Eqs. (3.31a) and (3.31b) for higher relative roughness values than 0.05 is not recommended. Note

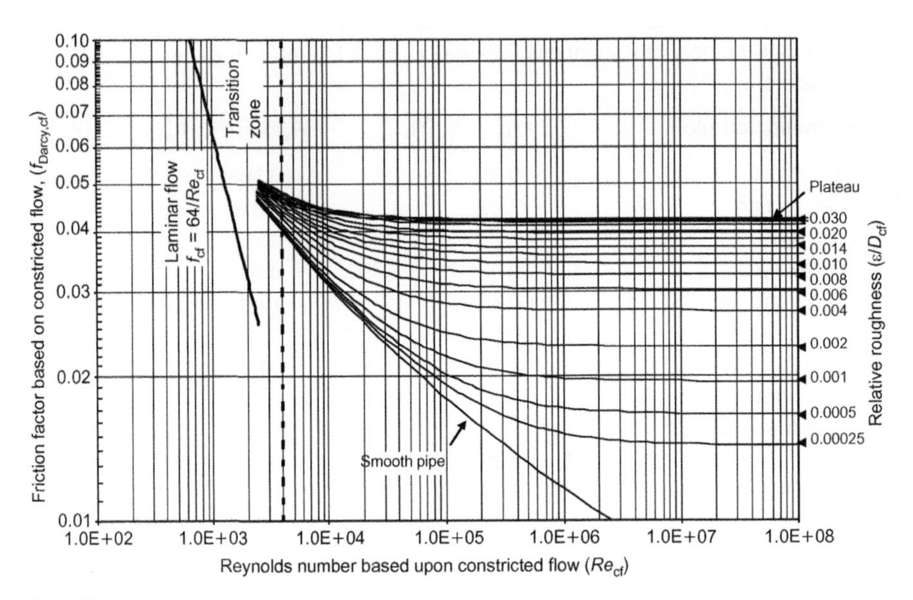

FIGURE 3.12

Darcy friction factor plot based on a constricted flow diameter.

Source: *From Kandlikar et al. (2005).*

that the friction factor and the geometrical and flow parameters are based on the constricted flow diameter, as given by the following equations:

$$D_{\mathrm{cf}} = D - 2\varepsilon \tag{3.32}$$

$$\Delta p = \frac{2 f_{\mathrm{cf}} \rho u_{m,\mathrm{cf}}^2 L}{D_{h,\mathrm{cf}}} \tag{3.33}$$

$$u_{m,\mathrm{cf}} = \dot{m}/A_{\mathrm{cf}} \tag{3.34}$$

$$Re_{\mathrm{cf}} = \frac{\rho u_{m,\mathrm{cf}} D_{h,\mathrm{cf}}}{\mu} \tag{3.35}$$

In the fully developed laminar flow region, the constricted friction factor is given by the following equation (Kandlikar et al., 2005):

Laminar region, $0 \leq \varepsilon/D_{\mathrm{h,cf}} \leq 0.15$

$$f_{\mathrm{cf}} = \frac{Po}{Re_{\mathrm{cf}}} \tag{3.36}$$

where the Poiseuille number Po is given by either Eq. (3.9) or (3.10) depending on the channel geometry. In the turbulent region, Kandlikar et al. (2005) derived the following expression for f_{cf} using the Colebrooke equation.

$$\frac{1}{(D_{cf}+2\varepsilon/D_{cf})^{2.5}f_{Darcy,cf}^{0.5}} = -2.0\log_{10}\left(\frac{(\varepsilon/D_{cf}+2\varepsilon)}{3.7}+\frac{2.51}{Re_{cf}(D_{cf}+2\varepsilon/D_{cf})^{1.5}f_{Darcy,cf}^{0.5}}\right) \tag{3.37}$$

Note that the equation given in Kandlikar et al. (2005) contains typographical errors, and the above equation is in the corrected form. However, since it requires an iterative calculation, an alternate form of friction factor suggested by Haaland (1983) is employed to obtain f_{cf} directly without an iterative procedure. In the fully developed turbulent region, $0 \le \varepsilon/D_{h,cf} < 0.03$, the friction factor is given by

$$f_{cf} = \frac{f_{Darcy,cf}}{4} = \frac{1}{4}\left\{-18\log_{10}\left[\left(\frac{1}{3.7(D_{cf}/\varepsilon+2)}\right)^{1.11}+\frac{6.9}{Re_{cf}\left(\frac{D_{cf}}{D_{cf}+2\varepsilon}\right)}\right]\right\}^{-2}\left[\frac{1}{1+\frac{2\varepsilon}{D_{cf}}}\right]^{5} \tag{3.38}$$

In the fully developed turbulent region, $0.03 \le \varepsilon/D_{h,cf} \le 0.05$ and $Re_{cf} < 10^8$, the friction factor is given by

$$f_{cf} = 0.0105 \tag{3.39}$$

Fully developed Darcy friction factors in the turbulent region for roughness $\varepsilon/D_{h,cf} < 0.05$ may be approximated to 0.042 (or Fanning friction factor of 0.0105), as indicated by the modified Moody diagram. Recently, Brackbill and Kandlikar (2010) studied the friction factors for rough channels in the transition and turbulent regions. They found that for structured two-dimensional roughness elements, such as grooves, the fully developed friction factors were reduced as the pitch increased. Eventually the friction factors approached the smooth channel values as the pitch approached infinity. The pitch is thus seen to affect the friction factor to approach the fully developed turbulent value of 0.042 at lower pitches and the smooth channel values at higher pitches. Further experimental work in this area is recommended.

For rectangular channels with all sides having a roughness ε, the mean flow velocity is calculated using the constricted flow area and the constricted channel dimensions of $a_{cf} = a - 2\varepsilon$ and $b_{cf} = b - 2\varepsilon$. The hydraulic diameter is calculated using the constricted channel dimensions.

Schmitt and Kandlikar (2005) conducted experiments to study the effect of a large relative roughness of $\varepsilon/D_{h,cf}$ up to 0.15 in smooth and artificially roughened rectangular minichannels. They introduced sawtooth roughness elements in a 10.3-mm-wide and 100-mm-long test channel. Cross-sectional views of the smooth channels and channels with sawtooth roughness elements are shown in Figure 3.13. The channel gap b_{cf} was varied to produce values of $\varepsilon/D_{h,cf}$ from 0.03 to 0.15. Differential pressure taps for measuring the local pressure drops were located at several locations along the flow length in the fully developed region.

The smooth channel results reported by Kandlikar et al. (2005) closely follow the classical laminar flow theory for friction factor. The experimental data from

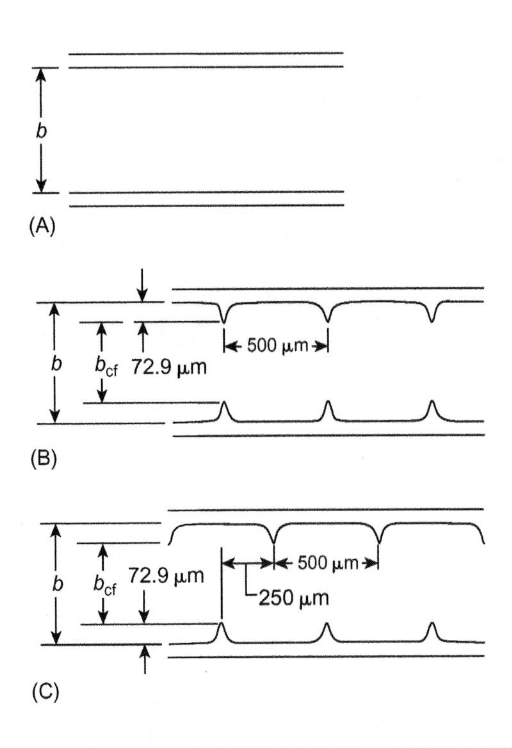

FIGURE 3.13

Roughness elements used by Schmitt and Kandlikar (2005) and Kandlikar et al. (2005): (A) smooth channel; (B) aligned sawtooth; and (C) offset sawtooth.

the two sawtooth structures are represented in Figure 3.14 for flow of water with $b = 500$ μm, resulting in a D_h of 953 μm. This plot uses the height b in calculating D_h and Re. The dashed line represents the $f\ Re = Constant$ line corresponding to the channel aspect ratio $\alpha_c = a/b$. It can be seen that the agreement in the low Reynolds number region corresponding to laminar flow is considerably off.

Figure 3.15 shows the same data used in Figure 3.14, but plotted with the constricted flow areas and using the constricted flow parameters given by Eqs. (3.32)–(3.35). The results shown in Figure 3.15 in the laminar region show good agreement with the accepted laminar flow theory ($f_{cf}Re_{cf} = Constant$).

Similar results were obtained for different gap sizes (yielding different relative roughness values) for both the offset sawtooth and aligned sawtooth geometries. Brackbill and Kandlikar (2010) tested the sawtooth roughness elements of different heights and different pitch to roughness height ratios (β) between 4 and 40, in the relative roughness range from 1.4% to 27.6%. Figure 3.16 shows the results for two roughness elements: (i) a roughness height of 99 μm and a pitch of 405 μm, and (ii) a roughness height of 52 μm and a pitch of 1008 μm. The gap was varied for both cases to yield a range of relative roughnesses shown in Figure 3.16. In the laminar region, although the friction factors were close to the

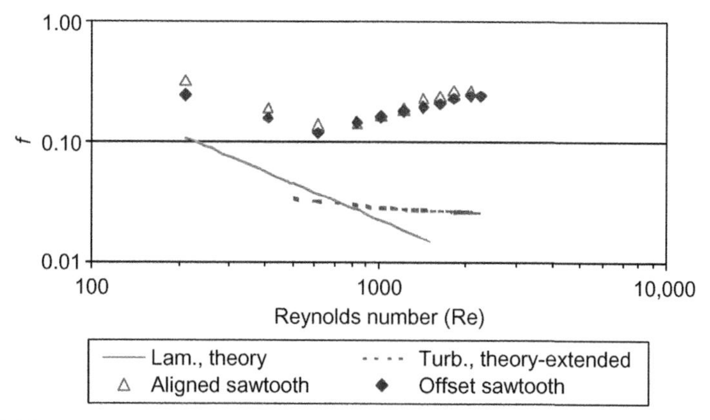

FIGURE 3.14

Fully developed friction factor versus Reynolds number, both based on hydraulic diameter for water flow. $D_h = 953$ μm, $b = 500$ μm, $b_{cf} = 354$ μm, $w = 10.03$ mm, $\varepsilon/D_h = 0.0735$.

Source: *From Kandlikar et al. (2005).*

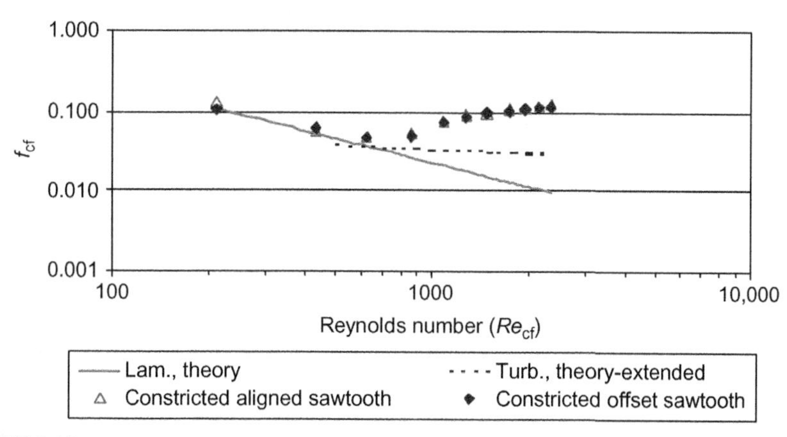

FIGURE 3.15

Fully developed friction factor versus Reynolds number, both based on constricted flow hydraulic diameter; water flow. $D_{h,cf} = 684$ μm, $b = 500$ μm, $b_{cf} = 354$ μm, $w = 10.03$ mm, $\varepsilon/D_{h,cf} = 0.1108$.

Source: *From Kandlikar et al. (2005).*

contricted model prediction, a clear effect of pitch was observed. The authors noted that for shorter pitches with pitch to roughness ratios of less than 5, the constricted model is able to predict the friction factor well. For higher pitch to roughness ratios, the friction factor approaches the smooth channel value corresponding to the base of the roughness elements.

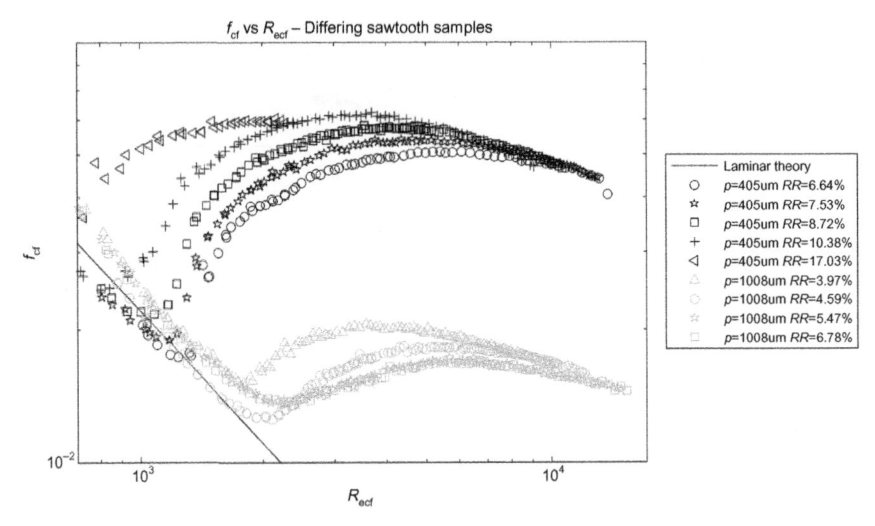

FIGURE 3.16

Constricted friction factor versus constricted Reynolds number plot with two sawtooth roughness elements and varying channel height in the laminar, transition, and turbulent regions. Upper set of points $\varepsilon_{FP} = 97\ \mu m$, $p = 405\ \mu m$, lower set of points $\varepsilon_{FP} = 5\ \mu m$, $p = 1008\ \mu m$.

Wall roughness region–based models: Koo and Kleinsstreuer (2004a,b) considered the roughness region as a porous medium layer at the wall and considered additional viscous forces due to the rough elements present at the wall. The viscous forces were modeled considering the permeability of the porous medium. Baviere et al. (2006) modified the porous layer model by using a discrete element approach to determine the frictional effects of the roughness elements. Subsequently, Gamrat et al. (2008) further extended this model into their rough-layer model (RLM) by considering a periodical distribution of discrete roughness elements on a smooth wall. The total flow resistance is calculated as the sum of the frictional resistance from the bottom of the smooth wall, frictional resistance from the top surface of the roughness elements, and the drag coefficient resulting from the roughness elements. They considered the effect of roughness geometry on the drag coefficient and refined the interface boundary condition between the porous medium and the bulk flow by accounting for the discontinuity at this interface. The analytical model was incorporated into their numerical modeling scheme and compared with the experimental results with a good degree of success for parallelopipedic, circular pin, and random roughness elements.

Lubrication model: Brackbill and Kandlikar (2010) applied lubrication approximation to the rough channels by neglecting the radial velocity component near the roughness elements on the wall and integrating the localized velocity profile. In the laminar region for relative roughness below 5%, the constricted model of Kandlikar et al. (2005) worked very well for lower values of pitch. The results indicated that the lubrication approximation also worked well for low values of roughness heights, below about 5%. The friction factor approached the smooth channel values as the pitch became larger.

Wagner and Kandlikar (2012) studied sinusoidal roughness elements by considering that the slope of the trajectory of the fluid elements in the vicinity of the roughness elements is small. In other words, the roughness elements provide smoothly varying roughness profiles. The periodic roughness element profiles were fitted with a cosine power function and introduced in the conventional laminar flow equations. The experimental results were compared with the model and it was noted that the effects of geometrical parameters such as pitch and roughness height both can be accounted for with this approach.

3.4.3 Roughness effect on the laminar-to-turbulent flow transition

The transition from laminar-to-turbulent flow has been reported to occur in microchannels at Reynolds numbers considerably below 2300. In many experiments, the transition has been mistakenly identified to occur early, based on experimental data uncorrected for the developing length (Steinke and Kandlikar, 2005a). Kandlikar et al. (2003) conducted experiments with stainless steel tubes and noted that early transition occurred for a 0.62 mm ID stainless steel tube with a surface roughness ε/D_h of 0.355%. Schmitt and Kandlikar (2005) conducted careful experiments with plain and sawtooth roughened channels with air and water. Their results for smooth rectangular channels showed a transition Reynolds number between 2000 and 2300, but for increasing relative roughness values there were decreasing transition Reynolds numbers, as seen in Figures 3.14 and 3.15. Figure 3.17 shows the transition Reynolds number as a function of the relative roughness (based on the constricted hydraulic diameter). The following equations are used to describe the roughness effects based on their experimental data.

Laminar-to-turbulent transition criteria:

$$\text{For } 0 < \varepsilon/D_{h,\mathrm{cf}} \leq 0.08: \quad Re_{t,\mathrm{cf}} = 2300 - 18,750(\varepsilon/D_{h,\mathrm{cf}}) \tag{3.40}$$

$$\text{For } 0.08 < \varepsilon/D_{h,\mathrm{cf}} \leq 0.15: \quad Re_{t,\mathrm{cf}} = 800 - 3270(\varepsilon/D_{h,\mathrm{cf}} - 0.08) \tag{3.41}$$

Brackbill and Kandlikar (2010) further modified Eq. (3.40) to match the transition Reynolds number for the base channel for a roughness height of zero (smooth channel), Re_0 as follows.

$$\text{For } 0 < \varepsilon/D_{h,\mathrm{cf}} \leq 0.08: \quad Re_{t,\mathrm{cf}} = Re_0 - \frac{(Re_0 - 800)}{0.08}(\varepsilon/D_{h,\mathrm{cf}}) \tag{3.42}$$

The experimental data on the transition Reynolds number by Brackbill and Kandlikar (2010) agrees well with the above model. At a relative roughness value of 27.6%, the turbulent transition was observed to occur at a Reynolds number of only 200.

The friction factor in the transition region for rough tubes with $\varepsilon/D_\mathrm{h,cf} < 0.05$ at a given Re_cf can be obtained by a linear interpolation of (i) the laminar friction

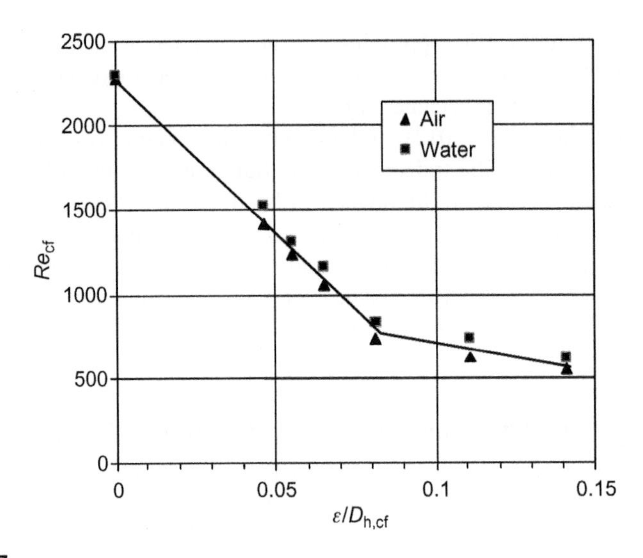

FIGURE 3.17

Transition Reynolds number variation with relative roughness based on constricted flow diameter, Eqs. (3.37) and (3.38) plotted along with data from Schmitt and Kandlikar (2005).

factor obtained from Eq. (3.36) at the transition Reynolds number, and (ii) the turbulent friction factor at $Re_{cf} = 2300$ given by Eq. (3.38) or (3.39) depending on the value of $\varepsilon/D_{h,cf}$. For $\varepsilon/D_{h,cf} > 0.05$ in the transition region, additional experimental data is needed before any recommendations can be made. Brackbill and Kandlikar (2010) present some of the constricted friction factor data in the transition region, as seen in Figure 3.16.

3.4.4 Developing flow in rough tubes

Developing flow in rough tubes has not been explored in the literature. As a preliminary estimate, the methods described in Section 3.2.3 for smooth tubes are recommended, introducing the constricted flow diameter in calculating the flow velocity, Reynolds number, and friction factor. Further research in this area is needed. Figure 3.16 shows the friction factor behavior in the transition region obtained by Brackbill and Kandlikar (2010) for two different roughness element pitches and varying gaps.

3.4.5 Turbulent flow in rough tubes

There are very few studies available on the effect of roughness on the turbulent flow friction factor. For uniform roughness beyond $\varepsilon/D_{cf} > 0.05$ in the fully developed turbulent flow regime, the friction factor based on constricted flow diameter is constant at 0.042, according to the modified Moody diagram given by

Kandlikar et al. (2005). For sawtooth roughness features, Brackbill and Kandlikar (2010) experimentally obtained the friction factors for two different pitches and a range of relative roughnesses. Figure 3.16 shows the values of constricted friction factors versus constricted Reynolds numbers for a rectangular channel with a roughness pitch of 405 and 1008 μm. The relative roughness was varied from 4% to 17%. It is seen that the friction factor in the transition region increases dramatically with a clear effect of relative roughness. However, the curves converge to a single line for each pitch. The asymptotic values decrease as the pitch increases and are below the value of 0.042 for the uniform roughness case.

3.5 Heat transfer in microchannels

3.5.1 Fully developed laminar flow

The Nusselt number in fully developed laminar flow is expected to be constant, as predicted by classical theory. However, there are a number of investigations reported in the literature that show a trend increasing with the Reynolds number in this range. This results from the experimental uncertainties, as discussed in Section 3.5.3.

The Nusselt number in the fully developed laminar flow is constant and depends on the channel geometry and the wall heat transfer boundary condition. Table 3.1 presents the Nusselt numbers for commonly used geometries under axially constant heat flux and circumferentially constant wall temperature (H1) boundary conditions with four-sided heating.

For a rectangular channel, the Nusselt number depends on the channel aspect ratio $\alpha_c = a/b$ and the wall boundary conditions. Three boundary conditions are identified in the literature; the Nusselt number for each one is given below.

Constant wall temperature, T-boundary condition:

$$Nu_T = 7.54(1 - 2.610\alpha_c + 4.970\alpha_c^2 - 5.119\alpha_c^3 + 2.702\alpha_c^4 - 0.548\alpha_c^5) \quad (3.43)$$

Constant circumferential wall temperature, uniform axial heat flux, H1 boundary condition:

$$\begin{aligned} Nu_{H1} &= 8.235(1 - 2.0421\alpha_c + 3.0853\alpha_c^2 - 2.4765\alpha_c^3 \\ &\quad + 1.0578\alpha_c^4 - 0.1861\alpha_c^5) \end{aligned} \quad (3.44)$$

Constant wall heat flux, both circumferentially and axially:

$$\begin{aligned} Nu_{H2} &= 8.235(1 - 10.6044\alpha_c + 61.1755\alpha_c^2 - 155.1803\alpha_c^3 \\ &\quad + 176.9203\alpha_c^4 - 72.9236\alpha_c^5) \end{aligned} \quad (3.45)$$

Dharaiya and Kandlikar (2012) numerically analyzed the rectangular geometry under the axially and circumferentially uniform heat flux (H2) boundary condition. The five conditions studied included heating from (i) all four walls, (ii) only three walls, (iii) only two opposite walls, (iv) only one wall, and (v) only two adjacent walls. Their results for the fully developed laminar flow under these boundary conditions are shown in Figure 3.18 and are also presented in Table 3.4. They also presented the following correlations based on their numerical results for the fully developed laminar flow Nusselt numbers for these five conditions.

For four walls heated under H2 boundary condition ($0.1 \le \alpha \le 1$), case 1:

$$Nu_{\mathrm{H2},f,d,4\ \mathrm{walls}} = [(0.3816)\alpha + 2.886] \tag{3.46}$$

Note that the aspect ratio is defined as the ratio of the smaller to larger side and is always between 0 and 1 for the four walls heated case.

For three walls heated under H2 boundary conditions ($0.1 \le \alpha \le 10$), case 2:

$$Nu_{\mathrm{H2},f,d,3\ \mathrm{walls}} = [(1.861 \times 10^{-3})\alpha^6 - (4.864 \times 10^{-2})\alpha^5 + (0.4477)\alpha^4 - (1.708)\alpha^3 + (2.403)\alpha^2 - (0.4319)\alpha + (2.464)] \tag{3.47}$$

For two opposite walls heated under H2 boundary conditions ($0.1 \le \alpha \le 10$), case 3:

$$Nu_{\mathrm{H2},f,d,2\ \mathrm{walls}} = [(8.464 \times 10^{-3})\alpha^4 - (0.1689)\alpha^3 + (1.145)\alpha^2 - (3.627)\alpha + (7.121)] \tag{3.48}$$

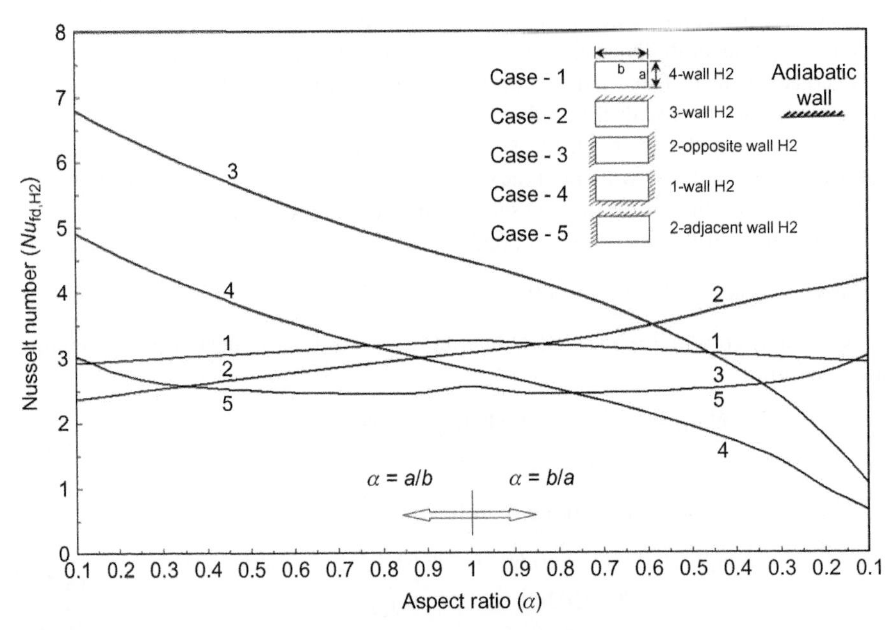

FIGURE 3.18

Nusselt numbers under H2 boundary conditions with heating on (1) all four walls, (2) three walls, (3) two opposing walls, (4) one wall, and (5) two adjacent walls. See Table 3.5 for further details on the wall configuration.

Source: *Adapted from Dharaiya and Kandlikar (2012).*

Table 3.4 Rectangular Ducts: $Nu_{H2,fd}$ for Fully Developed Laminar Flow, for One or More Walls Heated Under Uniform Heat Flux, H2 Boundary Conditions

Aspect Ratio, α (a/b)	Fully Developed Laminar Flow Nusselt Number, $Nu_{H2,fd}$				
	Four-Wall BC 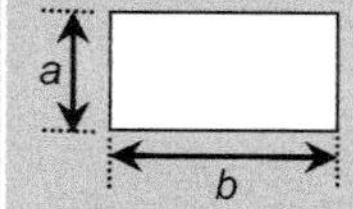	Three-Wall BC 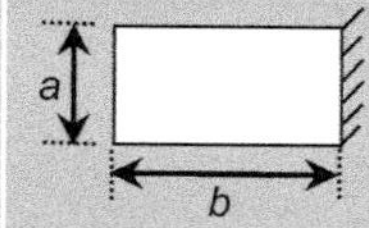	Two-Opposite Wall BC 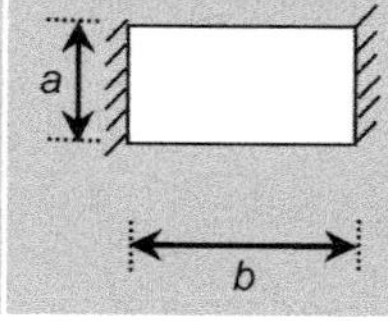	One-Wall BC 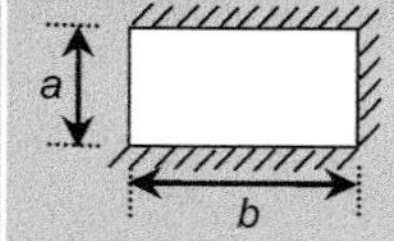	Two-Adjacent Wall BC 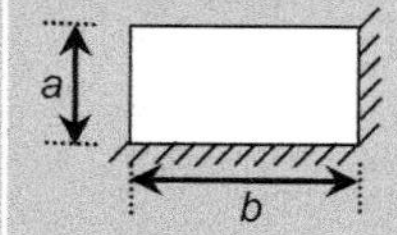
0.10	2.924	2.463	6.803	5.036	3.031
0.25	2.981	2.421	6.215	4.358	2.688
0.33	3.029	2.538	6.094	4.151	2.580
0.50	3.115	2.727	5.638	3.744	2.524
0.60	3.113	2.764	5.243	3.456	2.464
0.75	3.110	2.839	4.856	3.172	2.451
1	3.301	3.149	4.614	2.884	2.565
2	3.115	3.277	3.212	1.928	2.524
5	2.962	3.804	1.786	1.019	2.643
10	2.924	4.625	1.067	0.641	3.031

Source: *Adapted from Dharaiya and Kandlikar (2012).*

For one wall heated under H2 boundary conditions ($0.1 \le \alpha \le 10$), case 4:

$$Nu_{H2,f,d,2\ \text{walls}} = [(1.365 \times 10^{-2})\alpha^4 - (0.2563)\alpha^3 + (1.534)\alpha^2 - (3.825)\alpha + (5.301)] \quad (3.49)$$

For two adjacent walls heated under H2 boundary conditions ($0.1 \le \alpha \le 1$), case 5:

$$Nu_{H2,f,d,2\ \text{adjwalls}} = [(6.1571)\alpha^4 - (14.7340)\alpha^3 + (13.4030)\alpha^2 - (5.7474)\alpha + 3.4864] \quad (3.50)$$

Note that for the two adjacent wall heated case, the aspect ratio is defined as the ratio of the smaller to the larger side and is always between 0 and 1.

Cases 1 and 5 represent symmetric wall conditions, and the channel aspect ratio α varies between 0.1 and 1, while cases 2–4 are asymmetric and α varies between 0.1 and 10, as defined in Table 3.4.

In reality, all practical situations fall somewhere in the middle of these three boundary conditions. This becomes an especially important issue in the case of microchannels because of the difficulty in identifying a correct boundary condition with discretely spaced heat sources, and two-dimensional effects in the base and the fins.

The heating in microchannel geometrics generally comes from three sides, as a cover of glass or some other material is bonded on top of the microchannels to form the flow passages. The fully developed Nusselt numbers for both three- and four-side heated ducts have been compiled from various sources, including Wibulswas (1966) and Phillips (1987), and are given in Table 3.5. Note that the side with dimension a is not heated, and the channel aspect ratio is defined as $\alpha_c = a/b$.

3.5.2 Thermally developing flow

The thermal entry length is expressed by the following form for flow in ducts:

$$\frac{L_t}{D_h} = cRePr \quad (3.51)$$

For circular channels, the leading constant c in Eq. (3.51) is found to be 0.05, while for rectangular channels, the plots presented by Phillips (1987) suggest $c = 0.1$.

The local heat transfer in the developing region of a circular tube is given by the following equations (Shah and London, 1978):

$$Nu_x = 4.363 + 8.68(10^3\,x^*)^{-0.506} e^{-41x^*} \quad (3.52)$$

$$x^* = \frac{x/D_h}{RePr} \quad (3.53)$$

Table 3.5 Fully Developed Laminar Flow Nusselt Numbers Under H1 Boundary Conditions

$\alpha_c = a/b$	$Nu_{fd,3}$	$Nu_{fd,4}$
0	8.235	8.235
0.10	6.939	6.700
0.20	6.072	5.704
0.30	5.393	4.969
0.40	4.885	4.457
0.50	4.505	4.111
0.70	3.991	3.740
1.00	3.556	3.599
1.43	3.195	3.740
2.00	3.146	4.111
2.50	3.169	4.457
3.33	3.306	4.969
5.00	3.636	5.704
10.00	4.252	6.700
$>$ 10.00	5.385	8.235

a—unheated side in three-side heated case.
For intermediate values, use the curve-fit equations provided in Appendix A at the end of the chapter.

For rectangular channels with the four-side heating configuration, Nusselt numbers in the thermally developing region presented in Table 3.6 are derived from Phillips' (1987) work. For the three-side heating configuration, the following scheme is suggested by Phillips (1990).

Three-side heating, $\alpha_c \geq 0.1$ and $\alpha_c \leq 10$, use four-side heating table without any modification.

Three-side heating, $0.1 \leq \alpha_c \leq 10$:

$$Nu_{x,3}(x^*, \alpha_c) = Nu_{x,4}(x^*, \alpha_c)\frac{Nu_{fd,3}(x^* = x^*_{fd}, \alpha_c)}{Nu_{fd,4}(x^* = x^*_{fd}, \alpha_c)} \tag{3.54}$$

The subscripts x,3 and x,4 refer to the location at a distance x in the heated length for the three-sided and four-sided heating cases, respectively. The Nusselt numbers in the fully developed region for both heating configurations are obtained from Table 3.5; in the developing region Nusselt numbers for the four-sided heating are obtained from Table 3.6.

The use of fully developed hydrodynamic conditions in heat transfer analysis is reasonable for water. Garimella and Singhal (2004) noted that assuming fully developed hydrodynamic conditions and thermally developing conditions resulted in a satisfactory agreement with their data for microchannels.

Table 3.6 Thermal Entry Region Nusselt Numbers

x^*	$\alpha_c \leq 0.1$[a]	$Nu_{x,4}$				
		$\alpha_c = 0.25$	$\alpha_c = 0.333$	$\alpha_c = 0.5$	$\alpha_c = 1.0$	$\alpha_c \geq 10$[b]
0.0001	31.4	26.7	27.0	23.7	25.2	31.6
0.0025	11.9	10.4	9.9	9.2	8.9	11.2
0.005	10	8.44	8.02	7.46	7.1	9.0
0.00556	9.8	8.18	7.76	7.23	6.86	8.8
0.00625	9.5	7.92	7.5	6.96	6.6	8.5
0.00714	9.3	7.63	7.22	6.68	6.32	8.2
0.00833	9.1	7.32	6.92	6.37	6.02	7.9
0.01	8.8	7	6.57	6.05	5.69	7.49
0.0125	8.6	6.63	6.21	5.7	5.33	7.2
0.0167	8.5	6.26	5.82	5.28	4.91	6.7
0.025	8.4	5.87	5.39	4.84	4.45	6.2
0.033	8.3	5.77	5.17	4.61	4.18	5.9
0.05	8.25	5.62	5.00	4.38	3.91	5.55
0.1	8.24	5.45	4.85	4.22	3.71	5.4
1	8.23	5.35	4.77	4.11	3.6	5.38

$x^* = x/(Re\,PrD_h)$
[a]Parallel plates, both sides heated.
[b]Parallel plates, one side heated. For intermediate values, use the curve-fit equations provided in Appendix A.

For the H2 boundary conditions, Dharaiya and Kandlikar (2012) numerically analyzed the entry region problem. Their results for the five different wall heating configurations are given in Figures 3.19–3.23.

3.5.3 Agreement between theory and available experimental data on laminar flow heat transfer

Laminar flow heat transfer in microchannels has been studied by a number of researchers. Agreement with classical laminar flow theory is expected to hold, but the reported data show significant scatter due to difficulties encountered in making accurate measurements of local heat flux and temperature. Steinke and Kandlikar (2005b) reviewed the available data and presented a comprehensive table showing the range of parameters employed. The laminar flow heat transfer in the fully developed region is expected to be constant, but the data taken from the literature show a generally linear increase in Nusselt number with flow Reynolds number, as seen in Figure 3.24.

The main reasons for this discrepancy have been attributed to the following factors:

(i) *Entrance region effects:* The researchers have used the fully developed *theoretical* values for Nusselt numbers in their comparison with the

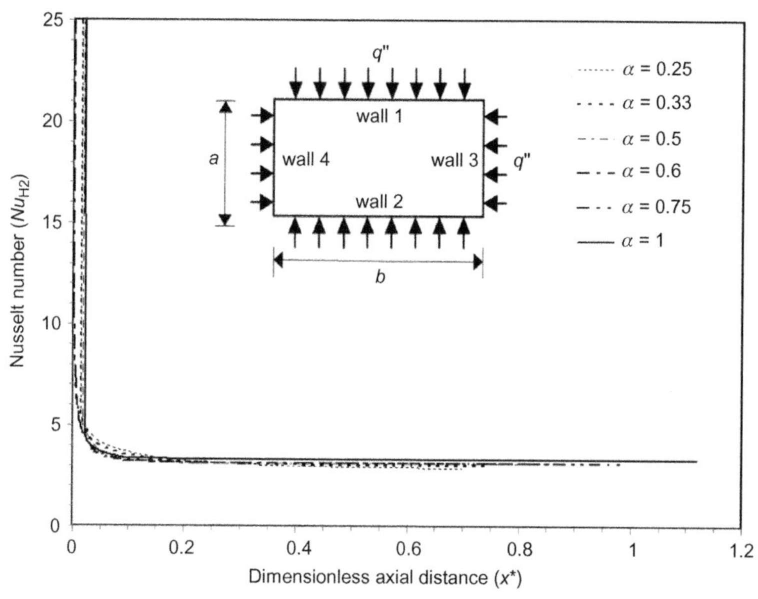

FIGURE 3.19

Local Nusselt number as a function of nondimensional length and channel aspect ratio for heating on all four walls, H2 boundary conditions.

Source: *Adapted from Dharaiya and Kandlikar (2012).*

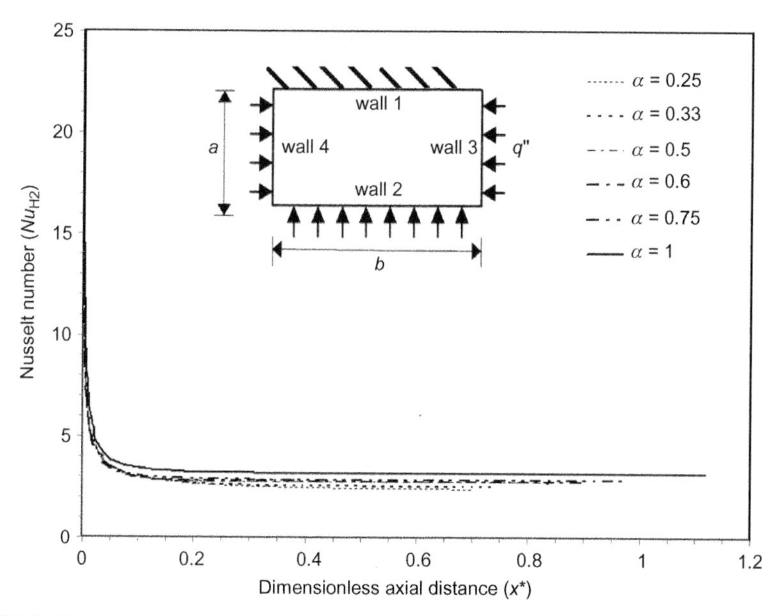

FIGURE 3.20

Local Nusselt number as a function of nondimensional length and channel aspect ratio for heating on three walls, H2 boundary conditions.

Source: *Adapted from Dharaiya and Kandlikar (2012).*

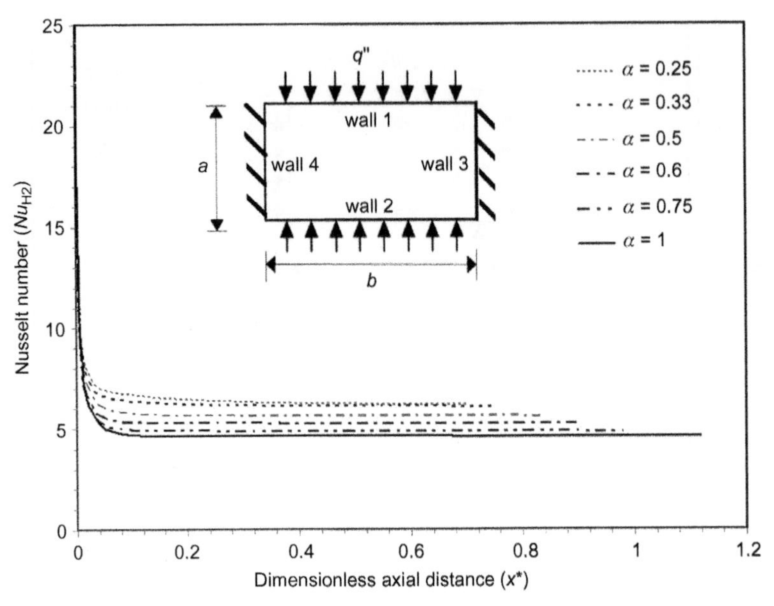

FIGURE 3.21

Local Nusselt number as a function of nondimensional length and channel aspect ratio for heating on two opposed walls, H2 boundary conditions.

Source: *Adapted from Dharaiya and Kandlikar (2012).*

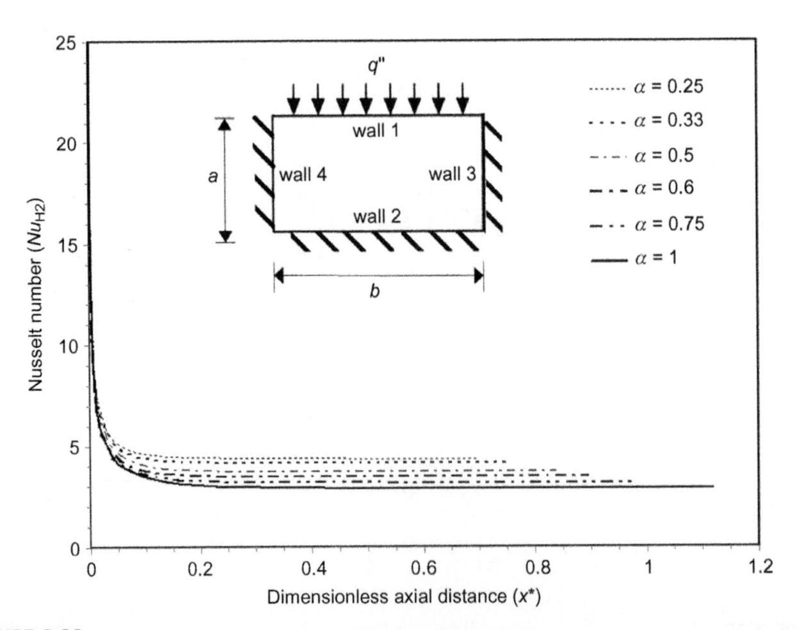

FIGURE 3.22

Local Nusselt number as a function of nondimensional length and channel aspect ratio for heating on one wall, H2 boundary conditions.

Source: *Adapted from Dharaiya and Kandlikar (2012).*

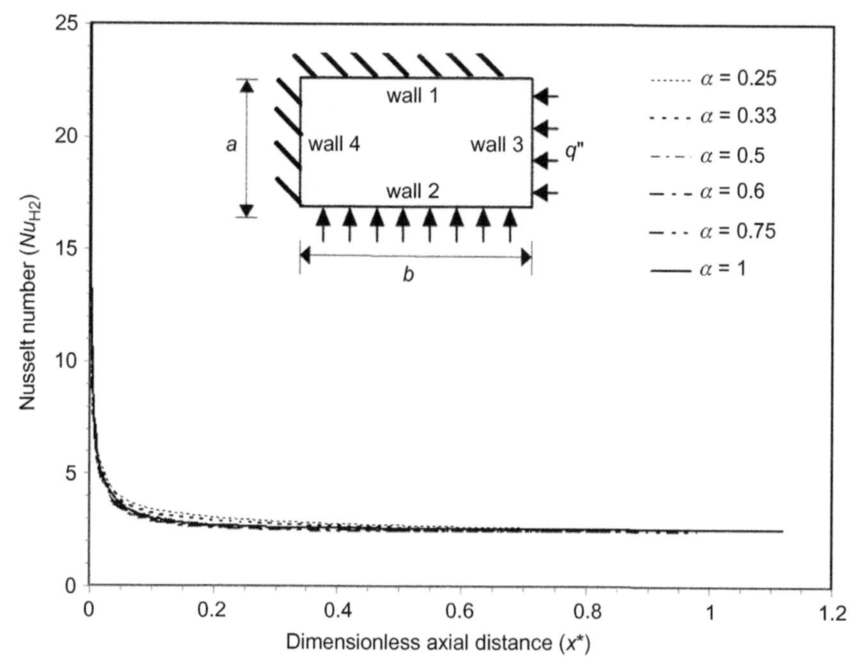

FIGURE 3.23

Local Nusselt number as a function of nondimensional length and channel aspect ratio for heating on two adjacent walls, H2 boundary conditions.

Source: *Adapted from Dharaiya and Kandlikar (2012).*

experimental data. Due to the relatively short lengths employed in microchannels, the influence of the entrance region cannot be neglected. The entrance region effects become more significant at higher Reynolds numbers, in part explaining the trend of increasing Nusselt number with Reynolds number, as seen in Figure 3.24.

(ii) *Uncertainties in experimental measurements:* The uncertainties in heat transfer measurements have been analyzed by a number of investigators, including Judy et al. (2002) and Steinke and Kandlikar (2005b). The following equation for the uncertainty in Nusselt number calculation from the uncertainties in the experimental measurements is presented by Steinke and Kandlikar for a rectangular channel of base width a and height b:

$$U_{Nu} = Nu \cdot \left[\left(\frac{U_{k_f}}{k_f}\right)^2 + \left(\frac{U_I}{I}\right)^2 + \left(\frac{U_V}{V}\right)^2 + 4\left(\frac{U_{T_s}}{T_s}\right)^2 + 2\left(\frac{U_{T_i}}{T_i}\right)^2 + 2\left(\frac{U_{T_o}}{T_o}\right)^2 + 3\left(\frac{U_L}{L}\right)^2 + 4\left(\frac{U_a}{a}\right)^2 + 5\left(\frac{U_b}{b}\right)^2 + 2\left(\frac{U_{\eta_f}}{\eta_f}\right)^2 \right]^{1/2} \tag{3.55}$$

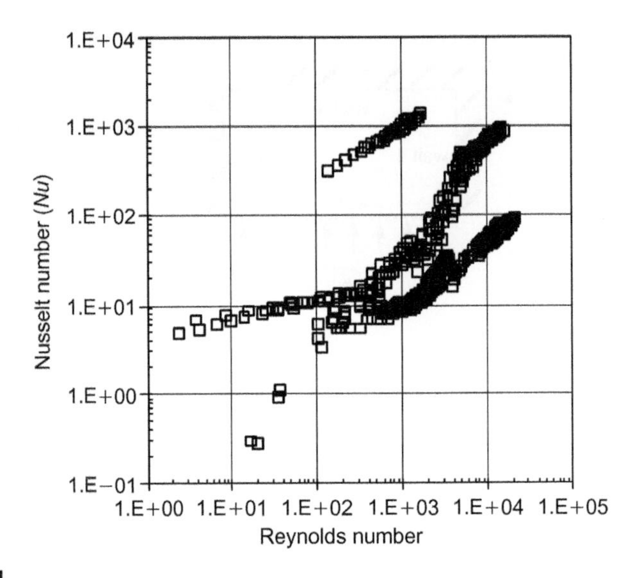

FIGURE 3.24

Selected experimental data for single-phase liquid flow in microchannels and minichannels, $D_h = 50\ \mu m$ to $600\ \mu m$.

where U = uncertainty, I = current, V = voltage, T_s = surface temperature, T_i = fluid inlet temperature, T_o = fluid outlet temperature, L = flow length, a and b = cross-section dimensions, and η_f = fin efficiency. The uncertainties in the measurement of the surface temperature and flow-channel dimensions play a critical role in the overall uncertainty. The importance of accurate geometrical measurement was emphasized in the friction factor estimation as well. Accurate surface temperature measurement poses a significant challenge due to the small dimensions of the test section. Use of silicon chips with integrated circuits to measure the temperatures is recommended.

Another factor that makes the temperature measurements critical in the overall uncertainty estimation is the small temperature difference between the surface and the fluid at the outlet. Since the heat transfer coefficients are very high in microchannels and the flow rate is relatively low (low Reynolds number), the outlet temperature in many experiments approaches the surface temperature. Proper experiments need to be designed to account for this effect.

(iii) *Ambiguity in the determination of the thermal boundary conditions:* Several experiments reported in the literature were conducted with microchannels or minichannels fabricated on copper or silicon substrates. The actual boundary conditions for these test sections are difficult to ascertain as they fall in between the constant temperature and constant heat flux boundary conditions. Furthermore, in many cases the heating is three-sided, with the side walls acting as fins. The fin efficiency effects also alter the heat flux and temperature distributions. A clear comparison is only possible after the conjugate heat transfer effects are incorporated into a detailed numerical simulation of the test section.

3.5.4 Heat transfer in the transition and turbulent flow regions

A detailed discussion was presented earlier on the critical Reynolds number for the laminar-to-turbulent transition. For smooth channels, the well-established criterion of $Re_c = 2300$ is expected to hold. The effect of roughness is described by Eqs. (3.42) and (3.41). Since the transition region is encountered in many minichannel and microchannel heat exchangers, there is a need to generate accurate experimental data for both smooth and rough tubes in this region.

Phillips (1990) suggests using the following equations in the developing turbulent region. For larger values of x, the influence of the term $[1 + (D_h/x)^{2/3}]$ reduces asymptotically to 1:

$$\text{For } 0.5 \le Pr \le 1.5: Nu = 0.0214[1.0 + (D_h/x)^{2/3}][Re^{0.8} - 100]Pr^{0.4} \tag{3.56}$$

$$\text{For } 1.5 \le Pr \le 1.5: Nu = 0.0214[1.0 + (D_h/x)^{2/3}][Re^{0.8} - 100]Pr^{0.4} \tag{3.57}$$

Further validation of these equations in microchannels is warranted. Adams et al. (1997) conducted experimental work in the turbulent region with flow of water in 0.76- and 0.109-mm diameter circular channels. Based on their data, they proposed the following equation, which matches the data by Yu et al. (1995) within $\pm$ 18.6%:

$$Nu = Nu_{Gn}(1 + F) \tag{3.58}$$

where

$$Nu_{Gn} = \frac{(f/8)(Re - 1000)Pr}{1 + 12.7(f/8)^{1/2}(Pr^{2/3} - 1)} \tag{3.59}$$

$$f = (1.82 \log(Re) - 1.64)^{-2} \tag{3.60}$$

$$F = C\,Re(1 - (D/D_o)^2) \tag{3.61}$$

Nu_{Gn} represents the Nusselt number predicted by Gnielinski's (1976) correlation. The least-squares fit to all the data sets studied by Adams et al. (1997) resulted in $C = 7.6 \times 10^{-5}$ and $D_o = 1.164$ mm.

Heat transfer coefficients in microchannels are very high due to their small hydraulic diameters. The high pressure gradients have led researchers to employ low flow rates. However, with the reduced flow rate, the ability of the fluid stream to carry the heat away for a given temperature rise becomes limited. In order to improve overall cooling performance, the following two options are available.

1. Reduce the flow length of the channels.
2. Increase the liquid flow rate.

As a result, employing multiple streams with short paths in a microchannel heat exchanger is recommended, similar to a split-flow arrangement, providing two streams, as will be discussed in Section 3.7. The reduced flow length will then enable the designer to employ higher flow rates under a given pressure drop limit. This scheme offers several advantages over a single-pass arrangement where the fluid traverses the entire length of the heat exchanger:

1. *The reduced flow length reduces the pressure drop*: the short flow length effectively reduces the overall pressure drop.
2. *Larger developing region*: the multiple inlets result in a larger channel area under developing conditions where the heat transfer is higher.
3. *Higher flow velocities*: some of the pressure drop reduction could be used to increase the flow velocity of individual streams. The possibility of employing turbulent flow should also be explored as the heat transfer coefficient is higher in this region.

3.5.5 Axial conduction effects

The high heat transfer rates associated with microchannels introduce a temperature gradient along the flow direction in the channel walls. This effect becomes more important at microscale as the channel wall dimensions become comparable to the channel dimensions, and the heat transfer in the walls cannot be neglected. Earlier researchers such as Guo and Li (2003) and Hetsroni et al. (2005) indicated that axial conduction effects may be among the reasons for discrepancies in predicting the heat transfer at microscale.

Maranzana et al. (2004) analyzed heat conduction effects and proposed a parameter M as the ratio of the wall conduction to fluid convection rates. An equation was derived to account for the axial conduction effects in reducing the heat exchanger efficiency. They also derived a set of equations, which were numerically solved to investigate these effects.

Lin and Kandlikar (2012a) considered the axial conduction effects to cause an increased heat transfer to the fluid near the entrance region, leading to a higher heat transfer rate and a higher fluid temperature at any section in the heat exchanger. They derived the following equation to determine the Nusselt number in the presence of axial conduction effects.

$$\frac{Nu_{\mathrm{ko}}}{Nu_{\mathrm{th}}} = \frac{1}{1 + 4(k_s A_{h,s} Nu_{\mathrm{th}}/k_{\mathrm{f}} A_{\mathrm{f}}(RePr)^2} \tag{3.62}$$

where Nu_{ko} is the Nusselt number neglecting the effect of axial conduction and Nu_{th} is the theoretical Nusselt number accounting for the heat conduction effect. The wall and fluid thermal conductivities are given by k_{s} and k_{f}, respectively, and $A_{\mathrm{h,s}}$ and A_{f} are the conduction heat transfer area in the wall and the flow cross-sectional area for fluid flow, respectively. This model was able to predict the heat

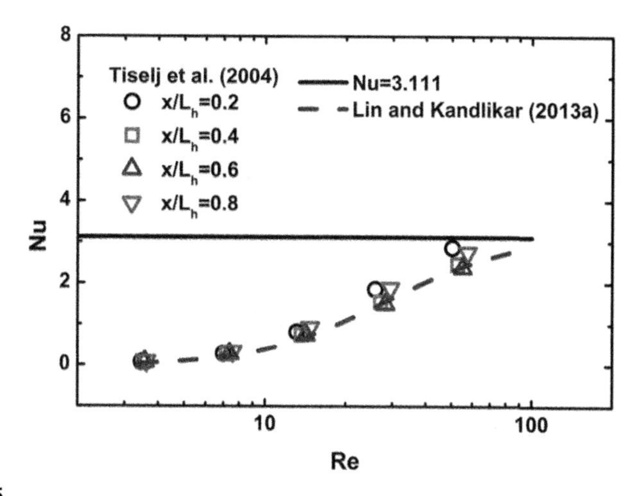

FIGURE 3.25

Comparison of the axial conduction model by Lin and Kandlikar (2012a) with the experimental data of Tiselj et al. (2004) for water flow in 160 μm diameter triangular silicon microchannels.

Source: *Adapted from Lin and Kandlikar (2012a).*

transfer coefficient data available in the literature quite well. Figure 3.25 shows a comparison between the model predictions with the data of Tiselj et al. (2004) for water flow in seventeen 160-μm triangular silicon microchannels for a Reynolds number between 3.2 and 64 for a heated length L_h of 10 mm. The silicon wafer was 530 μm thick and a Pyrex cover of 500 μm thickness was used as a cover plate. The axial conduction area was thus significantly higher than the fluid flow cross-sectional area, and the Reynolds number was quite low. Axial conduction effects are quite severe under these conditions.

Axial conduction effects become important when the wall cross-sectional area is comparable to the fluid flow area, increasing with an increase in the wall thermal conductivity and a reduction in the fluid Reynolds number. It can be seen from Figure 3.25 that the Nusselt number drops significantly below the theoretical laminar flow value at low Reynolds numbers employed in Tiselj et al.'s experiments. This effect becomes even more significant for gas flow due to the lower thermal conductivity of the fluid.

The effects of wall thermal conductivity on the ratio $Nu^* = Nu_{ko}/Nu_{th}$ are shown in Figure 3.26. The curves are plotted for different area ratios $A^* = A_{h,s}/A_f$ from 0.01 to 100. As the wall thermal conductivity and area ratio increase, the axial conduction effects become quite large and may not be negligible. The effect of cross-sectional area to flow area ratio and Reynolds number on Nu due to axial conduction is similarly shown in Figure 3.27.

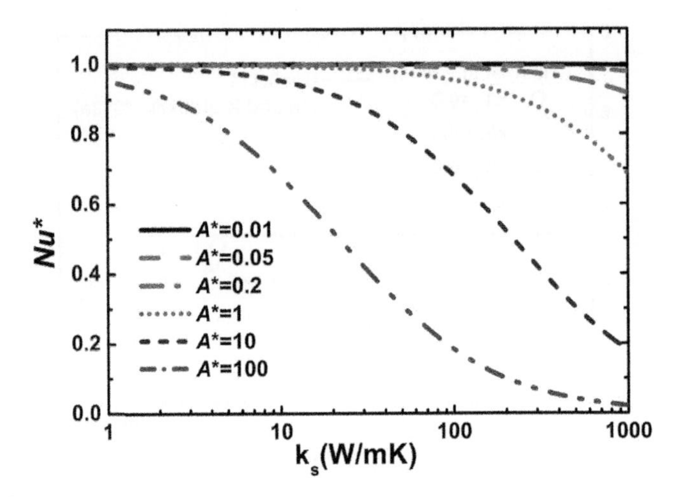

FIGURE 3.26

Effect of wall thermal conductivity and wall cross-sectional area on Nusselt number due to axial conduction during water flow in microchannels. $Nu^* = Nu_{ko}/Nu_{th}$ and $A^* = A_{h,s}/A_f$.

Source: *Adapted from Lin and Kandlikar (2012a).*

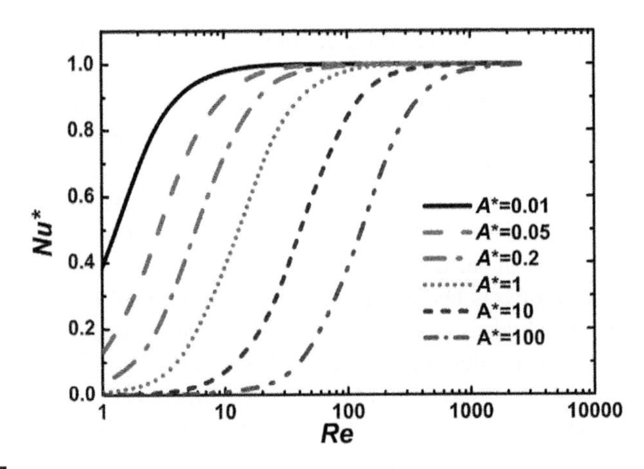

FIGURE 3.27

Effect of Reynolds number and wall cross-sectional area on Nusselt number due to axial conduction during water flow in silicon microchannels. $Nu^* = Nu_{ko}/Nu_{th}$ and $A^* = A_{h,s}/A_f$.

Source: *Adapted from Lin and Kandlikar (2012a).*

The axial conduction effects are particularly severe for gas flow. Yang et al. (2012) and Lin and Kandlikar (2013a,b) present the results of a detailed investigation on the effects of heat loss, viscous dissipation, and axial conduction on the heat transfer.

3.5.6 Variable property effects

The property ratio method is usually recommended in accounting for property variations due to temperature changes in heat exchanger flow passages. The following equations are recommended for liquids:

$$f/f_{cp} = [\mu_w/\mu_b]^M \tag{3.63}$$

$$Nu/Nu_{cp} = [\mu_w/\mu_b]^N \tag{3.64}$$

where the subscript cp refers to the constant property solution obtained from appropriate equations or correlations. For laminar flow, $M = 0.58$ and $N = -0.14$, and for turbulent flow, $M = 0.25$ and $N = -0.11$ (Kays and London, 1984).

3.6 Roughness effects on heat transfer in microchannels and minichannels

Recent studies on roughness effects presented in Section 3.4 on friction factor during laminar, transition, and turbulent flows clearly indicate a significant influence of roughness on fluid flow. Similar effects are to be expected in heat transfer.

Kandlikar et al. (2003) studied the heat transfer and pressure drop of laminar flow in smooth and rough stainless steel tubes of 1.067 and 0.62 mm ID. The surface roughness of the inner tube wall was changed by treating it with two different acid mixtures. The surface roughness actually went down after the acid treatment as the protruding peaks in the surface profile were smoothed out. The effect of changes in the relative roughness on pressure drop was minimal, but the heat transfer in the thermal entry region showed a distinct dependence on roughness. Figure 3.28 shows

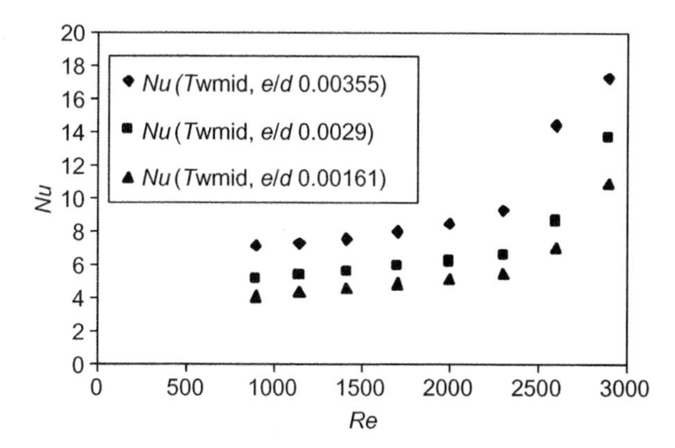

FIGURE 3.28

The effect of roughness on Nu in the thermal entrance region ($x = 52$ mm) of a 0.62 mm inner diameter tube with a fully developed velocity profile and a developing temperature profile (Kandlikar et al., 2003).

the local Nusselt number at a location 52 mm from the start of the heated length. The flow was hydrodynamically fully developed prior to entering the heated section.

It can be seen from Figure 3.28 that the Nusselt number increases even for a modest increase in the relative roughness from 0.161% to 0.355%. Also note that the location where the heat transfer measurements were taken was preceded by a tube length longer than the hydraulic entry length, which was 380 mm at $Re = 2300$. A number of researchers did not correctly identify the developing region and have mistakenly reported the increasing trend in the *Nu* versus *Re* plot as being in the fully developed region.

Dharaiya and Kandlikar (2013) numerically analyzed a two-dimensional sinusoidal roughness profile, shown in Figure 3.29. The width of the channel was 12.7 mm and the flow length was 114.3 mm. The simulations were validated with the laminar flow in smooth minichannels. They also studied the effects of abrupt and smooth entrances on friction factor and heat transfer in the entrance region, and found little difference between the two cases.

The effects of pitch and roughness height were systematically investigated by Dharaiya and Kandlikar (2013). The numerical results for rough channels were compared with the friction factors and Nusselt numbers obtained using the constricted flow diameter in fully developed laminar flow equations. Both friction factor and Nusselt number agreements were fairly good, within 10% for relative roughness values below 5% and pitch to roughness ratios from 2.5 to 25. The results indicated that the pitch to roughness ratio has a strong influence on Nusselt number enhancement. Very high enhancements of over 250% were noted for pitch to roughness ratios of 12.5 and 16; corresponding *Nu* were 19.3 and 17.2, respectively, for a separation of 550 μm and a pitch of 250 μm. The corresponding friction factor increase was only 30%. Figure 3.30 shows the velocity vectors from their simulation. The velocity vectors indicate that the reason for such high heat transfer enhancement with a relatively low friction factor increase was attributed to the smooth sinusoidal profile. Further details may be obtained

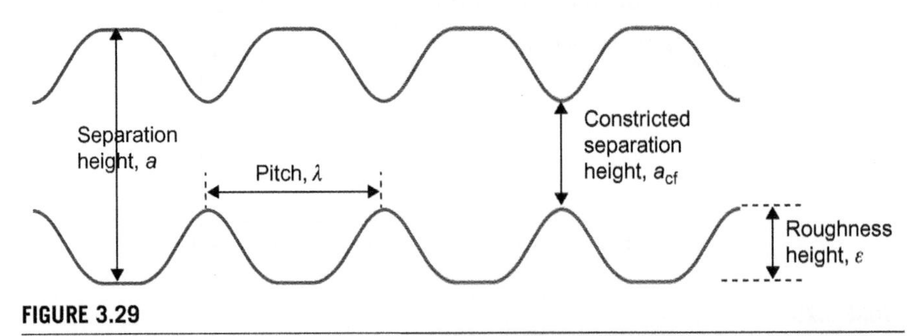

FIGURE 3.29

Two-dimensional roughness geometry analyzed by Dharaiya and Kandlikar (2013) with sinusoidal roughness profile.

Source: *Adapted from Dharaiya and Kandlikar (2013).*

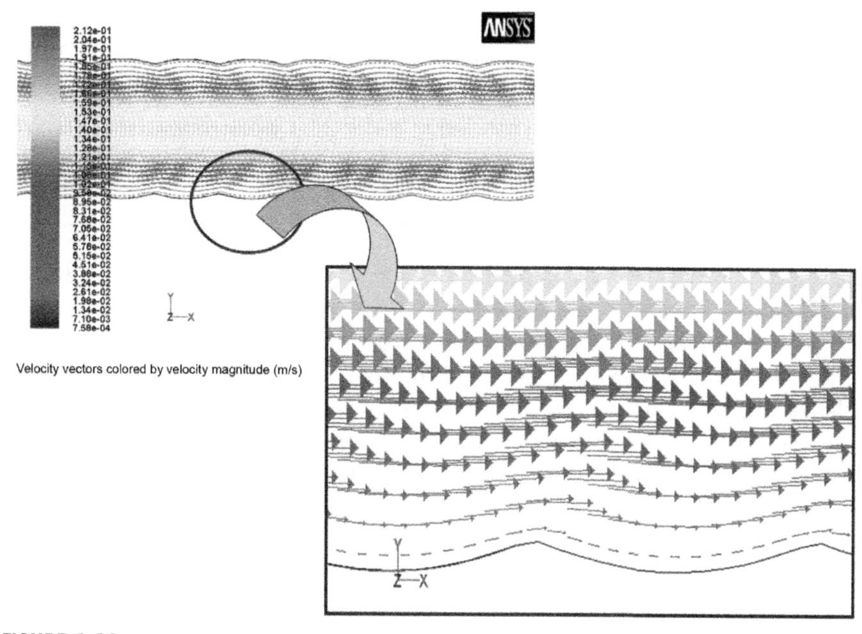

FIGURE 3.30

Velocity vectors for a roughness geometry with channel separation $a = 550\ \mu m$, pitch $\lambda = 250\ \mu m$, and roughness height $\varepsilon = 20\ \mu m$.

Source: *Adapted from Dharaiya and Kandlikar (2013).*

from Dharaiya and Kandlikar (2013). These model predictions were confirmed experimentally by Lin and Kandlikar (2012b).

Lin and Kandlikar (2012b) conducted a systematic experimental study as a complement to the numerical study by Dharaiya and Kandlikar (2013) discussed above. Eight sets of structured roughness geometries with sinusoidal profile were tested. The opposing walls of a 12.7-mm-wide channel were made of these roughness elements, and the gap was varied to achieve different relative roughness values. The height of the roughness elements varied from 18 to 96 μm, and the roughness pitch varied from 250 to 400 μm. The hydraulic diameter ranged from 0.71 to 1.87 mm, while the constricted diameter ranged from 0.68 to 1.76 mm.

The roughness elements were heated with silicone film heaters attached with adhesive on the back of the roughness elements. The experimental data indicated that the heat losses from the inlet and outlet ends played a significant role in evaluating the heat transfer coefficients. An elaborate scheme was applied to accurately estimate the heat losses in different sections. The heat transfer coefficients were obtained at 11 locations along the flow length after correcting for the heat losses and axial conduction effects. The sinusoidal elements tested provided considerable enhancement, with the maximum enhancement (a factor of 3.77 increase) was provided by a roughness element identified as B-1, with a pitch of

250 μm and a roughness height of 96 μm. The friction factor increase was also quite high, at 3.71. The second best performing roughness element was D-2 (*Nu* enhancement factor of 1.98, friction factor enhancement factor of only 1.03), with a pitch of 400 μm and a roughness height of 38 μm.

Sinusoidal roughness elements provided the highest heat transfer enhancement for a given friction factor enhancement. The authors defined an enhancement efficiency as

$$\eta = \frac{Nu/Nu_{\text{th,plain}}}{(f/f_{\text{th,plain}})^{1/3}} \tag{3.65}$$

The enhancement efficiency was 2.44 for the roughness element B-1 and 1.96 for the element D-2. These values were compared with some of the most efficient techniques reported in the literature. Figure 3.31 shows a comparison of these techniques. Wongcharee and Eiamsa-ard (2011) used alternate clockwise and counterclockwise twisted tapes with a heat transfer enhancement factor from 6 to 13, but the friction factor increase was from 8- to 15-fold. The porous medium studied by Huang et al. (2010) had a heat transfer enhancement factor of 5.5 to 4.5; the friction factor increased by a factor of 50–60. Similar observations were made by Krishna et al. (2009) with twisted tapes and by Akhavan-Behabadi et al. (2010) for coiled inserts.

The roughness profile results in an early transition to turbulent region. The enhancement in the turbulent region is also quite high, as seen from Figure 3.31. Roughness geometries with smoothly varying 2d profiles are thus seen as an effective way to enhance heat transfer at microscale.

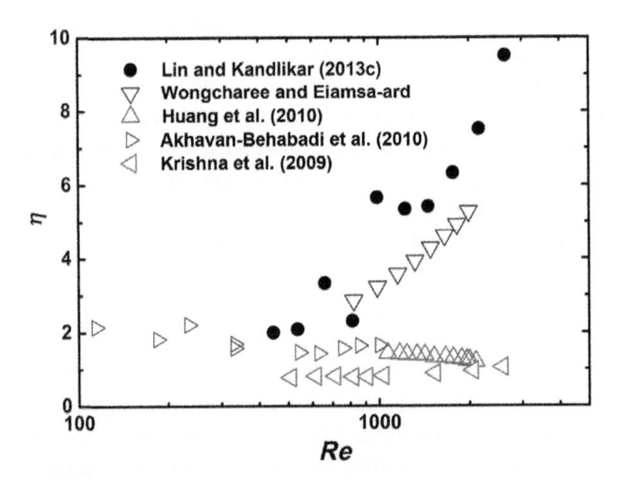

FIGURE 3.31

Comparison of enhancement factor η of B-1 with other available enhancement techniques from recent literature.

Source: *Adapted from Lin and Kandlikar (2012b).*

3.7 Heat transfer enhancement with nanofluids

Particles with at least one of their dimensions in the nanoscale range are termed nanoparticles. Nanofluids consist of these nanoparticles dispersed in a base liquid phase. Choi (1995) showed that the thermal conductivity of nanofluids is enhanced over the base fluid. The addition of nanoparticles was also reported to alter the specific heat of the mixture by Bergman (2009). This has been a very active topic of research in the last decade, but the effect of nanoparticles on the transport characteristics has not yet been clearly described, and their efficacy in enhancing single-phase heat transfer still remains an open question.

Chein and Huang (2005) performed numerical studies on heat transfer enhancement with Cu nanoparticles in water in various volumetric concentrations. Koo and Kleinstreuer (2005) also presented similar numerical studies. They considered two different microchannels with (i) 100 μm × 300 μm, and (ii) 57 μm × 365 μm channel cross-sectional dimensions. Their results showed that the pressure drop increased only marginally, while the thermal resistance of the heat sink dropped by as much as 50%. A number of researchers have reported analytical and numerical studies showing different levels of enhancement depending on the analytical models used in describing the thermal properties of the nanofluids.

Very few experimental studies on heat transfer with nanofluids have been reported in the literature. Wen and Ding (2004) studied the effect of nanofluids made of γ-Al_2O_3 particles in deionized water in a 4.5-mm-diameter and 970-mm-long copper tube in the laminar region. They found that the heat transfer coefficient enhancement was as much as 47% in the entrance region. This enhancement almost disappeared in the fully developed region. They postulated that the nanoparticles reduced the thermal boundary layer thickness in the entrance region. Nanoparticle migration within the boundary layer was proposed as an enhancement mechanism in nanofluids.

Lee and Mudawar (2007) conducted analytical and experimental studies in laminar and turbulent flow regions with Al_2O_3 nanoparticle suspensions in water. Their theoretical study showed that the friction factor was changed slightly, but the pressure drop increased because of the increased viscosity of the nanofluids. The experiments were performed with the nanofluid at 1% and 2% volume concentrations in deionized water. Thermal conductivity was measured to increase by 6% for the 2% concentration nanofluid. A test setup with 21 parallel channels with 215-μm-wide and 821-μm-deep grooves, with a hydraulic diameter of 341 μm over a 1 cm^2 area was used. The heat transfer enhancement was observed mainly in the entrance region, with only a modest improvement in the fully developed region. The enhancement increased with the concentration of nanoparticles. These observations are in agreement with the experimental results of Wen and Ding (2004) for a larger tube diameter of 4.5 mm.

Recently, Singh et al. (2011, 2012) conducted experimental and numerical investigations to study the hydrodynamics and thermal effects of nanofluids. They used alumina nanofluids in 0.25%, 0.5%, and 1% volume concentrations in three single microchannels of 130, 211, and 300 μm hydraulic diameter silicon microchannels. The cross-section of the channel was trapezoidal with side angles of 54.74°. The viscosity of the nanofluid reported in the literature was seen to have a significant scatter, varying as much as 50%. The friction factor trends are seen to depend on the channel size, shape, and nanoparticle concentration. An early transition was noticed with nanofluids, but the effect of channel roughness also seemed to play a role. The trends observed are quite different, and it is not clear whether the heat losses and the end effects were adequately considered in the single microchannel test section. The results by Jung et al. (2009) on single microchannel sections indicates a Nusselt number as low as 0.1, and validation with single-phase experiments with pure water is seen as essential before evaluating nanofluid trends.

3.8 Microchannel and minichannel geometry optimization

The first practical implementation of microchannels in silicon devices was demonstrated by Tuckerman and Pease (1981). They were able to dissipate 7.9 MW/m^2 with a maximum substrate temperature rise of 71°C and a pressure drop of 186 kPa. The allowable temperature differences and the pressure drops for cooling today's microprocessor chips have become significantly smaller, and many investigators have been working toward optimization of the channel geometrical configuration.

The application of microchannels to electronics cooling imposes severe design constraints on the system design. For a given heat dissipation rate, the flow rate, pressure drop, fluid temperature rise, and fluid inlet to surface temperature difference requirements necessitate optimization of the channel geometry. A number of investigators have studied the geometrical optimization of microchannel heat exchangers, including Phillips (1987), Harpole and Eninger (1991), Knight et al. (1992), Ryu et al. (2002), Bergles et al. (2003), and Kandlikar and Upadhye (2005). The results found by Kandlikar and Upadhye are presented below.

Kandlikar and Upadhye (2005) considered a microchannel system as shown in Figure 3.32. For a chip of width $W = 10$ mm and length $L = 10$ mm, they presented an analysis scheme for heat transfer and pressure drop by incorporating the entrance region effects. The number of channels was used as a parameter in developing the optimization scheme. The maximum chip temperature was set at 360 K while the fluid inlet temperature was set equal to 300 K. The channel depth was assumed to be 200 μm.

Figures 3.33 and 3.34 show the parametric plots resulting from the optimization program. The fin spacing ratio, defined as $\beta = s/a$, is plotted as a function of the number of channels on a 10-mm-wide chip with pressure drop, water flow rate, and fin thickness as parameters. A lower limiting value of approximately 40 μm was considered achievable for fin thickness using silicon microfabrication techniques. For a given pressure drop limit, the optimal number of channels and other channel geometrical parameters can be determined from this plot. Such plots could be generated for other system configurations as well.

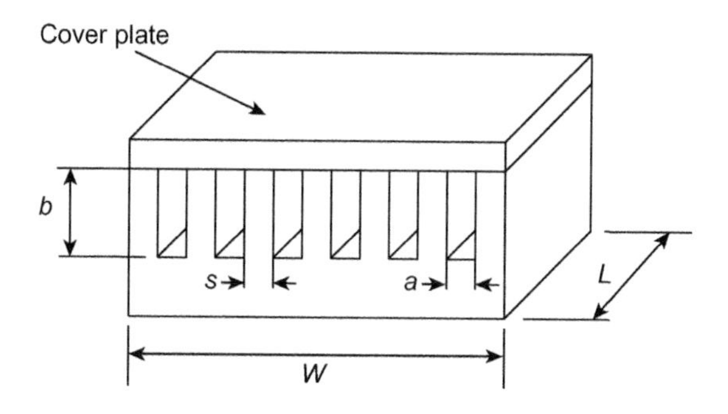

FIGURE 3.32

Microchannel geometry used in microprocessor heat sink channel size optimization.

Source: *From Kandlikar and Upadhye (2005).*

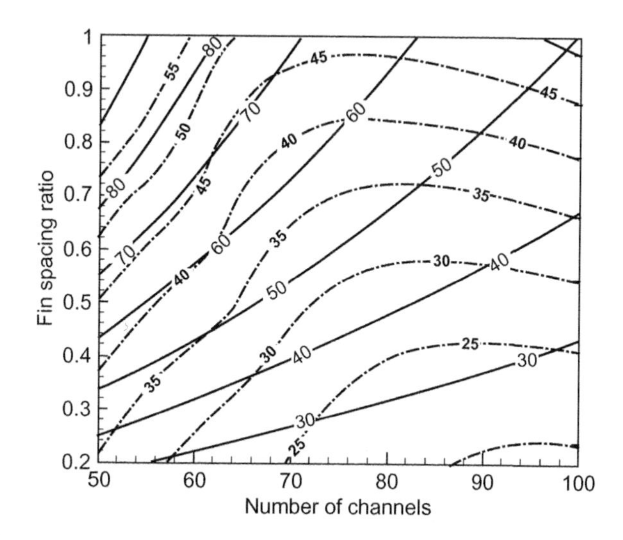

FIGURE 3.33

Contour plot of fin spacing ratio β versus the number of channels with pressure drop across them in kilopascal (dash-dot lines) and fin thickness in microns (solid lines) as parameters for water flow in plain rectangular microchannels in a single-pass arrangement at a heat flux of 3 MW/m^2.

Source: *From Kandlikar and Upadhye (2005).*

The effect of introducing offset strip-fins in the microchannel flow passages and a split-flow arrangement, as shown in Figures 3.35 and 3.36, respectively, was also analyzed by Kandlikar and Upadhye (2005). The resulting pressure drop versus the dissipated heat flux is shown in Figure 3.37. It can be seen that the pressure drop is reduced considerably in the split-flow arrangement.

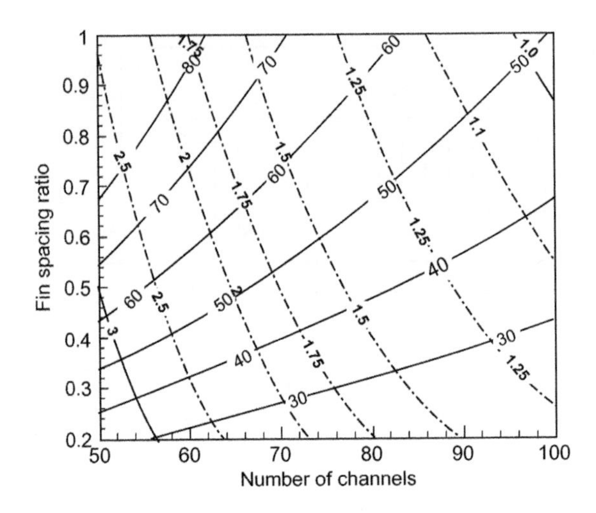

FIGURE 3.34

Contour plot of fin spacing ratio β versus the number of channels with water flow rate in 10^{-3} kg/s (dash-dot lines) and fin thickness in microns (solid lines) as parameters for water flow in plain rectangular microchannels in a single-pass arrangement at a heat flux of 3 MW/m^2.

Source: *From Kandlikar and Upadhye (2005).*

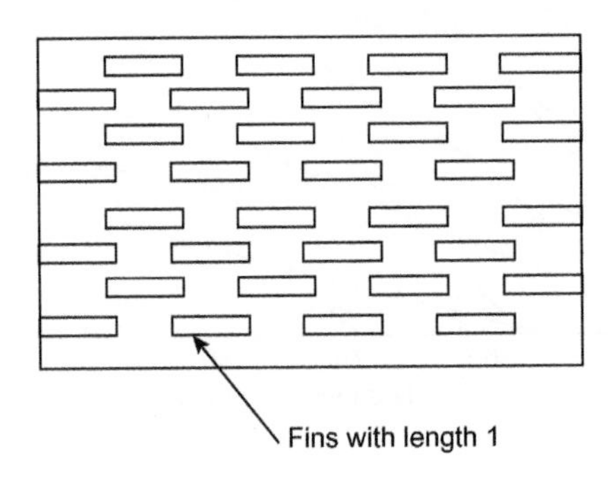

FIGURE 3.35

Offset strip-fins shown in the top view, with individual fin length l along the flow length.

Source: *From Kandlikar and Upadhye (2005).*

3.9 Enhanced microchannels

The use of enhanced microchannels was proposed by Kishimoto and Sasaki (1987). The need for higher heat transfer coefficients than those attainable with plain microchannels was identified by Kandlikar and Grande (2004), and some specific enhancement geometries were suggested by Steinke and Kandlikar (2004a,b). Kandlikar and Upadhye (2005) analyzed the enhanced-offset strip-fin geometry; the results were presented in the previous section.

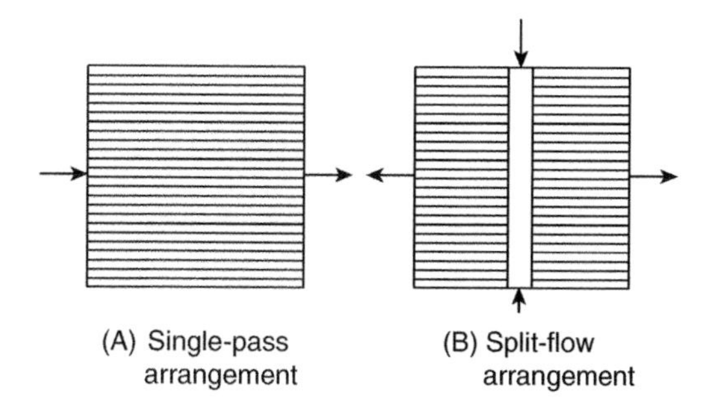

FIGURE 3.36

Schematics of single-pass and split-flow arrangements for fluid flow through microchannels: (A) single-pass arrangement and (B) split-flow arrangement.

Source: *From Kandlikar and Upadhye (2005).*

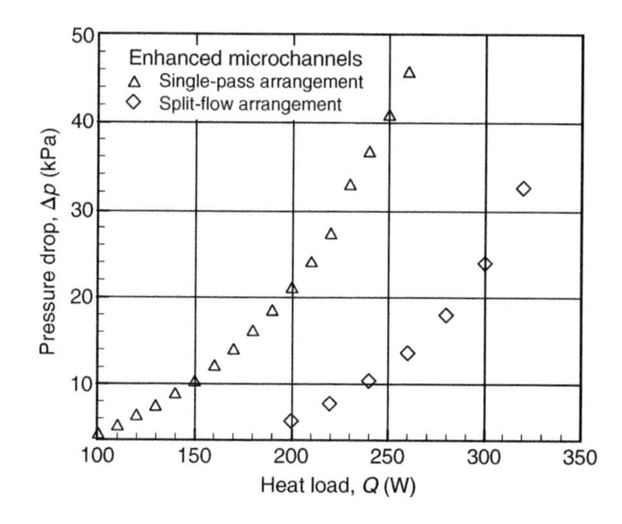

FIGURE 3.37

Comparison of pressure drops for the enhanced microchannels with offset strip-fins ($l = 0.5$ mm) in single-pass and split-flow arrangements on a 10 mm × 10 mm chip.

Source: *From Kandlikar and Upadhye (2005).*

Colgan et al. (2005) presented detailed experimental results comparing various offset fin geometries. A three-dimensional rendition is shown in Figure 3.38. There are several inlet and outlet manifolds, with short flow lengths through the enhanced structures.

The apparent friction factors and Nusselt numbers for plain and several enhanced configurations are shown in Figures 3.39 and 3.40. A plot of f_{app} versus Stanton number is shown in Figure 3.41. The relationship deviates somewhat from a linear behavior because the manifold effects are included in the data.

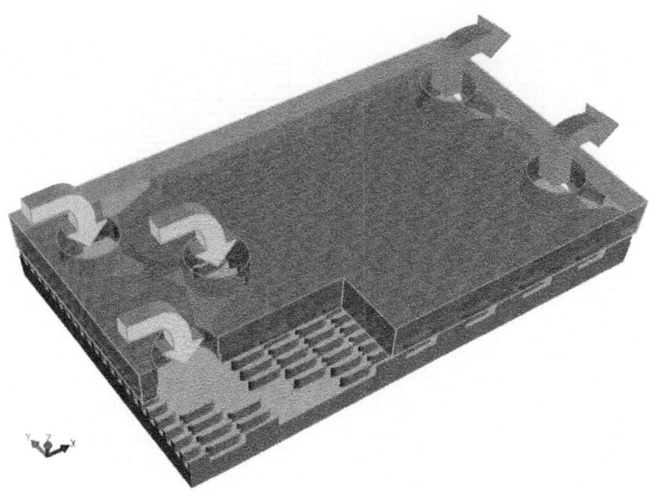

FIGURE 3.38

Three-dimensional rendition of the IBM enhanced silicon chip with short multiple fluid streams.

Source: *From Colgan et al. (2005).*

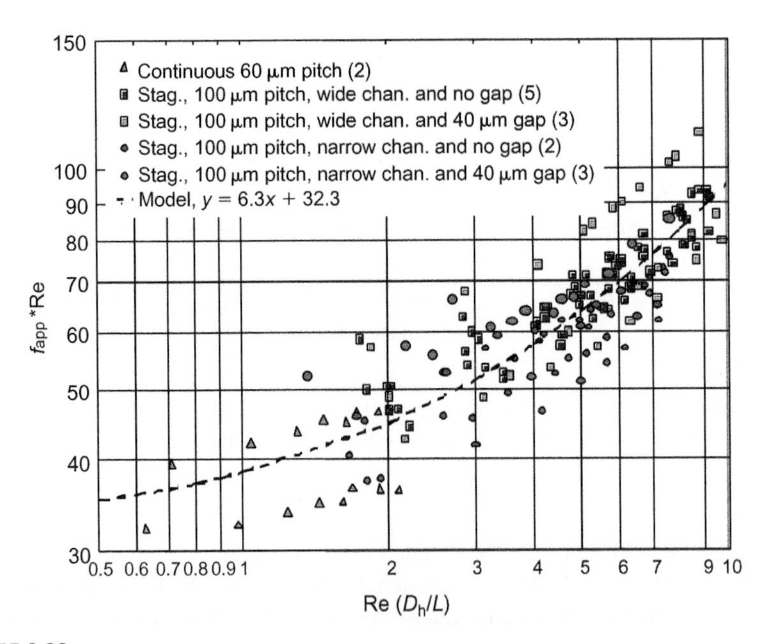

FIGURE 3.39

Apparent friction factor comparison of one plain and five enhanced microchannel configurations.

Source: *From Colgan et al. (2005).*

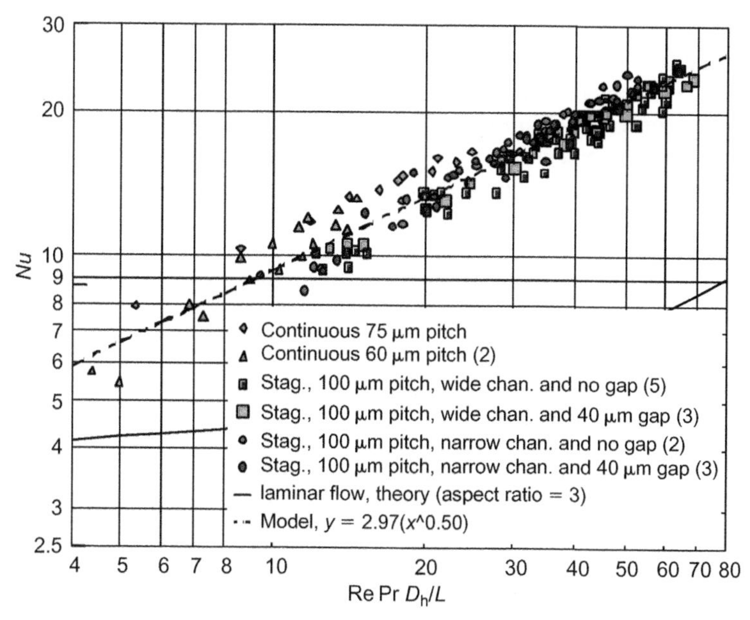

FIGURE 3.40

Nusselt number comparison of one plain and five enhanced microchannel configurations.

Source: *From Colgan et al. (2005).*

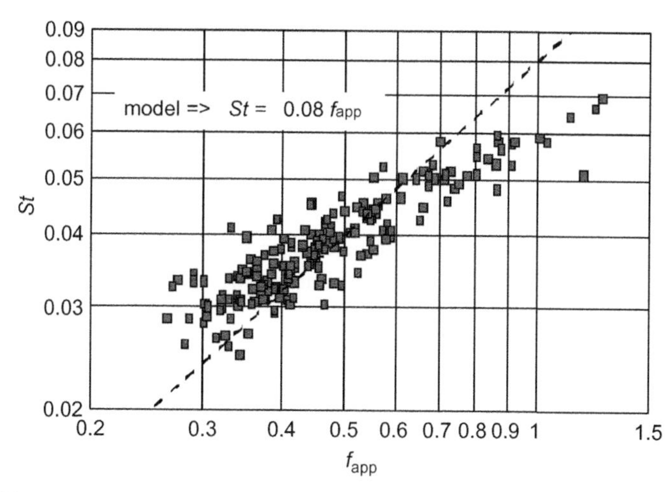

FIGURE 3.41

Stanton number and friction coefficient relationship for microchannels with different fin geometries.

Source: *From Colgan et al. (2005).*

Microfabrication technology opens up a whole new set of possibilities for incorporating enhancement structures derived from past experience in compact heat exchanger development. Different manufacturing constraints enter into play for microchannels on a copper substrate as compared to the silicon microfabrication technology. Novel developments are expected to arise as we continue forward in integrating the microfabrication technology using creative ways to achieve high-performance heat transfer devices. Steinke and Kandlikar (2004a,b) presented a good overview of single-phase enhancement techniques and illustrate various enhancement configurations that may be considered, especially with silicon microfabrication technology.

3.10 Solved examples

Example 3.1

Microchannels are directly etched into silicon in order to dissipate 100 W from a computer chip over an active surface area of 10 mm × 10 mm. The geometry may be assumed similar to Figure 3.32. Each of the parallel microchannels has a width $a = 50\ \mu m$, depth $b = 350\ \mu m$, and a spacing $s = 40\ \mu m$. The silicon thermal conductivity may be assumed to be $k = 180$ W/m K.

Assume a uniform heat load over the chip base surface and a water inlet temperature of 35°C. Assume one-dimensional steady state conduction in the chip substrate.

- **i.** Calculate the number of flow channels available for cooling.
- **ii.** Assuming the temperature rise of the water to be limited to 10°C, calculate the required mass flow rate of the water.
- **iii.** Calculate the flow Reynolds number using the mean water temperature for fluid properties.
- **iv.** Check whether the fully developed flow assumption is valid.
- **v.** Calculate the average heat transfer coefficient in the channels.
- **vi.** Calculate fin efficiency.
- **vii.** Assuming the heat transfer coefficient to be uniform over the microchannel surface, calculate the surface temperature at the base of the fin at the fluid inlet and fluid outlet sections.
- **viii.** Calculate the pressure drop in the core of the microchannel.
- **ix.** If two large reservoirs are used as the inlet and outlet manifolds, calculate the total pressure drop between the inlet and the outlet manifolds.

Solution

Properties: Water at 40°C

$\rho = 991.8$ kg/m^3; $\mu = 655 \times 10^{-6}$ N s/m^2; $c_p = 4179$ J/kg K; $k = 0.632$ W/m K; $Pr = 4.33$

(i) Number of flow channels (Answer: 111 channels).

Assumptions

(1) The first and the last channels are half the channel width from the chip edge.

The remaining width left for channels is: 10 mm−50 μm = 0.00995 m.
The pitch is: 50 μm + 40 μm = 90 μm = 90 × 10^{-6} m.
Assumption 1 requires that the pitch repeat up to the last channel, but not include the last channel.

Therefore, the number of channels is found by (0.00995 m − 50 × 10^{-6} m)/(90 × 10^{-6} m) = 110 channels.

But recall that that the above calculation was for a repeating pitch up to but not including the last channel, so the number of channels available for cooling is $n = 111$.

(ii) Required flow rate of water (Answer: 0.00239 kg/s or 143.6 ml/min, total).

The inlet water temperature is given as 35°C, and the stated assumption is that the water temperature rise is limited to 10°C. The mass flow rate can be calculated using the properties of water at the average temperature of 40°C by using the equation $q = \dot{m}c_{\mathrm{p}}\Delta T$; where q is the power dissipated from the chip, c_{p} is the specific heat of water at constant pressure, and ΔT is the temperature change of the water. The total mass flow rate is

$$\dot{m}_t = \frac{q}{c_{\mathrm{p}}\Delta T} = 100\ W/(4179\ \mathrm{J/kg\ K})(10\ \mathrm{K}) = 0.00239\ \mathrm{kg}/s = 143.6\ \mathrm{ml/min}$$

and the mass flow rate in one channel is given by

$$\dot{m}_{\mathrm{c}} = \frac{q}{nc_{\mathrm{p}}\Delta T} = 21.6 \times 10^{-6}\ \mathrm{kg}/s = 1.3\ \mathrm{mL/min}$$

(iii) Flow Reynolds number (Answer: *Re* = 164.7).

The flow Reynolds number using the mean water temperature (40°C) for fluid properties can be found using the hydraulic diameter D_{h} given by Eq. (3.7)

$$D_{\mathrm{h}} = \frac{2ab}{(a+b)} = 2(50 \times 10^{-6}\ \mathrm{m})(350 \times 10^{-6}\ \mathrm{m})/(50 \times 10^{-6}\ \mathrm{m} + 350 \times 10^{-6}\ \mathrm{m})$$
$$= 87.5 \times 10^{-6}\ \mathrm{m}$$

in the Reynolds number equation given by

$$Re = \frac{\rho u_{\mathrm{m}} D_{\mathrm{h}}}{\mu} = \frac{\rho u_{\mathrm{m}} A_{\mathrm{c}} D_{\mathrm{h}}}{A_{\mathrm{c}}\mu} = \frac{\dot{m}_{\mathrm{c}} D_{\mathrm{h}}}{A_{\mathrm{c}}\mu}$$
$$= \frac{(21.6 \times 10^{-6}\ \mathrm{kg/s})(87.5 \times 10^{-6}\ \mathrm{m})}{(17.5 \times 10^{-9}\ \mathrm{m}^2)(655 \times 10^{-6}\ \mathrm{N}\ s/\mathrm{m}^2)} = 164.7$$

This low Reynolds number clearly indicates laminar flow regime.

(iv) Fully developed assumption (Answer: fully developed assumption is valid—hydrodynamic entrance length = 0.72 mm and thermal entrance length = 6.25 mm).

The fully developed flow assumption is valid if the thermal and hydrodynamic entrance lengths are less than the channel length. For laminar flow, the hydrodynamic entrance length is given by Eq. (3.11) as

$$L_h = 0.05ReD_h = 0.05(164.7)(87.5 \times 10^{-6}\ m) = 720 \times 10^{-6}\ \mathrm{m}$$
$$= 0.72\ \mathrm{mm}$$

and the thermal entrance length is given by Eq. (3.51) as

$$L_t = 0.1RePrD_h = 0.1(164.7)(4.33)(87.5 \times 10^{-6}\ m) = 6.24 \times 10^{-3}\ \mathrm{m}$$
$$= 6.24\ \mathrm{mm}$$

Since the thermal and hydrodynamic entrance lengths are less than the channel length, the fully developed flow assumption is valid.

(v) Average heat transfer coefficient (Answer: 47.4 × 10³ W/m² K).

Assumptions

(1) Constant heat flux boundary condition.
(2) Three-sided heating condition.

The Nusselt number using the thermal conductivity of water at the average temperature (40°C) is given by $Nu = hD_h/k$. A value for the fully developed Nusselt number can be obtained from Table 3.5 using the aspect ratio of the channel $\alpha_c = a/b = 1/7$. By linear interpolation, the Nusselt number is $Nu_{fd,3} = 6.567$.

The average heat transfer coefficient is

$$\bar{h} = \frac{kNu}{D_h} = (0.632\ W/m\ \mathrm{K})(6.567)/87.5 \times 10^{-6}\ \mathrm{m} = 47.4 \times 10^3\ W/m^2\ \mathrm{K}$$

(vi) Fin efficiency (Answer: 67%).

Assumptions

(1) Adiabatic tip condition.

For an adiabatic tip, the fin efficiency equation is given by $\eta_f = \tanh(mb)/mb$ where b is the fin height

$$(350 \times 10^{-6}\ \mathrm{m})\ \text{and } m \text{ is defined as } m = (\bar{h}P/kA_c)^{1/2}$$

In the above equation, k is thermal conductivity of the fin material (given as $k = 180$ W/m K), A_c is the cross-sectional area of the fin $L \times s = (0.01\ \mathrm{m})(40 \times 10^{-6}\ \mathrm{m})$, and P is the perimeter of the fin which can be defined as $2L$ since the width of the fin is much smaller than its length. The term mb can be written as

$$mb = \left(\frac{\overline{h}2L}{ksL}\right)^{1/2} b = \left(\frac{2\overline{h}}{ks}\right)^{1/2} b$$

$$= \left(\frac{2(47.4 \times 10^3\ W/m^2\ K)}{(180\ W/m\ K)(40 \times 10^{-6}\ m)}\right)^{1/2} (350 \times 10^{-6}\ m) = 1.270$$

Therefore, the fin efficiency can be calculated as $\eta_f = \tanh(1.270)/1.270 = 0.672 = 67\%$.

(vii) Inlet and outlet surface temperatures at the base of the fin (Answer: $T_{s,i} = 36.3°C$ and $T_{s,o} = 48.6°C$).

Assumptions

(1) The heat flux is constant over the chip surface.

The surface heat flux considering the fin efficiency is given by

$$q'' = q/(2b\eta_f + a)nL$$

$$= \frac{100\ W}{[2(350 \times 10^{-6}\ m)(0.672) + 50 \times 10^{-6}\ m](111)(0.01\ m)}$$

$$= 173 \times 10^3\ W/m^2$$

The relationship between heat flux, heat transfer, and temperature difference is given by $q'' = h(T_s - T_f)$ where the subscripts s and f refer to the surface and fluid, respectively. The *local* heat transfer coefficients at the inlet and outlet of the microchannels are needed in order to compute the surface temperatures. Since the flow is developing at the entrance of the microchannel, Tables 3.5 and 3.6 are used with the channel aspect ratio and x^* for the inlet, where x^* is defined by Eq. (3.53). The entrance of the channel is assumed to begin at $x = 0.1$ mm from the edge of the chip, so x^* is

$$x^*_{in} = \frac{0.1 \times 10^{-3}\ m}{(87.5 \times 10^{-6}\ m)(164.7)(4.33)} = 1.603 \times 10^{-3}$$

Using a linear interpolation in Table 3.5, the fully developed Nusselt numbers for three-sided and four-sided heating are found to be 6.567 and 6.273, respectively. The thermal entry region Nusselt number for four-sided heating is found using Table 3.6 to be 18.42 for the inlet. Equation (3.54) is used to obtain the local Nusselt number at the entrance as

$$Nu_{x,3}(x^*, \alpha_c) = 18.42\frac{6.567}{6.273} = 19.28$$

Alternatively, equations given in Appendix A may be used for interpolating values from Tables 3.2, 3.5, and 3.6.

Using the thermal conductivity of water at the given inlet temperature (35°C), the local heat transfer coefficient at the inlet is found to be

$$h = \frac{kNu}{D_h} = (0.625\ W/\text{m K})(19.28)/87.5 \times 10^{-6}\ \text{m} = 47.9 \times 10^3\ W/\text{m}^2\ \text{K}$$

Since the flow is fully developed at the outlet, the three-sided Nusselt number from Table 3.5 is used with the thermal conductivity of water at the outlet temperature (45°C). The local heat transfer coefficient at the exit is found to be

$$h = \frac{kNu}{D_h} = (0.638\ W/\text{m K})(6.567)/87.5 \times 10^{-6}\ \text{m} = 47.9 \times 10^3\ W/\text{m}^2\ \text{K}$$

Using the value of the heat flux obtained above, the local heat transfer coefficients, and the given inlet and outlet water temperatures, the surface temperatures at the base of the fin at the fluid inlet and outlet are

$$T_s = \frac{q''}{h} + T_f = > T_{s,i} = 36.3°\text{C} \quad \text{and} \quad T_{s,o} = 48.6°\text{C}$$

(viii) Pressure drop in microchannel core (Answer: 43.6 kPa).

Assumptions

(1) The core of the microchannel includes pressure drop due to only frictional losses in the fully developed region and the loss due to the developing region.

Since the flow is fully developed at the exit, the pressure drop is defined by

$$\Delta p = \frac{2(fRe)\mu u_m L}{D_h^2} + K(\infty)\frac{\rho u_m^2}{2}$$

where $K(\infty)$ is the Hagenbach factor, which is defined by Eq. (3.18) as

$$K(\infty) = 0.6796 + 1.2197\alpha_c + 3.3089\alpha_c^2 + 9.5921\alpha_c^3 + 8.9089\alpha_c^4 + 2.9959\alpha_c^5 = 0.8969$$

The $f\,Re$ term is given by Eq. (3.10) as

$$fRe = 24(1 - 1.3553\alpha_c + 1.9467\alpha_c^2 - 1.7012\alpha_c^3 + 0.9564\alpha_c^4 - 0.2537\alpha_c^5) = 20.2$$

The pressure drop in the core of the microchannel is given by Eq. (3.14) as

$$\Delta p = \frac{2(20.2)(655 \times 10^{-6}\ \text{N s/m}^2)(1.24\ \text{m/s})(0.01\ \text{m})}{(87.5 \times 10^{-6}\ \text{m})^2} + (0.8969)\frac{(991.8\ \text{kg/m}^2)(1.24\ \text{m/s})^2}{2} = 43.6\ \text{kPa}$$

(ix) Total pressure drop (Answer: 45.0 kPa).

Assumptions

(1) Reservoirs are large so the Area of the reservoir ≫ the Area of microchannel.

The total pressure drop between the inlet an outlet manifolds would include the pressure drop calculated above (43.6 kPa) plus the minor losses at the entrance and exit.

The minor loss is defined by $\Delta p = K(\rho {u_m}^2/2)$ where K is a loss coefficient related to area changes at the entrance or exit. Taking the area ratio as zero, based on Assumption 1, for a laminar flow regime, the contraction and expansion loss coefficients K_c and K_e can be obtained from Figure 3.5B as $K_c = 0.8$ and $K_e = 1.0$.

The total pressure drop is obtained by

$$\Delta p = \frac{2(fRe)\mu u_m L}{D_h^2} + K(\infty)\frac{\rho u_m^2}{2} + K_c\frac{\rho u_m^2}{2} + K_e\frac{\rho u_m^2}{2} = 45.0\ \text{kPa}$$

The contraction and expansion coefficients were taken as the largest values on the chart in Figure 3.5B in order to design for the maximum expected pressure drop.

Example 3.2

A microchannel is etched in silicon. The microchannel surface is intentionally etched to provide an average roughness of 12 μm. The microchannel dimensions measured from the root of the roughness elements are: width = 200 μm, height = 200 μm, length = 10 mm. Water flows through the microchannels at a temperature of 300 K. Calculate the core frictional pressure drop when (i) $m = 90 \times 10^{-6}$ kg/s, and (ii) $\dot{m} = 180 \times 10^{-6}$ kg/s.

Schematic

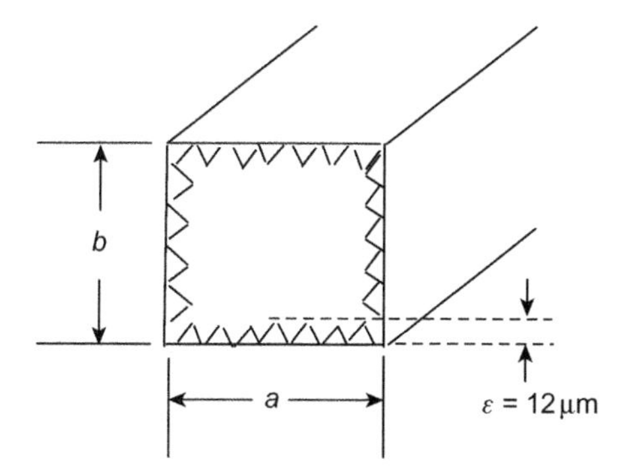

Solution

Properties: From Incropera et al. (2007), saturated water at 300 K: $\mu_f = 0.855 \times 10^{-3}$ N s/m^2, $\rho = 997$ kg/m^3, $c_{p,f} = 4179$ J/kg K, $k_f = 0.613$ W/m K.

(i) Core frictional pressure drop when $\dot{m} = 90 \times 10^{-6}$ kg/s (Answer: 29,365 Pa).

Calculate the hydraulic diameter using the constricted width (a_{cf}) and constricted height (b_{cf}) in Eq. (3.6)

$$D_{h,ci} = \frac{4A_c}{P_w} = a_{cf} = b_{cf} = 0.000176 \text{ m}$$

Calculate the Reynolds number of the fluid in a microchannel using Eq. (3.35):

$$Re_{cf} = \frac{\rho u_{m,cf} D_{h,cf}}{\mu} = \frac{(997.01 \text{ kg/m}^3)(2.91 \text{ m/s})(176 \times 10^{-6} \text{ m})}{(0.855 \times 10^{-3} \text{ N s}/m^2)} = 598$$

For fully developed laminar flow, the hydrodynamic entry length may be obtained using Eq. (3.11):

$$L_{h,cf} = 0.05 Re_{cf} D_{h,f} = 0.05(598)(0.176 \text{ mm}) = 5.26 \text{ mm}$$

Since $L > L_{h,cf}$, the fully developed flow assumption is valid.

The $f\,Re$ term can be obtained using Eq. (3.10):

$$fRe = 24(1 - 1.3553\alpha_c + 1.9467\alpha_c^2 - 1.7012\alpha_c^3 + 0.9564\alpha_c^4 - 0.2537\alpha_c^4)$$
$$= 14.23$$

The core frictional pressure drop can be calculated using Eq. (3.14):

$$\Delta p = \frac{2(fRe)\mu_{m,cf} L}{D_{h,cf}^2} + K(\infty)\frac{\rho u_{m,cf}^2}{2}$$

where $K(\infty)$ is given by Eq. (3.18).

$$K(\infty) = (0.6796 + 1.219\alpha_c + 3.3089\alpha_c^2 - 9.5921\alpha_c^3 + 8.9089\alpha_c^4 - 2.9959\alpha_c^5)$$
$$= 1.53$$

Hence

$$\Delta p = \frac{2(14.23)(0.855 \times 10^{-3} \text{ N s/m}^2)(2.91 \text{ m/s})(0.01 \text{ m})}{(176 \times 10^{-6} \text{ m})^2}$$
$$+ (1.53)\frac{(997.01 \text{ kg/m}^3)(2.91 \text{ m/s})^2}{2} = 29,356 \text{ Pa}$$

(ii) Core frictional pressure drop when $\dot{m} = 180 \times 10^{-6}$ kg/s (Answer: 68,694 Pa).

Calculate the Reynolds number of the fluid in a microchannel using Eq. (3.35):

$$Re_{cf} = \frac{\rho u_{m,cf} D_{h,cf}}{\mu} = \frac{(997.01 \text{ kg/m}^3)(5.83 \text{ m/s})(176 \times 10^{-6} \text{ m})}{(0.855 \times 10^{-3} \text{ N s/m}^2)} = 1200$$

For fully developed laminar flow, the hydrodynamic entry length may be obtained using Eq. (3.11):

$$L_{h,cf} = 0.05Re_{cf}D_{h,cf} = 0.05(1200)(0.176\ \text{mm}) = 10.5\ \text{mm}$$

Since $L < L_{h,cf}$, the fully developed flow assumption is not valid.

Calculate the core frictional pressure drop with the developing laminar flow assumption. The apparent friction factor can be obtained from Table 3.2 using x^+ calculated from Eq. 3.16:

$$x^+ = \frac{x/D_h}{Re_{cf}} = \frac{0.01\ \text{m}}{(176 \times 10^{-6}\ m)(1200)} = 0.0475$$

From Table 3.2, $f_{app}Re = 21.35$, so the core frictional pressure drop can be calculated using the following equation:

$$\Delta p = \frac{2(f_{app}Re)\mu_{m,cf}L}{D_{h,cf}^2}$$

$$= \frac{2(21.35)(0.855 \times 10^{-3}\ \text{N}\ s/m^2)(5.83\ m/s)(0.01\ \text{m})}{(176 \times 10^{-6}\ \text{m})^2} = 68,694\ \text{Pa}$$

Comments

(1) The total frictional pressure drop has to consider the minor losses because the actual Δp is higher than the core frictional pressure drop.

Example 3.3

Consider a copper minichannel heat sink with an area of 30 mm × 30 mm and relevant dimensions in Figure 3.14 with $a = 1$ mm, $b = 3$ mm, and $s = 1.5$ mm. The heat dissipation is 100 W and the water inlet temperature is 30°C. The maximum surface temperature in the heat sink is limited to 80°C. Calculate the water flow rate under these conditions. Also calculate the frictional pressure drop in the core. Assume a constant heat transfer coefficient corresponding to fully developed conditions and take the thermal conductivity of copper to be 400 W/m K.

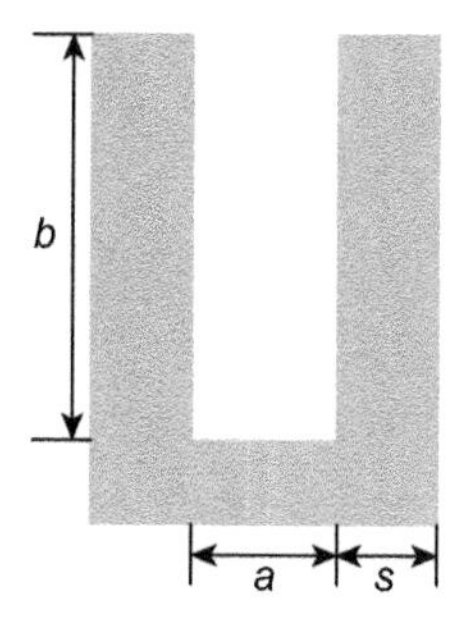

Assumptions

Laminar flow, constant heat flux equally distributed over the surface area, one-dimensional steady state conduction, and three-sided heated condition with an adiabatic tip for fin efficiency

$a = 1\ \text{mm} = 10^{-3}\ \text{m}$, $b = 3\ \text{mm} = 3 \times 10^{-3}\ \text{m}$, $s = 1.5\ \text{mm} = 1.5 \times 10^{-3}\ \text{m}$, $w = L = 30 \times 10^{-3}\ \text{m}$, $T_{f,i} = 30°\text{C}$, $T_{s,o} = 80°\text{C}$, $q = 100\ \text{W}$, $k_{Cu} = 400\ \text{W/m K}$, $A_c = a \times b = 3 \times 10^{-6}\ \text{m}^2$, $\alpha_c = a/b = 1/3$.

Solution

Equation (3.7) gives the hydraulic diameter:

$$D_h = \frac{2ab}{(a+b)} = \frac{2(10^{-3})(3 \times 10^{-3})}{(10^{-3} + 3 \times 10^{-3})} = 1.5 \times 10^{-3}\ \text{m}$$

The Nusselt number can be obtained using Table 3.5 for a three-sided heated channel. Through interpolation this gives

$$Nu = Nu_{fd,3} = 5.224$$

For calculating water properties, let's assume the water exit temperature is equal to the maximum surface temperature of the heat sink. Then the average water temperature is

$$T_{avg} = (30°\text{C} + 80°\text{C})/2 = 55°\text{C}$$

Properties of water at 55°C (from Incropera et al., 2007):

$$\rho = 985\ \text{kg/m}^3,\ c_p = 4.183\ \text{kJ/kg K},\ \mu = 505 \times 10^{-6}\ \text{kg/ms},\ k = 648 \times 10^{-3}\ \text{W/m K}$$

The average heat transfer coefficient is calculated using the Nusselt number, which is rearranged to give

$$\bar{h} = \frac{kNu}{D_h} = \frac{(0.648)(5.224)}{(1.5 \times 10^{-3})} = 2.26 \times 10^3\ \text{W/m}^2\ \text{K}$$

By relating the conductive and convective heat transfer for a constant heat flux, the average temperature difference between the surface and the fluid is given as

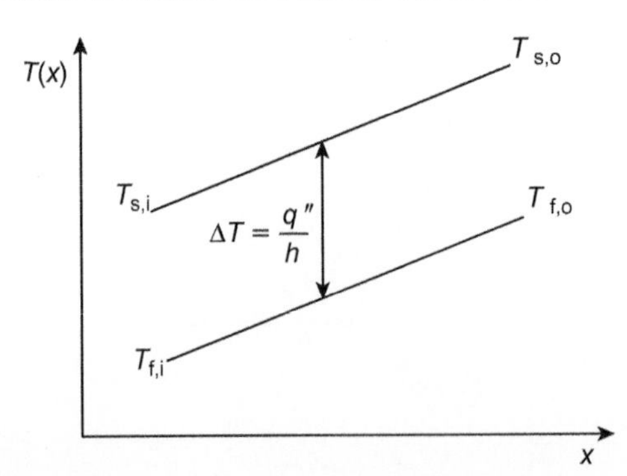

$$\Delta T = \frac{q}{h(2b\eta_f + a)nL}$$

The number of channels is $n = (\text{overall width/channel} + \text{fin}) = (w/a + s)$ $= (0.030/0.001 + 0.0015) = 12$

The fin efficiency is $\eta_f = (\tanh(mb)/mb)$

$$mb = \sqrt{\frac{2\bar{h}}{k_{\text{Cu}}s}}\, b = \sqrt{\frac{2(2.26 \times 10^3)}{(400)(1.5 \times 10^{-3})}} \times (3 \times 10^{-3}) = 0.260$$

$$\eta_f = \frac{\tanh(0.260)}{(0.260)} = 0.978$$

$$\Delta T = \frac{100}{(2.26 \times 10^3)[2(3 \times 10^{-3})(0.978) + 10^{-3}(12)(30 \times 10^{-3})]} = 17.9°\text{C}$$

The fluid outlet temperature is $T_{f,o} = T_{s,o} - \Delta T = 80°\text{C} - 17.9°\text{C} = 62.1°\text{C}$

The average water temperature is $T_{avg} = (30°\text{C} + 62.1°\text{C})/2 = 46.1°\text{C}$

T_{avg} is off by 9°. It is necessary to iterate again with the updated average temperature of 46°C.

Properties of water at 46°C (from Incropera et al., 2007):

$$\rho = 990\ \text{kg/m}^3,\ c_p = 4.180\ \text{kJ/kg K},\ \mu = 588 \times 10^{-6}\ \text{kg/ms},$$
$$k = 639 \times 10^{-3}\ W/m\ \text{K}$$

Following identical steps the new values are

$$\bar{h} = 2.23 \times 10^3\ W/\text{m}^2\ \text{K},\ mb = 0.258,\ \eta_f = 0.978,\ \Delta T = 18.2°\text{C},\ T_{f,o} = 61.8°\text{C}$$

$$T_{avg} = (30 + 61.8)/2 = 45.9°\text{C}$$

The calculated average temperature is approximately the same as the assumed value. No more iterations are necessary.

$$T_{s,i} = T_{f,i} + \Delta T = 30°\text{C} + 18.2°\text{C} = 48.2°\text{C}$$

$$q = \dot{m}c_p\Delta T_{\text{fluid}}$$

$$\dot{m}_t = \frac{q}{c_p\Delta T_{\text{fluid}}} = \frac{(100)}{(4180)(61.8 - 30)} = 752 \times 10^{-6}\ \text{kg/s}$$

$$\dot{Q}_c = \frac{\dot{m}_t(1/n)}{\rho} = \frac{(752 \times 10^{-6})(1/1/2)}{(990)} = 63.3 \times 10^{-9}\ \text{m}^3/\text{s} \quad \text{or} \quad 3.80\ \text{ml/min}$$

$$u_m = \frac{Q_c}{A_c} = \frac{(63.3 \times 10^{-9})}{(3 \times 10^{-6})} = 0.0211 \text{ m/s}$$

$$Re = \frac{D_h u_m \rho}{\mu} = \frac{(1.5 \times 10^{-3})(0.0211)(990)}{(588 \times 10^{-6})} = 53.3$$

Flow is laminar, so the original assumption is correct.

Equation (3.14) gives the pressure drop, $\Delta p = (2(fRe)\mu u_m L/D_h^2) + K(\infty)(\rho u_m^2/2)$

$f\,Re$ can be determined with Eq. (3.10):

$$fRe = 24(1 - 1.3553\alpha_c + 1.9467\alpha_c^2 - 1.7012\alpha_c^3 + 0.9564\alpha_c^4 - 0.2537\alpha_c^5) = 17.1$$

Hagenbach's factor comes from Eq. (3.18):

$$K(\infty) = 0.6796 - 1.2197\alpha_c + 3.3089\alpha_c^2 - 9.5921\alpha_c^3 + 8.9089\alpha_c^4 - 2.9959\alpha_c^5 = 1.20$$

The pressure drop in the core is therefore

$$\Delta p = \frac{2(17.1)(588 \times 10^{-6})(0.0211)(30 \times 10^{-3})}{(1.5 \times 10^{-3})^2} + (1.20)\frac{(990)(0.0211)^2}{2} = 5.92 \text{ Pa}$$

We need to calculate the hydrodynamic and thermal entrance lengths to justify fully developed conditions.

Using Eq. (3.11) we have $L_h = 0.05\ D_h Re = 0.05\ (1.5 \times 10^{-3})\ (53.3) = 0.00400 \text{ m} = 4.00 \text{ mm}$.

Similarly, Eq. (3.51) gives $L_t = 0.05\ D_h Re Pr = 0.05\ (1.5 \times 10^{-3})\ (53.3)\ (3.85) = 0.0154 \text{ m} = 15.4 \text{ mm}$.

This validates the fully developed assumption as given in the problem statement.

Comments

If the heat dissipation is increased to 200 W, the flow rate will be 18.4 ml/min. This increases the pressure drop to 38.8 Pa. This illustrates the performance hit for increased heat dissipation and the need to develop microchannel cooling under turbulent conditions where the heat transfer coefficient will be larger. For turbulent flow more heat can be dissipated with minimal performance loss in pressure drop.

3.11 Practice problems

Problem 3.1

Solve Example 3.1 with copper as the substrate, with a thermal conductivity of 380 W/m K.

Problem 3.2

For Example 3.1, plot the outlet surface temperature and pressure drop as a function of the channel depth b over a range of 200–600 μm.

Problem 3.3

For Example 3.1, plot the outlet surface temperature and pressure drop as a function of the channel width a over a range of 50–200 μm.

Problem 3.4

Solve Example 3.1 with a split-flow arrangement. Neglect the area reduction caused by the central manifold.

Problem 3.5

Solve Example 3.1 with copper substrate and a split-flow arrangement. Assume the channel width to be 100 μm and a depth of 800 μm. Neglect the area reduction caused by the central manifold.

Problem 3.6

For Example 3.3 with copper, plot the outlet surface temperature and the pressure drop as a function of the channel depth over a range of 200–2000 μm.

Problem 3.7

In Example 3.3, redesign the heat sink if the design heat load increases to 500 W. You may have to rework the channel geometry and flow arrangements.

Problem 3.8

Design a microchannel heat exchanger to dissipate 800 watts from a copper heat sink of 20 mm × 20 mm heated surface area. The inlet water temperature is 40°C, and the maximum surface temperature in the heat sink is desired to be below 60°C. Check your channel dimensions from a manufacturing standpoint (provide the manufacturing technique you will be implementing). Do not neglect the temperature drop occurring in the copper between the base of the channels and the bottom surface of the heat sink receiving heat. Show the details of the manifold design.

Problem 3.9

A copper minichannel heat sink has an area of 30 mm × 30 mm, and relevant dimensions in Figure 3.14 are $a = 1$ mm, $b = 3$ mm, and $s = 1.5$ mm. Calculate the maximum heat dissipation possible with an inlet water temperature of 30°C and the maximum surface temperature in the heat sink limited to 80°C. Calculate the water flow rate under these conditions. Also calculate the frictional pressure drop in the core under these conditions.

Problem 3.10

Design a minichannel heat exchanger to dissipate 5 kW of heat from a copper plate with a footprint of 10 cm × 12 cm. The plate surface temperature should not exceed 60°C, and the inlet design temperature for water is 35°C. Calculate the water flow rate, outlet water temperature, and core frictional pressure drop. Compare the performance of (a) straight once-through flow passages and (b) split-flow passages.

Problem 3.11

Design the heat exchanger for Problem 3.10 in aluminum, and compare its performance with a copper heat sink.

Appendix A

Table A.1 Curve-Fit Equations for Tables 3.2, 3.5, and 3.6

	Constants					
Equations	a	b	c	d	e	f
1	141.97	−7.0603	2603	1431.7	14364	−220.77
2	142.05	−5.4166	1481	1067.8	13177	−108.52
3	142.1	−7.3374	376.69	800.92	14010	−33.894
4	286.65	25.701	337.81	1091.5	26415	8.4098
5	8.2321	2.0263	1.2771	0.29805	2.2389	0.0065322
6	8.2313	1.9349	−2.295	0.92381	7.928	0.0033937
7	36.736	2254	17559	66172	555480	1212.6
8	30.354	1875.4	13842	154970	783440	−8015.1
9	31.297	2131.3	14867	144550	622440	−13297
10	28.315	3049	27038	472520	1783300	−35714
11	6.7702	−3.1702	0.4187	2.1555	2.76×10^{-6}	NA
12	9.1319	−3.7531	0.48222	2.5622	5.16×10^{-6}	NA

Equation form:
Equations 1–4: $y = (a + cx^{0.5} + ex/1 + bx^{0.5} + dx + fx^{1.5})$
Equations 5–10: $y = (a + cx + ex^2/1 + bx + dx^2 + fx^3)$
Equations 11–12: $y = a + bx + c(\ln x)^2 + d\ln x + ex^{-1.5}$

Equation variables:
Laminar flow friction factor in entrance region of rectangular ducts, Table 3.2:

y-variable—friction factor
x-variable—Equation 1: $\alpha_c = 1.0$
Equation 2: $\alpha_c = 0.5$
Equation 3: $\alpha_c = 0.2$
Equation 4: $0.1 = \alpha_c = 10$

Fully developed Nusselt number for three-sided heating, Table 3.5:

y-variable—Nusselt number for three-sided heating
x-variable—Equation 5: α_c

Fully developed Nusselt number for four-sided heating, Table 3.5:

y-variable—Nusselt number for four-sided heating
x-variable—Equation 6: α_c

Thermal entry region Nusselt numbers for four-sided heating, Table 3.6:

y-variable—Nusselt number for four-sided heating in the entry region
x-variable—Equation 7: $\alpha_c = 0.1$
Equation 8: $\alpha_c = 0.25$
Equation 9: $\alpha_c = 0.333$
Equation 10: $\alpha_c = 0.5$
Equation 11: $\alpha_c = 1.0$
Equation 12: $\alpha_c = 10$

References

Adams, T.M., Abdel-Khalik, S.I., Jeter, M., Qureshi, Z.H., 1997. An experimental investigation of single-phase forced convection in microchannels. Int. J. Heat Mass Transfer 41 (6–7), 851–857.

Akhavan-Behabadi, M.A., Kumar, R., Salimpour, M.R., Azimi, R., 2010. Pressure drop and heat transfer augmentation due to coiled wire inserts during laminar flow of oil inside a horizontal tube. Int. J. Therm. Sci. 49, 373–379.

ASME Standard B46.1-2002. Surface Texture (Surface Roughness, Waviness, and Lay). The American Society of Mechanical Engineers, An American National Standard (ASME B46.1-2002), pp. 1–98.

Baviere, R., Ayela, F., First local measurement in microchannels with integrated micromachined strain gauges. Paper No. ICMM2004-2406, Second International Conference on Microchannels and Minichannels, Rochester, NY, June 17–19 2004, pp. 221–228.

Baviere, R., Ayela, F., Le Person, S., Favre-Marinet, M., An experimental study of water flow in smooth and rough rectangular microchannels. Paper No. ICMM2004-2338, Second International Conference on Microchannels and Minichannels, Rochester, NY, June 17–19 2004, pp. 221–228.

Baviere, R., Gamrat, G., Favre-Marinet, M., Le Person, S., 2006. Modelling of laminar flows in rough-wall microchannels. J. Fluids Eng. 128 (4), 734–741.

Bergles, A.E., Lienhard, J.H.V, Kendall, G.E., Griffith, P., 2003. Boiling and condensation in small diameter channels. Heat Transfer Eng. 24, 18–40.

Bergman, T.L., 2009. Effect of reduced specific heats of nanofluids on single phase, laminar internal forced convection. Int. J. Heat Mass Transfer 52, 1240–1244.

Brackbill, T.P., Kandlikar, S.G., 2010. Application of lubrication theory and study of roughness pitch during laminar, transition and low Reynolds number turbulent flow at microscale. Heat Transfer Eng. 31 (8), 635–645.

Bucci, A., Celata, G.P., Cumo, M., Serra, E., Zummo, G., Water single-phase fluid flow and heat transfer in capillary tubes. Paper No. ICMM2004-2406, Second International Conference on Microchannels and Minichannels, Rochester, NY, 17–19 June 2004, pp. 221–228. International Conference on Microchannels and Minichannels. Paper # 1037, ASME, pp. 319–326.

Celata, G.P., Cumo, M., Guglielmi, M., Zummo, G., 2002. Experimental investigation of hydraulic and single phase heat transfer in 0.130 mm capillary tube. Microscale Thermophys. Eng. 6, 85–97.

Chen, R.Y., 1972. Flow in the entrance region at low Reynolds numbers. J. Fluids Eng. 95, 153–158.

Chein, R., Huang, G., 2005. Analysis of microchannel heat sink performance using nanofluids. Appl. Therm. Eng. 25, 3104–3114.

Choi, S.U.S., 1995, Enhancing thermal conductivity of fluids with nanoparticles, In: Developments and applications of non-Newtonian flows, Singer, D.A., and Wang, H.P. (Eds), FED231, ASME, New York, NY.

Colgan, E.G., Furman, B., Gaynes, M., Graham, W., LaBianca, N., Magerlein, J.H., et al., A practical implementation of silicon microchannel coolers for high power chips. Invited Paper presented at IEEE-Semi-Therm 21, San Jose, 15–17 March, 2005.

Costaschuk, D., Elsnab, J., Petersen, S., Klewicki, J.C., Ameel, T., 2007. Axial static pressure measurements of water flow in a rectangular microchannel. Exp. Fluids 43, 907–916.

Curr, R.M., Sharma, D., Tatchell, D.G., 1972. Numerical predictions of some three-dimensional boundary layers in ducts. Comput. Method Appl. Mech. Eng. 1, 143–158.

Darcy, H., 1857. Recherches experimentales relatives au movement de L'Eau dans les Tuyaux. Mallet-Bachelier, Paris, France.

Dharaiya, V.V., Kandlikar, S.G., 2012. Numerical investigation of heat transfer in rectangular microchannels under H2 boundary condition during developing and developed laminar flow. J. Heat Transfer 134 (2), 020911 (10 pages).

Dharaiya, V.V., Kandlikar, S.G., 2013. A numerical study on the effects of 2d structured sinusoidal elements on fluid flow and heat transfer at microscale. Int. J. Heat Mass Transfer 57, 190–201.

Fanning, T.A, 1886. Practical Treatise and Hydraulic and Water Supply Engineering. Van Nostrand (revised edition of 1877).

Gad-el-Hak, M., 1999. The fluid mechanics of microdevices. J. Fluids Eng. 121, 7–33.

Gamrat, G., Favre-Marinet, M., Asendrych, D., Numerical modeling of heat transfer in rectangular microchannels. Paper No. ICMM2004-2336, Second International Conference on Microchannels and Minichannels, 16–18 June 2004, Rochester, NY, pp. 205–212.

Gamrat, G., Favre-Marinet, M., Le Person, L., Baviere, R., Ayela, F., 2008. An experimental study of roughness effects on laminar flow in microchannels. J. Fluid Mech. 594, 399–423.

Garimella, S.V., Singhal, V., 2004. Single-phase flow and heat transport and pumping considerations in microchannel heat sinks. Heat Transfer Eng. 25 (1), 15–25.

Ghajar, A.J., Tang, C.C., Cook, W.L., 2010. Experimental investigation of friction factor in the transition region for water flow in minitubes and microtubes. Heat Transfer Eng. 31 (8), 646–657.

Gnielinski, V., 1976. New equations for heat and mass transfer in turbulent pipe and channel flow. Int. Chem. Eng. 16, 359–368.

Guo, Z.-Y., Li, Z.-X., 2003. Size effect on single-phase channel flow and heat transfer at microscale. Int. J. Heat Fluid Flow 24, 284–298.

Haaland, S.E., 1983. Simple and explicit formulas for the friction factor in turbulent flow. J. Fluids Eng. 105 (1), 89–90.

Harpole, G., Eninger, J.E., Micro-channel heat exchanger optimization. Proceedings—IEEE Semiconductor Thermal and Temperature Measurement Symposium, February, 1991, pp. 59–63.

Hartnett, J.P., Koh, J.C.Y., McComas, S.T., 1962. A comparison of predicted and measured friction factors for turbulent flow through rectangular ducts. J. Heat Transfer 84, 82–88.

Hetsroni, G., Mosyak, A., Pogrebnyak, E., Yarin, L.P., 2005. Heat transfer in micro-channels: comparison of experiments with theory and numerical results. Int. J. Heat Mass Transfer 48, 5580–5601.

Hornbeck, R.W., 1964. Laminar flow in the entrance region of a pipe. Appl. Sci. Res. 13, 224–232.

Huang, Z.F., Nakayama, A., Yang, K., Yang, C., Liu, W., 2010. Enhancing heat transfer in the core flow by using porous medium insert in a tube. Int. J. Heat Mass Transfer 53, 1164–1174.

Incropera, F.P., DeWitt, D.P., Bergman, T.L., Lavine, A.S., 2007. Introduction to Heat Transfer. John Wiley and Sons, Hoboken, NJ.

Jones Jr., O.C., 1976. An improvement in the calculation of turbulent friction in rectangular ducts. J. Fluids Eng. 98, 173–181.

Judy, J., Maynes, D., Webb, B.W., 2002. Characterization of frictional pressure drop for liquid flows through microchannels. Int. J. Heat Mass Transfer 45, 3477–3489.

Jung, J.-Y., Oh, H.-S., Kwak, H.-Y., 2009. Forced convective heat transfer of nanofluids in microchannels. Int. J. Heat Mass Transfer 52, 466–472.

Kakac, S., Shah, R.K., Aung, W., 1987. Handbook of Single-Phase Convective Heat Transfer. John Wiley and Sons, Inc., New York, NY.

Kandlikar, S.G., 2005. High heat flux removal with microchannels—a roadmap of challenges and opportunities. Heat Transfer Eng. 26 (8).

Kandlikar, S.G., Grande, W.J., 2004. Evaluation of single-phase flow in microchannels for high flux chip cooling—thermohydraulic performance enhancement and fabrication technology. Heat Transfer Eng. 25 (8), 5–16.

Kandlikar, S.G., Upadhye, H.R., Extending the heat flux limit with enhanced microchannels in direct single-phase cooling of computer chips. Invited Paper presented at IEEE-Semi-Therm 21, San Jose, 15–17 March 2005.

Kandlikar, S.G., Joshi, S., Tian, S., 2003. Effect of surface roughness on heat transfer and fluid flow characteristics at low Reynolds numbers in small diameter tubes. Heat Transfer Eng. 24 (3), 4–16.

Kandlikar, S.G., Schmitt, D., Carrano, A.L., Taylor, J.B., 2005. Characterization of surface roughness effects on pressure drop in single-phase flow in minichannels. Phys. Fluids 17 (10), Available from: http://dx.doi.org/10.1063/1.1896985, Article Number: 100606.

Kays, W.M., London, A.L., 1984. Compact Heat Exchangers. McGraw-Hill, New York, NY.

Kishimoto, T., Sasaki, S., 1987. Cooling characteristics of diamond-shaped interrupted cooling fins for high power LSI devices. Electron. Lett. 23 (9), 456–457.

Knight, R.W., Hall, D.J., Goodling, J.S., Jaeger, R.C., 1992. Heat sink optimization with application to microchannels. IEEE Trans. Comp. Hybrids Manuf. Technol. 15 (5), 832–842.

Koo, J., Kleinstreuer, C., Analyses of liquid flow in micro-conduits. Paper No. ICMM2004-2334, Second International Conference on Microchannels and Minichannels, Rochester, NY, 17–19 June 2004a, pp. 191–198.

Koo, J., Kleinstreuer, C., 2004b. Computational analysis of wall roughness effects for liquid flow in micro-conduits. J. Fluids Eng. 126 (1).

Koo, J., Kleinstreuer, C., 2005. Laminar nanofluid flow in microheat-sinks. Int. J. Heat Mass Transfer 48, 2652–2661.

Krishna, S.R., Pathipaka, G., Sivashanmugam, P., 2009. Heat transfer and pressure drop studies in a circular tube fitted with straight full twist. Exp. Therm. Fluid Sci. 33, 431–438.

Lee, J., Mudawar, I., 2007. Assessment of the effectiveness of nanofluids for single-phase and two-phase heat transfer in microchannels. Int. J. Heat Mass Transfer 50, 452–463.

Li, Z.X., Du, D.X., Guo, Z.Y. Experimental study on flow characteristics of liquid in circular microtubes. Proceeding of the International Conference on Heat Transfer and Transport Phenomena in Microscale, Banff, Canada, 15–20 October 2000, pp. 162–167.

Lin, T.-Y., Kandlikar, S.G., 2012a. A theoretical model for axial conduction effects during single-phase flow in microchannels. J. Heat Transfer 134 (2), . Available from: http://dx.doi.org/10.1115/1.4004936Article Number: 020902 (6 pages).

Lin, T-Y., Kandlikar, S.G., 2012b. An experimental investigation of structured roughness effect on heat transfer during single-phase liquid flow at microscale. J. Heat Transfer 134 (10), Available from: http://dx.doi.org/10.1115/1.4006844, Article Number: 101701 (9 pages).

Lin, T.-Y., Kandlikar, S.G., 2013a. Heat transfer investigation of air flow in microtubes I: effects of heat loss, viscous heating and axial conduction. J. Heat Transfer 135 (3), Available from: http://dx.doi.org/10.1115/1.4007876031703 (9 pages).

Lin, T.-Y., Kandlikar, S.G., 2013b. Heat transfer investigation of air flow in microtubes II: scale and axial conduction effects. J. Heat Transfer 135 (3), Available from: http://dx.doi.org/10.1115/1.4007877031704 (6 pages).

Maranzana, G., Perry, I., Maillet, D., 2004. Mini- and micro-channels: influence of axial conduction in the walls. Int. J. Heat Mass Transfer 47, 3993–4004.

Mises, R. v., 1914. Elemente der Technischen Hydrodynamik. B. G. Teubner, Leipzig.

Moody, L.F., 1944. Friction factors for pipe flow. ASME Trans. 66, 671–683.

Niklas, M. Favre-Marinet, M., Pressure losses in a network of triangular minichannels. Paper No. ICMM2003-1039, First International Conference on Microchannels and Minichannels, 24–25 April 2003, pp. 335–350.

Nikuradse, J., 1937. Strommungsgesetze in Rauen Rohren, VDI-Forschungsheft 361, Belige zu Forschung auf dem Gebiete des Ingenieurwesens Ausage B Band 4, July/August 1933. English Translation Laws of flow in rough pipes. NACA Tech. Mem. 1292.

Phillips, R.J., 1987. Forced Convection, Liquid Cooled, Microchannel Heat Sinks (MS thesis). Department of Mechanical Engineering, Massachusetts Institute of Technology, Cambridge, MA.

Phillips, R.J., 1990. Microchannel Heat Sinks, Advances in Thermal Modeling of Electronic Components and Systems. Hemisphere Publishing Corporation, New York, NY (Chapter 3).

Ryu, J.H., Choi, D.H., Kim, S.J., 2002. Numerical optimization of the thermal performance of a microchannel heat sink. Int. J. Heat Mass Transfer 45, 2823–2827.

Schmitt, D.J., Kandlikar, S.G., Effects of repeating microstructures on pressure drop in rectangular minichannels. Paper No. ICMM2005-75111, ASME, Third International Conference on Microchannels and Minichannels, Toronto, Canada, 13–15 June 2005.

Shah, R.K., London, A.L., 1978. Laminar Flow Forced Convection in Ducts, Supplement 1 to Advances in Heat Transfer. Academic Press, New York, NY.

Shen, P., Aliabadi, S.K., Abedi, J., A review of single-phase liquid flow and heat transfer in microchannels. Paper ICMM2004-2337, Second International Conference on Microchannels and Minichannels, 17–19 June 2004, pp. 213–220.

Singh, P.K., Harikrishna, P.V., Sundarajan, T., Das, S.K., 2011. Experimental and numerical investigation into the heat transfer study of nanofluids in microchannel. J. Heat Transfer 133, 121701 (9 pages).

Singh, P.K., Harikrishna, P.V., Sundarajan, T., Das, S.K., 2012. Experimental and numerical investigation into the hydrodynamics of nanofluids in microchannels. Int. J. Heat Mass Transfer 42, 174–186.

Steinke, M.E., 2005. Characterization of Single-Phase Fluid Flow and Heat Transfer in Plain and Enhanced Silicon Microchannels (Ph.D. thesis). Microsystems Engineering, Rochester Institute of Technology, Rochester, NY.

Steinke, M.E., Kandlikar, S.G., Single-phase enhancement techniques in microchannel flows. Paper No. ICMM2004-2328, Second International Conference on Microchannels and Minichannels, Rochester, NY, 17–19 June 2004a.

Steinke, M.E., Kandlikar, S.G., 2004b. Review of single-phase heat transfer enhancement techniques for application in microchannels, minichannels and microdevices. Int. J. Heat Technol. 22 (2), 3–11.

Steinke, M.E., Kandlikar, S.G., Single-phase liquid friction factors in microchannels. Paper No. ICMM2005-75112, Third International Conference on Microchannels and Minichannels, Toronto, Canada, 13–15 June 2005a.

Steinke, M.E., Kandlikar, S.G., Review of single-phase liquid heat transfer in microchannels. Paper No. ICMM2005-75114, ASME, Third International Conference on Microchannels and Minichannels, Toronto, Canada, 13–15 June 2005b.

Steinke, M.E., Kandlikar, S.G., Development of an experimental facility for investigating single-phase liquid flow in microchannels. Paper No. ICMM2005-75070, ASME, Third International Conference on Microchannels and Minichannels, Toronto, Canada, 13–15 June 2005c.

Szewczyk, H., 2008. Entrance and end effects in liquid flow in smooth capillary tubes. Chem. Process Eng. 29, 979–996.

Tiselj, I., Hetsroni, G., Mavko, B., Mosyak, A., Pogrebnyak, E., Segal, Z., 2004. Effect of axial conduction on the heat transfer in micro-channels. Int. J. Heat Mass Transfer 47, 2551–2565.

Tu, X. and Hrnjak, P., Experimental investigation of single-phase flow and pressure drop through rectangular microchannels. ASME, ICMM2003-1028, First International Conference on Microchannels and Minichannels, 24–25 April 2003, pp. 257–267.

Tuckerman, D.B., 1984. Heat Transfer Microstructures for Integrated Circuits (Ph.D. thesis). Stanford University, Stanford, CA.

Tuckerman, D.B., Pease, R.F.W., 1981. High performance heat sink for VLSI. IEEE Electron Dev. Lett. 2 (5), 126–129.

Wagner, R.N., Kandlikar, S.G., 2012. Effects of structures roughness on fluid flow at microscale level. Heat Transfer Eng. 33, 483–493.

Webb, R.L., Eckert, E.R.G., Goldstein, R.J., 1971. Heat transfer and friction in tubes with repeated-rib roughness. Int. J. Heat Mass Transfer 14, 601–617.

Wen, D.S., Ding, Y.L., 2004. Experimental investigation into convective heat transfer of nanofluids at the entrance region under laminar flow conditions. Int. J. Heat Mass Transfer 47 (24), 5181–5188.

Wen, D.S., Ding, Y.L., 2005. Effect of particle migration on heat transfer in suspensions of nanoparticles flowing through minichannels. Microfluid. Nanofluid. 1, 183–189.

Wibulswas, P., 1966. Laminar Flow Heat Transfer in Non-Circular Ducts (Ph.D. thesis). London University, London, UK.

Wongcharee, K., Eiamsa-ard, S., 2011. Friction and heat transfer characteristics of laminar swirl flow through the round tubes inserted with alternate clockwise and counter-clockwise twisted-tapes. Int. Commun. Heat Mass Transfer 38, 348–352.

Xiong, R., Chung, J., 2007. Flow characteristics of water in straight and serpentine micro-channels with miter bends. Exp. Therm. Fluid Sci. 31, 805–812.

Xu, B., Ooi, K.T., Wong, N.T., Choi, W.K., 2000. Experimental investigation of flow friction for liquid flow in microchannels. Int. Commun. Heat Mass Transfer 27 (8), 1165–1176.

Yang, C-Y., Lin, C-W., Lin, T-Y., Kandlikar, S.G., 2012. Heat transfer and friction characteristics of air flow in microtubes. Exp. Therm. Fluid Sci. 37, 12–18.

Yu, D., Warrington, R., Barron, R., Ameel, T., 1995. An experimental investigation of fluid flow and heat transfer in microtubes. Proc. ASME/JSME Therm. Eng. Conf. 1, 523–530.

CHAPTER

4 Single-Phase Electrokinetic Flow in Microchannels

Dongqing Li

Department of Mechanical and Mechatronics Engineering, University of Waterloo, West Waterloo, ON, Canada

4.1 Introduction

Understanding electrokinetic-driven liquid flow in microchannels is important for designing and controlling many lab-on-a-chip devices. Lab-on-a-chip devices are miniaturized biomedical or chemistry laboratories built on a small glass or plastic chip. Generally, a lab-on-a-chip device has a network of microchannels, electrodes, sensors, and electrical circuits. Electrodes are placed at strategic locations on the chip. Applying electrical fields along microchannels controls the liquid flow and other operations in the chip. These labs on a chip can duplicate the specialized functions of their room-sized counterparts, such as clinical diagnostics, DNA scanning, and electrophoretic separation. The advantages of these labs on a chip include dramatically reduced sample size, much shorter reaction and analysis time, high throughput, automation, and portability.

The key microfluidic functions required in various lab-on-a-chip devices include pumping, mixing, thermal cycling, dispensing, and separating. Most of these processes are electrokinetic processes. Basic understanding, modeling, and controlling of these key microfluidic functions/processes are essential to systematic design and operational control of lab-on-a-chip systems. Because all solid–liquid (aqueous solutions) interfaces carry electrostatic charge, there is an electrical double layer (EDL) field in the region close to the solid–liquid interface on the liquid side. Such an EDL field is responsible for at least two basic electrokinetic phenomena: electroosmosis and electrophoresis. Briefly, electroosmosis is the liquid motion in a microchannel caused by the interaction between the EDL at the liquid–channel wall interface with an electrical field applied tangentially to the wall. Electrophoresis is the motion of a charged particle relative to the surrounding liquid under an applied electrical field. Essentially, most on-chip electrokinetic microfluidic processes are realized using these two phenomena. This chapter will review basics of the EDL field and discuss some key on-chip microfluidic processes. A more comprehensive review of the electrokinetic-based microfluidic processes for lab-on-a-chip applications can be found elsewhere (Li, 2004).

Heat Transfer and Fluid Flow in Minichannels and Microchannels.
© 2014 Elsevier Ltd. All rights reserved.

4.2 Electrical double layer field

It is well known that most solid surfaces obtain a surface electric charge when they are brought into contact with a polar medium (e.g., aqueous solutions). This may be due to ionization, ion adsorption, or ion dissolution. If the liquid contains a certain amount of ions (for instance, an electrolyte solution or a liquid with impurities), the electrostatic charges on the solid surface will attract the counter-ions in the liquid. The rearrangement of the charges on the solid surface and the balancing charges in the liquid is called the EDL (Hunter, 1981; Lyklema, 1995). Immediately next to the solid surface, there is a layer of ions that are strongly attracted to the solid surface and are immobile. This layer is called the compact layer, normally about several Angstroms thick. Because of the electrostatic attraction, the counterion concentration near the solid surface is higher than that in the bulk liquid far away from the solid surface. The coions' concentration near the surface, however, is lower than that in the bulk liquid far away from the solid surface due to the electrical repulsion, so there is a net charge in the region close to the surface. From the compact layer to the uniform bulk liquid, the net charge density gradually reduces to zero. Ions in this region are affected less by the electrostatic interaction and are mobile. This region is called the diffuse layer of the EDL. The thickness of the diffuse layer is dependent on the bulk ionic concentration and electrical properties of the liquid, usually ranging from several nanometers for high ionic concentration solutions up to several microns for pure water and pure organic liquids. The boundary between the compact layer and the diffuse layer is usually referred to as the shear plane. The electrical potential at the solid–liquid surface is difficult to measure directly. The electrical potential at the shear plane is called the zeta potential, ς; it can be measured experimentally (Hunter, 1981; Lyklema, 1995). In practice, the zeta potential is used as an approximation to the potential at the solid–liquid interface.

The ion and electrical potential distributions in the EDL can be determined by solving the Poisson–Boltzmann equation (Hunter, 1981; Lyklema, 1995). According to the theory of electrostatics, the relationship between the electrical potential ψ and the local net charge density per unit volume ρ_e at any point in the solution is described by the Poisson equation

$$\nabla^2\psi = -\frac{\rho_e}{\varepsilon} \tag{4.1}$$

where ε is the dielectric constant of the solution. Assuming the equilibrium Boltzmann distribution equation is applicable, which implies uniform dielectric constant, the number concentration of the type-i ion is of the form

$$n_i = n_{io}\exp\left(-\frac{z_i e\psi}{k_b T}\right) \tag{4.2}$$

where n_{io} and z_i are the bulk ionic concentration and the valence of type-i ions, respectively, e is the charge of a proton, k_b is the Boltzmann constant, and T is the absolute temperature. For a symmetric electrolyte ($z_- = z_+ = z$) solution, the net volume charge density ρ_e is proportional to the concentration difference between symmetric cations and anions, via

$$\rho_e = ze(n_+ - n_-) = -2zen_o \sinh\left(\frac{ze\psi}{k_b T}\right) \tag{4.3}$$

Substituting Eq. (4.3) into the Poisson equation leads to the well-known Poisson–Boltzmann equation.

$$\nabla^2 \psi = \frac{2zen_o}{\varepsilon} \sinh\left(\frac{ze\psi}{k_b T}\right) \tag{4.4}$$

Solving the Poisson–Boltzmann equation with proper boundary conditions will determine the local EDL potential field ψ and hence, via Eq. (4.3), the local net charge density distribution.

4.3 Electroosmotic flow in microchannels

Consider a microchannel filled with an aqueous solution. There is an EDL field near the interface of the channel wall and the liquid. If an electric field is applied along the length of the channel, an electrical body force is exerted on the ions in the diffuse layer. In the diffuse layer of the EDL field, the net charge density, ρ_e, is not zero. The net transport of ions is the excess counterions. If the solid surface is negatively charged, the counterions are the positive ions. These excess counterions will move under the influence of the applied electrical field, pulling the liquid with them and resulting in electroosmotic flow. The liquid movement is carried through to the rest of the liquid in the channel by viscous forces. This electrokinetic process is called electroosmosis and was first introduced by Reuss (1809).

In most lab-on-a-chip applications, electroosmotic flow is preferred over pressure-driven flow. One reason is the plug-like velocity profile of electroosmotic flow. This means that fluid samples can be transported without dispersion caused by flow shear. Generally, pumping a liquid through a small microchannel requires applying a very large pressure difference depending on the flow rate. This is often impossible because of the limitations of the size and the mechanical strength of the microfluidic devices. Electroosmotic flow can generate the required flow rate even in very small microchannels without any applied pressure difference across the channel. Additionally, using electroosmotic flow to transport liquids in complicated microchannel networks does not require any external mechanical pump or moving parts; it can be easily realized by controlling the electrical fields via electrodes. Although high voltages are often required in electroosmotic flow, the required electrical power is very small due to the very low current involved, and it is generally safe. However, the heat produced in electroosmotic flow often presents problems in many applications where solutions of high electrolyte concentrations and long operation time are required.

Most channels in modern microfluidic devices and MEMS are made by micromachining technologies. The cross-section of these channels is close to a rectangular shape. In such a situation, the EDL field is two-dimensional and will affect the two-dimensional flow field in the rectangular microchannel. In order to control electroosmotic pumping as a means of transporting liquids in microstructures, we must understand the characteristics of electroosmotic flow in rectangular microchannels. In the following, we will review the modeling and the numerical simulation results of electroosmotic flow in rectangular microchannels (Arulanandam and Li, 2000a). The EDL field, the flow field, and the volumetric flow rate will be studied as functions of the zeta potential, the liquid properties, the channel geometry, and the applied electrical field.

Consider a rectangular microchannel of width $2W$, height $2H$, and length L, as illustrated in Figure 4.1. Because of the symmetry in the potential and velocity fields, the solution domain can be reduced to a quarter section of the channel (as shown by the shaded area in Figure 4.1).

The 2D EDL field can be described by the Poisson–Boltzmann equation:

$$\frac{\partial^2 \psi}{\partial y^2} + \frac{\partial^2 \psi}{\partial z^2} = \frac{2n_\infty ze}{\varepsilon\varepsilon_0} \sinh\left(-\frac{ze\psi}{k_b T}\right)$$

Along the planes of symmetry, the following symmetry boundary conditions apply:

$$\text{at} \quad y = 0 \quad \frac{\partial \psi}{\partial y} = 0 \quad \text{at} \quad z = 0 \quad \frac{\partial \psi}{\partial z} = 0$$

Along the surfaces of the solution domain, the potential is the zeta potential:

$$\text{at} \quad y = H \quad \psi = \zeta \quad \text{at} \quad z = W \quad \psi = \zeta$$

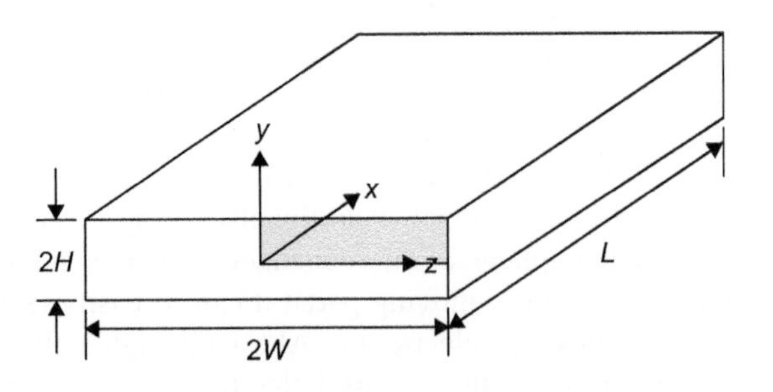

FIGURE 4.1

Geometry of microchannel. The shaded region indicates the computational domain.

The Poisson–Boltzmann equation and the boundary conditions can be transformed into nondimensional equations by introducing the following dimensionless variables:

$$y^* = \frac{y}{D_h} \quad z^* = \frac{z}{D_h} \quad \psi^* = \frac{ze\psi}{k_b T} \quad D_h = \frac{4A_{\text{cross-sectional}}}{P_{\text{wetted}}} = \left(\frac{4HW}{H+W}\right)$$

The nondimensional form of the Poisson–Boltzmann equation is given by

$$\frac{\partial^2 \psi^*}{\partial y^{*2}} + \frac{\partial^2 \psi^*}{\partial z^{*2}} = (\kappa D_h)^2 \sinh(\psi^*) \tag{4.5}$$

where κ, the Debye–Huckle parameter, is defined as follows:

$$\kappa = \left(\frac{2n_\infty z^2 e^2}{\varepsilon\varepsilon_0 k_b T}\right)^{1/2} \tag{4.6}$$

and $1/\kappa$ is the characteristic thickness of the EDL. The nondimensional parameter κD_h is a measure of the relative channel diameter, compared to the EDL thickness. κD_h is often referred to as the electrokinetic diameter. The corresponding nondimensional boundary conditions follow:

$$\text{at} \quad y^* = 0 \quad \frac{\partial \psi^*}{\partial y^*} = 0 \qquad \text{at} \quad z^* = 0 \quad \frac{\partial \psi^*}{\partial z^*} = 0$$

$$\text{at} \quad y^* = \frac{H}{D_h} \quad \psi^* = \zeta^* = \frac{ze\zeta}{k_b T} \quad \text{at} \quad z^* = \frac{W}{D_h} \quad \psi^* = \zeta^* = \frac{ze\zeta}{k_b T}$$

If we consider that the flow is steady, two-dimensional, and fully developed, and that there is no pressure gradient in the microchannel, the general equation of motion is given by a balance between the viscous or shear stresses in the fluid and the externally imposed electrical field force:

$$\mu\left(\frac{\partial^2 u}{\partial y^2} + \frac{\partial^2 u}{\partial z^2}\right) = F_x$$

where F_x is the electrical force per unit volume of the liquid, which is related to the electric field strength E_x and the local net charge density as follows:

$$F_x = \rho_e E_x$$

Therefore, the equation for the electroosmotic flow can be written as:

$$\mu\left(\frac{\partial^2 u}{\partial\, y^2} + \frac{\partial^2 u}{\partial\, z^2}\right) = \left(\frac{2n_\infty z\, e}{\varepsilon\varepsilon_0}\sinh\psi\right)E_x \tag{4.7}$$

The following boundary conditions apply along the planes of symmetry:

$$\text{at} \quad y = 0 \quad \frac{\partial u}{\partial y} = 0 \quad \text{at} \quad z = 0 \quad \frac{\partial u}{\partial z} = 0$$

Along the surface of shear (the surface of the solution domain), the velocity boundary conditions are given by

$$\text{at} \quad y = H \quad u = 0 \quad \text{at} \quad z = W \quad u = 0$$

The equation of motion can be nondimensionalized using the following additional transformations:

$$u^* = \frac{u}{U}$$

$$E_x^* = \frac{E_x L}{\zeta}$$

where U is a reference velocity and L is the distance between the two electrodes. The nondimensionalized equation of motion becomes

$$\frac{\partial^2 u^*}{\partial y^{*2}} + \frac{\partial^2 u^*}{\partial z^{*2}} = M E_x^* \sinh(\psi^*) \tag{4.8}$$

M is a new dimensionless group, which is a ratio of the electrical force to the frictional force per unit volume, given by

$$M = \frac{2 n_\infty z e \zeta D_h^2}{\mu U L}$$

The corresponding nondimensional boundary conditions are given by

$$\text{at} \quad y^* = 0 \quad \frac{\partial u^*}{\partial y^*} = 0 \quad \text{at} \quad z^* = 0 \quad \frac{\partial u^*}{\partial z^*} = 0$$

$$\text{at} \quad y^* = \frac{H}{D_h} \quad u^* = 0 \quad \text{at} \quad z^* = \frac{W}{D_h} \quad u^* = 0$$

Numerically solving Eqs. (4.5) and (4.8) with the boundary conditions will allow us to determine the EDL field and the electroosmotic flow field in such a rectangular microchannel. As an example, let's consider a KCl aqueous solution. At a concentration of 1×10^{-6} M, $\varepsilon = 80$, and $\mu = 0.90 \times 10^{-3}$ kg/(ms). An arbitrary reference velocity of $U = 1$ mm/s was used to nondimensionalize the velocity. According to experimental results (Mala et al., 1997), zeta potential values change from 100 to 200 mV, corresponding to three concentrations of the KCl solution, 1×10^{-6}, 1×10^{-5}, and 1×10^{-4} M. The hydraulic diameter of the channel varied from 12 to 250 μm, while the aspect ratio (H/W) varied from 1:4 to 1:1. Finally, the applied voltage difference ranged from 10 V to 10 kV.

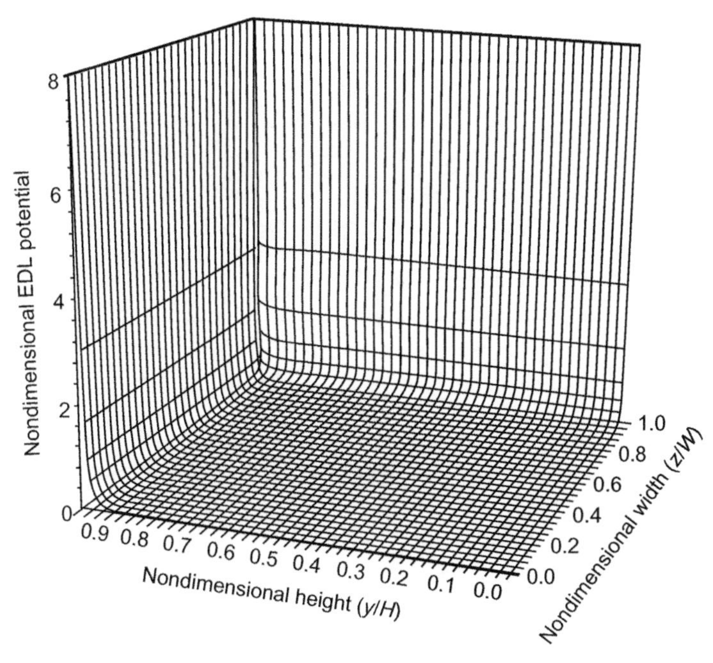

FIGURE 4.2

Nondimensional EDL potential profile in a quarter section of a rectangular microchannel, with $\kappa D_h = 79$, $\zeta^* = 8$, and $H{:}W = 2{:}3$.

The EDL potential distribution in the diffuse double layer region is shown in Figure 4.2. The nondimensional EDL potential profile across a quarter section of the rectangular channel exhibits characteristic behavior. The potential field drops off sharply very close to the wall. The region where the net charge density is not zero is limited to a small region close to the channel surface. Figure 4.3 shows the nondimensional electroosmotic velocity field for an applied potential difference of 1 kV/cm. The velocity field exhibits a profile similar to plug flow; however, in electroosmotic flow, the velocity increases rapidly from zero at the wall (shear plane) to a maximum velocity near the wall, and then gradually drops off to a slightly lower constant velocity that is maintained through most of the channel. This unique profile can be attributed to the fact that the externally imposed electrical field is driving the flow. In the mobile part of the EDL region very close to the wall, the larger electrical field force exerts a greater driving force on the fluid because of the presence of the net charge in the EDL region.

Variation of D_h affects the following nondimensional parameters: the electrokinetic diameter and the strength of the viscous forces in the ratio of electrical to viscous forces. The volumetric flow rate increases with approximately D_h^2, as seen

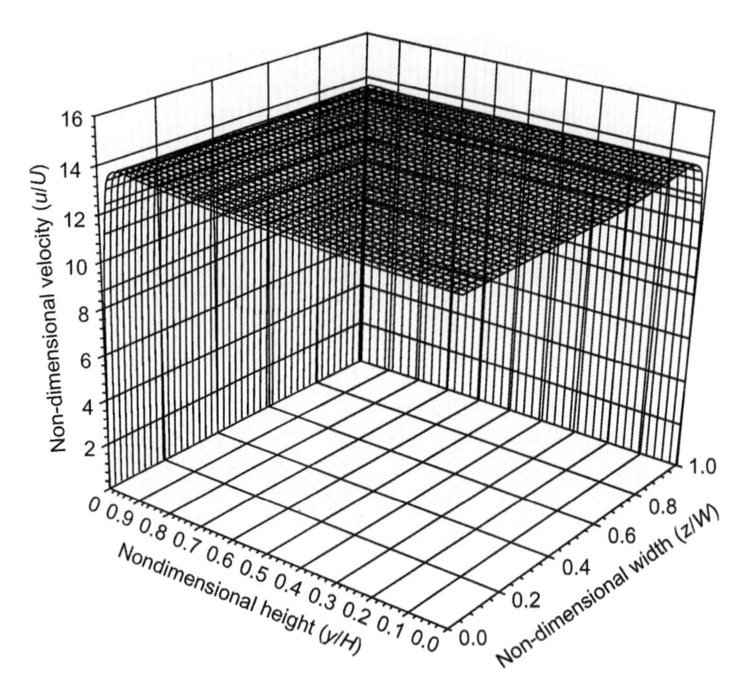

FIGURE 4.3

Nondimensional velocity field in a quarter section of a rectangular microchannel, with $\kappa D_h = 79$, $\zeta^* = 8$, $H/W = 2/3$, $E_x^* = 5000$, and $M = 2.22$.

in Figure 4.4. This is expected, since the cross-sectional area of the channel also increases proportionately to D_h^2. When larger pumping flow rates are desired, larger diameter channels would seem to be a better choice. However, there is no corresponding increase in the average velocity with increased hydraulic diameter. This is because of the nature of electroosmotic flow—the flow is generated by the motion of the net charge in the EDL region driven by an applied electrical field. When the double layer thickness ($1/\kappa$) is small, an analytical solution of the electroosmotic velocity can be derived from a one-dimensional channel system such as a cylindrical capillary with a circular cross-section, given by

$$v_{av} = -\frac{E_z \varepsilon_r \varepsilon_o \zeta}{\mu} \tag{4.9}$$

Equation (4.9) indicates that the electroosmotic flow velocity is linearly proportional to the applied electrical field strength and linearly proportional to the zeta potential. The negative sign indicates the flow direction and has to do with the sign of the ζ potential. If the ζ potential is negative (i.e., a negatively

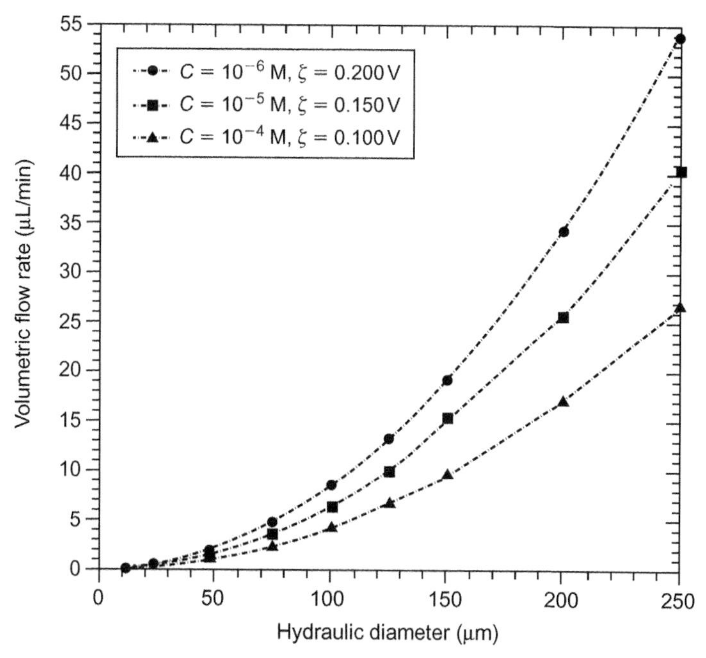

FIGURE 4.4

Variation of volumetric flow rate with hydraulic diameter for three different combinations of concentration and zeta potential, with $H/W = 2/3$ and $E_x = 1$ kV/cm.

charged wall surface), the excess counterions in the diffuse layer are positive, and therefore the electroosmotic flow in the microchannel is toward the negative electrode.

With a rectangular microchannel, not only the hydraulic diameter but also the channel shape will influence the velocity profile. This is because of the impact of the channel geometry on the EDL. Figure 4.5 shows the relationship between the aspect ratio (H/W) and the volumetric flow rate for a fixed hydraulic diameter. As the ratio of $H{:}W$ approaches 1:1 (for a square channel), the flow rate decreases. This is because of the larger role that corner effects have on the development of the EDL and the velocity profile in square channels.

From Eq. (4.6), it is clear that increasing the bulk ion concentration in the liquid results in an increase in κ or a decrease in the EDL thickness $1/\kappa$. Correspondingly, the EDL potential field falls off to zero more rapidly with distance; that is, the region influenced by the EDL is smaller. The ionic concentration effect on the velocity or the flow rate can be understood as follows. Since ionic concentration influences the zeta potential, as the ionic concentration is increased, the zeta potential decreases in value. As the zeta potential decreases, so does the electroosmotic flow velocity (Eq. (4.9)) and the volumetric flow rate.

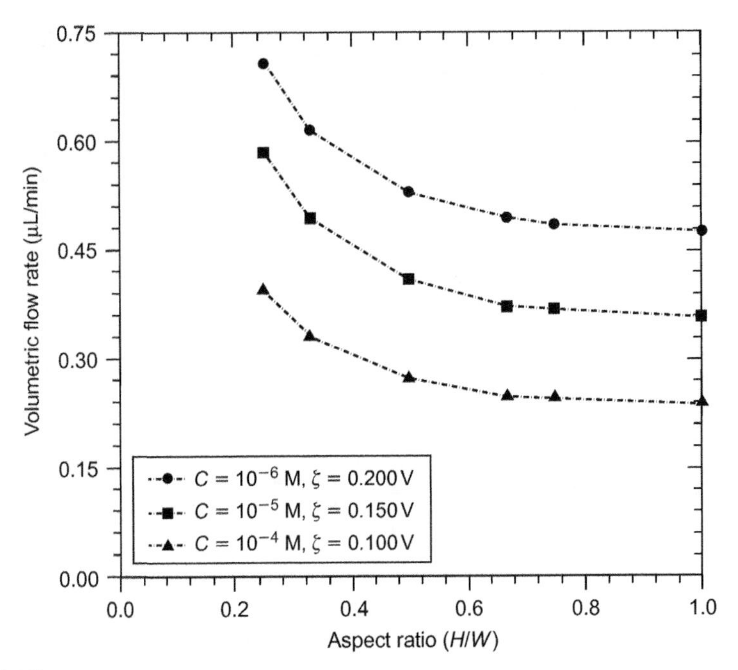

FIGURE 4.5

Variation of volumetric flow rate with aspect ratio for three different combinations of concentration and zeta potential, with $D_h = 24\ \mu m$ and $E_x = 1$ kV/cm. In this case, $z/W = 1.0$ represents the channel wall and $z/W = 0$ represents the center of the channel.

4.4 Experimental techniques for studying electroosmotic flow

In most electroosmotic flows in microchannels, the flow rates are very small (e.g., 0.1 μL/min) and the size of the microchannels is very small (e.g., 10–100 μm); it is extremely difficult to measure directly the flow rate or velocity of the electroosmotic flow in microchannels. To study liquid flow in microchannels, various microflow visualization methods have evolved. Microparticle image velocimetry (micro-PIV) is a method that was adapted from well-developed PIV techniques for flows in macro-sized systems (Herr et al., 2000; Selvaganapathy et al., 2002; Singh et al., 2001; Taylor and Yeung, 1993; Wereley and Meinhart, 2001). In the micro-PIV technique, the fluid motion is inferred from the motion of submicron tracer particles. To eliminate the effect of Brownian motion, temporal or spatial averaging must be employed. Particle affinities for other particles, channel walls, and free surfaces must also be considered. In electrokinetic flows, the electrophoretic motion of the tracer particles (relative to the bulk flow) is an additional consideration that must be taken. These are the disadvantages of the micro-PIV technique.

Dye-based microflow visualization methods have also evolved from their macro-sized counterparts. However, traditional mechanical dye injection techniques are difficult to apply to the microchannel flow systems. Specialized caged fluorescent dyes have been employed to facilitate the dye injection using selective light exposure (i.e., photo-injection of the dye). The photo-injection is accomplished by exposing an initially nonfluorescent solution seeded with caged fluorescent dye to a beam or a sheet of ultraviolet light. As a result of the ultraviolet exposure, caging groups are broken and fluorescent dye is released. Since the caged fluorescent dye method was first employed for flow tagging velocimetry in macro-sized flows in 1995 (Lempert et al., 1995), this technique has since been used to study a variety of liquid flow phenomena in microstructures (Dahm et al., 1992; Herr et al., 2000; Johnson et al., 2001; Molho et al., 2001; Paul et al., 1998; Sinton and Li, 2003a,b; Sinton et al., 2002, 2003a). The disadvantages of this technique are that it requires expensive specialized caged dye and extensive infrastructure to facilitate the photo-injection.

Recently, Sinton and Li (2003a) developed a microchannel flow visualization system and complementary analysis technique using caged fluorescent dyes. Both pressure-driven and electrokinetically driven velocity profiles determined by this technique compare well with analytical results and those of previous experimental studies. In particular, this method achieved a high degree of near-wall resolution. Generally, in the experiment, a caged fluorescent dye is dissolved in an aqueous solution in a capillary or microchannel. It should be noted that the caged dye cannot emit fluorescent light at this stage. Ultraviolet laser light is focused into a sheet crossing the capillary (perpendicular to the flow direction). The caged fluorescent dye molecules exposed to the UV light are uncaged and thus are able to shine. The resulting fluorescent dye is continuously excited by an argon laser and the emission light is transmitted through a laser-powered epi-illumination microscope. Full-frame images of the dye transport are recorded by a progressive scan CCD camera and saved automatically on the computer. In the numerical analysis, the images are processed and cross-stream velocity profiles are calculated based on tracking the dye concentration maxima through a sequence of several consecutive images. Several sequential images are used to improve the signal to noise ratio. Points of concentration maxima make convenient velocimetry markers as they are resistant to diffusion. In many ways, the presence of clearly definable, zero-concentration-gradient markers is a luxury afforded by the photo-injection process. The details of this technique can be found elsewhere (Sinton and Li, 2003a,b).

In an experimental study (Sinton and Li, 2003a), the CMNB-caged fluorescein with the sodium carbonate buffer and 102 μm ID glass capillaries were used. Images of the uncaged dye transport in four different electroosmotic flows are displayed in vertical sequence in Figure 4.6. The dye diffused symmetrically, as shown in Figure 4.6A. Image sequences given in Figure 4.6B–D were taken with voltages of

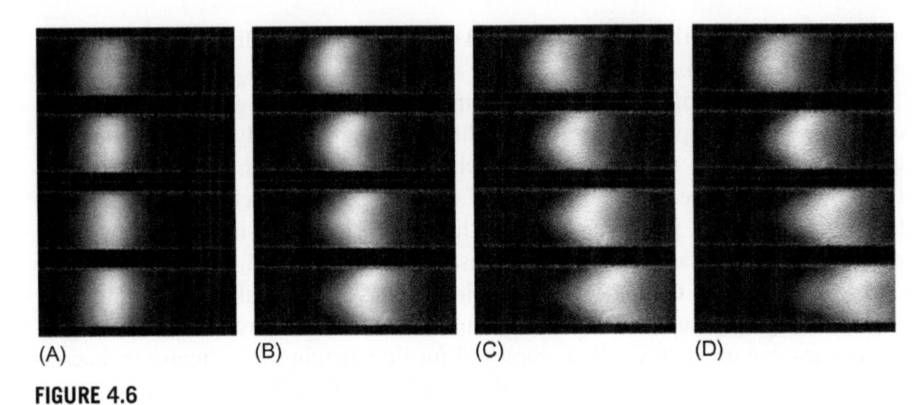

FIGURE 4.6

Images of the uncaged dye in electroosmotic flows through a 102 μm ID capillary at 133 msec intervals with applied electric field strength: (A) 0 V/0.14 m; (B) 1000 V/0.14 m; (C) 1500 V/0.14 m; and (D) 2000 V/0.14 m.

1000, 1500, and 2000, respectively (over the 14-cm length of capillary). The field was applied with the positive electrode at the left and the negative electrode at the right. The resulting plug-like motion of the dye is characteristic of electroosmotic flow in the presence of a negatively charged surface at high ionic concentration. The cup shape of the dye profile was observed in cases Figure 4.6B–D within the first 50 ms following the ultraviolet light exposure. This period corresponded to the uncaging time scale, in which the most significant rise in uncaged dye concentration occurs. Although the exact reason for the formation of this shape is unknown, it is likely that it was an artifact of the uncaging process in the presence of the electric field. Fortunately, however, the method is relatively insensitive to the shape of the dye concentration profile. Once formed, it is the transport of the maximum concentration profile that provides the velocity data. This also makes the method relatively insensitive to beam geometry and power intensity distribution.

Figure 4.7 shows velocity data for the four flows corresponding to the image sequences in Figure 4.6. Each velocity profile was calculated using an eight-image sequence and the numerical analysis technique described by Sinton and Li (2003a). The velocity profile resulting from no applied field, Figure 4.6A, corresponds closely to stagnation as expected. This run also serves to illustrate that, despite significant transport of dye due to diffusion, the analysis method is able to recover the underlying stagnant flow velocity. Although the other velocity profiles resemble that of classical electroosmotic flow (Wereley and Meinhart, 2001), a slight parabolic velocity deficit of approximately 4% was detected in all three flows. This was caused by a small back-pressure induced by the electroosmotic fluid motion (e.g., caused by the imperfectly leveled capillary along the length direction).

In additional to these PIV and dye-based techniques, electroosmotic flow velocity can be estimated indirectly by monitoring the electrical current change while one solution is replaced by another similar solution during electroosmosis

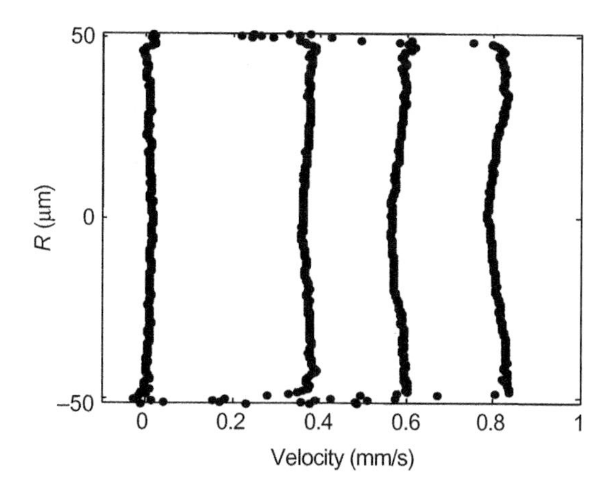

FIGURE 4.7

Plots of velocity data from four electroosmotic flow experiments through a 102 μm ID capillary with applied electric field strengths of: 0 V/0.14 m; 1000 V/0.14 m; 1500 V/0.14 m; and 2000 V/0.14 m (from left to right).

(Arulanandam and Li, 2000b; Ren et al., 2002; Sinton et al., 2002). In this method, a capillary tube is filled with an electrolyte solution, then brought into contact with another solution of the same electrolyte but with a slightly different ionic concentration. Once the two solutions are in contact, an electrical field is applied along the capillary in such a way that the second solution is pumped into the capillary and the first solution flows out of the capillary from the other end. As more and more of the second solution is pumped into the capillary and the first solution flows out of the capillary, the overall liquid conductivity in the capillary is changed, hence the electrical current through the capillary is changed. When the second solution completely replaces the first solution, the current will reach a constant value. Knowing the time required for this current change and the length of the capillary tube, the average electroosmotic flow velocity can be calculated by

$$u_{\mathrm{av,exp}} = \frac{L}{\Delta t} \tag{4.10}$$

where L is the length of the capillary and Δt is the time required for the higher (or lower)-concentration electrolyte solution to completely displace the lower (or higher)-concentration electrolyte solution in the capillary tube.

Figure 4.8 shows an example of the measured current–time relationship in a 10-cm-long glass capillary of 100 μm in diameter with a KCl solution under an applied electrical field of 350 V/cm. An example of measured electroosmotic velocity as a function of the applied electrical field is given in Figure 4.9. As seen in this figure, a linear relationship between the applied voltage and the

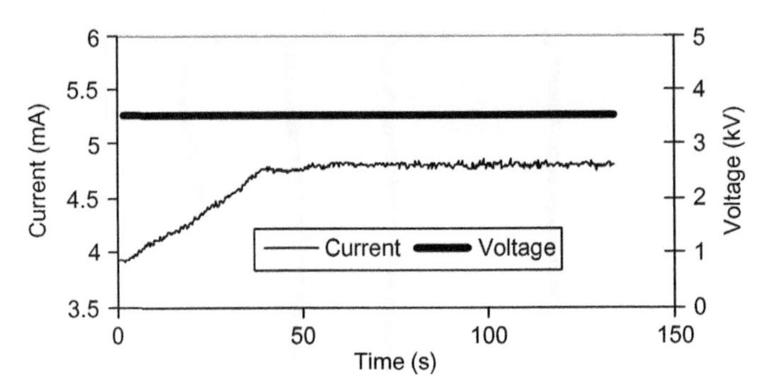

FIGURE 4.8

A typical result of current versus time. For the specific case of: capillary diameter $D = 100\ \mu m$, $E_x = 3500$ V/10 cm. KCl concentration in Reservoir 1 is $C_{95\%} = 0.95 \times 10^{-4}$ M. KCl concentration in Reservoir 2 is $C_{100\%} = 1 \times 10^{-4}$ M.

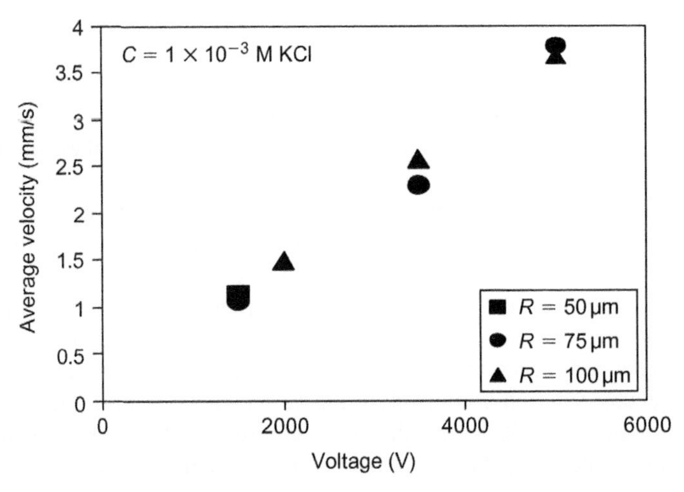

FIGURE 4.9

Average velocity versus applied voltage and capillary diameter, for a 10-cm-long polyamide-coated silica capillary tube.

average velocity is evident. In addition, it is clear that the average velocity is independent of the diameter of the capillary tubes.

However, in these experiments it is often difficult to determine the exact time required for one solution to completely replace another solution. This is because of the gradual changing of the current with time and small current fluctuations exist at both the beginning and the end of the replacing process. These make it very difficult to determine the exact beginning and ending time of the current change from the experimental results. Consequently, significant errors could be introduced in the average velocity determined in this way. It has been observed that, despite the curved beginning and ending sections, the measured current–time relationship is linear in most of the process, as long as the concentration difference is small. Therefore, a new method was developed to determine the average electroosmotic velocity by using the slope of the current–time relationship (Ren et al., 2002). When a high-concentration electrolyte solution gradually replaces another solution of the same electrolyte with a slightly lower concentration in a microchannel under an applied electric field, the current increases until the high-concentration solution completely replaces the low-concentration solution, at which time the current reaches a constant value. If the concentration difference is small, a linear relationship between the current and the time is observed. This may be understood in the following way. In such a process, because the concentration difference between these two solutions is small, e.g., 5%, the zeta potential, which determines the net charge density and hence the liquid flow, can be considered constant along the capillary. Consequently, the average velocity can be considered a constant during the replacing process. As one solution flows into the capillary and the other flows out of the capillary at essentially the same speed, and the two solutions have different electrical conductivities, the overall electrical resistance of the liquid in the capillary will change linearly, and hence the slope of current–time relationship is constant during this process.

The slope of the linear current–time relationship can be described as:

$$\text{slope} = \frac{\Delta I}{\Delta t} = \frac{\Delta(EA\lambda_b)}{\Delta t} \cdot \frac{L_{total}}{L_{total}} = \overline{u}_{ave} \frac{EA(\lambda_{b2} - \lambda_{b1})}{L_{total}} \tag{4.11}$$

where L_{total} is the total length of the capillary, $\overline{u}_{ave}$ is the average electroosmotic velocity, E is the applied electrical field strength, A is the cross-section area of the capillary, and $(\lambda_{b2} - \lambda_{b1})$ is the bulk conductivity difference between the high-concentration solution and the low-concentration solution. Using this equation, the average electroosmotic flow velocity can be evaluated from the slope of the measured current–time relationship. Figure 4.10 shows an example of the measured average electroosmotic velocity by the slope method.

In the straightforward current–time method, the total time required for one solution to completely replace another solution has to be found, which often involves some inaccuracy in choosing the beginning and the ending points of the replacment process. In this slope method, instead of identifying the total period of time, only a middle section of the current–time data is required to determine the slope of the current–time relationship. This method is easier and more accurate than finding the exact beginning and the ending points of the replacement process.

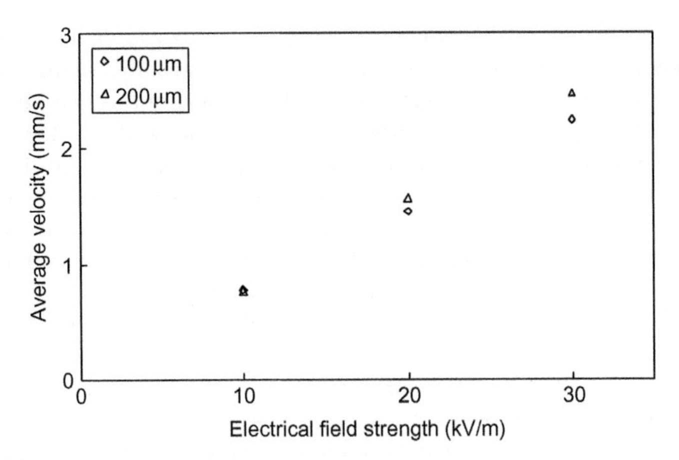

FIGURE 4.10

The average velocity versus the applied voltage and the capillary diameter, for 1×10^{-4} M KCl solution in a 10-cm-long capillary tube.

4.5 Electroosmotic flow in heterogeneous microchannels

Many microchannels don't have uniform surface properties. The surface heterogeneity may be caused by impurities of the materials, by adhesion of solutes (e.g., surfactants, proteins, and cells) of the solution, and by desired chemical surface modification. In the majority of lab-on-a-chip systems, electrokinetic means (i.e., electroosmotic flow, electrophoresis) are the preferred method of species transport. Although the surface heterogeneity has been long recognized as a problem leading to irregular flow patterns and nonuniform species transport, only recently have researchers begun to investigate the potential benefits that the presence of surface heterogeneity (nonuniform surface ζ potential or charge density) may have to offer. Early studies examining these effects were conducted by Anderson (1985), Ajdari (1995, 1996, 2001), and Ghosal (2002). In Ajdari's works it was predicted that the presence of surface heterogeneity could result in regions of bulk flow circulation, referred to as "tumbling" regions. This behavior was later observed in slit microchannels experimentally by Stroock et al. (2000), who found excellent agreement with their flow model. In another study, Herr et al. (2000) used a caged dye velocimetry technique to study electroosmotic flow in a capillary in the presence of heterogeneous surface properties and observed significant deviations from the classical plug-type velocity profile, an effect which was predicted by Keely et al. (1994). In another clever application, Johnson et al. (2001) used a UV excimer laser to introduce surface heterogeneity to the side wall of a polymeric microchannel and demonstrated how this could reduce sample band broadening around turns. Recently the use of surface heterogeneity to

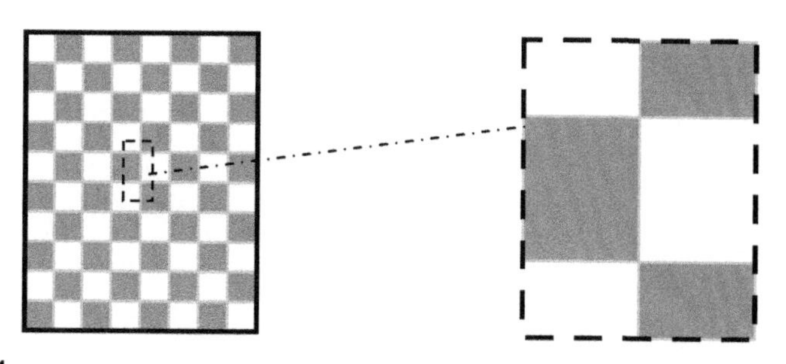

FIGURE 4.11

Periodic patchwise surface heterogeneity patterns. Dark regions represent heterogeneous patches; light regions are the homogeneous surface. The percent heterogeneous coverage is 50%.

increase the mixing efficiency of a T-shaped micromixer was proposed (Erickson and Li, 2002b). Here we wish to review a general theoretical analysis of three-dimensional electroosmotic flow structures in a slit microchannel with periodic, patchwise heterogeneous surface pattern (Erickson and Li, 2003a).

Let's consider the electroosmotically driven flow through a slit microchannel (i.e., a channel formed between two parallel plates) exhibiting a periodically repeating heterogeneous surface pattern, shown in Figure 4.11. Since the pattern is repeating, the computational domain is reduced to that over a single periodic cell, demonstrated by the dashed lines in Figure 4.11. To further minimize the size of the solution domain, it has been assumed that the heterogeneous surface pattern is symmetric about the channel mid-plane, resulting in the computational domain shown in Figure 4.12. As a result of these two simplifications, the inflow and outflow boundaries, surfaces 2 and 4 (as labeled in Figure 4.12 and referred to by the symbol Γ), represent periodic boundaries on the computational domain, while surface 3 at the channel mid-plane represents a symmetry boundary. From Figure 4.11 it can be seen that in all cases the surface pattern is symmetric about Γ_5 and Γ_6, and thus these surfaces also represent symmetry boundaries. For a general discussion of periodic boundary conditions, however, the reader is referred to a paper by Patankar et al. (1977).

Modeling the flow through this periodic unit requires a description of the ionic species distribution, the double layer potential, the flow field, and the applied potential. The divergence of the ion species flux (for simplicity here we consider a monovalent, symmetric electrolyte as our model species), often referred to as the Nernst–Planck conservation equations, is used to describe the positive and negative ion densities (given below in nondimensional form):

$$\tilde{\nabla}\cdot(-\tilde{\nabla}N^{+} - N^{+}\tilde{\nabla}\Phi + Pe^{+}N^{+}\mathbf{V}) = 0 \tag{4.12a}$$

$$\tilde{\nabla}\cdot(-\tilde{\nabla}N^{-} + N^{-}\tilde{\nabla}\Phi + Pe^{-}N^{-}\mathbf{V}) = 0 \tag{4.12b}$$

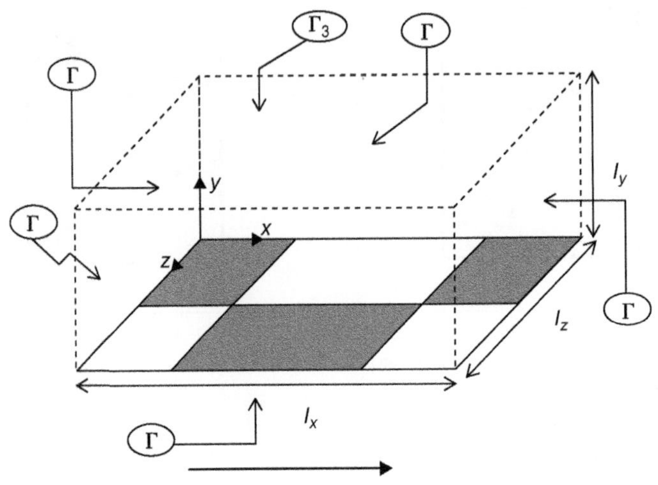

FIGURE 4.12

Domain for the periodical computational cell showing location of computational boundaries.

where $N^+$and N^- are the nondimensional positive and negative species concentrations ($N^+ = n^+/n_\infty$, $N^- = n^-/n_\infty$, where n_∞ is the bulk ionic concentration), Φ is the nondimensional electric field strength ($\Phi = e\phi/k_bT$, where e is the elemental charge, k_b is the Boltzmann constant, and T is the temperature in Kelvin), V is the nondimensional velocity ($V = v/v_o$, where v_o is a reference velocity), and the ~ sign signifies that the gradient operator has been nondimensionalized with respect to the channel half height (l_y in Figure 4.12). The Péclet numbers for the two species are given by $Pe^+ = v_o l_y/D^+$ and $\mathrm{Pe}^- = v_o l_y/D^-$, where $D^+$and D^- are the diffusion coefficients for the positively charged and negatively charged species, respectively.

Along the heterogeneous surface (Γ_1 in Figure 4.12) and symmetry boundaries of the computational domain (Γ_3 , Γ_5, and Γ_6), zero-flux boundary conditions are applied to both Eqs. (4.12a) and (4.12b) (see Erickson and Li, 2002a,b, for explicit equations). As mentioned earlier, along faces Γ_2 and Γ_4, periodic conditions are applied which take the form shown below,

$$N_2^+ = N_4^+, \text{ on } \Gamma_2, \Gamma_4, \; N_2^- = N_4^-, \text{ on } \Gamma_2, \Gamma_4$$

The velocity field is described by the Navier–Stokes equations for momentum, modified to account for the electrokinetic body force, and the continuity equation as shown below,

$$Re(\mathbf{V}\cdot\tilde{\nabla}\mathbf{V}) = -\tilde{\nabla}P + \tilde{\nabla}^2\mathbf{V} - F(N^+ - N^-)\tilde{\nabla}\Phi \qquad (4.13a)$$

$$\tilde{\nabla} \cdot \mathbf{V} = 0 \tag{4.13b}$$

where F is a nondimensional constant ($F = n_\infty l_y k_b T/\mu\, v_o$) which accounts for the electrokinetic body force, Re is the Reynolds number ($Re = \rho\, v_o l_y/\mu$, where ρ is the fluid density and μ is the viscosity), and P is the nondimensional pressure. Along the heterogeneous surface, Γ_1, a no-slip boundary condition is applied. At the upper symmetry surface, Γ_3, we enforce a zero-penetration condition for the y-component of velocity and zero-gradient conditions for the x- and z-terms, respectively. Similarly, a zero-penetration condition for the z-component of velocity is applied along Γ_5 and Γ_6, while zero-gradient conditions are enforced for the x- and y-components. As with the Nernst − Planck equations, periodic boundary conditions are applied along Γ_2 and Γ_4.

For the potential field, we choose to separate, without loss of generality, the total nondimensional electric field strength, Φ, into two components:

$$\Phi(X, Y, Z) = \Psi(X, Y, Z) + \mathrm{E}(X) \tag{4.14}$$

where the first component, $\Psi(X,Y,Z)$, describes the EDL field and the second component, $\mathrm{E}(X)$, represents the applied electric field. In this case the gradient of $\mathrm{E}(X)$ is of similar order of magnitude to the gradient of $\Psi(X,Y,Z)$ and thus cannot be decoupled to simplify the solution to the Poisson equation. As a result, for this case the Poisson equation has the form shown below,

$$\tilde{\nabla}^2 \Psi + d^2\mathrm{E}/dX^2 + K^2(N^+ - N^-) = 0 \tag{4.15a}$$

where K is the nondimensional double layer thickness ($K^2 = \kappa^2 l_y^{\,2}/2$, where $\kappa = (2n_\infty e^2/\varepsilon_w k_b T)^{1/2}$ is the Debye–Huckel parameter). Two boundary conditions commonly applied along the solid surface for the above equation are either an enforced potential gradient proportional to the surface charge density, or a fixed potential condition equivalent to the surface ζ potential. In this case we choose the former, resulting in the following boundary condition applied along the heterogeneous surface

$$\partial\Psi/\partial Y = -\Theta(X, Z) \text{ on } \Gamma_1 \tag{4.15b}$$

where Θ is the nondimensional surface charge density ($\Theta = \sigma l_y e/\varepsilon_w k_b T$, where σ is the surface charge density). Along the upper and side boundaries of the computational domain, Γ_3, Γ_5, and Γ_6, symmetry conditions are applied while periodic conditions are again applied at the inlet and outlet, Γ_2 and Γ_4.

In order to determine the applied electric field, $\mathrm{E}(X)$, we enforce a constant current condition at each cross-section along the x-axis, as below:

$$J_{\text{const}} = \int_0^{Z_{\max}} \int_0^{Y_{\max}} [J^+(X, Y, Z) - J^-(X, Y, Z)] \cdot n_x \, \mathrm{d}Y \, \mathrm{d}Z \tag{4.16a}$$

where n_x is a normal vector in the x-direction and J^+ and J^- are the nondimensional positive and negative current densities. Assuming monovalent ions for both positive and negative species (as was discussed earlier) and using the ionic species flux terms from Eqs. (4.12a) and (4.12b), condition Eq. (4.16a) reduces to the following:

$$J_{\text{const}} = \int_0^{Z_{\max}} \int_0^{Y_{\max}} \left(-\left[\frac{\partial N^+}{\partial X} - \frac{Pe^+}{Pe^-} \frac{\partial N^-}{\partial X} \right] - \left[N^+ + \frac{Pe^+}{Pe^-} N^- \right] \frac{\partial(\Psi + \text{E})}{\partial X} \right) \text{d}Y \, \text{d}Z + \int_0^{Z_{\max}} \int_0^{Y_{\max}} ([Pe^+(N^+ - N^-)V_x]) \text{d}Y \, \text{d}Z \tag{4.16b}$$

which is a balance between the conduction (term 2) and convection (term 3) currents with an additional term to account for any induced current due to concentration gradients. Solving Eq. (4.16b) for the applied potential gradient yields

$$\frac{\text{dE}}{\text{d}X} = \frac{\int_0^{Z_{\max}} \int_0^{Y_{\max}} \left(-\left[\frac{\partial N^+}{\partial X} - \frac{Pe^+}{Pe^-} \frac{\partial N^-}{\partial X} \right] - \left[N^+ + \frac{Pe^+}{Pe^-} N^- \right] \frac{\partial \Psi}{\partial X} \right) \text{d}Y \, \text{d}Z}{\int_0^{Z_{\max}} \int_0^{Y_{\max}} \left(N^+ + \frac{Pe^+}{Pe^-} N^- \right) \text{d}Y \, \text{d}Z} + \frac{\int_0^{Z_{\max}} \int_0^{Y_{\max}} \left([Pe^+(N^+ - N^-)V_x] \right) \text{d}Y \, \text{d}Z - J_{\text{const}}}{\int_0^{Z_{\max}} \int_0^{Y_{\max}} \left(N^+ + \frac{Pe^+}{Pe^-} N^- \right) \text{d}Y \, \text{d}Z} \tag{4.16c}$$

As an example, let's consider a $l_x = 50$ μm, $l_y = l_z = 25$ μm computational domain, as shown in Figure 4.12, containing a 10^{-5} M KCl solution and with an applied driving voltage of 500 V cm^{-1} and a $\sigma_{\text{homo}} = -4 \times 10^{-4}$ C/m^2. The numerical simulations of the above-described model revealed three distinct flow patterns, depending on the degree of surface heterogeneity, as shown in Figure 4.13. The gray scales in these figures represent the magnitude of the velocity perpendicular to the direction of the applied electric field ($v^2 = v_y^2 + v_z^2$) scaled by the maximum velocity in a homogeneous channel, which in this example is 1.7 mm/s.

At low degrees of surface heterogeneity ($\sigma_{\text{hetero}} \leq -2 \times 10^{-4}$ C/m^2) the streamline pattern shown in Figure 4.13A was obtained. As can be seen, a net counterclockwise flow perpendicular to the applied electric field is present at the first transition plane (i.e., at the initial discontinuity in the heterogeneous surface pattern) and a clockwise flow is present at the second transition plane. This flow circulation is a pressure-induced effect. It results from the transition from the higher local fluid velocity (particularly in the double layer) region over the homogeneous surface on the right-hand side at the entrance to the left-hand side after

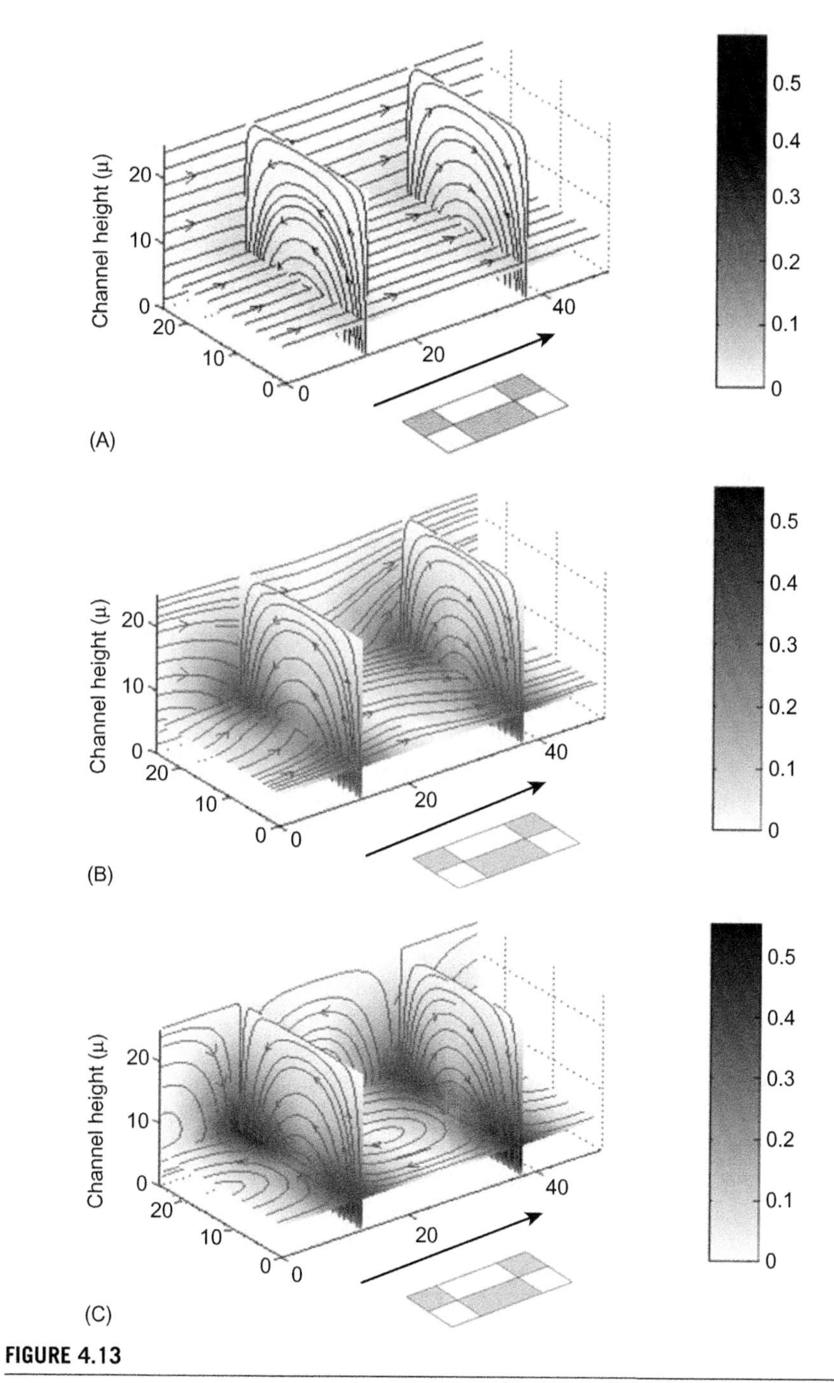

FIGURE 4.13

Electroosmotic flow streamlines over a patchwise heterogeneous surface pattern with $\sigma_{homo} = -4 \times 10^{-4}$ C/m^2 and (A) $\sigma_{hetero} = -2 \times 10^{-4}$ C/m^2 (B) $\sigma_{hetero} = +2 \times 10^{-4}$ C/m^2 (C) $\sigma_{hetero} = +4 \times 10^{-4}$ C/m^2. The gray scale represents magnitude of velocity perpendicular to the applied potential field, scaled by the maximum velocity in a homogeneous channel. Arrow represents direction of applied electric field.

the first transition plane (and vice versa at the second transition plane). To satisfy continuity, then, there must be a net flow from right to left at this point, which in this case takes the form of the circulation discussed above. The relatively straight streamlines parallel with the applied electric field indicate that at this level the heterogeneity is too weak to significantly disrupt the main flow.

While a similar circulation pattern at the transition planes was observed as the degree of surface heterogeneity was increased into the intermediate range (-2×10^{-4} C/m$^2 \leq \sigma_{hetero} \leq +2 \times 10^{-4}$ C/m^2), it is apparent from the darker contours shown in Figure 4.13B that the strength of the flow perpendicular to the applied electric field is significantly stronger, reaching nearly 50% of the velocity in the homogeneous channel. Unlike in the previous case it is now apparent that the streamlines parallel to the applied electric field are significantly distorted due to the much slower or even oppositely directed velocity over the heterogeneous patch.

At even higher degrees of heterogeneity ($+2 \times 10^{-4}$ C/m$^2 \leq \sigma_{hetero} \leq +4 \times 10^{-4}$ C/m^2), a third flow structure is observed in which a dominant circulatory flow pattern exists along all three coordinate axes, as shown in Figure 4.13C. This results in a negligible, or even nonexistent, bulk flow in the direction of the applied electric field (which is to be expected since the average surface charge density for these cases is very near zero). As indicated by the gray scales, the velocity perpendicular to the flow axis has again increased in magnitude, reaching a maximum at the edge of the double layer near the symmetry planes at the location where a step change in the surface charge density has been imposed.

It has also been shown (Erickson and Li, 2003a) that the induced three-dimensional flow patterns are limited to a layer near the surface with a thickness equivalent to the length scale of the heterogeneous patch. Additionally, the effect becomes smaller in magnitude and more localized as the average size of the heterogeneous region decreases. Moreover, it was demonstrated that while convective effects are small, the electrophoretic influence of the applied electric field could distort the net charge density field near the surface, resulting in a significant deviation from the traditional Poisson–Boltzmann double layer distribution. That is the reason why the Nernst–Planck conservation equations, instead of the Boltzmann distribution equation, were used in this study.

4.6 AC electroosmotic flow

Electroosmotic flow induced by unsteady applied electric fields or electroosmotic flow under an alternating electrical (AC) field has unique features and applications. Oddy et al. (2001) proposed and experimentally demonstrated a series of schemes for enhanced species mixing in microfluidic devices using AC electric fields. In addition, they presented an analytical flow field model, based on a surface slip condition approach, for an axially applied (i.e., along the flow axis)

AC electric field in an infinitely wide microchannel. Comprehensive models for such a slit channel have also been presented by Dutta and Beskok (2001), who developed an analytical model for an applied sinusoidal electric field, and by Söderman and Jönsson (1996), who examined the transient flow field caused by a series of different pulse designs. As an alternative to traditional DC electroosmosis, a series of novel techniques have been developed to generate bulk flow using AC fields. For example, Green et al. (2000) experimentally observed peak flow velocities on the order of hundreds of micrometers per second near a set of parallel electrodes subject to two AC fields, 180° out of phase with each other. The effect was subsequently modeled using a linear double layer analysis by Gonzàlez et al. (2000). Using a similar principle, both Brown et al. (2002) and Studer et al. (2002) presented microfluidic devices that incorporated arrays of nonuniformly sized embedded electrodes which, when subjected to an AC field, were able to generate a bulk fluid motion. Also using embedded electrodes, Selvaganapathy et al. (2002) demonstrated what they termed fr-EOF or bubble-free electroosmotic flow, in which a creative periodic waveform was used to yield a net bulk flow while electrolytic bubble formation was theoretically eliminated.

Recently, Erickson and Li (2003b) presented a combined theoretical and numerical approach to investigate the time-periodic electroosmotic flow in a rectangular microchannel. This work considers the time-periodic electroosmotic flow in a straight, rectangular microchannel. Assuming a fully developed flow field and considering the geometric symmetry, the analytical domain can be reduced to the upper left-hand quadrant of the channel cross-section, as shown in Figure 4.14. For pure electroosmotic flows (i.e., absent of any pressure

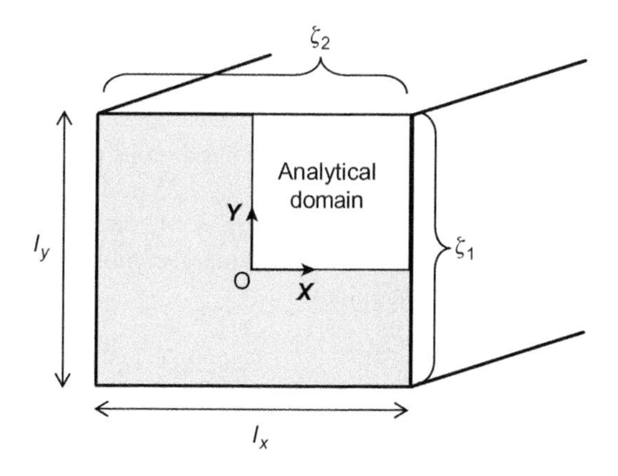

FIGURE 4.14

Illustration of the analytical domain in AC electroosmotic flow in a rectangular microchannel.

gradients) of incompressible liquids, the Navier–Stokes equations may take the following form:

$$\rho_f \frac{\partial v}{\partial t} = \mu \nabla^2 v + \rho_e E(\omega t) \tag{4.17}$$

where v is the flow velocity, t is time, ρ_f is the fluid density, ρ_e is the net charge density, μ is the fluid viscosity, and $E(\omega t)$ is a general time-periodic function with a frequency $\omega = 2\pi f$ and describes the applied electric field strength. To solve this flow equation, we must know the local net charge density ρ_e. This requires solving the EDL field. Introducing the nondimensional double layer potential, $\Psi = ze\psi/k_bT$, and nondimensional double layer thickness $K = D_h\kappa$ (where D_h is the hydraulic diameter of the channel, $D_h = 4l_xl_y/2(l_{x+}l_y)$, and κ is the Deybe–Hückel parameter), we obtain the nonlinear Poisson–Boltzmann distribution equation:

$$\tilde{\nabla}^2\Psi - K^2 \sinh(\Psi) = 0 \tag{4.18}$$

where the $\sim$ signifies that the spatial variables in the gradient operator have been nondimensionalized with respect to D_h (i.e., $X = x/D_h$, $Y = y/D_h$). Equation (4.18) is subject to the symmetry conditions along the channel center axes, i.e.,

$$\partial\Psi/\partial X = 0 \quad \text{at } X = 0 \quad \partial\Psi/\partial Y = 0 \quad \text{at } Y = 0$$

and the appropriate conditions at the channel walls,

$$\Psi = Z_1 \quad \text{at } X = L_x/2 \quad \Psi = Z_2 \quad \text{at } Y = L_y/2$$

where $Z_{1,2} = ze\zeta_{1,2}/k_bT$, $L_x = l_x/D_h$, and $L_y = l_y/D_h$.

Generally, for multidimensional space and complex geometry Eq. (4.18) must be solved numerically. However, under the condition that the double layer potential, $\Psi = ze\psi/k_bT$, is small, Eq. (4.18) can be linearized (the so-called the Deybe–Hückel approximation), yielding

$$\tilde{\nabla}^2\Psi - K^2\Psi = 0 \tag{4.19}$$

The Deybe–Hückel approximation is considered valid only when $|ze\zeta| < k_bT$, which in principle is limited to cases where $|\zeta| \leq 25$ mV at room temperature. For the rectangular geometry shown in Figure 4.14, an analytical solution to Eq. (4.19), subject to the aforementioned boundary conditions, can be obtained using a separation of variables technique:

$$\Psi(X,Y) = \sum_{n=1}^{\infty} \frac{4(-1)^{n+1}}{\pi(2n-1)} \left[Z_1 \cos(\lambda_n X) \frac{\cosh([\lambda_n{}^2 + K^2]^{1/2} Y)}{\cosh([\lambda_n{}^2 + K^2]^{1/2} L_Y)} \right] + \sum_{n=1}^{\infty} \frac{4(-1)^{n+1}}{\pi(2n-1)} \left[Z_2 \cos(\mu_n Y) \frac{\cosh([\mu_n^2 + K^2]^{1/2} X)}{\cosh([\mu_n^2 + K^2]^{1/2} L_X)} \right] \tag{4.20}$$

where λ_n and μ_n are the eigenvalues given by $\lambda_n = \pi(2n-1)/2L_X$ and $\mu_n = \pi(2n-1)/2L_Y$, respectively. Introducing the nondimensional time, $\theta = \mu\ t/\rho_f D_h^2$, and the nondimensional frequency, $\Omega = \rho_f D_h^2 \omega/\mu$, scaling the flow velocity by the electroosmotic slip velocity, $V = vze\mu/\varepsilon E_z k_b T$, where E_z is a constant equivalent to the strength of the applied electric field, and combining Eq. (4.17) with the Poisson equation:

$$\varepsilon \nabla^2 \psi = \rho_e$$

and Eq. (4.18) yields the nondimensional flow equation,

$$\frac{\partial V}{\partial \theta} = \tilde{\nabla}^2 V - K^2 \sinh(\Psi) F(\Omega\theta) \tag{4.21}$$

where F is a general periodic function of unit magnitude such that $E(\Omega\theta) = E_z F(\Omega\theta)$. Along the channel axes, Eq. (4.21) is subject to a symmetry boundary conditions:

$$\partial V/\partial X = 0 \quad \text{at } X = 0 \quad \partial V/\partial Y = 0 \quad \text{at } Y = 0$$

while no-slip conditions are applied along the solid channel walls:

$$V = 0 \quad \text{at } X = L_x/2 \quad \text{and} \quad Y = L_y/2.$$

In order to obtain an analytical solution to Eq. (4.21), the Deybe–Hückel approximation is implemented and $F(\Omega\theta)$ is chosen to be a sinusoid function, resulting in the following form of the equation:

$$\frac{\partial V}{\partial \theta} = \tilde{\nabla}^2 V - K^2 \Psi(X,Y)\sin(\Omega\theta) \tag{4.22}$$

Using a Green's function approach, an analytical solution to Eq. (4.22), subject to the homogeneous boundary conditions discussed above, can be obtained.

$$V(X,Y,\theta) = \frac{-16K^2}{L_X L_Y} \sum_{l=1}^{\infty}\sum_{m=1}^{\infty}\sum_{n=1}^{\infty} \frac{(-1)^{n+1}}{\pi(2n-1)} \cos(\lambda_l X)\cos(\mu_m Y)$$
$$\times \left[\frac{(\lambda_l^2 + \mu_m^2)\sin(\Omega\theta) - \Omega\cos(\Omega\theta) + \Omega\exp(-[\lambda_l^2 + \mu_m^2]\theta)}{(\lambda_l^2 + \mu_m^2)^2 + \Omega^2} \right] [f_{1mn} + f_{2ln}] \tag{4.23}$$

where f_{1mn} and f_{2ln} are given by

$$f_{1mn} = \left\{ \begin{matrix} 0 & n \neq m \\ \dfrac{Z_1 \mu_m (-1)^{m+1}}{\lambda_n^2 + \mu_m^2 + K^2} \left(\dfrac{L_X}{2} \right) & n = m \end{matrix} \right\}$$

$$f_{2ln} = \left\{ \begin{array}{cc} 0 & n \neq l \\ \dfrac{Z_2 \lambda_l (-1)^{l+1}}{\mu_n^2 + \lambda_l^2 + K^2} \left(\dfrac{L_Y}{2} \right) & n = l \end{array} \right\}$$

Equation (4.23) represents the full solution to the transient flow problem. The equation can be somewhat simplified in cases where the quasi-steady state time-periodic solution (i.e., after the influence of the initial conditions has dissipated) is of interest. As these initial effects are represented by the exponential term in Eq. (4.23), the quasi-steady state time-periodic solution has the form shown below:

$$V(X, Y, \theta) = \frac{-16K^2}{L_X L_Y} \sum_{l=1}^{\infty} \sum_{m=1}^{\infty} \sum_{n=1}^{\infty} \frac{(-1)^{n+1}}{\pi(2n-1)} \cos(\lambda_l X) \cos(\mu_m Y) \times \left[\frac{(\lambda_l^2 + \mu_m^2)\sin(\Omega\theta) - \Omega\cos(\Omega\theta)}{(\lambda_l^2 + \mu_m^2)^2 + \Omega^2} \right] [f_{1mn} + f_{2ln}] \tag{4.24}$$

In the above analytical description of the uniaxial electroosmotic flow in a rectangular microchannel, the governing parameter is Ω, which represents the ratio of the diffusion time scale ($t_{diff} = \rho_f D_h^2/\mu$) to the period of the applied electric field ($t_E = 1/\omega$). Figure 4.15 compares the time-periodic velocity profiles (as computed from Eq. (4.24)) in the upper left-hand quadrant of a square channel for two cases: (a) $\Omega = 31$ and (b) $\Omega = 625$. These two Ω values correspond to frequencies of 500 Hz and 10 kHz in a 100-μm square channel or, equivalently, a 100-μm and a 450-μm square channel at 500 Hz. To illustrate the essential features of the velocity profile, a relatively large double layer thickness has been used, $\kappa = 3 \times 10^6\ \mathrm{m}^{-1}$ (corresponding to a bulk ionic concentration $n_\infty = 10^{-6}$ M), and a uniform surface potential of $\zeta = -25$ mV was selected (within the bounds imposed by the Deybe–Hückel linearization).

From Figure 4.15, it is apparent that the application of the electrical body force results in a rapid acceleration of the fluid within the double layer. In the case where the diffusion time scale is much greater than the oscillation period (high Ω, Figure 4.15B), there is insufficient time for fluid momentum to diffuse far into the bulk flow, and thus while the fluid within the double layer oscillates rapidly the bulk fluid remains almost stationary. At $\Omega = 31$ there is more time for momentum diffusion from the double layer; however, the bulk fluid still lags behind the flow in the double layer (this out-of-phase behavior will be discussed shortly). Extrapolating from these results, when $\Omega < 1$, such that momentum diffusion is faster than the period of oscillation, the plug-type velocity profile characteristic of steady state electroosmotic flow would be expected at all times.

Another interesting feature of the velocity profiles shown in Figure 4.15 is the local velocity maximum observed near the corner (most clearly visible in the

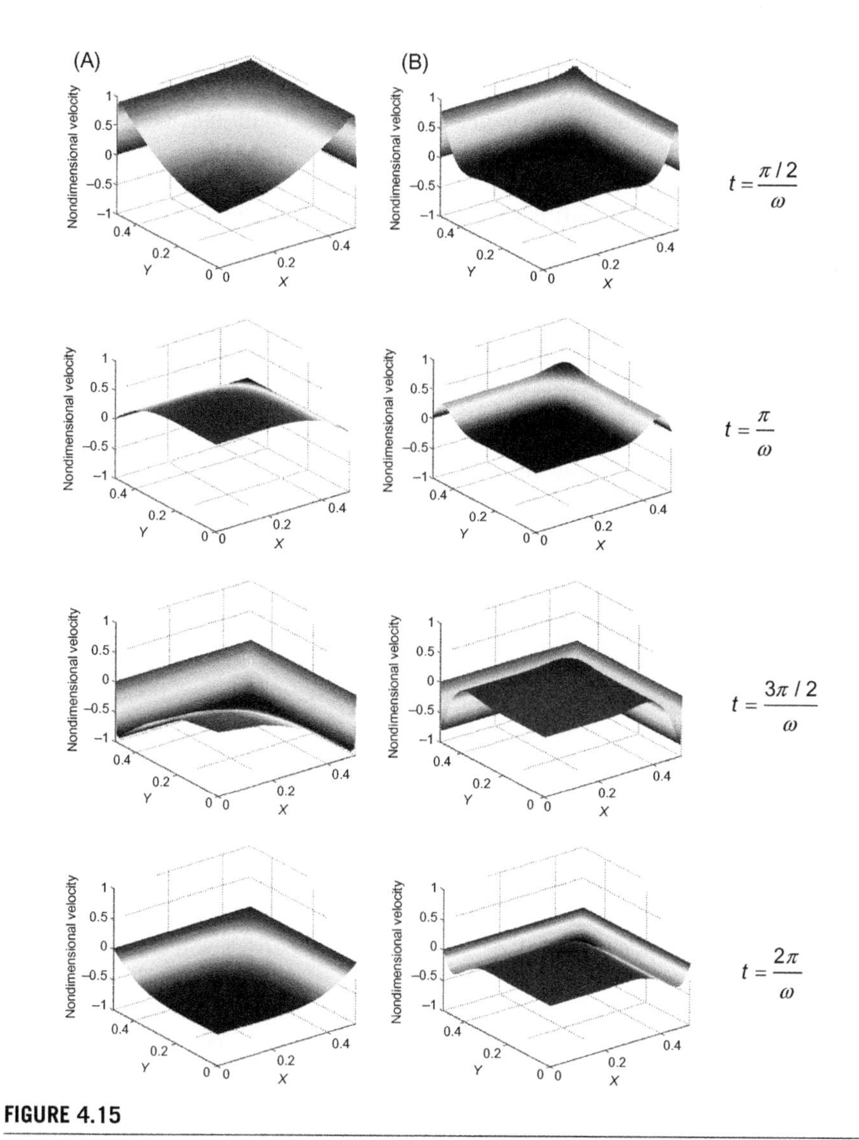

FIGURE 4.15

Steady state time-periodic electroosmotic velocity profiles in a square microchannel at (A) $\Omega = 30$ and (B) $\Omega = 625$ (equivalent to an applied electric field frequency of (A) 500 Hz and (B) 10 kHz in a 100 μm by 100 μm square channel).

$\Omega = 625$ case at $t = \pi/2\omega$ and $t = \pi/\omega$). The intersection of the two walls results in a region of double layer overlap and thus an increased net charge density over that region in the double layer. peak in the net charge density corresponds to a larger electrical body force and hence a local maximum electroosmotic flow velocity; it also increases the ratio of the electrical body force to the viscous retardation, allowing it to respond more rapidly to changes in the applied electric field.

The finite time required for momentum diffusion will inevitably result in some degree of phase shift between the applied electric field and the flow response in the channel. From Figure 4.15, however, it is apparent that within the limit of $\Omega > 1$, this phase shift is significantly different in the double layer region than in the bulk flow. it is apparent that the response of the fluid within the double layer to the AC field is essentially immediate; however, the bulk liquid lags behind the applied field by a phase shift depending on the Ω value. Additionally, while the velocity in the double layer reaches its steady state oscillation almost immediately, the bulk flow requires a period before the transient effects are dissipated. In Eq. (4.24), the out-of-phase cosine term ($\Omega\cos(\Omega\theta)$) is proportionally scaled by Ω; thus, as expected, when Ω is increased, the phase shift for both the double layer and bulk flow velocities is increased, as is the number of cycles required to reach the steady state. It is interesting to note the net positive velocity at the channel midpoint within the transient period before decaying into the steady state behavior. This is a result of the initial positive impulse given to the system when the electric field is first applied and is reflected by the exponential term in Eq. (4.23). The transient oscillations were observed to decay at an exponential rate, as expected from this transient term in Eq. (4.23). Similar to the out-of-phase cosine term, this exponential term is also proportionally scaled by the nondimensional frequency, suggesting that the effect of the initial impulse becomes more significant with increasing Ω.

The results presented above are limited to sinusoidal waveforms. To examine how the flow field will respond to different forms of periodic excitation, such as square (step) or triangular waveforms, numerical solutions to Eqs. (4.17) and (4.18) are required. Figure 4.15 shows that the particular waveform also has a significant effect on the transient response of the bulk fluid (here the channel midpoint is chosen as a representative point). As can be seen in Figure 4.16A for the $\Omega = 31$ case, a square wave excitation tends to produce higher velocities, whereas the triangular wave exhibits slightly smaller bulk velocities when compared with the sinusoidal waveform. As Ω is increased, the initial positive impulsive velocity is observed for both additional waveforms, as shown in Figure 4.16B. As expected, the fluid excited by the square waveform exhibits higher instantaneous velocities, which lead to an increase in the number of cycles required to reach the time-periodic quasi-steady-state oscillation.

4.7 Electrokinetic mixing

Let's consider a simple T-shaped microfluidic mixing system (Erickson and Li, 2002b). Without loss of generality, we will consider that two electrolyte solutions of the same flow rate enter a T-junction separately from two horizontal microchannels,

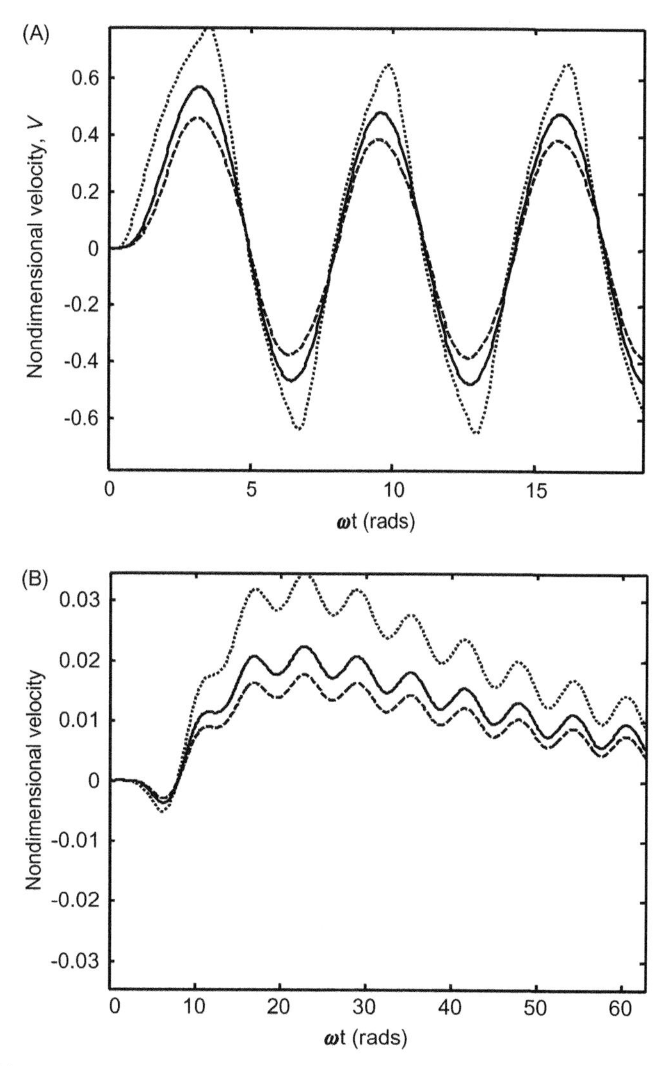

FIGURE 4.16

Transient stage velocity at channel midpoint for impulsively started flows using different waveforms at (A) $\Omega = 30$ and (B) $\Omega = 625$. Solid line represents sine wave, long dashed line represents triangular wave, and dotted line represents square wave.

and then start mixing while flowing along the vertical microchannel, as illustrated in Figure 4.17. The flow is generated by the applied electrical field via electrodes at the upstream and the downstream positions. This simple arrangement has been used for numerous applications, including the dilution of a sample in a buffer (Harrison et al., 1993), the development of complex species gradients (Dertinger et al., 2001;

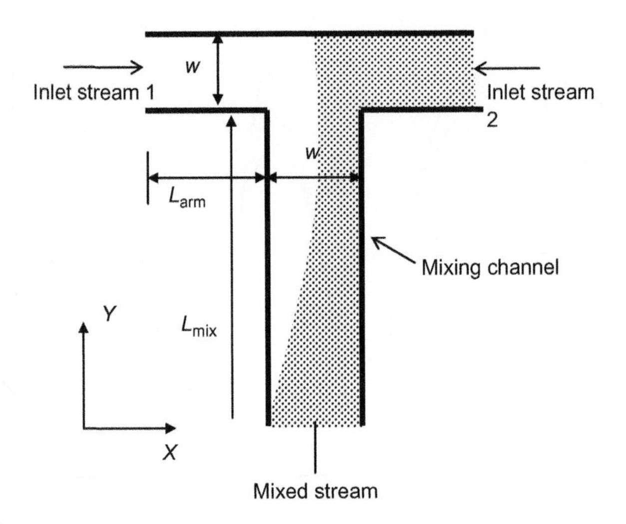

FIGURE 4.17

T-shaped micromixer formed by the intersection of two microchannels, showing a schematic of the mixing or dilution process.

Jeon et al., 2000), and measurement of the diffusion coefficient (Kamholz et al., 2001). Generally, most microfluidic mixing systems are limited to the low Reynolds number regime and thus species mixing is strongly diffusion dominated, as opposed to convection or turbulence dominated at higher Reynolds numbers. Consequently, mixing tends to be slow, occurring over relatively long distances and time frames. As an example, the concentration gradient generator presented by Dertinger et al. (2001) required a mixing channel length on the order of 9.25 mm for a 45 μm × 45 μm cross-sectional channel, or approximately 200 times the channel width, to achieve nearly complete mixing. Enhanced microfluidic mixing over a short flow distance is highly desirable for lab-on-a-chip applications. One possibility of achieving this is to utilize the local circulation flow caused by surface heterogeneous patches.

To model such an electroosmotic flow and mixing process, we need the following equations. The flow field is described by the Navier–Stokes equations and the continuity equation (given below in nondimensional form):

$$Re\left[\frac{\partial V}{\partial \tau} + (V\cdot\tilde{\nabla})V\right] = -\tilde{\nabla}P + \tilde{\nabla}^2 V \tag{4.25}$$

$$\tilde{\nabla}\cdot V = 0 \tag{4.26}$$

where V is the nondimensional velocity ($V = v/v_{eo}$, where v_{eo} is calculated using Eq. (4.27) given below), P is the nondimensional pressure, τ is the nondimensional time, and Re is the Reynolds number given by $Re = \rho v_{eo} L/\eta$, where L is a length scale taken as the channel width (w from Figure 4.17) in this case. The ~ symbol over the ∇ operator indicates the gradient with respect to the nondimensional coordinates ($X = x/w$, $Y = y/w$, and $Z = z/w$).

It should be noted that in order to simplify the numerical solution to the problem, we have treated the electroosmotic flow in the thin EDL as a slip flow velocity boundary condition, given by:

$$v_{eo} = \mu_{eo} \nabla \varphi = \left(\frac{\varepsilon_w \zeta}{\mu} \right) \nabla \varphi \tag{4.27}$$

where $\mu_{eo} = (\varepsilon_w \zeta / \mu)$ is the electroosmotic mobility, ε_w is the electrical permittivity of the solution, μ is the viscosity, ζ is the zeta potential of the channel wall, and ϕ is the applied electric field strength.

In general, the high voltage requirements limit most practical electroosmotically driven flows in microchannels to small Reynolds numbers; therefore, to simplify Eq. (4.25) we ignore transient and convective terms and limit ourselves to cases where $Re < 0.1$.

We consider the mixing of equal portions of two buffer solutions, one of which contains a species of interest at a concentration, c_o. Species transport by electrokinetic means is accomplished by three mechanisms—convection, diffusion and electrophoresis—and can be described by

$$Pe \left[\frac{\partial C}{\partial \tau} + \tilde{\nabla} \cdot (C(V + V_{ep})) \right] = \tilde{\nabla}^2 C \tag{4.28}$$

where C is the nondimensional species concentration ($C = c/c_o$, where c_o is the original concentration of the interested species in the buffer solution.), Pe is the Péclet number ($Pe = v_{eo} w/D$, where D is the diffusion coefficient), and V_{ep} is the nondimensional electrophoretic velocity equal to v_{ep}/v_{eo}, where v_{ep} is given by:

$$v_{ep} = \mu_{ep} \nabla \phi \tag{4.29}$$

and $\mu_{ep} = (\varepsilon_w \zeta_p / \eta)$ is the electrophoretic mobility (ε_w is the electrical permittivity of the solution, μ is the viscosity, ζ_p is the zeta potential of the charged molecules or particles to be mixed) (Hunter, 1981). As we are interested in the steady state solution, the transient term in Eq. (4.28) can be ignored.

The above-described model was solved numerically to investigate the formation of electroosmotically induced flow circulation regions near surface heterogeneities in a T-shaped micromixer and to determine the influence of these regions on the mixing effectiveness. In Figure 4.18 we compare the mid-plane flow field near the T-intersection of a homogeneous mixing channel with that of a mixing channel having a series of six asymmetrically distributed heterogeneous patches on the left and right channel walls. For clarity, the heterogeneous regions are marked as the crosshatched regions in this figure. The homogeneous channel surface

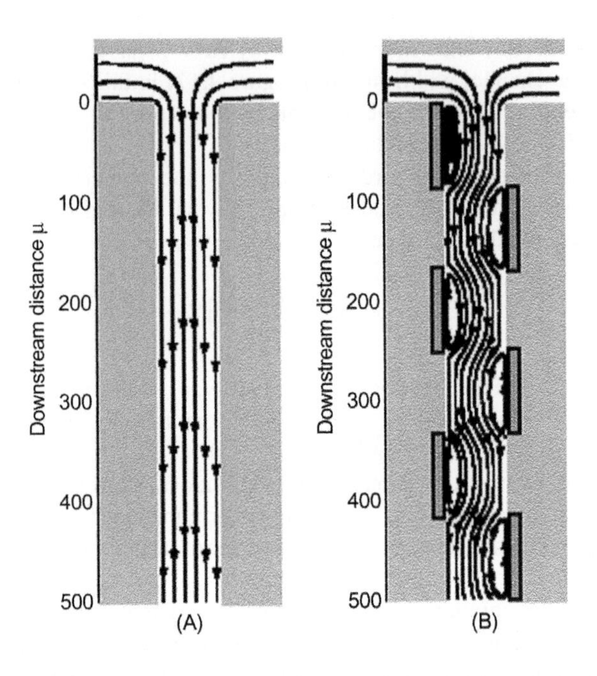

FIGURE 4.18

Electroosmotic streamlines at the mid-plane of a 50 μm T-shaped micromixer for (A) the homogeneous case with $\zeta = -42$ mV, (B) the heterogeneous case with six offset patches on the left and right channel walls. All heterogeneous patches are represented by the crosshatched regions and have a $\zeta = +42$ mV. The applied voltage is $\phi_{app} = 500$ V/cm.

has a ζ potential of -42 mV. A ζ potential of $\zeta = +42$ mV was assumed for the heterogeneous patches. Apparently, the channel with heterogeneous patches generates local flow circulations near the patches. These flow circulation zones are expected to enhance the mixing of the two streams.

Figure 4.19 compares the three-dimensional concentration fields of the homogeneous and heterogeneous mixing channel shown in Figure 4.18. In these figures a neutral mixing species (i.e., $\mu_{ep} = 0$, thereby ignoring any electrophoretic transport) with a diffusion coefficient $D = 3 \times 10^{-10}$ m^2/s is considered. While mixing in the homogeneous case is purely diffusive in nature, the presence of the asymmetric circulation regions, Figure 4.19B, enables enhanced mixing by convection.

Recently, a passive electrokinetic micromixer based on the use of surface charge heterogeneity was developed (Biddiss et al., 2004). The micromixer is a T-shaped microchannel structure (200 μm in width and approximately 8 μm in depth) made from polydimethylsiloxane (PDMS) and is sealed with a glass slide. Microchannels were fabricated using a rapid prototyping/soft lithography technique. The glass surface was covered by a PDMS mask with the desired

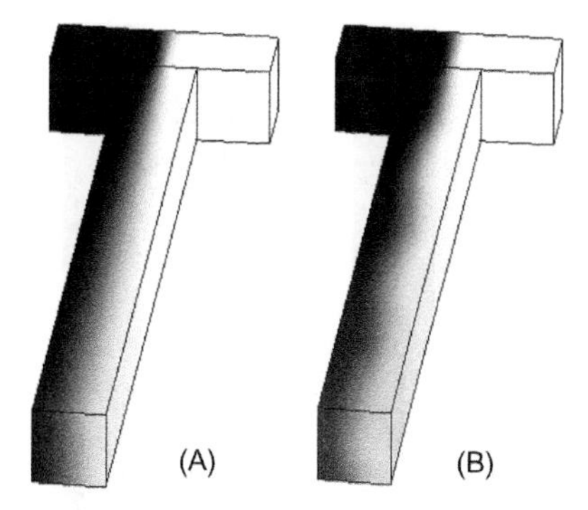

FIGURE 4.19

Three-dimensional species concentration field for a 50 μm × 50 μm T-shaped micromixer resulting from the flow fields shown in Figure 4.23: (A) homogeneous case and (B) heterogeneous case with offset patches. Species diffusivity is 3×10^{-10} m^2/s and zero electrophoretic mobility are assumed.

heterogeneous pattern, then treated with a polybrene solution. After removing the mask, the glass surface will have selective regions of positive surface charge while leaving the majority of the glass slide with its native negative charge (Biddiss et al., 2004). Finally, the PDMS plate (with the microchannel structure) will be bonded to the glass slide to form the sealed, T-shaped microchannel with heterogeneous patches on the mixing channel surface.

A micromixer consisting of six offset staggered patches (in the mixing channel) spanning 1.8 mm downstream and offset 10 μm from the channel centerline, with a width of 90 μm and a length of 300 μm, was analyzed experimentally. Mixing experiments were conducted at applied voltage potentials ranging between 70 V/cm and 555 V/cm; the corresponding Reynolds numbers range from 0.08 to 0.7 and Péclet numbers from 190 to 1500. The liquid is a 25 mM sodium carbonate/bicarbonate buffer. To visualize the mixing effects, 100 μM fluorescein was introduced through one inlet channel. As an example, Figure 4.20 shows the experimental images of the steady state flow for the homogenous and heterogeneous cases at 280 V/cm. The enhanced mixing effect is obvious.

This experimental study shows that the passive electrokinetic micromixer with an optimized arrangement of surface charge heterogeneities can increase flow narrowing and circulation, thereby increasing the diffusive flux and introducing an advective component of mixing. Mixing efficiencies were improved by 22–68% for voltages ranging from 70 to 555 V/cm. For producing a 95% mixture, this technology can reduce the required mixing channel length of up to 88% for flows with Péclet numbers between 190 and 1500 and Reynolds numbers between 0.08 and 0.7. In terms of required channel lengths, at 280 V/cm, a homogeneous microchannel would require a channel mixing length of 22 mm for reaching a

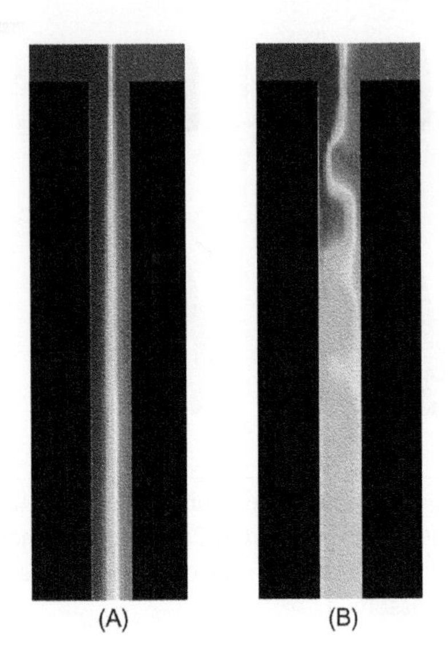

FIGURE 4.20

Images of steady state species concentration fields under an applied potential of 280 V/cm for (A) the homogeneous microchannel and (B) the heterogeneous microchannel with 6 offset staggered patches.

95% mixture. By implementing the developed micromixer, an 88% reduction in required channel length to 2.6 mm was experimentally demonstrated. Practical applications of reductions in required channel lengths include improvements in portability and shorter retention times, both of which are valuable advances applicable to many microfluidic devices.

4.8 Electrokinetic sample dispensing

An important component of many bio- or chemical lab chips is the microfluidic dispenser, which employs electroosmotic flow to dispense minute quantities (e.g., 300 pL) of samples for chemical and biomedical analysis. The precise control of the dispensed sample in microfluidic dispensers is key to the performance of these lab-on-a-chip devices.

Let's consider a microfluidic dispenser formed by two crossing microchannels, as shown in Figure 4.21 (Ren and Li, 2002). The depth and the width of all the channels are chosen to be 20 and 50 μm, respectively. There are four reservoirs connected to the four ends of the microchannels. Electrodes are inserted into these reservoirs to set up the electrical field across the channels. Initially, a sample solution (a buffer solution with sample species) is filled in Reservoir 1; the other reservoirs and the microchannels are filled with the pure buffer solution. When

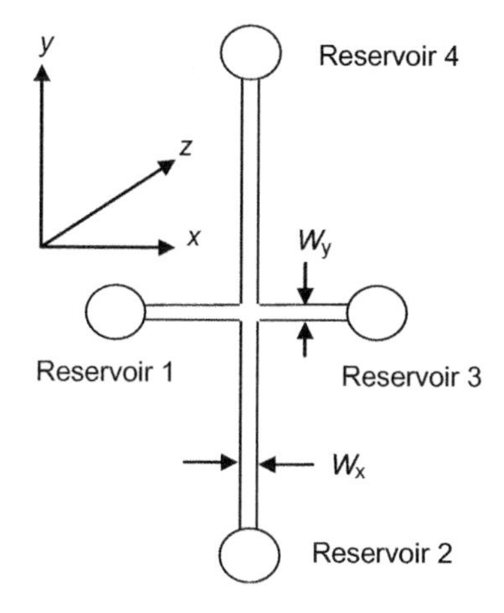

FIGURE 4.21

The schematic diagram of a crossing microchannel dispenser. W_x and W_y indicate the width of the microchannels.

the chosen electrical potentials are applied to the four reservoirs, the sample solution in Reservoir 1 will be driven to flow toward Reservoir 3, passing through the intersection of the cross-channels. This is the so-called loading process. After the loading process reaches a steady state, the sample solution loaded in the intersection will be "cut" or dispensed into the dispensing channel by the dispensing solution flowing from Reservoir 2 to Reservoir 4. This can be realized by adjusting the electrical potentials applied to these four reservoirs. This is the so-called dispensing process. The volume and the concentration of the dispensed sample are the key parameters of this dispensing process, and they depend on the applied electrical field, the flow field, and the concentration field during the loading and dispensing processes.

To model such a dispensing process, we must model the applied electrical field, the flow field, and the concentration field. To simplify the analysis, we consider this as a two-dimensional problem, i.e., ignoring the variation in the z-direction. The two-dimensional applied electrical potential in the liquid can be described by

$$\frac{\partial^2 \phi^*}{\partial x^{*2}} + \frac{\partial^2 \phi^*}{\partial y^{*2}} = 0$$

Here the nondimensional parameters are defined by

$$\phi^* = \frac{\phi}{\Phi}, \quad x^* = \frac{x}{h}, \quad y^* = \frac{y}{h}$$

where Φ is a reference electrical potential and h is the channel width, chosen as 50 μm. Boundary conditions are required to solve this equation. We impose the insulation condition to all the walls of microchannels and the specific nondimensional potential values to all the reservoirs. Once the electrical field in the dispenser is known, the local electric field strength can be calculated by

$$\vec{E} = -\vec{\nabla}\Phi$$

The electroosmotic flow field reaches steady state in milliseconds, much shorter than the characteristic time scales of the sample loading and sample dispensing. Therefore, the electroosmotic flow here is approximated as steady state. Furthermore, we consider a thin EDL, and use the slip flow boundary condition to represent the electroosmotic flow. The liquid flow field can thus be described by the following nondimensional momentum equation and the continuity equation:

$$u_{eo}^* \cdot \frac{\partial u_{eo}^*}{\partial x^*} + v_{eo}^* \cdot \frac{\partial u_{eo}^*}{\partial y^*} = -\frac{\partial P^*}{\partial x^*} + \frac{\partial^2 u_{eo}^*}{\partial x^{*2}} + \frac{\partial^2 u_{eo}^*}{\partial y^{*2}}$$

$$u_{eo}^* \cdot \frac{\partial v_{eo}^*}{\partial x^*} + v_{eo}^* \cdot \frac{\partial v_{eo}^*}{\partial y^*} = -\frac{\partial P^*}{\partial y^*} + \frac{\partial^2 v_{eo}^*}{\partial x^{*2}} + \frac{\partial^2 v_{eo}^*}{\partial y^{*2}}$$

$$\frac{\partial u_{eo}^*}{\partial x^*} + \frac{\partial v_{eo}^*}{\partial y^*} = 0$$

where u_{eo}, v_{eo} are the electroosmotic velocity component in the x- and y-direction, respectively, and nondimensionalized as follows:

$$P^* = \frac{P - Pa}{\rho(\nu/h)^2}, \quad u_{eo}^* = \frac{u_{eo} \cdot h}{\nu}, \quad v_{eo}^* = \frac{v_{eo} \cdot h}{\nu}$$

The slip velocity conditions are applied to the walls of the microchannels, the fully developed velocity profile is applied to all the interfaces between the microchannels and the reservoirs, and the pressures in the four reservoirs are assumed to be atmospheric pressure.

The distribution of the sample concentration can be described by the conservation law of mass, taking the following form:

$$\frac{\partial C_i^*}{\partial \tau} + (u_{eo}^* + u_{ep}^*)\frac{\partial C_i^*}{\partial x^*} + (v_{eo}^* + v_{ep}^*)\frac{\partial C_i^*}{\partial y^*} = \frac{D_i}{\nu}\left(\frac{\partial^2 C_i^*}{\partial x^{*2}} + \frac{\partial^2 C_i^*}{\partial y^{*2}}\right)$$

where C_i is the concentration of the ith species, u_{eo} and v_{eo} are the components of the electroosmotic velocity of the ith species, D_i is the diffusion coefficient of the ith species, and u_{epi} and v_{epi} are the components of the electrophoretic

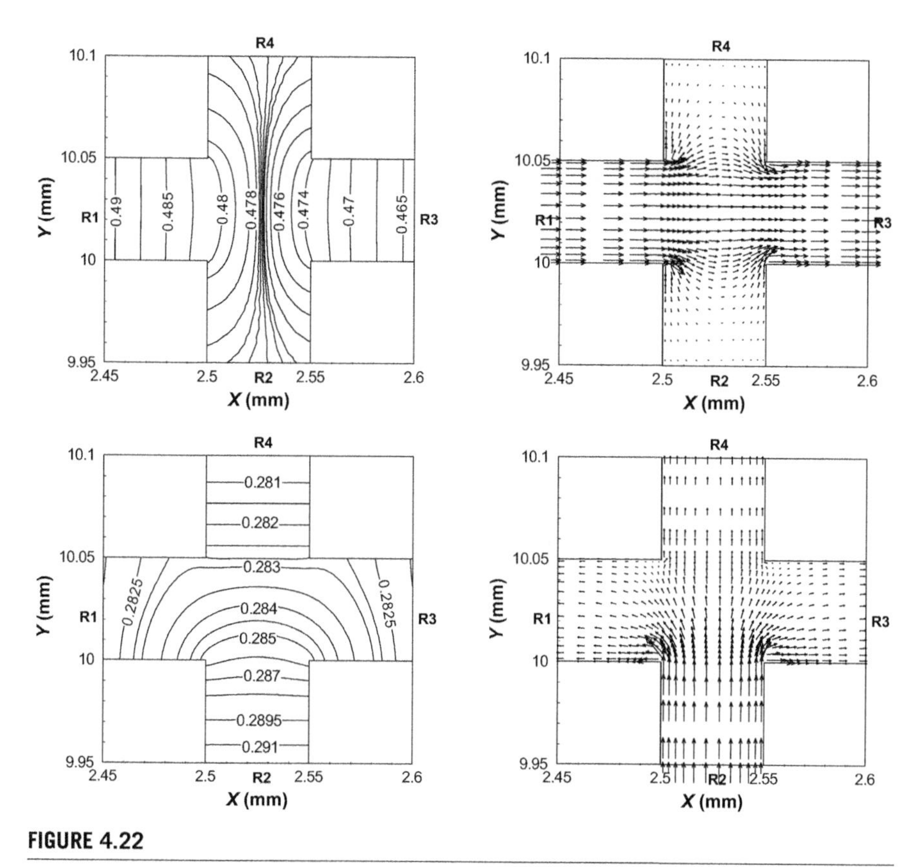

FIGURE 4.22

Examples of the applied electrical field (left) and the flow field (right) at the intersection of the microchannels in a loading process (top) and in a dispensing process (bottom).

velocity of the ith species given by $u_{epi} = E\mu_{epi}$, where μ_{epi} is the electrophoretic mobility. The nondimensional parameters in the above equation are defined by $C^* = C/\overline{C}$, and $\tau = t/(h^2/\nu)$, where $\overline{C}$ is a reference concentration.

Figure 4.22 shows the typical electrical field and flow field (computer simulated) for the loading and dispensing process, respectively. In this figure, the nondimensional applied electrical potentials are: $\phi^*(1) = 1.0$, $\phi^*(2) = 1.0$, $\phi^*(3) = 0.0$, $\phi^*(4) = 1.0$ for the loading process, and $\phi^*(1) = 0.2$, $\phi^*(2) = 2.0$, $\phi^*(3) = 0.2$, $\phi^*(4) = 0.0$ for the dispensing process, where $\phi^*(i)$ represents the nondimensional applied electrical potential to the ith reservoir. For this specific case, the electrical field and the flow field for the loading process are symmetric to the middle line of the horizontal channel, and the electrical field and the flow field for the dispensing process are symmetric to the middle line of the vertical channel.

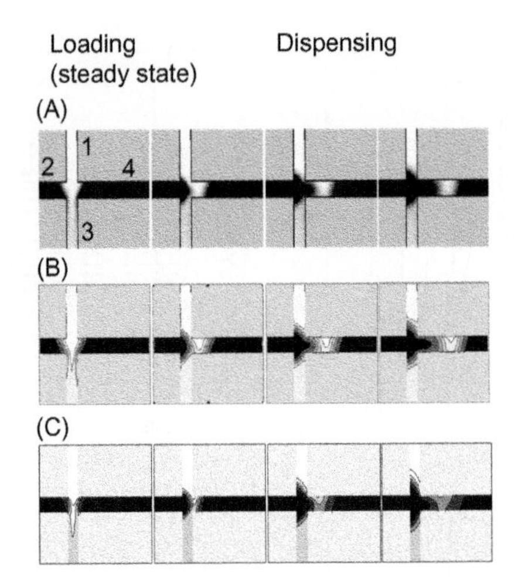

FIGURE 4.23

The loading and dispensing of a focused fluorescein sample: (A) processed images; (B) isoconcentration profiles at $0.1C_0$, $0.3C_0$, $0.5C_0$, $0.7C_0$, and $0.9C_0$, calculated from the images; and (C) corresponding isoconcentration profiles calculated through numerical simulation.

The electrokinetic dispensing processes of fluorescent dye samples were investigated experimentally (Sinton et al., 2003b,c,d). The measurements were conducted by using a fluorescent dye-based microfluidic visualization system. Figure 4.23 shows a sample dispensing process and the comparison of the dispensed sample concentration profile with the numerically simulated results. Both the theoretical studies and the experimental studies have demonstrated that the loading and dispensing of sub-nanoliter samples using a microfluidic crossing microchannel chip can be controlled electrokinetically. The ability to inject and transport large axial extent, concentration-dense samples was demonstrated. Both experimental and numerical results indicate the shape, cross-stream uniformity, and the axial extent of the samples were very sensitive to changes in the electric fields applied in the loading channel. In the dispensing process, larger samples were shown to disperse less than focused samples, maintaining more solution with the original sample concentration.

4.9 Electroosmotic flow with joule heating effects

Joule heating exists in the liquid when an axial electrical field is applied to generate the electrokinetic flow. This internal heat source can lead to a significant increase and nonuniformity of the liquid temperature. Consequently, the electrical field and the flow field are strongly affected via temperature-dependent electrical conductivity and viscosity of the liquid. the electroosmotic flow in a glass capillary that is suspended in the air and supported at the two ends by two liquid reservoirs. The set of equations governing the temperature, electric potential, and flow fields with the consideration of the Joule heating effects are summarized below.

Temperature field: As an axial electrical field is applied to produce the electroosmotic flow in a capillary, the electric current passing through the buffer solution results in Joule heating. This Joule heat is dissipated to the surroundings after conducting through the capillary wall. The general energy equation in the whole capillary takes the form

$$\rho C_p\left(\frac{\partial T}{\partial t} + \mathbf{u}\cdot\nabla T\right) = \nabla\cdot[k(T)\nabla T] + \sigma(T)\mathbf{E}\cdot\mathbf{E}$$

where ρ is the density of either the liquid or the solid, C_p the specific heat, T the absolute temperature, t the time, $\mathbf{u}$ the velocity vector, $k(T)$ the temperature-dependent thermal conductivity, $\sigma(T)$ the temperature-dependent electrical conductivity of the liquid, and $\mathbf{E}$ the externally applied electric field. Note that the terms containing $\mathbf{u}$ and $\sigma(T)$ vanish in the solid domain because there is neither liquid flow nor electric current in the capillary wall. The whole capillary system is initially at equilibrium with the ambient temperature T_0. Let us consider isothermal condition at both ends of the capillary, a symmetric condition with respect to the axis, and a convective boundary condition surrounding the capillary with a heat transfer coefficient $\bar{h}$. Note that although both reservoirs, particularly the downstream reservoir, receive heat from the liquid flowing in the capillary, this heat contribution is negligible as the reservoir volumes are of the order of microliters, sufficiently larger than the flow rate in the capillary.

Electric field: Because of the temperature dependence of liquid conductivity $\sigma(T)$, the electric field $\mathbf{E} = -\nabla\phi$ becomes nonuniform along the capillary, where the externally applied electric potential ϕ is determined by

$$\nabla\cdot[\sigma(T)\nabla\phi] = 0$$

In solving for ϕ, insulation conditions were imposed along the edges of the liquid domain. The potentials at the inlet and the outlet of the capillary were E_0L and 0, respectively, where E_0 represents the externally applied electric field and L is the length of the capillary.

Flow field: Under the condition of small Reynolds numbers, the liquid motion is governed by the Stokes and the continuity equations:

$$0 = -\nabla p + \nabla\cdot[\mu(T)\nabla\mathbf{u}] + \rho_e\mathbf{E}$$

$$\nabla \cdot \mathbf{u} = 0$$

where p is the hydrodynamic pressure and $\mu(T)$ is the temperature-dependent liquid viscosity. The net charge density ρ_e formed by the EDL is zero except in the thin EDL region adjacent to the capillary wall (the characteristic thickness is on the order of nanometers). Therefore, a slip boundary condition

$$U_{\text{wall}} = -\varepsilon(T)\varepsilon_0 \zeta E_z / \mu(T)$$

is applied at the capillary wall, dropping the electrical force term (the last term on the RHS) in the Stokes equation, and the solution of the EDL potential field was thus avoided. In the definition of U_{wall}, $\varepsilon(T)$ denotes the temperature-dependent dielectric constant of the liquid, ε_0 the permittivity of the vacuum, ζ the zeta potential of the capillary wall, and E_z the axial component of the local electrical field $\mathbf{E}$. Other boundary conditions for the flow field include fully developed velocity profiles at the ends of the capillary and a symmetric condition with respect to the axis.

In order to demonstrate the Joule heating effects on the electroosmotic flow, the following describes some experimental results (Xuan et al., 2004) conducted with a 10-cm-long fused-silica capillary (Polymicro Technologies Inc., Phoenix, AZ). The inner and outer diameters of the capillary are 200 and 320 μm, respectively. In order to make the heat transfer conditions uniform around the whole capillary, the capillary was suspended in air and supported only at the two ends by the liquid reservoirs, so that the condition of free air convection was realized. Two platinum electrodes were placed in the liquid in each of the reservoirs and connected to a high-voltage DC power source. Temperature-sensitive rhodamine B dye was used for the temperature measurement. Using this technique, the fluorescence intensity of rhodamine B dye is measured and its relative variation is then converted into the liquid temperature using the calibrated intensity versus temperature relationship. For the measurement of electroosmotic velocity in the capillary, the caged fluorescein dye method, as described previously, was used for the flow visualization in the capillary.

As can be seen in Figure 4.24, the numerically and experimentally predicted temperature distributions (15 s after electric fields were applied) agree well along the whole capillary. As numerically predicted, sharp temperature drops close to the two ends of the capillary and a high-temperature plateau in the main body of the capillary were observed. However, the numerical simulation tends to slightly overestimate the flow effect in the downstream region. In the presence of electroosmotic flow, the axial temperature profile of the liquid is inclined to the downstream. The higher the liquid flow is, the more significant the inclination becomes. The experimental results indicated a very small radial temperature difference in the capillary.

Furthermore, Figure 4.25 compares numerically calculated average velocities with those obtained from different measuring points. The electroosmotic velocities

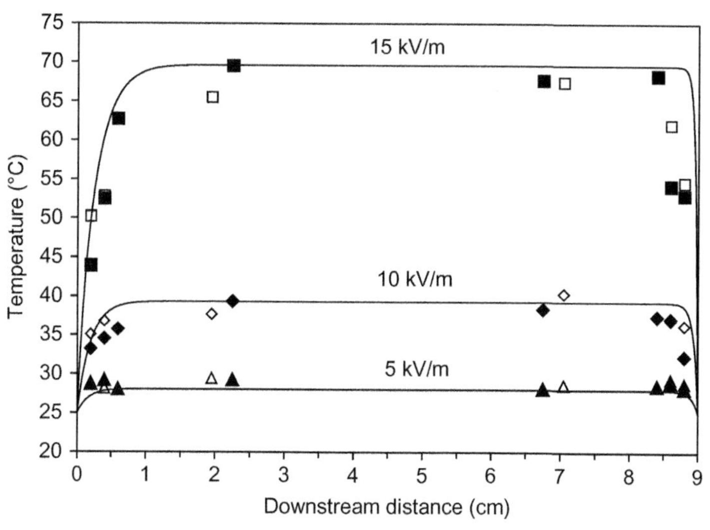

FIGURE 4.24

Comparison between numerically (solid lines) and experimentally (markers) obtained temperature distributions along the capillary axis, 15 s after the indicated electric fields were applied. Hollow and filled markers represent measurements on different days.

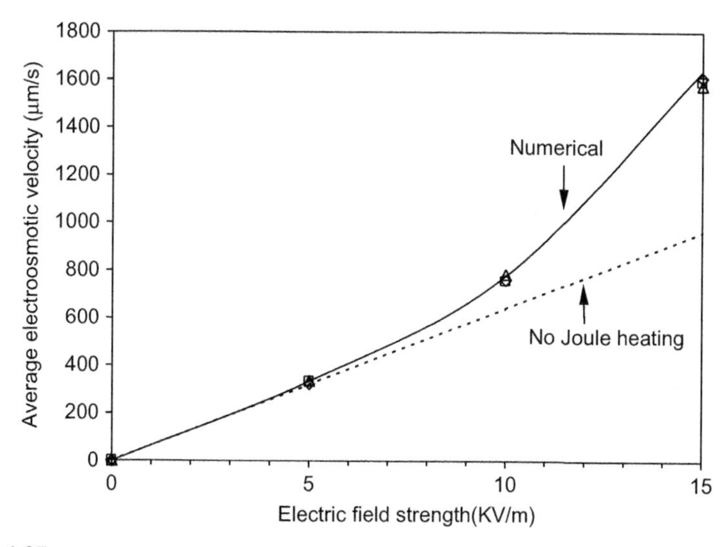

FIGURE 4.25

Comparison of numerically (solid line) and experimentally (markers) obtained average electroosmotic velocities at different electric fields. All data were extracted 15 s after the electric fields were applied. Dashed line gives the electroosmotic velocity in the absence of Joule heating effects.

in the absence of Joule heating effects are also shown in Figure 4.25. Note that the average velocity equals the volume flow rate divided by the cross-sectional area of the capillary. One can see that Joule heating effects significantly increase the average velocity (i.e., the flow rate) at high electric fields (for example, more than 50% increase at 15 kV/m, as demonstrated in Figure 4.25).

4.10 Practice problems

Problem 4.1

Model and simulate one-dimensional steady state electroosmotic flow in a circular capillary under an applied DC field. Plot and discuss the EDL field and flow field, effects of zeta potential, bulk ionic concentration, applied voltage, and diameter.

Problem 4.2

Model and simulate a one-dimensional steady state electroosmotic flow in a microchannel under an applied DC field and a small externally applied pressure difference along the channel. Plot and discuss the flow field in terms of the value and the direction of the applied pressure gradient, zeta potential, and the applied electrical field.

Problem 4.3

Model and simulate a one-dimensional electroosmotic flow in a circular capillary under an applied AC field. Plot and discuss the flow field, effects of the zeta potential, bulk ionic concentration, AC voltage and frequency, and diameter of the capillary.

Problem 4.4

Model and simulate a steady state electroosmotic flow in a microchannel with a 90-degree turn under an applied DC field. Plot and discuss the flow field in terms of the zeta potential, the channel size, and the applied electrical field.

Problem 4.5

Model and simulate electroosmotic flow in a converging–diverging circular microcapillary with a constant zeta potential, Plot the flow field (streamlines and velocity) and the induced pressure field, and discuss the effects of the applied field strength, dimensions of the converging–diverging microchannel, and the zeta potentials.

Problem 4.6

Model and simulate electroosmotic mixing of two aqueous solutions in a homogeneous T-shaped microchannel. Plot the flow field (streamlines and velocity) and the concentration field in the mixing channel, and discuss the effects of the applied field strengths at the two inlets, the zeta potential, and dimensions of the microchannel on the mixing length (i.e., the length required to have 99% mixing).

References

Ajdari, A., 1995. Electroosmosis on inhomogeneously charged surfaces. Phys. Rev. Lett. 75, 755–758.

Ajdari, A., 1996. Generation of transverse fluid currents and forces by an electric field: electro-osmosis on charge-modulated and undulated surfaces. Phys. Rev. E. 53, 4996–5005.

Ajdari, A., 2001. Transverse electrokinetic and microfluidic effects in micropatterned channels:lubrication analysis for slab geometries. Phys. Rev. E. 65, 016301.

Anderson, J.L., 1985. Effect of nonuniform zeta potential on particle movement in electricfields. J. Colloid Interface Sci. 105, 45–54.

Arulanandam, S., Li, D., 2000a. Liquid transport in rectangular microchannels by electroosmotic pumping. Colloids Surf. A 161, 89–102.

Arulanandam, S., Li, D., 2000b. Evaluation of zeta potential and surface conductance by measuring electro-osmotic current. J. Colloid Interface Sci. 225, 421–428.

Biddiss, E., Erickson, D., Li, D., 2004. Heterogeneous surface charge enhanced micro-mixing for electrokinetic flows. Anal. Chem. 76, 3208–3213.

Brown, A.B.D., Smith, C.G., Rennie, A.R., 2002. Pumping of water with an ac electric field applied to asymmetric pairs of microelectrodes. Phys. Rev. E. 63, 016305 1–8.

Dahm, W.J.A., Su, L.K., Southerland, K.B., 1992. A scalar imaging velocimetry technique for fully resolved four-dimensional vector velocity field measurement in turbulent flows. Phys. Fluids A 4, 2191–2206.

Dertinger, S.K.W., Chiu, D.T., Jeon, N.L., Whitesides, G.M., 2001. Generation of gradients having complex shapes using microfluidic networks. Anal. Chem. 73, 1240–1246.

Dutta, P., Beskok, A., 2001. Analytical solution of time periodic electroosmotic flows: analogies to stokes' second problem. Anal. Chem. 73, 5097–5102.

Erickson, D., Li, D., 2002a. Influence of surface heterogeneity on electrokinetically driven microfluidic mixing. Langmuir 18, 8949–8959.

Erickson, D., Li, D., 2002b. Microchannel flow with patch-wise and periodic surface heterogeneity. Langmuir 18, 1883–1892.

Erickson, D., Li, D., 2003a. Three dimensional structure of electroosmotic flows over periodically heterogeneous surface patterns. J. Phys. Chem. B. 107, 12212–12220.

Erickson, D., Li, D., 2003b. Analysis of AC electroosmotic flows in a rectangular microchannel. Langmuir 19, 5421–5430.

Ghosal, S., 2002. Lubrication theory for electro-osmotic flow in a microuidic channel of slowly varying cross-section and wall charge. J. Fluid Mech. 459, 103–128.

Gonzàlez, A., Ramos, A., Green, N.G., Castellanos, A., Morgan, H., 2000. Fluid flow induced by non-uniform ac electric fields in electrolytes on microelectrodes II: a linear double layer analysis. Phys. Rev. E. 61, 4019–4028.

Green, N.G., Ramos, A., Gonzàlez, A., Morgan, H., Castellanos, A., 2000. Fluid flow induced by non-uniform ac electric fields in electrolytes on microelectrodes I: experimental measurements. Phys. Rev. E. 61, 4011–4018.

Harrison, J.D., Fluri, K., Seiler, K., Fan, Z., Effenhauser, C., Manz, A., 1993. Micromachining a miniaturized capillary electrophoresis-based chemical analysis system on a chip. Science 261, 895–897.

Herr, A.E., Molho, J.I., Santiago, J.G., Mungal, M.G., Kenny, T.W., Garguilo, M.G., 2000. Electroosmotic capillary flow with nonuniform zeta potential. Anal. Chem. 72, 1053–1057.

Hunter, R.J., 1981. Zeta Potential in Colloid Science: Principle and Applications. Academic Press, London.

Jeon, N.L., Dertinger, S.K.W., Chiu, D.T., Choi, I.S., Stroock, A.D., Whitesides, G.M., 2000. Generation of solution and surface gradients using microfluidic systems. Langmuir 16, 8311–8316.

Johnson, T.J., Ross, D., Gaitan, M., Locascio, L.E., 2001. Laser modification of preformed polymer microchannels: application to reduce band broadening around turns subject to electrokinetic flow. Anal. Chem. 73, 3656–3661.

Kamholz, A.E., Schilling, E.A., Yager, P., 2001. Optical measurement of transverse molecular diffusion in a microchannel. Biophys. J. 80, 1967–1972.

Keely, C.A., van de Goor, T.A.A., McManigill, D., 1994. Modeling flow profiles and dispersion in capillary electrophoresis with nonuniform zeta potential. Anal. Chem. 66, 4236–4242.

Lempert, W.R., Magee, K., Ronney, P., Gee, K.R., Haugland, R.P., 1995. Low tagging velocimetry in incompressible flow using photo-activated nonintrusive tracking of molecular motion (PHANTOMM). Exp. Fluids 18, 249–257.

Li, D., 2004. Electrokinetics in Microfluidics. Academic Press, London.

Lyklema, J., 1995. Fundamentals of interface and colloid science, Solid–Liquid Interfaces, Vol. II. Academic Press, London.

Mala, G.M., Li, D., Werner, C., Jacobasch, H.J., Ning, Y.B., 1997. Flow characteristics of water through a microchannel between two parallel plates with electrokinetic effects. Int. J. Heat Fluid Flow 18, 489–496.

Meinhart, C.D., Wereley, S.T., Santiago, J.D., 1999. PIV measurements of a microchannel flow. Exp. Fluids 27, 414–419.

Molho, J.L., Herr, A.E., Mosier, B.P., Santiago, J.G., Kenny, T.W., Breenen, R.A., et al., 2001. Optimization of turn geometries for microchip electrophoresis. Anal. Chem. 73, 1350–1360.

Oddy, M.H., Santiago, J.G., Mikkelsen, J.C., 2001. Electrokinetic instability micromixing. Anal. Chem. 73, 5822–5832.

Patankar, S., Liu, C., Sparrow, E., 1977. Fully developed flow and heat transfer in ducts having streamwise-periodic variation of cross-sectional area. J. Heat Transfer 99, 180–186.

Paul, P.H., Garguilo, M.G., Rakestraw, D.J., 1998. Imaging of pressure- and electrokinetically driven flows through open capillaries. Anal. Chem. 70, 2459–2467.

Ren, L., Li, D., 2002. Theoretical studies of microfluidic dispensing processes. J. Colloid Interface Sci. 254, 384–395.

Ren, L., Escobedo, C., Li, D., 2002. Electro-osmotic flow in micro-capillary with one solution displacing another solution. J. Colloid Interface Sci. 250, 238–242.

Reuss, E.F., 1809. Memoires de la Societe Imperiale des Naturalistes de Moskou 2, 327.

Santiago, J.G., Wereley, S.T., Meinhart, C.D., Beebe, D.J., Adrian, R.J., 1998. A particle image velocimetry system for microfluidics. Exp. Fluids 25, 316–319.

Selvaganapathy, P., Ki, Y.-S.L., Renaud, P., Mastrangelo, C.H., 2002. Bubble-free electrokinetic pumping. J. Microelectromech. Syst. 11, 448–453.

Singh, A.K., Cummings, E.B., Throckmorton, D.J., 2001. Fluorescent liposome flow markers for microscale particle-image velocimetry. Anal. Chem. 73, 1057–1061.

Sinton, D., Li, D., 2003a. Caged-dye based microfluidic velocimetry with near-wall resolution. Int. J. Therm. Sci. 42, 847–855.

Sinton, D., Li, D., 2003b. Electroosmotic velocity profiles in microchannels. Colloids Surf. A 222, 273–283.

Sinton, D., Escobedo, C., Ren, L., Li, D., 2002. Direct and indirect electroosmotic flow velocity measurements in microchannels. J. Colloid Interface Sci. 254, 184–189.

Sinton, D., Erickson, D., Li, D., 2003a. Micro-bubble lensing induced photobleaching (μ-BLIP) with application to microflow visualization. Exp. Fluids 35, 178–187.

Sinton, D., Ren, L., Li, D., 2003b. Visualization and numerical simulation of microfluidic on-chip injection. J. Colloid Interface Sci. 260, 431–439.

Sinton, D., Ren, L., Li, D., 2003c. A dynamic loading method for controlling on-chip microfluidic chip sample injection. J. Colloid Interface Sci. 266, 448–456.

Sinton, D., Ren, L., Xuan, X., Li, D., 2003d. Effects of liquid conductivity differences on multi-component sample injection, pumping and stacking in microfluidic chips. Lab Chip 3, 173–179.

Söderman, O., Jönsson, B., 1996. Electro-osmosis: velocity profiles in different geometries with both temporal and spatial resolution. J. Chem. Phys. 105, 10300–10311.

Stroock, A.D., Weck, M., Chiu, D.T., Huck, W.T.S., Kenis, P.J.A., Ismagilov, R.F., et al., 2000. Patterning electro-osmotic flow with patterned surface charge. Phys. Rev. Lett. 84, 3314–3317.

Studer, V., Pépin, A., Chen, Y., Ajdari, A., 2002. Fabrication of microfluidic devices for AC electrokinetic fluid pumping. Microelectronic Eng. 61–62, 915–920.

Taylor, J.A., Yeung, E.S., 1993. Imaging of hydrodynamic and electrokinetic flow profiles in capillaries. Anal. Chem. 65, 2928–2932.

Wereley, S.T., Meinhart, C.D., 2001. Second-order accurate particle image velocimetry. Exp. Fluids 31, 258–268.

Xuan, X., Xu, B., Sinton, D., Li, D., 2004. Electroosmotic flow with Joule heating effects. Lab Chip 4, 230–236.

CHAPTER

Flow Boiling in Minichannels and Microchannels

5

Satish G. Kandlikar

Mechanical Engineering Department, Rochester Institute of Technology, Rochester, NY, USA

5.1 Introduction

Flow boiling in minichannels is of great interest in compact evaporator applications. Automotive air-conditioning evaporators use small passages with plate-fin heat exchangers. Extruded channels with passage diameters smaller than 1 mm are already being used in compact condenser applications. Developments in evaporators to this end are needed to overcome the practical barriers associated with flow boiling in minichannels and microchannels. Another application where flow boiling research is actively being pursued is in heat removal from high heat flux devices, such as computer chips, laser diodes, and other electronic devices and components.

Flow boiling is attractive over single-phase liquid cooling for two main reasons:

1. High heat transfer coefficient during flow boiling.
2. Higher heat removal capability for a given mass flow rate of the coolant.

Although the heat transfer coefficients are quite high in single-phase flow with small-diameter channels, flow boiling yields much higher values. For example, the single-phase heat transfer coefficient under laminar flow of water in a 200 μm square channel is around 10,000 W/m^2°C (Figure 1.2), whereas the flow boiling heat transfer coefficients can exceed 100,000 W/m^2°C (Steinke and Kandlikar, 2004). Thus, larger channel diameters can be implemented with flow boiling with comparable or even higher heat transfer coefficients than single-phase systems. This feature becomes especially important in light of the filtration requirements to keep the channels clean.

Another major advantage of flow boiling systems is the ability of the fluid to carry larger amounts of thermal energy through the latent heat of vaporization. With water, the latent heat is significantly higher (2257 kJ/kg) than its specific heat of 4.2 kJ/kg°C at 100°C. This feature is especially important for refrigerant systems. Although the latent heat of many potential refrigerants is around 150–300 kJ/kg at temperatures around 30–50°C, it still compares favorably with the single-phase cooling ability of water. However, the biggest advantage is that a

Heat Transfer and Fluid Flow in Minichannels and Microchannels.
© 2014 Elsevier Ltd. All rights reserved.

suitable refrigerant can be chosen to provide desirable evaporation temperatures, typically below 50°C, without employing a deep vacuum, as would be needed for flow boiling with water. Further research in identifying and developing specific refrigerants is needed. The specific desirable properties of a refrigerant for flow boiling application are identified by Kandlikar (2005) and are discussed in Section 5.10.

5.2 Nucleation in minichannels and microchannels

There are two ways in which flow boiling in small-diameter channels is expected to be implemented. They are:

1. Two-phase entry after a throttle valve.
2. Subcooled liquid entry into the channels.

The first mode is applicable to the evaporators used in refrigeration cycles. The throttle valve prior to the evaporator can be designed to provide subcooled liquid entry but, more commonly, a two-phase entry with quality between 0 and 0.1 is employed. Although this represents a more practical system, the difficulties encountered with proper liquid distribution in two-phase inlet headers have been a major obstacle to achieving stable operation. Even liquid distribution in the header provides a more uniform liquid flow through each channel in a parallel channel arrangement.

Subcooled liquid entry is an attractive option, since the higher heat transfer coefficients associated with subcooled flow boiling can be utilized. Such systems are essentially extensions of single-phase systems and rely largely on the temperature rise of the coolant in carrying the heat away.

In either of these systems, bubble nucleation is an important consideration. Even with a two-phase entry, it is expected that a slug flow pattern will prevail, and nucleation in liquid slugs will be important. With subcooled liquid entry, early nucleation is desirable to prevent the rapid bubble growth that has been observed by many investigators. An exhaustive review of this topic is provided by Kandlikar (2002a,b). A number of researchers have studied the flow boiling phenomena (Lazarek and Black, 1982; Cornwell and Kew, 1992; Kandlikar et al., 1995, 1997, 2001; Kuznetsov and Vitovsky, 1999; Cuta et al., 1996; Kew and Cornwell, 1996; Kandlikar and Spiesman, 1997; Kasza et al., 1997; Lin et al., 1998; 1999; Jiang et al., 1999; Kamidis and Ravigururajan, 1999; Lakshminarasimhan et al., 2000; Hetsroni et al., 2002).

The inception of nucleation plays an important role in flow boiling stability, as will be discussed in a later section. The nucleation criteria in narrow channels have been studied by a number of investigators, and it is generally believed that there are no significant differences from conventional theories for large-diameter tubes proposed by Bergles and Rohsenow (1962, 1964), Sato and Matsumura (1964), and Davis and Anderson (1966). These theories are extensions of the pool boiling nucleation models proposed by Hsu and Graham (1961) and Hsu (1962). Kenning and Cooper (1965) and Kandlikar et al. (1997) have suggested further modifications based on the local temperature field in the vicinity of a nucleating bubble under various flow conditions.

Consider subcooled liquid entering a small hydraulic diameter channel at an inlet temperature $T_{\mathrm{B,i}}$. Assuming (i) constant properties, (ii) uniform heat flux, and (iii) steady conditions, the bulk temperature $T_{\mathrm{B},z}$ along the flow length z is given by the following equation:

$$T_{\mathrm{B},z} = T_{\mathrm{B},i} + (q''Pz)/(\dot{m}c_{\mathrm{p}}) \tag{5.1a}$$

where the q'' is the heat flux, P is the heated perimeter, z is the heated length from the channel entrance, $\dot{m}$ is the mass flow rate through the channel, and c_{p} is the specific heat.

The wall temperature $T_{\mathrm{w},z}$ along the flow direction is related to the local bulk fluid temperature through the local heat transfer coefficient h_z.

$$T_{w,z} = T_{B,z} + q''/h_z \tag{5.1b}$$

The local heat transfer coefficient h_z is calculated with the single-phase liquid flow equations given in Chapter 3. For simplicity, the subscript z is not used in the subsequent equations. Considering the complexity in formulation (corner effects in rectangular channels, flow maldistribution in parallel channels, variation in local conditions, etc.), the equations for fully developed flow conditions are employed. For more accurate results, the equations presented in Chapter 3 should be employed in the developing region.

Small cavities on the heater surface trap vapor or gases and serve as nucleation sites. As the heater surface temperature exceeds the saturation temperature, a bubble may grow inside the cavity and appear at its mouth, as shown in Figure 5.1A. The force resulting from the difference in pressures between the outside liquid p_{L} and the inside vapor p_{V} is balanced by the surface tension forces. A force balance along a diametric plane through the bubble yields the following equation:

$$p_{\mathrm{v}} - p_{\mathrm{L}} = 2\sigma/r_{\mathrm{b}} \tag{5.2}$$

where σ is the surface tension and r_{b} is the bubble radius. Whether the bubble is able to nucleate and the cavity is able to act as a nucleation site depends on the local temperature field around the bubble. The local temperature in the liquid is evaluated by assuming a linear temperature gradient in a liquid sublayer of thickness $y = \delta_t$ from the temperature at the wall to the temperature in the bulk liquid. Equating the heat transfer rates obtained from the equivalent conduction and convection equations, the thickness δ_t is given by

$$\delta_t = k_{\mathrm{L}}/h \tag{5.3}$$

where k_{L} is the thermal conductivity of the liquid and h is the single-phase heat transfer coefficient in the liquid prior to nucleation. The heat transfer coefficient can be obtained from equations given in Chapter 3.

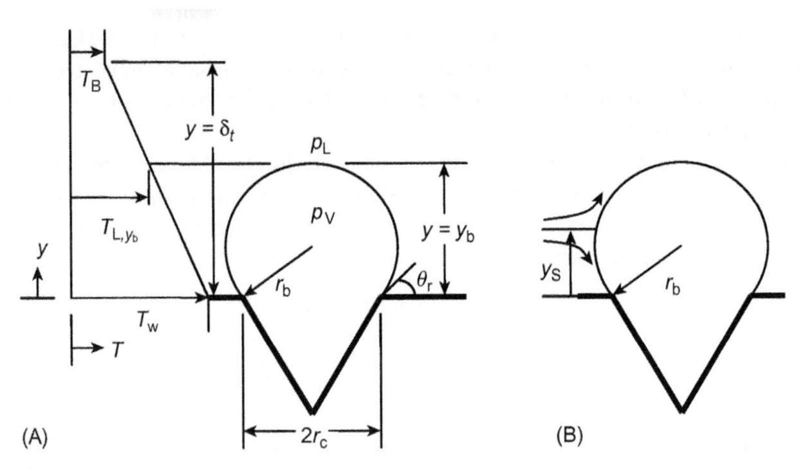

FIGURE 5.1

Schematic representation of (A) temperature and pressure around a nucleating bubble and (B) stagnation region in front of a bubble in the flow.

Source: *From Kandlikar et al. (1997).*

At a given location z, the temperature in the liquid at $y = y_b$ is obtained from the linear temperature profile as shown in Figure 5.1:

$$T_{L,y_b} = T_w - (y_b/\delta_t)(T_w - T_B) \tag{5.4}$$

where $T_{L,yb}$ is the liquid temperature at $y = y_b$ and T_w is the wall temperature. Neglecting the effect of interface curvature on the change in saturation temperature, and introducing the Clausius–Clapeyron equation, $dp/dT = h_{LV}/[T_{Sat}(v_V - v_L)]$, into Eq. (5.2b) to relate the pressure difference to the corresponding difference in saturation temperatures, the excess temperature needed to sustain a vapor bubble is given by

$$(p_V - p_L) = \frac{[T_{L,Sat}(p_V) - T_{Sat}]h_{LV}}{T_{Sat}(v_V - v_L)} \tag{5.5}$$

where $T_{L,Sat}(p_V)$ is the saturation temperature in K corresponding to the pressure p_V, T_{Sat} is the saturation temperature in K corresponding to the system pressure p_L, h_{LV} is the latent heat of vaporization at p_L, and v_V and v_L are the vapor- and liquid-specific volumes. Combining Eqs. (5.2) and (5.5), and assuming $v_V \gg v_L$, we get

$$T_{L,Sat}(pv) = T_{Sat} + \frac{2\sigma}{r_b}\frac{T_{Sat}}{\rho_v h_{LV}} \tag{5.6}$$

As a condition for nucleation, the liquid temperature $T_{L,yb}$ in Eq. (5.4) should be greater than $T_{L,Sat}(p_V)$, which represents the minimum temperature required at any point on the liquid–vapor interface to sustain the vapor bubble as given by

Eq. (5.6). Combining Eqs. (5.4) and (5.6) yields the condition for nucleating cavities of specific radii:

$$(y_b/\delta_t)(T_w - T_B) - (T_w - T_{Sat}) + \frac{2\sigma}{r_b}\frac{T_{Sat}}{\rho_V h_{LV}} = 0 \tag{5.7}$$

The liquid subcooling and wall superheat are defined as follows:

$$\Delta T_{Sub} = T_{Sat} - T_B \tag{5.8}$$

$$\Delta T_{Sat} = T_w - T_{Sat} \tag{5.9}$$

The bubble radius r_b and height y_b are related to the cavity mouth radius r_c through the receding contact angle θ_r as follows:

$$r_b = r_c/\sin\theta_r \tag{5.10}$$

$$y_b = r_b(1+\cos\theta_r) = r_c(1+\cos\theta_r)/\sin\theta_r \tag{5.11}$$

Substituting Eqs. (5.10) and (5.11) into Eq. (5.7), and solving the resulting quadratic equation for r_c, Davis and Anderson (1966) obtained the range of nucleation cavities given by

$$\{r_{c,min}, r_{c,max}\} = \frac{\delta_t \sin\theta_r}{2(1+\cos\theta_r)}\left(\frac{\Delta T_{Sat}}{\Delta T_{Sat}+\Delta T_{Sub}}\right) \times \left(1 \mp \sqrt{1 - \frac{8\sigma T_{Sat}(\Delta T_{Sat}+\Delta T_{Sub})(1+\cos\theta_r)}{\rho_V h_{LV}\delta_t \Delta T_{Sat}^2}}\right) \tag{5.12}$$

The minimum and maximum cavity radii $r_{c,min}$ and $r_{c,max}$ are obtained from the negative and positive signs of the radical in Eq. (5.12), respectively.

Different investigators have used different models to relate the bubble radius to the cavity radius and to the location where the liquid temperature T_L is determined. Hsu (1962) assumed $y_b = 1.6r_b$, which effectively translates into a receding contact angle of $\theta_r = 53.1°$. Substituting this value into Eq. (5.12), the range of cavities nucleating from Hsu's criterion is given by

$$\{r_{c,min}, r_{c,max}\} = \frac{\delta_t}{4}\left(\frac{\Delta T_{Sat}}{\Delta T_{Sat}+\Delta T_{Sub}}\right) \times \left(1 \mp \sqrt{1 - \frac{12.8\sigma T_{Sat}(\Delta T_{Sat}+\Delta T_{Sub})}{\rho_V h_{LV}\delta_t \Delta T_{Sat}^2}}\right) \tag{5.13}$$

Bergles and Rohsenow (1964) and Sato and Matsumura (1964) considered a hemispherical bubble at the nucleation inception with $y_b = r_b = r_c$. The resulting range of nucleating cavities is given by

$$\{r_{c,\min}, r_{c,\max}\} = \frac{\delta_t}{2}\left(\frac{\Delta T_{Sat}}{\Delta T_{Sat} + \Delta T_{Sub}}\right) \times \left(1 \mp \sqrt{1 - \frac{8\sigma T_{Sat}(\Delta T_{Sat} + \Delta T_{Sub})}{\rho v h_{LV} \delta_t \Delta T_{Sat}^2}}\right) \tag{5.14}$$

Kandlikar et al. (1997) analyzed the flow around a bubble and found that a stagnation point occurred at a certain distance y_S from the bubble base, as shown in Figure 5.1(b). For receding contact angles in the range of 20–60°, the location of the stagnation point was given by

$$y_S = 1.1 r_b = 1.1(r_c/\sin\theta_r) \tag{5.15}$$

Since a streamline farther away from the wall at this location would sweep over the bubble, as seen in Figure 5.1B, the temperature at $y = y_S$ was taken as the liquid temperature at $y = y_b$. The resulting range of nucleation cavities is then given by

$$\{r_{c,\min}, r_{c,\max}\} = \frac{\delta_t \sin\theta_r}{2.2}\left(\frac{\Delta T_{Sat}}{\Delta T_{Sat} + \Delta T_{Sub}}\right) \times \left(1 \mp \sqrt{1 - \frac{8.8\sigma T_{Sat}(\Delta T_{Sat} + \Delta T_{Sub})}{\rho v h_{LV} \delta_t \Delta T_{Sat}^2}}\right) \tag{5.16}$$

Note that there was a typographical error in the original publication by Kandlikar et al. (1997). The correct value for the constant is 8.8, as given in Eq. (5.16), though the actual graphs in the original publication were plotted using the correct value of 8.8.

The onset of nucleate boiling (ONB) is of particular interest in flow boiling. The radius $r_{c,crit}$ of the first cavity that will nucleate (if present) is obtained by setting the radical term in Eq. (5.16) to zero:

$$r_{c,crit} = \frac{\delta_t \sin\theta_r}{2.2}\left(\frac{\Delta T_{Sat}}{\Delta T_{Sat} + \Delta T_{Sub}}\right) \tag{5.17}$$

For a given heat flux, the wall superheat at the ONB, $\Delta T_{Sat,ONB}$, is given by

$$\Delta T_{Sat,ONB} = \sqrt{8.8 g \sigma T_{Sat} q''/(\rho_V h_{LV} k_L)} \tag{5.18}$$

If the local wall superheat at a given section is lower than that given by Eq. (5.18), nucleation will not occur. The local subcooling at the ONB can be determined from the following equation:

$$\Delta T_{Sub,ONB} = \frac{q''}{h} - \Delta T_{Sat,ONB} \tag{5.19}$$

In a channel with subcooled liquid entering, the local subcooling at the section where nucleation occurs is given by Eq. (5.19). If the subcooling is negative, this means that the local liquid is superheated and will cause extremely high bubble growth rates. Later it will be shown that such high rates result in reverse flow that leads to severe pressure drop fluctuations.

Figure 5.2 shows the comparison of different nucleation models with the experimental data by Kandlikar et al. (1997). A high-powered microscope and a high-speed camera were used to visualize the nucleation activity and measure the underlying cavity dimensions. Cavities were largely rectangular in shape, and the larger side of the opening was used in determining the cavity radius. The cavities nucleate at a certain minimum wall superheat and continue to nucleate for higher wall superheats. A majority of the data points shown in Figure 5.2 correspond to higher values of wall superheat than the minimum required for nucleation. It can be seen that all data points fall very close to or above the criterion given by Eq. (5.14).

As seen in Figure 5.2, the criterion by Davis and Anderson predicts higher wall superheats for larger cavities, whereas Bergles and Rohsenow's criterion allows the larger cavities to nucleate at lower wall superheats; Hsu's predictions are also quite close to the data. The criterion by Kandlikar et al. (1997) includes the contact angle effect.

The above analysis assumes the availability of cavities of all sizes on the heater surface. If cavities of radius $r_{c,crit}$ are not available, then higher superheats may be required to initiate nucleation on the existing cavities. The nucleation criteria plotted in Figure 5.2 indicate the superheat needed to activate cavities of specific radii. Alternatively, Eq. (5.7) along with Eqs. (5.3) and (5.10) may be

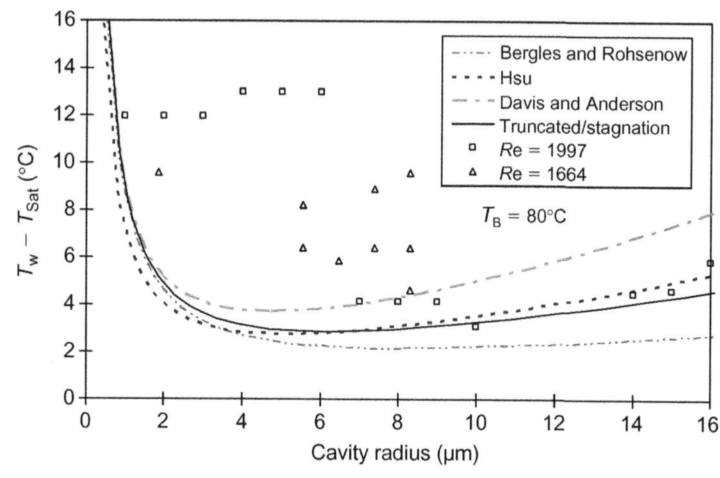

FIGURE 5.2

Comparison of different nucleation criteria against the experimental data taken with water at 1 atm. pressure in a 3 × 40 mm rectangular channel, $\theta_r = 40°$, truncated/stagnation model by Kandlikar et al. (1997).

employed to determine the local conditions required to initiate nucleation at a given cavity. The resulting equation for the wall superheat required to nucleate a given size cavity of radius r_c is given by

$$\Delta T_{Sat}|ONB \text{ at } r_c = \frac{1.1 r_c q''}{k_L \sin\theta_r} + \frac{2\sigma \sin\theta_r}{r_c} \frac{T_{Sat}}{\rho_V h_{LV}} \tag{5.20}$$

The cavity size used in Eq. (5.20) may be smaller or larger than $r_{c,crit}$ given by Eq. (5.20). Among the available cavities, the cavity with the smallest superheat requirement will nucleate first. Equation (5.19) can be employed to find the local bulk temperature at the nucleation location for a given heat flux. Again, if the local subcooling is negative, this means that the bulk liquid at this condition is superheated and will lead to severe instabilities, as will be discussed later.

The applicability of the above nucleation criteria to minichannels and microchannels is an open area. These equations were applied to design nucleation cavities in microchannels by Kandlikar et al. (2005), as will be discussed later in Section 5.6.

5.3 Nondimensional numbers used in microchannel flow boiling

Table 5.1 shows an overview of the nondimensional numbers relevant to two-phase flow and flow boiling as presented by Kandlikar (2004). The groups are classified as empirical- and theoretical-based.

With the exception of the empirically derived boiling number, the nondimensional numbers used in flow boiling applications have not incorporated the effect of

Table 5.1 Nondimensional Numbers in Flow Boiling

Nondimensional Number	Significance	Relevance to Microchannels
Groups based on empirical considerations		
Martinelli parameter, X $X^2 = \left(\frac{dp}{dz}\big|_F\right)_L / \left(\frac{dp}{dz}\big|_F\right)_V$	Ratio of frictional pressure drops with liquid and gas flow, successfully employed in two-phase pressure drop models	It is expected to be a useful parameter in microchannels as well
Convection number, Co $Co = [(1-x)/x]^{0.8}[\rho_V/\rho_L]^{0.5}$	Co is a modified Martinelli parameter, used in correlating flow boiling heat transfer data	Its direct usage beyond flow boiling correlations may be limited
Boiling number, Bo $Bo = \frac{q''}{G\, h_{LV}}$	Heat flux is nondimensionalized with mass flux and latent heat,	Since it combines two important flow parameters, q'' and G, it is used in

(Continued)

Table 5.1 (Continued)

Nondimensional Number	Significance	Relevance to Microchannels
	not based on fundamental considerations	empirical treatment of flow boiling
Groups based on fundamental considerations		
$K_1 = \left(\frac{q''}{G h_{LV}}\right)^2 \frac{\rho_L}{\rho_V}$	K_1 represents the ratio of evaporation momentum to inertia forces at the liquid–vapor interface	Kandlikar (2004) derived this number, which is applicable to flow boiling systems where surface tension forces are important
$K_2 = \left(\frac{q''}{h_{LV}}\right)^2 \frac{D}{\rho_V \sigma}$	K_2 represents the ratio of evaporation momentum to surface tension forces at the liquid–vapor interface	Kandlikar (2004) derived this number, which is applicable in modeling interface motion, such as in critical heat flux
Bond number, *Bo* $Bo = \frac{g(\rho_L - \rho_V)D^2}{\sigma}$	*Bo* represents the ratio of buoyancy force to surface tension force; used in droplet and spray applications	Since the effect of gravitational force is expected to be small, *Bo* is not expected to play an important role in microchannels
Eotvos number, *Eo* $Eo = \frac{g(\rho_L - \rho_V)L^2}{\sigma}$	*Eo* is similar to Bond number, except that the characteristic dimension L could be D_h or any other suitable parameter	Similar to *Bo*, *Eo* is not expected to be important in microchannels except at very low flow velocities and vapor fractions
Capillary number, *Ca* $Ca = \frac{\mu_V}{\sigma}$	*Ca* represents the ratio of viscous to surface tension forces, and is useful in bubble removal analysis	*Ca* is expected to play a critical role as both surface tension and viscous forces are important in microchannel flows
Ohnesorge number, *Z* $Z = \frac{\mu}{(\rho_L \sigma)^{1/2}}$	*Z* represents the ratio of viscous to the square root of inertia and surface tension forces, and is used in atomization studies	The combination of the three forces masks the individual forces, it may not be suitable in microchannel research
Weber number, *We* $We = \frac{LG^2}{\rho\sigma}$	*We* represents the ratio of the inertia to the surface tension forces; for flow in channels, D_h is used in place of L	*We* is useful in studying the relative effects of surface tension and inertia forces on flow patterns in microchannels
Jakob number, *Ja* $Ja = \frac{\rho_L c_{p,L} \Delta T}{\rho_V h_{LV}}$	*Ja* represents the ratio of the sensible heat required for reaching a saturation temperature to the latent heat	*Ja* may be used in studying liquid superheat prior to nucleation in microchannels and effect of subcooling

Source: *Adapted from Kandlikar (2004).*

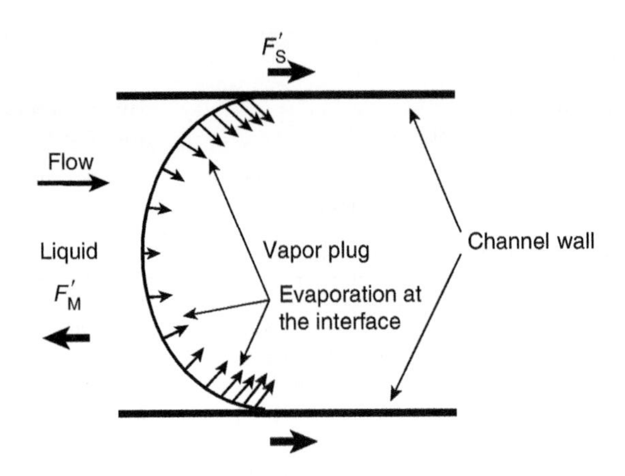

FIGURE 5.3

Schematic representation of evaporation momentum and surface tension forces on an evaporating interface in a microchannel or a minichannel.

Source: *From Kandlikar (2004).*

heat flux. This effect was recognized by Kandlikar (2004) as causing rapid interface movement during the highly efficient evaporation process occurring in microchannels. Figure 5.3 shows an evaporating interface occupying the entire channel. The change of phase from liquid to vapor is associated with a large momentum change due to the higher specific volume of the vapor phase. The resulting force is used in deriving two nondimensional groups K_1 and K_2, as shown in Table 5.1. The evaporation momentum force, the inertia force and the surface tension forces are primarily responsible for the two-phase flow characteristics and the interface shape and its motion during flow boiling in microchannels and minichannels. K_1 represents a modified boiling number with the incorporation of the liquid to vapor density ratio, while K_2 relates the evaporation momentum and surface tension forces. Future research work in this area is needed to utilize these numbers in modeling of the flow patterns and critical heat flux (CHF) phenomenon.

The Bond number compares the surface tension forces and the gravitational forces. Under flow boiling conditions in narrow channels, the influence of gravity is expected to be quite low. The Weber number and capillary numbers account for the surface tension, inertia, and viscous forces. These numbers are expected to be useful parameters in representing some of the complex features of the flow boiling phenomena.

5.4 Flow patterns, instabilities, and heat transfer mechanisms during flow boiling in minichannels and microchannels

A number of investigators have studied the flow patterns, pressure drop, and heat transfer characteristics of flow boiling in minichannels and microchannels.

Kandlikar (2002a) presented a comprehensive summary in tabular form. An abridged version of the table is reproduced in Table 5.2. The ranges of parameters investigated, along with some key results, are included in this table. Some of the researchers focused on obtaining heat transfer coefficient data. Table 5.3, derived from Steinke and Kandlikar (2004), gives the details of the experimental conditions of some of the studies reported in the literature. It may be noted that there are very few local data available that report local heat transfer coefficients.

The influence of surface tension forces becomes more dominant in small diameter channels, and this is reflected in the flow patterns observed in these channels. A comprehensive summary of adiabatic flow pattern studies is presented by Hewitt (2000). Kandlikar (2002a,b) presented an extensive summary of flow patterns and associated heat transfer during flow boiling in microchannels and minichannels. Although a number of investigators have conducted extensive studies on adiabatic two-phase flows with air–water mixtures, there are relatively few studies available on evaporating flows. Cornwell and Kew (1992) conducted experiments with R-113 flowing in 1.2 mm $\times$ 0.9 mm rectangular channels. They mainly observed three flow patterns, as shown in Figure 5.4 (isolated bubbles, confined bubbles, and high-quality annular flow). They observed the heat transfer to be strongly influenced by the heat flux, indicating the dominance of nucleate boiling in the isolated bubble region. The convective effects became important in other flow patterns. These flow patterns are observed in minichannels at relatively low heat flux conditions.

Mertz et al. (1996) conducted experiments in single and multiple channels with water and R-141b boiling in 1, 2 and 3 mm wide rectangular channels. They observed the presence of nucleate boiling, confined bubble flow, and annular flow. The bubble generation process was not a continuous process, and large pressure fluctuations were observed. Kasza et al. (1997) observed the presence of nucleate boiling on the channel wall similar to the pool boiling case. They also observed nucleation in the thin films surrounding a vapor core.

Bonjour and Lallemand (1998) reported flow patterns of R-113 boiling in a narrow space between two vertical surfaces. They noted that the Bond number effectively identifies the transition of flow patterns from conventional diameter tubes to minichannels. For smaller-diameter channels, however, the gravitational forces become less important and Bond number is not useful in modeling the flow characteristics. The presence of nucleation followed by bubble growth was visually observed by Kandlikar and Stumm (1995) and Kandlikar and Spiesman (1997).

The contribution of nucleate boiling to flow boiling heat transfer was clearly confirmed in the above studies, as well as in a number of other studies reported in the literature (Lazarek and Black, 1982; Wambsganss et al., 1993; Tran et al., 1996; Yan and Lin, 1998; Bao et al., 2000; Steinke and Kandlikar, 2004). However, for microchannels, it was also noted that the rapid evaporation and growth of a vapor bubble following nucleation caused major flow excursions, often resulting in reversed flow. Figure 5.5A shows a sequence of frames obtained by Kandlikar et al. (2005) for water boiling in 200-μm square parallel channels.

Table 5.2 Summary of Investigations on Evaporation in Minichannels and Microchannels

Author (year)	Fluid and Ranges of Parameters G (kg/m^2 s), q'' (kW/m^2)	Channel Shape, D_h (mm) Horizontal (Unless Otherwise Stated)	Flow Patterns	Remarks
Lazarek and Black (1982)	R-113, $G=125$–750, $q''=14$–380	Circular, $D=3.1$, $L=123$ and 246	Not observed	Subcooled and saturated data, h almost constant in the two-phase region, *dependent on* q''; behavior similar to large-diameter tubes
Cornwell and Kew (1992)	R-113, $G=124$–627, $q''=3$–33	Rectangular, 75 channels–each 1.2 mm × 0.9 mm, 36 channels–each 3.25 mm × 1.1 mm	Isolated bubble, confined bubble, annular slug	h was dependent on the flow pattern; isolated bubble region, $h \sim q^{0.7}$, lower q effect in confined bubble region, convection dominant in annular slug region
Moriyama and Inoue (1992)	R-113, $G=200$–1000, $q''=4$–30	Rectangular, 0.035–0.11 gap, $w=30$, $L=265$	Flattened bubbles, with coalescence, liquid strips/film	Data in narrow gaps obtained and correlated with an annular film flow model; nucleate boiling ignored, although h varied with q
Wambsganss et al. (1993)	R-113, $G=50$–100, $q''=8.8$–90.7	Circular, $D=2.92$ mm	Not reported	Except at the lowest heat and mass fluxes, both nucleate boiling and convective boiling components were present
Bowers and Mudawar (1994)	R-113, 0.28–1.1 ml/s, $q''=1000$–2000	Minichannels and microchannels, $D=2.54$ and 0.51	Not studied	Minichannels and microchannels compared. Minichannels are preferable unless liquid inventory or weight constraints are severe
Mertz et al. (1996)	Water/R-141b, $G=50$–300, $q''=3$–227	Rectangular, 1, 2, and 3 mm wide, aspect ratio up to 3	Nucleate boiling, confined bubble, and annular	Single- and multichannel test sections; flow boiling pulsations in multichannel, reverse flow, nucleate boiling dominant
Ravigururajan et al. (1996)	R-124, 0.6–5 ml/s, 20–400 W	270 μm wide, 1 mm deep, and 20.52 mm long	Not studied	Experiments were conducted over 0–0.9 quality and 5°C inlet subcooling; wall superheat from 0°C to 80°C

Tran et al. (1996)	R-12, $G = 44–832$, $q'' = 3.6–129$	Circular, $D = 2.46$; rectangular, $D_h = 2.4$	Not studied	Local h obtained up to $x = 0.94$; heat transfer in nucleate boiling dominant and convective boiling dominant regions obtained
Kasza et al. (1997)	Water, $G = 21$, $q'' = 110$	Rectangular, $2.5 \times 6.0 \times 500$	Bubbly, slug	Increased bubble activity on wall at nucleation sites in the thin liquid film responsible for high heat transfer
Tong et al. (1997)	Water, $G = 25–45 \times 10^3$, CHF 50–80 MW/m^2	Circular, $D = 1.05–2.44$	Not studied	Pressure drop measured in highly subcooled flow boiling, correlations presented for both single-phase and two-phase
Bonjour and Lallemand (1998)	R-113, $q'' = 0–20$	Rectangular, vertical 0.5–2 mm gap, 60 mm wide, and 120 mm long	Three flow patterns with nucleate boiling	Nucleate boiling with isolated bubbles, nucleate boiling with coalesced bubbles and partial dryout, transition criteria proposed
Peng and Wang (1998)	Water, ethanol, and mixtures	Rectangular, $a = 0.2–0.4$, $b = 0.1–0.3$, $L = 50$; triangular, $D_h = 0.2–0.6$, $L = 120$	Not observed	No bubbles observed, proposed a fictitious boiling model–did not use microscope/high-speed camera resulting in this erroneous conclusion
Kamidis and Ravigururajan (1999)	R-113, $Re = 190–1250$; 25–700 W	Circular, $D = 1.59$, 2.78, 3.97, 4.62	Not studied	Extremely high h, up to 11 kW/m^2C, were observed; fully developed subcooled boiling and CHF were obtained
Kuznetsov and Shamirzaev (1999)	R-318C, $G = 200–900$, $q'' = 2–110$	Annulus, 0.9 gap $\times$ 500	Confined bubble, cell, annular	Capillary forces important in flow patterns, thin film suppresses nucleation, leads to convective boiling
Lin et al. (1999)	R-141b, $G = 300–2000$, $q'' = 10–150$	Circular, $D = 1$	Not studied	Heat transfer coefficient obtained as a function of quality and heat flux; trends are similar to large-tube data

(Continued)

Table 5.2 (Continued)

Author (year)	Fluid and Ranges of Parameters G (kg/m^2 s), q'' (kW/m^2)	Channel Shape, D_h (mm) Horizontal (Unless Otherwise Stated)	Flow Patterns	Remarks
Downing et al. (2000)	R-113, ranges not clearly stated	Circular coils, $D_h = 0.23–1.86$, helix diameter $= 2.8–7.9$	Not studied	As the helical coil radius became smaller, pressure drop reduced – possibly due to rearrangement in flow patterns
Hetsroni et al. (2001)	Water, $Re = 20–70$, $q'' = 80–360$	Triangular, $\theta = 55°$, $n = 21, 26$, $D_h = 0.129–0.103$, $L = 15$	Periodic annular	Periodic annular flow observed in microchannels; significant enhancement noted in h during flow boiling
Kennedy et al. (2000)	Water, $G = 800–4500$, $q'' = 0–4000$	Circular, $D = 1.17$ and 1.45, $L = 160$	Not studied	q'' at the OFI* was 0.9 of q'' required for saturated vapor at exit; similarly, G at OFI was 1.1 times G for saturated exit vapor condition
Lakshminarasimhan et al. (2000)	R-11, $G = 60–4586$	Rectangular, $1 \times 20 \times 357$ mm	Boiling incipience observed through LCD	Boiling front observed in laminar flow, not visible in turbulent flow due to comparable h before and after; flow boiling data correlated by Kandlikar (1990) correlation
Kandlikar et al. (2001)	Water, $G = 80–560$	Rectangular, 16 channels, each 1 mm × 1 mm, $L = 60$ mm	High-speed photography	Flow oscillations and flow reversal linked to the severe pressure drop fluctuations, leading to flow reversal during boiling
Khodabandeh and Palm (2001)	R-134a/R-600a, G not measured, $q'' = 28–424$	Circular tube, 1.5 mm diameter	Not studied	h compared with 11 correlations. Mass flow rate not measured, assumed constant in all experiments—perhaps

				causing large discrepancies with correlations at higher h
Kim and Bang (2001)	R-22, $G = 384–570$, $q'' = 2–10$	Square tube, 1.66 mm; rectangular, 1.32 × 1.78	Flow pattern observed in rectangular chain	Heat transfer coefficient somewhat higher than correlation predictions; slug flow seen as the dominant flow pattern
Koo et al. (2001)	Water, 200 W heat sink	Parallel rectangular microchannels 50 μm × 25 μm	Thermal profile predicted on the chip and compared with experiments	Pressure drop using homogeneous flow model in good agreement with data; Kandlikar (1990) correlation predictions in good agreement with data
Lee and Lee (2001a,b)	R-113, $G = 50–200$, $q'' = 3–15$	Rectangular, 0.4–2 mm high, 20 mm wide	Not reported	Pressure drop correlated using Martinelli–Nelson parameter; heat transfer predicted well by Kandlikar (1990) correlation for film $Re > 200$; new correlation developed using film flow model for film $Re < 200$
Serizawa and Feng (2001)	Air–water, j_L = 0.003–17.52 m/s, $j_G = 0.0012–295.3$ m/s	Circular tubes, diameters of 50 μm for air–water and 25 μm for steam–water	Flow patterns identified over the ranges of flow rates studied	Two new flow patterns identified: liquid ring flow and liquid lump flow; steam–water ranges not given
Warrier et al. (2001)	FC-84, $G = 557–1600$, $q'' = 59.9$	Rectangular, dimensions not available, hydraulic diameter = 0.75 mm	Not studied	Overall pressure drop and local heat transfer coefficient determined; a constant value of $C = 38$ used in Eq. (1); heat transfer coefficient correlated as a function of boiling number alone

**OFI; Onset of flow instability.*

Source: *Adapted from Kandlikar (2002a).*

Table 5.3 Available Literature for Evaporation of Pure Liquid Flows in Parallel Minichannel and Microchannel Passages

Author (year)	Fluid	D_h (mm)	*Re*	*G* (kg/m² s)	(kW/m²)	Type*	Vis.**
Lazarek and Black (1982)	R-113	3.150	57–340	125–750	14–380	O	N
Moriyama and Inoue (1992)	R-113	0.140–0.438	107–854	200–1000	4.0–30	O/L	Y
Wambsganss et al. (1993)	R-113	2.920	313–2906	50–300	8.8–90.75	L	N
Bowers and Mudawar (1994)	R-113	2.540 and 0.510	14–1714	20–500	30–2000	O	N
Peng et al. (1994)	Water	0.133–0.343	200–2000	500–1626	–	O	N
Cuta et al. (1996)	R-124	0.850	100–570	32–184	1.0–400	O	N
Mertz et al. (1996)	Water	3.100	57–210	50–300	10–110	O	N
Ravigururajan et al. (1996)	R-124	0.425	217–626	142–411	5.0–25	L	N
Tran et al. (1996)	R-12	2.400–2.460	345–2906	44–832	3.6–129	L	N
Kew and Cornwell (1997)	R-141b	1.390–3.690	1373–5236	188–1480	9.7–90	L	Y
Ravigururajan (1998)	R-124	0.850	11,115–32,167	3583–10,369	20–700	L	N
Yan and Lin (1998)	R-134a	2.000	506–2025	50–900	5.0–20	L	N
Kamidis and Ravigururajan (1999)	R-113	1.540–4.620	190–1250	90–200	50–300	L	N
Lin et al. (1999)	R-141b	1.100	1591	568	–	L	N
Mudawar and Bowers (1999)	Water	0.902	16–49	2×10^4–1×10^5	1×10^3–2×10^5	O	N
Bao et al. (2000)	R-11/R-123	1.950	1200–4229	50–1800	5–200	L	N
Lakshminarasimhan et al. (2000)	R-11	3.810	1311–11,227	60–4586	7.34–37.9	O	Y
Jiang et al. (2001)	Water	0.026	1541–4811	2×10^4–5×10^4	–	O	Y
Kandlikar et al. (2001)	Water	1.000	100–556	28–48	1–150	O	Y
Kim and Bang (2001)	R-22	1.660	1883–2796	384–570	2.0–10	L	Y
Koizumi et al. (2001)	R-113	0.500–5.000	67–5398	100–800	1.0–110	O	Y

Lee and Lee (2001c)	R-113	1.569–7.273	220–1786	52–209	2.98–15.77	L	N
Lin et al. (2001)	R-141b	1.100	536–2955	50–3500	1–300	L	N
Hetsroni et al. (2002)	Vertrel XF	0.158	35–68	148–290	22.6–36	L	Y
Qu and Mudawar (2002)	Water	0.698	338–1001	135–402	1–1750	N	N
Warrier et al. (2002)	FC-84	0.750	440–1552	557–1600	1.0–50	L	N
Yen et al. (2002)	R-123	0.190	65–355	50–300	5.46–26.9	L	Y
Yu et al. (2002)	Water	2.980	534–1612	50–200	50–200	L	N
Zhang et al. (2002)	Water	0.060	127	590	2.2×10^4	O	N
Faulkner and Shekarriz (2003)	Water	1.846–3.428	20,551–41,188	3106–6225	250–2750	O	N
Hetsroni et al. (2003)	Water	0.103–0.161	8.0–42	51–500	80–220	O	Y
Kuznetsov et al. (2003)	R-21	1.810	148–247	30–50	3.0–25	L	N
Lee et al. (2003)	Water	0.036–0.041	22–51	170–341	0.2–301	O	Y
Lee and Garimella (2003)	Water	0.318–0.903	300–3500	260–1080	–	O	N
Molki et al. (2003)	R-134a	1.930	717–1614	100–225	14	L	N
Park et al. (2003)	R-22	1.660	1473–2947	300–600	10.0–20	L	N
Qu and Mudawar (2003)	Water	0.349	338–1001	135–402	10.0–1300	L	N
Wu and Cheng (2003)	Water	0.186	75–97	112	226	O	Y
Steinke and Kandlikar (2004)	Water	0.207	116–1218	157–1782	5–930	L	Y

**O = overall, L = local.*
***Visualization, Y = yes, N = no.*
Source: *Adapted from Steinke and Kandlikar (2004).*

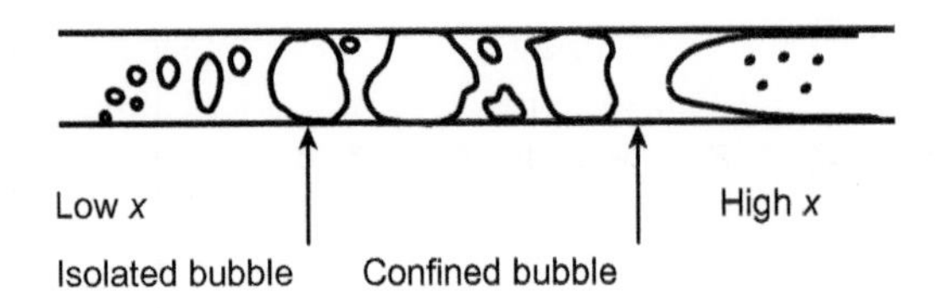

FIGURE 5.4

Flow patterns observed by Cornwell and Kew (1992) during flow boiling of R-113 in a 1.2 mm × 0.9 mm rectangular channel.

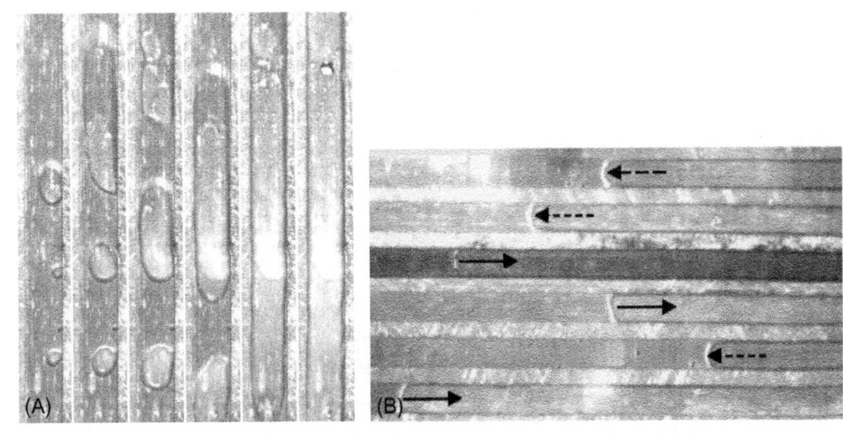

FIGURE 5.5

Flow patterns in 200 μm × 1054 μm parallel channels (A) expanding-bubble flow pattern (Kandlikar et al., 2005), six successive frames (from left to right) showing the progression of boiling in the same channel at 1 ms time intervals; and (B) a snapshot of interface movements in six parallel channels connected by common headers (Kandlikar and Balasubramanian, 2005).

The images are taken at intervals of 1 ms and the flow direction is upward. The growth of bubbles and their expansion against the flow direction is clearly seen. Figure 5.5B shows the movement of the liquid–vapor interface in both directions as obtained by Kandlikar and Balasubramanian (2005). Similar observations have been made by a number of investigators, including Mertz et al. (1996), Kennedy et al. (2000), Kandlikar et al. (2001), Kandlikar (2002a,b), Peles et al. (2001), Zhang et al. (2002), Hetsroni et al. (2001, 2003), Peles (2003), Brutin and Tadrist (2003), Steinke and Kandlikar (2004), and Balasubramanian and Kandlikar (2005).

Flow instability poses a major concern for flow boiling in minichannels and microchannels. A detailed description of flow boiling instabilities is provided by Kandlikar (2002a,b, 2005). Instabilities during flow boiling have been studied extensively in large-diameter tubes. Excursive (or Ledinegg) and parallel channel instabilities have been studied extensively in the literature. These instabilities are also present in small-diameter channels, as discussed by Bergles and Kandlikar (2005). Nucleation followed by an increase in the flow resistance due to two-phase flow in channels leads to a minimum in the pressure drop demand curve leading to instabilities. Parallel channel instabilities also occur at the minimum in the pressure drop demand curve. In a microchannel, in addition to these two instabilities, there is a phenomenon that comes into play due to rapid bubble growth rates, which causes instability and significant flow reversal problems. Large pressure fluctuations at high frequencies have been reported by a number of investigators including Kew and Cornwell (1996), Peles (2003), and Balasubramanian and Kandlikar (2005), among others.

The stability of the flow boiling process has been studied analytically using linear and nonlinear stability analyses. Some of the semi-empirical methods have yielded limited success (Peles et al., 2001; Brutin et al., 2002; Stoddard et al., 2002; Brutin and Tadrist, 2003). However, these models cannot be tested because of a lack of extensive experimental data under stable boiling conditions. It is expected that with the availability of such data sets, more rigorous models, similar to those available for flow boiling in conventional sized tubes, will be developed in the near future.

The pressure fluctuations associated with the flow boiling process were observed by many researchers, as described earlier. Figure 5.6 shows a typical plot of the pressure drop fluctuations obtained during flow boiling in minichannels by Balasubramanian and Kandlikar (2005). A fast Fourier transform analysis of the instantaneous pressure drop signal revealed that the dominant frequency was dependent on the wall temperature and was related to the nucleation activity within the channels. The growth rate of the bubbles was measured to be as high as 3.5 m/s. Figure 5.7 shows the growth rate of a slug in 1054 μm × 197 μm parallel channels. As the bubbles reach the opposite channel wall, the growth rate stabilizes until the rapid evaporation from the walls causes the growth rate to

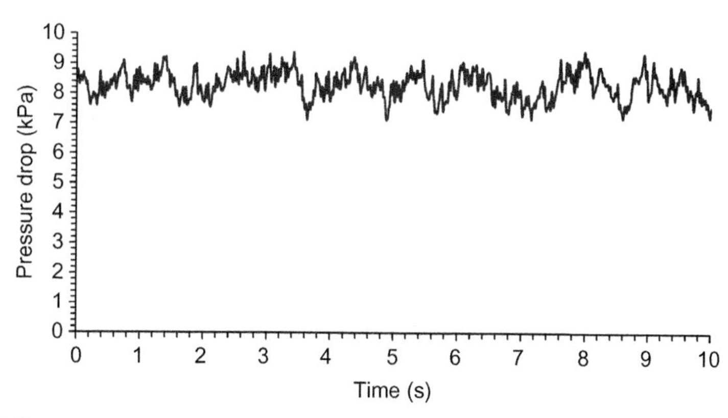

FIGURE 5.6

Pressure fluctuations observed during flow boiling of water in 1054 μm × 197 μm parallel minichannels.

Source: *From Balasubramanian and Kandlikar (2005).*

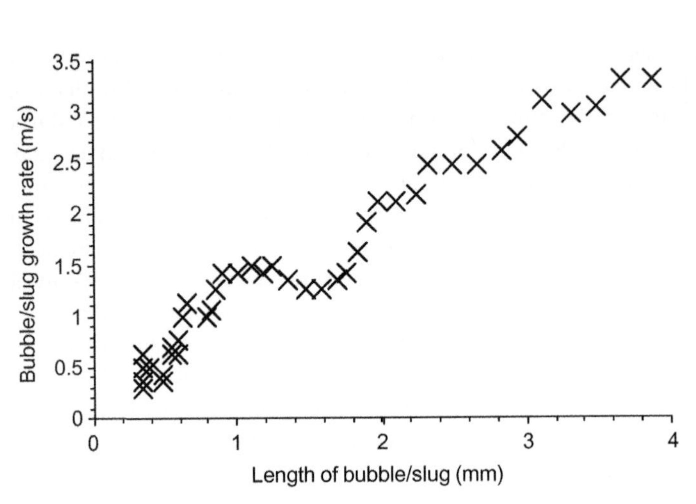

FIGURE 5.7

Bubble/slug growth rates during flow boiling of water in 1054 μm × 197 μm parallel minichannels.

Source: *From Balasubramanian and Kandlikar (2005).*

increase again. The numerical simulation by Mukherjee and Kandlikar (2005b) confirmed the bubble growth observed in Figure 5.7. Nucleation followed by rapid bubble growth is believed to be the cause of instabilities.

As a bubble grows under pool boiling conditions, the bubble growth rate is much higher immediately following its inception. The growth rate is initially proportional to time t and is controlled by the inertial forces. As the bubble grows, the growth rate slows down, following a $t^{1/2}$ trend in the thermally controlled region. In large-diameter channels (above 1–3 mm, depending on heat flux), the bubbles grow to sizes that are smaller than the channel diameter and leave the heater surface under inertia forces. In flow boiling in macrochannels, bubble growth is similar to that in pool boiling, except that the flow causes the bubbles to depart early, as shown in Figure 5.8. These departing bubbles contribute to the bubbly flow. As more bubbles are formed, they coalesce and develop into slug and annular flows, as shown in Figure 5.9.

In microchannels and minichannels, as a bubble nucleates and initially grows in the inertia-controlled region, it encounters the channel walls prior to entering the thermally controlled region found in conventional channels. The large surface area to fluid volume ratio in the channel causes the liquid to heat up rapidly. Thus, the bubble encounters a superheated liquid as it continues to grow and spreads over the other areas of the channel wall. The availability of heat from the superheated layer and from the channel walls causes a rapid expansion of the bubble, leading to the expanding-bubble flow pattern shown in Figure 5.8. The bubble occupies the entire channel cross-section and continues to grow as shown

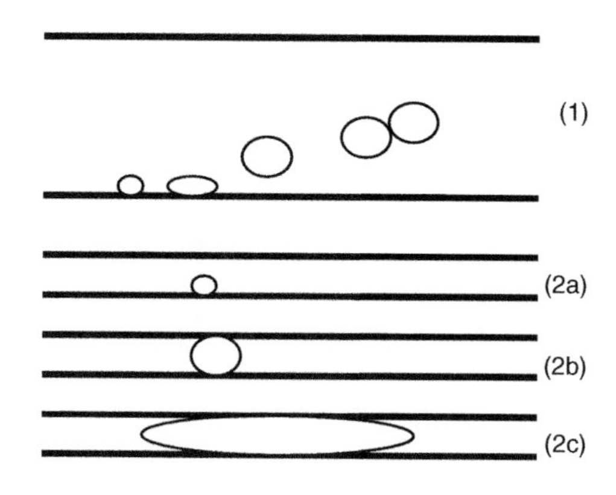

FIGURE 5.8

Schematic representation of bubble growth in (1) large-diameter tubes, and (2a–c) microchannels and minichannels.

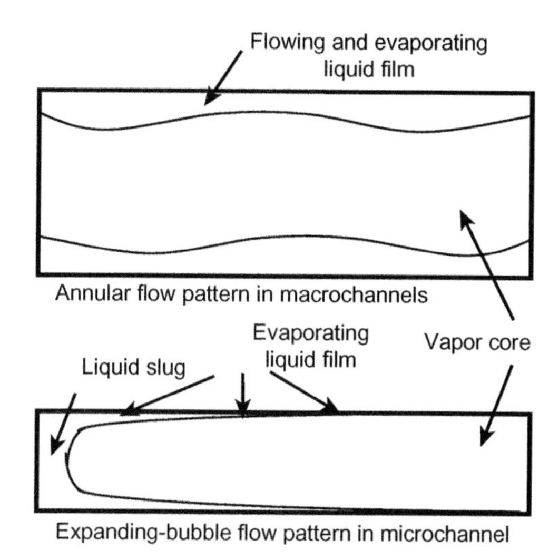

FIGURE 5.9

Comparison of the annular flow pattern in macrochannels and the expanding-bubble flow pattern in microchannels and minichannels.

in Figure 5.9. The expanding-bubble flow pattern differs from the annular flow pattern mainly in that the liquid on the wall acts similar to the film under a growing vapor bubble, rather than as a flowing film. This makes the heat transfer mechanism very similar to the nucleate boiling mechanism. A number of investigators have confirmed the strong heat flux dependence of the heat transfer coefficient during flow boiling in microchannels, which indicates the dominance of nucleate boiling.

FIGURE 5.10

The expanding-bubble flow pattern observed in a 197 μm × 1054 μm channel, using water at atmospheric pressure; the time duration between successive images is 0.16 ms; $G = 120\ \text{kg/m}^2\ \text{s}$, $q'' = 317\ \text{kW/m}^2$, $T_s = 110.9°\text{C}$; flow is from left to right.

Source: *From Kandlikar and Balasubramanian (2005).*

Figure 5.10 shows the expanding-bubble flow pattern observed during flow boiling in a rectangular minichannel by Kandlikar and Balasubramanian (2005). Note the rapid interface movement on the right side of the bubbles (downstream). The liquid film is essentially stationary and occasionally dries out before the upstream liquid slug flows through and rewets the surface. At other times, severe flow reversal was observed at the same site, and the bubble interface moved upstream rapidly.

Mukherjee and Kandlikar (2004) performed a numerical simulation of the bubble growth process in a microchannel evaporator. The bubble initially nucleated from one of the walls and then grew to occupy the entire channel cross-section. The bubble shapes were compared with the experimental observations by Balasubramanian and Kandlikar (2005) for the same geometry of 197 μm × 1054 μm with water in the expanding-bubble flow pattern region (Figure 5.11). Their agreement validates numerical simulation as a powerful tool for analyzing the flow boiling phenomenon.

Zhang et al. (2009) presented an analysis of the instability phenomenon. The stable operation was possible only when the demand-side pressure gradient curve intersected the supply side in the region of increasing pressure gradient with mass flow rate, as shown in Figure 5.12. The pressure drop for the all liquid and all vapor curves are obtained from the system characteristics. The supply curve is obtained from the pump characteristics. In the case of a constant pressure supply, the supply line becomes a horizontal line, while a positive displacement pump will introduce a vertical line. The system compressibility effects will introduce

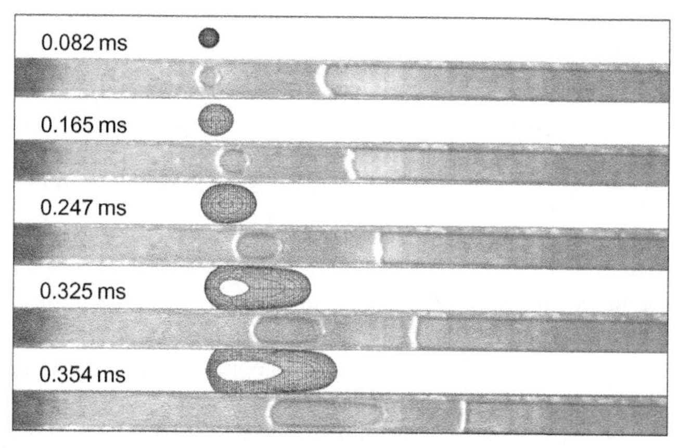

FIGURE 5.11

Comparison of bubble shapes obtained from numerical simulation by Mukherjee and Kandlikar (2004) and experimental observations by Balasubramanian and Kandlikar (2005).

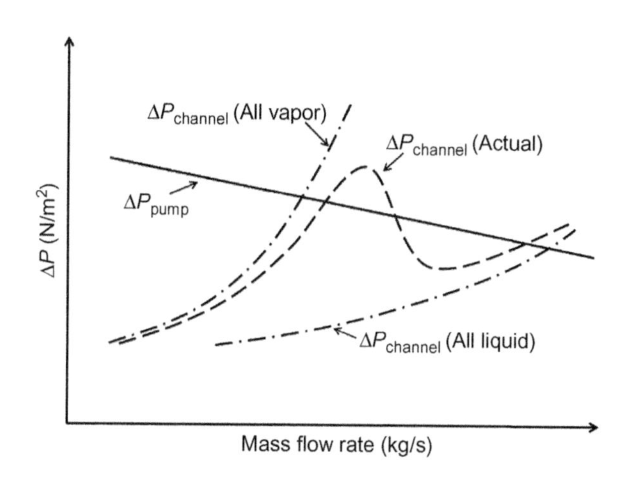

FIGURE 5.12.

Mass flow rate versus pressure drop, depicting channel demand and supply curves and the actual channel operating line for flow boiling in microchannels.

changes in the supply curve. As the flow rate is decreased, the channel pressure drop curve takes the shape shown in Figure 5.12. In the regions where the channel curve intersects with the pump curve in the negative slope region, the flow becomes unstable.

Zhang et al. (2009) presented the following equation for the demand-side (channel) pressure gradient by considering the single-phase and two-phase regions with subcooled liquid entry. The resulting equation is given as:

$$\frac{\partial(\Delta P)}{\partial G}=\frac{2Po\mu_1(L-z_s)}{\rho_1 D_h^2 x_0}\left(x_0-\frac{x_0^2}{2}+\frac{x_0^2}{2c}+\frac{5-5\exp(-319D_h)}{8\sqrt{c}}\right.$$
$$\times\left[\arcsin(2x_0-1)+\frac{\pi}{2}\right]+\frac{5-5\exp(-319D_h)}{16\sqrt{c}}$$
$$\left.\times\sin[2\arcsin(2x_0-1)]+\frac{2Po\mu_1 z_s}{\rho_1 D_h^2}+\frac{2G}{\rho_1}\left[\frac{x_0^2}{\alpha_0}\frac{\rho_1}{\rho_V}+\frac{(1-x_0)^2}{1-\alpha_0}-1\right]\right) \tag{5.21}$$

where G is the mass flux, c is the constant in the two-phase multiplier expression, ρ is density, and x is quality, D_h is hydraulic diameter, and Po is the Poiseuille number for the given channel cross-sectional geometry. A detailed treatment of flow boiling instabilities is presented by Peles (2012).

The instantaneous pressure spike occurring at the ONB in a microchannel was postulated to be a major reason for the flow instability by Kandlikar (2006). The instantaneous maximum pressure inside a nucleating bubble is postulated to correspond to the saturation pressure corresponding to the local wall temperature at the location, as shown in Figure 5.13. The location of ONB is identified with a nucleating bubble, and the pressure variation in the channel is depicted. Thus,

$$p_{V,max}=p_{Sat}|_{T_{w,ONB}} \tag{5.22}$$

The pressure difference between the $p_{V,max}$ and the pressure in the inlet manifold provides the driving force for the backflow in the channels and initiates the instability. The inertia force acting on the bubble also needs to be accounted for while determining the stability of the flow.

The intensity of the pressure spike depends on the local wall superheat. In the case of subcooled flow boiling, the ONB is delayed and large pressure spike may occur. This effect is explained in the solved Example 5.1 in Section 5.11.

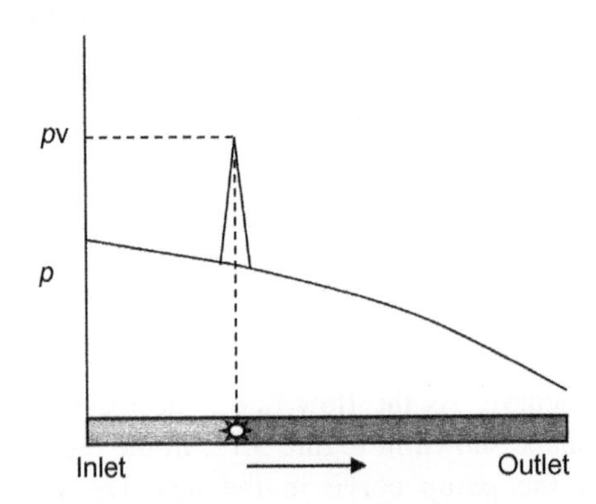

FIGURE 5.13.

Pressure spike in the bubble at the location of the ONB during flow boiling in a microchannel.

Source: *Adapted from Kandlikar (2006).*

5.5 Critical Heat Flux in microchannels

5.5.1 Comparison with pool boiling

The two-phase flow and local wall interactions during flow boiling set a limit for the maximum heat flux that can be dissipated in microchannels and minichannels. In electronics cooling applications, the inlet liquid is generally subcooled, while a two-phase mixture under saturated conditions may be introduced if a refrigeration loop is used to lower the temperature of the evaporating liquid. Thus CHF is of interest under both subcooled and saturated conditions.

Kandlikar (2001) modeled the CHF in pool boiling on the basis of the motion of the liquid–vapor–solid contact line on the heater surface. CHF was identified as the result of the interface motion caused by the evaporation momentum force at the evaporating interface near the heater surface. This force causes the "vapor-cutback" phenomenon, separating liquid from the heater surface by a film of vapor. High-speed images of the interface motion under such conditions were obtained by Kandlikar and Steinke (2002). This phenomenon is also responsible in restoring CHF conditions on subsequent rewetting of the heater surface. Extension of this pool boiling CHF model to microchannels and minichannels is expected to provide useful results.

Bergles (1963) and Bergles and Rohsenow (1964) provide extensive coverage on subcooled CHF in macrochannels. Under subcooled flow boiling conditions, the primary concern is the explosive growth of vapor bubbles upon nucleation. At the location where nucleation is initiated, the bulk liquid may have a low value of liquid subcooling. In some cases, the liquid may even be under superheated conditions. This situation arises when the proper nucleation sites are not available, and nucleation is initiated toward the exit end of the microchannels. This would result in an explosive bubble growth following nucleation, as shown in Figure 5.14A. Since the bubble growth is quite rapid, it often results in a reverse flow in the entire channel. When such bubble growth occurs near the channel entrance, the vapor is pushed back into the inlet plenum, as shown in Figure 5.14B. The vapor bubbles growing near the entrance region find the path

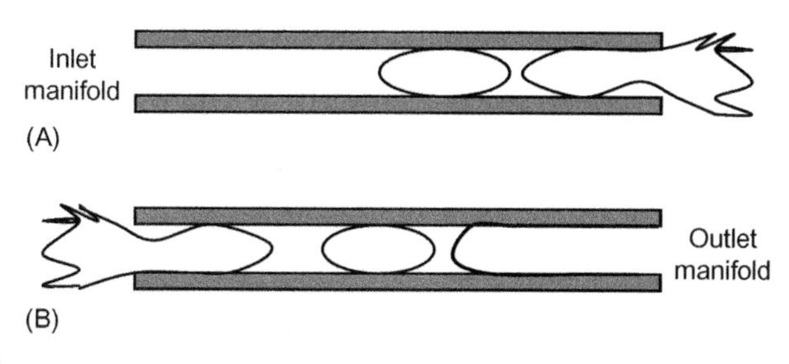

FIGURE 5.14

Reverse flow of vapor into the inlet manifold leading to early CHF.

of least resistance into the inlet plenum. The resulting instability leads to a CHF condition.

Unstable operating conditions are largely responsible for the low values of CHF reported in the literature. As reported in literature, the CHF under pool boiling conditions is around 1.2–1.6 MW/m^2, while the values reported for narrow channels are significantly lower. For example, Qu and Mudawar (2004) reported CHF data with a liquid subcooling between 70°C and 40°C, and their CHF values (based on the channel area) were only 316.2–519.7 kW/m^2. The main reason for such low values is believed to be the instabilities, which were reported by Qu and Mudawar in the same paper (see Figure 5.14). Bergles and Kandlikar (2005) reviewed the available CHF data and concluded that all the available data in the literature on microchannels suffers from this instability.

The effect of mass flux on CHF is seen to be quite significant. For example, Roach et al. (1999) obtained CHF data in 1.17- and 1.45-mm-diameter tubes and noted that the CHF increased from 860 kW/m^2 (for a 1.15-mm-diameter tube with $G = 246.6$ kg/m^2 s), to 3.699 MW/m^2 (for a 1.45-mm-diameter tube with $G = 1036.9$ kg/m^2 s). It is suspected that the higher flow rate results in a higher inertia force and induces a stabilization effect. This trend is supported by the experimental results obtained by Kamidis and Ravigururajan (1999) with R-113 in 1.59-, 2.78-, 3.97-, and 4.62-mm tubes, and by Yu et al. (2002) with water in a 2.98-mm-diameter tube.

Kandlikar (2010a) studied the scale effects on different forces applicable during flow boiling in microchannels. For a microchannel of diameter D, the forces per unit diameter, F'_i, F'_σ, F'_τ, F'_g, and F'_M, representing inertia, surface tension, shear, gravity, and evaporation momentum, respectively, were estimated as follows.

Inertia force: The inertia force acts over the entire channel due to the fluid velocity.

$$F'_{\mathrm{i}} \sim \rho V^2 \frac{D^2}{D} = \frac{G^2 D}{\rho} \tag{5.23}$$

where ρ is the density of the fluid (liquid prior to nucleation), V is the mean fluid velocity, and G is the mass flux.

Surface tension force: The surface tension force acts at the liquid–vapor–solid triple line.

$$F'_\sigma \sim \sigma \cos(\theta) D/D \sim \sigma \tag{5.24}$$

where σ is the surface tension of the liquid–vapor interface and θ is the contact angle of the liquid–vapor interface on the channel wall.

Shear force: Shear force arises due to the viscous effects at the wall.

$$F'_\tau \sim \frac{\mu V}{D} D^2/D = \mu V = \frac{\mu G}{\rho} \tag{5.25}$$

where μ is the fluid viscosity. Under two-phase conditions, the choice of fluid properties depends on the fluid that is in contact with the channel wall. Prior to nucleation, liquid properties are appropriate. In the two-phase region, liquid viscosity may still be employed due to wetter wall conditions, or a suitable averaging equation may be used between liquid and vapor phase properties.

Gravity (buoyancy) force: The buoyancy force results from the difference between the vapor and liquid densities, and is a body force, similar to the inertia force.

$$F'_{\mathrm{g}} \sim (\rho_{\mathrm{L}} - \rho_{\mathrm{V}})gD^3/D = (\rho_{\mathrm{L}} - \rho_{\mathrm{V}})gD^2 \tag{5.26}$$

where g is the acceleration due to gravity.

Evaporation momentum force: As the liquid evaporates, there is a force exerted at the evaporating interface due to the change in momentum caused by the density difference between liquid and vapor phases.

$$F'_{\mathrm{M}} \sim \left(\frac{q}{h_{\mathrm{fg}}}\right)^2 \frac{D}{\rho_{\mathrm{V}}} \tag{5.27}$$

To explore the scale effects at microscale, the variations of these forces were plotted as a function of diameter. Figures 5.15 shows these forces over a diameter range of 10 μm to 10 mm for flow boiling of water at a mass flux of 50 kg/m^{2}s and a heat flux of 1 MW/m^2. It can be seen that the surface tension forces become dominant at smaller diameters, while the shear force is relatively low due to low

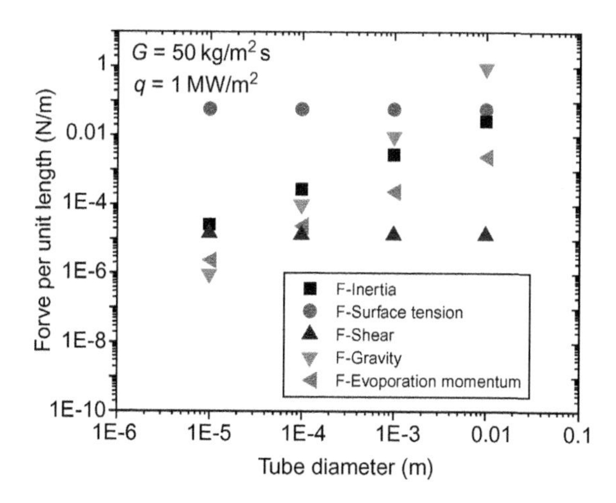

FIGURE 5.15

Scale effect of tube diameter on various forces during flow boiling, $G = 50$ kg/m^2 s, $q = 1$ MW/m^2.

Source: *Adapted from Kandlikar (2010a).*

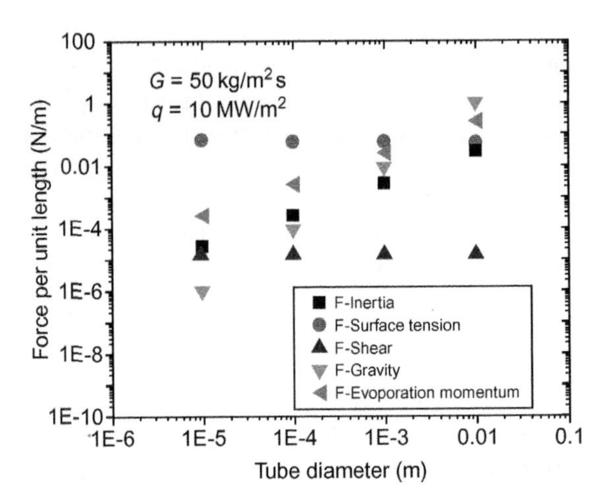

FIGURE 5.16

Scale effect of tube diameter on various forces during flow boiling, $G = 50$ kg/m^{2}s, $q = 10$ MW/m^2.

Source: *Adapted from Kandlikar (2010a).*

mass flux. The gravitational force remains quite low, and is insignificant below 100–200 μm diameters.

Figure 5.16 shows the variation of forces with diameter for the same conditions as Figure 5.15, but at a higher heat flux of 10 MW/m^2. The surface tension force still remains a major force, but the magnitude of the evaporation momentum force becomes quite large as compared to the inertia and shear forces. Thus, for microchannel flows, the evaporation momentum force becomes quite important. Since the effect of flow inertia and viscous forces become secondary, flow boiling in microchannels has similar characteristics to pool boiling (Kandlikar, 2010b).

A theoretical model for CHF based on the above scale analysis was developed by Kandlikar (2010c). This model is an extension of the pool boiling CHF model (Kandlikar, 2001). Figure 5.17 shows a schematic of the force balance conducted on an evaporating interface on the heater surface. The receding interface at the contact line represents the region where the liquid rewetting is prevented due to the interface being pulled away into the liquid. The retaining surface tension, inertia, and viscous forces are overcome by the evaporation momentum force at the onset of CHF condition. Performing the force balance, and setting this equality as the condition for CHF, the following equations were derived. The constants in the equations were obtained from 199 data points representing 13 experimental data sets using water, R-113, R-12, R-123, R-22, R-134a, R-236fa, and R-245fa. The diameter range covered was 10 μm to 3 mm. The absolute mean error with the correlation was 19.7%.

Three nondimensional groups were employed in the correlation. The nondimensional group K_2 represents the ratio of the evaporation momentum force to

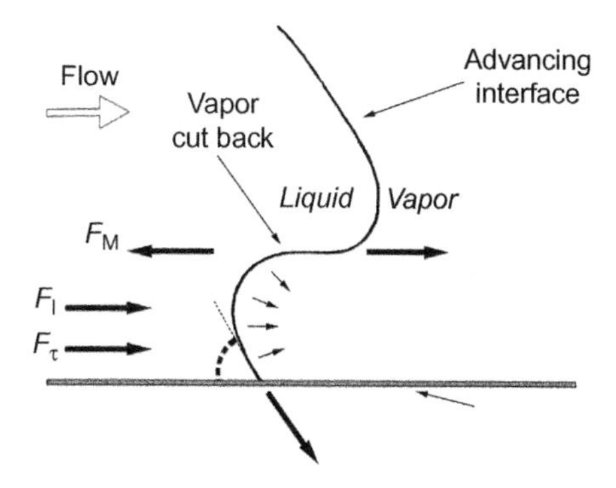

FIGURE 5.17

Schematic of a receding liquid–vapor interface at the CHF.

Source: *Adapted from Kandlikar (2010c).*

the surface tension force. The capillary number *Ca* represents the ratio of the viscous to surface tension forces, and the Weber number *We* represents the ratio of inertia to surface tension forces. The equations for the non-dimensional groups are as follows:

$$K_{2,CHF} = \left(\frac{q_{CHF}}{h_{fg}}\right)^2 \left(\frac{D_h}{\rho_V \sigma}\right) \tag{5.28}$$

$$We = \frac{G^2 D_h}{\rho_m \sigma} \tag{5.29}$$

$$Ca = \frac{\mu_L G}{\rho_L \sigma} \tag{5.30}$$

The entire region was subdivided into low inertia and high inertia regions based on the Weber number. Further, each region was subdivided into a low CHF region (LC) and a high CHF region (HC). The detailed set of equations is as follows.

Low Inertia Region, LIR: $We < 900$:

High CHF Subregion: LIR-HC—$L/D \leq 140$

$$K_{2,CHF} = a_1(1 + \cos\theta) + a_2 We(1 - x) + a_3 Ca(1 - x) \tag{5.31}$$

Low CHF Subregion: LIR-LC—$L/D \geq 230$

$$K_{2,CHF} = a_4[a_1(1 + \cos\theta) + a_2 We(1 - x) + a_3 Ca(1 - x)] \tag{5.32}$$

High Inertia Region, HIR: $We \geq 900$:

High CHF Subregion: HIR-HC—$L/D < 60$

$$K_{2,CHF} = a_1(1 + \cos\theta) + a_2 We(1 - x) + a_3 Ca(1 - x) \tag{5.33}$$

Low CHF Subregion: HIR-LC—$L/D \geq 100$

$$K_{2,CHF} = a_4[a_1(1 + \cos\theta) + a_2 We(1 - x) + a_3 Ca(1 - x)] \tag{5.34}$$

Since the changes between HC and LC subregions in both LIR and HIR regions are stepwise, as seen by the addition of the multiplier a_4 in Eq. (5.34) as compared to Eq. (5.33), CHF in the transition region cannot be interpolated. Additional experimental data is needed to further accurately define the transition criteria. It is noted that Eqs. (5.31) and (5.33) and Eqs. (5.32) and (5.34) are respectively identical, except that the transition criteria based on L/D values are different.

Improved Constant a_4 in the HIR-LC Region, HIR: $We \geq 900$:
Based on purely empirical considerations, Eq. (5.34) in the HIR-LC region is slightly modified to improve the agreement with the experimental data.

$$K_{2,CHF} = a_5\left(\frac{1}{WeCa}\right)^n [a_1(1 + \cos\theta) + a_2 We(1 - x) + a_3 Ca(1 - x)] \tag{5.35}$$

The use of the product $WeCa$ in the coefficient in Eq. (5.35) is purely empirical. It reflects some secondary effects that are correlated with this product in the available data sets. Since the improvement is small, caution is warranted until further testing is done with additional data.

The coefficients a_1–a_5 and n in Eqs. (5.31)–(5.35) are scaling parameters. Although they may be dependent on some of the system and operating parameters, they are assumed to be constants and are evaluated using available experimental data:

$$\begin{aligned} a_1 &= 1.03 \times 10^{-4} \\ a_2 &= 5.78 \times 10^{-5} \\ a_3 &= 0.783 \\ a_4 &= 0.125 \\ a_5 &= 0.14 \\ n &= 0.07 \end{aligned} \tag{5.36}$$

Table 5.4 shows a comparison of the predicted CHF values with the experimental data from different investigators.

The influence of tube diameter is somewhat confusing in light of the instabilities. In general, the CHF decreased with the tube diameter, and in many cases the reduction was rather dramatic (Qu and Mudawar, 2004). The presence of flow instability, especially in small-diameter tubes, needs to be addressed in obtaining reliable experimental data. Further research in this area is warranted.

Table 5.4 Comparison of the Present CHF Model with Experimental Data from Literature

Author (Year)/Fluid	x_{exit}	No. of Points	Abs. Mean Error, %	Mean Error, %	% Data in 30% Error Band	% Data in 50% Error Band	CHF Region
Kosar et al. (2009) water	0.003 to 0.046	15	23.5	7.5	40	70	HIR-LC
Kosar et al. (2005) water	0.47 to 0.89	8	15.6	− 10.7	100	100	LIR-HC
Kosar and Peles (2007a,b) R-123	0.003 to 0.046	30	16.7	− 11.9	80	100	LIR-HC
Kuan and Kandlikar (2008) water	0.387 to 0.776	6	12.8	9.4	83	100	LIR-LC
Kuan and Kandlikar (2008) R-123	0.857 to 0.927	6	17.5	17.5	67	100	LIR-HC
Roday and Jensen (2009) water	0.45 to 0.85	5	24.8	− 14.5	100	100	LIR-LC
Qu and Mudawar (2004) water, R-113	0.172 to 0.562	18	12.2	4.6	94	100	LIR-HC
Martin-Callizo et al. (2008) R-22, R-134a, R-256fa	0.78 to 0.98	11	31.1	28.3	55	91	LIR-LC
Inasaka and Nariai (1992) water	0.0025 to 0.039	4	16.6	− 9.3	75	100	HIR-HC
Roach et al. (1999) water	0.362 to 0.928	29	12.8	− 4.6	91	100	**HIR-LC***
Sumith et al. (2003) water	0.56 to 0.86	6	30.0	~0	33	60	LIR-HC
Cheng et al. (1997) R-12	0.003 to 0.59	38	18.	− 14.4	95	100	HIR-LC
Agostini et al. (2008) R-256fa	0.53 to ~1.00	23	25.0	− 22.0	72	86	**LIR-HC****
Overall	**0.003 to ~1.00**	**199**	**19.7**	**− 1.7**	**76**	**93**	

Notes: *LIR: low inertia region, HIR: high inertia region, HC: region with higher values of CHF, LC: region with lower values of CHF. Transition criterion between LIR and HIR is based on Weber number. LIR:* We < *900; HIR:* We ≥ *900. Transition criteria between HC and LC are based on the L/D ratio.*

**Two points fall in the LIR-LC region.*

***Two points fall in the HIR-HC region.*

Source: *Adapted from Kandlikar (2010c).*

5.6 Stabilization of flow boiling in microchannels

The reversed flow that leads to unstable operation poses a major concern in implementing flow boiling in practical applications. The rapid growth of a vapor bubble in a superheated liquid environment leads to flow reversal, which is identified as a major cause of instability.

The flow instability results from the reversed flow occurring in the parallel channels. Two methods for reducing the instabilities are discussed in this section.

5.6.1 Pressure drop element at the inlet to each channel

Parallel channels provide an effect similar to the upstream compressibility for each of the channels. Therefore, placing a flow restrictor in the flow loop prior to the inlet manifold will not reduce the instability arising in each channel. To reduce the intensity and occurrence of reverse flow, a pressure drop element (PDE) (essentially a flow constrictor or a length of reduced cross-sectional area) is placed at the entrance to each channel. This introduces an added resistance to fluid flow, but provides an effective way to reduce the flow instabilities arising from the reverse flow.

Mukherjee and Kandlikar (2004, 2005a,b) conducted an extensive numerical analysis of bubble growth in a microchannel in an effort to study the effect of the PDEs on flow stabilization. They utilized the level set method to define the interface and track its movement by applying conservation equations. Figure 5.18

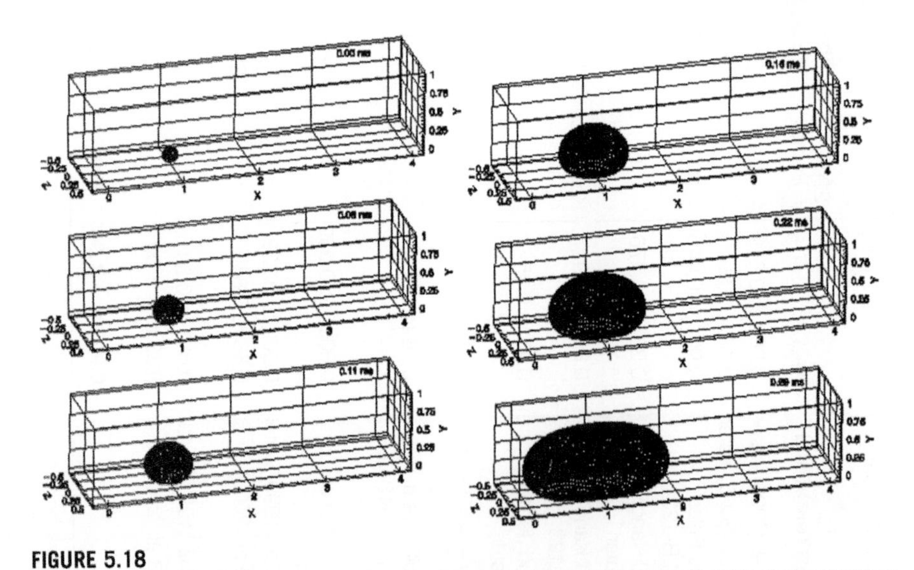

FIGURE 5.18

Simulation of bubble growth in a microchannel, with upstream to downstream flow resistance ratio $R = 1$.

Source: *From Mukherjee and Kandlikar (2005a,b).*

shows the consecutive frames of a bubble growing in a microchannel. The resistance to flow in both the upstream and downstream directions was the same. Bubble growth is seen to be slow at the beginning, but becomes more rapid after the bubble touches the other heated channel walls. This leads to an extension of the inertia-controlled region, where the heat transfer is more efficient since it does not depend on the diffusion of heat across a thin layer surrounding the liquid–vapor interface.

Mukherjee and Kandlikar (2005a,b) introduced a new parameter R to represent the upstream to downstream flow resistance ratio. In their simulation, they showed that increasing the upstream resistance reduced the intensity of the backflow. The results are shown in Figure 5.19. In Figure 5.19A the resistance is the same in both directions, while in Figure 5.19B the inlet to outlet ratio is 0.25, indicating a fourfold higher flow resistance in the backward direction. The resulting bubble growth in Figure 5.19B shows that reverse flow is completely eliminated in this case. As expected, the backflow characteristics were also found to be dependent on the heat flux and the local liquid superheat at the nucleation site.

5.6.2 Flow stabilization with nucleation cavities

Introducing artificial nucleation sites on the channel wall is another method for reducing instability. Introducing nucleation cavities of the right size would initiate nucleation before the liquid attains a high degree of superheat. Kandlikar et al. (2005) experimentally observed nucleation behavior with the introduction of artificial cavities into the microchannel. Figure 5.20 shows nucleation on these cavities much earlier, with significant reduction in reverse flow and instabilities.

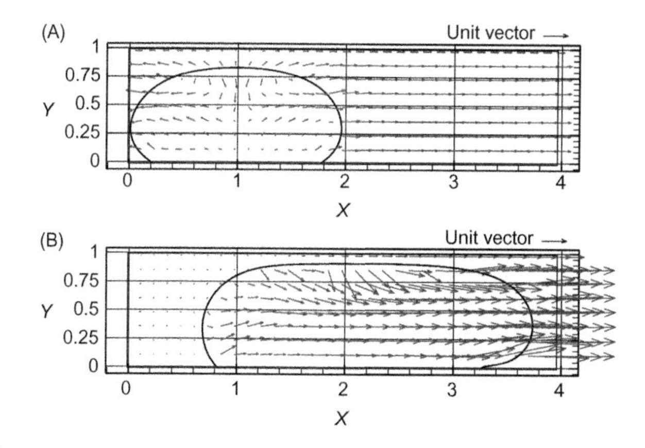

FIGURE 5.19

Bubble growth with different flow resistances on the upstream and downstream flow directions, upstream to downstream flow resistance ratio (A) $R=1$ and (B) $R=0.25$. Channel details: 200 μm square channel; X = channel length direction; Y = channel height direction.

Source: *From Mukherjee and Kandlikar (2005a,b).*

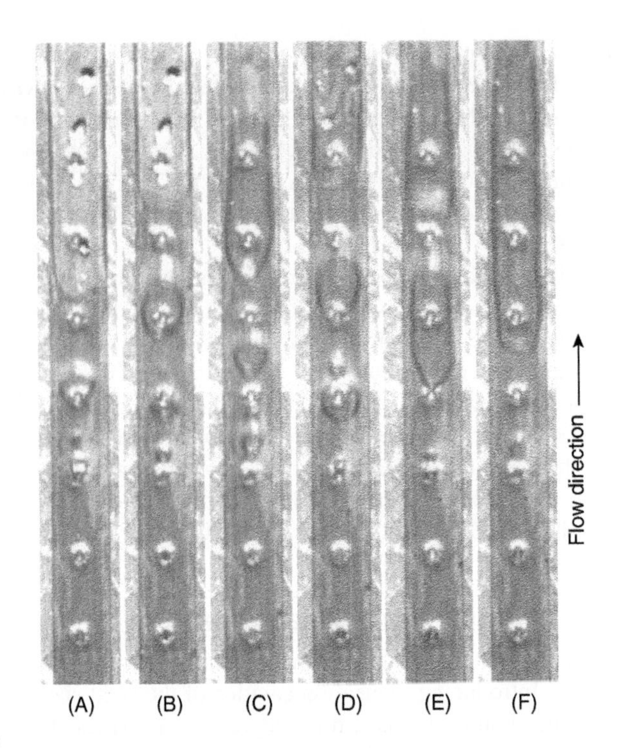

FIGURE 5.20

Stabilized flow with large fabricated nucleation sites; successive frames from (A) to (F), taken at 0.83 ms time intervals, illustrate stabilized flow in a single channel from a set of six parallel vertical microchannels; water, $G = 120$ kg/m^2 s, $q'' = 308$ kW/m^2, $T_s = 114$°C.

Source: *From Kandlikar et al. (2005).*

The flow pattern present after nucleation is initiated is an important feature of flow boiling in minichannels and microchannels. The bubbles expand to occupy the entire cross-section, with intermittent liquid slugs between successive expanding bubbles. Additional nucleation within the slugs further divides them. Figures 5.4, 5.5, and 5.10 show the two-phase structure, with a vapor core surrounded by a film of liquid. This flow pattern is similar to the annular flow pattern, but with an important distinction. In an annular flow pattern, liquid flows in the thin film surrounding the vapor core. The velocity profile and flow rate in the film of the classical annular flow are determined from the well-known triangular relationship between the wall shear stress, pressure drop, and liquid film flow rate.

The sizes of the nucleation cavities required to initiate nucleation are functions of local wall and bulk fluid temperatures, heat transfer coefficient, and heat flux. Examples 5.1 and 5.2 illustrate the size ranges of nucleating cavities under given flow and heat flux conditions.

Implementing cavities and PDEs together was found to be most effective in introducing the instabilities. The placement of nucleation cavities and the area

Table 5.5 Effect of Artificial Nucleation Sites and PDEs on Flow Boiling Stability in 1054 μm × 197 μm Channel (Kandlikar et al., 2005)

Case	Average Surface Temperature (°C)	Pressure Drop (kPa)	Pressure Fluctuation (± kPa)	Stability
Open header, 5–30 μm nucleation sites	113.4	12.8	2.3	Unstable
51% area PDEs, 5–30 μm nucleation sites	113.0	13.0	1.0	Partially stable
4% area PDEs, 5–30 μm nucleation sites	111.5	39.5	0.3	Completely stable

PDE: pressure drop element.

reduction at the inlet section are some of the design variables that need to be taken into account when designing a flow boiling system. Table 5.5 shows the effect of various configurations in reducing the flow boiling instabilities.

PDEs placed at the inlet of each channel provide the specified cross-sectional area for flow. For the cases investigated, the PDEs or artificial nucleation sites alone did not eliminate the instabilities, as noted from the pressure drop fluctuations. The 4% PDEs are able to eliminate the instabilities completely, but they introduce a very large pressure drop. The effect of a 51% area reduction in conjunction with the artificial nucleation sites is interesting to note. There was little pressure drop increase due to area reduction as compared to the fully open channel. It is therefore possible to arrive at an appropriate area reduction in conjunction with the artificial nucleation sites with a marginal increase in the pressure drop.

5.6.3 Flow stabilization with diverging microchannels

One way to ascertain that the demand curve has a positive slope is by increasing the flow cross-sectional area along the flow length so that dA/dx is positive in the flow direction. This will make the demand curve steeper than the supply curve and improve the stability of the flow boiling system. Mukherjee and Kandlikar (2005a,b, 2009) presented a diverging microchannel design that provided the desired cross-sectional area variation, as shown in Figure 5.21.

Lu and Pan (2011) implemented 10 diverging microchannels with a mean hydraulic diameter of 120 μm, as shown in Figure 5.22. Water was used as the working fluid. They added nucleation cavities in the microchannels in one sample. Their results indicated that the flow was stabilized, and a maximum dissipation rate of 48 W/cm^2 at a wall superheat of less than 15°C was obtained with the combination of expanding microchannels and the nucleation sites.

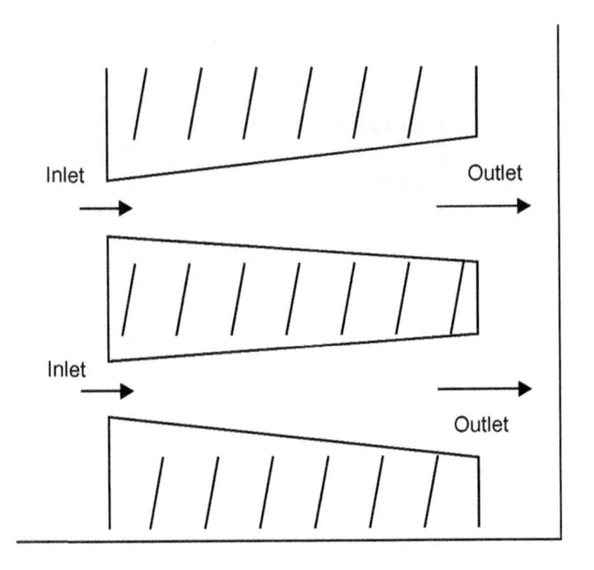

FIGURE 5.21

Diverging parallel microchannels to improve flow boiling stability.

Source: *From Mukherjee and Kandlikar (2005a,b, 2009).*

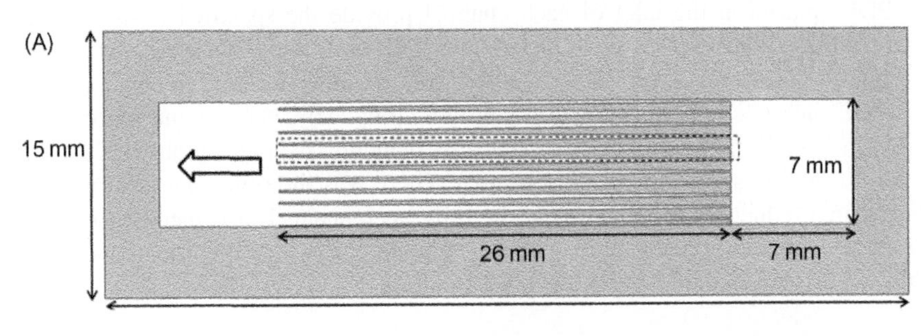

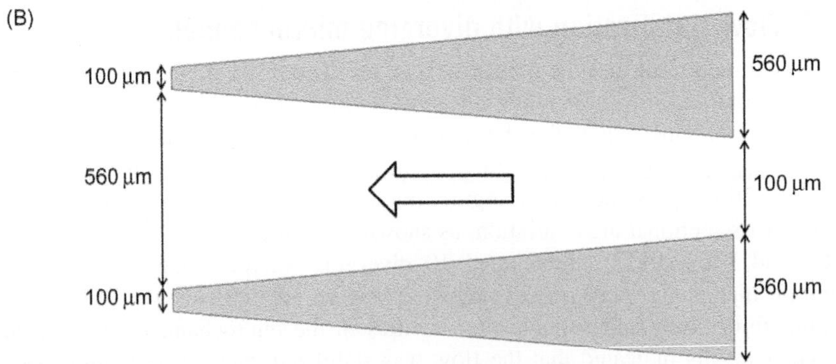

FIGURE 5.22

Details of the experimental setup with diverging microchannels: (A) channel and manifold configuration, (B) an individual channel.

Source: *Adapted from Lu and Pan (2011).*

Another recent study was conducted by Balasubramanian et al. (2011). Their test section had 40 microchannels of 300 μm channel width on a 25-mm square footprint area. After the first 15 mm length, alternate fins were removed, and another set of alternate fins were removed after 20 mm length. The flow cross-sectional area was thus increased from inlet to outlet. The geometry provided a more stable operation and a lower pressure drop. In spite of the reduced fin area, the heat transfer performance of the expanding geometry was better that the straight channels. They were able to dissipate a heat flux of 120 W/cm^2 with a heat transfer coefficient of around 24,000 W/m^2°C at a heat flux of 29 W/cm^2. The authors compared their results with a model by Kandlikar (2006) described by Eq. (5.22) and Figure 5.13. The diverging microchannels showed a smaller spike (higher ratio of inlet pressure to the maximum spike pressure) as compared to the straight microchannels for a mass flux of 100 kg/m^2 s, and the resulting pressure and temperature fluctuations were also reduced. At a mass flux of 133 kg/m^2 s, however, the ratio of the inlet pressure to the maximum spike pressure was the same for both straight and diverging microchannels, and the resulting pressure and wall temperature fluctuations were also observed to be similar in the two cases.

Miner et al. (2013) numerically analyzed the heat transfer in an expanding microchannel configuration using a separated flow model. They showed good agreement with the available experimental data, indicating the high heat flux dissipation potential of the expanding microchannels using increases in the width and height of the microchannel flow passages.

5.7 Predicting heat transfer in microchannels

Flow boiling heat transfer data were obtained by a number of investigators. Many of the researchers reported unstable operating conditions during their tests. Therefore, these data and the matching correlations should be used with some degree of caution. At the same time, it is recommended that experimental data under stable operating conditions be obtained by providing artificial nucleation sites and PDEs near the inlet inside each channel.

A good survey of available experimental data and correlations is provided by Qu and Mudawar (2004) and Steinke and Kandlikar (2004). Table 5.3 is adapted from Steinke and Kandlikar (2004); it lists the ranges of experimental data available in the literature. It can be seen that the fluids investigated include water, R-21, R-22, R-113, R-123, R-124, R-141b, FC-84, and Vertrel XF. The mass fluxes, liquid Reynolds numbers, and heat fluxes range from 20 to 6225 kg/m^2 s, 14 to 5236, and up to 2 MW/m^2, respectively. As the tube diameter decreases, the Reynolds number shifts toward a lower value.

In order to establish the parametric trends, local flow boiling heat transfer data are needed. Of the papers listed in Table 5.3, only those by Hetsroni et al. (2002),

Yen et al. (2002), and Steinke and Kandlikar (2004) deal with tubes around 200 μm and report local heat transfer data. Qu and Mudawar (2003) also report local data for parallel rectangular channels of 349 μm hydraulic diameter.

The experimental data of Bao et al. (2000) for a 1.95-mm-diameter tube indicate a strong presence of the nucleate boiling term, while Qu and Mudawar's data exhibit a strong influence of mass flux, indicating the presence of convective boiling. The decreasing trends in heat transfer coefficient with quality seen in Qu and Mudawar's data were also seen in Steinke and Kandlikar's (2004) data. A number of issues arise that make it difficult to assess the available experimental data accurately. Some of the factors are:

1. the presence of instabilities during the experiments, and
2. different ranges of parameters, especially heat flux and mass flux.

The influence of instabilities was discussed in earlier sections. The ranges of mass fluxes employed in small-diameter channels typically fall in the laminar range. The correlations developed for large-diameter tubes are in large part based on the turbulent flow conditions for Reynolds numbers based on all-liquid flow.

Some of the data available in the literature were seen to be correlated by the all-nucleate boiling-type correlations (Lazarek and Black, 1982; Tran et al., 1996). However, recent data reported by Qu and Mudawar (2004) has very large errors (36.2% and 98.8%). Similar observations can be made with the Steinke and Kandlikar (2004) data. Some of the more recent correlations by Yu et al. (2002) and Warrier et al. (2002) employ only the heat flux-dependent terms (similar to Lazarek and Black, 1982; and Tran et al, 1996) and correlate with Qu and Mudawar's (2003) data well. However, the presence of both nucleate boiling and convective boiling terms to a varying degree is reported by Kandlikar and Balasubramanian (2004). They recommend using the laminar flow equations for the all-liquid flow heat transfer coefficients in the Kandlikar (1990) correlation and cover a wide range of data sets.

The flow boiling correlation by Kandlikar (1990) utilizes the all-liquid flow, single-phase correlation. Since most of the available data were in the turbulent region, use of the Gnielinski correlation was recommended. However, later, Kandlikar and Steinke (2003) and Kandlikar and Balasubramanian (2004) introduced the laminar flow equation for laminar flow conditions based on Re_{LO}. The correlation based on the available data is given below:

For $Re_{LO} > 100$:

$$h_{TP} = \text{larger of} \begin{cases} h_{TP,NBD} \\ h_{TP,CBD} \end{cases} \tag{5.37}$$

$$h_{TP,NBD} = 0.6683 Co^{-0.2}(1-x)^{0.8} h_{LO} + 1058.0 Bo^{0.7}(1-x)^{0.8} F_{Fl} h_{LO} \tag{5.38}$$

$$h_{TP,CBD} = 1.136 Co^{-0.9}(1-x)^{0.8} h_{LO} + 667.2 Bo^{0.7}(1-x)^{0.8} F_{Fl} h_{LO} \tag{5.39}$$

where $Co = [(1-x)/x]^{0.8}(\rho_V/\rho_L)^{0.5}$ and $Bo = q''/Gh_{LV}$. The single-phase all-liquid flow heat transfer coefficient h_{LO} is given by:

$$\text{for } 10^4 \le Re_{LO} \le 5 \times 10^6 \quad h_{LO} = \frac{Re_{LO}Pr_L(f/2)(k_L/D)}{1 + 12.7(Pr_L^{2/3} - 1)(f/2)^{0.5}} \tag{5.40}$$

$$\text{for } 3000 \le Re_{LO} \le 10^4 \quad h_{LO} = \frac{(Re_{LO} - 1000)Pr_L(f/2)(k_L/D)}{1 + 12.7(Pr_L^{2/3} - 1)(f/2)^{0.5}} \tag{5.41}$$

$$\text{for } 100 \le Re_{LO} \le 1600 \quad h_{LO} = \frac{Nu_{LO}k}{D_h} \tag{5.42}$$

In the transition region between Reynolds numbers of 1600 and 3000, a linear interpolation is suggested for h_{LO}.

For Reynolds numbers below and equal to 100 ($Re \le 100$), the nucleate boiling mechanism governs, and the following Kandlikar Correlation is proposed:

For $Re_{LO} \le 100$:

$$h_{TP} = h_{TP,NBD} = 0.6683Co^{-0.2}(1-x)^{0.8}h_{LO} + 1058.0Bo^{0.7}(1-x)^{0.8}F_{Fl}h_{LO} \tag{5.43}$$

The single-phase all-liquid flow heat transfer coefficient h_{LO} in Eq. (5.43) is found from Eq. (5.42).

The fluid surface parameter F_{FL} in Eqs. (5.38–5.40) for different fluid surface combinations is given in Table 5.6. These values are for copper or brass surfaces. For stainless steel surfaces, use $F_{FL} = 1.0$ for all fluids. For silicon surfaces, no data are currently available. Use of the values listed in Table 5.6 for copper is suggested.

The above correlation scheme is based on the data available in the literature. It has to be recognized that all of the data suffer from the instability condition to some extent. It is expected that the correlation will undergo some changes as new data under stabilized flow conditions become available.

A comparison of the correlation scheme described in Eqs. (5.37)–(5.43) with some of the experimental data available in the literature is shown in Figures 5.23–5.25. The decreasing trend in the heat transfer coefficient with quality is evident in the data, indicating the dominance of the nucleate boiling mechanism. The complex nature of flow boiling in small-diameter channels, including liquid–vapor interactions, presence of expanding bubbles with thin evaporating film, nucleation of bubbles in the flow, as well as in the thin film, make it difficult to present a comprehensive analytical model to account for the heat transfer mechanisms during flow boiling. Further efforts with high-speed flow visualization techniques are recommended to provide fundamental information on this topic.

Table 5.6 Recommended F_{Fl} (Fluid Surface Parameter) Values in Flow Boiling Correlation by Kandlikar (1990, 1991)

Fluid	F_{Fl}
Water	1.00
R-11	1.30
R-12	1.50
R-13B1	1.31
R-22	2.20
R-113	1.30
R-114	1.24
R-134a	1.63
R-152a	1.10
R-32/R-132	3.30
R-141b	1.80
R-124	1.00
Kerosene	0.488
HFE 7000	2.0

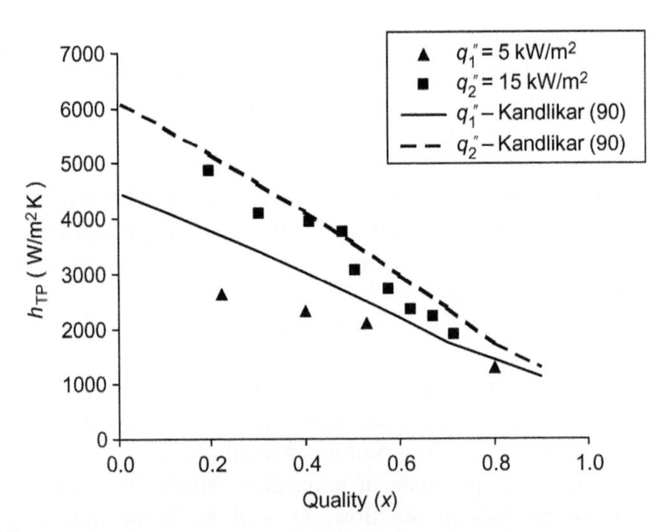

FIGURE 5.23

Yan and Lin (1998) data points for R-134a compared to the correlation by Kandlikar and Balasubramanian (2004) using the laminar single-phase equation; $D_h = 2$ mm, $G = 50$ kg/m^2 s, $q'' = 5$ and 15 kW/m^2, $Re_{LO} = 506$.

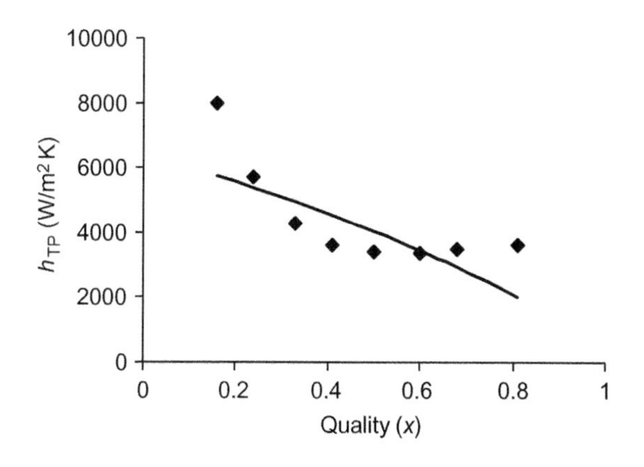

FIGURE 5.24

Yen et al. (2002) data points for HCFC 123 compared to the correlation by Kandlikar and Balasubramanian (2004) using the laminar single-phase flow equation; $D_h = 0.19$ mm, $G = 145$ kg/m^2 s, $q'' = 6.91$ kW/m^2, $p_{avg} = 151.8$ kPa, $Re_{LO} = 86$.

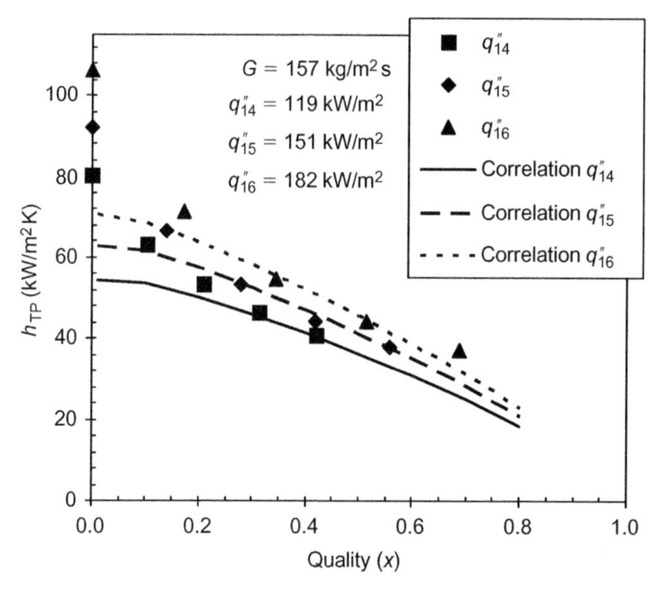

FIGURE 5.25

Steinke and Kandlikar (2004) data points compared to the correlation by Kandlikar and Balasubramanian (2004) using the laminar flow equation with the nucleate boiling dominant term only; $D_h = 207$ μm, $G = 157$ kg/m^2 s, $q'' = 119$, 151 and 182 kW/m^2, $Re_{LO} = 116$.

5.8 Pressure drop during flow boiling in microchannels and minichannels

The pressure drop in a microchannel or a minichannel heat exchanger is the sum of the following components:

$$\Delta p = \Delta p_{c} + \Delta p_{f,1\text{-}ph} + \Delta p_{f,tp} + \Delta p_{a} + \Delta p_{g} + \Delta p_{e} \tag{5.44}$$

where subscript c is the contraction at the entrance; f,1 − ph is the single-phase pressure loss due to friction, including entrance region effects; f,tp is the two-phase frictional pressure drop; a is the acceleration associated with evaporation; g is the gravitational; and e is the expansion at the outlet. Equations for calculating each of these terms are presented in the following sections.

5.8.1 Entrance and exit losses

The inlet to the microchannel may be single-phase liquid or a two-phase mixture. It is common to have liquid at the inlet when a liquid pump and a condenser are employed in the cooling system. When the refrigeration system forms an integral part of the cooling system, the refrigerant is throttled prior to entry into the microchannels. With the need to incorporate PDEs in each channel, liquid at the inlet in such cases is also possible.

The contraction losses in the single-phase liquid are covered in Chapter 3. The nature of the liquid's entry into the channels is another factor that needs to be taken into consideration. The channel floor may be flush with the manifold, or it may be shallower or deeper than the manifold. Lee and Kim (2003) used a micro-Particle Image Velocimetry (PIV) system to identify the entrance losses with sharp and smooth channel entrances.

For the two-phase entry and exit losses, Coleman (2003) recommends the following scheme proposed by Hewitt (2000).

The following equation is used to calculate the pressure loss due to a sudden contraction of a two-phase mixture using a separated flow model:

$$\Delta p_{c} = \frac{G^2}{2\rho_{L}}\left[\left(\frac{1}{C_{o}}-1\right)^2 + 1 - \frac{1}{\sigma_{c}^2}\right]\psi_{h} \tag{5.45}$$

where G is the mass flux, σ_{c} is the contraction area ratio (header to channel >1), C_{o} is the contraction coefficient given by:

$$C_{o} = \frac{1}{0.639(1-1/\sigma_{c})^{0.5}+1} \tag{5.46}$$

and ψ_{h} is the two-phase homogeneous flow multiplier given by:

$$\psi_{h} = [1 + x(\rho_{L}/\rho_{V} - 1)] \tag{5.47}$$

where x is the local quality.

The exit pressure loss is calculated from the homogeneous model:

$$\Delta p_e = G^2\sigma_e(1-\sigma_e)\psi_s \tag{5.48}$$

where σ_e is the area expansion ratio (channel to header <1) and ψ_s is the separated flow multiplier given by:

$$\psi_s = 1 + \left(\frac{\rho_L}{\rho_V} - 1\right)\left[0.25x(1-x) + x^2\right] \tag{5.49}$$

The frictional pressure drop in the single-phase region prior to nucleation is calculated from the equations presented in Chapter 3. In the two-phase region, the following equations may be used to calculate the frictional, acceleration, and gravity components of the pressure drop.

The local friction pressure gradient at any section is calculated with the following equation:

$$\left(\frac{dp_F}{dz}\right) = \left(\frac{dp_F}{dz}\right)_L \phi_L^2 \tag{5.50}$$

The two-phase multiplier $\phi_L{}^2$ is given by the following equation by Chisholm (1983):

$$\phi_L^2 1 = 1 + \frac{C}{X} + \frac{1}{X^2} \tag{5.51}$$

The value of the constant C depends on whether the individual phases are in the laminar or turbulent region. Chisholm recommended the following values of C:

$$\text{Both phases turbulent} \quad C = 21 \tag{5.52a}$$

$$\text{Laminar liquid, turbulent vapor} \quad C = 12 \tag{5.52b}$$

$$\text{Turbulent liquid, laminar vapor} \quad C = 10 \tag{5.52c}$$

$$\text{Both phases laminar} \quad C = 5 \tag{5.52d}$$

The Martinelli parameter X is given by the following equation:

$$X^2 = \left(\frac{dp_F}{dz}\right)_L \Big/ \left(\frac{dp_F}{dz}\right)_V \tag{5.53}$$

Mishima and Hibiki (1996) found that the constant C depends on the tube diameter, and recommended the following equation for C:

$$C = 21(1 - e^{-319D_h}) \tag{5.54}$$

where D_h is in meters. Mishima and Hibiki's correlation is used extensively and is recommended. English and Kandlikar (2005) found that their adiabatic air–water data in a 1-mm square channel was overpredicted using Eq. (5.54). Upon further investigation, they found that the diameter correction recommended by Mishima and Hibiki (1996) should be applied to C in Eq. (5.51). Accordingly,

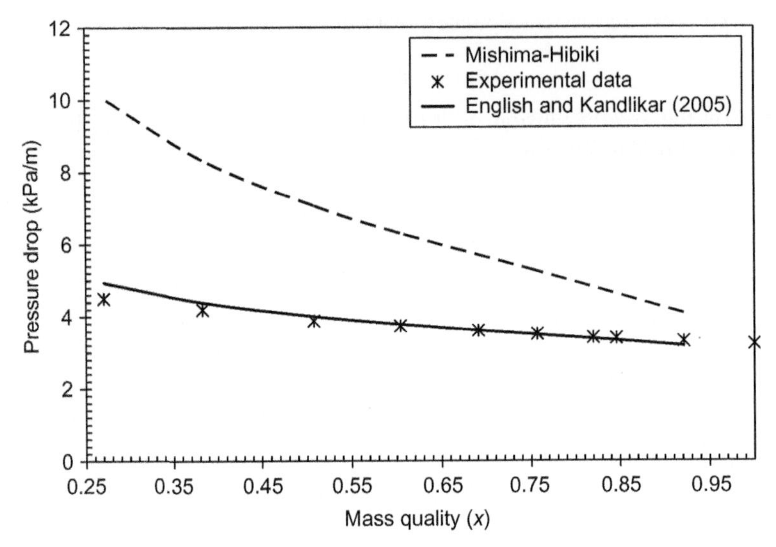

FIGURE 5.26

Comparison of experimental data from English and Kandlikar (2005) with their correlation and Mishima and Hibiki's correlation (1996).

the value of C will change depending on the laminar or turbulent flow of individual phases. Thus the following modified equation is recommended for frictional pressure drop calculation in microchannels and minichannels:

$$\phi_L^2 = 1 + \frac{C(1 - e^{-319D_h})}{X} + \frac{1}{X^2} \tag{5.55}$$

The value of C is obtained from Eqs. (5.52a)–(5.52d).

Figure 5.26 shows the agreement of experimental data obtained by English and Kandlikar (2005) during air–water flow in a 1-mm square channel. The average absolute deviation was found to be 3.5%, which was within the experimental uncertainties. Further validation of Eq. (5.55) is recommended.

The acceleration pressure drop is calculated from the following equation assuming homogeneous flow:

$$\Delta p_a = G^2 v_{LV} x_e \tag{5.56}$$

where v_{LV} is the difference between the specific volumes of vapor and liquid phases $= v_V - v_L$. The above equation assumes the inlet flow to be liquid only and exit quality to be x_e. For a two-phase inlet flow, the x_e should be replaced with the change in quality between the exit and inlet sections.

The gravitational pressure drop will be a very small component. It can be calculated from the following equation based on the homogeneous flow model:

$$\Delta p_g = \frac{g(\sin\beta)L}{v_{LV}x_e} \ln\left(1 + x_e \frac{v_{LV}}{v_L}\right) \tag{5.57}$$

where β is the angle made by the flow direction with the horizontal plane

Equation (5.57) also assumes a liquid inlet condition. In the case of two-phase inlet flow, the difference between the exit and inlet quality should be used in place of x_e.

5.9 Adiabatic two-phase flow

Adiabatic two-phase flow has been studied extensively in the literature, especially with air–water flow. For small-diameter channels, the work of Mishima and Hibiki (1996) is particularly noteworthy. Figure 5.27 shows their flow pattern map.

The effect of surfactants on the adiabatic two-phase flow pressure drop was studied by English and Kandlikar (2005) in a 1-mm square minichannel with air and water, with superficial gas and liquid velocities of 3.19–10 and 0.001–0.02 m/s, respectively. The resulting flow pattern was stratified (annular) in all cases, which is near the bottom right corner in Figure 5.27. The pressure drop was not affected by variation in surface tension from 0.034 to 0.073 N/m. The reason for this is believed to be the absence of a three-phase line (such as would exist with discrete droplets sliding on the wall). The Mishima and Hibiki (1996) correlation for pressure drop was modified as shown in Figure 5.26 using Eqs. (5.55) and (5.52).

5.10 Practical cooling systems with microchannels

Use of microchannels in a high heat flux cooling application using flow boiling systems was reviewed by Pokharna et al. (2004) for notebook computers and by Kandlikar (2005) for server applications. The major issues that need to be addressed before implementation of flow boiling becomes practical are listed as follows (Kandlikar, 2005).

High heat flux cooling systems with flow boiling have lagged behind the single-phase liquid cooled systems because of some operational challenges that still remain to be resolved. These are listed below:

1. Need for low-pressure water or a suitable refrigerant to match the saturation temperature requirement for electronics cooling.

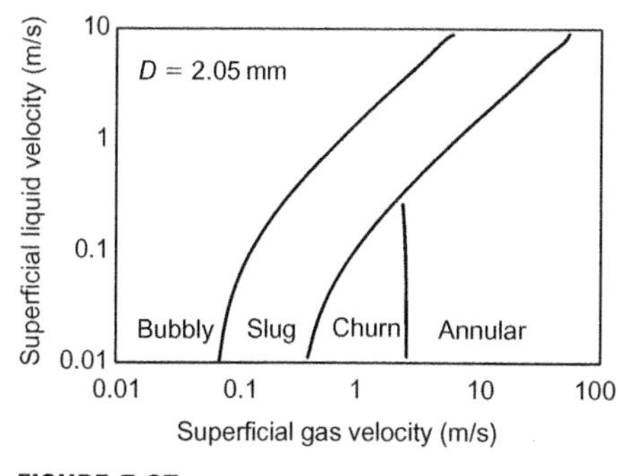

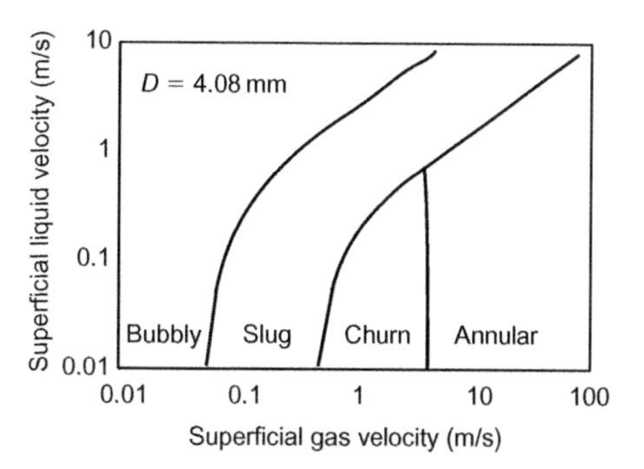

FIGURE 5.27

Flow pattern map for minichannels derived from Mishima and Hibiki (1996).

2. Unstable operation due to rapid bubble expansion and occasional flow reversal.
3. Unavailability of CHF data and a lack of fundamental understanding of the flow boiling phenomenon in microchannel passages.

The use of water is very attractive from a heat transfer performance perspective, but the positive pressure requirement may lead to a vacuum system, which is generally not desirable due to possible air leakage, which raises the saturation temperature in the system. Development of a refrigerant suitable for this application is recommended.

The desirable characteristics of an ideal refrigerant in a flow boiling system may be listed as (Kandlikar, 2005):

a. saturation pressure slightly above the atmospheric pressure at operating temperatures;
b. high latent heat of vaporization;
c. good heat transfer and pressure drop related properties (high liquid thermal conductivity, low liquid viscosity, low hysteresis for ONB);
d. high dielectric constant if applied directly into the chip;
e. compatible with silicon (for direct chip cooling) and copper (for heat sink applications), as well as other system components;
f. low leakage rates through pump seals;
g. chemical stability under system operating conditions;
h. low cost;
i. safe for human and material exposure under accidental leakages.

A fundamental understanding of the flow boiling phenomenon is slowly emerging, and efforts to stabilize flow using nucleating cavities and PDEs is expected to help in making flow boiling systems viable candidates for high heat flux cooling application.

Another application for flow boiling in narrow channels is in compact evaporators. The evaporators used in automotive applications currently do not employ some of the small channel configurations used in condenser application due to stability problems. Stabilized flow boiling, as described in Section 5.6, will enable practical implementation of microchannels and minichannels in a variety of compact evaporator applications. Some recent developments in this field will be presented in the remainder of this chapter.

5.11 Enhanced microchannel flow boiling systems

Flow boiling in microchannels suffers from three major drawbacks: (i) unstable operation, (ii) low CHF, and (iii) low heat transfer coefficient, except at the location of ONB during single-phase liquid at the inlet, and (iv) high pressure drop. The instability and the poor performance are interlinked, and the literature indicates that these issues represent inherent limitations of flow boiling in microchannels (Kandlikar, 2002a,b).

Several modifications have been proposed in the literature to overcome these limitations in an effort to enhance flow boiling heat transfer in recent years. Some of these enhancement techniques are introduced in the following sections.

5.11.1 Pin fins

Pin fin structures of different cross-sections have been used extensively in heat exchangers. A number of investigators have studied different fin arrangements with microscale flow passages between the fin arrays. Krishnamurthy and Peles (2008) studied flow boiling of water with a bank of staggered micro-pin fins, 250 μm long and 100 μm in diameter with pitch-diameter ratio of 1.5 for a mass fluxes ranging from 346 to 794 kg/m^2 s, and surface heat fluxes of 20–350 W/cm^2. The corresponding heat transfer coefficient was in the range of 60–75 kW/m^2°C. The single-phase pressure drop in the fin banks was exceedingly large as compared to the microchannel flows. Peles et al. (2005) reported a pressure drop of between 20 and 110 kPa for the single-phase flow of water at 3 ml/min in a 15 mm wide × 100 mm long pin fin section with a depth of 100 μm and a hydraulic diameter of 50 and 100 μm. The pressure drop during boiling was expected to be significantly higher. Kosar and Peles (2007a,b) reported a heat dissipation of 312 W/cm^2 in a 1.8-mm-wide channel fitted with 243-μm-deep hydrofoil-shaped pin fins with R-123. Although the heat transfer performance of the pin fin arrays was comparable with the microchannel performance, the high pressure drop poses a significant limitation.

Pin fin geometry was employed in a minichannel configuration by McNeil et al. (2010) in flow boiling with R-113. The pin fins were 1 mm square and 1 mm tall, placed in a 50-mm-wide and 50-mm-long channel. They dissipated a heat flux of 140 W/cm^2 with a pressure drop of 4–8 kPa for different pin fin configurations. The actual heat transfer coefficient over the fin surface was similar to that with a plain channel, between 5000 and 600 W/m^2°C, but the higher surface area of the pin fin allowed for a lower base temperature.

5.11.2 Microporous nanowire surfaces

Modifying the surface texture of the microchannel to improve the flow boiling performance has been studied by a number of researchers. Microporous surfaces were studied by Rainey et al. (2001) in a 10 mm × 10 mm × 1.5 mm high channel with flow boiling of FC-72. The microporous layer was made with aluminum particles and had cavities in the range of 0.1–1.0 μm and a layer thickness of 50 μm. The results indicated that the thermal resistance of the coating presented a significant barrier to heat transfer at high heat fluxes. Smaller thickness coatings are therefore desired. Recently, Sun et al. (2011) investigated flow boiling enhancement of FC-72 from microporous surfaces on copper substrate in minichannels of 10 mm width, 190 mm length, and varying heights of 0.25, 0.49, and 0.67 mm.

They studied the effect of structural parameters on the coating, such as particle size, pore size, coating thickness, and porosity, on boiling performance. The results were obtained at relatively low heat fluxes of below 15 W/cm^2. The porous coatings promoted nucleation, and significant reduction in the wall superheat was noted over a plain surface. At higher heat fluxes, the performance is expected to drop off considerably.

Khanikar et al. (2009) applied carbon nanotubes (CNT) to the bottom wall of a 10-mm-wide, 0.371-mm-tall, and 44.8-mm-long channel. Nucleation with water was initiated sooner than in a plain channel. The CNT-covered surface produced a large number of bubbles that transitioned the flow mainly into annular flow. At low mass fluxes, the CHF increased over a plain surface due to the increase in surface wettability, but the effect disappeared at higher heat fluxes, and the CHF was even lower. The CHF was, however, lower (30–40 W/cm^2) than with a plain copper surface under pool boiling ($\sim$120 W/cm^2).

Recently, Shenoy et al. (2011) applied multiwall CNT over a rectangular recess in a silicon wafer. The nanotubes were applied over the entire surface in one case, and were applied as cylindrical bundles serving as 500-μm-tall pin fins. They found heat transfer enhancement with subcooled liquid, but the testing was performed only at very low heat fluxes, below 45 W/cm^2.

The surface modification with nanostructures is being investigated by a few researchers, but the research is in its infancy. Integrating nanostructures in flow boiling channels is a very promising approach, and rapid development in this field is expected in the coming years.

5.11.3 Nanofluids

Nanofluids have been shown to improve CHF under pool boiling conditions due to deposits of the nanoparticles on the heater surface (Kim et al., 2007). As a bubble nucleates and evaporates, the local nanoparticle concentration increases, leading to their deposition in the vicinity of the nucleation cavities. Kim et al. (2008) conducted experiments with nanofluids of alumina particles under subcooled flow boiling in a large-diameter (8.7 mm) tube under vertical orientation. Pure water showed a CHF of 1.44 MW/m^2 at an inlet subcooling of 20°C, while the nanofluids with 0.01% by volume alumina nanoparticles resulted in a CHF of 3.25 MW/m^2 under a mass flux of 1500 kg/m^2 s. The nature of the CHF failure was also noted to be quite different. The CHF with pure water resulted in a catastrophic failure of the tube at the cross-section, while the CHF with the alumina nanoparticles resulted in a localized pinhole type failure. The higher wettability caused by nanoparticle deposits is believed to improve wettability and prevent the growth of local burnout at the CHF location. Although these results are for a macroscale tube, they are included here to illustrate the basic mechanism that may be affecting nanofluid behavior in minichannels and microchannels as well. In a subsequent paper, Kim et al. (2010) reported heat transfer coefficients for the same tests conducted by Kim et al. (2008). They observed no appreciable difference in the heat transfer coefficient between the nanofluids and pure water.

Boudouh et al. (2010) conducted experiments with copper–water nanofluids in 860-μm vertical channels under flow boiling conditions. They noted that the heat transfer coefficient and pressure drop both increased with the addition of three concentrations, 5 mg/L, 10 mg/L, and 50 mg/L. The heat transfer coefficient increased over the entire range of quality. The heat transfer coefficient with pure water was well correlated with the Kandlikar and Balasubramanian (2004) correlation given by Eqs. (5.37)–(5.43).

The increase in pressure drop with nanofluids observed by Boudouh et al. (2010) is somewhat surprising, but it may be caused by the more prominent role played by the bubbles, which are faced with a more hydrophilic surface with nanofluids under subcooled flow boiling conditions. Their two-phase friction pressure drop with pure water was well correlated with English and Kandlikar (2005) given by Eq. (5.55).

Henderson et al. (2010) found that direct dispersion of SiO_2 nanoparticles plays a critical role. When the particles were not well dispersed, the heat transfer coefficient decreased by as much as 55% in comparison to pure R-134a in a 7.9-mm inner diameter tube. Well-dispersed nanofluids containing polyester oil with CuO nanoparticles resulted in a 100% increase. The pressure drop increase was insignificant.

Vafaei and Wen (2010) conducted experiments with deionized water and alumina nanofluids in 510-μm-diameter microchannels under low mass flow rate conditions of 600–1650 kg/m^2 s. They found that CHF with nanofluids increased by 51% with 0.1% by volume of alumina nanoparticles. CHF increased with nanoparticle concentration from 0.001% to 0.1% by volume. They also noted that the pressure fluctuations were quite different with the nanofluids.

Ahn et al. (2012) modified the surface of a Zirlo tube used in nuclear applications. It was treated with anodic oxidation and resulted in improved wettability. This surface also exhibited up to 60% enhancement in CHF over a plain tube at a mass flux of 1500 kg/m^2 s. This further confirms that the surface wettability modification is the underlying reason for CHF enhancement with nanofluids.

Flow boiling with nanofluids results in the deposition of nanoparticles on the heater surface. This thin layer of nanoparticles changes the surface wettability of the channel walls. The higher wettability alters bubble behavior and enhances CHF. Since deposition of the nanoparticles depends on a number of factors, such as the size and dispersion of the nanoparticles, heat fluxes, nanoparticle–liquid interaction, concentration, duration of operation, and the base surface conditions, significant variations in the experimental results are expected from different sources. Providing a thin nanostructured layer on the heater surface by microfabrication techniques may be an alternate way to realize the same benefits. Lee and Mudawar (2007) observed that nanofluids offered marginal improvement in heat transfer, but the particles deposited in large clusters near the channel exit, causing catastrophic failure. In light of their findings, the long-term benefits on boiling performance need to be validated, and the effect of nanoparticles on the other system components needs to be carefully evaluated before their practical implementation.

5.12 Novel open microchannels with manifold

The four main issues that are currently inhibiting usage of microchannels in flow boiling application are:

1. Flow instability
2. Low heat transfer coefficient
3. Low CHF
4. High pressure drop.

Kandlikar et al. (2013) proposed a novel open microchannel design with a tapered manifold, as shown in Figure 5.28. The substrate has open microchannels, and a gap is introduced above it. This gap, called a manifold gap, has a taper with the gap increasing in the flow direction. The extra space available above the microchannels provides the additional cross-sectional area for the vapor generated along the flow length (Kandlikar et al., 2013). This helps in significantly reducing the pressure drop. The taper provides an expanding flow area configuration that is helpful in stabilizing flow. Further, providing separate pathways allows the vapor to flow in the manifold gap, while the liquid liquid prefers preferentially through the microchannels due to capillary forces.

Figure 5.29 shows the heat transfer performance of a microchannel 217 μm wide × 162 μm, with 160-μm-thick fins on a 10 mm × 10 mm copper substrate, with a uniform gap of 127 μm. It can be seen that the heat flux is insensitive to the mass flow rate, but drops off at higher flow rates. The tests were performed with degassed water at an inlet subcooling of 2–5°C. The performance of a tapered manifold was shown to be better than the uniform gap manifold.

It may be noted that none of the tested configurations reached CHF. The experiments were terminated because of the heater limitation in providing higher heat fluxes. The pressure drop was considerably lower, between 2 and 10 kPa with

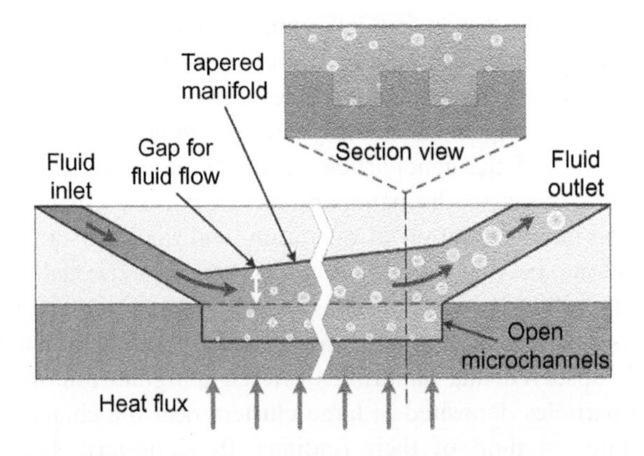

FIGURE 5.28

Schematic of the open microchannels with tapered manifold design.

Source: *Adapted from Kandlikar et al. (2013).*

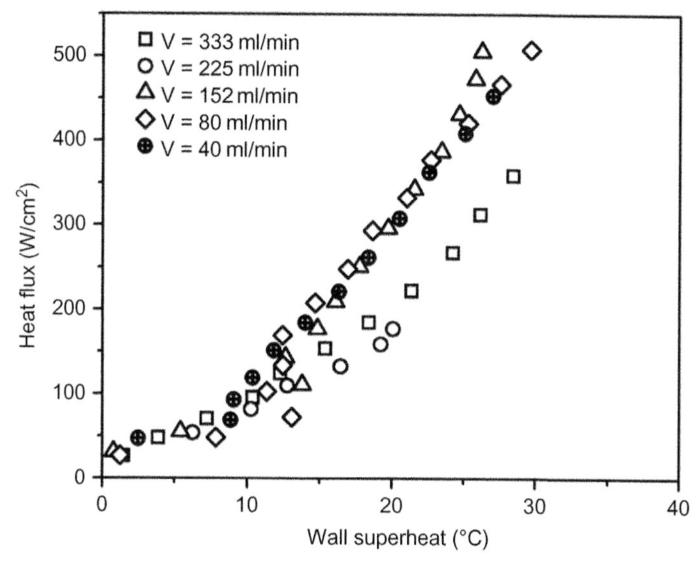

FIGURE 5.29

Heat flux versus wall superheat for an open microchannel (217 μm wide × 162 μm, with 160-μm-thick fins on a 10 mm × 10 mm copper substrate) with a manifold (Kandlikar et al., 2013).

the tapered design, while the pressure drop for a corresponding flow rate during flow boiling in the microchannels was in excess of 50 kPa at higher heat fluxes.

Introducing a slight taper was seen to improve heat transfer performance. However, the performance deteriorated with increasing taper beyond a certain limit. It appears that an optimum initial gap and taper exist for a given microchannel and manifold arrangement, the working fluid, and the operating conditions. It is expected that this configuration will provide the breakthrough in making the microchannels a viable option for high heat flux cooling. The pressure drop with this configuration was seen to be almost an order of magnitude lower as compared to the flow boiling in microchannels studied in the literature. The improvement in the heat transfer coefficient, significant enhancement in CHF (the setup dissipated a heat flux of over 500 W/cm^2 without reaching CHF), and a dramatic reduction in pressure drop make this configuration highly attractive. Further research into characterizing the heat transfer and pressure drop performance, and optimizing the geometrical parameters, are recommended.

5.13 Solved examples

Example 5.1

Water is used as the cooling liquid in a microchannel heat sink. The dimensions of one channel are $a = 1054$ μm × $b = 50$ μm, where a is the unheated length in the three-sided heating case. The inlet temperature of the water is 70°C, and the Reynolds number in the channel is 600.

1. Calculate the incipient boiling location and cavity radius and plot the wall superheat and liquid subcooling at the ONB versus the cavity radius r_c. Also plot the predicted heat transfer coefficient as a function of quality for the following values of heat flux:
   (i) $q'' = 50\ \text{kW/m}^2$; (ii) $q'' = 340\ \text{kW/m}^2$; (iii) $q'' = 1\ \text{MW/m}^2$.
2. Calculate the pressure drop in the heat exchanger core for a heat flux of $1\ \text{MW/m}^2$ and a channel length of 20 mm.

Assumptions

Fully developed laminar flow Nusselt number; three-sided heating; properties of water are at saturation temperature at 1 atm; receding contact angle is 40°.

Solution

Properties of water at 100°C: $h_{LV} = 2.26 \times 10^6$ J/kg, $i_L = 4.19 \times 10^5$ J/kg, $\rho_V = 0.596\ \text{kg/m}^3$, $v_L = 0.001044\ \text{m}^3\text{/kg}$, $v_g = 1.679\ \text{m}^3\text{/kg}$, $c_{p,L} = 4217$ J/kg K, $\mu_V = 1.20 \times 10^{-5}\ \text{N s/m}^2$, $\mu_L = 2.79 \times 10^{-4}\ \text{N s/m}^2$, $k_L = 0.68$ W/m K, $\sigma = 0.0589$ N/m, $Pr = 1.76$ (Incropera and DeWitt, 2002). Properties of water at 70°C: $i_L = 2.93 \times 10^5$ J/kg.

Part 1

(i) Heat flux is 50 kW/m².

The fully developed Nusselt number is found using Table 3.3 with aspect ratio $\alpha_c = a/b = 1054\ \mu\text{m}/50\ \mu\text{m} = 21.08$. Since the aspect ratio is greater than 10, the Nusselt number for three-sided heating is 5.385. The hydraulic diameter and heated perimeter are:

$$D_h = \frac{4A_c}{P_w} = \frac{2ab}{(a+b)} = \frac{2 \times 1054 \times 10^{-6} \times 50 \times 10^{-6}}{(1054 + 50) \times 10^{-6}} = 9.547 \times 10^{-6}$$

$$= 95.47\ \mu\text{m}$$

$$P = a + 2b = (2 \times 50 + 1054) \times 10^{-6} = 1.154 \times 10^{-3}\ \text{m} = 1154\ \mu m$$

The heat transfer coefficient is $h = k_L \text{Nu}/D_h = (0.68\ \text{W/m K})(5.385)/(9.55 \times 10^{-5}\ \text{m}) = 38{,}355\ \text{W/m}^2\ \text{K}$.

Under the fully developed assumption, a value for the temperature difference between the wall temperature and the bulk liquid temperature can be found using the equation $q'' = h(\Delta T) = h(T_w - T_B) = h(T_w - T_{Sat} + T_{Sat} - T_B) = h(\Delta T_{Sat} + \Delta T_{Sub})$. Therefore, the sum of ΔT_{Sat} and ΔT_{Sub} can be found for a given heat flux as $q''/h = (\Delta T_{Sat} + \Delta T_{Sub}) = (50{,}000\ \text{W/m}^2)/(38{,}355\ \text{W/m}^2\ \text{K}) = 1.30°\text{C}$.

The wall superheat at the critical cavity radius is found by setting the expression under the radical in Eq. (5.16) equal to zero.

$$\sqrt{1 - [8.8\sigma T_{Sat}(\Delta T_{Sat} + \Delta T_{Sub})]/(\rho_V h_{LV} \delta_t \Delta T_{Sat}^2)} = 0$$

Noting that $\delta_t = k_L/h$, the equation $q'' = h(\Delta T_{Sat} + \Delta T_{Sub})$ can be written as $q'' = (K_L/\delta_t)(\Delta T_{Sat} + \Delta T_{Sub})$.

Solving this equation for δ_t:

$$\delta_t = k_L(\Delta T_{Sat} + \Delta T_{Sub})/q'' = (0.68 \times 1.3)/50,000 = 1.77 \times 10^{-5}\ \text{m} = 17.7\ \mu\text{m}$$

and substituting it into the expression under the radical in Eq. (5.16), the value for ΔT_{Sat} at the critical r_c can be obtained as:

$$\Delta T_{Sat} = \sqrt{\frac{8.8\sigma T_{Sat} q''}{\rho_V h_{LV} k_L}} = \sqrt{\frac{8.8 \times 0.0589 \times 373.15 \times 50,000}{0.596 \times 2.26 \times 10^6 \times 0.68}} = 3.25°\text{C}$$

The cavity radius can be found by using Eq. (5.17):

$$r_{c,crit} = \left(\frac{\delta_t \sin\theta_r}{2.2}\right)\left(\frac{\Delta T_{Sat}}{\Delta T_{Sat} + \Delta T_{Sub}}\right) = \left(\frac{1.77 \times 10^{-5} \times \sin 40°}{2.2}\right)\frac{3.25}{1.3}$$
$$= 12.9 \times 10^{-6}\ \text{m} = 12.9\ \mu\text{m}$$

Using Eq. (5.1) and the definitions of wall superheat and liquid subcooling given by Eqs. (5.8) and (5.9), the location where ONB occurs is found:

$$z = \frac{(T_{B,z} - T_{B,i})\dot{m}c_p}{q''P} = \frac{(375.1 - 343.15) \times 92.4 \times 10^{-6} \times 4217}{50,000 \times 1.15 \times 10^{-3}}$$
$$= 0.216\ \text{m} = 21.6\ \text{m}$$

Realistically, this length is too large for a microchannel heat exchanger, and the actual length is expected to be much shorter. Also, the heat flux is very low for a microchannel heat exchanger. Therefore, it will be operating under single-phase liquid flow conditions throughout. Plotting Eq. (5.16) as a function of ΔT_{Sat} illustrates the nucleation criteria for the given heat flux. The wall superheat at ONB is plotted as a function of the nucleation cavity radius (ΔT_{ONB} versus r_c) by using Eq. (5.20). The liquid subcooling is also presented on this plot. Note that the negative values of ΔT_{Sub} indicate that the bulk is superheated at the point of nucleation. This will lead to significant flow instabilities as discussed in the Section 5.4.

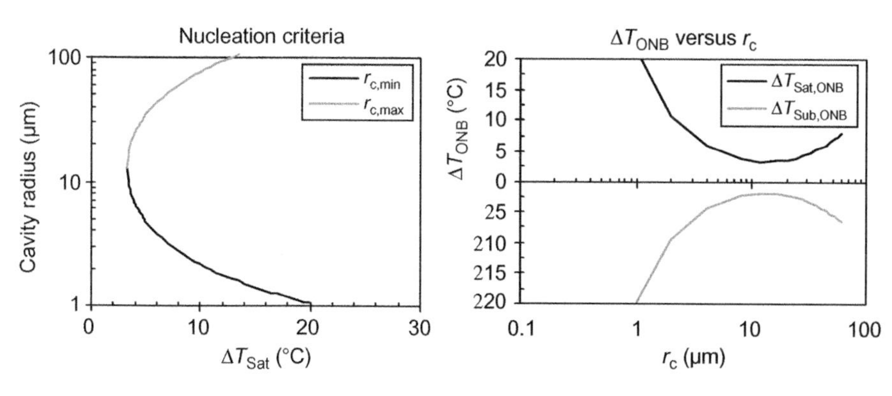

The boiling number is calculated from Table 5.1:

$$Bo = \frac{q''}{Gh_{LV}} = \frac{q''D_h}{h_{LV}Re\mu_L} = \frac{50,000 \times 9.547 \times 10^{-5}}{2.26 \times 10^6 \times 600 \times 2.79 \times 10^{-4}} = 1.262 \times 10^{-5}$$

For a flow of higher Reynolds number, the value of h_{LO} would not be the same as the single-phase value of h, as can be seen with Eqs. (5.25)–(5.27). As the Reynolds number is 600, the heat transfer coefficient h_{LO} is equal to the single-phase value of h by Eq. (5.42), $h_{LO} = h = 38{,}355$ W/m^2 K.

The convection number is a function of quality:

$$\begin{aligned} Co &= [(1-x)/x]^{0.8}[\rho_V/\rho_L]^{0.5} = [(1-x/x)]^{0.8}(0.596/957.9) \\ &= [(1-x)/x]^{0.8} \times 6.22 \times 10^{-4} \end{aligned}$$

From Eq. (5.37), the flow boiling h_{TP} is the larger of Eqs. (5.38) and (5.39), which can be plotted as functions of x as seen below using an F_{Fl} of 1:

$$h_{TP} = \text{larger of} \begin{cases} h_{TP,NBD} \\ h_{TP,CBD} \end{cases}$$

$$\begin{aligned} h_{TP,NBD} &= h_{LO}[0.6683Co^{-0.2}(1-x)^{0.8} + 1058.0Bo^{0.7}(1-x)^{0.8}F_{Fl}] \\ &= 38,355\{0.6683([(1-x)/x]^{0.8} \times 6.22 \times 10^{-4})^{-0.2}(1-x)^{0.8} \\ &\quad + 1058.0(1.262 \times 10^{-5})^{0.7}(1-x)^{0.8} \times 1\} \end{aligned}$$

$$\begin{aligned} h_{TP,CBD} &= h_{LO}[1.136Co^{0.9}(1-x)^{0.8} + 667.2Bo^{0.7}(1-x)^{0.8}F_{Fl}] \\ &= 38,355\{1.136([(1-x)/x]^{0.8} \times 6.22 \times 10^{-4})^{-0.9}(1-x)^{0.8} \\ &\quad + 667.2(1.262 \times 10^{-4})^{0.7}(1-x)^{0.8} \times 1 \end{aligned}$$

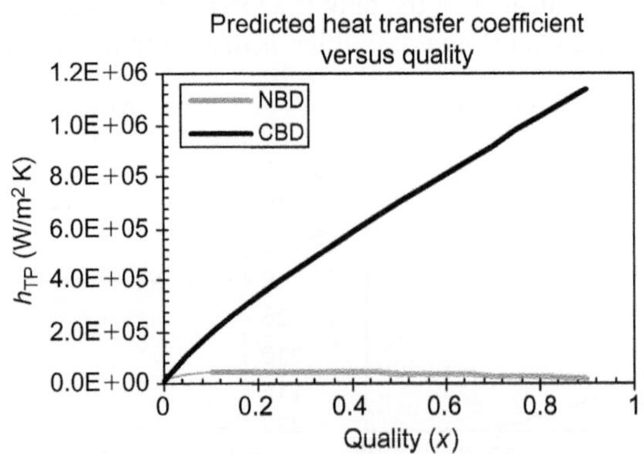

Since $h_{TP,NBD}$ yields higher values than $h_{TP,CBD}$, it represents the flow boiling thermal conditions and should be used for the present case. Note that h_{TP} increases with quality for this case.

(ii) Heat flux is 340 kW/m^2.

$a = 1054$ μm; $b = 50$ μm; $T_i = 70°C$; $Re = 600$; $\alpha_c = 21.08$; $h = 38{,}355$ W/m^2 K $(\Delta T_{Sat} + \Delta T_{Sub}) = 8.86°C$; $\Delta T_{Sat} = 8.48°C$; $r_{c,crit} = 5.0$ μm; $z = 2.9$ cm.

In plotting the heat transfer coefficient versus quality, it is seen that the nucleate boiling dominant prediction is actually higher than the convective dominant prediction for very low qualities (<0.1) and should be used in those cases. There is an increase in the nucleate boiling behavior in comparison to the lower heat flux case.

(iii) Heat flux is 1 MW/m^2.

$a = 1054$ μm; $b = 50$ μm; $T_i = 70°C$; $Re = 600$; $\alpha_c = 21.08$; $h = 38{,}355$ W/m^2 K $(\Delta T_{Sat} + \Delta T_{Sub}) = 26.1°C$; $\Delta T_{Sat} = 14.5°C$; $r_{c,crit} = 2.9$ μm; $z = 0.6$ cm.

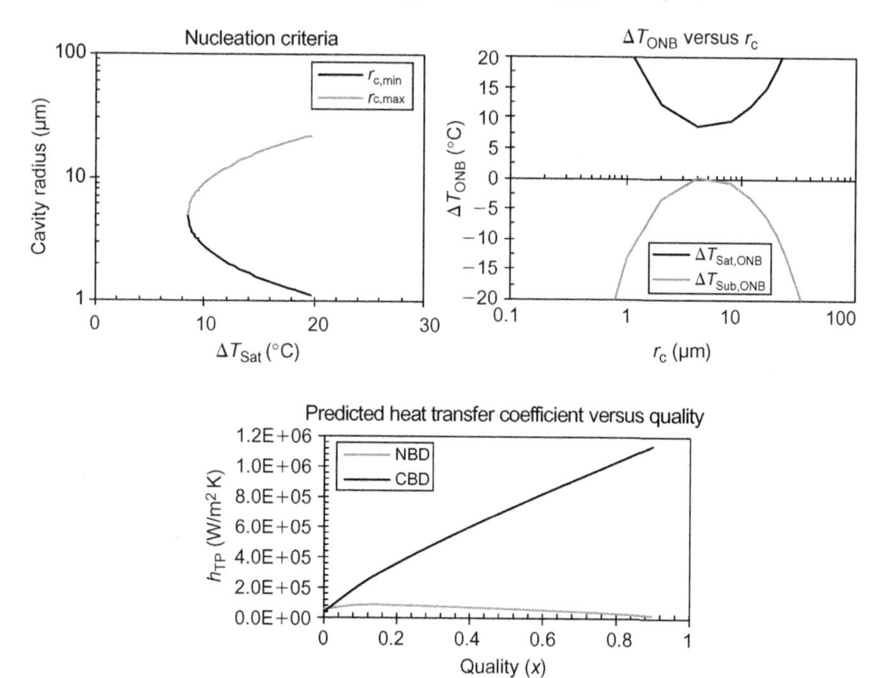

The nucleate boiling dominant prediction is higher than the convective boiling dominant for qualities lower than 0.4 and should be used for those cases. Note that a high wall superheat is needed to initiate nucleation (14.5°C). Also, the cavity sizes for nucleation become smaller as the heat flux increases. Thus a variety of different cavity sizes are needed on a surface to operate it at different heat flux conditions. Another point to note is that the saturation temperature of water at 1 atm is quite high for electronics cooling applications. Two options can be pursued: one is to use low-pressure steam, and the other is to use refrigerants. The next example illustrates the use of refrigerants.

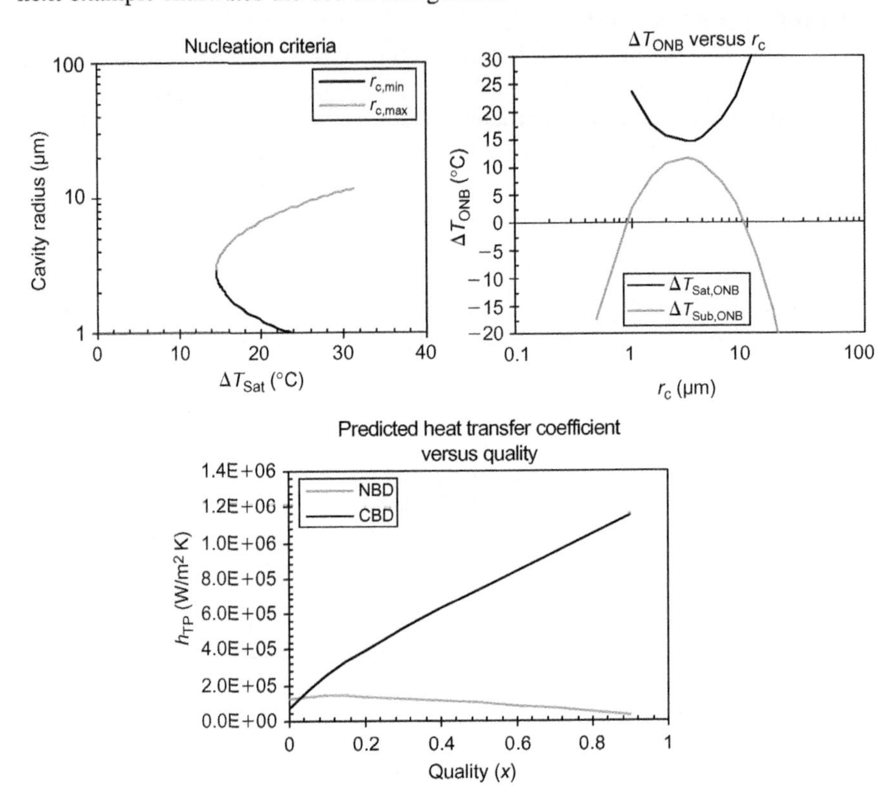

Part 2

From Part 1 (iii), it is found that the ONB occurs at 0.6 cm from the entrance end. Therefore, the flow is single-phase up to 0.6 cm at which point it becomes two-phase flow. However, the bulk is subcooled at this location and so Eq. (5.1) is used to determine the location where the saturation temperature is reached:

$$z = \frac{(T_{B,Z} - T_{B,i})\dot{m}c_p}{q''P} = \frac{(100 - 70) \times 92.4 \times 10^{-6} \times 4217}{1 \times 10^6 \times 1.15 \times 10^{-3}} = 0.0101 \text{ m} = 1.01 \text{ cm}$$

Using this approach simplifies the problem, but keep in mind that since ONB actually occurs before this location in the channel, the actual pressure drop in the core of the channel will be slightly higher due to the longer two-phase flow length.

The exit quality can be solved for by using the following equation to get $x_e = 0.055$, where A_T is the heated perimeter multiplied by the channel length:

$$q''A_T = \dot{m}[(i_{L,@T_{Sat}} + x_e h_{LV,@T_{Sat}}) - i_{L,@T_{B,in}}]$$

$$x_e = \left(\frac{q''A_T}{\dot{m}} + i_{L,@T_{B,in}} - i_{L,@T_{Sat}}\right) \Big/ h_{LV,@T_{Sat}}$$

$$= \left(\frac{10,00,000 \times 1.15 \times 10^{-3} \times 0.02}{9.24 \times 10^{-5}} + 2.93 \times 10^5 - 4.19 \times 10^5\right) \Big/ 2.26 \times 10^6$$

$$= 0.055$$

The total mass flux can be calculated from the Reynolds number and channel geometry:

$$G = Re\mu_L / D_h = (600 \times 2.79 \times 10^{-4})/(9.547 \times 10^{-5}) = 1753 \text{ kg/m}^2 \text{ s}$$

The superficial Reynolds numbers of the liquid and vapor phases in the two-phase flow region are found by using the following equations. The average quality over the channel length is taken as half the exit quality.

$$Re_V = \frac{G(x_e/2)D_h}{\mu_V} = \frac{1753 \times (0.055/2) \times 9.547 \times 10^{-5}}{1.2 \times 10^{-5}} = 384$$

$$Re_L = \frac{G(1-(x_e/2))D_h}{\mu_L} = \frac{1753(1-(0.055/2))9.547 \times 10^{-5}}{2.79 \times 10^{-4}} = 583$$

The single-phase friction factors f_L and f_V are obtained by using Eq. (3.10) with the Reynolds numbers calculated above.

$$\begin{aligned} Po = fRe &= 24(1 - 1.3553\alpha_c + 1.9467\alpha_c^2 - 1.7012\alpha_c^3 + 0.9564\alpha_c^4 - 0.2537\alpha_c^5) \\ &= 24(1 - 1.3553 \times 21.08 + 1.9467 \times 21.08^2 - 1.7012 \times 21.08^3 + 0.9564 \\ &\quad \times 21.08^4 - 0.2537 \times 21.08^5) \\ &= 22.56 \end{aligned}$$

The friction and acceleration pressure gradients in Eq. (5.53) are calculated as follows:

$$-\left(\frac{dp_F}{dz}\right)_L = \frac{2f_L G^2(1-(x_e/2))^2}{D_h \rho_L}$$

$$= \frac{2(22.56/583)1753^2(1-(0.055/2))^2}{9.547 \times 10^{-5} \times 957.9} = 2,458,697 \text{ Pa/m} = 2.46 \text{ MPa/m}$$

$$-\left(\frac{dp_F}{dz}\right)_V = \frac{2f_V G^2(x_e/2)^2}{D_h \rho_V} = \frac{2(22.56/384)1753^2(0.055/2)^2}{9.547 \times 10^{-5} \times 0.5956}$$

$$= 4,800,308 \text{ Pa/m} = 4.80 \text{ MPa/m}$$

$$X^2 = \left(\frac{dp_F}{dz}\right)_L \Big/ \left(\frac{dp_F}{dz}\right)_V = \frac{2.46}{4.80} = 0.5125$$

The two-phase multiplier is calculated with Eq. (5.40):

$$\phi_L^2 = 1 + \frac{C(1 - e^{-319D_h})}{X} + \frac{1}{X^2} = 1 + \frac{5(1 - e^{-319 \times 9.547 \times 10^{-5}})}{\sqrt{0.5125}} + \frac{1}{0.5125} = 3.16$$

and the two-phase pressure drop per unit length is calculated with Eq. (5.50):

$$\left(\frac{dp_F}{dz}\right)_{TP} = \left(\frac{dp_F}{dz}\right)_L \phi_L^2 = 2,458,69716 = 7,769,482 \text{ Pa/m} = 7.77 \text{ MPa/m}$$

The frictional two-phase pressure drop is found by multiplying the frictional pressure gradient by the two-phase flow length (0.02 m − 0.0101 m = 0.0099 m) to get 76.7 kPa.

The total pressure drop in the core is found by adding the pressure drop in the single-phase flow section length (0.0101 m) and the two-phase pressure drop:

$$\Delta p = \Delta p_{f,1-ph} + \Delta p_{f,tp} = 26,800 + 76,700 = 103,500 \text{ Pa} = 103.5 \text{ kPa}(15.0 \text{ psi})$$

Example 5.2

Microchannels are directly etched into silicon chips to dissipate a heat flux of 13,000 W/m^2 from a computer chip. The geometry may be assumed similar to Figure 3.19. Each of the parallel microchannels has a width $a = 200$ μm, height $b = 200$ μm, and length $L = 10$ mm. Refrigerant R-123 flows through the horizontal microchannels at an inlet temperature of $T_{B,i} = 293.15$ K. The heated perimeter $P = b + a + b = 600 \times 10^{-6}$ m, and the cross-sectional area $A_c = a \times b = 40 \times 10^{-9}$ m. Assume θ_r from Figure 5.1 is 20°, and $Re = 100$.

(i) Calculate the incipient boiling location and cavity radius?
(ii) Plot the wall superheat and liquid subcooling versus nucleating cavity radius for q" = 5, 13, and 30 kW/m^2.
(ii) Calculate the pressure drop in the test section.
(ii) Plot the predicted heat transfer coefficient as a function of quality.

Solution

Properties of R-123 at $T_{Sat} = 300.9$ K and 1 atm: $\mu_L = 404.2 \times 10^{-6}$ N s/m^2, $\mu_V = 10.8 \times 10^{-6}$ N s/m^2, $\rho_L = 1456.6$ kg/m^3, $\rho_V = 6.5$ kg/m^3, $c_{p,L} = 1023$ J/kg K, $k_L = 75.6 \times 10^{-3}$ W/m K, $k_V = 9.35 \times 10^{-3}$ W/m K, $\sigma_L = 14.8 \times 10^{-3}$ N/m, $h_{LV} = 170.19 \times 10^3$ J/kg, $i_L = 228 \times 10^3$ J/kg, $i_V = 398 \times 10^3$ J/kg.

(i) Calculate the incipient boiling location and cavity radius (Answers: $z = 0.584 \times 10^{-3}$ m, and $r_{c,crit} = 2.23 \times 10^{-6}$ m).
From Eq. (5.18):

$$\Delta T_{Sat,ONB} = \sqrt{8.8\sigma T_{Sat} q''/(\rho_V h_{LV} k_L)}$$
$$= \sqrt{8.8(14.8\times10^{-3})(300.9)(13{,}000)/(6.5)(170.19\times10^{3})(75.6\times10^{-3})}$$
$$= 2.47 \text{ K}$$

Calculate the hydraulic diameter:

$$D_h = \frac{4A_c}{P_w} = a = b = 200 \times 10^{-6} \text{ m}$$

From Table 3.3, $Nu_{fd,3} = 3.556$, and note that:

$$h = \frac{Nuk_L}{D_h} = \frac{(3.556)(0.0756)}{(200\times10^{-6})} = 1344 \text{ W/m}^2 \text{ K}$$

From Eq. (5.19):

$$\Delta T_{Sub,ONB} = \frac{q''}{h} - \Delta T_{Sat,ONB} = \frac{13{,}000}{1344} - 2.47 = 7.2 \text{ K}$$

From Eq. (5.8), T_B at the ONB is:

$$T_{B,ONB} = T_{Sat} - \Delta T_{Sub,ONB} = 300.9 - 7.2 = 293.7 \text{ K}$$

Calculate the flow velocity using the Reynolds number:

$$V = \frac{Re\mu_L}{\rho D_h} = \frac{(100)(404.2\times10^{-6})}{(1456.6)(200\times10^{-6})} = 0.139 \text{ m/s}$$

Calculate the mass flow rate:

$$\dot{m} = \rho V A_c = (1456.6)(0.139)(40\times10^{-9}) = 8.09\times10^{-6} \text{ kg/s}$$

Calculate the mass flux:

$$G = \frac{\dot{m}}{A_c} = \frac{8.09\times10^{-6}}{40\times10^{-9}} = 202 \text{ kg/m}^2 \text{ s}$$

The incipient boiling location can be calculated by rearranging Eq. (5.1):

$$z = (T_{B,ONB} - T_{B,i})\left(\frac{\dot{m}c_{p,L}}{q''P}\right)$$

$$= (293.7 - 293.15)\left(\frac{(8.09\times10^{-6})(1023)}{(13{,}000)(600\times10^{-6})}\right)$$

$$= 0.584\times10^{-3} \text{ m}$$

To find the cavity radius, substitute Eqs. (5.3) and (5.19) into Eq. (5.17); we get

$$r_{c,\mathrm{crit}} = \frac{k_L \sin\theta_r \Delta T_{\mathrm{Sat,ONB}}}{2.2q''} = \frac{(75.6\times10^{-3})(\sin 20^\circ)(2.47)}{2.2(13,000)} = 2.23\times10^{-6}\ \mathrm{m}$$

(ii) Plot the wall superheat and liquid subcooling versus nucleating cavity radius for q″ = 5, 13, and 30 kW/m².

Equations (5.19) and (5.20) are used to plot the following figures.

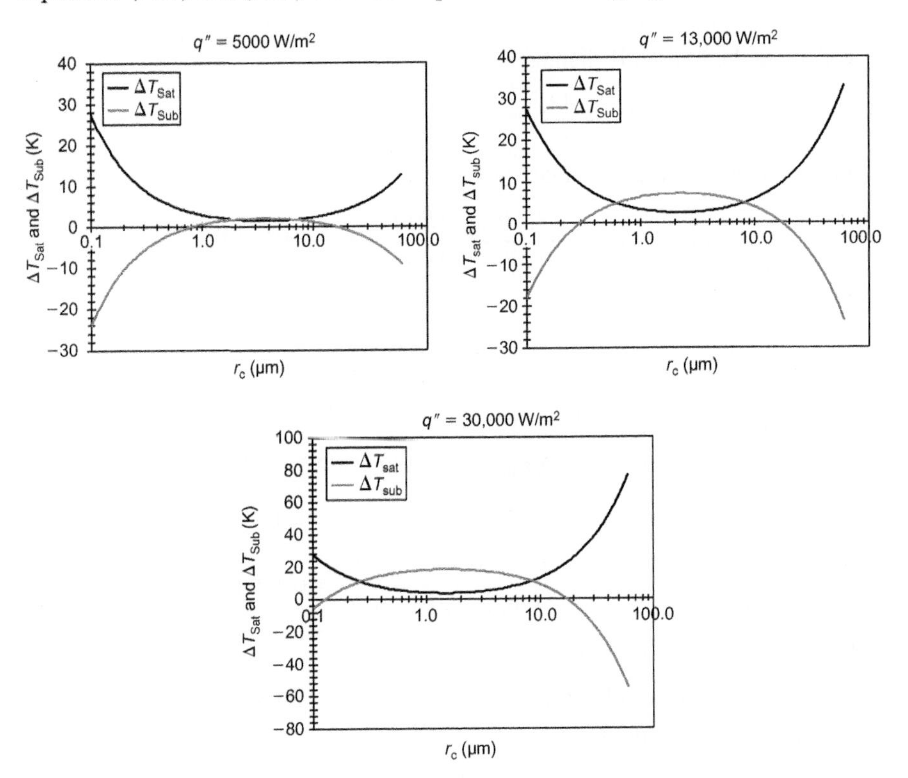

(iii) Calculate the pressure drop in the test section (Answer: 426 Pa).

For fully developed laminar flow, the hydrodynamic entry length may be obtained using Eq. (3.11):

$$L_h = 0.05ReD_h = 0.05(100)(0.200) = 1.0\ \mathrm{mm}$$

Since $L > L_h$, the fully developed flow assumption is valid.

From Eq. (5.1):

$$z = (T_{B,z} - T_{B,i})\left(\frac{\dot{m}c_{p,L}}{q''P}\right) = (300.97 - 293.15)\left(\frac{(8.09\times10^{-6})(1023)}{(13,000)(600\times10^{-6})}\right)$$

$$= 8.29\times10^{-3}\ \mathrm{m}$$

Note that z is the location where two-phase boiling begins. So we need to find the single-phase pressure drop until z and then add to that the value of the two-phase pressure drop from z to L.

The total pressure drop can be found using Eq. (5.44):

$$\Delta p = \Delta p_c + \Delta p_{f,1-ph} + \Delta p_{f,tp} + \Delta p_a + \Delta p_g + \Delta p_e$$

To find the single-phase pressure drop, the fRe term can be obtained using Eq. (3.10) and using an aspect ratio of 1:

$$\alpha_c = a/b = 200/200 = 1$$

$$\begin{aligned} fRe &= 24(1 - 1.3553\alpha_c + 1.9467\alpha_c^2 - 1.7012\alpha_c^3 + 0.9564\alpha_c^4 - 0.2537\alpha_c^5) \\ &= 24(1 - 1.3553 \times 1 + 1.9467 \times 1^2 - 1.7012 \times 1^3 + 0.9564 \times 1^4 - 0.2537 \times 1^5) \\ &= 14.23 \end{aligned}$$

The single-phase core frictional pressure drop can be calculated using Eq. (3.14):

$$\Delta p_{f,1-ph} = \frac{2(fRe)\mu_L U_m z}{D_h^2} + K(\infty) \cdot \frac{\rho_L U_m^2}{2}$$

where $K(\infty)$ is given by Eq. (3.18)

$$\begin{aligned} K(\infty) &= (0.6796 + 1.2197\alpha_c + 3.3089\alpha_c^2 - 9.5921\alpha_c^3) + 8.9089\alpha_c^4 - 2.9959\alpha_c^5) \\ &= 1.53 \end{aligned}$$

$$\begin{aligned} \Delta p_{f,1-ph} &= \frac{2(14.23)(0.404 \times 10^{-3})(0.139)(0.00829)}{(200 \times 10^{-6})^2} + (1.53)\frac{(1456.6)(0.139)^2}{2} \\ &= 353 \text{ Pa} \end{aligned}$$

Note that Δp_c, Δp_g, and Δp_e can be neglected because we are calculating the core pressure drop in horizontal microchannels.

Equations (9.1-12, 14, 35, 36, 37, 39, and 40) are from Chapter 9 of *Handbook of Phase Change* (Kandlikar et al., 1999). Assuming that the liquid pressure is 1 atm, then $i_{IN} = 228 \times 10^3$ J/kg. The heated area in the two-phase region is:

$$\begin{aligned} A_{h,tp} &= (b + a + b)(L - z) = (200 \times 10^{-6} + 200 \times 10^{-6} + 200 \times 10^{-6}) \\ &\quad \times (10 \times 10^{-3} - 8.29 \times 10^{-3}) \\ &= 1.026 \times 10^{-6} \text{ m}^2 \end{aligned}$$

From Eq. (9.1-12):

$$i_{TP} = i_{IN} + \frac{q'' A_{h,tp}}{G A_c} = 228 \times 10^3 + \frac{(13,000)(1.026 \times 10^{-6})}{(8.09 \times 10^{-6})} = 229.6 \times 10^3 \text{ J/kg}$$

From Eq. (9.1-14):

$$x_e = \frac{i_{TP} - i_L}{h_{LV}} = \frac{(229.6 \times 10^3 - 228 \times 10^3)}{(170.19 \times 10^3)} = 0.0094$$

Use the average thermodynamic quality between 0 and x_e, $x_{avg} = 0.0047$. From Eq. (9.1-39):

$$Re_V = \frac{GxD_h}{\mu_V} = \frac{(202)(0.0047)(200 \times 10^{-6})}{(10.8 \times 10^{-6})} = 17.58$$

From earlier, the vapor friction factor is:

$$f_V = \frac{14.23}{Re_V} = \frac{14.23}{17.58} = 0.809$$

From Eq. (9.1-35):

$$-\left(\frac{dp_F}{dz}\right)_V = \frac{2f_V G^2 x^2}{D_h \rho_V} = \frac{2(0.809)(202)^2(0.0047)^2}{(200 \times 10^{-6})(6.5)} = 1122 \text{ Pa/m}$$

From Eq. (9.1-40):

$$Re_L = \frac{G(1-x)}{\mu_L} = \frac{(202)(1-0.0047)(200 \times 10^{-6})}{(0.4042 \times 10^{-3})} = 9948$$

$$f_L = \frac{14.23}{Re_L} = \frac{14.23}{99.48} = 0.143$$

From Eq. (9.1-36):

$$-\left(\frac{dp_F}{dz}\right)_L = \frac{2f_L G^2(1-x)^2}{D_h \rho_L} = \frac{2(0.143)(202)^2(1-0.0047)^2}{(200 \times 10^{-6})(1453.6)} = 39,683 \text{ Pa/m}$$

From Eq. (5.53):

$$X^2 = \left(\frac{dp_F}{dz}\right)_L \bigg/ \left(\frac{dp_F}{dz}\right)_V = \frac{39683}{1122} = 35.37$$

Assuming both phases are laminar, from Eq. (5.52d), $C = 5$.
Using Eq. (5.55):

$$\phi_L^2 = 1 + \frac{C(1 - e^{-319D_h})}{X} + \frac{1}{X^2} = 1 + \frac{5(1 - e^{-319(200 \times 10^{-6})})}{\sqrt{35.37}} + \frac{1}{35.37} = 1.08$$

From Eq. (5.50):

$$\Delta p_{f,tp} = \left(\frac{dp_F}{dz}\right) = \left(\frac{dp_F}{dz}\right)_L \phi_L^2 = (39,683)(1.08) = 42,858 \text{ Pa/m}$$

The acceleration pressure drop is calculated from Eq. (5.56):

$$\Delta p_a = G^2 v_{LV} x_e = (202)^2(0.00069)(0.0094) = 0.26 \text{ Pa}$$

The value of the acceleration pressure drop is negligible, and hence the total pressure drop is:

$$\Delta p = \Delta p_{f,1-ph} + \Delta p_{f,tp} = (353) + (42,858)(0.01 - 0.00829) = 426 \text{ Pa}$$

(iv) Plot the predicted heat transfer coefficient as a function of quality.
From Table 5.1, the boiling number is:

$$Bo = \frac{q''}{Gh_{LV}} = \frac{(13,000)}{(202)(170.19 \times 10^3)} = 0.378 \times 10^{-3}$$

From Table 5.1, the convection number is:

$$Co = [(1-x)/x]^{0.8}[\rho_V/\rho_L]^{0.5} = [(1-x)/x]^{0.8}(6.5/1456.6)^{0.5}$$
$$= 0.0668[(1-x)/x]^{0.8}$$

For $Re_{LO} \leq 100$, from Eqs. (5.42) and (5.43):

$$h_{LO} = \frac{Nuk_L}{D_h} = \frac{(3.556)(0.0756)}{(200 \times 10^{-6})} = 1344 \text{ W/m}^2 \text{ K}$$

From table 3 in page 391 of *Handbook of Phase Change* (Kandlikar et al., 1999), assume $F_{Fl} = 1.3$.

For $Re_{LO} \leq 100$, $h_{TP} = h_{TP,NBD}$, and we can plot the predicted heat transfer coefficient as a function of quality using Eq. (5.43). This is very similar to the method followed in the previous example, but only the nucleate boiling dominant prediction is used from Eq. (5.43):

$$h_{TP} = h_{TP,NBD} = h_{LO}\{0.6683\ Co^{-0.2}(1-x)^{0.8} + 1058.0 Bo^{0.7}(1-x)^{0.8} F_{Fl}\}$$
$$= 1344\{0.6683(0.0668[(1-x)/x]^{0.8})^{-0.2}(1-x)^{0.8}$$
$$+ 1058.0(0.378 \times 10^{-3})^{0.7}(1-x)^{0.8}\ F_{Fl}\}$$

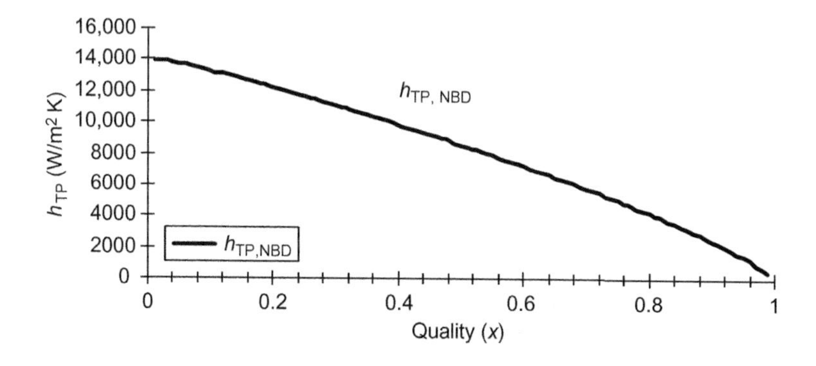

The total frictional pressure drop must include the minor losses because the actual Δp is higher than the core frictional pressure drop. Due to the laminar conditions and the dominance of nucleate boiling effects, note the decreasing trend in h_{TP} as a function of quality.

5.14 Practice problems

Problem 5.1

A microchannel with dimensions of 120 μm × 400 μm and 3 mm length operates at an inlet pressure of 60 kPa. The inlet liquid is at saturation temperature. A heat flux of 600 kW/m^2 is applied to the three sides of the channel walls:

- **i.** Calculate the mass flux to provide an exit quality of 0.1 from this channel. Assuming that cavities of all sizes are available, calculate the distance at which ONB occurs. What is the nucleating cavity diameter at this location?
- **ii.** Calculate the pressure drop in the channel.
- **iii.** Plot the variation of heat transfer coefficient with channel length assuming the evaporator pressure to be the mean of the inlet and exit pressures.

Problem 5.2

R-123 is used to cool a microchannel chip dissipating 400 W from a chip with 12 mm × 12 mm surface area. Assuming an allowable depth of 300 μm, design a cooling system to keep the channel wall temperature below 70°C. Calculate the associated pressure drop in the channel.

Problem 5.3

A copper heat sink has a 60 mm × 120 mm base. It is desired to dissipate 10 kW of heat while keeping the plate temperature below 130°C with a flow boiling system using water. Design a suitable cooling channel configuration and water flow rate to accomplish the design.

Problem 5.4

Flow boiling instability in microchannels arises due to rapid bubble growth, especially near the inlet of the channels. Identify the relevant parameters that affect the resulting instability and develop an analytical model to predict the instability condition.

References

Agostini, B., Revellin, R., Thome, J.R., Fabbri, M., Michel, B., Calmi, D., et al., 2008. High heat flux flow boiling in silicon multi-microchannels—part III: saturated critical heat flux of r236fa and two-phase pressure drops. Int. J. Heat Mass Transfer 51, 5426–5442.

Ahn, H.S., Kang, S.H., Lee, C., Kim, J., Kim, M.H., 2012. The effect of liquid spreading due to micro-structures of flow boiling critical heat flux. Int. J. Multiphase Flow 43, 1–12.

Balasubramanian, K., Lee, P.S., Jin, L.W., Chou, S.K., Teo, C.J., Gao, S., 2011. Experimental investigations of flow boiling in straight and expanding microchannels—a comparative study. Int. J. Therm. Sci. 50, 2413–2421.

Balasubramanian, P., Kandlikar, S.G., 2005. An experimental study of flow patterns, pressure drop and flow instabilities in parallel rectangular minichannels. Heat Transfer Eng. 26 (3), 20–27.

Bao, Z.Y., Fletcher, D.F., Haynes, B.S., 2000. Flow boiling heat transfer of Freon R11 and HCFC123 in narrow passages. Int. J. Heat Mass Transfer 43, 3347–3358.

Bergles, A.E., 1962. Forced Convection Surface Boiling Heat Transfer and Burnout in Tubes of Small Diameter (Doctoral Dissertation). Massachusetts Institute of Technology, Cambridge, MA.

Bergles, A.E., 1963. Subcooled Burnout in Tubes of Small Diameter, ASME Paper No. 63-WA-182.

Bergles, A.E., Kandlikar, S.G., 2005. On the nature of critical heat flux in microchannels. J. Heat Transfer 127 (**1**), 101–107.

Bergles, A.E., Rohsenow, W.M., 1962. Forced Convection Surface Boiling Heat Transfer and Burnout in Tubes of Small Diameter, M.I.T. Engineering Projects Laboratory Report No. DSR 8767-21.

Bergles, A.E., Rohsenow, W.M., 1964. The determination forced-convection surface boiling heat transfer. J. Heat Transfer 86, 365–372.

Bonjour, J., Lallemand, M., 1998. Flow patterns during boiling in a narrow space between two vertical surfaces. Int. J. Multiphase Flow 24, 947–960.

Boudouh, M., Gualous, H.L., Labachelerie, M.D., 2010. Local convective boiling heat transfer and pressure drop of nanofluids in narrow rectangular channels. App. Therm. Eng. 30, 2619–2631.

Bowers, M.B., Mudawar, I., 1994. High flux boiling in low flow rate, low pressure drop mini-channel and micro-channel heat sinks. Int. J. Heat Mass Transfer 37 (2), 321–334.

Brutin, D., Tadrist, F., 2003. Experimental study of unsteady convective boiling in heated minichannels. Int. J. Heat Mass Transfer 46, 2957–2965.

Brutin, D., Topin, F., Tadrist, L., 2002. Experimental study on two-phase flow and heat transfer in capillaries. In: Proceedings of the Twelfth International Heat Transfer Conference, France, pp. 789–784.

Cheng, X., Erbacher, F.J., Muller, U., Pang, F.G., 1997. Critical heat flux in uniformly heated vertical tubes. Int. J. Heat Mass Transfer 40, 2929–2939.

Chisholm, D., 1983. Two-Phase Flow in Pipelines and Heat Exchangers. Godwin, New York, NY.

Coleman, J.W., 2003. An experimentally validated model for two-phase sudden contraction pressure drop in microchannel tube headers. In: Proceedings of the First International Conference on Microchannels and Minichannels. ASME, Rochester, New York, NY, Paper No. ICMM2003-1026, 24–25 April, pp. 241–248.

Cornwell, K., Kew, P.A., 1992. Boiling in small parallel channels. In: Proceedings of CEC Conference on Energy Efficiency in Process Technology. Elsevier Applied Sciences, Athens, October 1992, Paper 22, pp. 624–638.

Cuta, J.M., McDonald, C.E., Shekarriz, A., 1996. Forced convection heat transfer in parallel channel array microchannel heat exchanger, In: Advances in Energy Efficiency, Heat and Mass Transfer Enhancement. ASME, New York, NY, In: ASME PID-ol. 2/HTD-ol. 338, pp. 17–22.

Davis, E.J., Anderson, G.H., 1966. The incipience of nucleate boiling in forced convection flow. AIChE J. 12 (4), 774–780.

Downing, R.S., Meinecke, J., Kojasoy, G., 2000. The effects of curvature on pressure drop for single and two-phase flow in miniature channels. In: Proceedings of NHTC2000: Thirty-fourth National Heat Transfer Conference, Paper No. NHTC2000-12100, 20–22 August, Pittsburgh, PA.

English, N.J., Kandlikar, S.G., 2005. An experimental investigation into the effect of surfactants on air-water two-phase flow in minichannels. In: Proceedings of the Third International Conference on Microchannels and Minichannels. ASME, Toronto, Canada, Paper No. ICMM2005-75110, 13–15 June, p. 8.

Faulkner, D.J., Shekarriz, R., 2003. Forced convective boiling in microchannels for kW/cm^2 electronics cooling. In: Proceedings of ASME Summer Heat Transfer Conference. ASME Publications, Las Vegas, NV, Paper No. HT2003-47160, 21–23 July.

Henderson, K., Park, Y.-G., Liu, L., Jacobi, A.M., 2010. Flow boiling heat transfer of R-134a-based nanofluids in a horizontal tube. Int. J. Heat Mass Transfer 53, 944–951.

Hetsroni, G., Segal, Z., Mosyak, A., 2001. Nonuniform temperature distribution in electronic devices cooled by flow in parallel microchannels. IEEE Trans. Compon. Packag. Manuf. Technol. 24 (1), 16–23.

Hetsroni, G., Mosyak, A., Segal, Z., Ziskind, G., 2002. A uniform temperature heat sink for cooling of electronic devices. Int. J. Heat Mass Transfer 45, 3275–3286.

Hetsroni, G., Klein, D., Mosyak, A., Segal, Z., Pogrebnyak, E., 2003. Convective boiling in parallel microchannels. In: Proceedings of the First International Conference on Microchannels and Minichannels. ASME, Rochester, New York, NY, Paper No. ICMM2003-1006, 24–25 April, pp. 59–68.

Hewitt, G.F., 2000. Fluid mechanics aspects of two-phase flow, Chapter 9. In: Kandlikar, S.G., Shoji, M., Dhir, V.K. (Eds.), Handbook of Boiling and Condensation. Taylor and Francis, New York, NY.

Hsu, Y.Y., 1962. On the size range of active nucleation cavities on a heating surface. J. Heat Transfer 84, 207–216.

Hsu, Y.Y., Graham, R., 1961. An Analytical and Experimental Study of the Thermal Boundary Layer and Ebullition Cycle in Nucleate Boiling, NASA TN-D-594.

Inasaka, F., Nariai, H., 1992. Critical heat flux of subcooled flow boiling in uniformly heated straight tubes. Fusion. Eng. Des. 19, 329–337.

Incropera, F.P., DeWitt, D.P., 2002. Fundamentals of Heat and Mass Transfer, fifth ed. John Wiley and Sons, New York, NY.

Jiang, L., Wong, M., Zohar, Y., 1999. Phase change in microchannel heat sinks with integrated temperature sensors. J. Microelectromech. Sys. 8, 358–365.

Jiang, L., Wong, M., Zohar, Y., 2001. Forced convection boiling in a microchannel heat sink. J. Microelectromech. Sys. 10 (1), 80–87.

Kamidis, D.E., Ravigururajan, T.S., 1999. Single and two-phase refrigerant flow in minichannels In: Proceedings of Thirty-third National Heat Transfer Conference, Paper No. NHTC99-5–17 August, ASME Publications, Albuquerque, NM.

Kandlikar, S.G., 1990. A general correlation for two-phase flow boiling heat transfer inside horizontal and vertical tubes. J. Heat Transfer 112, 219–228.

Kandlikar, S.G., 1991. A model for flow boiling heat transfer in augmented tubes and compact evaporators. J. Heat Transfer 113, 966–972.

Kandlikar, S.G., 2001. A theoretical model to predict pool boiling CHF incorporating effects of contact angle and orientation. J. Heat Transfer 123, 1071–1079.

Kandlikar, S.G., 2002a. Fundamental issues related to flow boiling in minichannels and microchannels. Exp. Therm. Fluid Sci. 26 (2–4), 389–407.

Kandlikar, S.G., 2002b. Two-phase flow patterns, pressure drop, and heat transfer during flow boiling in minichannel flow passages of compact heat evaporators. Heat Transfer Eng. 23 (5), 5–23.

Kandlikar, S.G., 2004. Heat transfer mechanisms during flow boiling in microchannels. J. Heat Transfer 126, 8–16.

Kandlikar, S.G., 2005. High flux heat removal with microchannels—a roadmap of opportunities and challenges. In: Proceedings of the Third International Conference on Microchannels and Minichannels. ASME, Toronto, Canada, Paper No. 75086, 13–15 June.

Kandlikar, S.G., 2006. Nucleation characteristics and stability considerations during flow boiling in microchannels. Exp. Therm. Fluid Sci. 30, 441–447.

Kandlikar, S.G., 2010a. Scale effects of flow boiling in microchannels: a fundamental perspective. Int. J. Therm. Sci. 49, 1073–1085.

Kandlikar, S.G., 2010b. Similarities and differences between flow boiling in microchannels and pool boiling. Heat Transfer Eng. 31 (3), 159–167.

Kandlikar, S.G., 2010c. A scale analysis based theoretical force balance model for critical heat flux (CHF) during saturated flow boiling in microchannels and minichannels. J. Heat Transfer 132 (8), 0181501-1–0181501-13.

Kandlikar, S.G., Balasubramanian, P., 2004. An extension of the flow boiling correlation to transition, laminar and deep laminar flows in minichannels and microchannels. Heat Transfer Eng. 25 (3), 86–93.

Kandlikar, S.G., Balasubramanian, P., 2005. Effect of gravitational orientation on flow boiling of water in 1054 × 197 micron parallel minichannels. J. Heat Transfer 127, 820–829.

Kandlikar, S.G., Spiesman, P.H., 1997. Effect of surface characteristics on flow boiling heat transfer. Paper Presented at the Engineering Foundation Conference on Convective and Pool Boiling, 18–25 May, Irsee, Germany.

Kandlikar, S.G., Steinke, M.E., 2002. Contact angles and interface behavior during rapid evaporation of liquid on a heated surface. Int. J. Heat Mass Transfer 45, 3771–3780.

Kandlikar, S.G., Steinke, M.E., 2003. Predicting heat transfer during flow boiling in minichannels and microchannels. ASHRAE Transfer 109 (1), 1–9.

Kandlikar, S.G., Stumm, B.J., 1995. A control volume approach for investigating forces on a departing bubble under subcooled flow boiling. J. Heat Transfer 117, 990–997.

Kandlikar, S.G., Cartwright, M.D., Mizo, V.R., 1995. A photographic study of nucleation characteristics of cavities in flow boiling, 30 May, 5 April, pp. 73–78. In: Proceedings of Convective Flow Boiling. Taylor and Francis, Banff, Alberta, Canada.

Kandlikar, S.G., Mizo, V.R., Cartwright, M.D., Ikenze, E., 1997. Bubble nucleation and growth characteristics in subcooled flow boiling of water, *ASME* In: Proceedings of the Thirty-Second National Heat Transfer Conference, **4**. HTD-ol. 342, pp. 11–18

Kandlikar, S.G., Dhir, V.K., Shoji, M., 1999. Handbook of Phase Change. Taylor and Francis, New York, NY.

Kandlikar, S.G., Steinke, M.E., Tian, S., Campbell, L.A., 2001. High-speed photographic observation of flow boiling of water in parallel mini-channels. Paper Presented at the ASME National Heat Transfer Conference, June, ASME.

Kandlikar, S.G., Kuan, W.K., Willistein, D.A., Borrelli, J., 2005. Stabilization of flow boiling in microchannels using pressure drop elements and fabricated nucleation sites. J. Heat Transfer 128, 389–396.

Kandlikar, S.G., Widger, T., Kalani, A., Mejia, V., 2013. J. Heat Transfer. Enhanced flow boiling over open microchannels with uniform and tapered gap manifolds (OMM) 135 (6), 061401 (9 pages).

Kasza, K.E., Didascalou, T., Wambsganss, M.W., 1997. Microscale flow visualization of nucleate boiling in small channels: mechanisms influencing heat transfer. In: Shah, R.K. (Ed.), Proceedings of the International Conference on Compact Heat Exchanges for the Process Industries. Begell House, Inc., New York, NY, pp. 343–352.

Kennedy, J.E., Roach Jr., G.M., Dowling, M.F., Abdel-Khalik, S.I., Ghiaasiaan, S.M., Jeter, S.M., et al., 2000. The onset of flow instability in uniformly heated horizontal microchannels. J. Heat Transfer 122 (1), 118–125.

Kenning, D.B.R., Cooper, M.G., 1965. Flow patterns near nuclei and the initiation of boiling during forced convection heat transfer. Paper Presented at the Symposium on Boiling Heat Transfer in Steam Generating Units and Heat Exchangers, Manchester, 15–16 September, London: IMechE.

Kew, P.A., Cornwell, K., 1996. On pressure drop fluctuations during boiling in narrow channels. In: Celata, G.P., Di Marco, P., Mariani, A. (Eds.), Second European Thermal Sciences and Fourteenth UIT National Heat Transfer Conference. Edizioni, Pisa, Italy, pp. 1323–1327.

Kew, P.A., Cornwell, K., 1997. Correlations for the prediction of boiling heat transfer in small-diameter channels. Appl. Therm. Eng. 17 (8–10), 705–715.

Khanikar, V., Mudawar, I., Fisher, T., 2009. Flow boiling in a micro-channel coated with carbon nanotubes. IEEE Trans. Compon. Packag. Technol. 32 (3), 639–649.

Khodabandeh, R., Palm, B., 2001. Heat transfer coefficient in the vertical narrow channels of the evaporator of an advanced thermosyphon loop, May 27–June 1. In: Proceedings of the Fourth International Conference on Multiphase Flow. ASME, New Orleans, LA.

Kim, J., Bang, K., 2001. Evaporation heat transfer of refrigerant R-22 in small hydraulic-diameter tubes. In: Michaelides, E.E. (Ed.), Proceedings of the Fourth International Conference on Multiphase Flow. ASME, New Orleans, LA, Paper no. 869, May 27–June 1.

Kim, S.J., Bang, I.C., Buongiorno, J., Hu, L.W., 2007. Surface wettability change during pool boiling of nanofluids and its effect on critical heat flux. Int. J. Heat Mass Transfer 50, 4105–4116.

Kim, S.J., McKrell, T., Buongiorno, J., 2008. Alumina nanoparticles enhance the flow boiling critical heat flux of water at low pressure. J. Heat Transfer 130, 044501, 3 pages.

Kim, S.J., McKrell, T., Buongiorno, J., Hu, L.-W., 2010. Subcooled flow boiling heat transfer of dilute alumina, zinc oxide, and diamond nanofluids at atmospheric pressure. Nucl. Eng. Des. 240, 1186–1194.

Koizumi, Y., Ohtake, H., Fujita, Y., 2001. Heat transfer and critical heat flux of forced flow boiling in vertical–narrow–annular passages In: Proceedings of the ASME International Mechanical Engineering Congress and Exposition. ASME, New York, NY Paper No. HTD-24219, 11–16.

Koo, J.M., Jiang, L., Zhang, L., Zhou, L., Banerjee, S.S., Kenny, T.W., et al., 2001. Modeling of two-phase microchannel heat sinks for VLSI chips In: Proceedings of the IEEE Fourteenth International MEMS Conference, January, IEEE, Interlaken, Switzerland.

Kosar, A., Peles, Y., 2007a. Critical heat flux of R-123 in silicon-based microchannels. J. Heat Transfer 129, 844–851.

Kosar, A., Peles, Y., 2007b. Boiling heat transfer in a hydrofoil-based pin fin heat sink. Int. J. Heat Mass Transfer 5 (5–6). 10.1016/j.ijheatmasstransfer.2006.07.032.

Kosar, A., Kuo, C.-J., Peles, Y., 2005. Boiling heat transfer in rectangular microchannels with reentrant cavities. Int. J. Heat Mass Transfer 48, 4867–4886.

Kosar, A., Peles, Y., Bergles, A.E., Cole, G.S., 2009. Experimental investigation of critical heat flux in microchannels for flow-field probes. In: Seventh International Conference on Nanochannels, Microchannels, and Minichannels, Paper No. ICNMM2009-82214, ASME, Pohang, Korea.

Krishnamurthy, S., Peles, Y., 2008. Flow boiling of water in a circular staggered micro-pin fin heat sink. Int. J. Heat Mass Transfer 51, 1349–1364.

Kuan, W.K., Kandlikar, S.G., 2008. Critical heat flux measurement and model for refrigerant-123 under stabilized flow conditions in microchannels. J. Heat Transfer 130 (3), Article Number 034503, 5 pages.

Kuznetsov, V.V., Shamirzaev, A.S., 1999. Two-phase flow pattern and flow boiling heat transfer in noncircular channel with a small gap. In: Celata, G.P., Shah, R.K. (Eds.), Two-Phase Flow Modelling and Experimentation. Edizioni, Pisa, Italy, pp. 249–253.

Kuznetsov, V.V., Vitovsky, O.V., 1999. Flow pattern of two-phase flow in vertical annuli and rectangular channel with narrow gap. In: Celata, G.P., Shah, R.K. (Eds.), Two-Phase Flow Modelling and Experimentation. Edizioni, Pisa, Italy.

Kuznetsov, V.V., Dimov, S.V., Houghton, P.A., Shamiraev, A.S., Sunder, S., 2003. Upflow boiling and condensation in rectangular minichannels, Paper No. ICMM2003-1087, 24–25 April .Proceedings of the First International Conference on Microchannels and Minichannels. ASME Publications, Rochester, New York, NY.

Lakshminarasimhan, M.S., Hollingsworth, D.K., Witte, L.C., 2000. Boiling incipience in narrow channels, In: Proceedings of the ASME Heat Transfer Division 2000, vol.4. ASME IMECE, pp. 55–63.

Lazarek, G.M., Black, S.H., 1982. Evaporative heat transfer, pressure drop and critical heat flux in a small diameter vertical tube with R-113. Int. J. Heat Mass Transfer 25 (7), 945–960.

Lee, H.J., Lee, S.Y., 2001a. Pressure drop correlations for two-phase flow within horizontal rectangular channels with small heights. Int. J. Multiphase Flow 27, 783–796.

Lee, H.J., Lee, S.Y., 2001b. Pressure drop and heat transfer characteristics of flow boiling in small rectangular horizontal channels. In: Proceedings of the International Conference on Multiphase Flow. ASME Fluid Mechanics Division, New Orleans, LA.
Lee, H.J., Lee, S.Y., 2001c. Heat transfer correlation for boiling flows in small rectangular horizontal channels with low aspect ratios. Int. J. Multiphase Flow 27 (823–827), 2043–2062.
Lee, J., Mudawar, I., 2007. Assessment of the effectiveness of nanofluids for single-phase and two-phase heat transfer in microchannels. Int. J. Heat Mass Transfer 50, 452–463.
Lee, P.C., Li, H.Y., Tseng, F.G., Pan, C., 2003. Nucleate boiling heat transfer in silicon-based micro-channels, In: Proceedings of ASME Summer Heat Transfer Conference. ASME Publications, Las Vegas, LA Paper No. HT2003-47220, July 21–23.
Lee, P.S., Garimella, S.V., 2003. Experimental investigation of heat transfer in microchannels In: Proceedings of ASME Summer Heat Transfer Conference. ASME Publications, Las Vegas, LA Paper No. HT2003-47293, 21–23 July.
Lee, S., Kim, G., 2003. Analysis of flow resistance inside microchannels with different inlet configurations using micro-PIV system. In: Proceedings of the First International Conference on Microchannels and Minichannels. ASME, Rochester, New York, NY, Paper No. ICMM2003-1108, 24–25 April, pp. 823–827.
Lin, S., Kew, P.A., Cornwell, K., 1998. Two-phase flow regimes and heat transfer in small tubes and channels. In: Proceedings of Eleventh International Heat Transfer Conference, vol. 2, Kyongju, Korea, pp. 45–50.
Lin, S., Kew, P.A., Cornwell, K., 1999. Two-phase evaporation in a 1 mm diameter tube. In: Proceedings of the Sixth UK Heat Transfer Conference, Edinburgh.
Lin, S., Kew, A., Cornwell, K., 2001. Flow boiling of refrigerant R141B in small tubes. Chem. Eng. Res. Des. 79 (4), 417–424.
Lu, C.T., Pan, C., 2011. Convective boiling in a parallel microchannel heat sink with a diverging cross section and artificial nucleation sites. Exp. Therm. Fluid Sci. 35 (5), 810–815.
Martin-Callizo, C., Rashid, A., Palm, B., 2008. Dryout heat flux in saturated flow boiling of refrigerants in a vertical uniformly heated microchannel. In: Proceedings of the 6th International Confernece on Nanochannels, Microchannels, and Minichannels. Darmstadt, Germany, June 23–25, 2008.
McNeil, D.A., Raeisi, A.H., Kew, P.A., Bobbili, P.R., 2010. A comparison of flow boiling heat-transfer in in-line mini pin fin and plane channel flow. App. Therm. Eng. 30, 2412–2424.
Mertz, R., Wein, A., Groll, M., 1996. Experimental investigation of flow boiling heat transfer in narrow channels. Calore Technol. 14 (2), 47–54.
Miner, M.J., Phelan, P.E., Odom, B.A., Ortiz, C.A., Prashar, R.S., Sherbeck, J.A., 2013. Optimized expanding microchannel geometry for flow boiling. J. Heat Transfer 135, 042901(8 pages).
Mishima, K., Hibiki, T., 1996. Some characteristics of air–water two-phase flow in small diameter vertical tubes. Int. J. Multiphase Flow 22 (4), 703–712.
Molki, M., Mahendra, P., Vengala, V., 2003. Flow boiling of R-134A in minichannels with transverse ribs, In: Proceedings of the First International Conference on Microchannels and Minichannels. ASME Publications, Rochester, New York, NY Paper no. ICMM2003-1074, 24–25 April.

Moriyama, K., Inoue, A., 1992. The thermodynamic characteristics of two-phase flow in extremely narrow channels (the frictional pressure drop and heat transfer of boiling two-phase flow, analytical model). Heat Transfer Jpn. Res. 21 (8), 838–856.

Mudawar, I., Bowers, M.B., 1999. Ultra-high critical heat flux (CHF) for subcooled water flow boiling-I: CHF data and parametric effects for small diameter tubes. Int. J. Heat Mass Transfer 42, 1405–1428.

Mukherjee, A., Kandlikar, S.G., 2005a. Numerical study of the effect of inlet constriction on flow boiling stability in microchannels. In: Proceedings of the Third International Conference on Microchannels and Minichannels. ASME, Toronto, Canada, Paper No. ICMM2005-75143, 13–15 June.

Mukherjee, A., Kandlikar, S.G., 2005b. Numerical simulation of growth of a vapor bubble during flow boiling of water in a microchannel. Proceedings of the Second International Conference on Microchannels and Minichannels. ASME, Rochester, New York, NY, Paper No. ICMM 2004-2382, pp. 565–572. (Also published in Microfluidics and Nanofluidics, 1(2), pp. 137–145.

Mukherjee, A., Kandlikar, S.G., 2009. The effect of inlet constriction on bubble growth during flow boiling in microchannels. Int. J. Heat Mass Transfer 52 (21–22), 5204–5212.

Park, K.S., Choo, W.H., Bang, K.H., 2003. Flow boiling heat transfer of R-22 in small-diameter horizontal round tubesIn: Proceedings of the First International Conference on Microchannels and Minichannels. ASME Publications, Rochester, New York, NY Paper no. ICMM2003-1073, 24–25 April.

Peles, Y., Contemporary Perspectives on Flow Boiling Instabilities in Microchannels and Minichannels, Begell House, 2012.

Peles, Y., 2003. Two-phase boiling flow in microchannels—instabilities issues and flow regime mapping. Proceedings of the First International Conference on Microchannels and Minichannels. ASME, Rochester, New York, NY, Paper No. 2003-1069, 24–25 April, pp. 559–566.

Peles, Y., Mishra, A., Kuo, C-J., Schneider, B., 2005. Forced convective heat transfer across a pin fin micro heat sink. Int. J. Heat and Mass Transfer 48 (17), 3615–3627.

Peles, Y., Yarin, L.P., Hetsroni, G., 2001. Steady and unsteady flow in a heated capillary. Int. J. Multiphase Flow 27 (4), 577–598.

Peng, X.F., Wang, B.X., 1998. Forced convection and boiling characteristics in microchannels. In: Heat Transfer 1998, In: Proceedings of Eleventh IHTC, 1. 23–28 August, Kyongju, Korea, pp. 371–390.

Peng, X.F., Peterson, G.P., Wang, B.X., 1994. Heat transfer characteristics of water flowing through microchannels. Exp. Heat Transfer 7 (4), 265–283.

Pokharna, H., Masahiro, K., DiStefanio, E., Mongia, R., Barry, J., Crowley, C.W., et al., 2004. Microchannel cooling in computing platforms: performance needs and challenges in implementation. In: Proceedings of the Second International Conference on Microchannels and Minichannels, Keynote Paper. ASME, Rochester, New York, NY, Paper No. ICMM2004-2325, 17–19 June, pp. 109–118.

Qu, W., Mudawar, I., 2002. Prediction and measurement of incipient boiling heat flux in micro-channel heat sinks. Int. J. Heat Mass Transfer 45, 3933–3945.

Qu, W., Mudawar, I., 2003. Flow boiling heat transfer in two-phase micro-channel heat sinks. Part-I. Experimental investigation and assessment of correlation methods. Int. J. Heat Mass Transfer 46, 2755–2771.

Qu, W., Mudawar, I., 2004. Measurement and correlation of critical heat flux in two-phase micro-channel heat sinks. Int. J. Heat Mass Transfer 47, 2045–2059.

Rainey, K.N., Li, G., You, S.M., 2001. Flow boiling heat transfer from plain and microporous coated surfaces in subcooled FC-72. J. Heat Transfer 123, 918–925.

Ravigururajan, T.S., 1998. Impact of channel geometry on two-phase flow heat transfer characteristics of refrigerants in microchannel heat exchangers. J. Heat Transfer 120, 485–491.

Ravigururajan, T.S., Cuta, J., McDonald, C.E., Drost, M.K., 1996. Effects of heat flux on two-phase flow characteristics of refrigerant flows in a micro-channel heat exchanger In: Proceedings of the National Heat Transfer Conference, vol.7. ASME pp. 167–178.

Roach, G.M., Abdel-Khalik, S.I., Ghiaasiaan, S.M., Dowling, M.F., Jeter, S.M., 1999. Low-flow critical heat flux in heated microchannels. Nucl. Sci. Eng. 131, 411–425.

Roday, A.P., Jensen, M.K., 2009. Study of the critical heat flux condition with water and R-123 during flow boiling in microtubes. Part I—experimental results and discussion of parametric trends. Int. J. Heat Mass Transfer 52, 3235–3249.

Sato, T., Matsumura, H., 1964. On the condition of incipient subcooled boiling with forced convection. Bull. JSME 7 (26), 392–398.

Serizawa, A., Feng, Z.P., 2001. Two-phase flow in microchannelsProceedings of the Fourth International Conference on Multiphase Flow. ASME, New Orleans, LA May 27–June 1.

Shenoy, S., Tullius, J.F., Bayazitoglu, Y., 2011. Minichannels with carbon nanotube structures surfaces for cooling applications. Int. J. Heat Mass Transfer 54, 5379–5385.

Steinke, M.E., Kandlikar, S.G., 2004. An experimental investigation of flow boiling characteristics of water in parallel microchannels. J. Heat Transfer 126 (4), 518–526.

Stoddard, R.M., Blasick, A.M., Ghiaasiaan, S.M., Abdel-Khalik, S.I., Jeter, S.M., Dowling, M.F., 2002. Onset of flow instability and critical heat flux in thin horizontal annuli. Exp. Therm. Fluid Sci. 26, 1–14.

Sumith, B., Kaminaga, F., Matsumura, K., 2003. Saturated boiling of water in a vertical small diameter tube. Exp. *Therm.* Fluid Sci. 27, 787–801.

Sun, Y., Zhang, L., Xu, H., Zhong, X., 2011. Flow boiling enhancement of FC-72 from microporous surfaces in minichannels. Exp. Therm. Fluid Sci. 35, 1418–1426.

Tong, W., Bergles, A.E., Jensen, M.K., 1997. Pressure drop with highly subcooled flow boiling in small-diameter tubes. Exp. Therm. Fluid Sci. 15, 202–212.

Tran, T.N., Wambsganss, M.W., France, D.M., 1996. Small circular- and rectangular-channel boiling with two refrigerants. Int. J. Multiphase Flow 22 (3), 485–498.

Vafaei, S., Wen, D., 2010. Critical heat flux (CHF) of subcooled flow boiling of alumina nanofluids in a horizontal microchannel. J. Heat Transfer 132, 102404, 7 pages.

Wambsganss, M.W., France, D.M., Jendrzejczyk, J.A., Tran, T.N., 1993. Boiling heat transfer in a horizontal small-diameter tube. J. Heat Transfer 115 (4), 963–972.

Warrier, G.R., Pan, T., Dhir, V.K., 2001. Heat transfer and pressure drop in narrow rectangular channels In: Proceedings of the Fourth International Conference on Multiphase Flow. ASME, New Orleans, LA May 27–June 1.

Warrier, G.R., Pan, T., Dhir, V.K., 2002. Heat transfer and pressure drop in narrow rectangular channels. Exp. Therm. Fluid Sci. 26, 53–64.

Wu, H.Y., Cheng, P., 2003. Liquid/two-phase/vapor alternating flow during boiling in microchannels at high heat flux. Int. Commun. Heat Mass Transfer 30 (3), 295–302.

Yan, Y., Lin, T., 1998. Evaporation heat transfer and pressure drop of refrigerant R-134a in a small pipe. Int. J. Heat Mass Transfer 42, 4183–4194.

Yen, T.H., Kasagi, N., Suzuki, Y., 2002. Forced convective boiling heat transfer in microtubes at low mass and heat fluxes. In: Symposium on Compact Heat Exchangers on the 60th Birthday of Ramesh K. Shah, August 24, Grenoble, France, pp. 401–406.

Yu, W., France, D.M., Wambsganss, M.W., Hull, J.R., 2002. Two-phase pressure drop, boiling heat transfer, and critical heat flux to water in a small-diameter horizontal tube. Int. J. Multiphase Flow 28, 927–941.

Zhang, L., Koo, J.-M., Jiang, L., Asheghi, M., 2002. Measurements and modeling of two-phase flow in microchannels with nearly constant heat flux boundary conditions. J. Microelectromech. Syst. 11 (1), 12–19.

Zhang, T., Tao, T., Chang, J.-Y., Peles, Y., Prashar, R., Jensen, M.K., et al., 2009. Ledinegg instability in microchannels. Int. J. Heat Mass Transfer 52, 5661–5674.

CHAPTER

Condensation in Minichannels and Microchannels

6

Srinivas Garimella
George W. Woodruff School of Mechanical Engineering, Georgia Institute of Technology, Atlanta, GA, USA

6.1 Introduction

Condensation in microchannels is increasingly being used to improve the efficiency of power generation, chemical processing, heating, and cooling systems. The high heat transfer coefficients characteristic of microchannel condensation yield compact heat transfer equipment, reducing cost and limiting the inventory of potentially toxic or environmentally harmful working fluids. Furthermore, microchannel geometries are better able to withstand the high operating pressure of newer synthetic (e.g., R-404A, R-410A) and natural (e.g., CO_2) working fluids. Ultimately, in these applications, microchannels are viewed as being responsible for the mitigation of ozone depletion by enabling the use of smaller amounts of environmentally harmful fluids, and also reducing greenhouse gas emissions by improving component and system energy efficiencies. Utilization in microsensors and micro-actuators, and in biological diagnostic devices are other candidate applications.

The large heat transfer coefficients in microchannels, and the large surface area-to-volume ratios they offer, have led to their use in compact condensers for the air-conditioning systems in automobiles for many years. The microchannels used in these condenser designs typically have hydraulic diameters in the 0.4–0.7 mm range, although an understanding of the fundamental phase-change phenomena in these heat exchangers is just beginning to emerge. Garimella and Wicht (1995) demonstrated that substantial reductions in condenser sizes can be achieved for small (~18 kW) condensers for ammonia–water absorption space-conditioning systems for residential application by using microchannel condensers instead of conventional round-tube, flat-fin designs. This compactness was attributed to large condensation heat transfer coefficients, high surface to volume ratios, and reduced air-side pressure drops due to the use of multilouver fins. Noting the potential enhancements due to the use of microchannels, Webb and

© 2014 Elsevier Ltd. All rights reserved.

Ermis (2001) investigated condensation of R-134a in extruded aluminum tubes with multiple parallel microchannels with and without microfins, and with hydraulic diameter $0.44 < D_H < 1.564$ mm. Their work documented that heat transfer coefficients and pressure gradients increase with decreasing hydraulic diameter. Webb and Lee (2001) investigated the use of these microchannel condensers as replacements for two-row, round-tube condensers with 7-mm-diameter tubes. They noted material cost reductions of up to 55% over the conventional geometries due to a combination of air-side and tube-side improvements. Other advantages such as decreased air-side pressure drop (due in part to form drag reduction over flat tubes) and material weight reductions were also noted due to the use of microchannel condensers, depending on the specific tube and fin geometry parameters. Jiang and Garimella (2001) demonstrated the implementation of microchannel tubes for residential air-conditioning systems. They compared conventional round-tube, flat-fin geometries to systems with air-coupled microchannel evaporators and condensers, as well as systems that coupled refrigerant heat exchange to intermediate glycol–water solution loops. Their work showed that indoor and outdoor units of air-coupled microchannel systems can be packaged in only one-half and one-third the space required for a conventional system, respectively. Furthermore, they showed that significant additional compactness can be achieved through hydronic coupling, because of the high heat transfer coefficients on both sides of the liquid-coupled heat exchangers, and the counterflow orientation, which is more suitable in the presence of a temperature glide. It was also shown that the refrigerant charge of the microchannel air-coupled system is 20% less than that of the round-tube heat pump. For the hydronically coupled system, the refrigerant charge is only 10% of the charge in the round-tube heat pump, yielding significant benefits pertaining to the reduction of the environmental impact of air-conditioning systems. These representative studies clearly demonstrate the wide-ranging benefits that can be obtained by the use of microchannels in phase-change applications.

Microchannel condensers used in the automotive industry are generally limited to a single synthetic working fluid (e.g., R-134a) operating in a narrow range of saturation temperatures (~30°C to 60°C). For this narrow band of operating conditions, empirical models could be developed and used for design with satisfactory results. More recently, microchannel condensers have been adopted or are being considered for other applications including stationary and transportation refrigeration, cooling, water heating, and energy recovery systems. Furthermore, the banning of chlorofluorocarbons (CFC) due to their ozone depletion potential and increased scrutiny of hydrochlorofluorocarbons (HCFC) and hydrofluorocarbons (HFC) due to their high global warming potential (GWP) has led to interest in new synthetic (R-1234yf, R-1234ze), natural (CO_2, NH_3, hydrocarbons), and mixtures (R-404A, R-407A, R-410A, NH_3/H_2O) as working fluids. These new fluids and the wider range of operating temperatures for the larger slate of applications lead to the need for addressing a much wider range of thermophysical properties and operating pressures.

An example of the magnitude of these differences is illustrated in Table 6.1. The variation in fluid properties suggests that empirical models developed over a narrow range of properties, geometries, and operating conditions may not be generally applicable. Thus, appropriate use of microchannel condensers for this wide range of applications requires that the science of obtaining a comprehensive understanding of phase change in these channels catch up to the art of the design and utilization of microchannels for achieving high heat transfer rates in such applications.

6.1.1 Defining microchannel condensation

Researchers have discussed the term "microchannel" at some length in the literature as well as elsewhere in this book. One extreme of the size range (submicron size channels) is where continuum phenomena do not apply, and properties such as viscosity, ice point, and others change due to molecular and electrokinetic forces, but not much information is available for that size range, and there are few practical phase-change applications at that range. However, such noncontinuum phenomena do not have to be present for there to be differences in flow and heat transfer phenomena from conventional, larger-diameter channels. One simple way of looking at microchannels is the existence of deviations in pressure gradients and heat transfer coefficients from widely available models, which is sufficient to warrant studies that understand and account for the different flow phenomena. In condensation applications, for example, the microchannel designation may be applied when the channel size leads to surface tension becoming the dominant force, which would result in differences from macrochannel phenomena. Essentially, when channel size affects flow, pressure gradients, and heat transfer in a manner that is not accounted for by models developed for the larger channels, it warrants a different treatment. The microchannel behavior, therefore, depends on phenomena that do not exist or are not prominent at the larger scales, or phenomena that are suppressed at the small scales. As these phenomena would depend on the liquid and vapor phase properties of the fluid under

Table 6.1 Comparison of Property Data

	Refrigerant Properties at $T_{sat}=50°C$ Air/water at P_{atm} and 20°C			
	P_{sat} (kPa)	ρ_L/ρ_V	μ_L/μ_V	$\sigma \cdot 10^{-3}$ (N m^{-1})
Air/water	101	845	48.2	38.1
R-134a	1320	16.6	10.7	4.9
R-404A	2313	6.6	5.5	1.6
Propane	1713	11.6	7.5	4.1
Ammonia	2033	35.8	15.7	14.3

consideration, a rigorous definition of the term microchannel would necessarily depend on the fluid in question. Serizawa et al. (2002), who conducted air–water and steam–water flow visualization studies in 50 μm channels, recommend the Laplace constant $L > D_H$ (also referred to as the capillary length scale) as the criterion for determining whether a channel should be considered to be a microchannel:

$$L = \sqrt{\frac{\sigma}{g(\rho_l - \rho_v)}} \tag{6.1}$$

The Laplace constant represents the ratio of surface tension and gravity forces, and has been interpreted as the dimension at which the influence of surface tension becomes more important than the stratifying effect of gravity (Fukano and Kariyasaki, 1993). This criterion implies that for a high surface tension fluid pair like air–water at standard temperature and pressure, a channel with $D_H < 2.7$ mm would be a microchannel, while for R-134a at 1500 kPa, the corresponding value is $D_H < 0.66$ mm. Another interpretation of a different form of this same quantity is provided in terms of the confinement of bubbles in channels (Kew and Cornwell, 1997). Thus, typically, microchannel two-phase flows are characterized by "confined" single elongated bubbles occupying much of the channel (as will be discussed at length in a subsequent section), whereas multiple bubbles may occupy the cross-section in macrochannels. Also, typical bubble sizes decrease as the pressure increases, affecting how the channel may be viewed in relation to the bubble size. Based on this interpretation, Kew and Cornwell (1997) recommend the use of the confinement number ($Co > 0.5$) as the transition criterion (for boiling applications) beyond which microchannel effects are present:

$$Co = \frac{\sqrt{\sigma/(g(\rho_l - \rho_v))}}{D_H} \tag{6.2}$$

Thus, $Co = L/D_H$, and the transition D_H value is 1.32 mm, that is, double the value recommended by Serizawa et al. (2002). The importance of fluid properties on microchannel behavior can be seen in Table 6.2, which shows the transition

Table 6.2 Microchannel Transition Criteria

	Transition Diameter (mm) $T_{sat} = 50°C$	
	$Co > 0.5$	$L > D_H$
R-134a	1.39	0.69
R-404A	0.91	0.46
Propane	2.01	1.01
Ammonia	3.26	1.63
Water	5.30	2.65

diameters for some common fluids at a saturation temperature of 50°C. Thus, the practical value of such categorizations is somewhat limited—even a change in operating temperature for the same fluid could affect whether the channel should be treated as a microchannel through the influence of temperature on surface tension and other properties. What is more important is to recognize and account for the respective flow phenomena in the development of pressure drop and heat transfer models. For example, substantial differences due to channel size have been documented by several researchers at hydraulic diameters of a few millimeters, as noted by Palm (2001).

It is also clear that the increase in surface area associated with the use of microchannels increases the importance of surface forces over that of body forces, as noted by Serizawa and Feng (2001). Thus, surface phenomena, and in some cases surface characteristics, become more prominent in these small channels, and the interactions between the fluid and wall increase in importance. This is another rationale for the commonly noted observation that in microchannels surface tension and viscous forces dominate over gravitational forces. As will also become clear in the following sections, neither the same measurement techniques nor the same modeling approaches used for the larger-scale channels are adequate for addressing these phenomena that are specific to microchannels.

6.1.2 Chapter organization and contents

The focus of this chapter is condensation in microchannels. Compared to single-phase flow and boiling in such channels, condensation has received less attention. In part this is because researchers have concentrated primarily on the removal of heat from small and inaccessible surfaces such as high-flux electronics components, using single-phase and boiling processes. Condensation has thus far been generally coupled with air: the dominant air-side resistance in such situations may be one reason for the relatively smaller attention to condensation. However, liquid coupling, condensation in heat pipes, fluid charge minimization, and other factors are making this process more prominent. The heat *removed* from the sources through boiling must also ultimately be *rejected*, requiring innovative compact designs and a better understanding of condensation in microchannels.

This chapter presents the available information in an emerging field for which directly relevant literature is limited; therefore, it is by no means a comprehensive treatment of all aspects of condensation in microchannels. Heat transfer coefficients and pressure drops for condensation inside microchannels are strongly dependent on the different flow patterns that are established as the fluid undergoes a transition from vapor to liquid. Methods to predict flow regimes as a function of flow-related parameters such as mass flux, quality, and fluid properties are therefore presented first. Where possible, the differences between the flow patterns in large channels and microchannels are discussed. This is followed by a brief discussion of void fraction in condensing flows. The void fraction is almost invariably required for the modeling of pressure drop and heat transfer in two-phase flow.

(Once again, it is noted that void fractions in microchannels have proved to be difficult to measure due to issues such as optical access, and the dynamic and sometimes indeterminate nature of the vapor–liquid interface.) Models for the prediction of pressure drops in the different flow regimes are then presented along with a comparison of the predicted pressure drops from the different models. This is followed by a presentation of the corresponding models for condensation heat transfer in microchannels. Finally, areas of critical research needs for furthering the understanding of condensation in microchannels are pointed out. At the time of this writing, there is ongoing research in this field at several laboratories around the world, and the information presented here will no doubt need to be updated on a continuous basis.

6.2 Flow regimes

The first step in understanding condensation in microchannels is the characterization of how the condensing fluid flows through the channels, because this forms the basis for the associated pressure drop and heat transfer. As the fluid condenses in the channel and progresses from the vapor phase toward the liquid phase over a range of qualities, different flow patterns are established at different regions of the condenser. Early attempts at flow regime mapping were conducted on relatively large tubes using adiabatic air–water or air–oil mixtures. Rouhani and Sohal (1983) provide a comprehensive review of experimental techniques for the development of two-phase flow regime maps in large diameter tubes. The progression of flow regimes, and in particular the transitions between the different flow mechanisms, is different in microchannels than in larger-diameter tubes, primarily due to differences in relative magnitudes of gravity, shear, viscous, and surface-tension forces. Similar differences exist between flow regimes and transitions in circular and noncircular microchannels. Therefore, extrapolation of large round-tube correlations to smaller diameters and noncircular geometries could introduce errors into pressure drop and heat transfer predictions.

Some insights from the large-diameter tube studies during adiabatic flow are important for understanding two-phase flow patterns during condensation in microchannels. The importance of the applicable flow regime in predicting two-phase pressure drop (and thus also heat transfer) was shown by Alves (1954) and Baker (1954). Baker (1954) also found that the onset of slug flow occurred at lower values of the Lockhart–Martinelli parameter, X, in larger pipes, and that the pipe diameter affects the two-phase multiplier. The importance of tube diameter was also recognized and quantified by Govier et al. (1957) and Govier and Short (1958), particularly for upward flow of air–water mixtures, even though their work was limited to a relatively large-diameter range (16.00–63.50 mm). Thus, the influence of diameter on flow patterns, even in the larger-diameter tubes, has been acknowledged for quite some time. Most of the early maps for large-diameter tubes were developed empirically; however, Taitel and Dukler (1976) devised a theoretical approach to flow regime mapping for air–water mixtures using a momentum balance on a stratified flow pattern. A set of four dimensionless parameters was used to define instability-driven transitions from the stratified to annular or intermittent flows, and to distinguish between stratified smooth and stratified wavy flows. This theoretical model showed good agreement with the experimentally developed flow regime map of Mandhane et al. (1974). However, despite its theoretical basis, the Taitel and Dukler (1976) map did not explicitly account for surface-tension forces.

6.2.1 Adiabatic air–water flow in microchannels

Unfortunately, the bulk of the literature on microchannel two-phase flow regimes is on adiabatic flows, in which air–water, nitrogen–water, and air–oil mixtures, typically in the absence of heat transfer, are used to simulate condensing flows. The obvious reason for this approach is the substantial simplification in experimental facilities that it offers. Thus, the requisite vapor–liquid "quality" can simply be "dialed in" by independently controlling the air and liquid flow rates entering the test section, whereas this must be achieved indirectly through upstream and downstream conditioning of a condensing fluid through pre- and post-heating/cooling, as appropriate. In addition, experiments on refrigerants and other working fluids require closed loops, whereas a nitrogen–water or air–water experiment can be conducted in a once-through manner, with the exiting fluid streams simply being exhausted and drained. Finally, several fluids (most refrigerants), at condensing temperatures of interest, are at pressures much higher than atmospheric, which presents challenges in providing optical access during the condensation process. On the other hand, air–water experiments are typically conducted at or near atmospheric pressure.

As was shown in Table 6.1, the properties of the air–water fluid pair are significantly different from those of refrigerant vapor–liquid mixtures; therefore, application of the results from air–water studies and extrapolation to condensing flows must be done with appropriate caution. Also, the absence of the heat of condensation during air–water experiments must be recognized when applying these studies to condensation in microchannels. Nevertheless, due to the preponderance of adiabatic air–water studies in the literature, and the relative scarcity of condensation studies, a discussion of the significant studies on adiabatic two-phase flow in small channels is provided here first, followed by studies on condensation (Table 6.3).

6.2.1.1 Circular and noncircular microchannels

Among the studies on flow regime maps for microchannels using adiabatic air–water mixtures, Suo and Griffith (1964) correlated the transition from elongated bubble to annular and bubbly flow in capillary tubes ($1.0 < D < 1.6$ mm) by using the average volumetric flows of the liquid and gas phases, and the velocity

Table 6.3 Summary of Flow Regime Studies

Investigators	Hydraulic Diameter	Fluids	Orientation/ Conditions	Range/ Applicability	Techniques, Basis, Observations
Adiabatic flow: conventional tubes					
Taitel and Dukler (1976)		Air–water			• Devised theoretical approach to flow regime mapping using a momentum balance on a stratified flow pattern
Mandhane et al. (1974)	12.7–165.1 mm	Air–water	Horizontal	$0.043 < j_G < 170.7$ m/s $0 < j_L < 7.315$ m/s	• "Best fit" map from available flow regime maps • Introduced physical property corrections
Adiabatic flow: small circular channels					
Suo and Griffith (1964)	1–6 mm	Air–water	Horizontal adiabatic	$\rho_1/\rho_g >> 1$ $\mu_1/\mu_g > 25$	• Transition from elongated bubble to annular and bubbly flow • Criterion to determine when buoyancy effects can be neglected
Barnea et al. (1983)	4–12 mm	Air–water	Horizontal	$T = 25°C$ $P_{exit} = 1$ atm $0.001 < j_L < 10$ m/s $0.01 < j_G < 100$ m/s	• Recommended Taitel and Dukler (1976) model for all transitions except stratified to nonstratified • Modified stratified-to-intermittent transition boundary of Taitel and Dukler (1976)
Damianides and Westwater (1988)	1–5 mm	Air–water	Horizontal	$P \sim 5$ atm $T \sim 15–20°C$ $0.001 < j_L < 10$ m/s $0.01 < j_G < 100$ m/s	• Documented effect of hydraulic diameter • Some agreement for 5 mm with Taitel and Dukler (1976), but poor agreement for 1 mm • 1-mm tube showed large intermittent flow region • Increasingly small stratified flow region in small-diameter tubes

Fukano et al. (1989)	1–4.9 mm	Air–water	Horizontal adiabatic	$P_{exit} = 1$ atm $0.04 < j_G < 40$ m/s $0.2 < j_L < 4$ m/s	• Results agreed with Barnea et al. (1983) but not with Damianides and Westwater (1988) • Mandhane et al. (1974) cannot predict transitions in small tubes • Taitel and Dukler (1976) and Weisman et al. (1979) not reliable for small diameters
Brauner and Maron (1992)	4–25.2 mm	Air–water	Horizontal	$0.01 < j_G < 40$ m/s $0.001 < j_L < 0.3$ m/s	• Incorporated the stabilizing effect of surface tension associated with practically finite wavelengths • Transition criteria based on Eotvos number
Ide et al. (1995)	0.5–6 mm	Air–water	Vertical upward		• Studied "liquid lumps" formed in vertically upward flow • Velocity characteristics of liquid lumps used to interpret flow pattern with corresponding maps by Fukano et al. (1989)
Mishima and co-workers (Mishima and Hibiki, 1996a,b; Mishima et al., 1997; Mishima and Hibiki, 1998)	1–4 mm	Air–water	Vertical upward	$0.0896 < j_G < 79.3$ m/s $0.0116 < j_L < 1.67$ m/s	• Neutron radiography for flow visualization and void fractions • Recommended transition criteria of Mishima and Ishii (1984)

(*Continued*)

Table 6.3 (Continued)

Investigators	Hydraulic Diameter	Fluids	Orientation/ Conditions	Range/ Applicability	Techniques, Basis, Observations
Coleman and Garimella (1999)	5.5-, 2.6-, 1.75-, 1.3-mm circular $D_H = 5.36$ mm ($\alpha = 0.725$) rectangular	Air–water	Horizontal	$0.1 < j_G < 100$ m/s $0.01 < j_L < 10$ m/s	• Diameter has significant effect on flow regime transitions • Intermittent flow regime increases in *small-diameter tubes* • Comparisons with the criteria of Damianides and Westwater (1988), Fukano et al. (1989), and Weisman et al. (1979)
Triplett et al. (1999b)	1.1- and 1.45-mm circular; semi-triangular 1.09 and 1.49 mm	Air–water	Horizontal adiabatic	$0.02 < j_G < 80$ m/s $0.02 < j_L < 8$ m/s	• Observed bubbly, churn, slug, slug-annular, and annular flow • No stratified flow • Maps agreed with Damianides and Westwater (1988), Fukano et al. (1989), and Fukano and Kariyasaki (1993) • Poor agreement with analytically derived criteria (Suo and Griffith, 1964; Taitel and Dukler, 1976)
Zhao and Bi (2001a)	Triangular channels $D_H = 2.886$, 1.443, and 0.866 mm	Air–water	Vertical upward	$0.1 < j_G < 100$ m/s; For $D_H = 2.886, 1.443$ $0.08 < j_L < 6$ m/s For $D_H = 0.866$ $0.1 < j_L < 10$ m/s	• Dispersed bubbly flow not observed in smallest channel, but capillary bubble flow pattern seen • Trends agreed well with several investigators (Barnea et al., 1983; Galbiati and Andreini, 1992; Mishima and Hibiki, 1996a,b; Ide et al., 1997), but not with Taitel and Dukler (1976) or Mishima and Ishii (1984)

Yang and Shieh (2001)	1, 2, and 3 mm	Air–water R-134a	Horizontal adiabatic	$T = 25–30°C$ $0.016 < j_G < 91.5$ m/s $0.006 < j_L < 2.1$ m/s $300 < G < 1600$ kg/m^2 s $T = 30°C$	• Transitions difficult to distinguish for air–water, but clear for R-134a • Lower σ of R-134a compared to air–water led to differences • R-134a transitions agreed with Hashizume (1983) and Wang et al. (1997a)
Tabatabai and Faghri (2001)	4.8–15.88-mm condensation, 12.3–1-mm air–water	Water, R-12, R-113, R-22, and R-134a	Horizontal		• Flow regime maps based on relative effects of surface tension, shear, and buoyancy forces • Better agreement with available data in shear-dominated regions • Predicts movement of boundaries with channel size
Adiabatic flow: narrow, high aspect ratio, rectangular channels					
Troniewski and Ulbrich (1984)	Rectangular $0.09 < \alpha < 10.10$ $7.45 < D_H < 13.1$	Air–water	Horizontal	$T = 25–35°C$ $0.6 < j_G < 43$ m/s $200 < j_G < 1600$ m/s	• Corrections to axes of Baker (1954) map based on single-phase velocity profiles
			vertical	$T = 25–35°C$ $0.7 < \mu_L < 40 \times 10^{-3}$ Pa s $995 < \rho_L < 1150$ kg/m^3	
Wambsganss et al. (1991, 1994)	Rectangular $\alpha = 6$, 0.167 $D_H = 5.44$ mm	Air–water	Horizontal adiabatic	$0.0001 < x < 1$ $50 < G < 2000$ kg/m^2 s	• Flow visualization and dynamic pressure measurements • Qualitative agreement with large circular channels but quantitative differences • Transition criteria for bubble or plug flow to slug flow based on root-mean-square pressure changes

(Continued)

Table 6.3 (Continued)

Investigators	Hydraulic Diameter	Fluids	Orientation/ Conditions	Range/ Applicability	Techniques, Basis, Observations
Xu and co-workers (Xu, 1999; Xu et al., 1999)	Rectangular gap = 0.3 0.6, and 1 mm	Air–water	Vertical adiabatic	$P_{exit} = 1$ atm $0.01 < j_G < 10$ m/s $0.01 < j_L < 10$ m/s	• Bubbly, slug, churn, and annular flow in 0.6- and 1-mm channels • Bubbly flow absent in 0.3-mm channel • Extended criteria of Mishima and Ishii (1984) for application to narrow rectangular geometry
Hibiki and Mishima (2001)	Rectangular channels with gap 0.3–17 mm	Air–water, steam–water	Vertical upward	Air–water data (from Sadatomi et al., 1982; Lowry and Kawaji, 1988; Ali et al., 1993; Mishima et al., 1993; Wilmarth and Ishii, 1994; Xu, 1999) Steam–water data (from Hosler, 1968)	• Criteria specifically for narrow rectangular channels based on Mishima and Ishii (1984) criteria • Collision/coalescence of bubbles increase when maximum distance between bubbles < projected bubble diameter
Adiabatic flow: $D \ll 1$ mm					
Kawahara et al. (2002)	100 μm	Nitrogen–water	Horizontal	$0.1 < j_G < 60$ m/s $0.02 < j_L < 4$ m/s	• Low *Re* at small D_H implies greater effects of wall shear and surface tension • Observed liquid slug, gas core with smooth-thin liquid film, gas core with smooth-thick liquid film, gas core with ring-shaped liquid film, and gas core with deformed interface

					• No gravitational effects, absence of bubbly flow (no bubble breakup due to absence of liquid phase turbulence at low *Re*) • Flow conditions determine likely, but not unique patterns • Map based on probability of occurrence of patterns • Four flow regimes (slug-ring, ring-slug, semi-annular, and multiple), which are combinations of flow mechanisms
Chung and Kawaji (2004)	530, 250, 100, and 50 μm	Nitrogen–water	Horizontal	$0.02 < j_G < 73$ m/s $0.01 < j_L < 5.77$ m/s	• Extended work of Kawahara et al. (2002) for effect of diameter • 530- and 250-μm flow similar to ~1-mm flow • Slug flow in small channels; absence of bubbly, churn, slug-annular, and annular flow
Chung et al. (2004)	96-μm square, 100-μm circular	Nitrogen–water	Horizontal	Circular data (from Chung and Kawaji, 2004)	• Negligible effect of channel shape
Serizawa et al. (2002)	20-, 25-, and 100-μm circular	Air–water, steam (only 50 μm)	Horizontal	$0.0012 < j_G < 295.3$ m/s $0.003 < j_L < 17.52$ m/s	• Results different from other investigators of $D \ll 1$-mm channels • Numerous terms to define flow patterns • Bubbly flow in 25-μm channel • High capillary pressure at small bubble interface keeps bubble spherical and prevents coalescence

(*Continued*)

Table 6.3 (Continued)

Investigators	Hydraulic Diameter	Fluids	Orientation/ Conditions	Range/ Applicability	Techniques, Basis, Observations
					• Slug flow due to entrance phenomenon, not bubble coalescence • Liquid ring flow due to rupture of liquid bridges between bubbles at high gas velocities • Surface tension prevents liquid slugs from spreading as film, leads to dry patches, liquid "lumps" • State that air–water and steam–water patterns in 25-μm channel similar; photographs presented seem dissimilar • Surprisingly state that flow map agrees with large channel Mandhane et al. (1974) map, but graphs indicate otherwise
Condensing flow					
Traviss and Rohsenow (1973)	8 mm	R-12	Horizontal condensing	$100 < G < 990$ kg/m^2 s $10 < T < 40.6$°C	• Disperse, annular, semi-annular, and slug flow observed • Froude number as basis for condensation mode • Von Karman velocity profile to describe film velocity
Breber et al. (1980)	4.8–50.8 mm	R-11, R-12, R-113, steam, *n*-pentane	Horizontal condensing	$108.2 < P < 1250$ kPa $17.6 < G < 990$ kg/m^2 s	• Taitel and Dukler (1976) map to develop simple criteria for condensation in horizontal tubes

					• Criteria based on ratio of shear to gravity forces • All map boundaries vertical and horizontal lines—easy to use • Defined transition regions, rather than abrupt transitions
Sardesai et al. (1981)	24.4 mm	R-113, steam, propane, methanol, *n*-pentane	Horizontal condensing		• Investigated annular-stratified/wavy transition • Taitel and Dukler (1976) annular-stratified/wavy transition criterion as basis to develop transition criterion from data
Soliman (1982, 1986)	$4.8 < D < 15.9$ mm	R-12, R-113, steam	Condensing		• Annular-wavy Froude number transition criterion based on data from different sources • Mist-annular Weber number transition criterion
Tandon et al. (1982, 1985)	$4.8 < D < 15.9$ mm	R-12, R-113, R-22	Condensing	$0.01 < j_B^* < 20$ $0.001 < 1 - \alpha/\alpha < 3$	• Simple transition criteria, good agreement with data for annular, semi-annular, and wavy flows
Hashizume and co-workers (Hashizume, 1983; Hashizume et al., 1985; Hashizume and Ogawa, 1987)	10 mm	R-12, R-22	Horizontal condensing	$570 < p < 1960$ kPa	• Refrigerant flow patterns different from air–water • Modified property corrections of Baker (1954) map proposed by Weisman et al. (1979) to make them applicable to R-12 and R-22

(*Continued*)

Table 6.3 (Continued)

Investigators	Hydraulic Diameter	Fluids	Orientation/ Conditions	Range/ Applicability	Techniques, Basis, Observations
Wang et al. (1997a,b)	6.5 mm	R-22, R-134a, R-407C	Horizontal	$50 < G < 700$ kg/m^2 s $T_{sat} = 2$, 6, and 20°C	• At low G, intermittent and stratified flow; at high G, annular • Agreed with modified Baker (1954) map of Hashizume and co-workers (Hashizume, 1983; Hashizume et al., 1985; Hashizume and Ogawa, 1987)
Dobson and co-workers (Dobson, 1994; Dobson et al., 1994; Dobson and Chato, 1998)	3.14, 4.6, and 7.04 mm	R-12, R-134a, R-22, R-32/ R-125	Horizontal condensing	$25 < G < 800$ kg/m^2 s $T = 35–60$°C	• Good agreement with Mandhane et al. (1974) map after axis transformation • Primary distinction between gravity and shear-controlled flows • Criteria based on modified Soliman number and mass flux
Coleman and Garimella (2000a,b, 2003), Garimella (2004)	Round: 4.91 mm; square: 4, 3, 2, 1 mm; rectangular: 4 × 6; 6 × 4; 2 × 4; 4 × 2 mm	R-134a	Horizontal condensing	$P = 1379–1724$ kPa $150 < G < 750$ kg/m^2 s	• Flow regimes subdivided into flow patterns, G-x based transition criteria • D_H more important than shape, aspect ratio for transition criteria • Intermittent and annular flow increases as D_H decreases • Wavy flow nonexistent in 1-mm channel
Cavallini et al. (2001, 2002b)	8 mm	R-22, R-134a, R-125, R-32, R-236ea, R-407C, R-410A	Condensation, horizontal tubes	$30 < T_{sat} < 50$°C $100 < G < 750$ kg/m^2 s	• Criteria similar to Breber et al. (1979, 1980) for primary flow regimes (annular, stratified, wavy, and slug) • Recommended Kosky and Staub (1971) model for annular flow

El Hajal et al. (2003)	$3.14 < D < 21.4$ mm	R-22, R-134a, R-125, R-32, R-236ea, R-410A	Horizontal condensing	$16 < G < 1532$ kg/m^2 s $0.02 < p_r < 0.8$ $76 < (We/Fr)_L < 884$	• Transitions between stratified, wavy, intermittent, annular, mist, and bubbly flow on *G-x*-coordinates, adapting boiling map of Kattan et al. (1998a,b, 1998c) with updates by Zurcher et al. (1999) • Void fraction for map construction deduced from heat transfer database, subsequently averaged with homogeneous value
Keinath and Garimella (2010)	Circular: $0.5 < D < 3.0$ mm	R-404A	Condensing	$200 < G < 800$ kg/m^2 s $30 < T_{sat} < 60°C$	• Identified similar trends in flow patterns to Coleman and Garimella (2000a,b, 2003) • Quantitative void fraction, bubble frequency, velocity, diameter, and length extracted from images
Kim et al. (2012)	Square: 1 mm	FC-72	Horizontal condensing	$68 < G < 367$ kg/m^2 s $T_{sat} = 60°C$	• Increased importance of intermittent flow at lower mass flux • No entrained liquid in annular and annular/wavy regime • Proposed dimensionless transition lines

of the bubbles. They also provided a criterion that determines when buoyancy effects can be neglected in terms of the parameter $\rho_l g D^2/\sigma$. However, it must be noted that they restrict their analysis to situations where $\rho_l/\rho_g \gg 1$ and $\mu_l/\mu_g < 25$. While they explicitly state these restrictions in their paper, several other air–water adiabatic flow papers do not establish such explicit restrictions; therefore, caution is advised when applying these air–water based criteria to refrigerant condensation. For example, for refrigerant R-134a condensing at 1500 kPa, the density ratio is 14, while the viscosity ratio is 9.7. For the air–water pair under standard conditions, the corresponding ratios are $\rho_l/\rho_g = 839$ and $\mu_l/\mu_g = 49$. Similarly, the surface tension of the air–water pair is 17 times higher than that of R-134a.

Barnea et al. (1983) classified two-phase flow patterns in small horizontal tubes ($4 < D < 12$ mm) into four major regimes (dispersed, annular, intermittent, and stratified). They found that all transitions except the stratified to nonstratified transition were satisfactorily described by the Taitel–Dukler (1976) model. Therefore, they modified the stratified-to-intermittent flow transition boundary predicted by Taitel and Dukler (1976) by accounting for surface tension in terms of the gas-phase height in the channel as follows:

$$h_G \leq \frac{\pi}{4}\sqrt{\frac{\sigma}{\rho g\left(1-\frac{\pi}{4}\right)}} \tag{6.3}$$

Even though the above equation includes surface tension and gravitational terms, they commented that for small channels, this condition is always satisfied, which leads to a further simplification of the criterion, namely, $h_G \leq (\pi/4)D$. They also provided different interpretations for the formation of intermittent flow from stratified flows. In large channels, they attribute the transition to the Kelvin–Helmholtz instability, whereas in small tubes with wetting liquid, wetting of the wall due to capillary forces causes the film to climb and form a bridge, which leads to intermittent flow.

Damianides and Westwater (1988) developed individual flow regime maps for air–water mixtures for $1 < D_H < 5$ mm to document the effect of diameter on flow regime transitions. They found some agreement between their results and the transition criteria of Taitel and Dukler (1976) for the 5-mm case, but the agreement was poor for 1-mm tubes. The 1-mm tube showed a large intermittent flow region, with the transition from annular flow attributed to the generation of roll waves. For the larger-diameter tubes, they reported the existence of an increasingly large region of stratified wavy flow, with the waves creeping up to form annular flow at larger gas velocities. The agglomeration of plugs of gas led to slug flow, rather than the growth of a finite amplitude wave, which meant that the Kelvin–Helmholtz instability could not be used to predict the onset of slugging. As observed by almost all investigators, they also stated the increasingly important role of surface tension at the smaller channel diameters. They noted that the intermittent-to-disperse flow transition occurred at larger liquid flow rates in the larger tubes, and that the intermittent-to-annular transition occurred at larger gas flow rates in smaller tubes. The stratified flow region was essentially nonexistent in the 1-mm tube.

Fukano et al. (1989) also investigated air–water flow patterns in tubes with $1 < D_H < 4.9$ mm and recorded bubbly, annular, plug, and slug flows, the latter two being combined into one intermittent region for pressure drop analyses. Their results agreed with the work of Barnea et al. (1983), and while there is some disagreement between the results reported by Fukano et al. (1989) and Damianides and Westwater (1988), these studies do point out that the flow regime map presented by Mandhane et al. (1974) cannot sufficiently predict the flow regime transitions in small-diameter tubes. Also, it appears that the theoretical predictions of Taitel and Dukler (1976) and the correlations presented by Weisman et al. (1979) are not reliable for small diameters.

Brauner and Maron (1992) noted that several previous researchers had based their flow regime transition analyses on single stability criteria of infinitely long waves. Therefore, they incorporated the stabilizing effects of surface tension associated with practically finite wavelengths, and were able to predict the transition from stratified flow over a wide range of channel sizes. This resulted in a transition criterion based on the Eötvos number:

$$Eö = \frac{(2\pi)^2 \sigma}{(\rho_1 - \rho_g) D^2 g} > 1 \quad (6.4)$$

Ide et al. (1995) studied the "liquid lumps" formed during vertical upward air–water flow in six tubes with $0.5 < D < 6.0$ mm, including the velocity of the lumps, the corresponding disturbance wave, and the base film. A momentum-averaged mean velocity of liquid lumps was used to characterize the flow for cases with large liquid lumps and many small waves. The liquid lump velocity characteristics were then used to interpret the flow patterns with corresponding maps (Fukano et al. 1989) for horizontal capillary tubes.

Mishima and Hibiki (1996b) investigated flow patterns for upward flow of air–water mixtures in vertical tubes (an orientation that is not particularly relevant for condensation applications) with $1 < D < 4$ mm. Some variants of flow patterns typically seen in larger tubes were observed, but the transitions were in reasonable agreement with Mishima and Ishii's (1984) transition criteria. Mishima et al. (1997) used neutron radiography, which uses the fact that thermal neutrons easily penetrate heavy materials such as metals, but are attenuated by light materials containing hydrogen (i.e., water) to nonintrusively investigate qualitative and quantitative two-phase flow phenomena. Thus, the incident neutron beam is attenuated in proportion to the liquid phase thickness, allowing projection of the image of the two-phase flow, which is then converted to an optical image by the scintillator. The image is intensified and enlarged, and detected by a high-speed video camera. Air–water vertical upward flow was visualized to identify the prevailing flow regime, rising velocity of bubbles, and wave height and interfacial area in annular flow. Representative images of slug flow before and after processing are shown in Figure 6.1. Void profiles and void fraction variations were also recorded, as discussed in a subsequent section. By utilizing attenuation characteristics of neutrons in materials, measurements of void profile and average void fraction were performed. This technique enables flow visualization in situations without optical access and where X-ray radiography is not applicable.

Coleman and Garimella (1999) investigated the effect of tube diameter and shape on flow patterns and flow regime transitions for air–water flow in circular tubes of 5.5-, 2.6-, 1.75-, and 1.3-mm diameter, and a rectangular tube with

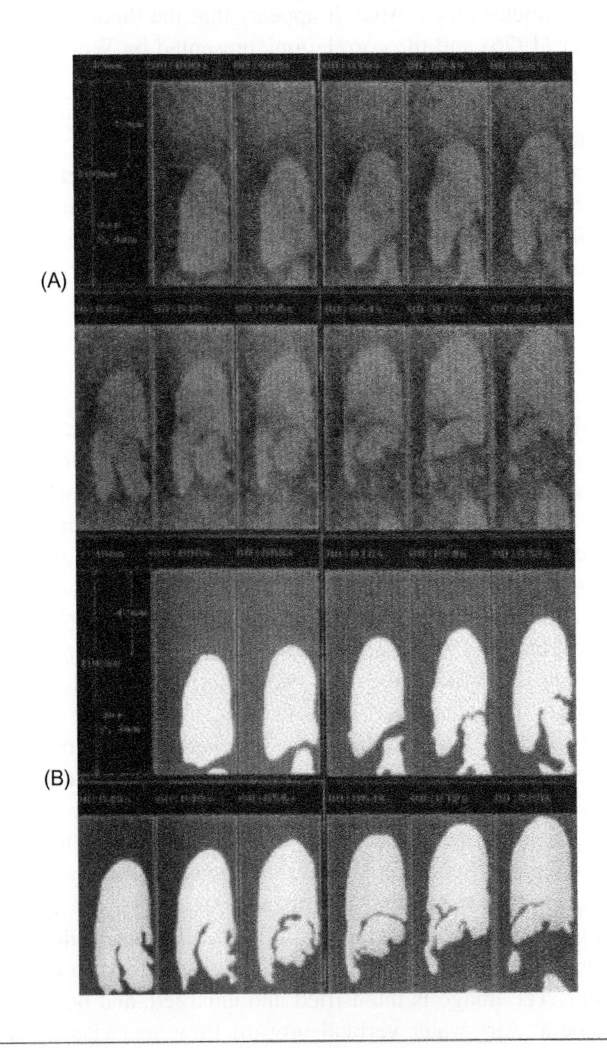

FIGURE 6.1

Raw and processed images of slug flow in vertical upward air–water flow.

Source: *Reprinted from Mishima et al. (1997) with permission from Elsevier.*

$D_H = 5.36$ mm and aspect ratio $\alpha = 0.725$. Gas and liquid superficial velocities were varied from $0.1 \text{ m/s} < U_{SG} < 100 \text{ m/s}$, and $0.01 \text{ m/s} < U_{SL} < 10.0 \text{ m/s}$, respectively. Bubble, dispersed, elongated bubble, slug, stratified, wavy, annular-wavy, and annular flow patterns were observed (Figure 6.2). This figure also shows their attempt to bring some uniformity to the terminology used to describe these flow mechanisms by the various investigators cited above. They divided the mechanisms into four major *regimes* (stratified, intermittent, annular, and dispersed) and subdivided these major regimes further into the applicable *patterns*. Thus, compared with the results of various investigators, the nomenclature for at least the major regimes would be somewhat uniform. The tube diameter had a significant effect on the transitions. As the tube diameter is decreased, the transition from the intermittent regime to the dispersed or bubbly regime occurred at progressively higher U_{SL}. Also, the transition from intermittent-to-annular flow occurs at a nearly constant U_{SL}, for a given tube, with the transition U_{SL} increasing as the diameter first decreases from 5.5 mm, but approaching a limiting value as the diameter decreases further to 1.75 and 1.30 mm. The size of the intermittent regime increases in small-diameter tubes. The primary difference in their results for round and rectangular channels of approximately the same D_H was that the rectangular tube showed a transition to disperse flows at higher U_{SL}, presumably because it is more difficult to dislodge the liquid from the corners of the rectangular tube due to the surface tension forces. Like other investigators, they found that very few points for the 5.5-mm tube were in the stratified regime, whereas the Taitel–Dukler (1976) criteria predicted the existence of a stratified region for these conditions. The agreement with the intermittent-to-disperse and the annular-to-disperse transitions was somewhat better. However, the assumptions inherent in the Taitel–Dukler analyses were not considered applicable to the small-diameter channels. The agreement with the data of Damianides and Westwater (1988) was much better. The transition to the dispersed region (including the effect of diameter) was also in good agreement with the results of Fukano et al. (1989). The transition correlations presented by Weisman et al. (1979) were only in agreement for the largest tube studied by Coleman and Garimella, with the agreement deteriorating as D_H decreased.

Triplett et al. (1999b) conducted an investigation similar to that of Coleman and Garimella (1999) of flow regimes in adiabatic air–water flows through circular microchannels of 1.1- and 1.45-mm diameter, and in semi-triangular microchannels with $D_H = 1.09$ and 1.49 mm. They identified bubbly, churn, slug, slug-annular, and annular flow patterns (stratified flow was not observed, as was the case with many other investigators), and plotted the data on gas and liquid superficial velocity axes. The progression between the respective regimes as the gas and liquid velocities are changed is shown in Figure 6.3. Bubbly flow undergoes a transition to slug flow when the gas superficial velocity U_{GS} increases due to the corresponding rise in the void fraction and the coalescence of bubbles. At high superficial liquid velocities U_{LS}, as the overall mass flux is increased, churn flow is established due to the disruption of slugs. Churn flow also occurred

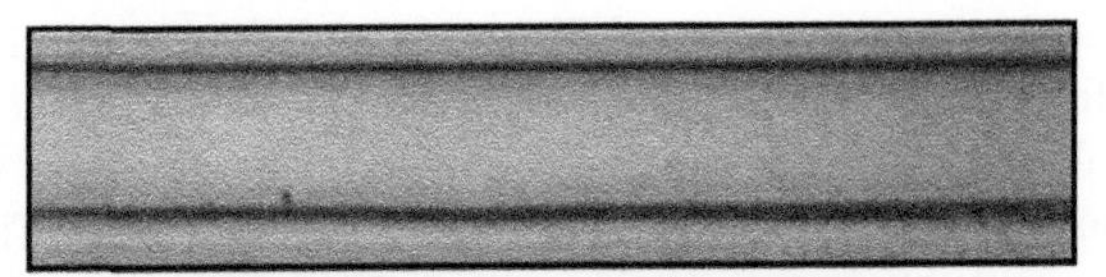

Stratified regime: wavy flow pattern

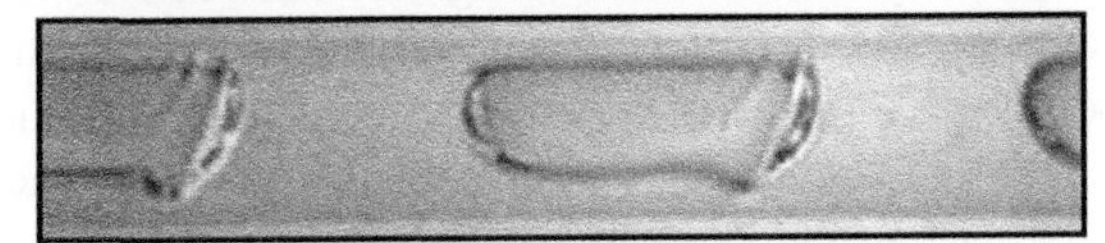

Intermittent regime: elongated bubble pattern

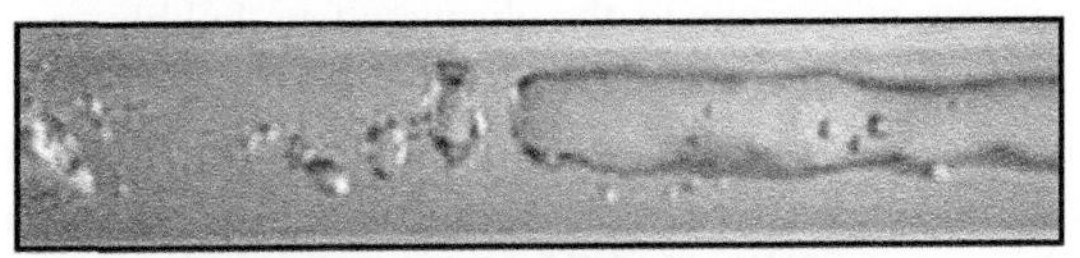

Intermittent regime: slug flow pattern

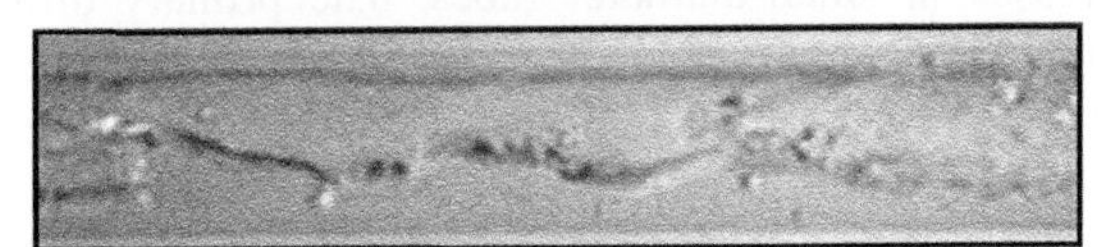

Annular regime: wavy–annular pattern

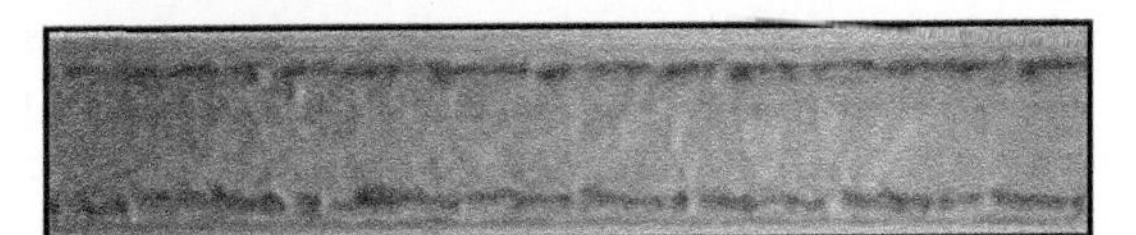

Annular regime: annular pattern

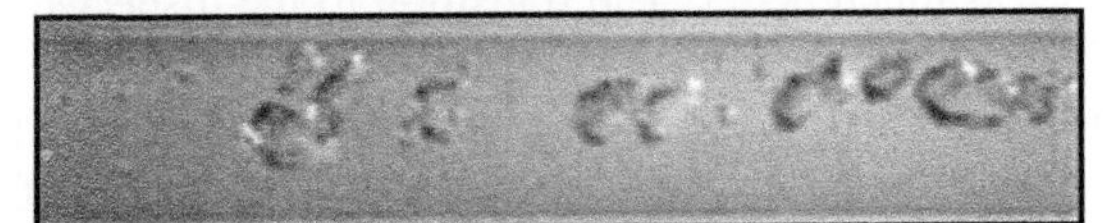

Dispersed regime: bubble flow pattern

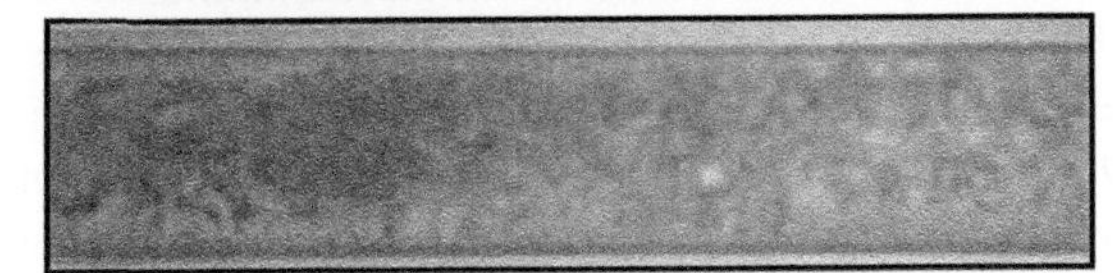

Dispersed regime: dispersed flow pattern

FIGURE 6.2

Air–water flow patterns in circular and rectangular tubes.

Source: *Reprinted from Coleman and Garimella (1999) with permission from Elsevier.*

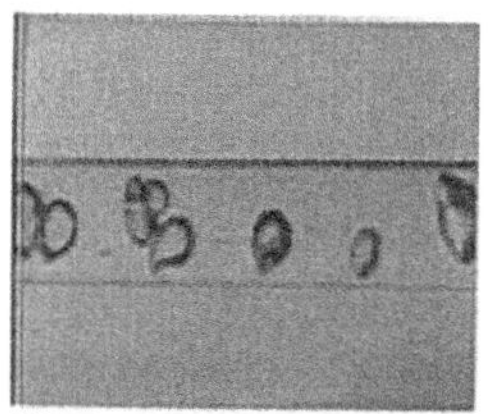

(A) U_{LS} = 3.021 m/s, U_{GS} = 0.083 m/s

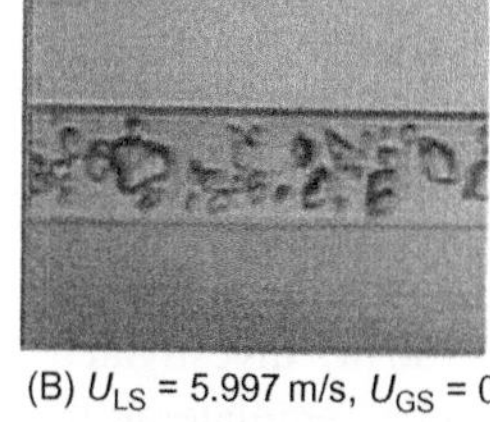

(B) U_{LS} = 5.997 m/s, U_{GS} = 0.396 m/s

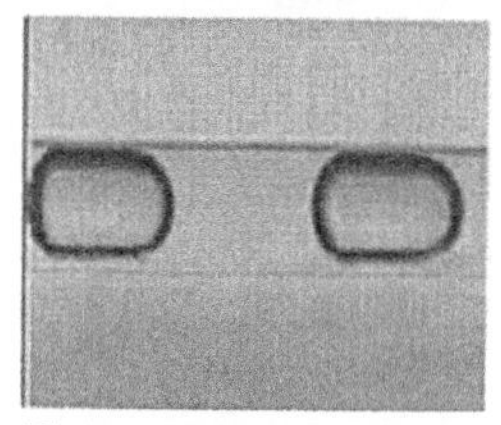

(C) U_{LS} = 0.213 m/s, U_{GS} = 0.154 m/s

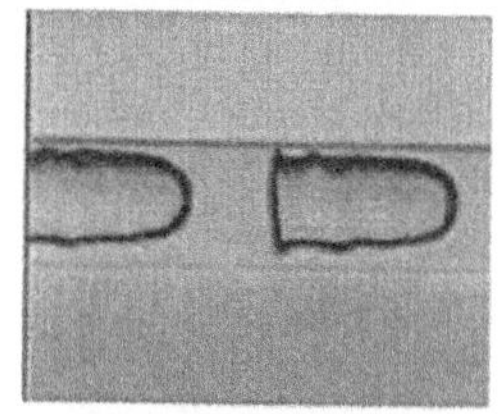

(D) U_{LS} = 0.608 m/s, U_{GS} = 0.498 m/s

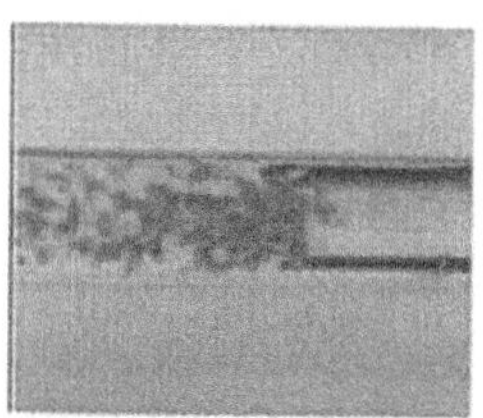

(E) U_{LS} = 0.661 m/s, U_{GS} = 6.183 m/s

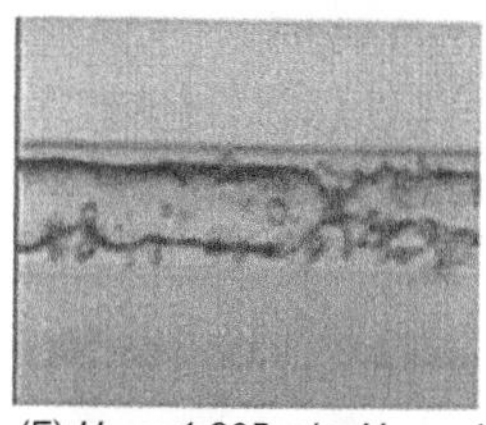

(F) U_{LS} = 1.205 m/s, U_{GS} = 4.631 m/s

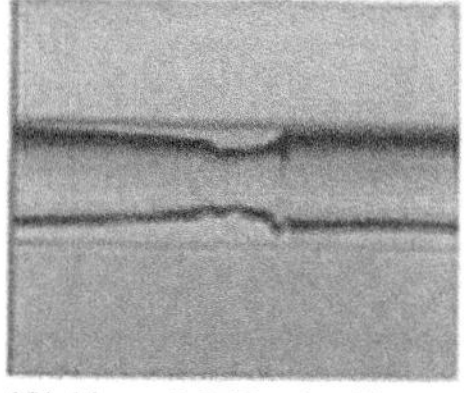

(G) U_{LS} = 0.043 m/s, U_{GS} = 4.040 m/s

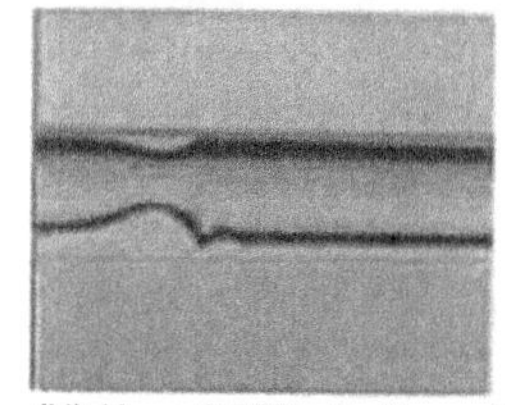

(H) U_{LS} = 0.082 m/s, U_{GS} = 6.163 m/s

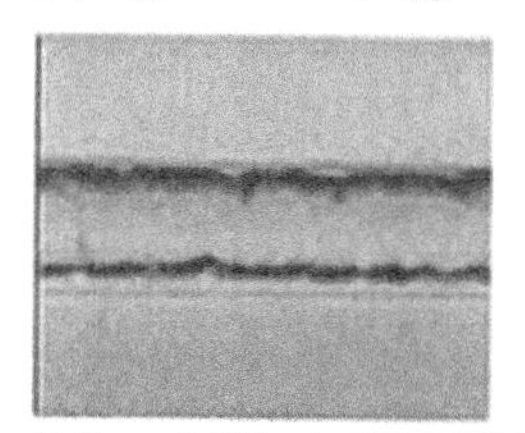

(I) U_{LS} = 0.082 m/s, U_{GS} = 73.30 m/s

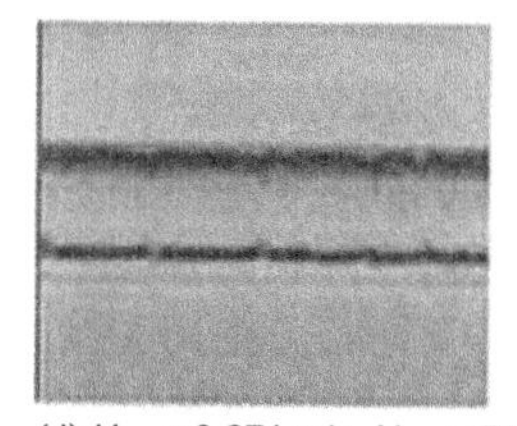

(J) U_{LS} = 0.271 m/s, U_{GS} = 70.42 m/s

FIGURE 6.3

Flow regimes for adiabatic air–water flow through 1.09-mm circular channels: (A) U_{LS} = 3.021 m/s, U_{GS} = 0.083 m/s; (B) U_{LS} = 5.997 m/s, U_{GS} = 0.396 m/s; (C) U_{LS} = 0.213 m/s, U_{GS} = 0.154 m/s; (D) U_{LS} = 0.608 m/s, U_{GS} = 0.498 m/s; (E) U_{LS} = 0.661 m/s, U_{GS} = 6.183 m/s; (F) U_{LS} = 1.205 m/s, U_{GS} = 4.631 m/s; (G) U_{LS} = 0.043 m/s, U_{GS} = 4.040 m/s; (H) U_{LS} = 0.082 m/s, U_{GS} = 6.163 m/s; (I) U_{LS} = 0.082 m/s, U_{GS} = 73.30 m/s; (J) U_{LS} = 0.271 m/s, U_{GS} = 70.42 m/s.

Source: *Reprinted from Triplett et al. (1999b) with permission from Elsevier.*

when large waves disrupted wavy-annular flows. Slug-annular flow is established at low U_{GS} and U_{LS}, and changes to annular flow when U_{GS} increases. The flow regime maps for all channels tested were substantially similar. Their maps agreed in general with the results of Damianides and Westwater (1988) and Fukano and Kariyasaki (1993), but the agreement with almost all analytically derived transition criteria (Suo and Griffith, 1964; Taitel and Dukler, 1976) was poor.

Zhao and Bi (2001a) investigated flow patterns for co-current upward air–water flow in vertical equilateral triangular channels with $D_H = 2.886$, 1.443, and 0.866 mm. The flow patterns encountered included dispersed bubbly flow, slug flow, churn flow, and annular flow for $D_H = 2.886$ and 1.443 mm. For $D_H = 0.866$ mm, dispersed bubbly flow was not observed, but a capillary bubbly flow pattern, consisting of a succession of ellipsoidal bubbles, spanned the cross-section of the channel (Figure 6.4) at low gas flow rates. Long slugs were seen at

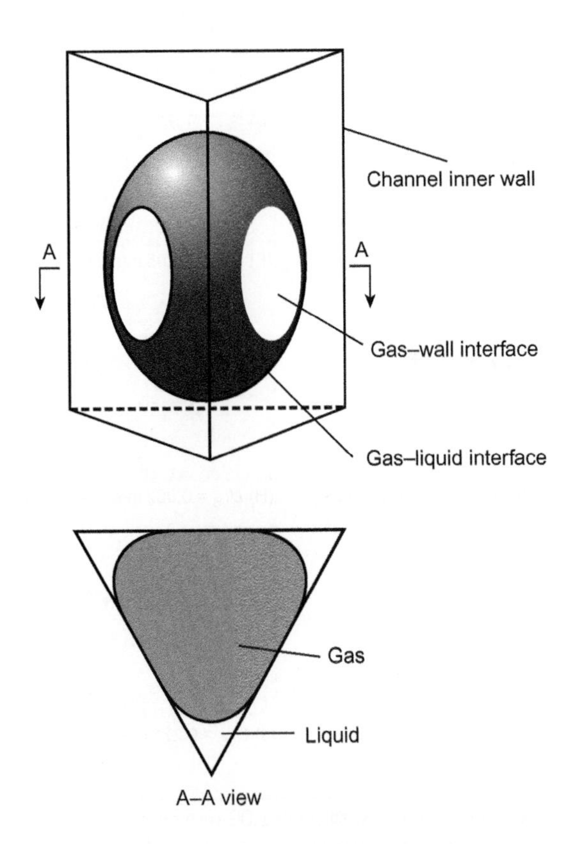

FIGURE 6.4

Capillary bubble in triangular channel.

Source: *Reprinted from Zhao and Bi (2001a) with permission from Elsevier.*

high superficial gas velocities in the slug flow regime. Churn and annular flows occurred at higher superficial gas velocities as the channel size decreased. These trends were in agreement with the results of other investigators (Barnea et al., 1983; Galbiati and Andreini, 1992; Ide et al., 1997; Mishima and Hibiki, 1996b) for small circular tubes. However, occurrence of the churn–annular transition line at higher gas velocities with decreasing channel diameter did not agree with the findings of Taitel et al. (1980) and Mishima and Ishii (1984), who predicted the opposite trend for conventional tubes. The significant discrepancies between the data and the Taitel et al. and Mishima–Ishii transition criteria were attributed to the large difference in channel size and the meniscus effect due to the triangular shape.

Yang and Shieh (2001) observed bubbly, plug, wavy, slug, annular, and dispersed regimes during air–water and R-134a (adiabatic) flow through 1-, 2-, and 3-mm horizontal tubes. Transitions between the regimes, especially slug-to-annular, were difficult to distinguish in air–water flow, while all the R-134a transitions were clear. In a manner much like the study by Coleman and Garimella (1999), they plotted their flow regime maps for the different tubes versus the transition lines of Taitel and Dukler (1976), Barnea et al. (1983), and Damianides and Westwater (1988). The results were in generally good agreement with those of the latter two investigators, but as pointed out by numerous other investigators, the stratified–intermittent transition predicted by Taitel and Dukler (1976) was not seen, due to the inapplicability of the basis for this transition to the smaller tubes. The slug-annular transition for R-134a occurred at lower gas velocities, while the intermittent–bubbly transition occurred at higher liquid velocities, both attributed to the lower surface tension of R-134a compared to the air–water pair. The R-134a transitions were in good agreement with the results of Wang et al. (1997a) and Hashizume (1983), although these two investigations focused on lower mass fluxes than those studied by Yang and Shieh (2001).

Tabatabai and Faghri (2001) developed a flow regime map for microchannels with particular attention to the influence of surface tension. They noted that in two-phase flow, ripples are generated on the annular layer with an increase in gas-phase velocity, which in turn leads to the formation of collars and bridges. They distinguish slug, plug, and bubble regimes based on the size and gap between the bridges. As with other investigators, they state that in large tubes, gravitational forces pull the liquid film down, preventing bridge formation. Their map is based on the relative effects of surface tension, shear, and buoyancy forces. The map uses the ratio of pressure difference due to surface tension and that due to shear forces as one axis, while the ratio of superficial velocities V_{GS}/V_{IS} forms the other axis. The void fraction model of Smith (1969) is used to compute the required vapor or gas-core flow area and interfacial surface area. The three terms in the force balance per unit length of the region of interest are given by

$$\frac{F_{\text{surface tension}}}{L} = 2\sigma(1-\alpha)^{0.5} \tag{6.5}$$

$$\frac{F_{\text{shear}}}{L} = \pi D \alpha f 1 \left(\frac{\rho_1 u_1^2}{4} \right) \tag{6.6}$$

$$\frac{F_{\text{buoyancy}}}{L} = g(\rho_1 - \rho_g) A_g \tag{6.7}$$

where the void fraction α appears in each of the three equations above (indirectly through the gas-phase area in the buoyancy term). This establishes a condition for the surface tension forces being dominant:

$$|F_{\text{surface tension}}| > |F_{\text{shear}}| + |F_{\text{buoyancy}}| \tag{6.8}$$

which then yields a criterion for the transition from surface-tension–dominated regions to shear-dominated regions:

$$u_1 < B = \left[\frac{2\sigma(1-\alpha)^{0.5} - g(\rho_1 - \rho_g)}{(\pi D \alpha \rho_1 f_1)/4} \right]^{0.5} \tag{6.9}$$

Relating the actual phase velocities to the corresponding superficial velocities results in the following form of the transition criterion:

$$\frac{V_{SG}}{V_{SL}} > \frac{V_{SG}}{(1-\alpha)B} \tag{6.10}$$

For the annular-to-intermittent transition, they used the following criterion for the liquid volume fraction m proposed by Griffith and Lee (1964), which is based on collars forming in annular flow due to instabilities at a wavelength of $5.5 \times D$, with the collar ultimately forming bridges and slugs:

$$m = \frac{(\pi r^2) 2r - (4/3)\pi r^3}{\pi r^2 (5.5) 2r} = 0.06 \tag{6.11}$$

where the numerator is the total liquid volume per slug and the denominator is the total volume of the slug at the most unstable wavelength of 5.5. Tabatabai and Faghri compared this flow regime map with data in the literature for condensation of steam and refrigerants in tubes with $D < 4.8$ mm, and for air–water flow in smaller-diameter tubes, and stated that better agreement was obtained in the surface-tension–dominated regions, and also that the map is able to predict the movement of the boundaries with channel size.

6.2.1.2 Narrow, high aspect ratio, rectangular channels

Most of the research on two-phase flow in small hydraulic diameter rectangular channels uses tubes of either small ($\alpha < 0.50$) or large ($\alpha > 2.0$) aspect ratios (Hosler ,1967; Jones and Zuber, 1975; Lowry and Kawaji, 1988; Wilmarth and Ishii, 1994). In a study on flow regime maps for $0.125 < \alpha < 0.50$ and $11.30 < D_H < 33.90$ mm, Richardson (1959) showed that the smaller aspect ratio suppressed the stratified and wavy flow regimes and promoted the onset of elongated bubble and slug flows. Troniewski and Ulbrich (1984) proposed corrections to the axes of the Baker (1954) map based on the single-phase velocity profiles in rectangular channels for horizontal and vertical channels with $0.09 < \alpha < 10.10$ and $7.45 < D_H < 13.10$ mm. Lowry and Kawaji (1988) studied rectangular geometries with $D_H < 2.0$ mm and $40 < \alpha < 60$ in vertical upward flows and concluded that the Taitel and Dukler (1976) model was not valid for narrow channel flow. Wambsganss et al. (1991) reported flow patterns and flow regime transitions in a single rectangular channel with aspect ratios of 6.0 and 0.167 and $D_H = 5.45$ mm through flow visualization and dynamic pressure measurements. They found qualitative agreement with the corresponding flow regime maps for circular channels and larger rectangular channels, but noted quantitative differences between those maps and their results for the relatively smaller rectangular channels. Wambsganss et al. (1994) extended this work to develop criteria for transition from bubble or plug flow to slug flow based on root-mean-square pressure changes. The transition was identified from a distinct change in the slope or a local peak in the pressure drop versus Martinelli parameter X or quality x-graph.

Xu (1999) investigated adiabatic air–water flow in vertical rectangular channels with gaps of 0.3, 0.6, and 1.0 mm. The flow regimes (bubbly, slug, churn, and annular) in the 1- and 0.6-mm channels were found to be similar to those reported in other studies. At smaller gaps, the bubbly–slug, slug–churn, and churn–annular flow transitions occurred at lower superficial gas velocities, due to the squeezing of the gas phase along the width of the narrow channels, which was believed to facilitate these transitions. In the 0.3-mm channel, bubbly flow was absent, but cap-bubbly flow was seen, characterized by two-dimensional (2-D) semicircular tops and flat bottoms. In addition, the presence of droplets together with slug and annular flows (attributed to the increased influence of surface tension and shear) led him to add slug-droplet and annular-droplet regimes to the others found in the bigger channels. Xu et al. (1999) later extended the transition criteria developed by Mishima and Ishii (1984) for application to the narrow rectangular geometry. Although this orientation and the corresponding physical considerations are not particularly relevant to condensation, their approach can be modified to yield analogous transition criteria for some of the transitions encountered in condensing flows. They based the bubbly-to-slug transition on the increasing probability of bubble collision and coalescence, which would lead to slug flow, with this transition occurring in the range of void fractions $0.1 < \alpha < 0.3$, as suggested by Mishima and Ishii (1984). The slug–churn transition was also adapted from the work of Mishima and Ishii (1984), based on the flow becoming unstable when two neighboring slugs start to touch each other as the channel mean void fraction increases. The transition to annular flow was assumed to occur somewhat arbitrarily for void fractions >0.75. They justified this based on the findings of Armand (1946), who found that the void fraction is related to the volumetric quality β through a simple linear relationship $\alpha = 0.833\beta$ for $\beta < 0.9$, but increased sharply when $\beta < 0.9$, signifying annular flow due to the uninterrupted gas core. Although these investigators applied some reasonable physical insights to the development of the transition criteria, the agreement was only acceptable with the 1-mm channel data, yielding poor agreement with the 0.6- and 0.3-mm data.

Using much the same approach as Xu et al. (1999), Hibiki and Mishima (2001) adapted Mishima and Ishii's (1984) model for round tubes to develop a flow regime map and transition criteria specifically for narrow rectangular cross-sections with vertical upflow. Their map consisted of bubbly, slug, churn, and annular flows. The bubbly-to-slug transition was based on fluctuations of bubbles distributed in a square lattice pattern. They postulated that collisions and coalescences of bubbles increase significantly when the maximum distance between bubbles is less than the projected diameter of a flat bubble. The bubble spacing was related to void fraction, which yielded the transition criterion in terms of the void fraction. The slug-to-churn transition was based on the pressure gradients around the bubble and in the liquid film surrounding it, while the churn flow-to-annular flow criterion was based on flow reversal considerations or the destruction of liquid slugs due to deformation or entrainment. Their transition criteria were in reasonable agreement with data for air–water flows in rectangular channels with gaps of 0.3–17 mm, and with some high–pressure water boiling data.

6.2.1.3 Channels with $D \ll 1$ mm

Flow regimes in two-phase flow through microchannels with $D \ll 1$ mm have only recently been investigated. Among the first groups of investigators of such flows are Feng and Serizawa (1999, 2000) and Serizawa et al. (2002), who studied air–water and steam–water flow through 25- and 50-μm channels. Their use of the term "liquid ring flow" for such small microchannels has since been used by other investigators in several different forms, as discussed throughout this section. Essentially, as the gas flow rate increases, the liquid bridge between adjacent gas slugs in slug flow becomes unstable, and transitions to this so-called liquid ring flow pattern. Kawahara et al. (2002) then conducted a comprehensive study of flow regimes, void fraction, and pressure drop for horizontal nitrogen–water flow in 100-μm circular channels. These authors contend that, in comparison with flow in channels in the 1-mm diameter range, the order of magnitude reduction in channel diameter results in lower Reynolds numbers and greater effects of wall shear and surface tension, which are manifested in different types of flow patterns. Although they stated the need for refractive index matching to remove the effects of optical distortion arising due to the small radius of curvature, they chose to conduct the experiments without such apparatus. This is because they felt that the near-wall phenomena, although distorted, could be viewed in more detail without optical correction, and the effects of distortion could be accounted for during the analysis phase. They observed five different flow mechanisms: liquid alone (liquid slug), gas core with smooth-thin liquid film, gas core with smooth-thick liquid film, gas core with ring-shaped liquid film, and gas core with a deformed interface. These flow regimes are shown in Figure 6.5 for various combinations of liquid and

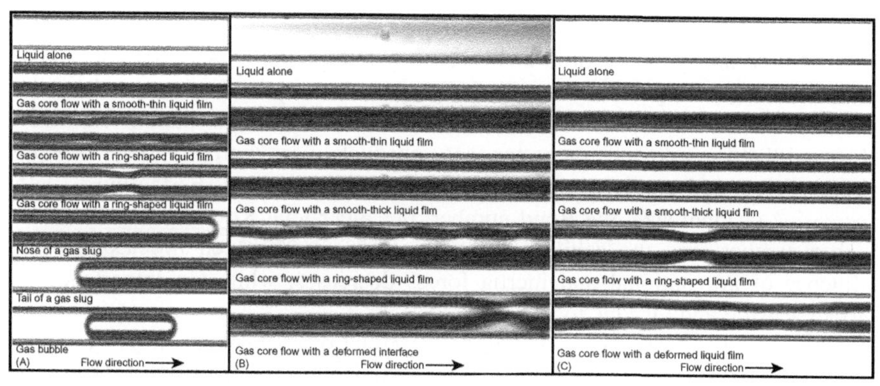

FIGURE 6.5

Air–water flow regimes in 100-μm channels: (A) $U_L = 0.15$ m/s, $U_G = 6.8$ m/s; (B) $U_L = 0.56$ m/s, $U_G = 20.3$ m/s; (C) $U_L = 3.96$ m/s, $U_G = 19.0$ m/s.

Source: *Reprinted from Kawahara et al. (2002) with permission from Elsevier.*

gas velocities. The axisymmetric (nonstratified) flow patterns clearly demonstrated the absence of gravitational effects. Similarly, the absence of small, dispersed bubbles (bubbly flow) was attributed to the liquid phase *Re* being very low; thus, no bubble breakup induced by liquid phase turbulence could occur. They also recognized that a given combination of liquid and gas-phase velocities does not always result in a unique flow pattern—it simply establishes the more likely patterns. Other investigators such as Coleman and Garimella (2000a, 2003) have addressed this issue by defining "overlap zones" in their flow regime maps, in which the flow could transition from one regime to another at the same conditions. Therefore, they developed a flow regime map based on the probability of occurrence of the different patterns at the different conditions. These probability analyses led them to define four flow regimes (slug-ring, ring-slug, semi-annular, and multiple), which were combinations of the flow mechanisms mentioned above. At low liquid flow rates, flows with a gas core surrounded by a smooth-thin liquid film occurred at low gas flow rates, while the gas core was accompanied with a ring-shaped liquid film at high gas flow rates. At high liquid flow rates, the gas core was surrounded by a thick liquid film, or the gas core moved through a deformed film in serpentine manner. At a combination of high gas and liquid flow rates, several combinations of flows occurred to yield the "multiple" flow pattern.

Chung and Kawaji (2004) extended the work of Kawahara et al. (2002) to investigate the effect of channel diameter, by conducting experiments on nitrogen–water flow through 530-, 250-, 100-, and 50-μm channels. They found that the two-phase flow characteristics in the 530- and 250-μm channels were similar to those reported for channels of ~1 mm diameter, for example, by Triplett et al. (1999b). For channels smaller than these, only slug flow was observed, with the absence of bubbly, churn, slug-annular, and annular flow attributed to the greater viscous and surface tension effects. From a combination of flow visualization and void fraction and pressure drop analyses, they interpreted slug flow as the rapid motion of a gas slug through a relatively quiescent liquid film retarded by the wall. Based on the values of the pertinent dimensionless parameters, they observed that the relevant forces for these channels were inertia, surface tension, viscous force, and gravity, in decreasing order of importance. Furthermore, with decreasing channel size, the Bond number, the superficial Reynolds numbers, the Weber number, and the capillary number all decrease, which implies that the influence of gravitational and inertia forces decreases, while the importance of surface tension and viscous forces increases.

Chung et al. (2004) also investigated the effect of channel shape on flow patterns, void fraction, and pressure drop by conducting similar tests on a 96-μm square, and a 100-μm circular, microchannel. The flow regime maps for the two channels were substantially similar, except that the square channel did not exhibit the ring-slug flow pattern that was present in the circular channel. They reasoned that whereas in the circular channel, the liquid film must increase in thickness or develop into thicker rings periodically, in the square channel, it accumulates in the corners of the square channel, which was also supported by the analyses of Kolb and Cerro (1991, 1993). Kawahara et al. (2005) subsequently also investigated water–nitrogen and ethanol–water/nitrogen mixtures in 50-, 75-, 100-, and 251-μm circular channels to demonstrate that fluid properties have little effect on flow patterns in these channels.

Serizawa et al. (2002) also conducted a study similar to that of Kawahara et al. (2002) on air–water flow in circular tubes of 20-, 25-, and 100-μm diameter, and steam–water flow in a 50-μm circular tube. Representative flow patterns recorded by them are shown in Figure 6.6. It must be noted that they develop a particularly imaginative and varied set of terms that lead to further proliferation (and perhaps confusion) about the descriptors for the observed flow mechanisms. Thus, the terms they use include: dispersed bubbly flow, gas slug flow, liquid ring flow, liquid lump flow, skewed barbecue (Yakitori) shaped flow, annular flow, frothy or wispy annular flow, rivulet flow, and liquid droplets flow. Unlike the findings of Kawahara et al. (2002), they observed bubbly flow in the 25-μm channel, often bubbles almost the size of the channel, closely followed by much smaller, finely dispersed bubbles. They noted that the capillary pressure at the small 5-μm bubble interface was as high as 0.3 bar, which kept the bubble from distorting from a spherical shape or from coalescing. They explained that slug flow occurs not due to bubble coalescence, but rather because of an entrance phenomenon, whereby gas at a high flow rate enters the channel, but the gas bubble velocity is not high enough to overcome the surface tension of the liquid bridge. This is an explanation quite different from that of most other investigators, and requires further study. They also state that surface tension prevents the liquid slug from spreading as a film, and coupled with the high pressure of the gas bubble, which pushes it to expand across the entire cross-section, leads to dry spots and patches. Liquid ring flow originates from liquid bridges between slugs that are

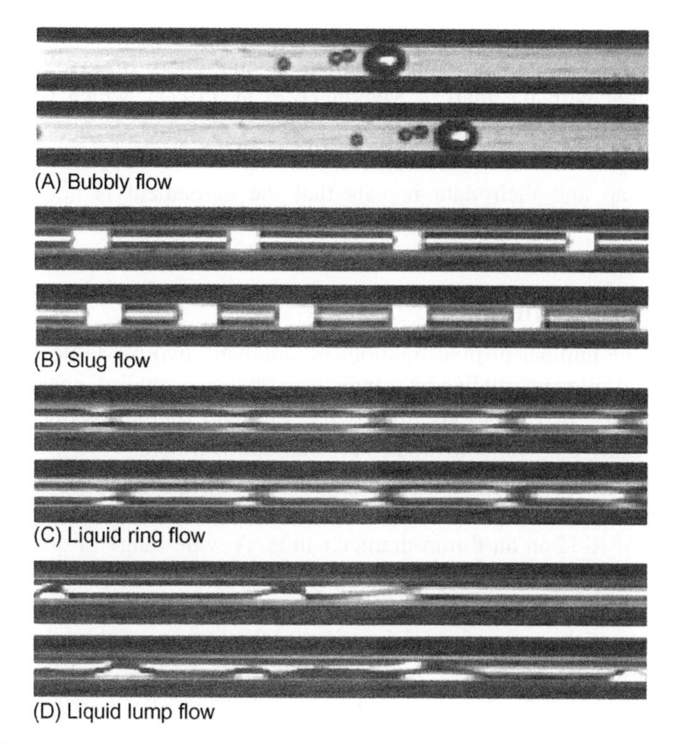

FIGURE 6.6

Air–water flow patterns in 25-μm circular channel: (A) bubbly flow, (B) slug flow, (C) liquid ring flow, and (D) liquid lump flow.

Source: *Reprinted from Serizawa et al. (2002) with permission from Elsevier.*

ruptured when the gas velocity is very high. Liquid ring motion is governed by a balance between viscous forces at the wall and shear at the gas interface: viscous forces dominate up to a certain threshold thickness, beyond which the shear from the gas core pushes the liquid ring. Liquid lumps are formed at even higher gas velocities, when liquid is entrained and oscillates, but surface tension prevents it from spreading as a film. Liquid lumps are considered by them to be "partially continuous films" rather than the term rivulet that is often used in the literature. They also present steam–water flow regimes that include bubbly, slug, liquid ring, and liquid droplet flows, and state that these flow mechanisms are similar to those shown for air–water in a 25-μm tube (Figure 6.6). However, given the degree of distinction they make in this work between small variations in flow mechanisms, it does not appear that the steam–water and air–water flow patterns are all that similar. Several of their photographs show large gas bubbles appearing together with adjoining small bubbles, and detailed explanations are provided based on the prominence of surface effects in small channels.

Given the above emphasis on the uniqueness of microchannel two-phase flow, one of their more surprising findings is that their flow regime map is deemed to agree with the map of Mandhane et al. (1974), which is one of the more widely quoted flow maps for *much larger* conventional channels. No real explanation is given for this stated agreement with such a map, although a quick look at the map and their data reveals that the agreement is not as good as they state.

6.2.2 Condensing flow

Compared to the number of investigations on adiabatic two-phase flows, there are few particularly relevant studies on condensing flows in small-diameter channels. A few of the key investigations toward the low end of the hydraulic diameter range from among the available studies are summarized here.

One of the earlier, well-cited investigations of flow patterns during condensation of refrigerants was conducted by Traviss and Rohsenow (1973), who studied condensation of R-12 in an 8-mm-diameter tube. A wide range of qualities, mass fluxes ($100 < G < 990$ kg/m^2 s), and saturation temperatures ($10°C < T < 40.6°C$) was tested. Disperse, annular, semi-annular (referred to by many as wavy, with a thin liquid film at the top and a thicker film at the bottom of the tube), and slug flows were observed. Their main focus, as is the case with most of the studies on tubes with $D < \sim 5$ mm, was the transition between shear-dominated annular flow and gravity-dominated stratified flow. They used the Froude number $Fr_1^2 = \overline{V}_1^2/g\delta$, where $\overline{V}_1$ is the average velocity of the liquid film and δ is the film thickness, as the basis for determining the applicable mode of condensation. They used the von Karman universal velocity profile to describe the film velocity. Using this velocity profile, and expressing the wall shear stress in terms of the two-phase frictional pressure drop (which was computed using a two-phase multiplier to the vapor-phase pressure drop), they developed expressions for the film Reynolds number $Re_1 = G(1 - x)D/\mu_1$ in terms of Fr_1, the Galileo number $Ga_1 = gD^3/v_1{}^2$, and the Martinelli parameter X_{tt}. By plotting their data on a Re_1 versus X_{tt} graph, they proposed that the annular to semi-annular transition occurs at a constant Fr_1 of 45. They further state that this boundary may be used to distinguish between nonstratified disperse and annular flows and "stratified" semi-annular, wavy, slug and plug flows. This development is widely available in the relevant textbooks, and the details are not presented here.

Breber et al. (1980) used the Taitel–Dukler (1976) map discussed in the previous section to develop simple transition criteria for condensation in horizontal tubes, once again basing them on the ratio of shear forces to gravity forces on the liquid film, which are quantified through the Wallis dimensionless gas velocity j_g^*:

$$J_g^* = \frac{G_t x}{\sqrt{Dg\rho_v(\rho_1 - \rho_v)}} \tag{6.12}$$

in addition to the ratio of liquid volume to vapor volume. In the absence of a reliable way to predict the slip between the phases, the Martinelli parameter X is used to quantify this:

$$X = \sqrt{\frac{\Delta P_l}{\Delta P_v}} = \left(\frac{1-x}{x}\right)^{0.9}\left(\frac{\rho_v}{\rho_l}\right)^{0.5}\left(\frac{\mu_l}{\mu_v}\right)^{0.1} \tag{6.13}$$

They applied these criteria to data from a variety of studies on fluids such as R-12, R-113, steam, and n-pentane flowing through $4.8 < D < 22.0$-mm tubes at $108 < p < 1249$ kPa and $18 < G < 990$ kg/m^2 s. On the basis of the comparisons between these data and the aforementioned coordinate axes, they recommended the following simplified transition criteria:

Annular flow	$j_g^* > 1.5;\ X < 1.0$
Wavy/stratified flow	$j_g^* < 0.5;\ X < 1.0$
Slug (intermittent) flow	$j_g^* < 1.5;\ X > 1.5$
Disperse (bubbly) flow	$j_g^* > 1.5;\ X > 1.5$

One advantage of these criteria is the simplicity they afford: the transition lines appear as horizontal or vertical lines, helping in the clear bracketing of the various regimes. The existence of a transition band between the respective regimes also allows for a more realistic accounting of overlap zones through linear interpolation of heat transfer coefficients and other performance parameters. The authors do note, however, that its predictive capabilities are the weakest for the wavy–slug transition (which happens to be of considerable interest in small-diameter and microchannels).

Sardesai et al. (1981) also investigated the annular-to-stratified/wavy transition for a variety of fluids (R-113, steam, propanol, methanol, and n-pentane) condensing in horizontal 24.4-mm-diameter tubes. They instrumented their test section with thermocouples around the circumference on the tube at several axial locations to determine the circumferential variation of heat transfer coefficients, which enabled them to compute the ratio of the heat transfer coefficient at the top and bottom of the tube. They used the Taitel–Dukler (1976) annular-stratified/wavy transition criterion (representing a ratio of the axial shear force to the circumferential gravity force in the liquid film) as the basis for developing their experimentally derived transition criterion. Thus, they used the data to find a multiplier β to this transition line $\beta(F, X) = \text{constant}$, which would result in a line parallel to the Taitel–Dukler transition line. The data showed that the heat transfer coefficient ratio stayed close to unity (signifying axisymmetric annular flow) down to $\beta = 1.75$, below which it decreased rapidly, signifying gravity-driven condensation. An empirical fit derived from the square root of the two-phase frictional multiplier of Chisholm and Sutherland (1969) was used such that $\phi_g{}^2 F = 1$ is used as the upper limit of gravity-controlled condensation, while $\phi_g{}^2 F = 1.75$ is used as the lower limit for annular flow, with ${\phi_g}^2 = 0.7X^2 + 2X + 0.85$. Here, F is the modified Froude number given by

$$F = \sqrt{\frac{\rho_g}{\rho_1 - \rho_g} \frac{U_{GS}}{\sqrt{Dg}}} \tag{6.14}$$

and U_{GS} is the superficial gas-phase velocity.

Soliman (1982) also proposed an annular-wavy transition criterion based on data from different sources that included R-12, R-113, and steam flowing through $4.8 < D < 15.9$-mm tubes for $28°C < T < 110°C$. During the progression from vapor to liquid, the increase in the liquid volume and the significance of gravity, coupled with the decreasing velocity, and thus the inertial effects, were quantified based on the decrease in the Froude number. Using the liquid film thickness expressions of Kosky (1971), the Martinelli parameter X_{tt}, and the Azer et al. (1972) equation $\phi_G = 1 + 1.09{X_{tt}}^{0.039}$, he developed the following expressions for the film Reynolds number:

$$\begin{aligned} Re_1 &= 10.18Fr^{0.625}Ga^{0.313}(\phi_G/X_{tt})^{-0.938} \quad \text{for } Re_1 \leq 1250 \\ Re_1 &= 0.79Fr^{0.962}Ga^{0.481}(\phi_G/X_{tt})^{-1.442} \quad \text{for } Re_1 \leq 1250 \end{aligned} \tag{6.15}$$

The data from the various sources were then used to determine that the annular-to-wavy and intermittent transition occurs when $Fr = 7$ in the above expressions, irrespective of the diameter or fluid under consideration.

Subsequently, Soliman (1986) also developed a correlation for the mist-annular transition. Mist flow occurs in the entrance region of the condenser, with the liquid phase flowing as entrained droplets without any visible liquid film at the wall. He used the large discrepancies between experimentally determined heat transfer and pressure drop values and predictions typically using annular flow models as the rationale to propose that mist flow must be treated separately. He reasoned that entrainment occurs due to the inertia of the vapor phase ($\rho_G {V_G}^2$) shearing droplets from the surface of the liquid film, while viscous ($\mu_L V_L \delta$) and surface tension (σ/D) forces tend to stabilize the liquid film, with the balance between them being represented by the Weber number, for which the following expressions were derived:

$$\begin{aligned} We &= 2.45Re_{GS}^{0.64}\left(\frac{\mu_G^2}{\rho_G \sigma D}\right)^{0.3}\phi_G^{0.4} \quad \text{for} \quad Re_{LS} \leq 1250 \\ We &= 0.85Re_{Re}^{0.79}\left(\frac{\mu_G^2}{\rho_G \sigma D}\right)^{0.3}\left[\left(\frac{\mu_G}{\mu_L}\right)^2\left(\frac{\rho_L}{\rho_G}\right)\right]^{0.084}(X_{tt}/\phi_G^{2.55})^{0.157} \quad \text{for} \quad Re_{LS} > 1250 \end{aligned} \tag{6.16}$$

Based on the above definitions and comparison with the database mentioned above, the following criteria were established:

$$\begin{aligned} We &< 20 \text{ always annular} \\ We &> 30 \text{ always mist} \end{aligned} \tag{6.17}$$

Tandon et al. (1982) also used data for R-12 and R-113 from several sources for the range $4.8 < D < 15.9$ mm, and proposed a flow regime map with dimensionless gas velocity j_G^* as the ordinate and $(1-\alpha)/\alpha$ as the abscissa, which resulted in simple transition equations and good agreement with the data for annular, semi-annular, and wavy flows. The correlation by Smith (1969) was used to compute the void fraction α. Similar to the Breber et al. (1980) criteria, their transition criteria, given below, also result in horizontal and vertical transition lines on the map. Thus,

$$\begin{aligned}
&\text{Spray} \quad 6 < J_G^* \text{ and } \frac{1-\alpha}{\alpha} \leq 0.5 \\
&\text{Annular/semi-annular} \quad 1 \leq J_G^* \leq 6 \text{ and } \frac{1-\alpha}{\alpha} \leq 0.5 \\
&\text{Wavy} \quad J_G^* \leq 1 \text{ and } \frac{1-\alpha}{\alpha} \leq 0.5 \\
&\text{Slug} \quad 0.01 \leq J_G^* \leq 0.5 \text{ and } \frac{1-\alpha}{\alpha} \leq 0.5 \\
&\text{Plug} \quad J_G^* \leq 0.01 \text{ and } \frac{1-\alpha}{\alpha} \leq 0.5
\end{aligned} \tag{6.18}$$

Subsequently, Tandon et al. (1985) also demonstrated good agreement with their own data on condensation of R-12, R-22, and mixtures of the two fluids flowing through 10-mm tubes.

Hashizume and co-workers (Hashizume, 1983; Hashizume and Ogawa, 1987; Hashizume et al., 1985) conducted a three-part study on flow patterns, void fraction, and pressure drop for refrigerant (R-12 and R-22) flows in 10-mm-diameter horizontal tubes. Tests were conducted at saturation pressures in the range $570 < p < 1960$ kPa, corresponding to saturation temperatures in the range $20°\text{C} < T < 72°\text{C}$ for R-12 and $4°\text{C} < T \leq 50°\text{C}$ for R-22, and it was shown that these flow patterns for refrigerants are quite different from those for the air–water system. The flows were analyzed using simplified models for annular and stratified flow, and the velocity profiles in the liquid and gas phases were described using the Prandtl mixing length. They modified the property corrections to the Baker map (1954) proposed by Weisman et al. (1979) (specifically, they decreased the surface tension exponent in $\psi = (\sigma_W/\sigma)^1\,[(\mu_L/\mu_W)(\rho_W/\rho_L)^2]^{1/3}$ from 1 to 0.25) to make those maps applicable for R-12 and R-22.

Wang et al. (1997a,b) investigated two-phase flow patterns for refrigerants R-22, R-134a, and R-407C in a 6.5-mm circular tube over the mass flux range $50 < G < 700$ kg/m^2 s, for saturation temperatures of 2°C, 6°C, and 20°C. They noted, as did Wambsganss et al. (1991), that prior flow pattern studies had focused on diameters in the range $9.5 \leq D \leq 75.0$ mm, and had been conducted at high mass fluxes ($G > 300$ kg/m^2 s), which are not particularly applicable to refrigeration and air-conditioning applications. At $G = 100\ \text{kg/m}^2\ \text{s}$, plug, slug, and stratified flow patterns were observed, with no occurrence of annular flow. As the mass flux increased to 200 and $400\ \text{kg/m}^2\ \text{s}$, annular flow appeared, and dominated the map at higher mass fluxes. They found the transitions in their study to be in good agreement with the modified Baker map of Hashizume (Hashizume, 1983; Hashizume and Ogawa, 1987; Hashizume et al., 1985), with the transitions occurring at somewhat lower mass fluxes, which they attributed to the smaller diameter used in their study. The slug-to-wavy and wavy-to-annular flow regime transitions for the refrigerant blend R-407C occurred at higher G and x than for the pure fluids. The higher phase densities and the consequent lower gas- and liquid-phase velocities for R-407C compared to R-22, and the higher liquid-phase viscosity of the constituents of R-407C were assumed to be the cause for this lagging behavior of the refrigerant flow transitions.

One of the more exhaustive studies of refrigerant condensation in small-diameter ($D = 3.14$, 4.6, and 7.04 mm) tubes was conducted by Dobson and Chato (1998) using refrigerants R-12, R-22, R-134a, and two different compositions of a near-azeotropic blend of R-32/R-125 over the mass flux range $25 < G < 800\ \text{kg/m}^2\ \text{s}$. The authors provide a good overview of the various studies on condensation, including a step-by-step description of the progression of flows from one regime to another as the quality changes at different mass fluxes. At the lowest mass flux of $25\ \text{kg/m}^2\ \text{s}$, the flow was entirely smooth stratified for all qualities, while wavy flow was observed at $75\ \text{kg/m}^2\ \text{s}$. As the mass flux was increased to $150\text{–}300\ \text{kg/m}^2\ \text{s}$, condensation occurred over annular, wavy-annular, wavy, and slug flow as the quality decreased. Fluid properties and tube diameter affected the flow transitions the most at these mass fluxes. At the highest mass fluxes (500, 650, and $800\ \text{kg/m}^2\ \text{s}$), the flow progressed from annular-mist (indicating entrainment) at high qualities to annular, wavy-annular, and slug flow with decreasing quality. With tests conducted at $T = 35$ and 45°C, they commented that the decreasing difference in vapor and liquid properties with an approach toward the critical pressure (increasing reduced pressure) explained the changes in the flow patterns. Thus, annular flow was seen over a larger portion of the map at lower reduced pressures; for example, annular flow was established in R-134a at 35°C at qualities as low as 15% at $G = 600\ \text{kg/m}^2\ \text{s}$. Also, the R-32/R-125, blend showed transitions to annular flow at much higher qualities than the other fluids, with a significant amount of stratification seen at the high temperature case even at $G = 600\ \text{kg/m}^2\ \text{s}$. It was also shown that the slug flow region increases at higher reduced pressures. As the tube diameter decreased, the transition from wavy flow to wavy-annular flow, and from wavy-annular to annular flow moved to lower qualities. They found very good agreement with the Mandhane et al. (1974) map after correcting the superficial vapor velocity with the factor $\sqrt{\rho_g/\rho_a}$ to account for the fact that the Mandhane et al. map was based primarily on air–water data, where the gas-phase density is drastically lower than refrigerant vapor density. This correction accounts for the different gas-phase kinetic energies better than the original map. Since their stated objective was to primarily develop flow-regime-based heat transfer correlations, they divided the above-mentioned flow regimes into gravity-dominated and shear-controlled regimes (as done by the many other researchers cited above), and did not provide detailed transition criteria between each of the flow regimes observed. Instead, in their prior work (Dobson, 1994), they established that the wavy-annular to annular flow transition would occur at a Soliman-modified Froude number Fr_{so} of 18, instead of $Fr_{so} = 7$, as recommended by Soliman (1982). In the heat transfer correlations also developed in this work, and to be discussed in a subsequent section of this chapter, they recommend that the shear-controlled annular flow correlation be used for $G \geq 500$ kg/m^2 s for all x, while for $G < 500$ kg/m^2 s, the annular flow correlation should be used for $Fr_{so} > 20$, and the gravity-driven correlation should be used for $Fr_{so} < 20$.

The studies by Coleman and Garimella (2000a,b, 2003) and Garimella (2004) were the first investigations that specifically addressed all the aspects of the subject at hand. Thus, several studies cited above either report detailed flow regimes in microchannels as small as 25 μm, but for adiabatic air–water flow, or those that study flow regimes in condensing flows are typically restricted to $D > \sim$ 5 mm, with the exception of the one 3.14-mm tube studied by Dobson and Chato (1998) and discussed above. The work of Coleman and Garimella simultaneously addressed: (a) small diameters, (b) condensing instead of adiabatic flow, (c) flow visualization during condensation, (d) temperatures representative of practical condensation processes, and (e) a wide range of mass fluxes and qualities. They conducted flow visualization studies for refrigerant R-134a in nine different tubes of round, square, and rectangular cross-sections with $1 < D_H < 4.91$ mm over the mass flux range $150 < G < 750$ kg/m^2 s and quality range $0 < x < 1$. The channels studied by them are shown in Figure 6.7. In their studies, refrigerant of a desired quality was supplied to the test section by precondensing a superheated vapor. Visualization was conducted in a counterflow heat exchanger, with refrigerant flowing through an inner glass tube of the cross-section of interest, and air flowing through the space between this inner tube and another transparent outer Plexiglas tube, thus enabling visualization of the respective flow regimes. Heat transfer between the cold air and refrigerant causes condensation. Compressed air flowing in the annulus provided a low differential pressure for the glass microchannel, making it possible to conduct tests at saturation pressures as high as 1379–1724 kPa. Post-condensers/subcoolers completed the closed loop facility. They recorded flow patterns for each mass flux at nominally 5–10% quality increments for a total of about 50 data per tube, to provide a fine resolution in the variation of flow patterns along the condensation process.

Four major flow *regimes*, including annular, intermittent, wavy, and dispersed flow, were identified, with the regimes further subdivided into flow *patterns* (Figure 6.8). In the annular flow regime, the vapor flows in the core of the tube with a few entrained liquid droplets, while liquid flows along the circumference of the tube wall. The flow patterns within this regime (mist, annular ring, wave

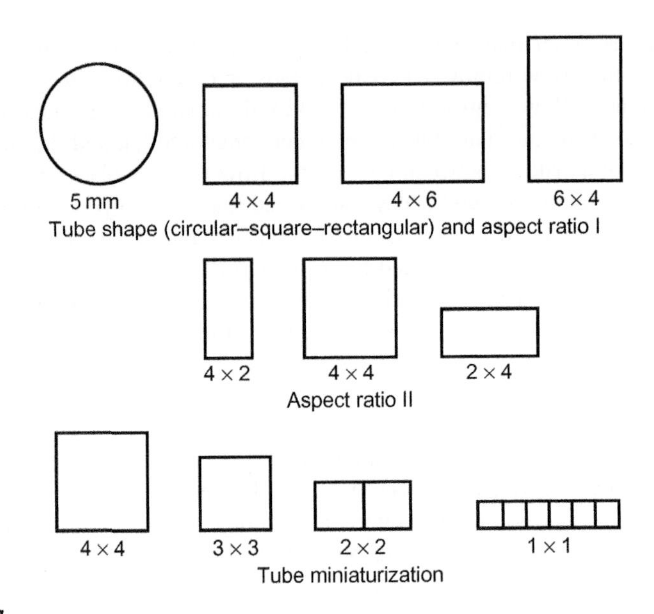

FIGURE 6.7

Flow visualization geometries of Coleman and Garimella (2000a,b, 2003).

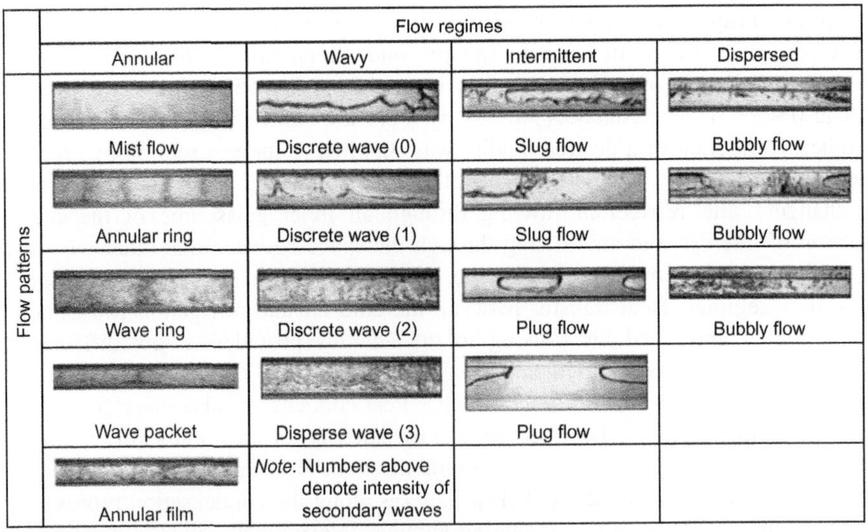

	Flow regimes			
	Annular	Wavy	Intermittent	Dispersed
Flow patterns	Mist flow	Discrete wave (0)	Slug flow	Bubbly flow
	Annular ring	Discrete wave (1)	Slug flow	Bubbly flow
	Wave ring	Discrete wave (2)	Plug flow	Bubbly flow
	Wave packet	Disperse wave (3)	Plug flow	
	Annular film	*Note*: Numbers above denote intensity of secondary waves		

FIGURE 6.8

Condensation flow regimes and patterns ($1 < D_H < 4.9$ mm).

Source: *Reprinted from Coleman and Garimella (2003) with permission from Elsevier and International Institute of Refrigeration (iifiir@iifiir.org or www.iifiir.org).*

ring, wave packet, and annular film) show the varying influences of gravity and shear forces as the mass flux and quality changes. Flow patterns with a significant influence of gravity (vapor flowing above the liquid, or a noticeable difference in film thickness at the top and bottom of the tube) and a wavy interface were assigned to the wavy flow regime. The waves at the liquid–vapor interface are caused by interfacial shear between the two phases moving at different velocities. Thus, this regime was subdivided into discrete waves of larger structure moving along the phase interface, and disperse waves with a large range of amplitudes and wavelengths superimposed upon one another, as shown in Figure 6.8. The other flow regimes shown in the figure have already been described in detail in this chapter. In some cases, the flow mechanisms corresponded to more than one flow regime, typically indicating a transition between the respective regimes. A typical flow regime map for the 4.91-mm circular tube plotted using the mass flux G and quality x-coordinates is shown in Figure 6.9. A major portion of this map is occupied by the wavy flow regime, with a small region where the plug, slug, and discrete wave flow patterns coexist. The waves become increasingly disperse as the quality and mass flux is increased (shown by the arrow in Figure 6.9). The approximate demarcation between discrete and disperse waves is shown by the dashed line in this figure, although this transition occurs gradually.

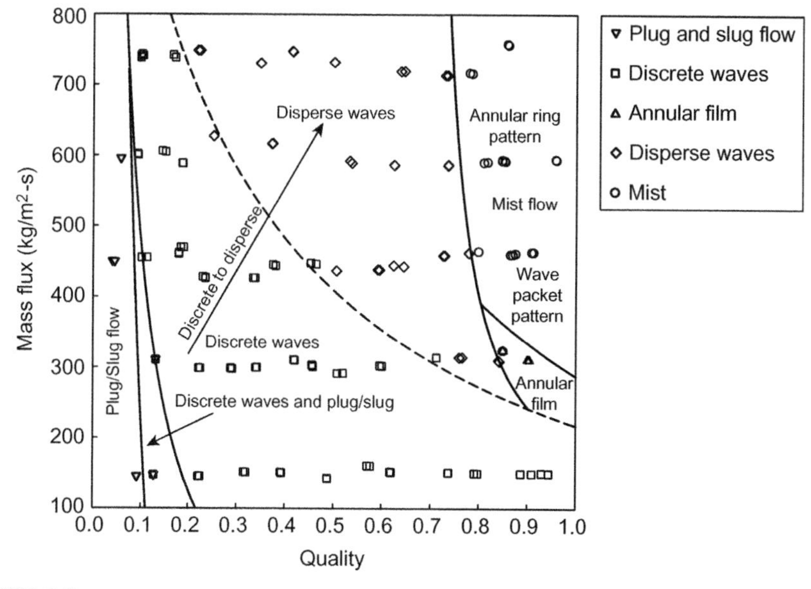

FIGURE 6.9

Condensation flow regime map (4.91 mm).

Source: *Reprinted from Coleman and Garimella (2003) with permission from Elsevier and International Institute of Refrigeration (iifiir@iifiir.org or www.iifiir.org).*

The effect of hydraulic diameter on the flow regime maps is shown in Figures. 6.10 and 6.11. Figure 6.10, which depicts the transition from the intermittent regime for the four square tubes investigated, shows that the size of the intermittent regime increases as D_H decreases, with this effect being greater at the lower mass fluxes. The large increase in the size of the intermittent regime in the smaller hydraulic diameter tubes is because surface tension achieves a greater significance in comparison with gravitational forces at these dimensions. This also occurs because, in square channels, it is easier for the liquid to be held in the sharp corners, counteracting to some extent the effects of gravity. This facilitates plug and slug flow at higher qualities as the hydraulic diameter is decreased. Figure 6.10 shows that the 4-mm tube map is dominated by the wavy flow regime (with an absence of the annular film flow pattern). As D_H is decreased, the annular flow regime appears and occupies an increasing portion of the map. Thus, for the 4-mm tube, the effects of gravity dominate, resulting in most of the flow regime map being covered by the wavy flow regime. As the hydraulic diameter decreases, the effects of surface tension increasingly counteract the effects of gravity, promoting and extending the size of the annular film flow pattern region instead of the more stratified wavy flow regime. Thus, as D_H decreases, the wavy flow regime is increasingly replaced by the annular flow regime, and is nonexistent in the $D_H = 1$-mm tube. Furthermore, surface tension stabilizes the waves, which leads to more discrete waves at small diameters.

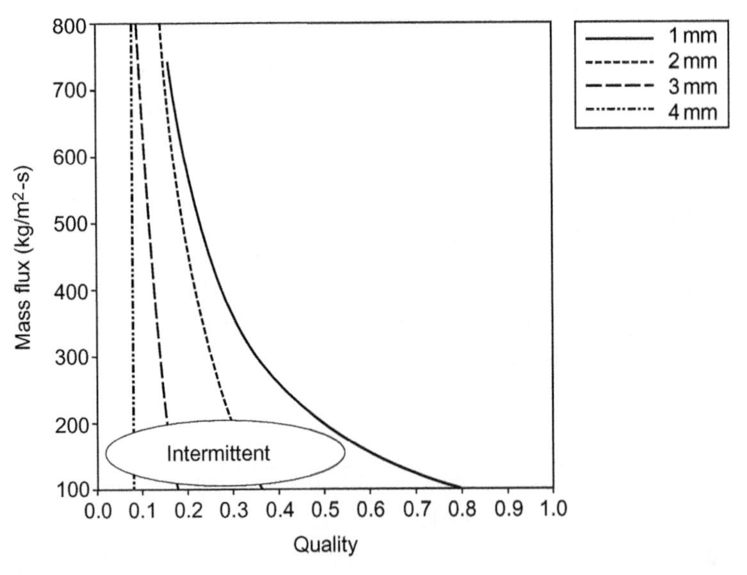

FIGURE 6.10

Effect of D_H on intermittent flow regime.

Source: *From Garimella (2004).*

The effect of tube shape was investigated using maps for circular, square, and rectangular tubes of approximately the same $D_H \sim 4-4.9$ mm (Figure 6.12). This figure shows that the intermittent regime is larger in the round tube than in the square tube at lower mass fluxes and approximately the same at higher mass fluxes. The extent of the intermittent regime for the rectangular tubes is in between that of the circular and square tubes. The wavy flow regime is also larger in the round tube. The square and rectangular channels help liquid retention in the corners and along the entire circumference of the tube leading to annular flow, rather than preferentially at the bottom of the tube as would be the case in the wavy flow regime. In the 4 × 6-mm and 6 × 4-mm tubes, the larger aspect ratio results in a slight increase in the size of the intermittent regime at the lower mass fluxes and a small reduction in the size of this regime at the higher mass fluxes. However, these effects are small, and it can be concluded that this transition line is only weakly dependent on the aspect ratio. The smaller aspect ratio also results in a larger annular film flow pattern region, which is to be expected because of the reduced influence of gravity for the tubes with the smaller height. Discrete waves are more prevalent in the round tube compared to the square and rectangular tubes. It was also found that the smaller aspect ratio results in a smaller wavy flow regime but with a larger fraction of discrete waves. At the higher mass fluxes, the effect of the aspect ratio is negligible. The increase in the annular film flow pattern region was more pronounced in the smaller hydraulic diameter tubes

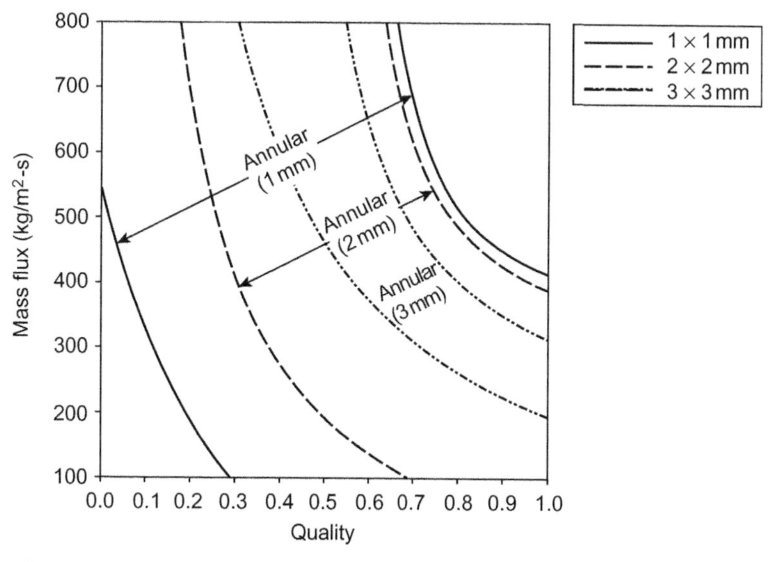

FIGURE 6.11

Effect of D_H on annular flow regime.

Source: *From Garimella (2004).*

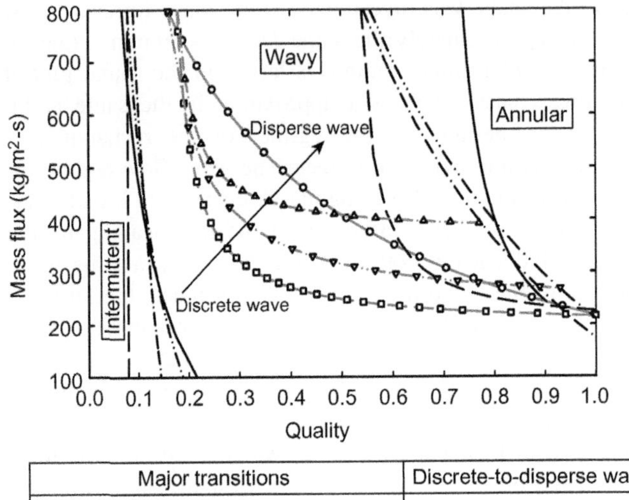

Major transitions	Discrete-to-disperse wave
Round (4.91 mm)	
Square (4 × 4 mm)	
Rectangular 4 (H) × 6 (W) mm	
Rectangular 6 (H) × 4 (W) mm	

FIGURE 6.12

Effect of tube shape on flow regime transitions.

Source: *From Garimella (2004).*

tested, perhaps due to the greater influence of surface tension. Thus, while tube shape had some effect on the transitions between the various flow regimes, the influence of hydraulic diameter was found to be far more significant.

The study of Coleman and Garimella (2000a,b, 2003) provided qualitative insights into flow regimes and patterns. Keinath and Garimella (2010) extended the study using improved video imaging and analysis to record similar flow regimes for the condensation of high–pressure refrigerant R-410A, and also obtained quantitative information for such flows in small-diameter channels. Experiments were conducted with R-404A in round tubes ($0.5 < D < 3$ mm) at high reduced pressure ($0.38 < P_r < 0.77$) and mass fluxes ranging from 200 to 800 kg/m^2 s. They identified the same major flow regimes as those defined by Coleman and Garimella (Figure 6.8), and the trends in the flow regime transitions for R-404A were similar to those observed for R-134a. In addition to qualitatively determining the local flow regimes and patterns for condensing R-404A, Keinath and Garimella (2010) were also able to obtain quantitative parameters, including bubble dimensions and velocity, slug frequency and local temporally and spatially varying void fraction, which is discussed in the following section.

A two-part study on flow regimes and heat transfer during condensing flows in channels with $3.14 < D < 21$ mm was conducted by El Hajal et al. (2003) and

Thome et al. (2003). Although this work is primarily on large channels, the comprehensive nature of the database used for model development warrants discussion here. However, due to the highly coupled nature of the flow regime (El Hajal et al., 2003) and heat transfer parts (Thome et al., 2003), it is presented in a subsequent section on heat transfer.

Kim et al. (2012) conducted experiments on condensing FC-72 in parallel, square microchannels ($D_H = 1$ mm) formed in a copper plate. Data were obtained at a nominal saturation temperature of 60°C and mass fluxes ranging from 68 to 367 kg/m^2 s. Using high-speed video imaging, they were able to capture and classify five distinct flow regimes: (1) annular smooth, (2) annular wavy, (3) transition, (4) slug, and (5) bubbly. Consistent with the observations of Coleman and Garimella, they noted a complete absence of gravity-dominated stratified flow and an increase in the range of qualities over which the slug (intermittent) flow existed as mass flux decreased. They observed a lack of entrained liquid in the annular smooth or annular wavy regime, which they note is in contrast to observations in microchannel boiling experiments. Furthermore, they note that entrained droplets in microchannel boiling flows (Qu and Mudawar, 2003) have a significant influence on pressure drop and heat transfer. Thus, while the working fluid and tube geometry may be the same between condensation and boiling experiments, the boundary condition has an important effect on the flow regime and the associated heat transfer and pressure drop modeling. They noted fair agreement with their data and the transition criteria of Triplett et al. (1999b) and Chung and Kawaji (2004), although exact correspondence between qualitative definitions of flow regimes between different investigators remains elusive. Despite the fair agreement, they highlighted the weakness of using dimensional axis to present flow regime data (e.g., mass flux versus superficial velocity). In part two (Kim and Mudawar, 2012) of the study, they propose a set of transition lines as a function of a modified Weber number and the turbulent–turbulent Martinelli parameter.

Nema et al. (2013) also recognized the shortcomings of dimensional axes for mapping microchannel condensation flow regimes, and developed transition criteria based on dimensional analysis of the condensation data of Coleman and Garimella (2000a,b, 2003). Criteria were developed for transitions from annular or wavy regime to intermittent, discrete-to-disperse wave, intermittent or wavy-to-annular film flow, intermittent to dispersed flow, and transition to mist flow. The use of dimensionless parameters allows for wider application of the criteria while simultaneously accounting for the effects of mass flux, tube size, and fluid properties. As tube size decreases, the most important flow regime transition is from intermittent flow to either wavy or annular flow. As noted by Coleman and Garimella (2003) and Kim et al. (2012), direct transition from annular-to-intermittent flow is observed for the smallest tubes ($D_H = 1.0$ mm), with no wavy flow regime present. As tube size increases, the relative size of the intermittent flow regime decreases. Thus, the transition to intermittent flow can occur from either annular or wavy flow, depending on the relative influence of surface tension and gravity forces. To account for the relative contribution of each force, Nema et al. (2013) defined a critical Bond number.

$$Bo_{crit} = \frac{1}{((\rho_L/\rho_L - \rho_V) - (\pi/4))} \tag{6.19}$$

When the Bond number is less than Bo_{crit}, surface tension forces dominate, and liquid slugs form as soon as the minimum liquid volume fraction to form them is present. Surface tension pulls the condensate film and forms a bridge across the tube. This is similar to the mechanism proposed by Barnea et al. (1983) and Tabatabi and Faghri (2001). For $Bo > Bo_{crit}$, the liquid volume fraction at which slugs form is predicted to increase with increasing tube size, approaching a volume of 50% for sufficiently large tubes. A 50% volume fraction corresponds to a Martinelli parameter of 1.6, equivalent to the transition criteria established by Taitel and Dukler (1976), thus smoothing the transition between microchannel and conventional channel behavior.

6.2.3 Summary observations and recommendations

The above discussion shows that numerous studies have been conducted to understand two-phase flow in small-diameter channels. The preponderance of studies is on adiabatic air–water flows. While these flows have historically provided some understanding of condensation phenomena, many of the discrepancies between the behavior observed in these studies and the actual flow regimes during the condensation process can be attributed to the substantially different properties of the air–water pair compared to refrigerant vapor–liquid pair. It also appears that video recording technology is affording investigators the ability to capture every nuance of two-phase flow in great detail. High-speed video presents the investigator with a plethora of images separated in time by very small increments. While this is certainly advantageous in understanding the intricacies of two-phase flow, it has led to the further proliferation of creative, confusing, and vague definitions of flow patterns. The number of flow patterns cited in this chapter alone would approach about 50! There are many *definitions* with relatively less *distinction*. Flow patterns from different investigators can be as easily thought to agree as not, depending on the interpretation of these terms. As one who has himself pored over numerous frames of high-speed video, it is easy to understand how each researcher attempts to capture every nuance of a dynamic bubble or a wave in a description of the pattern. It is not clear whether such detailed descriptions have led to a significant advance in the quantification of the relevant physical phenomena that are responsible for these mechanisms in the first place. These minute distinctions between flow patterns would only be useful if they lead to quantitative models of essential features such as interfacial area, shear, slip, void fraction, momentum coupling, and, ultimately, heat transfer and pressure drop.

The summary evaluation of the above-mentioned studies is as follows:

1. Adiabatic flow patterns are now available for a wide range of microchannels, mostly using air or nitrogen–water mixtures, all the way down to 25 μm. Extrapolation of air–water flow patterns to condensing refrigerant applications must be done with caution, however.
2. Many of the studies on small circular and narrow rectangular channels were conducted on co- or countercurrent vertical upflow, which is an orientation seldom used in condensation.
3. It has now been very well documented that gravitational forces have little significance in the small channels ($D > \sim 1$ mm) of interest in this book. This has been validated through observations of symmetric films, bubbles and slugs, and also flow patterns independent of channel orientation.
4. In the channels with $1 > D > 10$ mm, the flow patterns can broadly be classified as disperse, annular, wavy, and intermittent. Most of the phase-change studies in this size range have focused on channels with $D > 3$ mm, and the primary focus of most of these investigations has been to understand gravity-dominated versus shear-controlled condensation.
5. In channels with $D > 1$ mm, it appears that the primary patterns are some kind of slug/plug flow, a transition region designated as liquid ring by some investigators, and annular flow (with some mist flow at high qualities and mass fluxes).
6. Maps and transition criteria such as those by Taitel–Dukler (1976) are tempting to use because of their analytical basis; however, rarely do they predict condensation phenomena in microchannels—the original investigators perhaps never intended for them to be applied in this manner.
7. For the ~5-mm channels, simple criteria such as those by Breber et al. (1980), Tandon et al. (1982), and Soliman (1982, 1986) may suffice, with minor modifications to match the application under consideration to distinguish between gravity and shear-controlled condensation. The more recent descriptive reporting of flow patterns by Dobson and Chato (1998) may also be used to obtain qualitative insights.
8. In addition, transition criteria that are applicable over a wide range of fluid properties, operating conditions, and channel sizes must be further developed based on the underlying forces and validated by the detailed experimentally obtained phase-change maps such as those of Coleman and Garimella (2000a,b, 2003), Keinath and Garimella (2010), and Kim et al. (2012).
9. Similarly, investigators of microchannels with $D << 1$ mm should also test condensation processes to supplement the information gathered from adiabatic flow visualization studies.

The assignment of flow regimes for nine representative cases (combinations of mass fluxes and qualities) for condensation of refrigerant R-134a in a 1-mm diameter channel using several of the prominent transition criteria is illustrated in Example 6.1.

6.3 Void fraction

An important parameter for the calculation of heat transfer coefficients, and especially pressure drops, is the void fraction in two-phase flow. For an annular flow situation, knowledge of the void fraction allows an estimate of the liquid condensate film thickness, through which the latent heat of condensation must be transported. El Hajal et al. (2003) showed that condensation heat transfer is highly sensitive to void fraction, particularly at high quality, when liquid inventory is small. The void fraction also determines the amount of fluid to be charged into a closed loop system such as a refrigeration cycle. Void fraction is simply the ratio of the cross-sectional area A_g occupied by the gas to the total cross-sectional area of the channel (Table 6.4):

$$\alpha = \frac{A_g}{A} \tag{6.20}$$

The calculation of this parameter is an essential step and provides closure to the set of equations that must be solved to estimate the pressure drop and heat transfer in two-phase flow. While the vapor quality x is a thermodynamic quantity determined by the state of the two-phase mixture, the void fraction depends on the characteristics of the flow, and has been the subject of several investigations. Like the flow regime studies discussed above, most existing void fraction investigations have focused on larger tubes with adiabatic air–water flows. There are relatively fewer studies conducted in microchannels, and there is little information on void fraction during condensation. Winkler et al. (2012b) presented a comprehensive review of experimental and modeling techniques to obtain void fractions, with a focus on condensation in small channels. Winkler et al. show that void fraction has been measured with intrusive and nonintrusive means, including the use of shut-off valves, where the relative volume of liquid and vapor are manually measured, electrical conductance or capacitance methods (Elkow and Rezkallah, 1996, 1997a,b), neutron radiography (Hibiki and Mishima, 1996; Hibiki et al., 1997; Mishima and Hibiki, 1998; Mishima et al., 1997) or direct visualization through high-speed recording. Direct visualization is most commonly used for evaluating void fraction in mini- and microchannels. Furthermore, they noted that many void fraction models developed for larger channels (Armand, 1946) are used by researchers in microchannel applications, irrespective of the conditions for which the model was originally validated. Again, it is also common to apply models developed with air–water or steam–water data to fluids with vastly different properties. The validity of extrapolating these findings to two-phase refrigerant flows has not been established, and has been questioned by researchers including Yashar et al. (2001) and Winkler et al. (2012b).

The slip, that is, the ratio of the velocities of the two phases, is required to enable computation of the void fraction. The simplest assumption that can be made is of homogeneous flow, where it is assumed that the vapor- and liquid-phase velocities are equal, and that the two phases behave like a single, uniform

Table 6.4 Summary of Void Fraction Studies

Investigator	Hydraulic Diameter	Fluids	Orientation	Range/Applicability	Techniques, Basis, Observations
Conventional tubes					
Zivi (1964)	Several	Steam–water	Horizontal, vertical		• Analytical void fraction model (assuming annular flow), minimizes rate of energy dissipation • Three models: (I) without liquid entrainment and wall friction, (II) with wall friction, and (III) with liquid entrainment • Data show wall friction effect ≪ effect of liquid entrainment • As p increases, model (I) approaches homogeneous model • Simplest model (I) widely used, does not rely on data
Yashar et al. (2001)	• Microfin tubes: 0.2-mm fin height • 7.3, 8.9 mm (evaporation) • 8.9 mm (condensation)	R-134a, R-410A	Horizontal	$75 < G < 700$ kg/m^2 s $0.05 < x < 0.8$	• Measurements in condensing microfin tubes (uncertainty 10%) • Similar trends for microfin and smooth tubes • Lower void fractions compared to evaporation, for same G, x • No effect of fin arrangement (helix angle)
El Hajal et al. (2003)	8 mm	R-22, R-134a, R-236ea, R-125, R-32, R-410A	Horizontal condensing	$65 < G < 750$ kg/m^2 s $0.22 < ps < 3.15$ MPa $0.15 < x < 0.88$	• Used earlier boiling maps (Kattan et al., 1998a,b,c; Zurcher et al., 1999) as basis • Deduced void fractions for annular flow from heat transfer database (Cavallini et al., 2000, 2002a) and film thickness considerations • Logarithmic mean of homogeneous and Steiner (1993) models

(Continued)

Table 6.4 (Continued)

Investigator	Hydraulic Diameter	Fluids	Orientation	Range/Applicability	Techniques, Basis, Observations
Koyama et al. (2004)	7.52 mm (smooth) 8.86 mm (microfin)	R-134a	Horizontal adiabatic	$0.01 < x < 0.96$ $p = 0.8$, 1.2 MPa $G = 125$, 250 kg/m^2 s (smooth) $G = 90$, 180 kg/m^2 s (microfin)	• Void fraction increases as pressure decreases (smooth and microfin) • Smooth-tube data agree with Smith's (1969) and Baroczy's (1965) models • Microfin tube data lower than smooth tube values • Results different from Yashar et al. (2001), who reported similar values for both tube types • Yashar et al. (2001) correlation overpredicts microfin tube data • α in microfin tubes increases with increasing G, at low p, x • Model for stratified annular and annular flow; reasonable agreement with data, especially at high x
Mini- and microchannel adiabatic void fraction					
Armand (1946)		Air–water	Horizontal	$\beta < 0.9$ atmospheric pressure	• Empirical correlation
Kariyasaki et al. (1991)	1, 2.4, 4.9 mm	Air–water	Horizontal	$0.1 < j_G < 25$ m/s $0.03 < j_L < 2$ m/s	• Numerous curve-fits to match data without physical considerations
Mishima and Hibiki (1996b)	1–4 mm	Air–water	Vertical	$0.0896 < j_G < 79.3$ m/s $0.0116 < j_L < 1.67$ m/s	• Image processing from neutron radiography for nonintrusively measuring void fraction • Correlated void fraction using drift-flux model with empirical parameters for vertical bubbly and slug flow • Distribution parameter function of D, larger values for small channels than reported by Ishii (1977) for larger tubes • Agreed with Kariyasaki et al. (1992) correlation

Hibiki and Mishima (1996)	1–4 mm	Air–water	Vertical	$j < 19$ m/s	• Best agreement with 1-mm data • Technique also works for annular flow at high α in larger tubes
Hibiki et al. (1997)	3.9 mm	Air–water	Vertical upward	$j_G = 0.131$ m/s $j_L = 0.0707$ m/s	• Correlation for radial void fraction profile for vertical slug flow
Mishima et al. (1997)	2.4 mm	Air–water	Vertical upward	$j < 8$ m/s	• Neutron radiography image processing for variation of void fraction along channel length for slug flow • Rise of bubble tracked using high void fraction region • Merging of void fraction peaks indicated bubble coalescence
Kureta et al. (2001; 2003)	3, 5 mm	Steam–water	Vertical upward	$240 < G < 2000$ kg/m^2 s	• Instantaneous and time-averaged 2-D α for subcooled boiling (at low α)
Triplett et al. (1999a)	1.1, 1.45 mm	Air–water	Horizontal adiabatic	$0.02 < j_G < 80$ m/s $0.02 < j_L < 8$ m/s	• Volumetric void fraction from still photographs (rotational symmetry assumed, uncertainty ~15%) • Void fraction increases with increasing gas velocity and decreases with increasing superficial liquid velocity • Homogeneous model best for slug and bubble flow, Premoli et al. (1971), Chexal and co-workers (Chexal and Lellouche, 1986; Chexal et al., 1992, 1997), Lockhart–Martinelli (1949), and Baroczy (1965) overpredict for annular and churn flow
Serizawa et al. (2002)	20, 25, 100, 50 μm	Air–water (steam–water)	Horizontal	$0.0012 < j_G < 295.3$ m/s $0.003 < j_L < 17.52$ m/s	• α in bubbly and slug flow from video analysis • Follows Armand (1946) and Kariyasaki et al. (1991) correlations

(Continued)

Table 6.4 (Continued)

Investigator	Hydraulic Diameter	Fluids	Orientation	Range/Applicability	Techniques, Basis, Observations
Kawahara et al. (2002)	100 μm	Nitrogen–water	Horizontal	$0.1 < j_G < 60$ m/s $0.02 < j_L < 4$ m/s	• Time-averaged void fraction and curve-fit from video analysis • α increases with increasing homogenous void fraction. Very low α for $\beta < 0.8$, rapid increase for $\beta < 0.8$ • Different from Serizawa et al. (2002), who found linear correlation to homogeneous α for similar diameters • Low α at high gas flow rates indicates large slip ratios and weak momentum coupling between phases
Chung and Kawaji (2004)	50, 100, 250, 530 μm	Nitrogen–water	Horizontal	$T = 22.9–29.9°C$ $0.02 < j_G < 72.98$ m/s $0.01 < j_L < 5.77$ m/s	• 530 μm: homogeneous model; 250 μm: Armand (1946) correlation; 50, 100 μm: Kawahara et al. (2002) correlation
Chung et al. (2004)	96-μm square	Nitrogen–water	Horizontal adiabatic	$0.000310 < Bo <$ 0.000313 $0.00010 < We_{ls} < 25$ $0.000019 < We_{GS} < 9$ $1 < Re_{LS} < 438$ $1 < Re_{GS} < 612$	• Slightly different constants for 50-μm channel • Kawahara et al. (2002) correlation holds for 96-μm square channel
Kawahara et al. (2005)	50, 75, 100, 251 μm	Water–nitrogen, ethanol–water/nitrogen	Horizontal	$0.08 < j_G < 70$ m/s $0.02 < j_L < 4.4$ m/s	• No effects of fluid properties • 50-, 75-, 100 μm data follow Kawahara et al. (2002) correlation, 251 μm data follow linear Armand (1946) correlation • Used to demarcate micro- and minichannels (100–251 μm)

Mini- and microchannel condensation void fraction					
Winkler et al. (2012a)	Square, rectangular, circular: $2 < D_H < 4.91$ mm	R-134a	Horizontal condensing	$150 < G < 750$ kg/m^2 s $T_{sat} \approx 52°C$	• Determined void fraction in intermittent and wavy regimes of Coleman and Garimella (2000a,b, 2003) data • Minimal influence of D_H or G on void fraction in wavy flow regime
Keinath and Garimella (2011)	Circular: $0.5 < D < 3.0$ mm	R-404A	Horizontal condensing	$200 < G < 800$ kg/m^2 s $30 < T_{sat} < 60°C$	• Improved optics allowed void fraction measurement in annular flow. • Negligible influence of G or D_H on *bulk average* void fraction • Significant effect of D_H on distribution of liquid/vapor

mixture that has representative average properties, which yields the following expression:

$$\alpha = \frac{x\rho_v}{(1-x)/\rho_l + x/\rho_v} \tag{6.21}$$

While the homogeneous model is in fact quite arbitrary, and serves as a starting point for the more representative models of void fraction, it has gained some acceptance for the modeling of microchannels. Other void fraction models for conventional channels are readily available in textbooks such as those by Carey (2008), Collier and Thome (1994), and Hewitt et al. (1994). While the details of the flow patterns must be taken into account to obtain models based on the physical phenomena, several of the more widely used correlations often assume annular flow and relate the void fraction to a parameter that resembles the Martinelli parameter X, often X_{tt} (which assumes turbulent flow in both phases). Incorporating the appropriate expressions for the pressure drops in the two phases, these expressions can be represented by the following generic expression (Butterworth, 1975):

$$\alpha = \left[1 + B_B\left(\frac{1-x}{x}\right)^{n1}\left(\frac{\rho_v}{\rho_l}\right)^{n2}\left(\frac{\mu_l}{\mu_v}\right)^{n3}\right]^{-1} \tag{6.22}$$

A convenient tabulation of the leading coefficient B_B and the exponents $n_1 \ldots n_3$ for this model appears in Carey (2008) for the Zivi (1964) entropy minimization model, the Wallis (1969) separated cylinder model, the Lockhart–Martinelli (1949) model, the Thom (1964) correlation, and the Baroczy (1965) correlation.

6.3.1 Void fraction in adiabatic flow through mini- and microchannels

For the small-diameter range, Kariyasaki et al. (1991) measured void fraction in 1-, 2.4-, and 4.9-mm tubes and proposed several empirical curve-fits to the measured void fractions in small regions as a function of the homogeneous volume fraction:

$$\begin{aligned}
&\beta < \beta_A: && \alpha = \beta\\
&\beta_B < \beta < 0.6 \text{ (intermittent flow)}: && \alpha = 0.833\beta\\
&\beta_B < \beta,\ 0.6 < \beta < 0.95 \text{ (intermittent flow)}: && \alpha = 0.69\beta + 0.0858\\
&\beta_B < \beta,\ 0.95 < \beta \text{ (annular and intermittent flow)}: && \alpha = 0.83\log(1-\beta) + 0.633
\end{aligned} \tag{6.23}$$

where β_A and β_B are also empirically derived transition points. While the above correlation matched their data, it appears to have been derived simply based on statistical curve-fits to follow the data closely, and the trends exhibited show numerous unrealistic discontinuities, changes in slope and inflections for which no physical rationale is provided. The first intermittent flow correlation above is the so-called Armand (1946) correlation.

Mishima and Hibiki (1996b) used neutron radiography to conduct flow visualization studies on air–water flows in 1–4-mm-diameter tubes. The void fraction was correlated well by the drift-flux model with a new equation for the distribution parameter as a function of inner diameter. The drift-flux model was used to correlate the void fraction as follows:

$$u_G = j_G/\alpha = C_o j + V_{GJ} \tag{6.24}$$

where u_G is the gas velocity, j is the mixture volumetric flux, $j = j_G + j_L$, C_o, is the distribution parameter, and V_{Gj} is the drift velocity. For each regime (bubbly, slug, annular, and churn), the required distribution parameter was taken from Ishii (1977). However, in accordance with the findings of several other researchers (Gibson, 1913; Kariyasaki et al., 1992; Tung and Parlange, 1976; Zukoski, 1966), the drift velocity in slug and bubbly flows was assumed to be zero, and the distribution parameter was fit to the following equation based on the data:

$$C_o = 1.2 + 0.510_e^{-0.691D} \tag{6.25}$$

where the tube diameter D is in millimeters. These findings were in good agreement with the results of Kariyasaki et al. (1992), who modeled their data in terms of the volumetric void fraction $\beta = j_G/(j_G + j_L)$. The distribution parameters resulting from the data and the above equation for the small-diameter channels are larger than those reported by Ishii (1977) for larger channels. Mishima et al. (1993) also found this increase in C_o for narrow rectangular channels, and attributed this trend to the centralized void profile and the laminarization of flow in these smaller and narrower channels. Hibiki and Mishima (1996) developed an approximate method for the processing of images obtained from neutron radiography, which was then used for nonintrusively measuring void fractions in air–water flows in tubes with $1 < D < 4$ mm. The agreement between the data and the predictions of the radiography technique was best at 1 mm, but the technique was deemed to be acceptable for annular flows with large void fractions in the larger tubes also. Hibiki et al. (1997) then used the neutron radiography method to measure radial void fraction distributions in vertical upward air–water flow in a 3.9-mm channel. An assumed power law or saddle-shaped void fraction profile was subsequently verified using measured integrated variations in the void fraction profile along the channel axis. A triple-concentric-tube apparatus with some channels filled with water or gas in different combinations provided baseline attenuations and ensured calibration of the experimental results. The method was demonstrated by developing the following equation for void fraction in slug flow:

$$\alpha_r = 0.8372 + \left[1 - \left(\frac{r}{r_w}\right)^{7.316}\right] \tag{6.26}$$

The variation in void fraction along channel length in slug flow was determined by Mishima et al. (1997), who used the images shown in the previous section on flow patterns (Figure 6.1) to compute the void profile across the channel at intervals of 6.45 mm (Figure 6.13). This figure shows that the high void fraction region moves upward with the rise of the slug bubble. Also, the merging of two peaks into one peak is illustrated, corresponding to the coalescence of two bubbles. Unesaki et al. (1998) later conducted Monte Carlo simulations for the triple-concentric-tube apparatus described above to further validate the technique.

This group (Kureta et al., 2001, 2003) also measured instantaneous and time-averaged void fractions for subcooled boiling of water (at low void fractions) in narrow rectangular channels (3- and 5-mm gap) using neutron radiography in the mass flux range $240 < G < 2000$ kg/m^2 s. Although their primary objective was to determine the point of net vapor generation and its effect on critical heat flux, their technique and results demonstrate the feasibility of nonintrusively obtaining temporally (0.89 ms) and spatially (2-D) resolved void fractions in two-phase flows in minichannels. A representative distribution of instantaneous void fractions obtained by them is shown in Figure 6.14. Such temporal resolution is particularly helpful for flows more prevalent in minichannels (such as slug flows) with large temporal variations in void fraction. Figure 6.15 shows void fraction profiles computed by averaging instantaneous values over 2 s (2250 images).

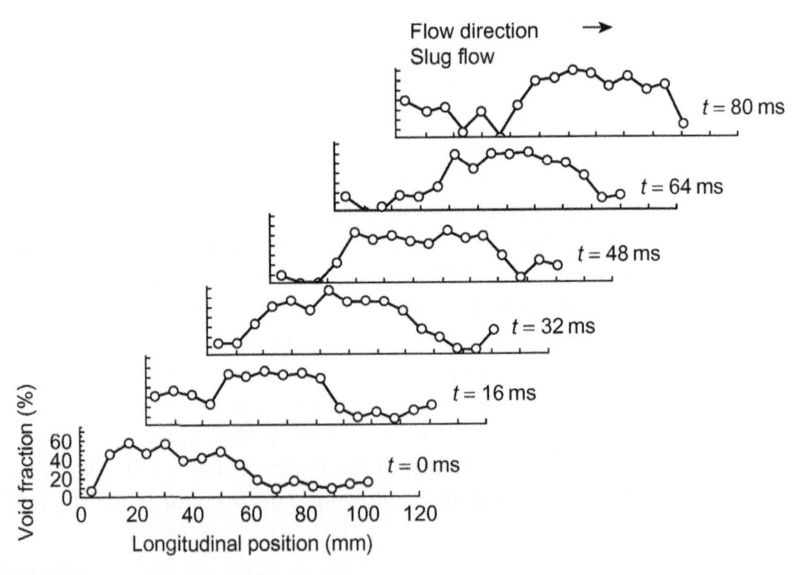

FIGURE 6.13

Progression of void fraction profiles with time.

Source: *Reprinted from Mishima et al. (1997) with permission from Elsevier.*

Such techniques could also be applied to gain detailed insights into condensing flows.

Triplett et al. (1999a,b) used still photographs of flow patterns in 1.1- and 1.45-mm horizontal circular channels (although, as noted in the previous section,

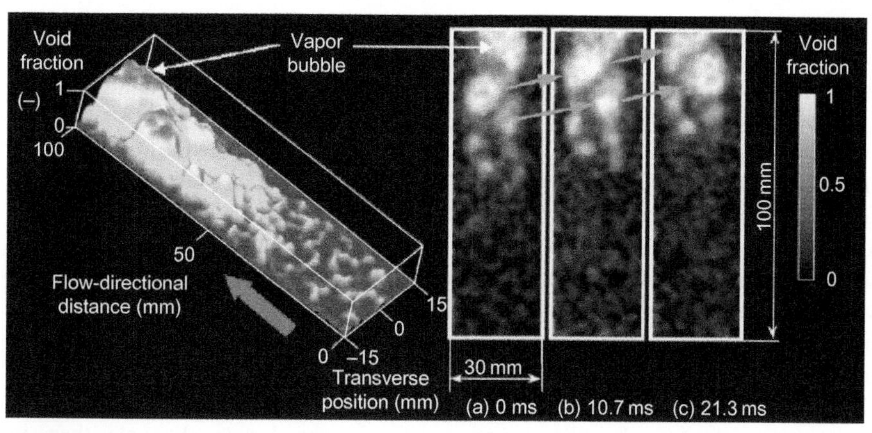

FIGURE 6.14

Instantaneous 2-D void fraction profiles in narrow rectangular channels.

Source: *Reprinted from Kureta et al., 2003 with permission from Elsevier.*

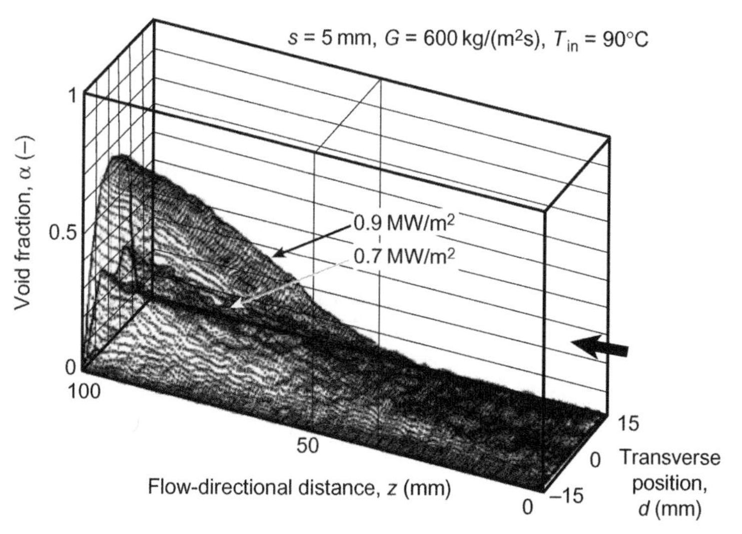

FIGURE 6.15

Time-averaged 2-D void fraction in narrow rectangular channels.

Source: *Reprinted from Kureta et al., 2003 with permission from Elsevier.*

flow patterns were also recorded for semi-triangular channels) to obtain the volumetric void fraction. Bubbles in the bubbly flow pattern were assumed to be spheres or ellipsoids, while the bubbles in slug flow were assumed to consist of cylinders and spherical segments. Annular flow void fractions were estimated by averaging cylindrical segments of the vapor core. The estimated uncertainties in these void fractions were relatively high, 15%. Also, they were unable to reliably measure void fractions in slug-annular and churn flow, and assigned a void fraction of 0.5 to the portion of churn flow in which the gas phase was dispersed. In general, they found that the void fraction increases with increasing gas superficial velocity, U_{GS} at a constant liquid superficial velocity, U_{LS}, and decreases with increasing U_{LS} for constant U_{GS}. They compared their results with the void fraction correlations of Butterworth (1975), Premoli et al. (1971), and the correlation of Chexal and co-workers (Chexal and Lellouche, 1986; Chexal et al., 1992, 1996), which is based on the drift-flux model of Zuber and Findley (1965). They found that the homogeneous flow model was the best predictor of the measured void fractions in bubbly and slug flow, that is, at low U_{GS} values. For annular and churn flow, however, the homogeneous model and other empirical models significantly overpredicted the experimental values. They believe that this overprediction by the empirical correlations, typically based on annular flow in large channels, is due to the slip being greater in the larger channels than in microchannels.

Serizawa et al. (2002) studied air–water flow in microchannels (20, 25, and 100 μm) and steam–water flow in a 50-μm tube. Although a variety of flow regimes (discussed in the previous section) were identified by them, cross-sectional-averaged void fractions were calculated from high-speed video pictures only for the bubbly and slug flow regimes (liquid ring flow was not included for void fraction analysis). The Armand (1946) correlation $\alpha = 0.833\beta$, discussed above in connection with the work of Kariyasaki et al. (1991), was recommended for these regimes.

Kawahara et al. (2002) measured void fractions for nitrogen–water flow in a 100-μm tube by analyzing video frames and assigning a void fraction of 0 to the liquid slug, and a void fraction of 1 to a frame with a gas core surrounded by a smooth-thin liquid film or ring-shaped liquid film. Time-averaged void fractions were then calculated over several frames. For high liquid flow rates, gas-core flow with a thick liquid film was assigned a void fraction ($0 < \alpha < 1$), thus accounting for a finite film thickness surrounding the core in such flows. They found that the void fraction is not strongly dependent on j_L, and developed the following empirical fit for the void fraction in terms of the homogeneous void fraction:

$$\alpha = \frac{0.03\beta^{0.5}}{1 - 0.97\beta^{0.5}} \tag{6.27}$$

This variation of void fraction (Figure 6.16) is significantly different from the linear relationship with the homogeneous void fraction described above.

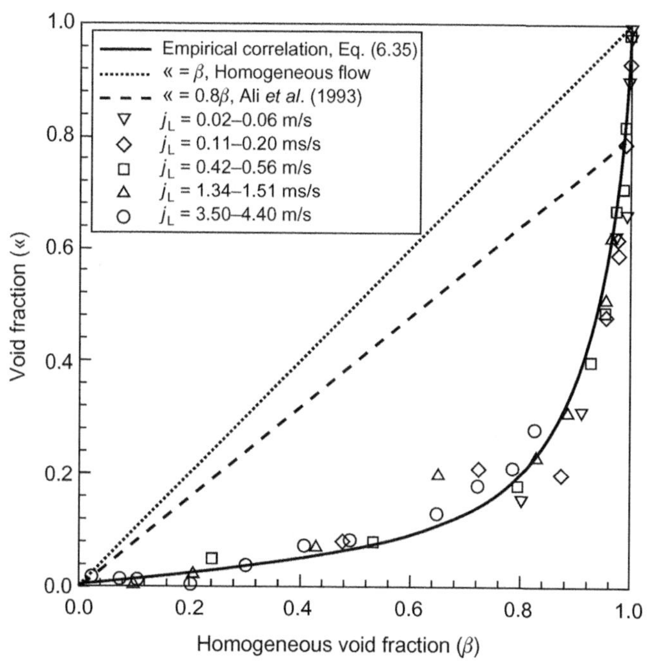

FIGURE 6.16

Void fractions for air–water flow in 100-μm channels.

Source: *Reprinted from Kawahara et al. (2002) with permission from Elsevier.*

Thus, void fractions remain low even at high values β ($\beta < 0.8$), and increase steeply for $0.8 < \beta < 1$. They attributed this phenomenon to the observation that single-phase liquid flow occurred most frequently at high liquid flow rates for all gas flow rates, and also at low liquid and gas flow rates. Wall shear and surface tension resulted in channel bridging under such conditions, and the associated low average void fractions. At low liquid and high gas flow rates, gas-core flow surrounded by a liquid film with weak momentum coupling was seen. While the liquid flow was governed by wall shear and surface tension, the gas phase flowed without significant resistance due to the undeformed shape characteristic of the flow in such small channels observed by them. The nonlinear dependence of the area-based void fraction on the volumetric void fraction implies that slip ratios are much higher (as high as 16) in these channels compared to the much lower values seen in channels with $D_H > 1$ mm. It should be noted that these results contradict the results of Serizawa et al. (2002), who reported the adequacy of the Armand correlation for a similar range of D_H.

Chung and Kawaji (2004) investigated the effect of channel diameter on void fraction for the flow of nitrogen–water mixtures in circular 530-, 250-, 100-, and 50-μm channels. For the larger channels (530 and 250 μm), multiple images at each data point were analyzed by estimating the gas fraction volume as a combination of symmetrical shapes. For the smaller channels, the void fraction was estimated as described above in Kawahara et al. (2002). They found that the void fraction for the 530-μm channel was predicted well by the homogeneous flow model, and the void fraction for the 250-μm channel agreed well with the Armand-type correlation. However, the 100- and 50-μm channel void fractions followed the slow linear increase up to $\beta = 0.8$, followed by a steep exponential increase captured by Eq. (6.27) above that was developed by them earlier (Kawahara et al., 2002). (Based on this latter work, however, they modified the leading constants for the 50-μm channel to 0.02 and 0.98 in the numerator and denominator, respectively.) They attributed the departure from homogeneous flow at high volumetric void fractions to a larger slip between the phases. This modified equation indicates a further decrease from the homogeneous model, even compared to the values for the 100-μm channel. They argued that in the larger (mini) channels, the wavy and deformed gas–liquid interface leads to more momentum exchange between the gas and liquid phases, which slows the flow of the gas bubbles, keeping the void fraction high. On the other hand, according to them, the lack of this momentum exchange in the microchannels leads to a lower void fraction. Chung et al. (2004) later confirmed that the coefficients in the void fraction correlation for circular 100-μm channels ($C_1 = 0.03$ and $C_2 = 0.97$) would apply to square channels with $D_H = 96$ μm.

Kawahara et al. (2005) further investigated the effect of fluid properties and diameter on void fraction. They tested the two-phase flow of water–nitrogen and ethanol–water/nitrogen gas mixtures (of varying ethanol concentrations) in 50-, 75-, 100-, and 251-μm circular channels. Their results showed that the void fraction was not sensitive to fluid properties over the range of diameters studied. The void fraction data for the 50-, 75-, and 100-μm channels agreed with the correlation of Kawahara et al. (2002), but the data for the 251-μm diameter channel followed the homogeneous flow type model of Armand (1946). They interpreted these results to imply that the boundary between "microchannels" and "minichannels" would lie between 100 and 251 μm.

6.3.2 Void fraction in condensing flow through mini- and microchannels

Few experiments have been conducted on void fraction during condensation in microchannels. For optical visualization, as in the flow regime studies discussed above, the flow must be viewed through a transparent test section that can withstand the high pressures characteristic of condensing flows at most common operating conditions of interest. Using the flow visualization data of Coleman and Garimella (2003), Winkler et al. (2012a) extracted quantitative void fraction information. To obtain such information from video frames, it is necessary to (1) identify the liquid–vapor interface, and (2) infer 3-D structure from 2-D images through simplifying assumptions. Due to limitations on resolution of the data, particularly at the small tube diameters, the study focused on determining void fraction in the intermittent and wavy flow regimes in the data of Coleman and Garimella. These regimes corresponded to the entire quality range for $G = 150\ \mathrm{kg\ m^{-2}\ s^{-1}}$ data, and quality up to 0.5 ($300\ \mathrm{kg\ m^{-2}\ s^{-1}}$), 0.3 ($450\ \mathrm{kg\ m^{-2}\ s^{-1}}$), 0.15 ($600\ \mathrm{kg\ m^{-2}\ s^{-1}}$), and 0.10 ($750\ \mathrm{kg\ m^{-2}\ s^{-1}}$) for the other mass fluxes. The video recordings of Coleman and Garimella provided a 2-D view of a longitudinal section of the flow field. This view was extrapolated to a cross-sectional view of intermittent flow (Figure 6.17) or wavy flow (Figure 6.18), depending on how the flow was originally classified by Coleman and Garimella (2000a,b, 2003).To analyze the results for data in transitional flow, Winkler et al. (2012a) developed a statistical approach which considered the standard deviation of the liquid height and the number of mean crossings in each analyzed frame. It was assumed that a video frame corresponding to the intermittent

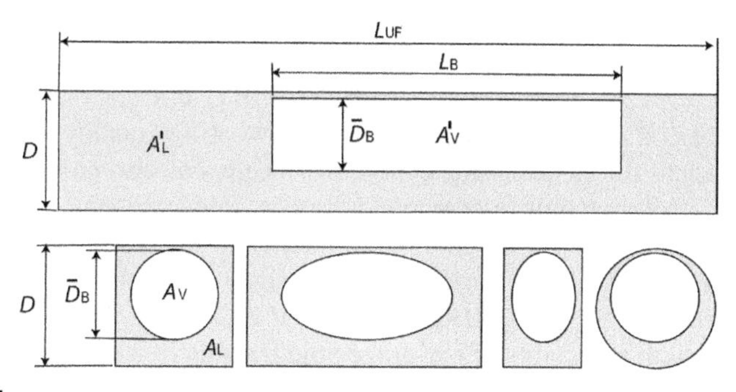

FIGURE 6.17

Schematic of the longitudinal section (top) and the cross-section (bottom) for intermittent flow in differently shaped tubes and channels.

Source: *Winkler et al. (2012a)*

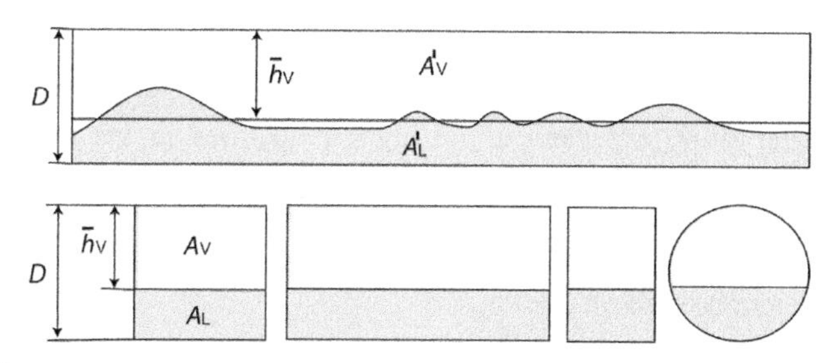

FIGURE 6.18

Schematic of the longitudinal section (top) and the cross-section (bottom) for wavy flow in differently shaped tubes and channels.

Source: *Winkler et al. (2012a)*

regime would have a small standard deviation and many mean crossings (representative of a smooth bubble–liquid film interface), while wavy flow would exhibit higher standard deviations characteristic of interfacial waves. For the wavy regime, no clear influence of hydraulic diameter or mass flux on the calculated void fraction was observed. Trends in the intermittent and intermittent/wavy overlap flow regimes were less clear due to limited data and difficulty in distinguishing between flow regimes. However, despite some of these limitations, this study was one of the first to directly measure void fraction for condensing refrigerants in channels with hydraulic diameter as small as $D_H = 2$ mm.

Keinath and Garimella (2010, 2011) improved on the Winkler et al. study with better optical resolution and image processing, analysis of tubes with hydraulic diameter as small as $D_H = 0.5$ mm, and for a wider range of reduced pressures ($0.38 < P_r < 0.77$) of condensing R-404A. By spanning such a wide range of reduced pressures, they were able to assess the influence of fluid properties on void fraction. Experiments were conducted at mass fluxes ranging from 200 to 800 kg/m^2 s in round tubes with $D = 0.5$, 1.0, and 3.0 mm. The improved resolution allowed direct measurement of void fraction in the annular flow regime. Like Winkler et al. (2012a), Keinath and Garimella (2011) did not find a significant influence of mass flux on void fraction, particularly at low quality. This observation was consistent for all saturation conditions. There was also no observed effect of hydraulic diameter on bulk average void fraction; however, they noted a significant effect of diameter on the distribution of liquid and vapor within the flow. This can be observed in Figure 6.19, which shows images and a plot of axial and bulk average void fraction for $G = 200$ kg/m^2 s flow at $T_{sat} = 30$°C in tubes with $D = 1$ and 3 mm. For both tubes, the bulk average void fraction is the same, even though one image clearly shows intermittent flow and the other wavy flow. Thus, basing pressure drop and heat transfer models only on calculated bulk average void fraction is inadequate; different interfacial areas, even at the same void fraction, would lead to different pressure drops and heat and mass transfer coefficients.

Keinath and Garimella (2011) compared the calculated void fractions for 142 data points to the homogenous model and the correlations of Armand (1946), Baroczy (1965), and El Hajal et al. (2003). There was reasonable agreement between existing models and the measured void fractions, with agreement best at high qualities. This is not surprising, as most void fraction correlations are developed for annular flow, expected at the highest qualities in the small channels under investigation. However, none of the models performed particularly well in the intermittent or intermittent/wavy flow regime. This further highlights the need for models developed specifically for microchannels, where intermittent flow is expected to be increasingly important.

6.3.3 Summary observations and recommendations

The above discussion has shown that void fraction models for conventional tubes have been developed on the basis of an assumption of homogeneous flow, or as a

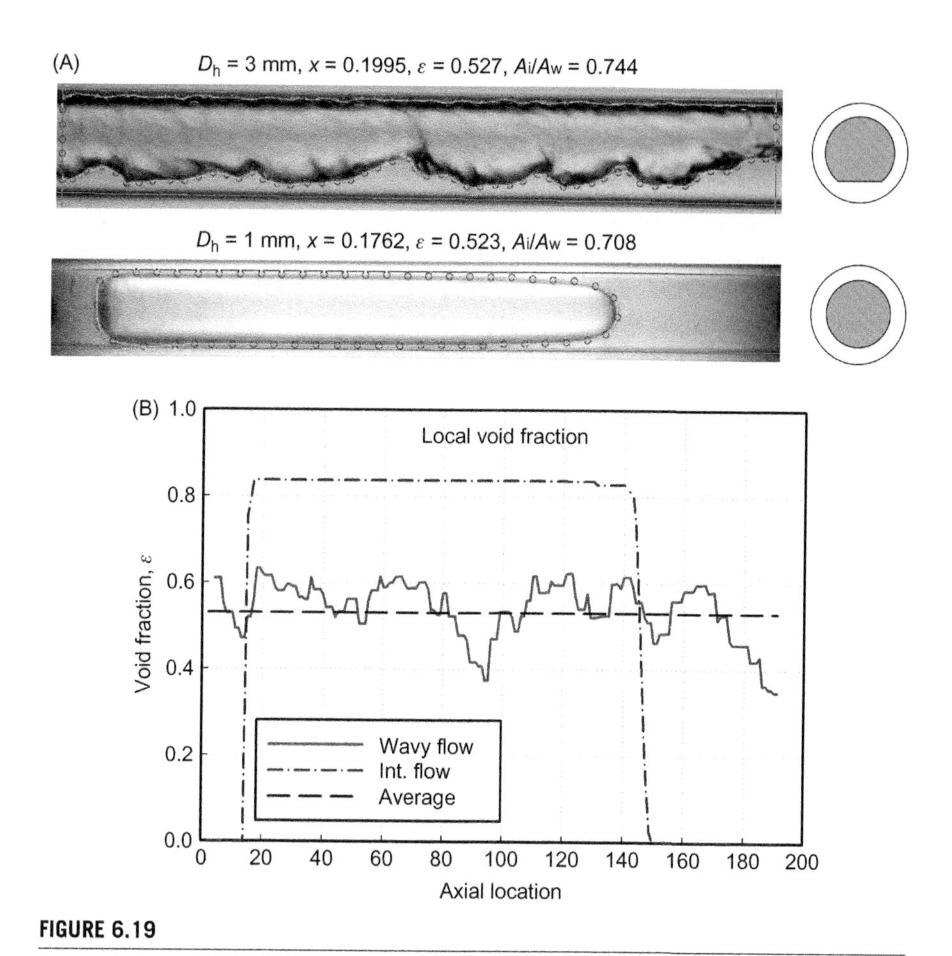

FIGURE 6.19

Comparison of vapor–liquid distribution on (A) frame and (B) local basis for $T_{sat} = 30°C$, $G = 200$ kg/m^2 s.

function of slip ratio, the Lockhart–Martinelli parameter, and those that incorporate some dependence on the mass flux. Many of these models assume annular flow regardless of the flow conditions. Also, several of these models were developed using data from air–water adiabatic flow, or evaporating flows, although in practice they are routinely used for condensation also. Yashar et al. (2001) acknowledged the higher vapor phase densities in condensing flows compared to air–water or evaporating flows, and pointed out that this leads to lower vapor velocities, more stratified flows, and a dependence of the void fraction on mass flux. This group (Yasher et al., 1998) also used a Froude rate parameter and developed a simple model for the changing void fraction from stratified to annular flows by accounting for the ratio of the vapor kinetic energy to the gravitational drag, combined with the Lockhart–Martinelli parameter, which accounts for the ratio of the viscous drag to the vapor kinetic energy. Linked shutoff valves were used by this group as well as by Koyama et al. (2004) to measure the vapor trapped in the test section, which yields the volumetric void fraction, but not the cross-sectional void fraction.

Mishima, Hibiki, and co-workers (Hibiki and Mishima, 1996; Hibiki et al., 1997; Kureta et al., 2001, 2003; Mishima and Hibiki, 1996a,b; Mishima et al., 1993, 1997) have used neutron radiography to measure the distribution of void fractions in 2-D in small channels, as well as the temporal evolution of the void fraction, for example, as the vapor and liquid phases in slug flow pass through the region of interest. Triplett et al. (1999a,b) have used image analysis to measure void fractions in these channels, while Rezkallah and co-workers (Clarke and Rezkallah, 2001; Elkow and Rezkallah, 1997a,b) have focused on nonintrusive measurement of void fraction using capacitance sensors in microgravity environments. Their work has also identified the use of time traces and PDF of the void fraction as signatures for the different kinds of flow regimes and transitions between them. Kawaji, Kawahara, and co-workers (Chung and Kawaji, 2004; Chung et al., 2004; Kawahara et al., 2002, 2005) have used image analysis to obtain void fraction models for the <1-mm range. Their work shows that the void fraction for channels of ~ 1-mm diameter follows the homogeneous flow model, while smaller channels ($D < 250\ \mu m$) show much lower void fractions at volumetric void fraction $\beta < 0.8$, beyond which the void fraction sharply rises to the homogeneous value. Serizawa et al. (2002), on the other hand, found a variant of the homogeneous flow model to apply to smaller geometries.

As was the case for the identification of flow regimes and transition criteria, much of the work on void fractions has been conducted using adiabatic air–water flow. There are very few studies on void fraction for condensing flows in microchannels. Clearly, much additional research on void fractions in different flow regimes during condensation in microchannels is required.

Void fractions computed for representative conditions for R-134a condensing in a 1-mm tube at a pressure of 1500 kPa, at different mass fluxes and qualities using a wide variety of correlations, are illustrated in Example 6.2. A comparison of the resulting void fractions is shown in Figure 6.20.

6.4 Pressure drop

As was noted in the preceding sections, there is relatively little information on flow regimes and void fraction in condensing flows through microchannels. For pressure drop, most of the available work at the microscale has been conducted in small tubes with adiabatic air–water or nitrogen–water mixtures. Recently, there have been experiments conducted with condensing refrigerants in microchannels. In general, models for pressure drop in microchannels have either adopted a flow-regime-based approach (i.e., separate models for annular and intermittent flow)

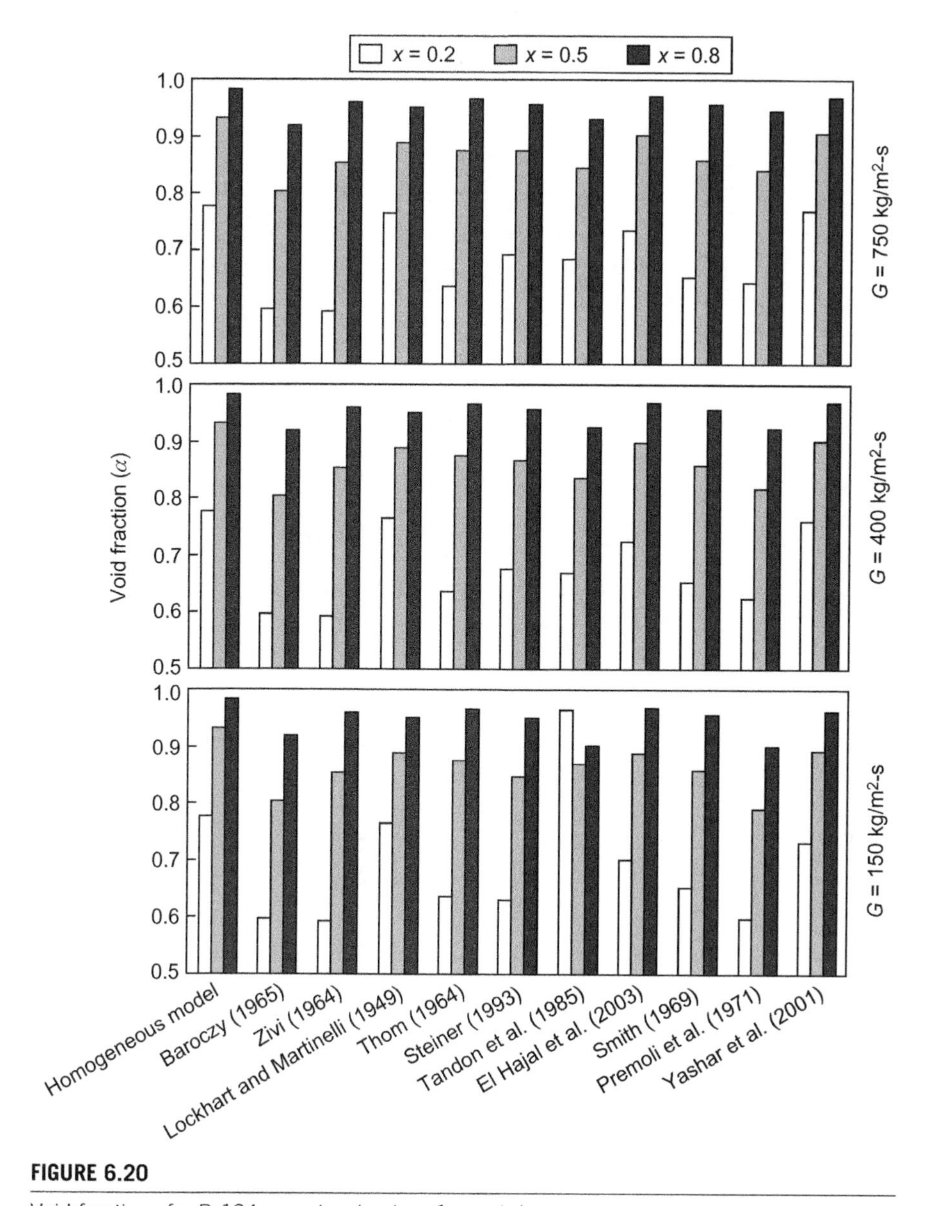

FIGURE 6.20

Void fractions for R-134a condensing in a 1-mm tube.

or empirically modified the two-phase multiplier from classical correlations in an effort to capture microchannel phenomena. Therefore, the available relevant information on pressure drops in condensing flows through relatively small channels and primarily adiabatic flows through microchannels is presented here (Table 6.5).

Table 6.5 Summary of Pressure Drop Studies

Investigator	Hydraulic Diameter (mm)	Fluids	Orientation/ Conditions	Range/ Applicability	Techniques, Basis, Observations
Classical correlations					
Lockhart and Martinelli (1949)	1.5–26 mm	Air and benzene, kerosene, water, and various oils	Adiabatic		• ΔP correlated based on whether gas and liquid phases are laminar or turbulent • ΔP related in terms of two-phase multiplier ϕ^2 to corresponding liquid- or gas-phase pressure drop • Chisholm (1967) correlated ϕ_L^2 terms of X_{tt}
Chisholm (1973)		Steam			• Modified procedure and equation developed by Baroczy (Baroczy and Sanders, 1961; Baroczy, 1966) to develop correlation based on liquid only ΔP
Friedel (1979)	$D > 1$ mm		Adiabatic	$\mu_1/\mu_v < 1000$	• Developed ϕ_{LO}^2 correlating from database of 25,000 points • Includes surface tension effects
Condensation or adiabatic liquid–vapor studies ($\sim 2 < D_H < \sim 10$ mm)					
Hashizume et al. (1985)	10 mm	R-12, R-22	Horizontal condensing	$570 < p < 1960$ kPa	• ΔP expressions for annular and stratified flow, liquid- and gas-phase velocity in terms of Prandtl mixing length • Stratified flow modeled as flow between parallel plates

Wang et al. (1997b)	6.5 mm	R-22, R-134a, R-407C	Horizontal adiabatic	$50 < G < 700$ kg/m^2 s $T = 2°C$, 6°C, and 20°C	• Modified Chisholm (1967)C for ϕ_L^2 for $G < 200$ kg/m^2 s • Empirical correlation for ϕ_G^2 for $G < 200$ kg/m^2 s
Hurlburt and Newell (1999)	3–10 mm	R-22, R-134a (Dobson, 1994), R-11, R-12, R-22 (Sacks, 1975)	Horizontal	$0.2 < x < 0.9$ $200 < G < 650$ kg/m^2 s	• Scaling equation to predict ΔP at new condition from available information at a different condition • Assumed bulk of condensation occurs in annular flow • Turbulent vapor and liquid films, law of wall assumed
Chen et al. (2001)	1.02, 3.17, 5.05, 7.02 mm	Air–water	Horizontal	Room temperature $50 < G < 3000$ kg/m^2 s $0.001 < x < 0.9$	• Accounted for increased σ influence, decreased g influence • Poor agreement with Chisholm (1967), Friedel (1979, 1980), and homogeneous flow models
	3.17, 5.05, 7.02, 9 mm	R-410A		5°C and 15°C $50 < G < 600$ kg/m^2 s $0.1 < x < 0.9$	• Modified homogeneous model to include *Bo* and *We* to account for effect of surface tension and mass flux
Wilson et al. (2003)	$D_H = 7.79$, 6.37, 4.40, and 1.84	R-134a, R-410A	Horizontal condensing	$T = 35°C$ $75 < G < 400$ kg/m^2 s	• Flattened round smooth, axial, and helical microfin tubes • ΔP increases as tube approaches rectangular shape
Souza et al. (1993)	10.9 mm	R-12, R-134a	Horizontal	$P_r = 0.07–0.12$ $200 < G < 600$ kg/m^2 s $0.1 < x < 0.9$	• New correlation for ϕ_{LO}^2 • Demonstrated effect of oil in refrigerant on pressure drop

(*Continued*)

Table 6.5 (Continued)

Investigator	Hydraulic Diameter (mm)	Fluids	Orientation/ Conditions	Range/ Applicability	Techniques, Basis, Observations
Cavallini et al. (2001, 2002a)	8 mm	R-22, R-134a, R-125, R-32, R-236ea, R-407C, R-410A	Horizontal tubes	$30 < T_{sat} < 50°C$ $100 < G < 750$ kg/m^2 s	• Modified Friedel (1979) correlation to apply only to annular flow, whereas it had originally been intended for annular and stratified regimes
Mini- and microchannel adiabatic studies					
Fukano et al. (1989)	1, 2.4, and 4.9 mm	Air–water	Horizontal isothermal	$P_{exit} = 1$ atm $0.04 < j_G < 40$ $0.2 < j_L < 4$	• *Correlation* for ΔP in bubbly, slug, plug, and annular flow • Proposed a new slug flow model accounting for slug/bubble velocity ratios for different gas- and liquid-phase regimes, and for expansion losses as liquid flowed from the annular film surrounding the gas bubble into liquid slug region
Mishima and Hibiki (1996a,b)	1–4 mm	Air–water	Vertical upward	$0.0896 < j_G < 79.3$ m/s $0.0116 < j_L < 1.67$ m/s	• C in Chisholm (1967) correlation decreases for small D, proposed new correlation for C based on D_H
Triplett et al. (1999a)	Circular 1.1 and 1.45 mm, semi-triangular 1.09 and 1.49 mm	Air–water	Horizontal adiabatic	$0.02 < j_G < 80$ m/s $0.02 < j_L < 8$ m/s	• Homogeneous model predicted bubbly and slug flow data well at high Re_L • Poor agreement in slug-annular and annular region, also in slug flow at very low Re_L • Friedel (1979) correlation resulted in large deviations

Zhao and Bi (2001b)	Equilateral triangular channels $D_H = 0.866$, 1.443, and 2.886 mm	Air–water	Vertical upward co-current	$0.1 < j_G < 100$ m/s For $D_H = 2.886$ 1.443 mm $0.08 < j_L < 6$ m/s For $D_H = 0.866$ mm $0.1 < j_L < 10$ m/s	• Frictional ΔP correlated using Lockhart and Martinelli (1949) approach
Lee and Lee (2001)	Rectangular 20 mm wide Gap 0.4–4 mm	Air–water	Horizontal	$175 < Re_{LO} < 17{,}700$ $0.303 < X < 79.4$	• Commented on weaker coupling between phases in smaller channels, correlated parameter C in the two-phase multiplier (Chisholm, 1967) in terms of nondimensional parameters to include surface tension effects
Kawahara et al. (2002)	100 μm	Nitrogen–water	Horizontal	$0.1 < j_G < 60$ m/s $0.02 < j_L < 4$ m/s	• Homogeneous correlations did not predict data well • Lockhart and Martinelli (1949) approach with Chisholm (1967) coefficient significantly overpredicted data • Proposed new values for parameter C based on data
Chung and Kawaji (2004)	530, 250,100, and 50 μm	Nitrogen–water	Horizontal	$T = 22.9$–29.9°C $0.02 < j_G < 72.98$ m/s $0.01 < j_L < 5.77$ m/s	• Effect of diameter investigated • New slug flow model based on work of Garimella et al. (2002, 2003a,b, 2005) for 50- and 100-μm tubes

(*Continued*)

Table 6.5 (Continued)

Investigator	Hydraulic Diameter (mm)	Fluids	Orientation/ Conditions	Range/ Applicability	Techniques, Basis, Observations
Adiabatic two-phase refrigerant studies					
Yang and Webb (1996b)	Rectangular plain: $D_H = 2.64$ Microfin: $D_H = 1.56$	R-12	Horizontal adiabatic	$400 < G < 1400$ kg/m^2 s $0.1 < x < 0.9$	• ΔP in microfin tube higher than in plain tube • Equivalent *Re*-based model for ΔP • Concluded surface tension does not affect ΔP
Zhang and Webb (2001)	3.25 and 6.25 mm	R-134a, R-22	Adiabatic	$T = 20–65°C$	• Friedel (1979, 1980) correlation did not predict data well
	Multiport extruded Al $D_H = 2.13$ mm	R-404A		$200 < G < 1000$ kg/m^2 s	• ΔP function of reduced pressure rather than density or viscosity ratio; proposed new correlation for ϕ_{LO}^2
Ribatski et al. (2006)	$0.05 < D_H < 3$ mm	Refrigerants	Adiabatic and boiling	$23 < G < 6000$ kg/m^2 s	• Correlation of Müller-Steinhagen and Heck (1986) predicted ΔP data best • Poorest agreement at $x > 0.6$
Revellin and Thome (2007)		R-134a, R-245fa	Adiabatic		• Observed laminar, transition, and turbulent zone while plotting f_{TP} versus Re_{TP} • Proposed new correlation as function of two-phase *Re* number
Cioncolini et al. (2009)	$0.52 < D_H < 25$ mm	Air–water, refrigerants	Adiabatic		• Empirical correlation as function of We_V and Re_L

Mini- and microchannel condensation studies					
Yan and Lin (1999)	2 mm	R-134a	Horizontal	$T = 40–50°C$	• Used Akers et al. (1959) equivalent mass velocity concept to correlate condensation ΔP
Garimella et al. (2002, 2003a,b, 2005)	0.5–4.91 mm	R-134a	Horizontal	$T \sim 52°C$ $0 < x < 1$ $150 < G < 750$ kg/m^2 s	• Flow-regime-based models for intermittent and annular/mist/disperse flow regimes • Intermittent flow model treats ΔP as combination of ΔP due to liquid slug, film–bubble interface, and transitions between slug and bubble, slug frequency deduced from ΔP data • Annular flow model: interfacial friction factor derived from measured pressure drops correlated with Re_L, y_L, X, and surface tension parameter • Combined model ensures smooth transitions across regimes
Andresen (2007)	Circular $0.76 < D < 9.4$ mm	R-404A, R-410A	Horizontal condensing		• Near-critical condensation • Proposed new model as function of confinement number
Agarwal (2006)	Rectangular: $0.10 < D_H < 0.16$ mm	R-134a	Horizontal condensing		• Intermittent model for all data, with annular flow being continuous vapor bubble • New model for number of unit cells

6.4.1 Classical correlations

Correlations developed for large tubes often serve as the starting point for microchannel pressure drop models. Pressure drops in conventional channels have long been calculated using three well-known correlations by Lockhart and Martinelli (1949), Chisholm (1973), and Friedel (1979), sometimes with modifications to account for the specific geometry or flow conditions under investigation. The Lockhart–Martinelli (1949) correlation was based on adiabatic flow of air and benzene, kerosene, water, and various oils flowing through 1.5–26-mm pipes, and the pressure drops were correlated based on whether the individual liquid and gas phases were considered to be in laminar or turbulent flow. As many of the separated flow correlations for small channels are also based on this basic methodology, a brief overview is presented here. The two-phase pressure drop is expressed in terms of two-phase multipliers to the corresponding single-phase liquid or gas-phase pressure drop:

$$\phi_{\mathrm{L}}^2 = \frac{(\mathrm{d}P_F/dz)}{(\mathrm{d}P_F/dz)_{\mathrm{L}}} \tag{6.28}$$

$$\phi_{\mathrm{G}}^2 = \frac{(\mathrm{d}P_F/dz)}{(\mathrm{d}P_F/dz)_{\mathrm{G}}} \tag{6.29}$$

$$\phi_{\mathrm{LO}}^2 = \frac{(\mathrm{d}P_F/dz)}{(\mathrm{d}P_F/dz)_{\mathrm{LO}}} \tag{6.30}$$

where ϕ^2 is the two-phase multiplier, the subscripts L and G refer to the flow of the liquid and gas phases flowing through the whole channel, and LO refers to the entire fluid flow occurring as a liquid phase through the channel. These multipliers are in turn most often correlated in terms of the Martinelli parameter $X = [((\mathrm{d}P_F/dz)_{\mathrm{L}})/((\mathrm{d}P_F/dz)_{\mathrm{G}})]$ (which has been referred to in the previous sections). Chisholm (1967) developed the following correlations for the two-phase multipliers of Lockhart and Martinelli:

$$\phi_{\mathrm{L}}^2 = 1 + \frac{C}{X} + \frac{1}{X^2} \tag{6.31}$$

$$\phi_{\mathrm{G}}^2 = 1 + CX + X^2 \tag{6.32}$$

where C depends on the flow regime of the liquid and gas phases. While these (Chisholm, 1967; Lockhart and Martinelli, 1949) correlations have shown considerable deviations from the data for small channels with phase-change flows, they continue to be the basis for many of the more recent correlations, as will be seen later in this section.

Chisholm (1973) modified the procedure and equations developed by Baroczy (Baroczy, 1966; Baroczy and Sanders, 1961) that account for fluid properties, quality, and mass flux based on steam, water–air, and mercury–nitrogen data to develop the following correlation:

$$\phi_{\mathrm{LO}}^2 = 1 + (Y^2 - 1)[Bx^{(2-n)/2}(1-x)^{(2-n)/2} + x^{2-n}] \tag{6.33}$$

where n is the exponent for Reynolds number in the turbulent single-phase friction factor correlation, for example, $n = 0.5$ for the Blasius equation. The parameter Y is the Chisholm parameter:

$$Y = \left[\frac{(dP_F/dz)_{GO}}{(dP_F/dz)_{LO}}\right]^{0.5} \tag{6.34}$$

and B is given by:

$$\begin{aligned} B &= \frac{55}{G^{0.5}} \quad 0 < Y < 9.5 \\ &= \frac{520}{(YG^{0.5})} \quad 9.5 < Y < 28 \\ &= \frac{15,000}{(Y^2G^{0.5})} \quad 28 < Y \end{aligned} \tag{6.35}$$

Friedel (1979) developed the following correlation based on a database of 25,000 points for adiabatic flow through channels with $D > 1$ mm:

$$\phi_{LO}^2 = E + \frac{0.324FH}{Fr^{0.045}We^{0.035}} \tag{6.36}$$

where

$$E = (1-x)^2 + x^2\left(\frac{\rho_L f_{GO}}{\rho_G f_{LO}}\right) \tag{6.37}$$

$$F = x^{0.78}(1-x)^{0.24} \tag{6.38}$$

$$H = \left(\frac{\rho_L}{\rho_G}\right)^{0.91}\left(\frac{\mu_G}{\mu_L}\right)^{0.19}\left(1-\frac{\mu_G}{\mu_L}\right)^{0.7} \tag{6.39}$$

and $Fr = G^2/gD\alpha^2{}_{TP}$ and $We = G^2D/\rho_{TP}\alpha$, and f_{LO} and f_{GO} are the single-phase friction factors for the total fluid flow occurring as liquid and gas, respectively. The two-phase mixture density is calculated as follows:

$$\rho_{TP} = \left(\frac{x}{\rho_G} + \frac{1-x}{\rho_L}\right)^{-1} \tag{6.40}$$

The above correlation is the more widely used correlation for vertical upward and horizontal flow, and is recommended by Hewitt et al. (1994) for situations where surface tension data are available, and by Hetsroni (1982) for $\mu_l/\mu_v < 1000$.

While these correlations have been used widely, their predictive capabilities are often not particularly good, primarily because investigators have tried to use them beyond their originally intended range of applicability, and also because these models do not account for the flow regimes that are established over these diverse ranges of conditions and geometries.

6.4.2 Condensation or adiabatic liquid–vapor flows for $\sim 2 < D_H < \sim 10$ mm

Hashizume et al. (1985) conducted flow visualization and pressure drop experiments on refrigerants R-12 and R-22 flowing through horizontal 10-mm tubes over the range 20°C < T < 50°C. They developed pressure drop expressions for annular and stratified flow, describing the liquid- and gas-phase velocity profiles in terms of the Prandtl mixing length. The pressure drop was modeled as $dP/dL = (4(Re^+)^2\mu^2_L/\rho_L D^3)$, where Re^+ is the friction Reynolds number. Stratified flow was modeled as flow between parallel plates with the liquid pool forming one of the parallel plates, with an equivalent gap between the plates derived in terms of the tube diameter, quality, and mass flux. Transition (wavy) region pressure drops were calculated by interpolating between the annular and stratified regions, with the corresponding transitions established by the modified Baker map developed by them (Hashizume, 1983). They later (Hashizume and Ogawa, 1987) tried to extend this work to air–water and steam–water flows, but found that the best agreement was with refrigerant data, and not with air–water or steam–water mixtures.

Wang et al. (1997b) measured pressure drops during the adiabatic flow of refrigerants R-22, R-134a, and R-407C in a 6.5-mm tube for the mass flux range $50 < G < 700$ kg/m^2 s. They computed two-phase multipliers ϕ^2_G analogous to the Lockhart–Martinelli (1949) correlation from the measured pressure drops, and tried to obtain the best fit for the C parameter in Chisholm's (1967) equation for the multiplier, because they found that C was strongly dependent on the flow pattern. For R-22 and R-134a, the intermittent flow data were predicted well with $C = 5$. For $G < 200$ kg/m^2 s, wavy-annular flow was observed with the two phases being turbulent, and the multipliers did not depend on the mass flux. For $G = 50$ and 100 kg/m^2 s, there was a pronounced influence of the mass flux. They stated that for $G = 100$ kg/m^2 s, the flow is not shear dominated; thus the waves do not reach the top of the tubes, and the pressure drop would therefore depend on the wetted perimeter, which in turn depends on the mass flux. However, for $G < 200$ kg/m^2 s, even though the flow continues to be wavy (with a thicker film at the bottom), the waves do reach the top of the tube and render the tube completely wet, which makes the wetted perimeter relatively constant as the mass flux varies. Based on these considerations, they developed the following equations for the two-phase multiplier:

For $G > 200$ kg/m^2 s:

$$\phi^2_G = 1 + 9.4X^{0.62} + 0.564X^{2.45} \tag{6.41}$$

and for $G < 200$ kg/m^2 s, the C parameter in the Chisholm (1967) correlation for the Lockhart–Martinelli (1949) multiplier was modified to include a mass flux dependence as follows:

$$C = 4.566 \times 10^{-6} X^{0.128} Re_{LO}^{0.938} \left(\frac{\rho_L}{\rho_G}\right)^{-2.15} \left(\frac{\mu_L}{\mu_G}\right)^{5.1} \tag{6.42}$$

Hurlburt and Newell (1999) developed scaling equations for condensing flows that would enable the prediction of void fraction, pressure drop, and heat transfer for a refrigerant at a given condition and tube diameter from the available results for another similar fluid (R-11, R-12, R-22, and R-134a) operating at a different condition in the diameter range 3–10 mm. To achieve this, they assumed that the bulk of the condensation process occurs under annular flow conditions, and that even though in reality the condensing film is nonuniform around the circumference, an equivalent average film thickness would suffice for the prediction of shear stress. They further considered that the primary resistance to heat transfer occurs in the viscous and buffer layers of the film, regardless of the actual thickness. The approach used for modeling the film and core flows was similar to that of Rohsenow et al. (Bae et al., 1968; Traviss et al., 1973), assuming turbulent vapor and liquid films and law-of-the-wall universal velocity and temperature profiles. (The liquid film was assumed to be turbulent even at low *Re* ($Re > 240$) due to interactions with the turbulent vapor core, based on the work of Carpenter and Colburn (1951).) They used the interfacial shear stress, rather than the void fraction, as an input to determine the liquid film thickness, which enabled them to capture the dependence of void fraction on mass flux. They used void fraction, pressure drop, and heat transfer data of Sacks (1975) for R-11, R-12, and R-22, and the R-22 and R-134a data of Dobson (1994) to validate their model. Using these considerations, they developed the following scaling equation for pressure drop prediction:

$$\Delta P_2/\Delta P_1 = (\rho_{g2}/\rho_{g1})^{-0.75}(\rho_{L2}/\rho_{L1})^{-0.25}(\mu_{g2}/\mu_{g1})^{-0.15}(\mu_{L2}/\mu_{L1})^{0.125} \times (D_2/D_1)^{-1.0}(G_2/G_1)^2(x_2/x_1)^{1.65}\left(\frac{1-x_2}{1-x_1}\right)^{0.38} \tag{6.43}$$

This equation predicts a five-fold increase in pressure drop between R-22 and R-11, primarily due to the drastic difference in saturation pressure at 40°C (959 kPa for R-22, 175 kPa for R-11), which leads to a much higher vapor-phase density for R-22 (65.7 kg/m^3) than for R-11 (9.7 kg/m^3). This approach was considered to be valid for the quality range $0.2 < x < 0.8$, where annular flow prevails in general. Although there were some disagreements between the trends predicted by this equation and the experimental data, particularly in the effect of mass flux, the approach provides some general guidance for the modeling of annular flows.

Chen et al. (2001) attempted to account for the increased influence of surface tension and the decreased influence of gravity in tubes with $D < 10$ mm for fluids encompassing a wide range of properties, air–water (in 1.02-, 3.17-, 5.05-, 7.02 mm tubes) and R-410A (in 3.17-, 5.05-, 7.02-, 9.0 mm tubes). The air–water tests were conducted at room temperature, while the R-410A tests were conducted at 5°C and 15°C, somewhat lower than typical condensation temperatures of interest. Poor agreement was found between their pressure drop data and the predictions of the Chisholm (1967), Friedel (1979), and homogeneous flow (with average viscosity; Beattie and Whalley, 1982) models. Therefore, they modified the homogeneous flow model by including the Bond number and the Weber number to account for the effects of surface tension and mass flux:

$$\frac{dP}{dz} = \left.\frac{dP}{dz}\right|_{\text{hom}} \Omega_{\text{hom}}$$
$$\Omega_{\text{hom}} = \begin{cases} 1 + (0.2 - 0.9\exp(-Bo)) & Bo < 2.5 \\ 1 + (We^{0.2}/\exp(Bo^{0.3})) - 0.9\exp(-Bo) & Bo \geq 2.5 \end{cases} \tag{6.44}$$

where $We = G^2D/\sigma\rho_m$ and $Bo = g(\rho_L - \rho_G)\,((d/2)^2/\sigma)$. They also developed a similar modification to the Friedel (1979) correlation using the rationale that, when used for small tubes, this correlation does not emphasize surface tension (We) enough, and may emphasize gravity (Fr) too much. The resulting modification is as follows:

$$\frac{dP}{dz} = \left.\frac{dP}{dz}\right|_{\text{Friedel}} \Omega$$
$$\Omega = \begin{cases} \dfrac{0.0333Re_{LO}^{0.45}}{Re_G^{0.09}(1 + 0.4\exp(-Bo))} & Bo < 2.5 \\[2ex] \dfrac{We^{0.2}}{(2.5 + 0.06Bo)} & Bo \geq 2.5 \end{cases} \tag{6.45}$$

These modifications improved the predictive capabilities of the homogeneous and Friedel correlations from deviations of 53.7% and 218.0% to 30.9% and 19.8%, respectively. In addition, these equations also agreed reasonably well with the pressure drop data of Hashizume (1983) for R-12 and R-22 in 10-mm tubes.

Wilson et al. (2003) studied the effect of progressively flattening 8.91-mm round smooth tubes and tubes with axial and helical microfin tubes on void fraction, pressure drop, and heat transfer during condensation of refrigerants R-134a and R-410A at 35°C. The hydraulic diameters of the tubes tested were 7.79, 6.37, 4.40, and 1.84 mm, with the microfin tubes having 60 fins of 0.2 mm height for a surface area increase of 60% over a smooth tube. The pressure drop at a given mass flux and quality increased as the tube approached a rectangular shape. They found that the following circular tube liquid-only two-phase multiplier correlation of Jung and Radermacher (1989) predicted their results within 40%:

$$\phi_{LO}^2 = 12.82X_{tt}^{-1.47}(1-x)^{1.8} \tag{6.46}$$

Similarly, the following annular and stratified flow correlation of Souza et al. (1993) with the D_H of the flattened tube as the characteristic dimension, also predicted their data within 40%:

$$\begin{aligned}
&\phi_L^2 = 1.376 + C_1 X_{tt}^{-C_2}\\
&Fr_l < 0.7:\\
&C_1 = 4.172 + 5.48 Fr_l - 1.564 Fr_l^2\\
&C_2 = 1.773 - 0.169 Fr_l\\
&Fr_l > 0.7:\\
&C_1 = 7.242; \quad C_2 = 1.655
\end{aligned} \tag{6.47}$$

where $Fr_l = G/(\rho_L \sqrt{gD})$. The corresponding single-phase friction factor was calculated using the Colebrook (1939) equation and an expression for the relative roughness of the microfin tubes developed by Cavallini et al. (2000). Wilson et al. (2003) deemed this 40% agreement to be adequate and did not develop a new correlation specifically for flat (flattened) tubes.

6.4.3 Adiabatic flows through mini- and microchannels

6.4.3.1 Adiabatic air–water flows

Fukano et al. (1989) proposed correlations for pressure drop in bubbly, slug, plug, and annular flow based on experiments with air–water flow in 1-, 2.4-, and 4.9-mm tubes. In slug flow, sequential photographs showed that the velocity of the large gas bubble can be correlated as $u_S = 1.2(j_G + j_L)$, where $j_G + j_L$ is the liquid slug velocity. From this, the relative velocity between the gas bubble and the liquid in the slug is given by $\mu_r = 0.2(j_G + j_L)$. From photographic observations, it was also established that the liquid slug length is given by:

$$L_l/(L_l + L_g) = K j_L/(j_L + j_G) \tag{6.48}$$

where the constant of proportionality K varied from 0.9 at 1 mm to 0.72 at 4.9 mm. Assuming that, in slug and plug flows, the pressure drop occurs in the liquid slug only, while in annular and bubbly regions it occurs over the entire length of the channel, they developed the following equations for the two-phase multiplier:

$$\phi_L^2 \propto K\left(\frac{\rho_T}{\rho_L}\right)\left(\frac{\nu_L}{\nu_L}\right)n\left(\frac{j_G + j_L}{j_L}\right)^{n+c+1} Re_L^{n-m} \tag{6.49}$$

where $c = 0$ for intermittent flow and $c = 1$ for annular and bubbly flows, subscript T refers to the two-phase mixture, and m and n refer to the Re exponent in the single-phase friction factor for the liquid and two-phase flows, respectively, dependent on whether the flow is laminar or turbulent. As the experimental ϕ_L^2 values were larger than those predicted by this treatment, they proposed a new slug flow model that accounts for the expansion losses (loss coefficient = 1) as the liquid flowed from the annular film surrounding the gas bubble into the liquid slug region. This additional contribution of the expansion losses was shown to be significant, and led to much better agreement between the measured and predicted values. The analysis also showed that for a 2.4-mm channel, the expansion losses could in fact be larger than the frictional losses in the slug. The expansion losses were found to first increase with increasing j_G because the liquid slug length decreases, presumably leading to more expansion events; but further increases in j_G make the expansion losses less significant because at very high j_G, the frictional losses increase rapidly.

Mishima and Hibiki (1996b) (whose work was discussed in the previous sections on flow regimes and void fraction) measured frictional pressure drops in air–water flows through 1–4-mm tubes. By comparing their results with the Lockhart–Martinelli (1949) correlation, they noticed that the parameter C in Chisholm's (1967) curve-fit [Eq. (6.31) above] to the multiplier decreased with a decrease in tube diameter. Similar trends were also observed by Sugawara et al. (1967) for air–water flow in horizontal 0.7–9.1-mm round tubes, Ungar and Cornwell (1992) for the flow of ammonia in horizontal 1.46–3.15-mm round tubes, Mishima et al. (1993) for vertical upward flow of air–water in rectangular 1.07×40-, 2.45×40-, and 5.00×40-mm channels, Sadatomi et al. (1982) for vertical upward flow of air–water through rectangular $(7-17) \times 50$- and 7×20.6-mm channels, and Moriyama et al. (1992a,b) for the flow of R-113–nitrogen through horizontal rectangular $(0.007-0.098) \times 30$-mm channels. Including the data from these investigators and their own data, they developed the following equation for the parameter C:

$$C = 21(1 - \exp(-0.319 D_H)) \quad (6.50)$$

where D_H is in millimeters. They state that this equation is valid for vertical and horizontal round tubes as well as rectangular ducts (although the predictions did not agree well with the ammonia vapor data of Ungar and Cornwell, 1992).

It should be noted that several investigators have shown that the homogeneous flow model is reasonably successful in predicting pressure drop during adiabatic flow and boiling in channels with a diameter of a few millimeters. These include the previously mentioned work of Ungar and Cornwell (1992) on the flow of ammonia in horizontal 1.46–3.15-mm round tubes, and Kureta et al. (1998) on the boiling of water in 2–6-mm channels at atmospheric pressure over a wide range of mass fluxes. Similarly, Triplett et al. (1999a) found that this model was able to predict some of their data on adiabatic flow of air–water mixtures through 1.1- and 1.45-mm circular, and 1.09- and 1.49-mm semi-triangular channels. In particular, the homogeneous model predicted the data in the bubbly and slug flow regimes well at high Re_L. The agreement in the slug-annular and annular flow regions was poor, as was the agreement with the slug flow data at very low Re_L.

Also, the Friedel (1979) correlation resulted in larger deviations from the data, especially at low Re_L values. This same group (Ekberg et al. 1999) also investigated air–water flow in two narrow, concentric annuli with inner and outer diameters of 6.6 and 8.6 mm for one annulus, and 33.2 and 35.2 mm for the second annulus, respectively. The measured pressure drops did not show particularly good agreement with any of the available models, and even the Friedel correlation, which they state provided the best agreement, had deviations of over 50%.

Zhao and Bi (2001b) measured pressure drops for upward co-current air–water flow through equilateral triangular channels with $D_H = 0.866$, 1.443, and 2.886 mm. They found that the void fraction could be simply represented by $\alpha = 0.838\beta$, which is very similar to the Armand (1946) correlation and was used to compute acceleration losses in two-phase flow. The frictional component was then correlated using the Lockhart–Martinelli (1949) approach. For this, they computed the required single-phase friction factor by replacing the laminar and turbulent friction factors in the comprehensive laminar-transition-turbulent correlation of Churchill (1977) with the corresponding expressions for triangular channels. This modification yielded two-phase multipliers that were between the values predicted by the Chisholm (1967) expression using $C = 5$ and $C = 20$.

Lee and Lee (2001) investigated pressure drop for air–water flow through 20-mm-wide horizontal rectangular channels with gaps of 0.4–4 mm. They recalled that the original basis for the Chisholm (1967) correlation of the two-phase multiplier in the Lockhart–Martinelli (1949) correlation consisted of accounting for the pressure drop due to (a) gas phase, (b) liquid phase, and (c) interactions between the two phases. The parameter C in Chisholm's correlation represents this interaction term, which in turn depends on the flow regimes of the two phases. Therefore, they proposed different values for C, accounting for the gap size as well as the phase flow rates. They reasoned that as the gap size decreases, the flow tends more and more to plug and slug flow, with an increasing effect of surface tension due to the curved gas/liquid interface at the edge of the bubble. In such surface-tension-dominated flows, they stated that of the several dimensionless parameters identified by Suo and Griffith (1964) for capillary tubes, the Reynolds number of the liquid slug, $Re = (\rho_L j D_H)/\mu_L$, the ratio of viscous and surface tension effects, $\psi = \mu_L j/\sigma$, and a combination of parameters independent of the liquid slug velocity, $\lambda = \mu_L^2/(\rho_L \sigma D_h)$ were the relevant ones. Lee and Lee then used their data to obtain individual values of the constant A and exponents q, r, and s for each combination of liquid and gas flow regimes in the following equation for the parameter C in the two-phase multiplier:

$$C = A\lambda^q \psi^r Re^s{}_{LO} \tag{6.51}$$

Regression analysis showed that the dependence on parameters λ and ψ above, that is, surface tension effects, was only significant when both phases were laminar, with C being simply a function of Re_{LO} when either or both phases were in the turbulent regime. This approach predicted their data within $\pm 10\%$, and also predicted the data of Wambsganss et al. (1992) and Mishima et al. (1993) within

15% and 20%, respectively. This correlation is valid for $175 < Re_{LO} < 17700$ and $0.303 < X < 79.4$.

Kawahara et al. (2002) measured pressure drops for air–water flow through a 100-μm circular tube (the corresponding results on flow patterns and void fractions were discussed in previous sections). Test section entrance (contraction) losses, and acceleration losses due to the reduction in density across the test section were removed from the measured pressure drop in a manner similar to that discussed in connection with the work of Garimella et al. (2005), below. The void fraction required for the acceleration losses was evaluated using Eq. (6.27) developed from their image analysis work. The contributions of these contraction and acceleration losses to the total pressure drop were found to be 0.05–9% and 0–4.5%, respectively, of the measured pressure drop. Although their flow pattern and void fraction results showed that the flow was not well represented by a homogeneous model, based on the success of others in using this model for small channels (Triplett et al., 1999b; Ungar and Cornwell, 1992), they compared their results with a homogeneous model using the homogeneous density and homogeneous viscosity models of Owens (1961), McAdams (1954), Cicchitti et al. (1960), Dukler et al. (1964), Beattie and Whalley (1982), and Lin et al. (1991). They found that none of these models were able to predict their data adequately (the Dukler et al. (1964) model, $\mu_H = \beta\mu_G + (1 - \beta)\mu_L$, provided somewhat reasonable predictions). When a Lockhart–Martinelli (1949) type two-phase multiplier approach was used, the Chisholm (1967) coefficient of $C = 5$ significantly overpredicted the data, while a C value of 0.66 from Mishima and Hibiki's (1996b) work resulted in a 10% overprediction, and agreement within ±10% was achieved using Lee and Lee's (2001) correlation for C. Therefore, they conducted a regression analysis of the data to obtain a value of $C = 0.24$, which also yielded a ±10% agreement. In later work on a 96-μm square channel (Chung et al., 2004), they noted that channel shape did not have any appreciable effect on two-phase pressure drop, and provided revised values of $C = 0.22$ for circular channels and $C = 0.12$ for square channels for the C coefficient in Chisholm's (1967) model.

Chung and Kawaji (2004) also investigated the effects of channel diameter on pressure drop during adiabatic flow of nitrogen–water through 530-, 250-, 100-, and 50-μm circular channels. In this study, they found that the Dukler et al. (1964) viscosity model (when used assuming homogeneous flow) did not predict the pressure drops for the 530- and 250-μm channels, although the agreement was within ±20% for the 100- and 50-μm data. Beattie and Whalley's (1982) mixture viscosity model showed the opposite trends, succeeding somewhat in predicting the 530- and 250-μm channel data, but significantly overpredicting the 100- and 50-μm channel data. They attributed this to the fact that Beattie and Whalley's (1982) model yields a higher mixture viscosity than that of Dukler et al. (1964), resulting in better agreement with the larger channel data because of the additional mixing losses in the bubbly and churn flows in these channels. When predominated by the more laminar slug flow with weak momentum coupling of the phases, however, as is the case in the smaller channels, the lower viscosities predicted by Dukler et al. (1964) model are more appropriate. When attempting the use of a two-phase multiplier approach as discussed above, they noted a mass flux effect on the friction multiplier for the 530- and 250-μm channels. The data at different mass fluxes for the 100- and 50-μm channels, however, were predicted well by a single C value ($C = 0.22$ for 100 μm, $C = 0.15$ for 50 μm). This decrease in C value (based on the above discussion of Lee and Lee's (2001) work) with a decrease in channel diameter implies a decrease in momentum coupling of the two phases, that is, completely separated laminar flow of gas and liquid, as also noted by Ali et al. (1993).

Based on the somewhat inadequate predictions of the homogeneous and two-phase multiplier approaches, they adapted the intermittent flow model of Garimella et al. (2002) (discussed below) to calculate the pressure drops in the 50- and 100-μm channels, which have predominantly intermittent flows. They also used the unit-cell concept consisting of gas bubbles surrounded by a liquid film and flowing through slower moving liquid slugs. Although, in accordance with the model of Garimella et al. (2002), they represented the total pressure drop as the summation of the frictional pressure drop in each of these regions, they ignored the mixing losses at the transitions between these regions. The rationale used was that in these smaller channels, the liquid flow is typically laminar, with suppressed mixing leading to negligibly small transition losses. For the single-phase liquid region, they computed the average liquid velocity $U_L = j_L/(1-\alpha)$, which yielded $Re_L = (\rho_L U_L D)/\mu_L$. Laminar or turbulent single-phase friction factors were calculated based on Re_L. To compute the pressure drop in the film/bubble region, they assumed the average velocity of the gas phase and the gas bubble to be equal, $U_G = U_B$, where $U_G = j_G/\alpha$. This assumption is different from the assumption by Garimella et al. (2002) that $U_B = 1.2U_S$, which, according to Chung and Kawaji (2004), is equivalent to assuming that the Armand (1946) correlation for void fraction approximately holds. Instead, they used the nonlinear expression for void fraction derived by them and discussed here in the previous section. Defining the bubble Re_B in terms of the relative velocity between the bubble and the interface, the computation of the pressure drop in the film–bubble region develops much like that in Garimella et al. (2002), with an assumption that $D_B = 0.9D$. The interface velocity, the film–bubble pressure gradient, and the bubble Reynolds number are then obtained iteratively. To calculate the relative lengths of the bubble and slug regions, they relied on the void fraction measured by them. Thus, treating the bubble as a cylinder with flat ends, the bubble length is given by:

$$\frac{L_B}{L_{UC}} = \alpha\left(\frac{D}{D_B}\right)^2 \tag{6.52}$$

Finally, the total pressure drop in the two regions is given by:

$$\left(\frac{dP_f}{dz}\right)_{TP} = \left(\frac{dP}{dz}\right)_{F/B}\frac{L_B}{L_{UC}} + \left(\frac{dP}{dz}\right)_L\frac{L_L}{L_{UC}} \tag{6.53}$$

The model was shown to agree better with the 50- and 100-μm channel pressure drop data than the homogeneous and two-phase multiplier approaches. Also, the predictions of the 50-μm channel data were better than those of the 100-μm data, presumably because the smaller channel exclusively exhibits slug flow, while the larger channel may have some regions of other kinds of flow. This also points out that this model should not be applied to larger ($D_H > 100$ μm) channels where the flow is not exclusively intermittent.

6.4.3.2 Adiabatic two-phase refrigerant flows

Yang and Webb (1996b) measured pressure drops in single- and adiabatic two-phase flows of refrigerant R-12 in rectangular plain and microfin tubes with $D_H = 2.64$ and 1.56 mm, respectively. The measurements were conducted at somewhat higher mass fluxes ($400 < G < 1400$ kg/m^2 s) than those of interest for refrigeration and air-conditioning applications. Their single-phase friction factors were correlated by $f_{smooth} = 0.0676 Re_{D_H}^{-0.22}$ and $f_{microfin} = 0.0814 Re_{D_H}^{-0.22}$. The pressure gradient in the microfin tube was higher than that of the plain tube. They tried to model the two-phase pressure drops using the Lockhart–Martinelli (1949) approach and curve-fitting the two-phase multiplier ϕ_v^2, but this did not lead to a satisfactory correlation. The equivalent mass velocity concept of Akers et al. (1959) was then attempted to develop a correlation. Thus, the single-phase friction factor is first calculated using the aforementioned equations at the Reynolds number Re_{Dh} for the liquid phase flowing alone. The equivalent Reynolds number for two-phase flow is then calculated as follows:

$$Re_{eq} = \frac{G_{eq} D_H}{\mu_l}; \quad G_{eq} = G\left[(1-x) + x\left(\frac{\rho_l}{\rho_v}\right)^{1/2}\right] \tag{6.54}$$

The equivalent friction factor is therefore based on an equivalent all-liquid flow that yields the same frictional pressure drop as the two-phase flow. Based on their data, the two-phase friction factor at this equivalent Reynolds number was fit to the following equation:

$$\frac{f}{f_l} = 0.435\, Re_{eq}^{0.12} \tag{6.55}$$

Finally, the two-phase pressure drop is calculated as follows:

$$\Delta P_f = f \frac{Re_{eq}^2 \mu_l^2}{2\rho_l} \frac{4L}{D_H^3} \tag{6.56}$$

This approach was able to predict both plain and microfin tube data within ±20%. It should be noted that this approach collapses both the single- and two-phase pressure gradients for each tube on the same line, when plotted against the Re_{eq}. From their data, Yang and Webb (1996b) also concluded that surface tension did not play a role in determining the pressure drop in these tubes, although in related work (Yang and Webb, 1996a) they found surface tension to be significant for the heat transfer coefficient, especially at low G and high x.

Zhang and Webb (2001) measured adiabatic two-phase pressure drops for R-134a, R-22, and R-404A flowing through 3.25- and 6.25-mm circular tubes and through a multiport extruded aluminum tube with a hydraulic diameter of 2.13 mm. They noted that the Friedel (1979) correlation, although recommended widely for larger tubes, was developed from a database with $D > 4$ mm, and therefore did not predict their data well, especially at the higher reduced pressures ($T_{sat} = 65°C$). Since the dependence of the Friedel correlation on Weber number and Froude number was weak, they decided not to include them as variables while correlating their data. Also, based on the work of Wadekar (1990) for convective boiling, they postulated that the pressure drop would be a function of the reduced pressure rather than the density and viscosity ratios. Using these considerations, and additional data from a previous study on multiport tubes with round 1.45- and 0.96-mm ports, and 1.33-mm rectangular ports, they developed the following modified version of the Friedel correlation:

$$\phi_{LO}^2 = (1-x)^2 + 2.87x^2(p/p_{crit})^{-1} + 1.68x^{0.8}(1-x)^{0.25}(p/p_{crit})^{-1.64} \quad (6.57)$$

Ribatski et al. (2006) conducted a comprehensive review of heat transfer and pressure drop in microchannels ($0.05 < D_H < 3$ mm) consisting of eight fluids, mass fluxes from 23 to 6000 kg/m^2 s, and vapor qualities up to 1. The reviewed studies included adiabatic and evaporation studies. They found that the pressure drop correlation of Müller-Steinhagen and Heck (1986), developed for large channels, predicted the data best, followed by the method of Mishima and Hibiki (1996b). However, even the best correlation predicted less than 50% of the data within $\pm 30\%$. They noted particularly poor agreement at higher quality (>0.6), where annular flow is expected to dominate. Revellin and Thome (2007) conducted adiabatic pressure drop experiments with R-134a and R-245fa in circular microchannels ($0.509 < D_H < 0.790$ mm). They observed laminar, transition, and turbulent zones when plotting two-phase friction factor versus two-phase Reynolds number (using the two-phase viscosity model of McAdams et al., 1942). They proposed a new correlation for two-phase friction factor as a function of two-phase Reynolds number. However, the correlations are specific to the tube diameters investigated and not easily extrapolated. Additionally, they note that there is still poor agreement in the transition and laminar regimes, which tend to correspond to the intermittent flow regime.

Cioncolini et al. (2009) conducted a review of annular flow pressure drop data from micro- to macrochannel size ($0.52 < D_H < 25$ mm) for fluids including air–water and refrigerants. The method of Lombardi and Carsana (1992) was found to predict the data best over the entire set of conditions, with a mean average deviation of 15.9% for macroscale tubes and 18.9% for the microscale tubes. They then proposed an empirical correlation (Eq. (6.58)) as a function of vapor core Weber number and liquid-phase Reynolds number. The droplet-laden vapor core Weber number and the film Reynolds number are defined as function of the fraction of liquid entrained in the core from the correlation of Oliemans et al. (1986).

$$f_{TP} = 0.172 \cdot We_C^{-0.372} \quad \text{for } Bo \geq 4$$
$$f_{TP} = 0.0196 \cdot We_C^{-0.372} \cdot Re_l^{0.318} \quad \text{for } Bo < 4 \tag{6.58}$$

6.4.4 Condensing flows through mini- and microchannels

Yan and Lin (1999) used the same Akers et al. (1959) equivalent mass velocity concept that was used by Yang and Webb (1996b), discussed in the previous section, to correlate condensation pressure drop of R-134a in a 2-mm circular tube. They arranged 28 such copper tubes in parallel, held between copper blocks, to which were soldered similar cooling channels to form a crossflow heat exchanger that served as the test section. The resulting friction factor correlation was as follows:

$$f = 498.3Re_{eq}^{-1.074} \tag{6.59}$$

It should be noted that this is considerably different from the results of Yang and Webb cited above; however, Yan and Lin's single-phase friction factors were significantly higher than the predictions of the Blasius equation. They attributed this to the influence of entrance lengths and tube roughness, but this discrepancy may explain the substantial differences between the two studies.

Despite being developed from boiling data, the Tran et al. (2000) correlation has often been applied to condensation pressure drop. The correlation was developed for pressure drop during boiling of refrigerants in circular (2.46 and 2.92 mm) and rectangular (4.06 × 1.7 mm) channels. Their tests did not record local pressure drops—the pressure drop was measured from the inlet at nominally zero quality to a varying exit quality. For the tests with large exit qualities, therefore, the measured pressure drops could be the result of considerably varying local pressure drops as the flow regimes change along the test section. Although they found that the trends in the data (in terms of the effect of mass flux, quality, and saturation pressure) were similar to those found in large tubes (Eckels et al., 1994), the large tube correlations (Chisholm 1967, 1973; Friedel, 1979; Jung and Radermacher, 1989; Souza and Pimenta, 1995) typically underpredicted the data, with the discrepancy increasing at large qualities and mass fluxes. The reason for the higher-pressure drop in small tubes was thought to be the fact that coalesced bubbles in small channels are confined, elongated, and slide over a thin liquid film, whereas in large tubes, the bubbles may grow and flow unrestricted through the tubes. In view of this, they proposed a modified version of the Chisholm (1973) correlation that accounted for the role of surface tension through the confinement number introduced by Cornwell and Kew (1993) as follows:

$$\Delta P_f = \Delta p_{fLO}\{1 + (4.3Y^2 - 1)[N_{conf}x^{0.875}(1-x)^{0.875} + x^{1.75}]\} \tag{6.60}$$

where Y is the ratio of the gas- and liquid-only pressure drops.

Garimella et al. developed experimentally validated models for pressure drops during condensation of refrigerant R-134a in intermittent flow through circular (Garimella et al., 2002) and noncircular (Garimella et al., 2003b) microchannels (Figure 6.21) with $0.4 < D_H < 4.9$ mm. In addition, they developed a

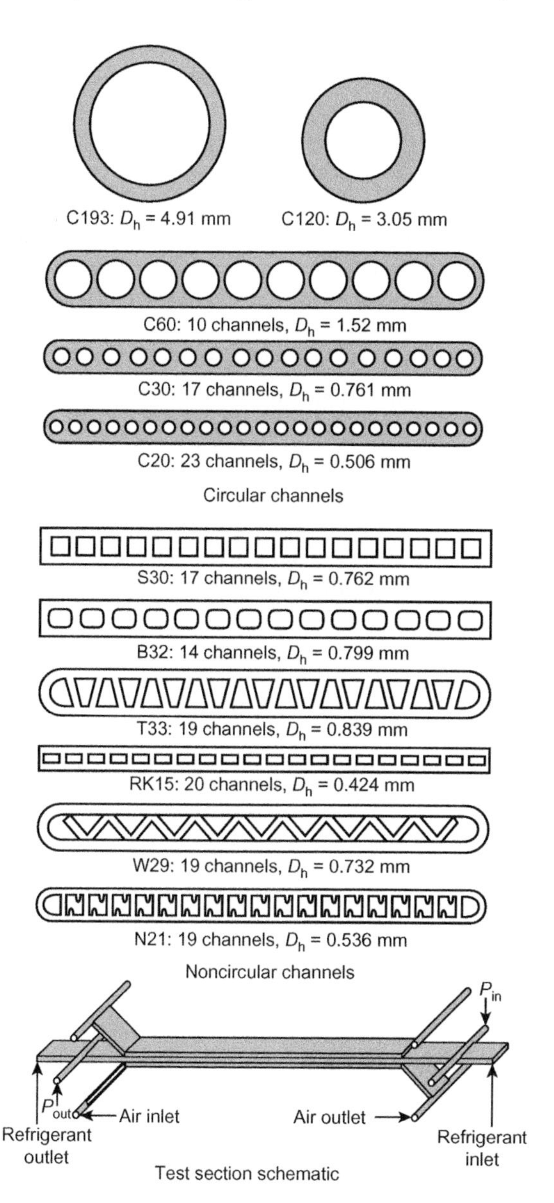

FIGURE 6.21

Refrigerant condensation geometries of Garimella.

Source: *From Garimella et al. (2005).*

model for condensation pressure drop in annular flow, and further extended it to a comprehensive multi-regime pressure drop model (Garimella et al., 2005) for microchannels for the mass flux range $150 < G < 750\ \mathrm{kg/m^2\,s}$. For $D_H < 3.05$ mm, they used flat tubes with multiple extruded parallel channels to ensure accurately measurable flow rates and heat balances. Three such tubes were brazed together (Figure 6.21), with refrigerant flowing through the center tube, and coolant (air) flowing in counterflow through the top and bottom tubes. The low thermal capacity and heat transfer coefficients of air maintained small changes in quality in the test section, which in turn enabled the measurement of the pressure drop variation as a function of quality with high resolution. The measurement techniques were first verified by conducting single-phase pressure drop measurements for each tube in the laminar and turbulent regimes for both the superheated vapor and subcooled liquid cases. The single-phase pressure drops were in excellent agreement with the values predicted by the Churchill (1977) correlation.

The measured pressure drop included expansion and contraction losses due to the headers at both ends of the test section, and the pressure change due to deceleration caused by the changing vapor fraction as condensation takes place, and was represented as follows:

$$\Delta P_{\mathrm{measured}} = \Delta P_{\mathrm{frict}} + \Delta P_{\mathrm{exp+contr}} + \Delta P_{\mathrm{decel}} \tag{6.61}$$

The pressure drop due to contraction was estimated using a homogeneous flow model recommended by Hewitt et al. (1994):

$$\Delta P_{\mathrm{contr}} = \frac{G^2}{2\rho_{\mathrm{L}}}\left[\left(\frac{1}{C_c} - 1\right)^2 + 1 - \gamma^2\right]\left[1 + x\left(\frac{\rho_{\mathrm{L}}}{\rho_{\mathrm{G}}} - 1\right)\right] \tag{6.62}$$

where γ is the area ratio ($A_{\text{test-section}}/A_{\text{header}}$) and C_c is the coefficient of contraction, which in turn is a function of this area ratio as given by Chisholm (1983):

$$C_c = \frac{1}{0.639(1-\gamma)^{0.5} + 1} \tag{6.63}$$

For the expansion into the header from the test section, the following separated flow model recommended by Hewitt et al. (1994) was used:

$$\Delta P_{\mathrm{expansion}} = \frac{G^2\gamma(1-\gamma)\psi_{\mathrm{S}}}{\rho_{\mathrm{L}}} \tag{6.64}$$

where ψ_{S}, the separated flow multiplier, is also a function of the phase densities and the quality. These estimates were validated using pressure drop measurements on a "near-zero" length test section, in which the measured pressure drops (with small frictional pressure drops) showed excellent agreement with the contraction/expansion contributions calculated as shown above. Contraction and expansion loss contributions were less than 5% of the total measured pressure drop for all their data. The pressure change due to acceleration (deceleration) of the fluid (due to the change in quality across the test section) was estimated as follows (Carey, 2008):

$$\Delta P_{\text{acceleration}} = \left[\frac{G^2x^2}{\rho_V\alpha} + \frac{G^2(1-x)^2}{\rho_L(1-\alpha)}\right]_{x=\text{Xout}} - \left[\frac{G^2x^2}{\rho_V\alpha} + \frac{G^2(1-x)^2}{\rho_L(1-\alpha)}\right]_{x=\text{Xin}} \tag{6.65}$$

where the void fraction was evaluated using the Baroczy (1965) correlation. For almost all their data, the deceleration term was extremely small compared to the overall pressure drop. These estimates were also corroborated by adiabatic flow and condensation tests at the same nominal conditions. The residual frictional component of the two-phase pressure drop, which generally was at least an order of magnitude larger than these minor losses, was used for developing condensation pressure drop models for the respective flow regimes.

To develop flow-regime-based pressure drop models, transition from the intermittent to the other flow regimes (based on their earlier flow visualization work) was determined as follows:

$$x \leq \frac{a}{G+b} \tag{6.66}$$

where a and b are geometry-dependent constants given by:

$$a_I = 69.57 + 22.60 \times \exp(0.259 \times D_H) \tag{6.67}$$

$$b_I = -59.99 + 176.8 \times \exp(0.383 \times D_H) \tag{6.68}$$

where D_H is in millimeters. On a mass flux versus quality map, these transition lines appear in the lower left corner, as shown in Figure 6.22. For pressure drop model development, they broadly categorized the flow into the primary regimes,

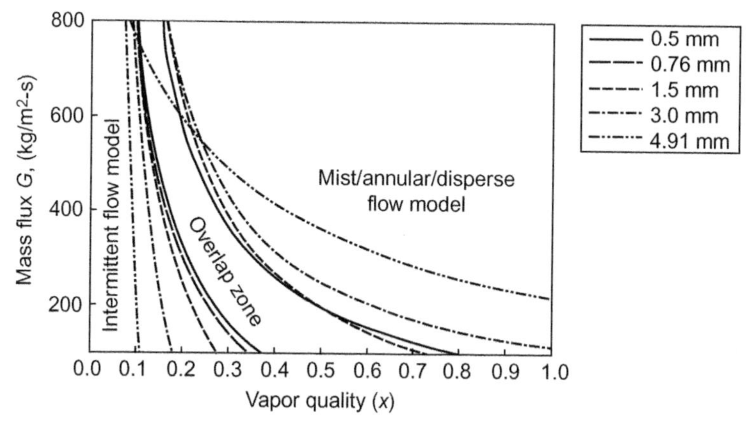

FIGURE 6.22

Flow regime assignment.

Source: *From Garimella et al. (2005).*

intermittent (lower left) and annular film/mist/disperse patterns of the annular flow regime (upper right). The very few data that were in the wavy regime in these small channels were incorporated into the annular or intermittent regimes, as appropriate. The overlap region in Figure 6.22 accounts for the gradual transition between the primary regimes, during which the flow switches back and forth between the respective regimes. For transition from other flow regimes to the annular flow regime, Garimella and Fronk (2012) reported the following criteria:

$$G \geq a_A + \frac{b_A}{x} \tag{6.69}$$

The coefficients a_A and b_A are functions of the hydraulic diameter, and are determined from a linear interpolation of experimental data given in Table 6.6. Pressure drop models for the intermittent, annular, and transition regimes are described below.

The unit cell for the development of the intermittent model is shown in Figure 6.23, in which the vapor phase travels as long solitary bubbles surrounded

Table 6.6 Annular Flow Transition Criteria

D_H (m)	a_A	b_A
0.00493	−21	135.2
0.00305	−21	135.2
0.002	−140.3	165.2
0.001	−56.3	125.4
0.0001	−56.3	125.4

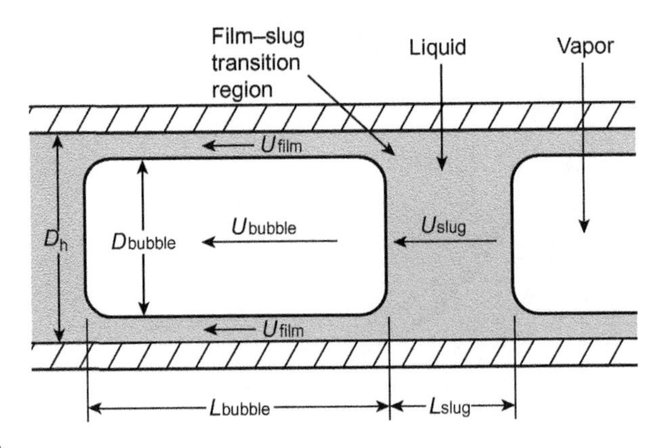

FIGURE 6.23

Unit cell for intermittent flow.

Source: *Reprinted from Garimella et al. (2002) with permission from the American Society of Mechanical Engineers.*

by an annular liquid film and separated by liquid slugs. Based on the recorded flow patterns, it was assumed that the bubble is cylindrical, and that there is no entrainment of vapor in the slug, or liquid in the bubble. Further, for any given condition, it was assumed that the length/frequency/speed of bubbles/slugs is constant, with no bubble coalescence, and a smooth bubble/film interface. The total pressure drop for this flow pattern includes contributions from: the liquid slug, the vapor bubble, and the flow of liquid between the film and slug as follows:

$$\Delta P_{\text{total}} = \Delta P_{\text{slug}} + \Delta P_{f/b} + \Delta P_{\text{transitions}} \tag{6.70}$$

A simple control volume analysis (Garimella et al., 2002, 2003b) similar to that performed by Suo and Griffith (1964) was used to calculate the liquid slug velocity from the superficial phase velocities, in turn calculated from the overall mass flux and the quality. Thus, $U_{\text{slug}} = j_{\text{V}} + j_{\text{L}}$, which yielded the slug Reynolds number $Re_{\text{slug}} = (\rho_{\text{L}} U_{\text{slug}} D_{\text{H}})/\mu_{\text{L}}$. This Re_{slug} was used to calculate the pressure gradient in the slug from an appropriate single-phase friction factor correlation from the literature. The results of several investigations (Dukler and Hubbard, 1975; Fukano et al., 1989; Suo and Griffith, 1964) discussed in this chapter suggested that $U_{\text{bubble}} = 1.2U_{\text{slug}}$. Unlike air–water studies, where many of these investigators have neglected the pressure drop in the film–bubble region, since the ratio of the liquid to vapor densities for R-134a is much lower (approximately 16:1 versus 850:1 for air–water), as is the ratio of liquid and vapor viscosities (approximately 10:1 versus 50:1), the contribution of the film/bubble region was not neglected. Based on the data, the film was assumed to be laminar, while the gas bubble was turbulent. The respective Reynolds numbers were defined as follows:

$$Re_{\text{film}} = \frac{\rho_{\text{L}} U_{\text{film}} (D_h - D_{\text{bubble}})}{\mu_{\text{L}}} \qquad Re_{\text{bubble}} = \frac{\rho_{\text{V}} (U_{\text{bubble}} - U_{\text{interface}}) D_{\text{bubble}}}{\mu_{\text{V}}} \tag{6.71}$$

The film flow was assumed to be driven by the combination of the pressure gradient in the film/bubble region and shear at the film/bubble interface. The velocity profile for combined Couette–Poiseuille flow through an annulus where the inner surface moves at the interface velocity, $U_{\text{interface}}$, is represented by the superposition of the pressure-driven and the shear-driven components:

$$u_{\text{film}}(r) = \frac{-(\mathrm{d}P/dx)_{f/b}}{4\mu_{\text{L}}} \left[R_{\text{tube}}^2 - r^2 - (R_{\text{tube}}^2 - R_{\text{bubble}}^2) \frac{\text{In}(R_{\text{tube}}/r)}{\text{In}(R_{\text{tube}}/R_{\text{bubble}})} \right] + U_{\text{interface}} \frac{\text{In}(R_{\text{tube}}/r)}{\text{In}(R_{\text{tube}}/R_{\text{bubble}})} \tag{6.72}$$

The bubble velocity, assuming a power law profile ($n = 7$), is given by:

$$u_{\text{bubble}}(r) = \frac{(n+1)(2n+1)}{2n^2} (U_{\text{bubble}} - U_{\text{interface}}) \left(1 - \frac{r}{R_{\text{bubble}}}\right)^{1/n} + U_{\text{interface}} \tag{6.73}$$

A shear balance at the film–bubble interface yields the interface velocity:

$$U_{\text{interface}} = \frac{-(\mathrm{d}P/dx)_{f/b}}{4\mu_{\text{L}}}(R_{\text{tube}}^2 - R_{\text{bubble}}^2) \tag{6.74}$$

With a parabolic velocity profile in the film, the average film velocity can also be calculated. Also, with the film pressure gradient calculated using a single-phase friction factor for the gas phase, continuity yields:

$$U_{\text{slug}} = U_{\text{bubble}}\left(\frac{R_{\text{bubble}}}{R_{\text{tube}}}\right)^2 + U_{\text{film}}\left(1 - \left(\frac{R_{\text{bubble}}}{R_{\text{tube}}}\right)^2\right) \tag{6.75}$$

These calculations yield $0.899 < R_{\text{bubble}}/R_{\text{tube}} < 0.911$; that is, the bubble diameter is ~90% of the tube diameter. With the pressure drop/length known in the slug and the film/bubble regions, the frictional pressure drop was calculated using the relationship proposed by Fukano et al. (1989) for the length of the slug (Eq. (6.76)):

$$\Delta P_{\text{friction only}} = L_{\text{tube}}\left[\left(\frac{\mathrm{d}P}{\mathrm{d}x}\right)_{f/b}\left(1 - \frac{L_{\text{slug}}}{L_{\text{slug}} + L_{\text{bubble}}}\right) + \left(\frac{\mathrm{d}P}{\mathrm{d}x}\right)_{\text{slug}}\left(\frac{L_{\text{slug}}}{L_{\text{slug}} + L_{\text{bubble}}}\right)\right] \tag{6.76}$$

The remaining contribution to the total pressure drop is the loss associated with the flow of liquid between the film and the slug. As the film moves slowly, the front of the liquid slug is constantly taking up fluid from the film. A pressure loss is associated with the acceleration and subsequent mixing of this liquid, with the total pressure loss from these transitions given by $\Delta P_{\text{film/slug transitions}} = N_{\text{UC}} \times \Delta P_{\text{one transition}}$, where N_{UC} is the number of cells per unit length. The pressure drop for one transition is based on the force to accelerate the portion of the liquid film overtaken by the rear of the bubble from the average film velocity to the average slug velocity (Dukler and Hubbard, 1975):

$$\Delta P_{\text{one transtion}} = \rho_{\text{L}}\left(1 - \left(\frac{R_{\text{bubble}}}{R_{\text{tube}}}\right)^2\right)(U_{\text{slug}} - U_{\text{film}})(U_{\text{bubble}} - U_{\text{film}}) \tag{6.77}$$

These components of the total pressure drop are shown below:

$$\frac{\Delta P}{L} = \left(\frac{\mathrm{d}P}{\mathrm{d}x}\right)_{\text{film bubble}}\left(\frac{L_{\text{bubble}}}{L_{\text{UC}}}\right) + \left(\frac{\mathrm{d}P}{\mathrm{d}x}\right)_{\text{slug}}\left(\frac{L_{\text{slug}}}{L_{\text{UC}}}\right) + \Delta P_{\text{one transition}}\left(\frac{N_{\text{UC}}}{L}\right) \tag{6.78}$$

A depiction of these various contributions to the measured pressure drop for circular channels is shown in Figure 6.24. The number of unit cells per unit length

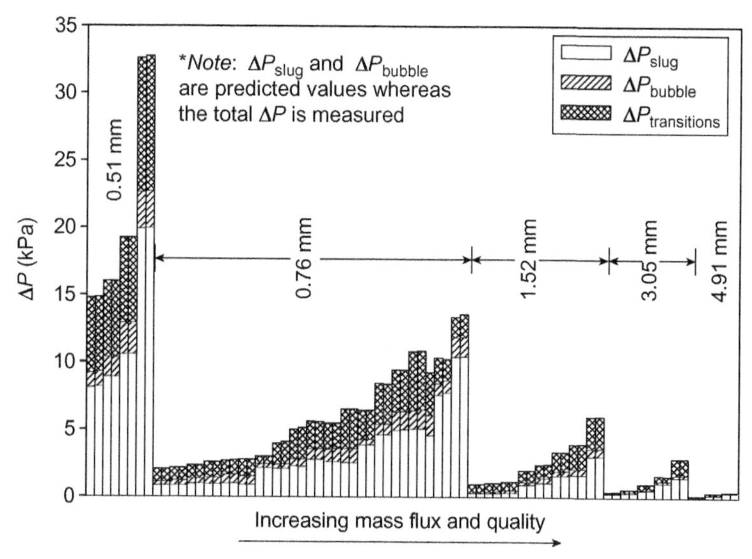

FIGURE 6.24

Contribution of each pressure drop mechanism to total pressure drop.

Source: *Reprinted from Garimella et al. (2002) with permission from the American Society of Mechanical Engineers.*

required for closure of the model was computed from the slug frequency w, which was correlated based on slug Re and D_H:

$$a(Re_{\text{slug}})^b = \omega \frac{D_h}{U_{\text{bubble}}} = D_H \left(\frac{N_{UC}}{L_{\text{tube}}}\right) = \left(\frac{D_H}{L_{UC}}\right) \tag{6.79}$$

The coefficients a and b were fit using the difference between the measured pressure drop and the pressure drop calculated as described above for the slug and bubble/film regions, that is, the net pressure drop due to transitions. The correlation yielded $a = 2.437$, $b = -0.560$ for both circular and noncircular (except triangular, for which different coefficients were necessary, as discussed in Garimella et al., 2003b) channels. For circular channels (0.5–4.91 mm), the predicted pressure drops were on average within $\pm$ 13.5% of the measured values.

The intermittent model was also extended to the discrete wave flow region by Garimella et al. (2005). This is because as the progression from the intermittent region to the discrete wave region occurs, the gas bubbles start disappearing, to be replaced by stratified, well-defined liquid and vapor layers. Within the discrete wave region, traversing from the intermittent flow boundary toward the annular flow boundary, the bubbles disappear completely, with the number of unit cells per unit length approaching zero. Based on this conceptualization of the intermittent and discrete flow regions, the slug frequency model developed by

Garimella et al. (2002, 2003b) for intermittent flow was modified to include data from the discrete wave flow region, resulting in the following combined model for the two regions:

$$N_{UC}\left(\frac{D_H}{L_{tube}}\right) = \left(\frac{D_H}{L_{UC}}\right) = 1.573(Re_{slug})^{-0.507} \tag{6.80}$$

Garimella et al. (2005) also developed pressure drop models for annular flow condensation in microchannels, assuming equal pressure gradients in the liquid and gas core at any cross-section, uniform film thickness, and no entrainment of the liquid in the gas core. The measured pressure drops were used to compute the Darcy form of the interfacial friction factor to represent the interfacial shear stress as follows:

$$\frac{\Delta P}{L} = \frac{1}{2} \times f_i \rho_g V_g^2 \times \frac{1}{D_i} \tag{6.81}$$

The above equation uses the interface diameter, D_i. This same expression can be represented in terms of the more convenient tube diameter, D, by using the Baroczy (1965) void fraction model:

$$\frac{\Delta P}{L} = \frac{1}{2} \times f_i \frac{G^2 \times x^2}{\rho_g \times \alpha^{2.5}} \times \frac{1}{D} \tag{6.82}$$

The ratio of this interfacial friction factor (obtained from the experimental data) to the corresponding liquid-phase Darcy friction factor was then computed and correlated in terms of the Martinelli parameter and the surface tension parameter $\psi = j_L \mu_L/\sigma$ introduced by Lee and Lee (2001):

$$\frac{f_i}{f_l} = A \times X^a Re_l^b \psi^c \tag{6.83}$$

where $j_L = ((G(1-x))/(\rho_l(1-\alpha)))$ is the mean liquid velocity. The phase Reynolds numbers required above and for the Martinelli parameter were defined in terms of the flow areas of each phase:

$$Re_l = \frac{GD(1-x)}{(1+\sqrt{\alpha})\mu_l} Re_g = \frac{GxD}{\mu_g\sqrt{\alpha}} \tag{6.84}$$

The friction factors were computed using $f = 64/Re$ for $Re < 2100$ and the Blasius expression $f = 0.316 \times Re^{-0.25}$ for $Re > 3400$. Regression analysis on data grouped into two regions based on Re_l yielded:

$$\text{Laminar region}(Re_l < 2100): A = 1.308 \times 10^{-3} \quad a = 0.427 \quad b = 0.930 \quad c = -0.021 \tag{6.85}$$

$$\text{Turbulent region}(Re_l < 3400): A = 25.64 \quad a = 0.532 \quad b = -0.327 \quad c = 0.021 \tag{6.86}$$

Interpolations based on G and x are used to compute the pressure drop for the transition region. This model predicted 87% of the data within ±20%. These predictions include not only the annular flow region, but also the mist and disperse-wave flow data, although the model development uses a physical representation where the liquid forms an annular film around a gas core. Garimella et al. (2005) explained this based on the fact that the annular flow regime increases in extent for smaller tubes, causing several data classified into the wavy and mist flow regions using $D_H = 1$ mm transition criteria to be in fact in the annular flow regime. Representative pressure drop trends predicted by this model are shown in Figure 6.25.

Agarwal (2006) extended the multi-regime pressure drop model concept to channels with $D_H < 0.16$ mm. Experiments were conducted in rectangular microchannels ($0.1 < D_H < 0.16$, $1 < AR < 4$) with condensing R-134a. Like the Garimella et al. (2005) model, he considered pressure drop contributions from the vapor bubble, liquid slug, and film/bubble transition sections. Thus, the total pressure drop was modeled using Eq. (6.78), as before. The primary difference for these much smaller D_H channels was that Agarwal's model used an intermittent flow approach for the entire data set, where the annular flow regime is treated as intermittent flow with an infinitely long bubble or with negligible slug length compared to the bubble length. He proposed a new model for the number of unit cells as a function of slug Reynolds number, quality, density ratio, and channel aspect ratio:

$$N_{UC} \cdot \frac{D}{L_{tube}} = (2.8 \cdot e^{0.4 \cdot AR}) \cdot Re_{slug}^{-0.35} \cdot \left(\frac{1-x}{x}\right)^{0.46} \cdot \left(\frac{\rho_V}{\rho_L}\right)^{0.868} \tag{6.87}$$

The most interesting aspect of the model is that it achieves a smooth transition from intermittent-to-annular flow, where the length of the bubble region increases to infinity as the flow becomes more annular. This direct transition from annular to intermittent is unique to microchannel flows. To quantify the progression from intermittent-to-annular flow, he proposed a new term, the annular flow factor (AFF), shown in Eq. (6.88).

$$AFF = \sqrt{\left(1 - \frac{L_{slug}}{L_{slug} + L_{bubble}}\right) \cdot \frac{1}{N_{UC}}} \tag{6.88}$$

The two parameters that play an important role in determining whether the flow tends to intermittent or annular flow are the number of unit cells (N_{UC}) and the slug length ratio (SLR). Both the N_{UC} and SLR decrease with increasing quality and decreasing saturation temperature due to an increase in void fraction. AFF = 1 represents fully annular flow, while AFF = 0 represents fully intermittent (slug) flow.

Andresen (2007) conducted near-critical-pressure ($0.8 < P_r < 1.0$) condensation pressure drop experiments with R-404A and R-410A in tubes ranging in diameter

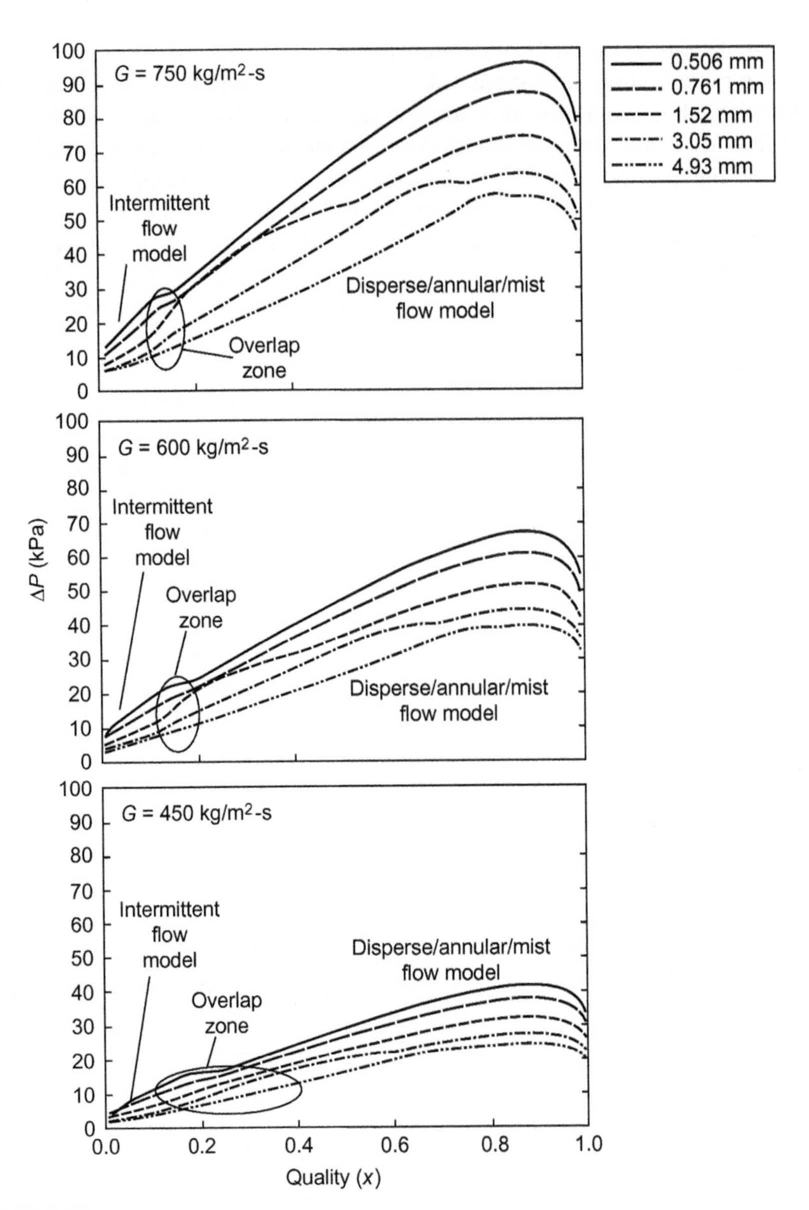

FIGURE 6.25

Predictions of Garimella et al. (2005) multiple-flow regime model for refrigerant condensation pressure drops in microchannels.

Source: *From Garimella et al. (2005).*

from $0.76 < D < 9.4$ mm. He found that the model of Garimella et al. (2005) was not able to accurately account for the small vapor–liquid property differences in the near-critical region. Thus, he proposed a continuous model for the Chisholm parameter C as a function of liquid Reynolds number and confinement number.

$$\begin{aligned} C &= a \cdot Re_{\mathrm{L}}^{b} \cdot N_{\mathrm{conf}}^{c} \\ a &= 24 \\ b &= -0.3 \\ c &= -0.4 \end{aligned} \tag{6.89}$$

The confinement number is intended to capture the transition from gravity to surface-tension-dominated forces as the channel size decreases. The two-phase pressure drop multiplier could be then be calculated from the following equation, where X is the Martinelli parameter:

$$\Phi_{\mathrm{L}}^{2} = 1 + \frac{C}{X} + \frac{1}{X^{2}} \tag{6.90}$$

Additional pressure drop studies on condensation of synthetic and natural refrigerants in mini- and microchannels have been conducted by Cavallini, Del Col, and co-workers (2005a,b,2006b, 2009); Del Col et al., 2010. They first presented an empirical correlation similar in form to the Friedel (1979) correlation, for pressure gradient in annular flow of refrigerants in minichannels. They attempted to account for liquid entrained in the vapor core through the entrainment ratio proposed by Paleev and Filippovich:

$$\begin{aligned} E &= 0.015 + 0.44 \cdot \log\left((\rho_{\mathrm{GC}}/\rho_{\mathrm{L}}) \cdot \left(\frac{\mu_{\mathrm{L}} \cdot j_{\mathrm{V}}}{\sigma} \right)^{2} \cdot 10^{4} \right) \\ \rho_{\mathrm{GC}} &= \rho_{\mathrm{V}} \left[1 + (1-x) \cdot \frac{E}{x} \right] \end{aligned} \tag{6.91}$$

Data from several studies on condensation of refrigerants in mini- and microchannels were then correlated, resulting in the expression in Eq. (6.92) for the two-phase multiplier as a function of quality, fluid properties, and reduced pressure.

$$\begin{aligned} \Phi_{\mathrm{LO}}^{2} &= Z + 3.595 \cdot F \cdot H \cdot (1-E)^{W} \\ Z &= (1-x)^{2} + x^{2} \left(\frac{\rho_{\mathrm{L}}}{\rho_{\mathrm{V}}} \right) \cdot \left(\frac{\mu_{\mathrm{V}}}{\mu_{\mathrm{L}}} \right)^{0.2} \\ F &= x^{0.9525} \cdot (1-x)^{0.414} \\ H &= \left(\frac{\rho_{\mathrm{L}}}{\rho_{\mathrm{V}}} \right)^{1.132} \cdot \left(\frac{\mu_{\mathrm{V}}}{\mu_{\mathrm{L}}} \right)^{0.44} \left(1 - \frac{\mu_{\mathrm{V}}}{\mu_{\mathrm{L}}} \right)^{3.542} \\ W &= 1.398 \cdot P_{\mathrm{r}} \end{aligned} \tag{6.92}$$

They investigated adiabatic frictional pressure drop in a rectangular multiport minichannel tube ($D_H = 1.4$ mm, $L = 1.13$ m) (Cavallini et al., 2006b). Data were obtained for mass fluxes ranging from 200 to 1400 kg/m^2 s at a saturation temperature of 40°C. The refrigerants were selected to span a range of reduced pressures, high ($P_r = 0.49$ for R-410a), medium ($P_r = 0.25$ for R-134a), and low ($P_r = 0.14$ for R-236ea). They found that most correlations were not able to predict the frictional pressure gradient for this range of fluids and operating conditions for the multiport tube. However, they showed that the model given in Eqs. (6.90) and (6.91) was able to predict the data over the reduced pressure range from 0.2 to 0.5 satisfactorily. More recently, they have obtained additional data for R-134a (Cavallini et al., 2009) and R-1234yf (Del Col et al., 2010) in a single circular channel with $D = 0.96$ mm at mass fluxes ranging from 200 to 1000 kg/m^2 s.

6.4.5 Summary observations and recommendations

The available information on pressure drop in two-phase flows was discussed above. One overall observation is that most of the studies continue to rely on the classical works, or modifications of Lockhart and Martinelli (1949), Chisholm (1967), and Friedel (1979) for the evaluation of separated flow pressure drops. As in the case of flow regime and void fraction evaluation, most of the work in the small channels has been on adiabatic flows of air–water mixtures. It appears that the original two-phase multipliers proposed in these studies have been modified by the latter investigators (specifically the C coefficient in the Chisholm (1967) model of the Lockhart–Martinelli (1949) multiplier) to match the data on the smaller channels. This indicates that of the liquid-phase, gas-phase, and gas–liquid interaction parameters, the gas–liquid interaction parameter has required the most modification to make it suitable for small channels. Examples include the work of Mishima and Hibiki (1996a) and Lee and Lee (2001). As the channel approaches the smaller dimensions, Kawahara et al. (2002), Chung and Kawaji (2004), and others have found that the pressure drop predictions in the 530- and 250-μm channels and those for 100- and 50-μm channels required different mixture viscosity models to make the homogeneous flow models work. The standard two-phase flow multiplier approach with much smaller values of the C coefficient also yielded some success. However, it seems that for the plug/slug flows in the small channels, an approach that recognizes the intermittent nature of the flow, and accounts for the pressure drops in the liquid slug, the gas bubble, and, in some cases, the mixing losses between the bubbles and slugs, as proposed by Garimella et al. (2002, 2003b) and later adapted by Chung and Kawaji (2004) is the best approach. Furthermore, the vastly different properties of different fluids and operating conditions (i.e., high versus low reduced pressure) need to be better accounted for in a general model for predicting pressure drop. In addition, accounting for the range of flow regimes encountered in microchannels, that is, intermittent, annular, and to a certain extent mist flow, has just recently been attempted. One multi-region model that accounts for the flow mechanisms in each of these regimes and results in smooth transitions between these regimes for condensing flows in the $0.4 < D_H < 5$ mm range is the model by Garimella et al. (2005, 2006) and for channels with $D_H < 0.4$ mm by Agarwal (2006).

The various techniques for predicting pressure drops in microchannels are illustrated in Example 6.3.

6.5 Heat transfer coefficients

In the above sections, the available information on flow regimes, void fractions, and pressure drop in two-phase flow was presented, including adiabatic flow of air–water mixtures and condensing flows. This is because the large body of work on adiabatic two-phase flows can be used as a starting point to understand the corresponding phenomena during phase change. However, for the evaluation of heat transfer coefficients, only the literature on condensing flows is relevant. The measurement and understanding of heat transfer in really small channels ($D_H < 1$ mm) has proved to be particularly challenging in view of the confined spaces for measurement probes, the small flow rates in microchannels that correspond to small heat transfer rates that are therefore difficult to measure accurately, and the large heat transfer coefficients that lead to very small temperature differences that are also difficult to measure. There would be little work to report, indeed, if the scope is restricted to channels with $D_H < 1$ mm. Therefore, in this section, an overview of the classical models of condensation is presented first to serve as a background for the work on the relatively smaller channels. Gravity-driven condensation is expected to be less important at the microscale. Thus, the focus will be on shear-driven and multi-regime models in macrochannels, followed by a presentation of the limited information on channels in the 400 μm to 3 mm range, which have loosely been referred to as microchannels by the air-conditioning and refrigeration industry. The focus here is on condensation inside *horizontal* tubes (Table 6.7).

6.5.1 Conventional channel models and correlations

6.5.1.1 Shear-driven condensation

Carpenter and Colburn (1951) obtained data for local and average heat transfer coefficients for the condensation of steam, methanol, ethanol, toluene, and trichloroethylene inside a vertical 11.66-mm ID tube. The data were compared with the Nusselt falling film equation, which underpredicted the values due to the influence of vapor shear. Carpenter and Colburn stated that the thin condensate layer for annular flow might become turbulent at very low liquid Reynolds numbers (~240). They considered vapor shear, gravity, and the momentum change due to condensation acting upon this condensate layer. The major thermal resistance was considered to be due to the laminar sublayer. Using the von Karman universal

Table 6.7 Summary of Heat Transfer Studies

Investigator	Hydraulic Diameter (mm)	Fluids	Orientation/ Condition	Range/Applicability	Techniques, Basis, Observations
Conventional channel models and correlations: shear-driven condensation					
Carpenter and Colburn (1951)	11.66 mm	Steam, methanol, ethanol, toluene, trichloroethylene	Vertical		• Data for local and average h • Thin condensate film becomes turbulent at $Re_1 \sim 240$ • Considered vapor shear, gravity, and momentum change due to condensation acting upon condensate layer • Major thermal resistance in laminar sublayer • Von Karman velocity profile for liquid film to obtain local h as function of film thickness and sum of forces on condensate • Condensation divided into: laminar condensate entrance region, turbulent condensate, laminar vapor exit (Nusselt analysis)
Akers et al. (1959)				$Re_l > 5000$ $Re_v > \left(\frac{\mu_v}{\mu_1}\right)\left(\frac{\rho_1}{\rho_v}\right)^{0.5} > 20,000$	• Equivalent all-liquid mass flux to replace vapor core, and provided same h for condensing annular flow • Enables single-phase flow treatment
Akers and Rosson (1960)	15.88 mm	Methanol, R-12	Horizontal	$1000 < \frac{DG_v}{\mu_1}\left(\frac{\rho_1}{\rho_v}\right)^{0.5}$	• Three different experimentally validated correlations for "semi-stratified" flow, laminar, and turbulent annular flow • Multipliers to single-phase Nu based on data

Soliman et al. (1968)	$7.44 < D < 11.66$ mm	Steam, R-113, ethanol, methanol, toluene, trichloroethylene, R-22	Horizontal vertical downward	$1 < Pr < 10$ $6 < U_v < 300$ m/s $0.03 < x < 0.99$	• Modifications to Carpenter and Colburn (1951) shear stress • Wall shear stress combination of friction, momentum, and gravity contributions • Wall shear stress used to develop model for h for annular flow • Parametric analyses to highlight importance of different forces
Traviss et al. (1973)	8-mm tube	R-12, R-22	Horizontal	$T = 25–58.3°C$ $161 < G < 1533$ kg/m^2 s	• Heat–momentum analogy and von Karman velocity profile in liquid film to correlate Nu in annular flow • Frictional term from two-phase multiplier • Three-region turbulent film model
Shah (1979)	$7 < D < 40$ mm	R-11, R-12, R-22, R-113, methanol, ethanol, benzene, toluene, trichloroethylene	Horizontal, inclined and vertical tubes, annulus	$11 < G < 211$ kg/m^2 s $21 < T < 310°C$ $3 < U_v < 300$ m/s $0.002 < P_r < 0.44$ $1 < P_n < 13$ $Re_1 > 350$	• Widely used general purpose empirical correlation • Large database from 21 sources • Basis: condensation h similar to evaporative h in absence of nucleate boiling • Two-phase multiplier in terms of p_r and x
Soliman (1986)	$7.4 < D < 12.7$ mm	Steam, R-113, R-12	Horizontal vertical	$21 < T < 310°C$ $80 < G < 1610$ kg/m^2 s $0.20 < x < 0.95$ $8 < We < 140$ $1 < P_n < 7.7$	• Correlation for annular-mist flow • Heat–momentum analogy for single-phase turbulent flow used as starting point • Homogeneous flow in thermodynamic equilibrium Mixture viscosity basis

(Continued)

Table 6.7 (Continued)

Investigator	Hydraulic Diameter (mm)	Fluids	Orientation/ Condition	Range/Applicability	Techniques, Basis, Observations
Chen et al. (1987)			Co-current annular film in vertical and horizontal tube		• Film condensation correlation based on analytical and empirical approach to asymptotically combine values at limits • Interfacial shear, interface waviness, and turbulent transport • Included effects of interfacial shear, interface waviness, and turbulent transport • Average and local *Nu* from asymptotic limits
Tandon et al. (1995)	10 mm	R-12, R-22	Horizontal	$20 < T < 40°C$ $175 < G < 560$ kg/m^2 s	• At same G_v, h for R-22 higher than for R-12 • h trends slope change at $Re_v \sim 3 \times 10^4$; based on transition, proposed wavy, and annular/semi-annular flow correlations
Chitti and Anand (1995, 1996)	8 mm	R-22, R-32/R-125 with oil	Horizontal	$32.2 < T < 51.7°C$ $148 < G < 450$ kg/m^2 s	• Analytical model for annular condensation • Used Prandtl mixing length theory with van Driest's (1956) hypothesis (law of wall) to obtain ε_m, $Pr_t = 0.9$ • h for R-32/R-125 15% higher than for R-22, effect of oil minimal

Moser et al. (1998)	$3.14 < D < 20$ mm	R-22, R-134a, R-410A, R-12, R-125, R-11			• Akers et al. (1959) G_{eq} model consistently underpredicts data • Akers assumptions of constant and equal f_v, f_1 not correct, also ΔT of Akers et al. not correct • New Re_{eq} with Friedel (1979) ϕ_{lo}^2 boundary-layer analysis to obtain equivalent Nu
Hurlburt and Newell (1999)	3–10 mm	R-22, R-134a, R-11, R-12, R-22	Horizontal	$0.2 < x < 0.9$ $200 < G <$ 650 kg/m^2 s	• Scaling equations to translate between α, ΔP, h for different refrigerants at different conditions and geometries
Conventional channel models and correlations: multi-regime condensation					
Jaster and Kosky (1976)	12.5 mm	Steam	Horizontal	$2 \times 10^4 < P < 1.7 \times 10^5$ $12.6 < G < 145$ kg/m^2 s	• Algebraic expression for film thickness facilitates computations • Stress ratio to distinguish between annular/stratified regimes • Heat–momentum transfer analogy (annular flow) with two-phase multiplier for shear stress • For stratified flows, related liquid pool angle to void fraction • Linear interpolation for transition region
Breber et al. (1979; 1980)	4.8–50.8 mm	R-11, R-12, R-113, steam, *k*-pentane	Horizontal	$108.2 < P < 1250$ kPa $17.6 < G < 990$ kg/m^2 s	• Forced-convective correlation with two-phase multiplier for annular, intermittent, and bubbly flow • Nusselt-type equation for stratified wavy • Interpolations in transition zones

(Continued)

Table 6.7 (Continued)

Investigator	Hydraulic Diameter (mm)	Fluids	Orientation/ Condition	Range/Applicability	Techniques, Basis, Observations
Nitheanandan et al. (1990)	7.4–15.9 mm	R-12, R-113, steam		Data from various sources	• Evaluation of correlations in literature • Mist flow ($We \geq 40$, $Fr \geq 7$): Soliman (1986) • Annular flow ($We < 40$, $Fr \geq 7$): Shah (1979) • Wavy flow ($Fr < 7$): Akers and Rosson (1960)
Dobson et al. (1994)	4.57 mm	R-12, R-134a	Horizontal	$75 < G < 500$ kg/m^2 s $T = 35°C$ and $60°C$	• Annular flow: h increased with G, x due to increased shear and thinning of liquid film—used two-phase multiplier • Wavy flow: h independent of G, slight increase with x—Chato (1962) with leading constant changed to function of X_{tt}
Dobson and Chato (1998)	31.4, 4.6, and 7.04 mm	R-12, R-134a, R-22, near-azeotropic blends of R-32/R-125	Horizontal	$25 < G < 800$ kg/m^2 s $T = 35–60°C$	• Extended Dobson et al. (1994)—additional tubes, fluids • Fluid properties effect not very significant • No significant effect of diameter reduction • Wavy and annular correlations updated, forced convection in liquid pool and film condensation at the top, void fraction for pool depth estimate • Fr_{so} and G used to separate regimes

Wilson et al. (2003)	$D = 7.79$, 6.37, 4.40, and 1.84	R-134a, R-410A	Horizontal	$T = 35°C$ $75 < G < 400$ kg/m^2 s	• Progressively flattened 8.91-mm round smooth/microfin tubes • h increases with flattening, h highest for helical microfins • No predictive models or correlations
Cavallini et al. (2001, 2002a)	8 mm	R-22, R-134a, R-125, R-32, R-236ea, R-407C, R-410A	Horizontal	$30 < T < 50°C$ $100 < G < 750$ kg/m^2 s	• Breber et al. (1979, 1980) type approach for k recommendations • Own data and database from others used for correlations • Kosky and Staub (1971) for annular h with modified Friedel (1979) for shear stress • Stratified: film condensation and forced convection in pool • Interpolations for other regimes, bubbly treated as annular • Mass flux criterion for wavy–slug transition
Thome et al. (2003)	$3.1 < D < 21.4$ mm	R-11, R-12, R-113, R-32/R-125, propane, *a*-butane, iso-butane, butane, propylene	Horizontal	$24 < G < 1022$ kg/m^2 s $0.02 < Pr < 0.8$ $0.03 < x < 0.97$	• Curve-fitting from large database • h data used to refine void fraction model, which is used to predict transitions, in turn used to predict tuned blend of stratified-convective heat transfer coefficients • Stratified pool spread out as partial annular film • Includes effect of interfacial waves on heat transfer • Smoothly varying h across regimes through interpolations • No new intermittent, bubbly, or mist models; treated as annular

(*Continued*)

Table 6.7 (Continued)

Investigator	Hydraulic Diameter (mm)	Fluids	Orientation/ Condition	Range/Applicability	Techniques, Basis, Observations
Goto et al. (2003)	Helical, herringbone grooved tubes, 8.00 mm	R-410A, R-22	Horizontal	$30 < T < 40°C$ $130 < G < 400$ kg/m^2 s	• h tests in complete vapor compression system • h of herringbone grooved tube $\sim 2 \times$ helical grooved tubes • Modified Koyama and Yu's (1996) correlation for stratified and annular condensation, new expressions for annular flow part
Condensation in small channels					
Yang and Webb (1996a)	Rectangular plain $D_H = 2.64$, microfin tubes $D_H = 1.56$ mm	R-12	Horizontal	$T = 65°C$ $400 < G < 1400$ kg/m^2 s	• Shah (1979) significantly overpredicted data, Akers et al. (1959) better agreement, except at high G • Enhancement due to microfins decreased with increasing G • h showed heat flux dependence • Surface tension drainage rationale for microfin enhancement • Yang and Webb (1996b) concluded surface tension did not play role in ΔP in these same tubes
Yang and Webb (1997)	Extruded microchannels ($D_H = 1.41$ and 1.56 mm) with microfins	R-12, R-134a			• $G = 400$ kg/m^2 s and $x > 0.5$, surface tension contribution equals and exceeds vapor shear term • $G = 1400$ kg/m^2 s, surface tension contribution very small • h model based on shear and surface tension drainage (flooded and unflooded parts) • Small fin tip radius enhanced h, large inter-fin drainage area activated surface tension effect at lower x

Kim et al. (2003)	$D_H = 1.41$ smooth, 1.56 mm microfin	R-22, R-410A	Horizontal	$T = 45°C$ $200 < G < 600$ kg/m^2 s	• In smooth tubes, R-410A h slightly higher than R-22, opposite true for microfin tubes • Many qualitative explanations for varying enhancement trends • Recommend Moser et al. (1998) model with modified two-phase multiplier for smooth tubes, and Yang and Webb's (1997) model with minor modifications for microfin tubes
Yan and Lin (1999)	2 mm	R-134a	Horizontal	$T = 40–50°C$	• Condensation h higher at lower T_{sat}, especially at higher x • h decreased significantly as q'' increased, particularly at high x; some unusual trends in data • Re_{eq} based h correlation
Wang et al. (2002)	Rectangular $D_H = 1.46$ mm (1.50 × 1.40 mm)	R-134a	Horizontal	$T = 61.5–66°C$ $75 < G < 750$ kg/m^2 s	• Large variations in x across test section • Stratification seen even at such small D_H • Akers et al. (1959) Re_{eq} agreed with annular data, Jaster and Kosky (1976) agreed best with stratified flow data • Breber et al. (1980) and Soliman (1982, 1986) transition criteria • Curve-fits for two-phase multiplier, dimensionless temperature • Stratified and annular flow correlations with interpolations

(*Continued*)

Table 6.7 (Continued)

Investigator	Hydraulic Diameter (mm)	Fluids	Orientation/ Condition	Range/Applicability	Techniques, Basis, Observations
Koyama et al. (2003b)	Multiport extruded Al tubes	R-134a	Horizontal	$T = 60°C$ $100 < G < 700$ kg/m^2 s	• Local h measured every 75 mm of using heat flux sensors • Combination of convective and film condensation terms; annular and stratified terms from Haraguchi et al. (1994a,b), two-phase multiplier replaced by Mishima and Hibiki (1996a, b)
Wang, Rose, and co-workers (Wang and Rose, 2004; Wang et al., 2004)	Triangular (Wang and Rose, 2004) $D_H = 0.577$ mm Square (Wang et al., 2004)	R-134a	Horizontal	$T = 50°C$ $100 < G < 1300$ kg/m^2 s	• Wang and Rose (2004) numerical analyses to predict varying condensate flow pattern across cross-section and length • Wang et al. (2004) model for film condensation of R-134a in square, horizontal, 1-mm microchannels • Surface tension, gravity, and shear terms can be turned on or off to demonstrate individual effects
Cavallini et al. (2005)	Multiple parallel 1.4-mm channels	R-134a, R-410A		$T = 40°C$ $200 < G < 1000$ kg/m^2 s	• h from T_{wall} measurements, available models (Akers et al., 1959; Moser et al., 1998; Zhang and Webb, 2001; Cavallini et al., 2002a; Wang et al., 2002; Koyama et al., 2003a) underestimate results • Mist flow in experiments might lead to high h

Kim and Shin (2005)	Circular and square channels $0.5 < D_H < 1$ mm	R-134a	Horizontal	$T = 40°C$ $100 < G < 600$ kg/m^2 s	• Technique matched T_{out} of electrically heated air stream with similar air stream heated by condensing refrigerant R-134a • Measured small, local condensation Q • Most available models and correlations (Akers et al., 1959; Soliman et al., 1968; Traviss et al., 1973; Cavallini and Zecchin, 1974; Shah, 1979; Dobson, 1994; Moser et al., 1998) underpredict data at low G • At lower G, square channels had higher h than circular channels; reverse was true for high G
Bandhauer et al. (2006) Agarwal et al. (2010)	$0.4 < D_H < 4.9$ mm	R-134a	Horizontal	$150 < G < 750$ kg/m^2 s	• Thermal amplification technique for accurate h measurement • h model for circular and noncircular tubes based on Traviss et al. (1973) boundary-layer analyses • Two-region dimensionless film temperature • Interfacial shear stress from models developed specifically for microchannels • Addresses annular, mist, and disperse wave regimes
Andresen (2007)	$0.76 < D < 9.4$ mm	R-404A, R-410A	Horizontal	$200 < G < 800$ kg/m^2 s $0.8 < P_r < 1.0$	• Near-critical data not predicted well • Model developed for wavy, annular, and annular-wavy regime
Kim and Mudawar (2012)	Square: $D_H = 1$ mm	FC-72	Horizontal	$68 < G < 367$ kg/m^2 s $T_{sat} = 60°C$	• Good agreement with macrochannel models • Proposed annular flow model with three-sided cooling from boundary-layer analysis

velocity profile for the liquid film, a relationship for the local condensation heat transfer coefficient was derived, which was a function of the film thickness (i.e., the laminar sublayer thickness), fluid properties, and the sum of the forces acting on the condensate. The gravity force was deemed to be unimportant, which is also the case for horizontal annular flow. The vapor friction force was determined from an equivalent vapor friction factor, which was higher than for single-phase flow due to the presence of the liquid film. They also developed a separate correlation for a laminar condensate film with a linear velocity profile. Hence, the condensation process was divided into three different regions: laminar condensate entrance region, turbulent condensate, and laminar vapor exit region (where a Nusselt analysis was more appropriate).

As discussed in the previous section, Akers et al. (1959) defined an all-liquid flow rate that provided the same heat transfer coefficient for condensing annular flow. This all-liquid flow rate was expressed by an "equivalent" mass flux, which was used to define an equivalent Reynolds number. This equivalent Reynolds number was substituted in a single-phase heat transfer equation, which was a function of the liquid Prandtl and Reynolds numbers, to predict the two-phase condensation Nusselt number. Thus, the equivalent liquid mass flux G_{eq} is composed of the actual liquid condensate flux G_l, in addition to a liquid flux G'_l that replaces the vapor core mass flux and produces the same interfacial shear stress as the vapor core; $G_{eq} = G_l + G'_l$. By equating the shear stress due to the new liquid flux G'_l and the original vapor core flux G_v, they obtained:

$$G'_l = G_v \sqrt{\frac{\rho_l}{\rho_v}} \sqrt{\frac{f_v}{f_l}} \tag{6.93}$$

At this point, they assumed that f_l and f_v would be constant due to a presumed fully rough interface, and also that $f_l = f_v$, which resulted in $G_{eq} = G_l + G_v \sqrt{\rho_l / \rho_v}$. This G_{eq} was used to define Re_{eq}, which, when substituted into a typical single-phase turbulent heat transfer equation, yields:

$$Nu = 0.0265 Re_{eq}^{0.8} Pr^{1/3} \tag{6.94}$$

This equation is typically recommended for $Re_l > 5000$ and $Re_v(\mu_v/\mu_l)(\rho_l/\rho_v)^{0.5} > 20{,}000$.

Akers and Rosson (1960) developed three different experimentally validated heat transfer correlations for condensation inside horizontal tubes for "semi-stratified" flow (annular condensation and run down superimposed on stratified flow), laminar annular flow, and turbulent annular flow. Local values of the condensation heat transfer coefficient were determined experimentally for methanol and R-12 flowing inside a 15.88-mm horizontal tube. For $Re_l < 5000$, the heat transfer coefficient was a function of the Prandtl number, vapor Reynolds number, and wall temperature difference. For this region, the constant multiplier for the Nusselt number was determined from the experimental data. This constant and the exponent for Re_v were different for Re_v above and below 20,000, the resulting expressions being:

$$1000 < \frac{DG_v}{\mu_l}\left(\frac{\rho_l}{\rho_v}\right)^{0.5} < 20{,}000 \quad \frac{hD}{k_l} = 13.8Pr_l^{1/3}\left(\frac{h_{fg}}{C_p\Delta T}\right)^{1/6}\left[\frac{DG_v}{\mu_l}\left(\frac{\rho_l}{\rho_v}\right)^{0.5}\right]^{0.2}$$

$$20{,}000 < \frac{DG_v}{\mu_l}\left(\frac{\rho_l}{\rho_v}\right)^{0.5} < 100{,}000 \quad \frac{hD}{k_l} = 0.1Pr_l^{1/3}\left(\frac{h_{fg}}{C_p\Delta T}\right)^{1/6}\left[\frac{DG_v}{\mu_l}\left(\frac{\rho_l}{\rho_v}\right)^{0.5}\right]^{2/3} \tag{6.95}$$

For turbulent liquid films, the correlation developed by Akers et al. (1959), which neglected the wall temperature difference, was employed.

Ananiev et al. (1961) suggested that the two-phase heat transfer coefficient can be related to the corresponding single-phase heat transfer coefficient with the entire mass flux flowing as a liquid, $h = h_o\sqrt{\rho_l/\rho_m}$, where the mixture density is given by $1/\rho_m = (1/\rho 1)(1-x) + (1/\rho_v)^x$ Boyko and Kruzhilin (1967) used this approach to correlate steam condensation data for $10 < D < 17$ mm from Miropolsky (1962), with a Dittus–Boelter type single-phase heat transfer coefficient proposed by Mikheev (1956) of the form $(h_oD/k) = CRe^{0.8}(Pr_b^{0.43}/Pr_w)^{0.25}$. Different leading constants C were required for the single-phase heat transfer coefficient to fit the data from tubes of different materials.

Soliman et al. (1968) developed a model for predicting the condensation heat transfer coefficient for annular flow. In this work, they note that Carpenter and Colburn (1951) did not appropriately account for the momentum contribution to the wall shear stress. They evaluated the wall shear stress as a combination of friction, momentum, and gravity contributions, and used the resulting expression to evaluate the heat transfer coefficient, much like the approach used by Carpenter and Colburn (1951). Soliman et al. (1968) performed momentum balances around the vapor core and the liquid film. To obtain the shear stress at the vapor–liquid interface, they used the Lockhart–Martinelli (1949) two-phase multiplier, representing it as $\phi_v = 1 + 2.85X_{tt}{}^{0.523}$, yielding

$$\tau_i = \frac{D}{4}\left(-\frac{dP}{dz}\right)_F\left(-\frac{dP}{dz}\right)_F = \phi_g^2\left(-\frac{dP}{dz}\right)_v \tag{6.96}$$

in the usual manner. The momentum term accounts for the slowing down of the condensing vapor as it converts into the liquid phase. Depending on whether the liquid film is laminar or turbulent, the interface velocity is related to the mean film velocity as $U_{li} = \beta U_l$, where $\beta = 2$ for a laminar film, and $\beta = 1.25$ for a turbulent film. The shear due to momentum change is then represented as follows:

$$\tau_m = \frac{D}{4}\left(\frac{G^2}{\rho_v}\right)\left(\frac{dx}{dz}\right)\sum_{n=1}^{5} a_n\left(\frac{\rho_v}{\rho_l}\right)^{n/3} \tag{6.97}$$

The respective coefficients $a_1 \ldots a_5$ in terms of the local quality and the velocity ratio β are available in their paper as well as in Carey (2008). For the general case of an inclined tube, the shear due to gravity is given by:

$$\tau_g = \frac{D}{4}(1-\alpha)(\rho_l - \rho_g)g \sin\theta \tag{6.98}$$

where the void fraction is calculated using the Zivi (1964) correlation. The three shear stress terms are summed as follows to yield the wall shear stress:

$$\tau_w = \tau_i + \tau_m + \tau_g \tag{6.99}$$

Parametric analyses on the various components of the shear term were also conducted to determine their relative significance along the condensation path. The friction term dominated at high and intermediate x, and progressively decreased toward the end of the condenser due to the decreasing vapor velocities. The gravity term is negligible at high x, and increases with increasing film thickness as condensation proceeds. The momentum term consists of the contribution due to the momentum added to the liquid film by the condensing vapor (positive term) and the reverse shear stress at the wall due to momentum recovery of the vapor (only significant at low void fractions). Momentum effects are significant for high liquid to gas density ratios and become important at low x in the absence of a gravity field. These considerations were also used to predict the onset of flooding in countercurrent gas–liquid flow. They commented that while some investigators may have obtained reasonable agreement between their heat transfer data and models even when ignoring or incorrectly accounting for momentum effects, this may simply be due to the relative unimportance of these terms under the tested conditions. Due to the variations of the constituent terms in the overall shear stress, however, these terms must be properly accounted for, so that the model can be valid over a wide range of conditions. With the wall shear evaluated, the following expression for the heat transfer coefficient was developed using the data from several investigators:

$$\frac{h\mu_l}{k_l \rho_l^{1/2}} = 0.036 Pr_l^{0.65} \tau_w^{1/2} \tag{6.100}$$

The above equation is valid for $7.44 < D < 11.66$ mm, the fluids steam, R-113, ethanol, methanol, toluene, trichloroethylene, and R-22, $1 < Pr < 10$, $6 < U_v < 300$ m/s, $0.03 < x < 0.99$, for horizontal as well as vertical downward condensation when annular flow prevails.

Traviss et al. (1973) used the heat–momentum analogy and the von Karman universal velocity distribution in the liquid film to develop a correlation for the Nusselt number in annular flow condensation. The turbulent vapor core prompted the assumption that the vapor core and vapor–liquid interface temperatures were the same. The pressure gradient was considered to be due to friction, gravity, and momentum change. In a manner analogous to that used by Soliman et al. (1968), they computed the frictional term using the Lockhart–Martinelli (1949) two-phase multiplier approach with the expression for ϕ_g developed by Soliman et al. (1968).

They also used Zivi's (1964) void fraction model to compute the gravitational term. For the momentum term, assuming that the liquid film was thin compared to the tube length, flat-plate flow was used. The ratio of the interface velocity to the average liquid film velocity, β (defined above in the discussion of the work of Soliman et al., 1968), was obtained from the universal velocity profile as a function of the nondimensional film thickness. The eddy diffusivity ratio (i.e., the turbulent Prandtl number) was assumed to be unity. Using the assumed liquid velocity profile, a relationship for the condensation heat transfer coefficient was determined as a function of the turbulent film thickness. They then derived a relationship for the liquid Reynolds number as a function of this film thickness. By arguing that the interfacial shear to wall shear ratio was approximately unity, the final relationship for the condensation heat transfer coefficient was developed as follows:

$$\frac{hD}{k_1} = \frac{0.15 Pr_1 Re_1^{0.9}}{F_T}\left(\frac{1}{X_{tt}} + \frac{2.85}{X_{tt}^{0.476}}\right) \quad Re_1 = \frac{G(1-x)D}{\mu_1} \tag{6.101}$$

where F_T is given as follows:

$$\begin{aligned} F_T &= 5Pr_1 + 5\ln(1+5Pr_1) + 2.5\ln(0.0031 Re_1^{0.812}) \quad Re_1 > 1125 \\ &= 5Pr_1 + 5\ln[1 + Pr_1(0.0964 Pe_1^{0.585} - 1)] 50 < \quad Re_1 < 1125 \\ &= 0.707 Pr_1 Pe_1^{0.5} \quad Re_1 < 50 \end{aligned} \tag{6.102}$$

The results were compared with experiments conducted on R-12 and R-22 condensing in an 8-mm tube for $161 < G < 1533\ \text{kg/m}^2\,\text{s}$. Agreement with the experimental data was good for qualities as low as 0.1, and the quality range from 0 to 0.1 (presumed by them to be slug flow) was predicted well by conducting a linear interpolation between this model and a single-phase heat transfer correlation. When the turbulent Martinelli parameter was below 0.155, the correlation produced good results. However, for Martinelli parameters above 0.155, most probably a mist flow (high quality and mass flux) condition, the experimental data were underpredicted by the correlation. A correction factor was proposed to improve predictions in this area.

One of the most widely used general purpose condensation correlations, due to the large database from 21 investigators used for its development, and also its ease of use, is the Shah (1979) correlation. Shah reasoned that in the absence of nucleate boiling, condensation heat transfer should be similar to evaporative heat transfer when the tube is completely wet, and extended the correlation developed previously (Shah, 1976) for evaporation to condensation as follows:

$$\begin{aligned} \frac{h}{h_{lo}} &= (1-x)^{0.8} + \frac{3.8x^{0.76}(1-x)^{0.04}}{(P/P_{crit})^{0.38}} \\ h_{lo} &= 0.023\left(\frac{k_1}{D}\right)\left(\frac{GD}{\mu_1}\right)0.8 Pr_1^{0.4} \end{aligned} \tag{6.103}$$

He validated the correlation using data for water, R-11, R-12, R-22, R-113, methanol, ethanol, benzene, toluene, and trichloroethylene condensing inside horizontal, inclined, and vertical tubes, as well as an annulus. The mean deviation between the predictions and the 474 data points used for correlation development was 17%. Some noticeable discrepancies included the underprediction of the data at high qualities (85–100%), which could be due to entrance or entrainment effects. The operating conditions included $11 < G < 211$ kg/m^2 s, $21°C < T_{sat} < 310°C$, $3 < U_v < 300$ m/s, reduced pressure from 0.002 to 0.44, and $1 < Pr_l < 13$, for tube diameters between 7 and 40 mm. Shah recommended that the correlation should be used only for $Re_l > 350$ due to the lack of lower Re_l data for the development of the correlation.

Soliman (1986) developed a heat transfer correlation for condensation in annular-mist flow. The annular-mist transition criterion developed by him was discussed in a previous section of this chapter. He stated that previous approaches for modeling mist flow heat transfer had primarily tried to modify annular flow models, with limited success. Therefore, he developed a model specifically for this regime, using data from several investigators, selected because of the existence of mist flow. Thus, the selected data included data for steam, R-113, and R-12, $7.4 < D < 12.7$ mm, horizontal and vertical orientations, $21°C < T_{sat} < 310°C$, $80 < G < 1610$ kg/m^2 s, $0.20 < x < 0.95$, $8 < We < 140$, and $1 < Pr_l < 7.7$. Of the five data sets chosen, one was in the annular flow regime, one was in annular and mist flow, and three were predominantly in mist flow, facilitating the development of this correlation. He found that although the Akers et al. (1959) correlation was successful in predicting the data for $We > 20$, the data were seriously underpredicted for $We > 30$. The Traviss et al. (1973) correlation was also found to seriously underpredict the data. Based on these observations, Soliman assumed that the flow would occur as a homogeneous mixture in thermodynamic equilibrium. Vapor shear would entrain the condensate into the bulk flow, leaving alternate dry spots and thin ridges of liquid at the wall. Heat transfer to this alternately dry or thin-film covered wall should therefore result in higher heat transfer coefficients than those predicted by annular flow models. The analogy between heat and momentum transfer for single-phase turbulent flow was used as a starting point. Thereafter, the effect of wall temperature difference was accounted for, which resulted in the following correlation (valid for $We > \sim 30-35$):

$$Nu = 0.00345 Re_m^{0.9} \left(\frac{\mu_G h_{LG}}{k_G (T_{sat} - T_w)} \right)^{1/3} \tag{6.104}$$

where the mixture Reynolds number is given by $Re_m = GD/\mu_m$, and the mixture viscosity is given by $1/\mu_m = x/\mu_G + (1 - x)/\mu_L$.

Chen et al. (1987) developed a film condensation heat transfer correlation based on analytical and empirical approaches. They noted that at moderate film Re, waves on the film increase the surface area, which enhances heat transfer above the Nusselt predictions. At higher film Re ($Re > 1800$), the enhancement was attributed to turbulent transport in the condensate film. At high vapor velocities, as noted by others, they state that vapor shear modifies the condensate film, making it thinner in co-current flow, and thicker in countercurrent flow. Thus, in this work, they included the effects of interfacial shear, interface waviness, and turbulent transport. They considered the limiting cases of film condensation and interpolated between these limits to obtain a more general correlation. For laminar film condensation in a quiescent vapor, they accounted for the effect of waves at the interface by using Chun and Seban's (1971) expression $Nu_1 = 0.823\ Re_x^{-0.22}$. For turbulent films, they used the data of Blangetti and Schlunder (1978) for steam condensation in a vertical tube, which leads to increasing Nu at increasing Re due to higher turbulent intensities, to express the Nusselt number as $Nu_T = 0.00402\ Re_x^{0.4} Pr^{0.65}$. For condensation with high interfacial shear stress, they neglected the gravitational and momentum terms at high qualities to develop a simple relationship assuming that the wall shear stress was equal to the interfacial shear stress. Thus, the equation developed by Soliman et al. (1968) was simplified to $Nu_i = 0.036 Pr^{0.65} \tau_i^{*1/2}$, where $\tau_i^* = \tau_i / \rho_L (g\nu_L)^{2/3}$ for vertical as well as horizontal film condensation. Having developed expressions for individual regimes, they combined these using the technique of Churchill and Usagi (1974). They first developed an expression for gravity-dominated film condensation in a quiescent vapor by combining the wavy-laminar and turbulent expressions discussed above as follows:

$$Nu_O = [(Nu_L^{n_1}) + (Nu_T^{n_1})]^{1/n_1} \tag{6.105}$$

The resulting Nu_o for quiescent vapor condensation was combined with Nu_i for high interfacial shear condensation in like manner to yield:

$$Nu_x = [(Nu_o^{n_2}) + (Nu_i^{n_2})]^{1/n_2} \tag{6.106}$$

Using the data of Blangetti and Schlunder for downflow condensation of steam in a 30-mm vertical tube over a wide range of film Re and vapor velocities, they determined that $n_1 = 6$ and $n_2 = 2$, which yields the following correlation for local film condensation heat transfer:

$$Nu_x = \frac{h_x}{k_L}\left(\frac{\nu_L}{g}\right)^{1/3} = \left[\left(0.31 Re_x^{-1.32} + \frac{Re_x^{2.4} Pr^{3.9}}{2.37 \times 10^{14}}\right)^{1/3} + \frac{Pr^{1.3}}{771.6}\tau_i^*\right]^{1/2} \tag{6.107}$$

Based on the findings of Chen et al. (1984), they assumed that the interfacial shear for a turbulent vapor core would be the same for condensing and adiabatic flows, and used the shear stress relationship of Dukler (1960) to compute the Nusselt number above:

$$\tau_i^* = A(Re_T - Re_x)^{1.4} Re_x^{0.4} A = \frac{0.252 \mu_L^{1.177} \mu_g^{0.156}}{D^2 g^{2/3} \rho_L^{0.553} \rho_g^{0.78}} \tag{6.108}$$

where Re_T is the film Re if the entire vapor flow condenses, and the parameter A accounts for gravitational and viscous forces. The resulting Nusselt number is valid for co-current annular film condensation in vertical tubes. For horizontal annular film condensation, the gravity terms are neglected to yield a simplified version of the Nusselt number equation as follows:

$$Nu_x = 0.036 Pr^{0.65} A^{0.5} (Re_T - Re_x)^{0.7} Re_x^{0.2} \tag{6.109}$$

Chen et al. cautioned that these equations would not apply at the very entrance of the tube when high vapor velocities could cause significant entrainment and mist flow, or toward the end of the condensation process, where stratified and slug flows would prevail. They also developed expressions for the average Nusselt numbers for film condensation using the corresponding limiting expressions as the basis (rather than integrating the local Nu_x), which resulted in the following:

$$\overline{Nu} = \left(Re_T^{-0.44} + \frac{Re_T^{0.8} Pr^{1.3}}{1.718 \times 10^5} + \frac{A Pr^{1.3} Re_T^{1.8}}{2075.3} \right)^{1/2} \tag{6.110}$$

for vertical co-current condensation, and

$$\overline{Nu} = 0.036 Pr^{0.65} \overline{\tau}_i^{*\,1/2}$$
$$\overline{\tau}_i^* = A \frac{Re_L^2}{Re_T^{0.2}} \left(\frac{Re_L}{Re_T} \right)^{-1.6} \left(1.25 + 0.39 \frac{Re_L}{Re_T} \right)^{-2} \tag{6.111}$$

for horizontal condensation. Using similar considerations, Chen et al. also developed expressions for the Nusselt number in countercurrent annular flow (reflux condensation), details of which are available in their paper.

Tandon et al. (1995) conducted condensation experiments for R-12 and R-22 flowing through a 10-mm horizontal tube, primarily in the wavy flow regime. They combined these data with previously reported data in the annular and semi-annular regimes for an overall range of $20°C < T_{sat} < 40°C$ and $175 < G < 560\ kg/m^2\,s$ to propose modifications to the Akers–Rosson (1960) correlations. They found that at the same vapor mass flux, the heat transfer coefficient for R-22 was higher than that for R-12, primarily due to the higher thermal conductivity and latent heat of R-22. They noted that the heat transfer coefficient trends showed a change in slope at $Re_v \sim 3 \times 10^4$, indicating a transition from wavy to annular or semi-annular flow. This prompted them to propose the following correlations for shear-controlled and gravity-controlled flow regimes:

$$Re_v > 3 \times 10^4 \quad Nu = 0.084 Pr_L^{1/3} \left(\frac{h_{fg}}{C_p \Delta T} \right)^{1/6} Re_v^{0.67}$$
$$Re_v > 3 \times 10^4 \quad Nu = 23.1 Pr_L^{1/3} \left(\frac{h_{fg}}{C_p \Delta T} \right)^{1/6} Re_v^{1/8} \tag{6.112}$$

Chitti and Anand (1995) developed an analytical model for annular flow condensation in smooth horizontal round tubes. The results were compared with data for R-22 condensing inside an 8-mm tube. Their model used the Prandtl mixing length theory with van Driest's hypothesis (1956) (for the law of the wall) to obtain the momentum diffusivity, and $Pr_t = 0.9$. Thus, the computation of the shear stress distribution, pressure drop, and void fraction was avoided, and the heat flux was evaluated through continuity in the radial direction, with the film thickness being calculated iteratively. For closure, the friction velocity is calculated through the use of the empirical Lockhart–Martinelli (1949) two-phase multiplier ϕ^2_{vv} from Azer et al. (1972). The model showed reasonable agreement with their data, with a mean deviation of 15.3%, about the same as the agreement achieved with the Traviss et al. (1973) model and the Shah (1979) correlation. Agreement with the equivalent mass flux-type models of Akers et al. (1959) and Boyko and Kruzhilin (1967), however, was not particularly good, and they attributed this to the lack of a strong physical basis for these latter models. Chitti and Anand (1996) later extended the experiments to R-32/R-125 mixtures (50/50 by weight) with oil concentrations of 0%, 2.6%, and 5.35%, and pure R-22 condensing in the 8-mm horizontal tube, primarily in the annular regime based on the Taitel–Dukler (1976) map. They found a minimal effect ($\sim$10%) of condensing temperature on the heat transfer coefficient because the operating conditions (24–38°C) were well below the critical temperature. Heat transfer coefficients for R-32/R-125 mixtures were about 15% higher than those for R-22, and the effect of polyolester oil was also found to be minimal, within the uncertainty estimates.

Moser et al. (1998) noted that the Akers et al. (1959) equivalent mass velocity model discussed above had been shown to consistently underpredict the data of several investigators. They stated that this was primarily due to two assumptions in their model that could not be adequately supported: (1) the assumption of constant and equal friction factors for the vapor core and the liquid film, because the latter would most certainly be affected by the Re_l; (2) the driving temperature differences for single-phase $(T_b - T_w)$ and condensing flows $(T\delta - T_w)$ are completely different, something that the Akers et al. model does not take into account. To correct the first problem, they related the friction in the vapor and liquid phases through the two-phase multiplier concept, rather than assuming the friction factors to be equal. This led to the following definition of the equivalent Reynolds number: $Re_{eq} = \phi_{lo}^{8/7}$. To evaluate ϕ^2_{lo} they recommended the use of the Friedel (1979) correlation. For the driving temperature difference, they related the single-phase and condensing ΔT_s as follows: $F = (T_b - T_w)/(T\delta - T_w)$. Using a boundary-layer analysis similar to that of Traviss et al. (1973), along with the single-phase ΔT being determined by the Petukhov (1970) equation, the correction factor F was expressed as follows:

$$F = 1.31(R^+)^{C_1} Re_l^{C_2} Pr_l^{-0.185} \quad C_1 = 0.126 Pr_L^{-0.448} \quad C_2 = -0.113 Pr_l^{-0.563} \tag{6.113}$$

The dimensionless pipe radius was obtained through the friction velocity and the use of the Blasius friction factor to yield $R^+ = 0.0994Re_{eq}^{7/8}$. The final Nusselt number equation was as follows:

$$Nu = \frac{hD}{k_l} = \frac{0.0994^{C_1} Re_l^{C_2} Re_{eq}^{1+0.875C_1} Pr_l^{0.815}}{(1.58 \ln Re_{eq} - 3.28)(2.58 \ln Re_{eq} + 13.7Pr_l^{2/3} - 19.1)} \tag{6.114}$$

This model was shown to predict data from a variety of sources for $3.14 < D < 20$ mm with a mean deviation of 13.6%, about as well as the empirical Shah (1979) correlation, and better than the Traviss et al. (1973) correlation. In general, the trend was toward underprediction of data, which they stated could be corrected by the application of better pressure drop models for the evaluation of the two-phase multiplier required for this model. In addition, accounting for entrainment and stratified annular film would yield better predictions.

It was mentioned in the previous section that Hurlburt and Newell (1999) developed scaling equations for condensing flows that would enable the prediction of void fraction, pressure drop, and heat transfer for a refrigerant at a given condition and tube diameter from the available results for another similar fluid (R-11, R-12, R-22, and R-134a) operating at a different condition in the diameter range 3–10 mm. They developed an equivalent average film thickness model assuming annular flow, which enabled the prediction of shear stress. Like several other researchers, they assumed that the primary resistance to heat transfer occurs in the viscous and buffer layers of the film, regardless of the actual thickness. Film and core flows were modeled after the approach of Rohsenow et al. (1957; Traviss et al., 1973), assuming turbulent vapor and liquid films, and a law-of-the-wall universal velocity and temperature profiles. They used data from Sacks (1975) for R-11, R-12, and R-22, and the R-22 and R-134a data of Dobson (1994) to validate their model. Using these considerations, they developed the following scaling equation for heat transfer prediction:

$$\begin{aligned} h_2/h_1 = {} & (\rho_{g_2}/\rho_{g_1})^{-0.375}(\rho_{L_2}/\rho_{L_1})^{0.375}(\mu_{g_2}/\mu_{g_1})^{-0.75}(\mu_{L_2}/\mu_{L_1})^{0.062} \\ & \times (C_{pL_2}/C_{pL_1})(Pr_{L_2}/Pr_{L_1})^{-0.25}(G_2/G_1)(x_2/x_1)^{0.82}\left(\frac{1-x_2}{1-x_1}\right)^{0.19} \end{aligned} \tag{6.115}$$

Upon comparison with the data of Dobson (1994), they commented that diameter effects may not be fully accounted for with the above approach, and perhaps the above equation should also have included the factor $(D_2/D_1)^{-0.15}$. Also, the fluid property exponents in this equation were different from those found in the literature; specifically, the thermal conductivity exponent varied from 0.35 for Chen et al. (1987), to 0.38 for Shah (1979), and 0.65 for Cavallini and Zecchin (1974), while the above equation shows an exponent of 0.25 through the Prandtl number. They noted that several properties together determine the thicknesses of the various layers; therefore, differences in the dependence on an individual property may not be too significant.

6.5.1.2 Multi-regime condensation

For films under the simultaneous influence of shear and gravity, Kosky (1971) provided a simple algebraic equation for the convenient evaluation of the nondimensional film thickness to facilitate closed-form computations. Thus, the work of Dukler (1960) (considered accurate close to the wall and for thick films) and Kutateladze (1963) (considered accurate for thick films, $\delta^+ > \sim 10$) were combined to obtain the following simplified expressions: $\delta^+ = (Re_L/2)^{1/2}$ for $\delta^+ < 25$, and $\delta^+ = 0.0504 Re_L{}^{7/8}$ for $\delta^+ > 25$. Jaster and Kosky (1976) performed experiments on condensing steam inside a 12.5-mm horizontal tube. Coupled with the experimental results from Kosky and Staub (1971), the mass flux ranged from 12.6 to 145 kg/m^2 s. A stress ratio $F = \tau_w/(\rho_L g \delta)$ (axial shear-to-gravity) was used to divide the data into annular ($F > 29$), transition ($29 \geq F \geq 5$), and stratified ($F < 5$) regimes. The heat and momentum transfer analogy was applied to determine the annular flow heat transfer, with the two-phase gas multiplier ϕ_g^2 used to determine the shear stress and the nondimensional film thickness obtained from Kosky (1971). For stratified flow, they related the half-angle subtended by the liquid pool β to the void fraction α using Zivi's (1964) model approximately as $\cos(\beta) = 2\alpha - 1$, which resulted in a modified version of the Nusselt condensation equation with the leading constant of 0.725 and a multiplier of $\alpha^{3/4}$. In the transition regime, a linear interpolation between the two was suggested:

$$Nu_{tr} = Nu_{an} + \frac{F - 29}{24}(Nu_{an} - Nu_{str}) \tag{6.116}$$

It was stated in a previous section that Breber et al. (1979, 1980) developed a four flow zone map based on data for a variety of fluids for diameters as small as 4.8 mm. The corresponding transition criteria were listed with that discussion. For Zone I (annular flow, $j_g^* > 1.5$, $X < 1.0$), they recommend convective-type correlations:

$$Re_l > 1500 \quad h_l = 0.024\left(\frac{k_l}{D}\right) Re_l^{0.8} Pr_l^{1/3} (\mu_l/\mu_w)^{0.14}$$

$$Re_l < 1500 \quad h_l = 1.86\left(\frac{k_l}{D}\right)\left(\frac{Re_l Pr_l D}{L}\right)^{1/3} (\mu_l/\mu_w)^{0.14} \tag{6.117}$$

$$h_{annular} = h_l(\phi_l^2)^{0.45}$$

For Zone II (wavy or stratified flow, $j_g^* < 0.5$, $X < 1.0$), they recommend a Nusselt-type correlation:

$$h_{stratified} = F_g \left[\frac{k_l^3 \rho_l (\rho_l - \rho_v) g h_{lv}}{4\mu_l (T_{sat} - T_{wall}) D}\right]^{1/4} \tag{6.118}$$

where the factor F_g is a function of the liquid volume fraction and the vapor Reynolds number, but may be approximated by $F_g = 0.79$ as suggested by

Palen et al. (1979). For Zone III (intermittent flow, $j_g^* < 1.5$, $X > 1.1$), in view of the lack of appropriate models, they suggest the use of the annular flow correlation, because in their view, the convective mechanism governs, and also because it approaches the single-phase convection limit as $x \rightarrow 0$. For Zone IV (bubble flow, $j_g^* > 1.5$, $X > 1.1$), once again they suggest the convective equation presented above. For transitions between Zones I and II, they recommend interpolation between $h_{annular}$ and $h_{stratified}$ based on j_g^*, while for transitions between Zones II and III, they recommend interpolation based on X.

Nitheanandan et al. (1990) evaluated the available condensation correlations using condensation data from several sources for flow inside horizontal tubes with the intention of developing a more accurate approach spanning multiple-flow regimes for the design of condensers. The heat transfer database included R-12, R-113, and steam condensing inside tubes with diameters ranging from 7.4 to 15.9 mm, divided into three regimes: wavy, annular, and mist flow. Following the work of Soliman (1982, 1986), they recommended the use of the following heat transfer correlations for the respective regimes:

Mist flow ($We \geq 40$, $Fr \geq 7$) Soliman (1986)
Annular flow ($We < 40$, $Fr \geq 7$) Shah (1979)
Wavy flow ($Fr < 7$) Akers and Rosson (1960)

It should be noted that the $We = 40$ criterion to determine the mist-annular transition was different from the value of 30 recommended earlier by Soliman (1986). Similar recommendations were also made for condensation in vertical tubes based on the Hewitt and Roberts (1969) flow regime map.

Dobson (1994) investigated condensation of R-12 and R-134a in a 4.57-mm horizontal tube at $75 < G < 500$ kg/m^2 s at 35°C and 60°C saturation temperature. They found that the $Fr = 7$ criterion proposed by Soliman (1982) was adequate for establishing the wavy to wavy-annular flow transition. In addition, they introduced $Fr = 18$ as the transition criterion for the wavy-annular to annular transition, and also stated that the $We = 30$ criterion of Soliman (1986) was appropriate for the annular-to-mist flow transition. Having thus divided their data into gravity- and shear-driven regions, they interpreted several trends in heat transfer coefficients. In annular flow, the heat transfer coefficient increased with mass flux and quality, due to the increased shear and the thinning of the liquid film. The heat transfer coefficients for R-134a were about 20% higher than those for R-12 at constant mass flux. In wavy flow, the heat transfer coefficient was independent of mass flux, and showed a slight increase with quality. As the primary heat transfer in this region occurs in the upper portion due to Nusselt-type condensation, with very little heat transfer through the liquid pool, the slight increase in heat transfer coefficients with quality was attributed to the thinning of the pool and a smaller pool depth at the higher qualities. In wavy flow, the R-134a heat transfer coefficients were about 10% higher than those for R-12. The heat transfer coefficients for R-134a were about 15% higher at 35°C than at 60°C because of the decrease in the ratio of the densities of the vapor and liquid phases, which leads to lower slip ratios, and also because of a decrease in the thermal conductivity at the higher temperature. To correlate their data in the wavy flow regime, they employed the Chato (1962) type correlation, but changed the leading coefficient from 0.555 to a function of the Lockhart–Martinelli parameter to account for the change in void fraction with quality, yielding:

$$Nu_{\text{gravity}} = \frac{0.375}{X_{\text{tt}}^{0.23}}\left[\frac{g\rho_{\text{l}}(\rho_{\text{l}}-\rho_{\text{v}})D^{3}h'_{\text{lv}}}{\mu_{\text{l}}(T_{\text{sat}}-T_{\text{wall}})k_{\text{l}}}\right]^{0.25} \tag{6.119}$$

For annular flow, they used the familiar two-phase multiplier approach:

$$Nu_{\text{annular}} = 0.023Re_{\text{L}}^{0.80}Pr_{\text{L}}^{0.3}\left(\frac{2.61}{X_{\text{tt}}^{0.805}}\right) \tag{6.120}$$

They also found that the annular flow heat transfer coefficients predicted by this equation are quite similar to those for convective evaporation found by Wattelet et al. (1994), implying similar heat transfer mechanisms for the two convective phase-change processes. Finally, they stated that several widely used correlations in the literature (Cavallini and Zecchin, 1974; Shah, 1979; Traviss et al., 1973) overpredicted their data. Subsequently, Dobson and Chato (1998) extended this work to 3.14- and 7.04-mm tubes, and also to the fluids R-22 and mixtures of R-32/R-125. Data from these latter two tubes were used for the development of updated heat transfer correlations. The only discernible effect of tube diameter in their work was that transition to annular flow happened at higher mass fluxes and qualities in the larger tubes, although they were able to predict this using the large-diameter correlations. Thus, surface tension effects were not seen within this diameter range. Due to a variety of compensating influences such as thermophysical properties, different reduced pressures, and different void fractions, the effect of the particular fluid used on the heat transfer coefficient was at most 10% in the wavy regime, with a slightly better performance of R-134a in annular flow. In developing the revised correlations, they noted that the boundary-layer analyses used by several investigators, including primarily Traviss et al. (1973), could be shown to be similar in basis to the two-phase multiplier approach used by others. Noting also that the primary thermal resistance in annular flow occurs in the laminar and buffer layers (even the presence of waves at the interface or the varying film thickness around the circumference would not significantly affect the near-wall behavior), they did not find it necessary to include a multi-region model of the liquid film resistance. The role of entrainment was also not found to be very significant in affecting heat transfer coefficients, because they found that rarely did true mist flow without a thin annular film coating the wall exist. With these considerations, the following annular flow correlation was proposed:

$$Nu_{\text{annular}} = 0.023Re_{\text{L}}^{0.80}Pr_{\text{L}}^{0.3}\left(1+\frac{2.22}{X_{\text{tt}}^{0.89}}\right) \tag{6.121}$$

For the stratified wavy flow region, they accounted for the film condensation at the top of the tube and forced convection in the liquid pool as follows:

$$Nu_{wavy} = \frac{0.23Re_{vo}^{0.12}}{1 + 1.11X_{tt}^{0.58}}\left[\frac{GaPr_l}{Ja_l}\right]^{0.25} + (1 - \theta_l/\pi)Nu_{forced} \tag{6.122}$$

where the forced convection term is given by:

$$Nu_{forced} = 0.0195Re_L^{0.80}Pr_L^{0.4}\sqrt{1.376 + \frac{c_1}{X_{tt}^{c_2}}} \tag{6.123}$$

The constants c_1 and c_2 in the above equation are given as follows: for $0 < Fr_l \leq 0.7$:

$$c_1 = 4.172 + 5.48Fr_l - 1.564Fr_l^2 \quad c_2 = 1.773 - 0.169Fr_l \tag{6.124}$$

and for $Fr_l > 0.7$, $c_1 = 7.242$, $c_2 = 1.655$. The angle from the top of the tube to the liquid pool level, θ_l, was approximated as follows:

$$1 - \frac{\theta_l}{\pi} \simeq \frac{\arccos(2\alpha - 1)}{\pi} \tag{6.125}$$

The void fraction required for the above equation was calculated using Zivi's (1964) correlation. They recommended that the above equations should be used as follows:

$$\begin{aligned} &G \geq 500\ \text{kg/m}^2\ \text{s}\ Nu = Nu_{annular} \\ &G < 500\ \text{kg/m}^2\ \text{s}\ Nu = Nu_{annular} \quad \text{for } Fr_{so} > 20 \\ &\qquad\qquad Nu = Nu_{wavy} \quad \text{for } Fr_{so} < 20 \end{aligned} \tag{6.126}$$

It should be noted that the single-value demarcation on the basis of G shown above could result in sharp discontinuities between the heat transfer coefficient predictions in the neighborhood of $G = 500$ kg/m^2 s as the annular or wavy correlations are used. The more gradual transition between wavy, wavy-annular, annular, and mist flows in their flow visualization work may have warranted correspondingly gradual transitions between heat transfer coefficient models. Also, they cautioned that for the small-diameter (3.14 mm) tube, the data had large uncertainties due to difficulties in measuring small heat transfer rates accurately, which led to relatively larger deviations from the above correlations.

Wilson et al. (2003) reported heat transfer coefficients for 7.79-, 6.37-, 4.40-, and 1.84-mm tubes made by progressively flattening 8.91-mm round smooth tubes and tubes with axial and helical microfin tubes. Refrigerants R-134a and R-410A were condensed at 35°C in these tubes at $75 < G < 400$ kg/m^2 s. Heat transfer coefficients were seen to increase over corresponding smooth-tube values as the tube was flattened. (It should be noted that their smooth-tube condensation heat transfer coefficients were consistently lower by 20–30% than the Dobson and Chato (1998) predictions, which is somewhat surprising because these experiments were conducted by researchers from the same group; the Dobson and Chato predictions were therefore artificially increased by the factor required to match the data. Since the enhancements observed were primarily in the 1.5–3.5 range, presumably including the 60% enhancement due to surface area increases in the microfinned tubes, these discrepancies could affect the conclusions considerably.) The enhancement was the highest for the flattened tubes with 18° helix angle microfins. Tentative explanations for these trends were provided including: potential early transition from stratified to annular flow, "modifications to the flow field without altering flow field configuration," formation of different flow field configurations, and others. No predictive models or correlations were proposed, but it was noted that flattening to 5-mm height would lead to about a 10% reduction in condenser size, 70% increase in pressure drop, and 40% reduction in refrigerant charge for the example case studied. A similar study on condensation of R-404A in 9.52-mm OD tubes at 40°C and $200 < G < 600\ \mathrm{kg/m^2\,s}$ by Infante Ferreira et al. (2003) also showed enhancements of 1.8–2.4 in microfin and cross-hatched tubes.

In a set of related papers, Cavallini et al. (2001, 2002a) obtained heat transfer and pressure drop data for a variety of refrigerants and blends (R-22, R-134a, R-125, R-32, R-236ea, R-407C, and R-410A) condensing in smooth 8-mm horizontal tubes, and developed multiple-flow regime models to predict their own data as well as data from other investigators. The experiments covered the temperature range $30 < T_{sat} < 50°C$, and $100 < G < 750\ \mathrm{kg/m^2\,s}$. When comparing the available correlations against these data, they noted that the data were often outside the recommended range of applicability for many of these correlations, especially for the newer high–pressure refrigerants such as R-125, R-32, and R-410A. The predictions of some of the empirical correlations were also not satisfactory within the stated range of applicability. They systematically plotted the various correlations in the literature in different graphs for low and high–pressure refrigerants, adhering to the recommended ranges of applicability to demonstrate the inadequacy of most of these correlations as general purpose predictive tools. Therefore, they tried to develop new procedures patterned after the approach of Breber et al. (1979, 1980) that would span the primary flow regimes encountered in condensation in horizontal tubes; that is, annular, stratified, wavy, and slug. For annular flow, for which they chose $j_g^* > 2.5$; $X_{tt} < 1.6$ as the transition criterion, they recommended the use of the Kosky and Staub (1971) model, which relates the heat transfer coefficient to the frictional pressure gradient through the interfacial shear stress τ. However, to compute the necessary frictional pressure gradient for this model, they modified the Friedel (1979) correlation to apply only to annular flow, whereas it had originally been intended for annular and stratified regimes. This restricted version was intended to match the single-regime data better. Their modified annular flow model is given by the following equations:

$$h_{annular} = \frac{\rho_L C_{pL} (\tau/\rho_L)^{0.5}}{T^+} \tag{6.127}$$

where the dimensionless temperature is given by the following:

$$\begin{aligned} T^+ &= \delta^+ Pr_L & \delta^+ \le 5 \\ T^+ &= 5\{Pr_L + \ln[1 + Pr_L(\delta^+/5 - 1)]\} & 5 < \delta^+ < 30 \\ T^+ &= 5[Pr_L + \ln(1 + 5Pr_L) + 0.495 \ln(\delta^+/30)] & \delta^+ \ge 30 \end{aligned} \tag{6.128}$$

with

$$\delta^+ = (Re_L/2)^{0.5} \text{ for } Re_L \le 1145; \quad \delta^+ = 0.0504 Re_L^{7/8} \text{ for } Re_L > 1145 \tag{6.129}$$

The shear stress is evaluated from the two-phase pressure drop in the usual manner as follows:

$$\tau = \frac{(\mathrm{d}P/dz)_f}{4}; \quad \text{where } (\mathrm{d}P/dz)_f = \phi_{LO}^2 (\mathrm{d}P/dz)_{LO} = \phi_{LO}^2 \; {}^2 f_{LO}\left(\frac{G^2}{D\rho_L}\right) \tag{6.130}$$

The two-phase multiplier contains several modifications from the original Friedel (1979) model; therefore, it is presented completely below:

$$\phi_{LO}^2 = E + \frac{1.262FH}{We^{0.1458}} \tag{6.131}$$

The parameters E, F, and H are given by:

$$\begin{aligned} E &= (1-x)^2 + x^2 \frac{\rho_L f_{GO}}{\rho_G f_{LO}} \\ F &= x^{0.6978} \\ H &= \left(\frac{\rho_L}{\rho_G}\right)^{0.3278} \left(\frac{\mu_G}{\mu_L}\right)^{-1.181} \left(1 - \frac{\mu_G}{\mu_L}\right)^{3.477} \end{aligned} \tag{6.132}$$

The Weber number is defined based on the total mass flux and the gas-phase density instead of the mixture density, $We = G^2 D/\rho G\sigma$. For $j_g^* < 2.5$; $X_{tt} < 1.6$, stratification begins, and they compute the heat transfer coefficient in this region using a combination of the annular model evaluated at the $j_g^* = 2.5$ boundary and a stratified heat transfer model. It should be noted that throughout the region $0 < j_g^* < 2.5$; $X_{tt} < 1.6$, a progressively decreasing contribution of the annular term is included; that is, the flow is never treated as exclusively stratified. Therefore they refer to this region as "annular-stratified transition and stratified region." The interpolation formula recommended is as follows:

$$h_{transition} = (h_{annular, j_g^*=2.5} - h_{stratified})(j_g^*/2.5) + h_{stratified} \tag{6.133}$$

where the stratified flow contribution is evaluated in the usual manner as a combination of film condensation at the top and forced convection in the liquid pool:

$$h_{stratified} = 0.725\left\{1 + 0.82\left[\frac{1-x}{x}\right]^{0.268}\right\}^{-1}\left[\frac{k_L^3 \rho_L(\rho_L - \rho_G) g h_{LG}}{\mu_L D(T_{sat} - T_{wall})}\right]^{0.25} \tag{6.134}$$

where the forced convection heat transfer coefficient in the liquid pool is given by:

$$h_L = 0.023\frac{k_L}{D}\left(\frac{G(1-x)D}{\mu_L}\right)^{0.8} Pr_L^{0.4} \tag{6.135}$$

and the liquid pool angle θ is defined using the Zivi (1964) void fraction model as follows:

$$1 - \frac{\theta}{\pi} = \frac{\arccos(2\alpha - 1)}{\pi} \tag{6.136}$$

Cavallini et al. commented that the liquid pool heat transfer term is significant at high values of reduced pressure. For conditions where $j_g^* < 2.5$; $X_{tt} < 1.6$, transition from stratified-to-intermittent flows starts to occur. This transition, however, occurs at a varying value of X_{tt}, and is defined by a transition mass flux, $G > G_W$, defined by Rabas and Arman (2000) as the mass flux required to fill the tube when being discharged to a gas-filled space. For slug flow, they used data from Dobson and Chato (1998) and Tang (1997) to curve-fit a single-phase heat transfer correlation. Smooth transition between the heat transfer coefficients in stratified and slug flows is provided once again by an interpolation technique. Thus, the stratified-slug transition and slug flow region heat transfer coefficient is given by:

$$h_{stratified-slug} = h_{LO} + \left(\frac{x}{x_{X_{tt}} = 1.6}\right)(h_{X_{tt}} = 1.6 - h_{LO}) \tag{6.137}$$

The transition quality can be simply calculated from the definition of X_{tt} as follows:

$$x_{X_{tt}} = 1.6 = \frac{(\mu_L/\mu_G)^{1/9}(\rho_G/\rho_L)^{5/9}}{1.686 + (\mu_L/\mu_G)^{1/9}(\rho_G/\rho_L)^{5/9}} \tag{6.138}$$

The liquid-only (h_{LO}) and liquid-phase ($h_{X_{tt}} = 1.6$) heat transfer coefficients required above are calculated using the Dittus–Boelter (1930) equation at Re_{LO} and Re_L (at the x for $X_{tt} = 1.6$), respectively.

There is also a possibility for an annular-to-slug transition across $j_g^* = 2.5$ at high values of X_{tt}; however, in common practice, this rarely occurs because it requires $G > 1000$ kg/m^2 s and very low qualities for most refrigerants. However, in such instances, they recommend a linear interpolation between $h_{annular}$ and h_{LO} based on the vapor quality. At high values of j_g^* and X_{tt}, especially at high reduced pressures, bubbly flow is encountered, and once again, in the absence of reliable correlations in this region, they recommend the use of the annular flow correlation. This multi-region map is recommended for the refrigerants listed above, and for the following conditions:

$$3 < D < 21 \text{ mm} \quad p_r < 0.75 \quad \rho_L/\rho_G > 4 \tag{6.139}$$

This composite model was shown to predict their own data and data from several other investigators with an average absolute deviation of about 10%. A chart for the use of the above heat transfer coefficient correlations is provided in Figure 6.26. For the prediction of pressure drops, Cavallini et al. recommend the use of the modified Friedel two-phase multiplier ϕ_{LO}^2 developed by them for annular flow when $j_g > 2.5$; for $j_g < 2.5$, the original Friedel (1979) model is recommended.

El Hajal et al. (2003) and Thome et al. (2003) conducted another detailed exercise (similar to that of Cavallini et al., 2001, 2002a, described above) of developing flow-regime-based heat transfer correlations over a wide range of conditions by fitting data from several different investigators for many different fluids. They patterned this work after the similar work of Kattan et al. (1998a,b,c) on flow boiling. Their basic premise is that void fraction is the most important variable in determining flow regimes, pressure drop, and heat transfer. However, reliable void fraction data were not available, especially at high reduced pressures. Therefore, they first used the Steiner (1993) horizontal tube version of the Rouhani–Axelsson (1970) drift-flux void fraction map because it includes mass flux and surface tension effects:

$$\alpha = \frac{x}{\rho_V}\left([1+0.12(1-x)]\left[\frac{x}{\rho_V}+\frac{1-x}{\rho_L}\right]+\frac{1.18(1-x)[g\sigma(\rho_L-\rho_V)]^{0.25}}{G\rho_L^{0.5}}\right)^{-1} \tag{6.140}$$

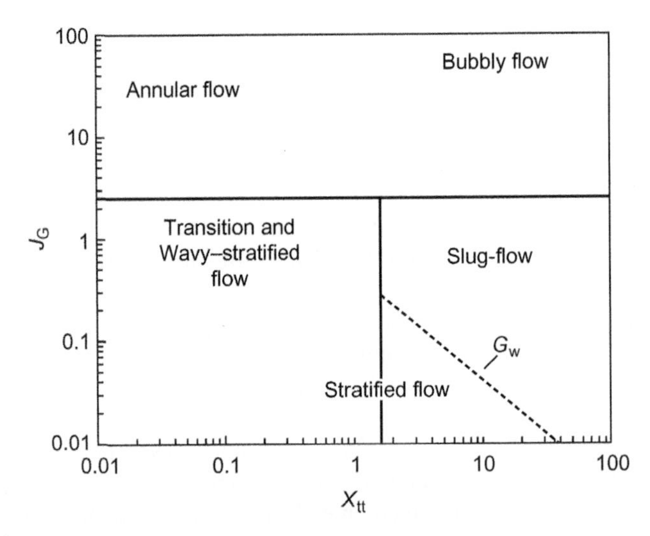

FIGURE 6.26

Illustration of Cavallini et al. (2002a) model implementation.

Source: *From Cavallini et al. (2002a) with permission from the American Society of Heating, Refrigerating and Air-Conditioning Engineers.*

The void fraction predicted by this model does not approach the homogeneous void fraction as the critical pressure is approached. Therefore, they set about trying to improve upon this model by indirect means. They stated that turbulent annular flow condensation heat transfer data could in fact be used to deduce the applicable void fraction. This is because the corresponding liquid film heat transfer coefficient may be represented as $h = cRe_L^n Pr_L^{0.5}(k_L/\delta)$, where the film thickness δ is simply by definition $\delta = D(1 - \alpha)/4$. Substituting this expression back into the heat transfer coefficient equation, it is clear that $h\alpha(1 - x)^n/(1 - \alpha)$. Since void fractions in annular flow are very large, the denominator in this expression, representative of the liquid film thickness, is a small quantity that affects the heat transfer coefficient significantly. It is this sensitivity of the heat transfer coefficient to void fraction that they sought to exploit in deducing the void fraction from available heat transfer data. Thus, they used the annular flow data from Cavallini et al. (2001, 2002a) to statistically curve-fit the leading constant and exponent for Re_L in the heat transfer expression, and assumed the Pr exponent to be 0.5 instead of 0.4. Although reasonable prediction was achieved, the prediction accuracy seemed to show a strong saturation pressure dependency. Therefore, they decided, somewhat arbitrarily, that the actual void fraction would in fact lie somewhere between the Axelsson–Rouhani prediction and the homogeneous value. They then chose a logarithmic mean of these two values to represent the actual void fraction:

$$\alpha = \frac{\alpha_{\text{homogeneous}} - \alpha_{\text{Axelsson-Rouhani}}}{\ln(\alpha_{\text{homogeneous}}/\alpha_{\text{Axelsson-Rouhani}})} \tag{6.141}$$

This formulation was then used to curve-fit the heat transfer data again, which demonstrated an improvement in the prediction of the heat transfer data over a larger range of reduced pressures, which they took to be a validation of the void fraction curve-fit. Once this was established, they plotted the transitions between stratified, wavy, intermittent, annular, mist, and bubbly flow on G-x-coordinates, adapting the transition expressions presented by Kattan et al. (1998a,b,c; with updates by Zurcher et al., 1999) for boiling to the condensing situation. For example, allowance is made for the fact that dryout does not occur in condensing flows; thus they remove that portion of the wavy transition line and arbitrarily extend it from the minima of that transition to the $x = 1$ location. Various liquid–vapor cross-sectional areas and perimeters are derived from the void fraction to enable plotting of the transition lines. A representative map constructed in this manner is shown in Figure 6.27. It should be noted that because the void fraction is dependent on the mass flux, strictly speaking, these maps have to be redrawn for each mass flux under consideration, even though g is in fact one of the coordinates. However, El Hajal et al. (2003) state that the transition lines are not very sensitive to the mass flux, and a first approximation may be achieved by choosing a representative mass flux to plot the transition lines. They demonstrate qualitative agreement with several of the widely used flow regime maps discussed in one of the previous sections of this chapter. Quite a bit of the agreement or

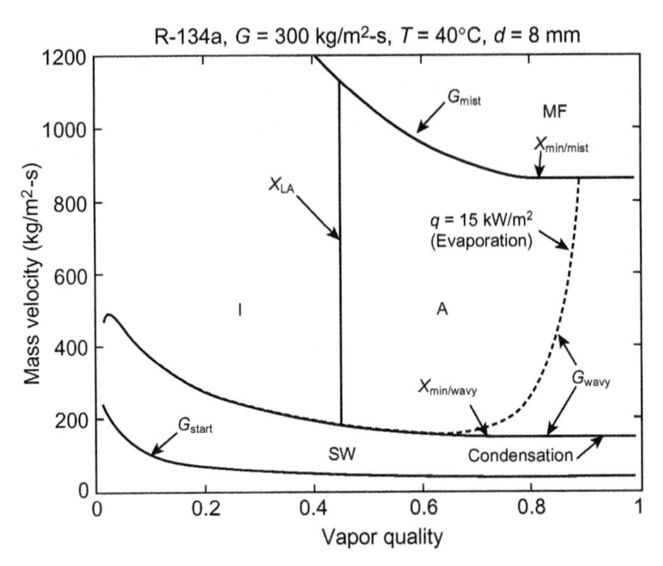

FIGURE 6.27

Flow regime map of El Hajal et al. (2003).

Source: *Reprinted from El Hajal et al. (2003) with permission from Elsevier.*

otherwise is, however, subject to interpretation and definition of the respective flow regimes used by the various investigators. It is surprising, however, that Figure 6.27 shows such a large intermittent regime for an 8-mm tube, even at qualities as high as about 45% and mass fluxes as high as 1000 kg/m^2 s and above. It is unusual to see intermittent slugs and plugs at such high, G, x values in such large tubes. They state that the maps are applicable for $16 < G < 1532$ kg/m^2 s, $3.14 < D < 21.4$ mm, $0.02 < p_r < 0.8$, $76 < (We/Fr)_L < 884$, and for a variety of refrigerants and blends. It should be noted, however, that the applicability of the maps down to 3.14 mm may need additional validation. This is because very few of the data on which the maps are based were for such small tubes, the vast majority of the data being for approximately 8-mm tubes. Also, for example, Dobson and Chato's (1998) work, from which much of the 3.14-mm data came, explicitly stated that these data were associated with large uncertainties.

For the development of the heat transfer model, Thome et al. (2003) start with an annular flow formulation. The turbulent annular flow forced-convective heat transfer equation discussed in El Hajal et al. (2003) is applied around the perimeter for annular flow. (But it should be noted that near the entrance of the condenser, the liquid fraction is very low, and the film may in fact be laminar.) For stratified flows, instead of treating the liquid phase as a pool with a flat upper surface at the bottom of the tube, they redistribute it as an annular ring of thickness δ, occupying the portion of the tube circumference that would yield a liquid phase cross-section equivalent to that of the stratified pool. The liquid phase cross-section that should be redistributed is in turn determined from the void fraction correlation described in the companion paper. The liquid film thickness in the film condensation region at the top is deemed negligible in this reapportioning of areas. This transformation allows them to use a gravity-driven Nusselt (1916) film condensation expression (with an assumption of zero vapor shear) in the upper part of the tube, coupled with the forced-convective annular flow expression for the lower portion of the circumference for what would otherwise be a liquid pool. The rest of the model is simply a weighted average of these different contributions based on the fraction of the circumference occupied by the "stratified" portion:

$$h = \frac{h_{\text{falling-film}} r\theta + h_{\text{convective}}(2\pi - \theta)}{2\pi r} \tag{6.142}$$

Like most other models discussed in this section, all of the multiple regimes identified in the first part of the study are handled either as (a) fully annular forced-convective, or (b) consisting of varying combinations of upper gravity-driven and lower forced-convective terms. It is stated that intermittent flow is very complex, and is therefore assumed to be predicted by annular flow equations. Similarly, mist flow is handled as annular flow, assuming that the liquid inventory entrained in the vapor phase can be viewed as an unsteady annular film. Bubbly flow is not modeled in this work, partly because it is not commonly encountered. The novel manner of handling the stratified pool does, however, yield smoother transitions between the heat transfer coefficient predictions across transitions. Further, the degree of stratification is itself obtained through quadratic interpolation across the flow regime transition lines, ensuring a smooth introduction of the stratification component into the heat transfer coefficient:

$$\theta = \theta_{\text{strat}}\left(\frac{G_{\text{wavy}} - G}{G_{\text{wavy}} - G_{\text{strat}}}\right) \tag{6.143}$$

Another innovation introduced is the accounting of the effects of interfacial waves, using concepts of the "most dangerous" wavelength from stability considerations, and developing an enhancement factor due to interfacial roughness in terms of the slip ratio and the roughness amplitude:

$$f_{\text{i}} = 1 + \left(\frac{u_{\text{V}}}{u_{\text{L}}}\right)^{j}\left(\frac{(\rho_{\text{L}} - \rho_{\text{V}})g\delta^{2}}{\sigma}\right)^{k}\left(\frac{G}{D_{\text{strat}}}\right) \tag{6.144}$$

The exponents j and k are picked to be ½ and ¼ without discussion or substantiation from physical principles. The presence of the surface tension in the denominator correctly implies a decrease in the intensity (damping) of the interfacial waves at high surface tension values. In addition, the last term in the equation ensures that the waves are progressively damped out upon approach to fully stratified conditions, although why this term should be a linear decrease (evidenced by the exponent of 1 for this term) and not raised to another exponent is not made clear. The authors present numerous graphs to demonstrate that the predictions are not biased toward underprediction or overprediction with respect to the independent variables. Their model predicts the data from the large database over a wide range of fluids, operating conditions, and diameters well (with the caution stated above about the applicability to the 3.14-mm diameter), with 85% of the refrigerant heat transfer data predicted within 20%. It is interesting to note that this is about the same level of predictive accuracy as the earlier multi-flow regime correlations proposed by Cavallini et al. (2002a). This model does offer some new approaches—heat transfer data are used to refine a void fraction model, which is then used to predict flow regime transitions, which are in turn used to predict a "tuned blend" of stratified-convective heat transfer coefficients. It achieves smoothly varying heat transfer coefficients across multiple regimes (Figure 6.28) with good accuracy, using some first principles concepts and several key assumptions and empirical constants or curve-fitted terms at appropriate locations for closure. Additional substantiation of the use of an average of homogeneous and Steiner (1993)/Rouhani–Axelsson (1970) void fractions, and the physical basis for extensions of transition lines determined from boiling considerations would lend further credence to the model. Measurement, rather than deduction from heat transfer data, of the fundamental underlying quantity (void fraction), which the authors credit for much of the success of this approach,

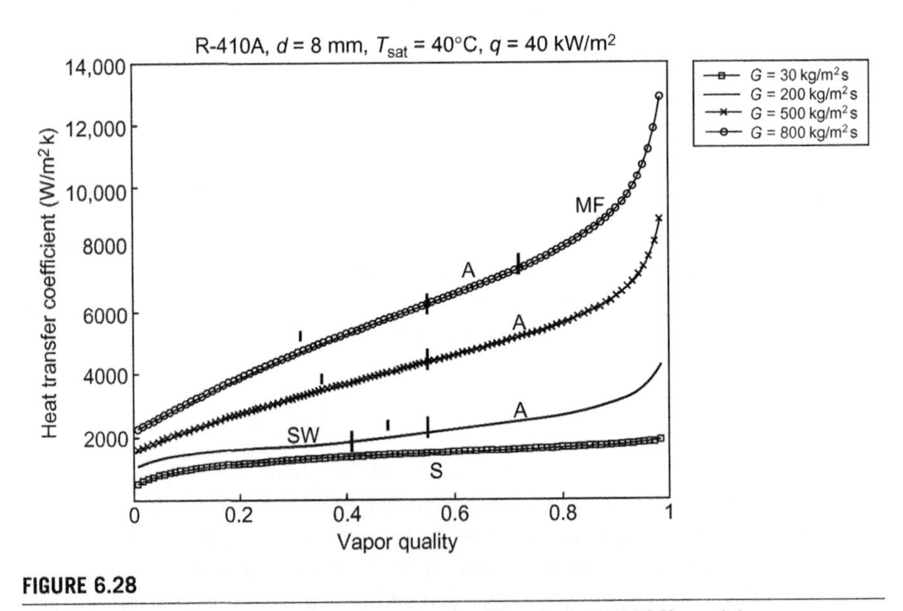

FIGURE 6.28

Heat transfer coefficient trends predicted by the Thome et al. (2003) model.

Source: *Reprinted from Thome et al. (2003) with permission from Elsevier.*

would also enhance its predictive abilities. This would be particularly important if the model were to be extended in the future to microchannels.

Goto et al. (2003) measured heat transfer coefficients for the condensation of R-410A and R-22 inside five different helical and herringbone internally grooved tubes of about 8.00-mm OD. Local heat transfer coefficients were measured for $130 < G < 400$ kg/m^2 s and $T_{sat} = 30$°C and 40°C by dividing a 1-m-long test section into 10 small sections. Unlike most other studies discussed in this chapter, the heat transfer tests were conducted in a complete vapor compression system. The reported uncertainties in the heat transfer coefficients were very high, about 40%. Heat transfer coefficients of the herringbone grooved tube were found to be about twice as large as those for the helical grooved tubes. Using Koyama and Yu's (1996) $Nu = (Nu^2_{annular} + Nu^2_{stratified})^{0.5}$ correlation for stratified and annular condensation as a starting point, they developed new expressions for the annular flow portion based on their data. A turbulent liquid film basis, $Nu = \mathrm{Re}^*_\mathrm{L}\mathrm{Pr}_\mathrm{L}/T^*_i$, was used, with the liquid Reynolds number $Re^*_\mathrm{L} = (\rho_\mathrm{L}\sqrt{\tau_\mathrm{w}/\rho_\mathrm{L}}D)/\mu_\mathrm{L}$ computed from the friction velocity with the two-phase multiplier of Goto et al. (2001), $\phi_\mathrm{v} = 1 + 1.64X_{tt}^{0.79}$, for the computation of the wall shear stress. This approach resulted in following correlation for helical and cross microfins:

$$Nu_{annular} = 0.743\sqrt{f}\left(\frac{\phi_\mathrm{V}}{X_{tt}}\right)\left(\frac{\mu_\mathrm{L}}{\mu_\mathrm{V}}\right)^{0.1}\left(\frac{x}{1-x}\right)^{0.1} Re_\mathrm{L}^{0.7} \tag{6.145}$$

while herringbone microfin tubes required a leading constant of 2.34 and a *Re* exponent of 0.62. These correlations predicted the data within 20%. Individual equations were reported for each tube for the friction factor required in the above equation.

6.5.2 Condensation in small channels

There has been a slow progression toward the measurement and modeling of heat transfer coefficients in small channels, and unlike the studies on flow regimes, void fraction, and pressure drop cited in previous sections, there are few studies that address $D_\mathrm{H} \leq 1$ mm.

Yang and Webb (1996a) measured heat transfer in single- and two-phase flow of refrigerant R-12 at 65°C in rectangular plain and microfin tubes with $D_\mathrm{H} = 2.64$ and 1.56 mm, respectively, using the modified Wilson plot technique. The companion pressure drop study (Yang and Webb, 1996b) was discussed in the previous section. The measurements were conducted at somewhat higher mass fluxes ($400 < G < 1400$ kg/m^2 s) than those of interest for refrigeration and air-conditioning applications. They found that the Shah (1979) correlation significantly overpredicted the data, whereas the Akers et al. (1959) correlation showed better agreement, except at high mass fluxes, where it also overpredicted the data. The microfin tube heat transfer coefficients showed steeper slopes than those for the plain tubes when plotted against quality, especially at the lower mass fluxes.

The heat transfer coefficients also showed a heat flux dependence ($h \alpha q''^{0.2}$); such heat flux dependence is typically seen in stratified flows (which is certainly not the case at such high mass fluxes) and in boiling. They explained this dependence based on the work of Soliman et al. (1968), who argued that the momentum contribution will cause an increase in heat transfer coefficient when q'' is increased. Their results also showed that the heat transfer enhancement due to the microfins decreased with increasing mass flux. By plotting the heat transfer data against the equivalent mass velocity Reynolds number proposed by Akers et al. (1959), they showed that while data for different mass fluxes and qualities collapsed to a single curve for plain tubes, this only occurred at low qualities for the microfin tube. This led them to explain the different trends on the basis of the drainage of the liquid film by the microfins. They reasoned that at high qualities, the microfins are not fully submerged in liquid; therefore, they act like Gregorig (1962) surfaces, helping to drain the film to the base of the fins, thus increasing heat transfer over the plain tube values. At low qualities, however, the fins are submerged; therefore, the drainage mechanism is not active. According to them, this mechanism is particularly important at low mass fluxes, because at high mass fluxes, vapor shear exerts the dominant influence on heat transfer. It should be noted that in the companion work, Yang and Webb (1996b) had concluded that surface tension did not play a role in determining the pressure drop in these same tubes.

Yang and Webb (1997) continued this work to develop a heat transfer model for condensation of R-12 and R-134a in extruded microchannels ($D_H = 1.41$ and 1.56 mm) with microfins (0.2 and 0.3 mm deep). The data for these channels were reported in their earlier papers cited above. They used the observations discussed above about the contribution of vapor shear and surface tension to represent the heat transfer coefficient as follows:

$$h = h_u \frac{A_u}{A} + h_f \frac{A_f}{A} \tag{6.146}$$

where the subscripts u and f refer to unflooded and flooded portions of the channel, respectively. They calculated the surface and cross-sectional areas of the liquid in the drainage regions using the geometric features of the microfin and the Zivi (1964) void fraction model to estimate the liquid volume occupying the drainage region. For flooded conditions (e.g., at low vapor qualities) they used a slightly modified version of the Akers et al. (1959) model (which corresponds to their experimentally derived single-phase Nusselt number $Nu_{D_H} = 0.10 Re_{D_H}^{0.73} Pr_L^{1/3}$) to compute the shear-driven heat transfer coefficient. When the microfins were unflooded, they assumed a thin laminar film draining on the walls of the microfin and computed the pressure gradients in the axial and microfin wall directions due to vapor shear and surface tension, respectively. These perpendicular vapor shear (sh) and surface tension (st) stresses were combined to yield the wall shear stress, which was then related to the unflooded heat transfer coefficient, $h_u^2 = h_{sh}^2 + h_{st}^2$. Relating the surface-tension-driving force to the microfin profile and the Weber number $We = (G_{eq}^2 D_H)/(\sigma_1 \rho)$, they developed the following expression for the surface tension contribution to the heat transfer coefficient:

$$h_{st} = C \frac{\tau_z - \tau_i}{dp/dz} k_1 \frac{d(1/r)}{ds} \frac{Re_{eq} Pr_1^{1/3}}{We} \tag{6.147}$$

where z refers to the axial direction, subscript i refers to the interface, and C was derived through regression of their data to be 0.0703. The relative fractions of the surface tension and vapor shear contributions to the heat transfer coefficient were plotted to show that at low mass flux ($G = 400$ kg/m^2 s) and $x > 0.5$, the surface tension contribution could equal and exceed the vapor shear term. At high mass fluxes ($G = 1400$ kg/m^2 s), the surface tension contribution was very small. A small fin tip radius enhanced the heat transfer, while a large inter-fin drainage area allowed the surface tension effect to be activated at lower qualities.

Kim et al. (2003) conducted a study similar to that of Yang and Webb (1996a, b, 1997) above, on similar tubes ($D_H = 1.41$ smooth, 1.56-mm microfin with 39% greater surface area than the smooth tube), with a similar experimental setup using R-22 and R-410A for $200 < G < 600$ kg/m^2 s at 45°C. Representative tube cross-sections and the test section used for the experiments are shown in Figure 6.29. To test the tubes, they formed a jacket around the channel with a 1-mm gap. Since the Wilson plot technique was used, it was important to minimize the coolant-side resistance (an issue not recognized or stated by investigators often enough.) In their tests, the coolant side constituted approximately one-third of total resistance. Further decreases could not be achieved because the required higher coolant flow rates would result in an extremely small coolant temperature rise, which would significantly decrease the accuracy of heat duty measurement. At the low mass flux conditions, the heat duties were measured using a 1.3°C coolant temperature rise. For R-22, the microfin tubes yielded higher heat transfer coefficients than the smooth tube, with the enhancement decreasing to 1 at a mass flux of 600 kg/m^2 s. For R-410A, the microfin tube heat transfer coefficients were similarly higher than those of the smooth tubes at low mass fluxes, but in fact decreased to values lower than those for the smooth tube at $G = 600$ kg/m^2 s. Using the rationale provided by Carnavos (1979, 1980) for single-phase performance in finned tubes, they explained this decrease in heat transfer by postulating that fins reduce the inter-fin velocity, thus decreasing heat transfer. They further pointed out that in condensing flows, the surface tension drainage force that is prominent at low mass fluxes compensates for the decrease in velocity, whereas at high mass flux, the velocity reduction effect dominates. No detailed validation of this hypothesis is provided. They found also that in smooth tubes, R-410A heat transfer coefficients were slightly higher than those of R-22, while the opposite was true for microfin tubes. In smooth tubes, the higher R-410A heat transfer coefficients were attributed to its larger thermal conductivity and smaller viscosity. In microfin tubes, however, they noted that the Weber number of R-22 was 2.7 times lower than that of R-410A at the test conditions, leading to greater surface tension drainage and heat transfer. The two times larger vapor to liquid

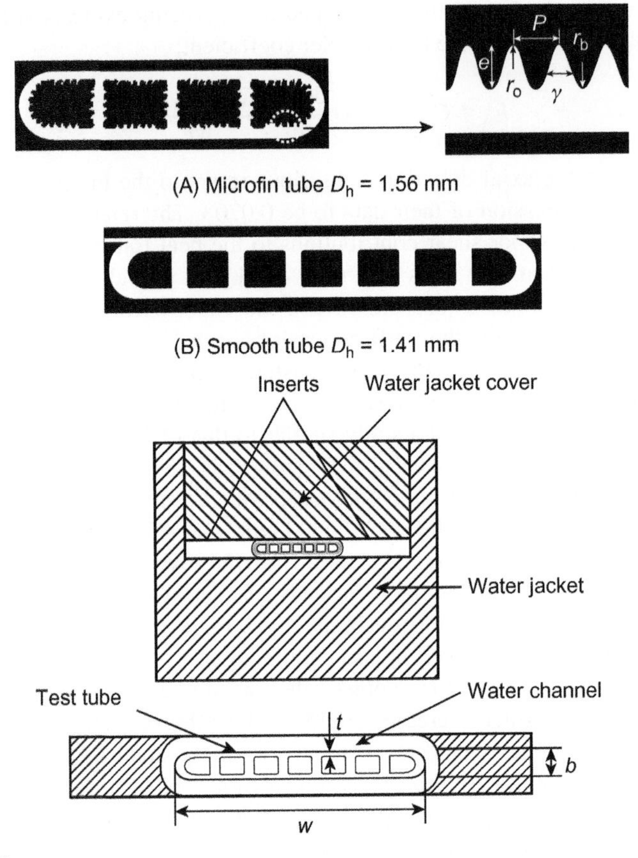

FIGURE 6.29

Representative plain and microfin extruded channels and test section used by Yang and Webb (1996a) and Kim et al. (2003): (A) microfin tube $D_h = 1.56$ mm and (B) smooth tube $D_h = 1.41$ mm

Source: *Reprinted from Kim et al. (2003) with permission from Elsevier and International Institute of Refrigeration (iifiir@iifiir.org or www.iifiir.org).*

volume ratio of R-22 was also believed to have resulted in a larger fraction of the fin surface being exposed to the vapor, thus enhancing heat transfer. Several other plausible explanations in terms of the fluid properties are provided, although additional validation would be necessary to confirm these explanations of the small differences in heat transfer coefficients, especially in view of the uncertainties in the measurement of heat transfer coefficients at these small heat transfer rates. Based on these trends, they recommended the use of Moser et al.'s (1998) model with a modified two-phase multiplier for the smooth tubes, and Yang and Webb's (1997) model with minor modifications for the microfin tubes.

Yan and Lin (1999) measured heat transfer coefficients during condensation of R-134a through a bank of 2-mm tubes arranged in parallel. Their pressure drop results were discussed in the previous section. Copper plates, 5 mm thick, were soldered onto the bank of refrigerant tubes, and coolant tubes were in turn soldered onto the copper plates to provide a crossflow cooling orientation. The authors state that the tubes and the plate had good thermal contact, eliminating the need for avoiding gaps between the tubes and the plates. However, they provide no substantiation that the thermal resistance from the refrigerant to the coolant through the refrigerant and coolant tubes, the gaps between the plates and the tubes, and the two 5-mm-thick copper plates was either accounted for properly, or was in fact small in magnitude. The issue of maldistribution through the 28 parallel channels was also not discussed in detail, except to state that the inlet and outlet headers were appropriately designed. Their single-phase friction factors were significantly higher than the predictions of the Blasius equation, and the single-phase heat transfer coefficients were also higher than the predictions of the Gnielinski (1976) correlations. Entrance length effects and tube roughness may have accounted for some of this, according to the authors. Condensation heat transfer coefficients were higher at the lower saturation temperatures, especially at the higher qualities. They attributed this to the lower thermal conductivity of R-134a at the higher temperatures. The heat transfer coefficients also decreased significantly as the heat flux was increased, particularly at the higher qualities. This is not typical of condensation heat transfer, where any heat flux (or ΔT) dependence is usually seen at the lower qualities in gravity-dominated flow. The graphs they presented to demonstrate the effect of mass flux show inflection points where the heat transfer coefficient first increased sharply as quality was increased from around 10% to 20%, followed by a flattening with further increases in quality at the lower mass fluxes, finally again increasing with quality. Adequate explanations for these trends are not provided. They proposed the following correlation for their data:

$$\frac{hD}{k_l}Pr_l^{-0.33}Bo^{0.3}Re = 6.48Re_{eq}^{1.04} \quad Bo = \frac{q''}{h_{fg}G} \tag{6.148}$$

It is not clear why the Reynolds number as well as the equivalent Reynolds number according to the Akers et al. (1959) model were required to correlate the data.

Rectangular channels with $D_H = 1.46$ mm (1.50×1.40 mm) were investigated by Wang et al. (2002) for the condensation of R-134a at 61.5–66°C over the mass flux range 75–750 kg/m^2 s. The experiments were conducted in 610-mm-long tubes with finned 10 multiport channels cooled by air in crossflow. Flow visualization experiments were also conducted by replacing one aluminum wall of the tube with a glass window to help determine the applicable flow regimes. The authors report inlet qualities to the test section, which were controlled over a wide range; however, the quality change across the test section was substantial, indicating that for many of the tests, the experiments did not yield local heat transfer coefficients over small increments of quality. From the combination of flow visualization and heat transfer experiments, they note that even in a channel this small, stratified (wavy) flows were seen at the low mass fluxes. The heat transfer coefficients were insensitive to quality at the low mass fluxes. At the high mass fluxes, the dependence on quality was strong, signifying annular flow. Of the available heat transfer correlations, the Akers et al. (1959) equivalent *Re* correlation was found to agree most with their data in the annular flow regime, while the Jaster and Kosky (1976) correlation agreed best with their stratified flow data. However, they noted that the Re_{eq} concept of Akers et al. (1959) does not explicitly account for the applicable flow regime, because Re_{eq} can be the same for different flow regimes. They attempted to improve the predictions by developing their own stratified and annular flow correlations. They separated the data into stratified and annular regimes using the Breber et al. (1980) and Soliman (1968, 1986) transition criteria. Their stratified-annular transition occurred at a lower gas superficial velocity ($j_g^* = 0.24$ m/s) than that predicted by the Breber et al. map, because the corners of the rectangular channel aided the formation of annular flow. This earlier transition to annular flows was also observed with respect to the Soliman map, while the annular-mist transition was not observed. For annular flow, they conducted a boundary-layer analysis similar to Traviss et al. (1973), and through the analysis of their data, proposed the following curve-fits for the two-phase multiplier, and unlike most other researchers, for the dimensionless temperature as well:

$$\phi_v = \sqrt{1.376 + 8X_{tt}^{1.655}} \quad T_\delta^+ = 5.4269\left(\frac{Re_l}{x}\right)^{0.2208} \tag{6.149}$$

The resulting annular flow correlation was as follows:

$$Nu_{annular} = 0.0274 Pr_l Re_l^{0.6792} x^{0.2208}\left(\frac{1.376 + 8X_{tt}^{1.655}}{X_{tt}^2}\right)^{0.5} \tag{6.150}$$

For stratified flows, they used the Chato (1962) correlation for the film condensation portion and the Dittus–Boelter (1930) correlation for forced convection in the liquid pool, combining the two as follows:

$$Nu_{stratified} = \alpha Nu_{film} + (1 - \alpha) Nu_{convection} \tag{6.151}$$

Therefore, they essentially used the Zivi (1964) void fraction as a measure of the liquid pool height in this rectangular channel. In addition, for design purposes, especially since their condensation tests were conducted over large quality changes, they proposed a weighted average between these two correlations using the quality at Soliman Froude number $Fr_{so} = 8$ as the transition between the regimes.

Koyama et al. (2003b) conducted a study on the condensation of R-134a in two multiport extruded aluminum tubes with eight 1.11-mm channels, and nineteen 0.80-mm channels, respectively. Local heat transfer coefficients were measured at every 75 mm of the 600-mm-long cooling section using heat flux sensors. They acknowledged that it is very difficult to accurately measure local heat transfer coefficients in such small channels using the temperature rise in the coolant or a Wilson plot method due to the inaccuracies associated with the measurement of low heat transfer rates and small temperature differences. The tests were conducted over the range $100 < G < 700$ kg/m^2 s at 60°C. For correlating their data, they used a combination of convective and film condensation terms to yield $Nu = (Nu^2_{annular} + Nu^2_{stratified})^{0.5}$. The individual annular and stratified terms were obtained from the work of Haraguchi et al. (1994b) except that their two-phase multiplier was replaced by the multiplier proposed by Mishima and Hibiki (1996b). They stated that this was an improvement over several other correlations; however, the graphs of predicted and experimental Nusselt numbers show consistent and large (as high as 80%) overpredictions of all their data at almost all conditions.

In some recent studies, Wang, Rose, and co-workers have developed analytical approaches for addressing condensation heat transfer in triangular (Wang and Rose, 2004) and square (Wang and Rose, 2005; Wang et al., 2004) microchannels. For film condensation of R-134a in horizontal 1-mm triangular channels, Wang and Rose (2004) assumed that the condensate was in laminar flow, and developed one of the first models that accounts for surface tension, shear stress, and gravity. They were able to predict and substantiate the varying condensate flow pattern across the cross-section as well as along the length of the channel. Thus, they were able to model the corresponding variations in heat transfer coefficient. Using a similar approach, Wang et al. (2004) developed a model for film condensation of R-134a in square, horizontal, 1-mm microchannels. As these papers account for the three primary governing influences in microchannel condensation, they provide a good start for the modeling of phenomena specific to microchannels. Furthermore, the respective forces can be "switched on or off" to understand their respective significances (Figure 6.30).

In another recent study, Cavallini et al. (2005a) conducted measurements of heat transfer coefficients and pressure drops during condensation of R-134a and R-410A inside multiple parallel 1.4-mm D_H channels. The test section was divided into three separate segments to provide quasi-local pressure drops and heat transfer coefficients. They deduced the frictional pressure drop from the measured drop in saturation temperature and found good agreement between the data for R-134a and the correlations of Friedel (1979), Zhang and Webb (2001), Mishima and Hibiki (1996a,b), and Müller-Steinhagen and Heck (1986). All of these correlations were found to overpredict the R-410A data, however. Their heat transfer coefficients were obtained from wall temperature measurements, and they found that the available models in the literature (Akers et al., 1959; Cavallini et al., 2002a; Koyama et al., 2003a; Moser et al., 1998; Wang et al., 2002; Zhang and Webb, 2001) underestimated their results, particularly at high values of mass flux. They attributed these differences to the much higher gas velocities in their experiments where mist flow might prevail, whereas the available correlations are primarily for annular flow in larger-diameter tubes. Later, Cavallini et al. (2006a) presented a new flow-regime-based ($\Delta T = (T_{sat} - T_{wall})$-dependent and ΔT-independent flow regimes) model developed from an extensive database of condensation heat transfer research

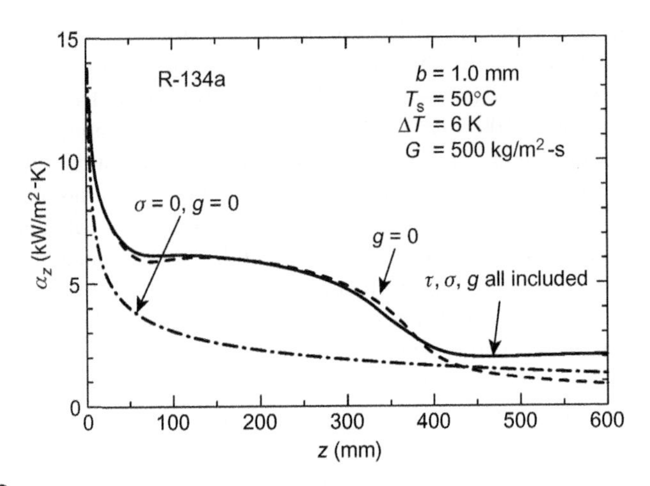

FIGURE 6.30

Variation of mean heat transfer coefficient along a square 1-mm channel.

Source: *Reprinted from Wang and Rose, 2005*

in tubes with $D > 3$ mm. Despite being comprised of data for large tubes, Del Col et al. (2010) found that the model predicted condensation data for R-1234yf in circular channels ($D = 0.96$ mm) within $\pm 15\%$.

Kim and Shin (2005) used a technique that matched the outlet temperature of an electrically heated air stream with that of a similar air stream heated by condensing refrigerant R-134a to measure small, local condensation heat transfer rates. Circular and square channels with $0.5 < D_H < 1$ mm were tested for the mass flux range $100 < G < 600$ kg/m^2 s at 40°C. For circular and square channels, they also found that most of the available models and correlations discussed above (Akers et al., 1959; Cavallini and Zecchin, 1974; Dobson, 1994; Moser et al., 1998; Shah, 1979; Soliman et al., 1968; Traviss et al., 1973) underpredict their data at the low mass fluxes, which they deemed to be the important range for engineering applications. The agreement with these correlations improved somewhat at the higher mass fluxes. Although they found no significant effect of the heat flux, they noted that at lower mass fluxes, square channels had higher heat transfer coefficients than those for circular channels, whereas the reverse was true for high mass fluxes. No satisfactory explanation was provided for these trends.

Bandhauer et al. (2006) and Agarwal et al. (2010) presented experimental results and new models for condensation of R-134a in circular ($0.506 \leq D \leq 1.524$ mm) and noncircular ($0.424 \leq D_H \leq 0.839$ mm) microchannels. Data were obtained for mass fluxes ranging from 150 to 750 kg/m^2 s. Experiments were conducted on the facility used for the pressure drop experiments of Garimella et al. (2002, 2003b, 2005), described above. They specifically addressed the problems in heat transfer coefficient determination due to the high heat transfer coefficients and low mass flow rates in microchannels. For the small Δx required for local measurements, the heat duties at the mass fluxes of interest are relatively small, and to ensure reasonable accuracies in heat duty measurement, the coolant inlet-to-outlet temperature difference must be increased to minimize uncertainties. However, since this requires low coolant flow rates, the coolant-side thermal resistance becomes dominant, making it difficult to deduce the refrigerant-side resistance from the measured U_A. These conflicting requirements for the accurate measurement of heat duty and the refrigerant heat transfer coefficients were resolved by developing a thermal amplification technique (Figure 6.31) that decoupled these two issues. Thus, the test section was cooled using water flowing in a closed (primary) loop at a high flow rate to ensure that the condensation side presented the governing thermal resistance. Heat exchange between this primary loop and a secondary cooling water stream at a much lower flow rate was used to

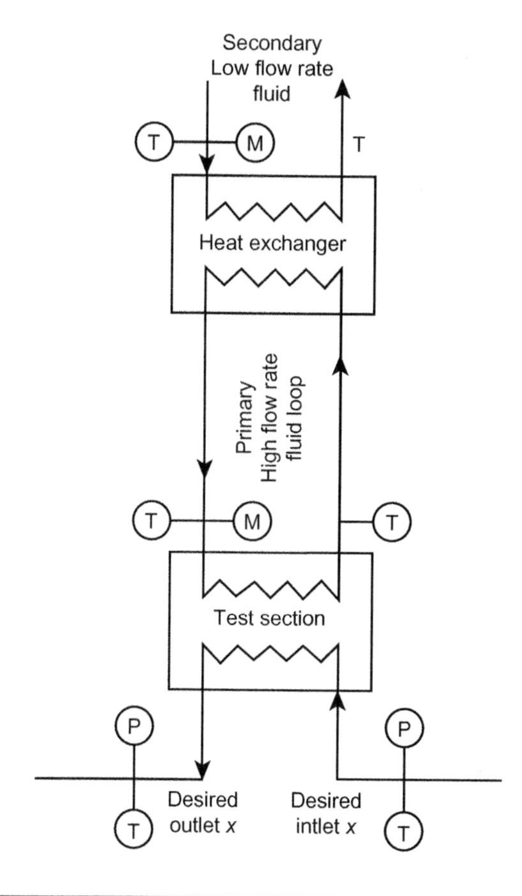

FIGURE 6.31

Thermal amplification technique.

Source: *Reprinted from Bandhauer et al. (2006).*

obtain a large temperature difference, which was in turn used to measure the condensation duty. The secondary coolant flow rate was adjusted as the test conditions change to maintain a reasonable ΔT and also small condensation duties in the test section. By ensuring that the pump heat dissipation in the primary loop and the ambient heat loss were small fractions of the condensation load, sensitivity to these losses and gains was minimized. Local heat transfer coefficients were therefore measured accurately in small increments for the entire saturated vapor–liquid region. Additional details of this thermal amplification technique are provided in Garimella and Bandhauer (2001). The thermal amplification resulted in uncertainties typically as low as $\pm 2\%$ in the measurement of the secondary loop heat duty, even at the extremely small heat transfer rates under consideration. Combining the errors in the secondary loop duty, the pump heat addition, and the ambient heat loss, the local condensation duty in these small channels was typically known to within a maximum uncertainty of $\pm 10\%$. The large coolant flow rate and the enhancement in surface area (indirect area of about 4.7 times the direct area) provided by the coolant port walls on both sides of the microchannel tube resulted in high refrigerant to coolant resistance ratios (between 5 and 30). With this high resistance ratio, even an uncertainty of $\pm 25\%$ in the tube-side heat transfer coefficient did not appreciably affect the refrigerant-side heat transfer coefficient, ensuring low uncertainties. Representative heat transfer coefficients reported by Bandhauer et al. (2006) are shown in Figure 6.32. In general, the heat transfer coefficient increases with decreasing diameter, with the effect of diameter becoming more significant at the higher qualities ($\sim x > 0.45$) and mass fluxes. Thus,

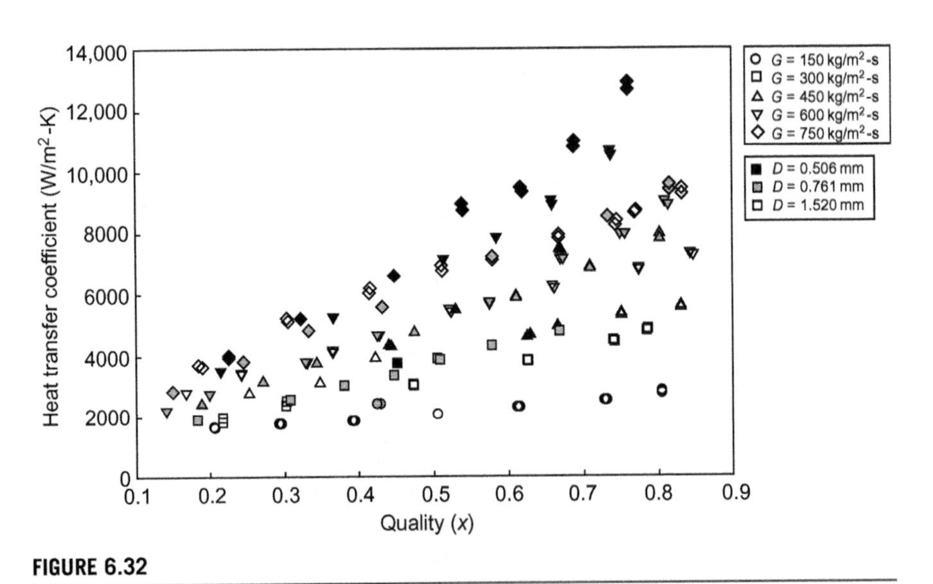

FIGURE 6.32

Microchannel condensation heat transfer coefficients.

Source: *From Bandhauer et al. (2006).*

h increases between 10% and 40% for $x > 0.45$ as D_H is reduced from 1.524 to 0.506 mm, with the effect of D_H being more significant for a decrease in D_H from 0.761 to 0.506 mm. Gravity-driven correlations were found to be poor predictors of the data, with the Chato (1962) correlation underpredicting the data considerably, the Jaster and Kosky (1976) correlation resulting in significant overprediction, and the Rosson and Meyers (1965) and Dobson and Chato (1998) correlations being somewhat better—these latter correlations account for heat transfer through the bottom of the tube. This is not surprising because these models are not appropriate for Garimella and Bandhauer's data, which were demonstrated to be not in the stratified wavy regimes (Coleman and Garimella, 2000a,b, 2003). Among correlations that use a two-phase multiplier approach, the Moser et al. correlation predicted the data reasonably well (14% average deviation), with the Shah (1979) and Cavallini and Zecchin (1974) correlations resulting in some overprediction. Homogeneous flow correlations of Boyko and Kruzhilin (1967) and the mist flow correlation of Soliman (1986) were better at predicting the data, with the Soliman correlation leading to some overprediction. Shear-dominated annular flow correlations using boundary-layer treatments also showed reasonable agreement: Soliman et al. (1968) (18% average deviation), Chen and Kocamustafaogullari (1987) (13% average deviation). However, in spite of the theoretical basis of the Traviss et al. (1973) correlation, as reported by several investigators, this model showed the largest average deviation (38%) among shear-driven treatments.

Bandhauer et al. (2006) noted that during the condensation process, the flow changes from mist flow (where applicable) to annular film flow to intermittent, with large overlaps in the types of flow resulting in transition flow (intermittent/annular, intermittent/annular/mist, and annular/mist). Based on the work of Coleman and Garimella (2000a,b, 2003), wavy flow was not expected in any of the channels considered. Due to the large overlap regions between these flows, with the bulk of the data being in transition between annular flow and other regimes, they developed a heat transfer model based on annular flow considerations. They noted that many of the available shear-driven models, though sound in formulation, led to poor predictions because of the inadequate calculation of shear stresses using pressure drop models that were not applicable to microchannels. Thus, their model is based on boundary-layer analyses, but with the requisite shear stress being calculated from the pressure drop models of Garimella et al. (2005) developed specifically for microchannels. Since, in annular flow, the condensate layer might become turbulent at very low Re_l (~ 240) (Carpenter and Colburn, 1951) due to the turbulent vapor core and the destabilizing effect of condensation (Schlichting and Gersten, 2000), a model based on turbulent parameters with appropriate modifications for the conditions under study was developed. The friction velocity $u^* = \sqrt{\tau_i/\rho_l}$ was expressed in terms of the interfacial shear stress, rather than the commonly used wall shear stress. The turbulent dimensionless temperature was defined as follows:

$$T^+ = \frac{\rho_l \times C_{pl} \times u^*}{q''}(T_i - T_w) \tag{6.152}$$

The shear stress and heat flux were expressed in the usual manner:

$$\tau = (\mu + \rho \times \varepsilon_\mathrm{m})\frac{du}{dy} \tag{6.153}$$

$$q'' = -(k + \varepsilon_\mathrm{h} \times \rho \times C_p)\frac{dT}{dy} \tag{6.154}$$

Assuming that the all of the heat is transferred in the liquid film results in the following expression for the heat transfer coefficient, where the interface is at the saturation temperature:

$$h = \frac{q''}{(T_\mathrm{sat} - T_\mathrm{w})} = \frac{\rho_\mathrm{l} \times C_{p_\mathrm{l}} \times u^*}{T^+} \tag{6.155}$$

The heat flux above yields the dimensionless temperature gradient:

$$\frac{dT^+}{dy^+} = \left(\frac{1}{Pr_\mathrm{l}} + \frac{\rho_\mathrm{l} \times \varepsilon_h}{\mu_\mathrm{l}}\right)^{-1} \tag{6.156}$$

To integrate this, the turbulent film thickness $\delta = (1 - \sqrt{\alpha})D/2$ is determined using the Baroczy (1965) model, which is different from the manner in which Traviss et al. (1973) determined film thickness. The dimensionless turbulent film thickness is then $\delta^+ = (\delta\rho^*_{\mathrm{l}u})/\mu_\mathrm{l}$. Also, $\tau = (1 - y/K)\tau_\mathrm{w}$ yields the following equation:

$$\frac{\rho_\mathrm{l}\varepsilon_\mathrm{m}}{\mu_\mathrm{l}} = \frac{1 - y^+/R^+}{du^+/dy^+} - 1 \tag{6.157}$$

In a manner analogous to the development by Traviss et al. (1973), with the assumptions of small film thickness compared to the tube radius, and $\varepsilon_\mathrm{m} \cong \varepsilon_\mathrm{h}$, the following simplified *two-region* turbulent dimensionless temperature expressions were proposed: for $Re_\mathrm{l} < 2100$,

$$T^+ = 5Pr_\mathrm{l} + 5\ln\left[Pr_\mathrm{l}\left(\frac{\delta^+}{5} - 1\right) + 1\right] \tag{6.158}$$

for $Re_\mathrm{l} > 2100$,

$$T^+ = 5Pr_\mathrm{l} + 5\ln(5Pr_\mathrm{l} + 1) + \int_{30}^{\delta+} \frac{dy^+}{((1/Pr_\mathrm{l}) - 1) + (y^+/5)(1 - (y^+/R^+))} \tag{6.159}$$

The only unknown quantity, the interfacial shear stress for determining u^*, is obtained from the annular flow portion of the multiple-flow regime pressure drop model by Garimella et al. (2005), discussed in the previous section. Thus, the interfacial friction factor (and then the interfacial shear stress) is computed from the corresponding liquid-phase Re and friction factor, the Martinelli parameter, and the surface tension parameter: $f_\mathrm{i}/f_\mathrm{l} = AX^a Re_\mathrm{l}{}^b \psi^c$. Additional details are available in the previous section. In summary, to obtain the heat transfer coefficient, the interfacial shear stress is first calculated using the pressure drop model. The resulting shear stress is used to compute the friction velocity u^* and the dimensionless film thickness δ^+. The dimensionless temperature T^+ is then calculated, which yields the heat transfer coefficient. This model predicted 86% of the data within $\pm 20\%$, with an average absolute deviation of 10%. A more explicit accounting of the heat transfer in the liquid slugs, vapor bubbles, and the film–bubble interface in intermittent flow, and of the entrainment of liquid into the vapor core in mist flow would improve the predictions further. The data and the predictions at representative mass fluxes for each tube are shown in Figure 6.33. The heat transfer coefficient increases with an increase in mass flux and quality, and with a decrease in the tube diameter. As the liquid film becomes thinner with increasing vapor quality, the heat transfer coefficient increases. The steep slope at the higher qualities represents an approach to a vanishingly thin film. However, it must be noted that the actual behavior in this region will be dependent on the inlet conditions of a condenser, including the inlet superheat, the coolant temperature, and the consequent wall subcooling; these phenomena are not accounted for in this model. Although the model was developed based on an annular flow mechanism, it appears to predict the heat transfer coefficients in the mist and mist-annular overlap regions in the high-mass flux and high-quality cases adequately. At the extreme end of the graph ($x > 80\%$), the flow may be exclusively in the mist flow region. For a given flow rate, the quality at which the mist flow regime occurs could decrease with decreasing diameter due to increased interfacial shedding. As liquid entrainment increases, the heat transfer coefficient will increase due to the thinning of the liquid film (Soliman, 1986). Although liquid entrainment and mist formation are not explicitly accounted for in the heat transfer part of this model, the trends in the data (the steeper slope in the data at the higher qualities, particularly for small D_H) can be interpreted on this basis.

In addition to the pressure drop experiments described previously, Andresen (2007) conducted near-critical condensation heat transfer experiments with R-404A and R-410A in horizontal tubes with diameters ranging from 0.76 to 9.4 mm. Using an experimental technique similar to that used by Bandhauer et al. (2006), he measured local heat transfer coefficients in small quality increments for mass fluxes ranging from 200 to 800 kg/m^2 s. As with the pressure drop results, he found that existing correlations failed to accurately predict heat transfer in this region. Thus, he introduced a multi-regime model for heat transfer in the wavy, annular, and annular/wavy regimes. For microchannels, the annular flow regime is expected to be more prevalent. In this region, he proposed an empirical model where the two-phase multiplier closely resembles the Martinelli parameter.

$$Nu_{\mathrm{annular}} = 0.0133 \cdot Re_{\mathrm{L}}^{4/5} \cdot Pr_{\mathrm{L}}^{1/3} \cdot \left[1 + \left(\frac{x}{1-x}\right)^{0.80} \left(\frac{\rho_l}{\rho_{\mathrm{v}}}\right)^{0.88}\right] \tag{6.160}$$

Kim and Mudawar (2012) reported heat transfer results and proposed a new correlation for FC-72 condensing in parallel square ($D_H = 1$ mm) microchannels.

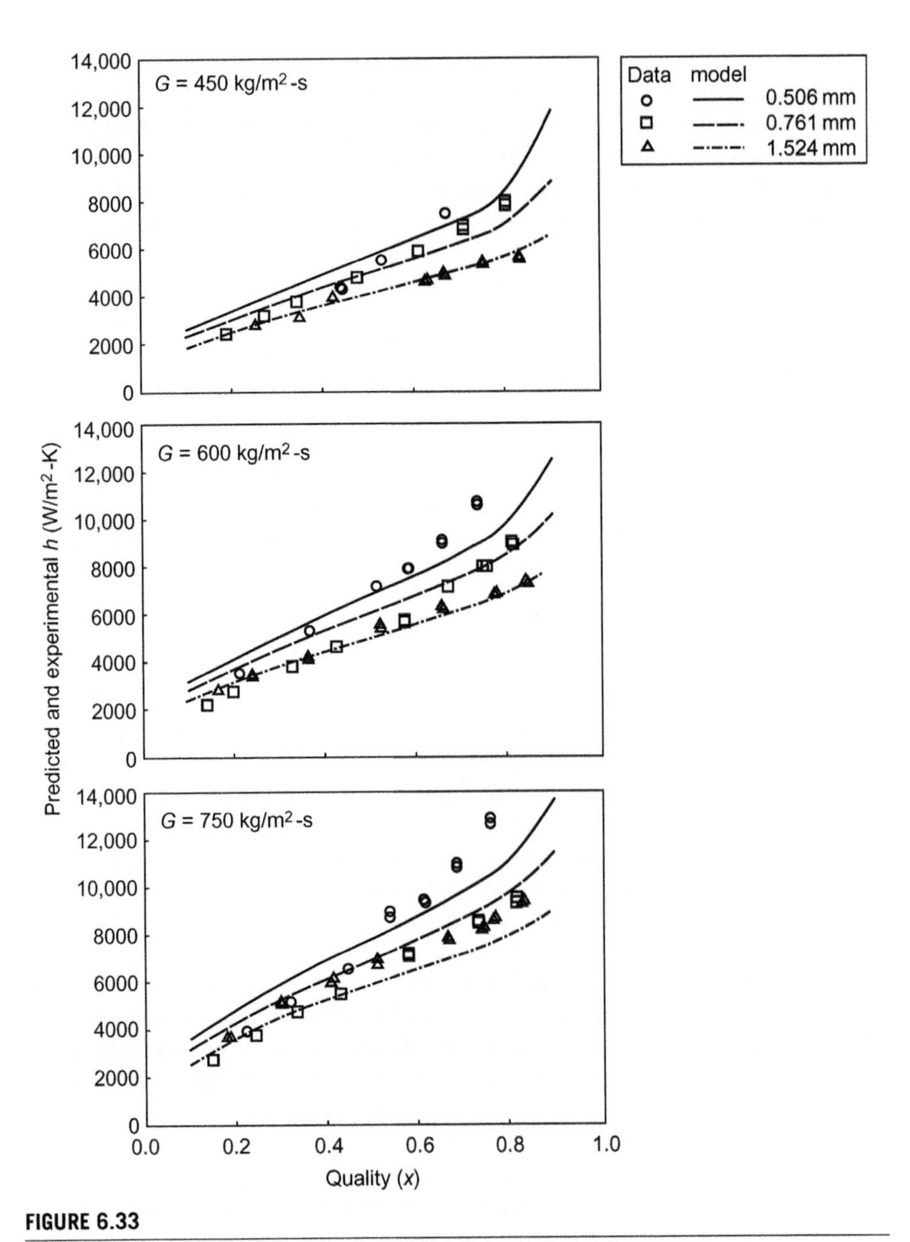

FIGURE 6.33

Microchannel condensation heat transfer coefficient predictions.

Source: *From Bandhauer et al. (2006).*

Part one of the study (Kim et al., 2012) was discussed previously in the flow regime section. Because the test section had a transparent top cover for visual access, they assumed the condensing flow was cooled only on three sides. They then applied a correction for three-sided cooling to the calculated values to allow comparison with conventional condensation heat transfer models. They reported good agreement with the conventional tube models of Cavallini and Zecchin (1974), Shah (1979), and Dobson and Chato (1998). They noted that the relatively low surface tension of FC-72 may lead to good agreement with macrochannel models, particularly in annular flow. They then proposed a new model for annular flow in microchannels with three-sided cooling derived from a boundary-layer analysis.

Recently, new numerical modeling studies have been reported by Nebuloni and Thome (2010, 2012) and Da Riva and Del Col (2012). Nebuloni and Thome (2010, 2012) investigated laminar annular condensation in microchannels with varying internal shape using a finite volume formulation of the energy equations in the liquid phase. They showed that as D_H decreases, the effect of axial conduction on local tube wall temperature and h can be significant. Da Riva and Del Col (2012) simulated condensation of R-134a in a 1-mm diameter circular channel using a volume of fluid method to assess the influence of surface tension. For the circular channel, they found condensation to be gravity-dominated. This is in contrast to the other numerical studies (Wang and Rose, 2005, 2011) and Nebuloni and Thome (2010, 2012), which show a surface tension effect when condensation occurs in geometries with sharp corners.

6.5.3 Summary observations and recommendations

The above discussion of the available literature on condensation heat transfer shows that much of the available information is on tubes larger than about 7 mm. In these tubes, heat transfer models have treated the multitude of flow regimes identified by many investigators and discussed in the previous sections as either stratified or annular flow regimes. Thus, almost all the available models in essence still use only gravity- or shear-dominated approaches. What differs from investigator to investigator is the specific analysis that addresses these two modes. In models of gravity-dominated condensation, for example, the earliest models were patterned very closely after the Nusselt condensation analysis. They then progressed from neglecting the heat transfer in the liquid pool to accounting for it in some manner, usually through the application of single-phase liquid forced-convective heat transfer correlations. The apportioning of film condensation and forced convection regions within this stratified flow has also differed in minor details. Most of these models use some void fraction model, usually the one by Zivi (1964), to determine the liquid and vapor fractions of the cross-section (which is usually quantified as the liquid pool angle), which then allows them to predict the surface area over which these two submodels are applied. In many cases, however, the void fraction models were originally based on annular flow data or analyses. A few models have also attempted to account for vapor shear on the falling-film condensation portion of the tube.

In annular flow models, the technique of relating the interfacial shear (through a boundary-layer analysis that usually uses the von Karman velocity profile) to the heat transfer across the liquid film has been used widely, starting with the work of Carpenter and Colburn (1951), with refinements and improvements to the contribution of various forces to the interfacial shear, for example, by Soliman et al. (1968), Traviss et al. (1973), and others. There have also been some attempts such as the work of Chen et al. (1987) to develop general purpose annular flow correlations starting with asymptotic limits, and blending them through some simple combination of the terms at the respective limits. Another approach has been to directly apply a multiplication factor to a single-phase liquid or vapor phase heat transfer correlation, with the multiplier determined using adaptations of the Lockhart–Martinelli (1949) and Friedel (1979) pressure drop multipliers. This approach is in fact analogous to the boundary-layer shear-stress analysis used by many investigators. The only essential difference is in where the multiplier is applied: the first type of models use the multiplier first for shear stress determination, and calculate the heat transfer coefficient directly thereafter; the second type calculate the single-phase heat transfer coefficient directly, and apply the multiplier thereafter. An explanation of the similarity of these two approaches is provided in Dobson and Chato (1998). One correlation that continues to be used frequently is the purely empirical Shah (1979) correlation, because of its simplicity, the wide range of data that were utilized in its development, and its comparatively good predictions for annular flows. Another model that has been used widely until recently is the Akers et al. (1959) technique of determining an equivalent mass flux that would provide the same shear as the two-phase flow, thus replacing the vapor core with an additional liquid flow rate, and then treating the combined flux as being in single-phase flow. At the time of its development, this approach was perhaps motivated by the desire to facilitate computation of heat transfer coefficients in a manner more familiar to most designers (i.e., single-phase analysis). However, recent papers have shown that the predictions of these models are not very good, and also that the appropriate friction factors and driving temperature difference are not applied when transforming the two-phase flow to an equivalent single-phase flow. Although corrected versions (Moser et al., 1998) of this model are now available, implementing the corrected versions renders them as involved as the boundary-layer analyses, and does not seem to offer any additional ease of use.

In recent years, large experimental efforts or analyses of data from multiple researchers have been undertaken to develop models that address the entire process of condensation from inlet to outlet, spanning a wide range of mass fluxes, diameters, and fluids. Representative models include Dobson and Chato (1998), Cavallini et al. (2002a), and Thome et al. (2003). While these studies have yielded smoother, and in general more accurate, predictions over a wide range of conditions, they too still group the multitude of flow regimes into only stratified/wavy and annular flows. This is in stark contrast to the flow visualization studies discussed in previous sections, where the tendency is more toward identifying and categorizing every single nuance of the flow structure. Important flow regimes (and increasingly so with the progression toward smaller-diameter channels) such as intermittent and mist flow have, until now, not been successfully modeled, except as unsubstantiated extensions of annular flow. A quasi-homogeneous model of Soliman (1986) is one of the few examples of a model dedicated to one such flow regime (mist).

The availability of reliable local condensation heat transfer measurements and models in microchannels of $D<3$ mm is especially limited. (The wealth of gravity-dominated condensation models in large tubes are of little relevance to condensation in microchannels.) A group of researchers led by Webb (Webb and Ermis, 2001; Yang and Webb, 1996a,b, 1997) has made some progress in measuring heat transfer coefficients in extruded aluminum tubes with multiple parallel ports of $D_H<3$ mm, although the mass fluxes they have focused on are at the high end of the range of interest for common refrigeration and air-conditioning applications. In these papers, they have attempted several different approaches to modeling the heat transfer coefficients including typical shear stress models and equivalent mass flux models. A reliable model that predicts and explains the variety of trends seen in these results has, however, yet to be developed. While surface tension has been mentioned routinely as an important and governing parameter in microchannel flows, there is almost no model that explicitly tracks surface tension forces on a first principles basis from pressure gradient to heat transfer evaluation. At best, some investigators such as Bandhauer et al. (2006) and Agarwal et al. (2010) have indirectly accounted for surface tension through a surface tension parameter in the pressure drop model used for the shear stress calculation, which has yielded accurate microchannel heat transfer predictions over a wide range of conditions. Also, Yang and Webb (1997) are among the few researchers who explicitly account for surface tension forces in microchannels (with microfins) by computing the drainage of the liquid film from the microfin tips and the associated heat transfer enhancement when the fin tips are not flooded. The analytical treatments proposed recently by Wang and co-workers (Wang and Rose 2005, 2011; Wang et al., 2004) for microchannels with $D\sim1$ mm that account for the combined influence of surface tension, shear, and gravity as condensation proceeds also hold promise.

The use of representative heat transfer correlations and models discussed in this section is illustrated in Example 6.4.

6.6 Conclusions

This chapter has provided an overview of the available literature that is relevant to the modeling of condensation in microchannels. Microchannel definitions and governing influences, flow regimes and transition criteria, void fraction correlations, pressure drop, and heat transfer were addressed. Based on the information presented, it can be said that the building blocks for the modeling of condensation in microchannels are slowly becoming available, although the bulk of the available information is on larger channels. These macrochannel studies and models serve as a starting point for corresponding studies on microchannels. By far the most progress on addressing microchannel flows has been in the area of flow regime determination: adiabatic and condensing flows have been documented in detail in minichannels, whereas in channels with $D_H < \sim 3$ mm, there are few studies of condensing flows, almost all the research being on adiabatic two-phase flow. It is not clear whether these results for adiabatic flows can be reliably extrapolated to refrigerants with lower surface tension, larger vapor densities, and other property differences. Void fractions required for closure of most heat transfer and pressure drop models have primarily been for annular flows, with recent advances yielding some empirical correlations based on video analysis for adiabatic flows in channels of $D_H \approx 1$ mm and lower. Models for pressure drops during condensation are primarily available for channels with $D_H >$ about 3 mm, and they continue to use two-phase multiplier approaches, with some modifications to the multipliers to match the data. A few flow-regime-based pressure drop models for condensing flows in microchannels are beginning to emerge.

Heat transfer coefficients for condensation present additional challenges that must be addressed with great care when designing and running experiments. The widely used Wilson plot techniques that were ubiquitously used for large channels are limited in their utility for small channels. Controlled experiments that can measure heat transfer rates on the order of a few watts for local condensation ($\Delta x \approx 0.05$) in microchannels are particularly challenging. Innovative techniques, including nonintrusive measurements, and attention to the many sources of uncertainty are essential to ensure providing an accurate database for which models can be developed. As stated in this chapter, this challenge of small heat duties is coupled with the large condensing heat transfer coefficients, which imply that heat is often being transferred across $\Delta T < \sim 0.5°C$. Any appreciable uncertainties in temperature measurement would render the resulting heat transfer coefficients unreliable and inaccurate. On the analysis front, models for heat transfer coefficients reflect the cumulative effects of reliable or unreliable establishment of flow regimes, modeling of void fractions, and computation of shear stress through pressure drop, and, of course, uncertainties in the measurement of heat transfer coefficients. Thus, models that account for the underlying phenomena, including surface tension considerations unique to microchannels, are particularly important.

Research needs also extend to practical issues such as maldistribution of refrigerant flows through multiple channels—many innovative designs fail or are infeasible because adequate consideration is not given to establishing uniform distribution of two-phase fluid through parallel channels. Similarly, models for "minor" losses through expansions, bends, and contractions should be developed to provide the designer a complete set of tools.

Environmental and other concerns will continue to change the refrigerants that are acceptable and allowable for use in practical air-conditioning and refrigeration systems. Therefore, the models developed must be robust enough to be used for a variety of fluids, including refrigerant blends. If the chosen refrigerants are zeotropic blends, techniques for handling the additional mass transfer resistance must either be validated using large channel methodologies as a basis, or new modeling techniques should be developed. The acceptable synthetic refrigerant blends and natural refrigerants such as CO_2 are increasingly higher-pressure refrigerants; therefore, the effect of the approach to critical pressure on condensation phenomena requires special attention. Also, fluids such as CO_2 and steam have high thermal conductivities, specific heats, and latent heat of condensation, which make measurement of the resulting higher heat transfer coefficients particularly challenging.

These are but a few of the foreseeable research challenges and needs; given the state of infancy of understanding and modeling condensation in microchannels, a considerable amount of careful effort is required to obtain a comprehensive understanding. The achievement of such a fundamental understanding of condensation at the microscale will yield far-reaching benefits not only for the air-conditioning and refrigeration industries, but also for other as yet untapped applications such as portable personal cooling devices, hazardous duty, high ambient air-conditioning, sensors, and medical/surgical devices, to name a few.

Example 6.1 Flow regime determination

Determine the applicable flow regime for the flow of refrigerant R-134a in a 1-mm diameter tube at a pressure of 1500 kPa, at mass fluxes of 150, 400, and 750 kg/m^2 s, and qualities of 0.2, 0.5, and 0.8. Use the flow regime maps and transition criteria of Sardesai et al. (1981), Tandon et al. (1982), Dobson and Chato (1998), Breber et al. (1980), Soliman (1982, 1986), Coleman and Garimella (2000a,b, 2003), and Cavallini et al. (2002a), and compare and comment on the flow regimes predicted by each model.

Refrigerant properties

Saturation temperature $T_{sat} = 55.21°C$
Surface tension $\sigma = 0.00427$ N/m

Liquid Phase	Vapor Phase
$\rho_l = 1077$ kg/m^3	$\rho_v = 76.5$ kg/m^3
$\mu_l = 1.321 \times 10^{-4}$ kg/m s	$\mu_v = 1.357 \times 10^{-5}$ kg/m s

For $G = 400$ kg/m^2 s and $x = 0.5$:

Liquid Reynolds number,

$$Re_l = \frac{GD(1-x)}{\mu_l} = \frac{400 \times 0.001 \times (1-0.5)}{1.321 \times 10^{-4}} = 1514$$

Vapor Reynolds number,

$$Re_v = \frac{GDx}{\mu_v} = \frac{400 \times 0.001 \times 0.5}{1.357 \times 10^{-5}} = 14743$$

Vapor-only Reynolds number,

$$Re_{vo} = \frac{GD}{\mu_v} = \frac{400 \times 0.001}{1.357 \times 10^{-5}} = 29486$$

Using the Churchill (1977) correlation for the single-phase friction factors, assuming a smooth tube, $f_l = 0.0423$, and $f_v = 0.0279$. The corresponding single-phase pressure drops are:

$$\left(\frac{dP}{dz}\right)_l = \frac{f_l G^2 (1-x)^2}{2D\rho_l} = \frac{0.0423 \times 400^2 (1-0.5)^2}{2 \times 0.001 \times 1077} = 784.6 \text{ Pa}/m$$

$$\left(\frac{dP}{dz}\right)_v = \frac{f_v G^2 x^2}{2D\rho_v} = \frac{0.0279 \times 400^2 \times 0.5^2}{2 \times 0.001 \times 76.5} = 7304 \text{ Pa}/m$$

The predictions from each model listed above are computed as follows.

Sardesai et al. (1981)

Martinelli parameter,

$$X = \left[\frac{(dP/dz)_l}{(dP/dz)_v}\right]^{1/2} = \left[\frac{784.6}{7304}\right]^{1/2} = 0.3278$$

Superficial gas velocity,

$$j_v = \frac{xG}{\rho_v} = \frac{0.5 \times 400}{76.5} = 2.615 \; m/s$$

Modified Froude number:

$$F = \sqrt{\frac{\rho_v}{\rho_l - \rho_v}} \times \frac{j_v}{\sqrt{Dg}} = \sqrt{\frac{76.5}{1077 - 76.5}} \times \frac{2.615}{\sqrt{0.001 \times 9.81}} = 7.298$$

Function $\beta = (0.7 X^2 + 2X + 0.85)F = (0.7 \times 0.3278^2 + 2 \times 0.3278 + 0.85) \times 7.298 = 11.54$.

The corresponding values for the other conditions are calculated in a similar manner, and the regimes are assigned as follows:

$\beta < 1.75$: Stratified and stratified/wavy flow
$\beta \geq 1.75$: Annular flow

The resulting flow regimes are listed and plotted below.

G	x	X	β	Flow Regime
150	0.2	1.631	6.538	Annular
150	0.5	0.4666	5.297	Annular
150	0.8	0.01974	5.571	Annular
400	0.2	1.024	10.61	Annular
400	0.5	0.3278	11.54	Annular
400	0.8	0.1375	13.29	Annular
750	0.2	1.212	23.55	Annular
750	0.5	0.3527	22.47	Annular
750	0.8	0.1082	23.53	Annular

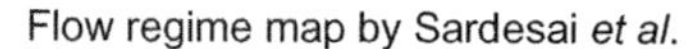

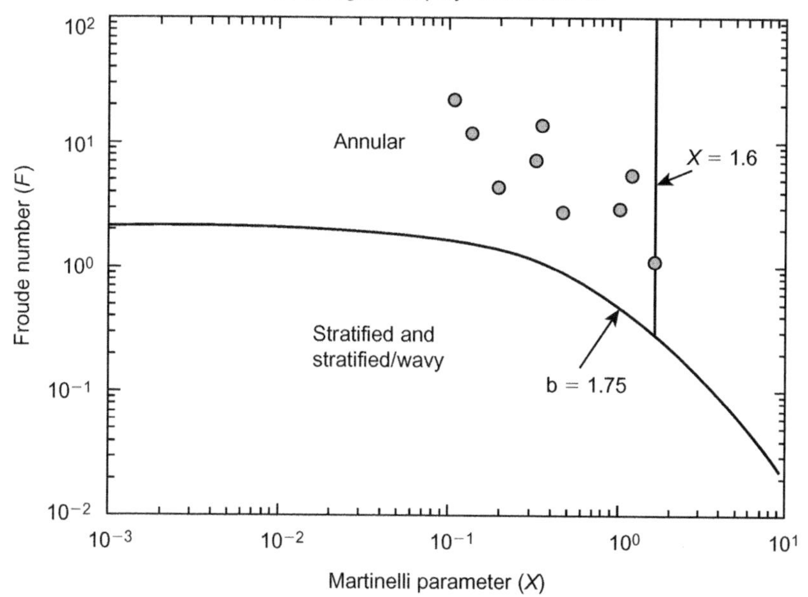

Tandon et al. (1982)

For the sample condition with $G = 400\ \text{kg/m}^2\ \text{s}$ and $x = 0.5$, the required parameters are calculated below.

Dimensionless gas velocity,

$$j_g^* = \frac{Gx}{\sqrt{Dg\rho_v(\rho_l - \rho_v)}} = \frac{400 \times 0.5}{\sqrt{0.001 \times 9.81 \times 76.5(1077 - 76.5)}} = 7.298$$

Void fraction,

$$\alpha = \left\{1 + \frac{\rho_v}{\rho_l}\left(\frac{1-x}{x}\right)\left[0.4 + 0.6\sqrt{\frac{(\rho_l/\rho_v) + 0.4(1-x/x)}{1+0.4(1-x/x)}}\right]\right\}^{-1}$$

$$\Rightarrow \alpha = \left\{1 + \frac{76.5}{1077}\left(\frac{1-0.5}{0.5}\right)\left[0.4 + 0.6\sqrt{\frac{(1077/76.5) + 0.4(1-(0.5/0.5))}{1+0.4(1-(0.5/0.5))}}\right]\right\}^{-1} = 0.858$$

$$\Rightarrow \frac{1-\alpha}{\alpha} = 0.1654$$

In a similar manner, these parameters are computed for each of the conditions, and the flow regimes assigned according to the transition criteria listed in Section 6.2.2.

G	x	j_g^*	$1-\alpha/\alpha$	Flow Regime
150	0.2	1.095	0.5322	Undesignated
150	0.5	2.737	0.1654	Annular and semi-annular
150	0.8	4.379	0.04535	Annular and semi-annular
400	0.2	2.919	0.5322	Undesignated
400	0.5	7.298	0.1654	Spray
400	0.8	11.68	0.04535	Spray
750	0.2	5.473	0.5322	Undesignated
750	0.5	13.68	0.1654	Spray
750	0.8	21.89	0.04535	Spray

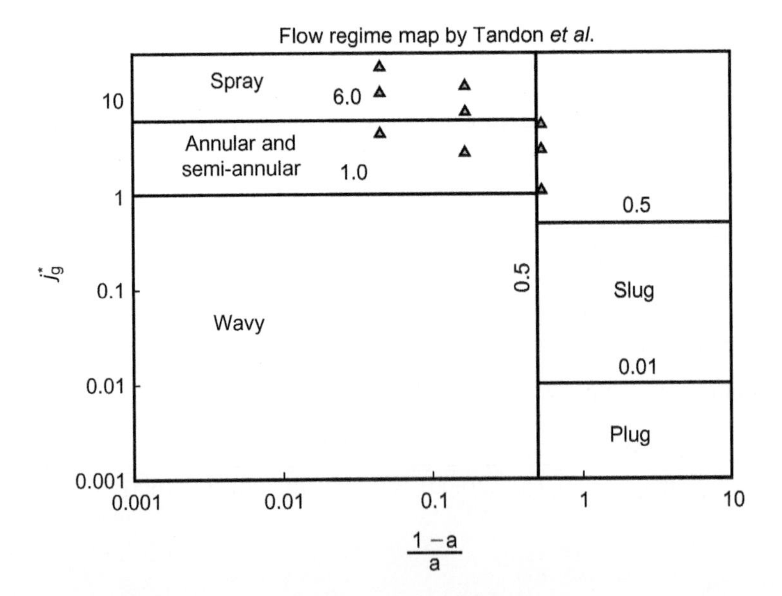

Dobson and Chato (1998)

For the sample condition $G = 400\ \mathrm{kg/m^2\,s}$ and $x = 0.5$, the required parameters are as follows.

$$X_{tt} = \left(\frac{1-x}{x}\right)^{0.9}\left(\frac{\rho_v}{\rho_l}\right)^{0.5}\left(\frac{\mu_l}{\mu_g}\right)^{0.1}$$

$$\Rightarrow X_{tt} = \left(\frac{1-0.5}{0.5}\right)^{0.9}\left(\frac{76.5}{1077}\right)^{0.5}\left(\frac{1.321\times10^{-4}}{1.357\times10^{-5}}\right)^{0.1} - 0.3346$$

Void fraction,

$$\alpha = \left[1+\frac{1-x}{x}\left(\frac{\rho_v}{\rho_l}\right)^{2/3}\right]^{-1}$$

$$\Rightarrow \alpha = \left[1+\frac{1-0.5}{0.5}\left(\frac{76.5}{1077}\right)^{2/3}\right]^{-1} = 0.8536$$

Galileo number,

$$Ga = g\rho_l(\rho_l-\rho_v)\frac{(\sqrt{\alpha}D)^3}{\mu_l^2}$$

$$\Rightarrow Ga = 9.81\times1077(1077-76.5)\frac{(\sqrt{0.8536}\times0.001)^3}{(1.321\times10^{-4})^2} = 478,182$$

$$Fr_{so} = \begin{cases} 0.025\times Re_l^{1.59}\left(\dfrac{1+1.09X_{tt}^{0.039}}{X_{tt}}\right)^{1.5}\dfrac{1}{Ga^{0.5}} & \text{for } Re_l \le 1250 \\ 1.26\times Re_l^{1.04}\left(\dfrac{1+1.09X_{tt}^{0.039}}{X_{tt}}\right)^{1.5}\dfrac{1}{Ga^{0.5}} & \text{for } Re_l > 1250 \end{cases}$$

Therefore for $Re_l = 1514$,

$$Fr_{so} = 1.26\times Re_l^{1.04}\left(\frac{1+1.09X_{tt}^{0.039}}{X_{tt}}\right)^{1.5}\frac{1}{Ga^{0.5}}$$

$$\Rightarrow Fr_{so} = 1.26\times1514^{1.04}\left(\frac{1+1.09\times0.3346^{0.039}}{0.3346}\right)^{1.5}\frac{1}{478,182^{0.5}} = 55.86$$

In a similar manner, these parameters are computed for each of the conditions, and the flow regimes assigned according to the transition criteria based on Fr_{so} and G listed in Section 6.2.2. The respective flow regimes are as follows.

G	x	Fr_{so}	Flow Regime
150	0.2	5.797	Wavy
150	0.5	13.08	Wavy
150	0.8	17.49	Wavy
400	0.2	19.12	Wavy
400	0.5	55.86	Annular
400	0.8	83.17	Annular
750	0.2	36.77	Annular
750	0.5	107.4	Annular
750	0.8	226	Annular

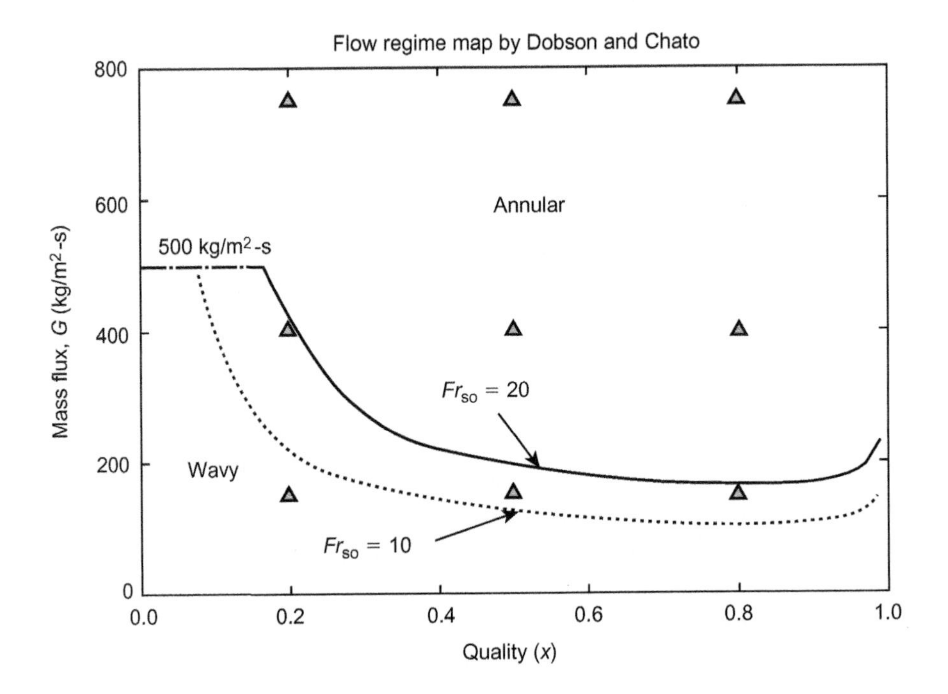

Breber et al. (1980)

For the sample condition $G = 400\ \text{kg/m}^2\ \text{s}$ and $x = 0.5$, the dimensionless gas velocity, $j_g^* = 7.298$, and the Martinelli parameter $X_{tt} = 0.3346$ from the calculations above. The respective flow regimes are assigned based on the specific values of these parameters as shown below.

G	x	X	j_g^*	Flow Regime
150	0.2	1.165	1.095	Transition between all four regimes
150	0.5	0.3346	2.737	Annular and mist-annular
150	0.8	0.0961	4.379	Annular and mist-annular
400	0.2	1.165	2.919	Transition from annular and mist-annular flow regime to bubble flow regime
400	0.5	0.3346	7.298	Annular and mist-annular
400	0.8	0.0961	11.68	Annular and mist-annular
750	0.2	1.165	5.473	Transition from annular and mist-annular flow regime to bubble flow regime
750	0.5	0.3346	13.68	Annular and mist-annular
750	0.8	0.0961	21.89	Annular and mist-annular

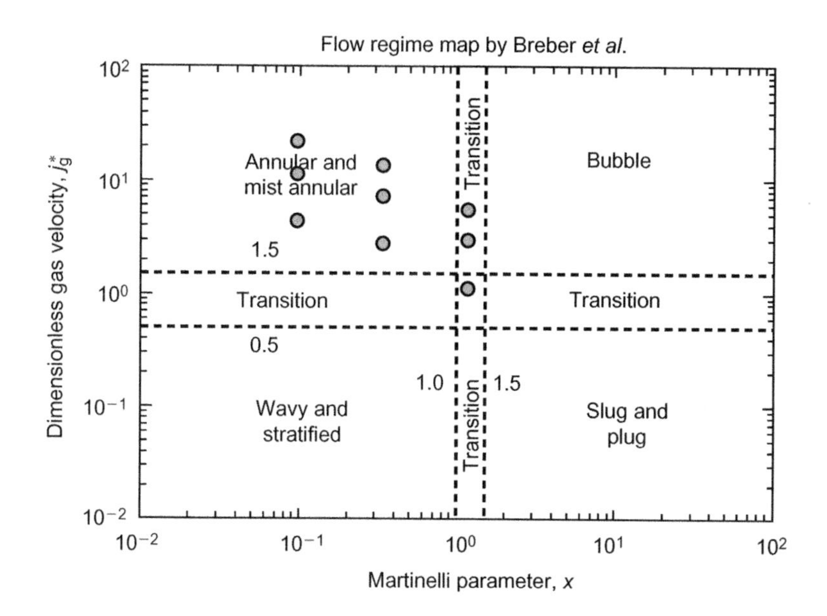

Soliman (1982, 1986)

The annular-to-wavy and intermittent transition (Soliman, 1982) was established based on the Froude number and the annular-to-mist transition (Soliman, 1986) was based on the modified Weber number. The Froude number is calculated using the following equations:

$$Re_l = 10.18\,Fr^{0.625}Ga^{0.313}(\phi_v/X_{tt})^{-0.938} \quad \text{for} \quad Re_l \le 1250$$
$$Re_l = 0.79\,Fr^{0.962}Ga^{0.481}(\phi_v/X_{tt})^{-1.442} \quad \text{for} \quad Re_l > 1250$$

From the above calculations, for $G = 400\ \text{kg/m}^2\ \text{s}$ and $x = 0.5$, liquid Reynolds number $Re_1 = 1514$, and Martinelli parameter $X_{tt} = 0.3346$. The Galileo number for this model is defined as $Ga = gD^3(\rho_1/\mu_1)^2$. Thus,

$$Ga = 9.81 \times 0.001^3 \left(\frac{1077}{1.321 \times 10^{-4}} \right)^2 = 652,650\sqrt{a^2 + b^2}$$

The curve-fit for the square root of the two-phase multiplier, is given by:

$$\phi_v = 1 + 1.09X_{tt}^{0.039} = 1 + 1.09 \times 0.3346^{0.039} = 2.044$$

The values are substituted into the Reynolds number equation to yield:

$$1514 = 0.79Fr^{0.962} \times 652650^{0.481} \left(\frac{2.044}{0.3346} \right)^{-1.442}; \quad Fr = 48.22$$

The modified Weber number is given by:

$$We = \begin{cases} 2.45Re_v^{0.64} \left(\dfrac{\mu_v^2}{\rho_v \sigma D} \right)^{0.3} \phi_v^{-0.4} & \text{for } Re_1 \leq 1250 \\ 0.85Re_v^{0.79} \left(\dfrac{\mu_v^2}{\rho_v \sigma D} \right)^{0.3} \left[\left(\dfrac{\mu_v}{\mu_1} \right)^2 \left(\dfrac{\rho_1}{\rho_v} \right) \right]^{0.084} (X_{tt}/\phi_v^{2.55})^{0.157} & \text{for } Re_1 > 1250 \end{cases}$$

For the present condition, therefore:

$$\begin{aligned} We &= 0.85 \times 14743^{0.79} \left(\frac{(1.357 \times 10^{-5})^2}{76.5 \times 4.27 \times 10^{-3} \times 0.001} \right)^{0.3} \\ &\quad \times \left[\left(\frac{1.357 \times 10^{-5}}{1.321 \times 10^{-4}} \right)^2 \left(\frac{1077}{76.5} \right) \right]^{0.084} \left(\frac{0.3346}{2.044^{2.55}} \right)^{0.157} \\ &= 12 \end{aligned}$$

Now,

$Fr < 7$ Wavy and intermittent
$Fr > 7$ and $We < 20$ Always annular
$Fr > 7$ and $We > 30$ Always mist

Thus, at $G = 400\ \text{kg/m}^2\ \text{s}$ and $x = 0.5$, flow regime is annular. The flow regimes for the other points are shown below.

G	x	Fr	We	Flow Regime
150	0.2	3.908	3.36	Wavy and intermittent
150	0.5	11.54	6.10	Annular
150	0.8	16.7	8.324	Annular
400	0.2	12.58	7.008	Annular
400	0.5	48.22	12.00	Annular
400	0.8	80.20	15.59	Annular
750	0.2	24.17	11.52	Annular
750	0.5	92.68	19.72	Annular
750	0.8	219.3	23.32	Annular-mist transition

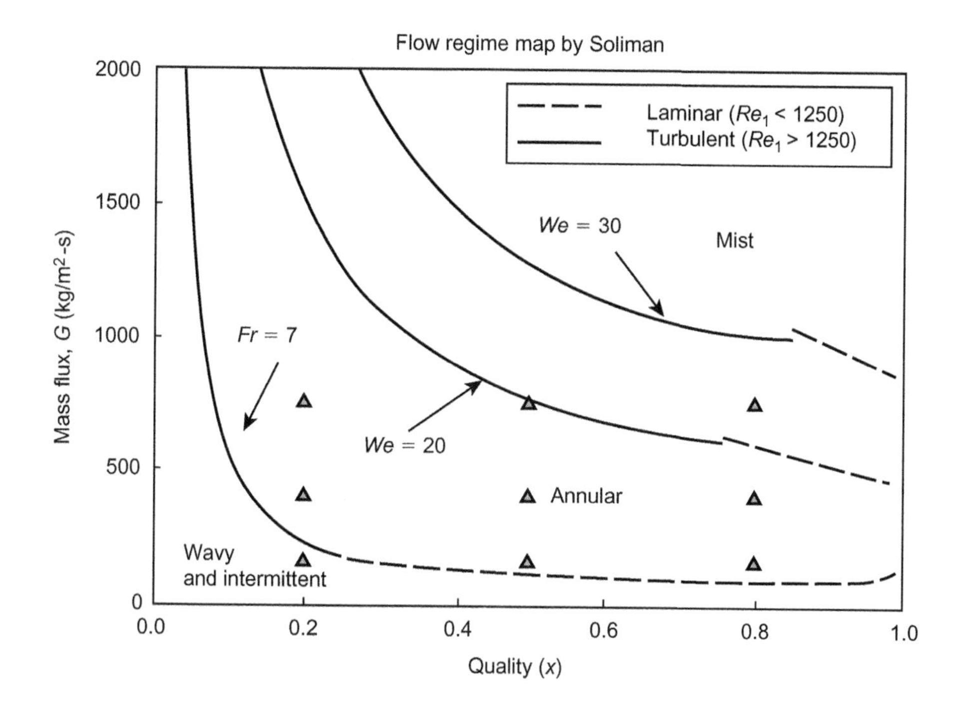

Coleman and Garimella (2000a,b, 2003)

Simple algebraic curve-fits to their G-x transition lines are given below.

Intermittent flow to intermittent and annular film flow:

$$G = \frac{547.7 - 1227x}{1 + 2.856x}$$

Intermittent and annular film flow to annular film flow:

$$G = -56.13 + 125.4/x$$

Annular film flow to annular film and mist flow:

$$G = 206.1 + 85.8/x$$

Annular film and mist flow to mist flow:

$$G = \frac{207.2 - 527.5x}{1 - 1.774x}$$

Dispersed bubble flow:

$$G = 1376 - 97.1/x$$

The respective lines and the conditions of interest are shown in the graph below. Thus, the flow regimes are assigned as follows:

G	x	Flow Regime
150	0.2	Intermittent flow
150	0.5	Intermittent and annular film flow
150	0.8	Annular film flow
400	0.2	Intermittent and annular film flow
400	0.5	Annular film and mist flow
400	0.8	Annular film and mist flow
750	0.2	Annular film and mist flow
750	0.5	Annular film and mist flow
750	0.8	Mist flow

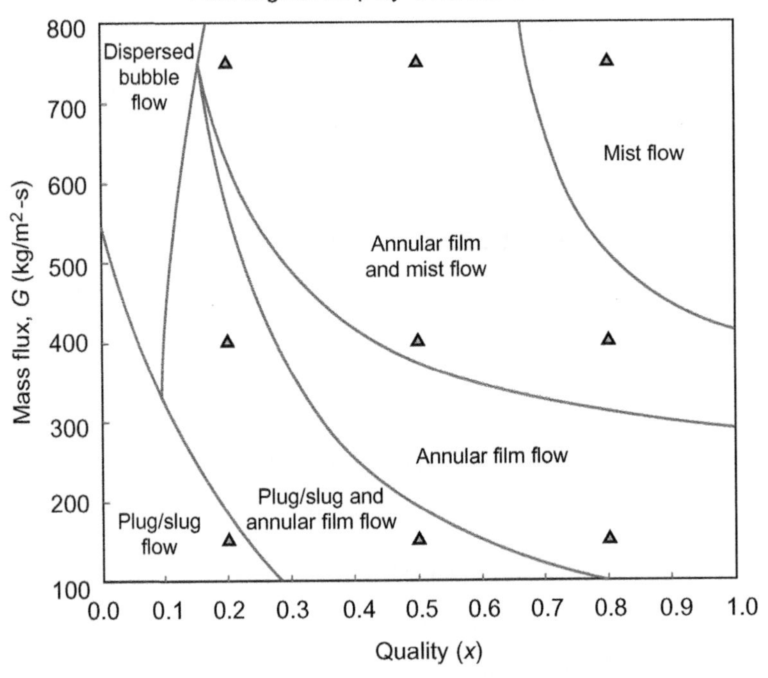

Cavallini et al. (2002a)

The Cavallini et al. map is similar to the Breber et al. map, being based on the Martinelli parameter and the dimensionless gas velocity, both of which were calculated above. The flow regime assignment for all the conditions of interest is shown below, and also plotted on their map.

G	x	X_{tt}	f_g^*	Flow Regime
150	0.2	1.165	1.095	Annular-stratified flow transition and stratified flow
150	0.5	0.3346	2.737	Annular
150	0.8	0.0961	4.379	Annular
400	0.2	1.165	2.919	Annular
400	0.5	0.3346	7.298	Annular
400	0.8	0.0961	11.68	Annular
750	0.2	1.165	5.473	Annular
750	0.5	0.3346	13.68	Annular
750	0.8	0.0961	21.89	Annular

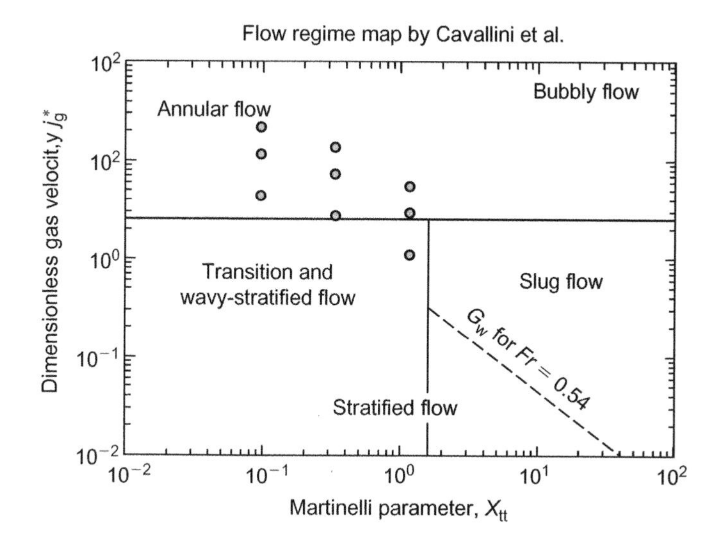

A comparison of the predictions of each of these maps is shown below.

This comparison shows that the different investigators predict a wide range of regimes for the conditions of interest in this example. The bulk of the points are in annular flow according to most models and transition criteria, although the criteria of Dobson and Chato predict wavy flow for all the $G = 150$ kg/m^2 s conditions, as well as for the low quality, $G = 400$ kg/m^2 s case. It should be noted that in Dobson and Chato's paper, a discussion appears about the $10 < Fr < 20$ region being some form of a transition region between wavy and annular flow, although in their recommended criteria, $Fr = 20$ is stated as a unique transition line. Tandon et al., Breber et al., and Coleman and Garimella predict some form of mist flow, either in combination with annular flow or by itself, at the higher mass fluxes and qualities. The similarities in the bases between the Breber et al. and

Cavallini et al. maps are also evident in this table. Of these maps, only the Coleman and Garimella criteria are reported to apply for condensation in tubes as small as 1 mm, while the Dobson and Chato criteria apply down to $D = 3.14$ mm, and many of the other criteria were for tubes with $D > 4.8$ mm, with most being primarily based on tubes with $D \leq 8$ mm. Flow regime studies for the channels much smaller than 1 mm were not represented in the above illustration because they are all for adiabatic air–water or nitrogen–water flow, and many of them do not propose explicit transition equations.

Example 6.2 Void fraction calculation

Compute the void fraction for refrigerant R-134a flowing through a 1-mm tube at a pressure of 1500 kPa, at mass fluxes 150, 400, and 750 kg/m^2 s and qualities of 0.2, 0.5, and 0.8. Use the homogeneous model and correlations by Kawahara et al. (2002), Baroczy (1965), Zivi (1964), Lockhart and Martinelli (1949), Thom (1964), Steiner (1993), El Hajal et al. (2003), Smith (1969), Premoli et al. (1971), and Yashar et al. (2001). Compare the values predicted by each model and comment.

Refrigerant properties

Saturation temperature: $T_{sat} = 55.21°C$
Surface tension: $\sigma = 4.27 \times 10^{-3}$ N/m

Representative calculations are shown here for the $G = 400$ kg/m^2 s, $x = 0.5$ case.

Liquid Phase	Vapor Phase
$\rho_l = 1077$ kg/m^3	$\rho_v = 76.5$ kg/m^3
$\mu_l = 1.321 \times 10^{-4}$ kg/m s	$\mu_v = 1.357 \times 10^{-5}$ kg/m s

Homogeneous model

The void fraction is given by:

$$\alpha = \left[1 + \frac{1-x}{x}\left(\frac{\rho_v}{\rho_l}\right)\right]^{-1} = \left[1 + \frac{1-0.5}{0.5}\left(\frac{76.5}{1077}\right)\right]^{-1} = 0.9337$$

Some investigators, for example, Triplett et al. (1999a), recommend the homogeneous model for calculation of void fraction in channels with $D_H \sim 1$ mm.

Kawahara et al. (2002)

Kawahara et al. (2002) recommend the following correlation for void fraction in tubes of diameters 100 and 50 μm.

$$\alpha = \frac{C_1 \alpha_h^{0.5}}{1 - C_2 \alpha_h^{0.5}}$$

where α_h is the homogeneous void fraction

Condition	Sardesai et al.	Tandon et al.	Dobson and Chato	Breber et al.	Soliman	Cavallini et al.	Coleman and Garimella
$G = 150$ kg/m^2 s $x = 0.2$	Annular	–	Wavy	Transition between all four regimes	Wavy and intermittent	Annular-stratified flow transition and stratified flow	Plug/slug flow
$G = 150$ kg/m^2 s $x = 0.5$	Annular	Annular and semi-annular	Wavy	Annular and mist annular	Annular	Annular	Plug/slug and annular film flow
$G = 150$ kg/m^2 s $x = 0.8$	Annular	Annular and semi-annular	Wavy	Annular and mist annular	Annular	Annular	Annular film flow
$G = 400$ kg/m^2 s $x = 0.2$	Annular	–	Wavy	Transition from annular and mist-annular flow regime to bubble flow regime	Annular	Annular	Plug/slug and annular film flow
$G = 400$ kg/m^2 s $x = 0.5$	Annular	Spray	Annular	Annular and mist-annular	Annular	Annular	Annular film and mist flow
$G = 400$ kg/m^2 s $x = 0.8$	Annular	Spray	Annular	Annular and mist-annular	Annular	Annular	Annular film and mist flow
$G = 750$ kg/m^2 s $x = 0.2$	Annular	–	Annular	Transition from annular and mist-annular flow regime to bubble flow regime	Annular	Annular	Annular film and mist flow
$G = 750$ kg/m^2 s $x = 0.5$	Annular	Spray	Annular	Annular and mist-annular	Annular	Annular	Annular film and mist flow
$G = 750$ kg/m^2 s $x = 0.8$	Annular	Spray	Annular	Annular and mist-annular	Transition	Annular	Mist flow

$$C_1 = \begin{cases} 0.03 & \text{For } D = 100\ \mu m \\ 0.02 & \text{For } D = 50\ \mu m \end{cases} \quad \text{and } C_2 = \begin{cases} 0.97 & \text{For } D = 100\ \mu m \\ 0.98 & \text{For } D = 50\ \mu m \end{cases}$$

Based on experimental data, they recommend that for $D > 250\ \mu\text{m}$, the void fraction be computed using the homogeneous model. Therefore, for this case ($D = 1$ mm), the void fraction is 0.9337.

Baroczy (1965)

The Baroczy void fraction is calculated as follows:

$$\alpha = \left[1 + \left(\frac{1-x}{x}\right)^{0.74}\left(\frac{\rho_v}{\rho_l}\right)^{0.65}\left(\frac{\mu_l}{\mu_v}\right)^{0.13}\right]^{-1}$$

$$\Rightarrow \alpha = \left[1 + \left(\frac{1-0.5}{0.5}\right)^{0.74}\left(\frac{76.5}{1077}\right)^{0.65}\left(\frac{1.321 \times 10^{-5}}{1.357 \times 10^{-5}}\right)^{0.13}\right]^{-1} = 0.8059$$

Koyama et al. (2004) recommend this correlation for smooth tubes, but mention that it does not predict the void fraction in microfin tubes well.

Zivi (1964)

This widely used correlation is given by:

$$\alpha = \left[1 + \frac{1-x}{x}\left(\frac{\rho_v}{\rho_l}\right)^{2/3}\right]^{-1}$$

$$\Rightarrow \alpha = \left[1 + \frac{1-0.5}{0.5}\left(\frac{76.5}{1077}\right)^{2/3} = 0.8536\right]^{-1}$$

Lockhart and Martinelli (1949)

This correlation is given by:

$$\alpha = \left[1 + 0.28\left(\frac{1-x}{x}\right)^{0.64}\left(\frac{\rho_v}{\rho_l}\right)^{0.36}\left(\frac{\mu_l}{\mu_v}\right)^{0.07}\right]^{-1}$$

$$\Rightarrow \alpha = \left[1 + 0.28\left(\frac{1-0.5}{0.5}\right)^{0.64}\left(\frac{76.5}{1077}\right)^{0.36}\left(\frac{1.321 \times 10^{-4}}{1.357 \times 10^{-5}}\right)^{0.07}\right]^{-1} = 0.8875$$

Based on experimental data for air–water flow in 6.6- and 33.2-mm diameter tubes, Ekberg et al. (1999) recommend the Lockhart and Martinelli correlation for void fraction calculations.

Thom (1964)

The calculation is as follows:

$$\alpha = \left[1 + \left(\frac{1-x}{x}\right)\left(\frac{\rho_v}{\rho_l}\right)^{0.89}\left(\frac{\mu_l}{\mu_v}\right)^{0.18}\right]^{-1}$$

$$\Rightarrow \alpha = \left[1 + \left(\frac{1-0.5}{0.5}\right)\left(\frac{76.5}{1077}\right)^{0.89}\left(\frac{1.321\times10^{-4}}{1.357\times10^{-5}}\right)^{0.18}\right]^{-1} = 0.8748$$

Steiner (1993)

$$\alpha = \frac{x}{\rho_v}\left([1+0.12(1-x)]\left[\frac{x}{\rho_v} + \frac{1-x}{\rho_l}\right] + \frac{1.18(1-x)[g\times\sigma(\rho_l-\rho_v)]^{0.25}}{G\times\rho_l^{0.5}}\right)^{-1}$$

$$\Rightarrow \alpha = \frac{0.5}{76.5}\left([1+0.12(1-0.5)]\left[\frac{0.5}{76.5} + \frac{1-0.5}{1077}\right]\right.$$

$$\left. + \frac{1.18(1-0.5)[9.81\times4.27\times10^{-3}\times(1077-76.5)]^{0.25}}{400\times1077^{0.5}}\right)^{-1}$$

The resulting void fraction is $\alpha = 0.8675$.

El Hajal et al. (2003)

El Hajal et al. (2003) proposed the following correlation, which uses the void fraction calculated by the homogeneous model and the correlation by Steiner (1993).

$$\alpha = \frac{\alpha_h - \alpha_{Steiner}}{\ln(\alpha_h/\alpha_{Steiner})}$$

where α_h and $\alpha_{Steiner}$ are the void fractions calculated by the homogeneous model and the correlation by Steiner (1993). In the present case, $\alpha_h = 0.9337$ and $\alpha_{Steiner} = 0.8675$. Therefore,

$$\alpha = \frac{0.9337 - 0.8675}{\ln(0.9337/0.8675)} = 0.9002$$

Smith (1969)

This correlation is given by:

$$\alpha = \left\{ 1 + \left(\frac{\rho_v}{\rho_l}\right) \left(\frac{1-x}{x}\right) \times \left[K + (1-K)\sqrt{\frac{(\rho_l/\rho_v) + K((1-x)/x)}{1+K((1-x)/x)}} \right] \right\}^{-1}$$

Smith found good agreement with experimental data (air–water and steam–water, 6–38-mm diameter tubes) for $K = 0.4$. In the present case:

$$\alpha = \left\{ 1 + \left(\frac{76.5}{1077}\right) \left(\frac{1-0.5}{0.5}\right) \times \left[0.4 + (1-0.4)\sqrt{\frac{(1077/76.5) + 0.4((1-0.5)/0.5)}{1+0.4((1-0.5)/0.5)}} \right] \right\}^{-1} = 0.858$$

Tabatabai and Faghri (2001) found the best agreement with their data for 4.8–15.88-mm tubes (refrigerants) and 1–12.3-mm tubes (air–water) with the Smith and Lockhart–Martinelli correlations. They preferred the Smith correlation for its wide range of applicability.

Premoli et al. (1971)

The liquid-only Reynolds number is given by:

$$Re_{lo} = \frac{GD}{\mu_l} = \frac{400 \times 1 \times 10^{-3}}{1.321 \times 10^{-4}} = 3028$$

The liquid-only Weber number is given by:

$$We_{lo} = \frac{G^2 D}{\sigma \rho_l} = \frac{400^2 \times 1 \times 10^{-3}}{0.00427 \times 1077} = 34.78$$

The parameters E_1 and E_2 are calculated as follows:

$$E_1 = 1.578 Re_{lo}^{-0.19} \left(\frac{\rho_l}{\rho_v}\right)^{0.22} = 1.578 \times 3028^{-0.19} \times \left(\frac{1077}{76.5}\right)^{0.22} = 0.6157$$

$$E_2 = 0.0273 We_{lo} Re_{lo}^{-0.51} \left(\frac{\rho_l}{\rho_v}\right)^{-0.08} = 0.0273 \times 34.78 \times 3028^{-0.51} \times \left(\frac{1077}{76.5}\right)^{-0.08} = 0.01289$$

The parameter y is a function of the homogeneous void fraction:

$$y = \frac{\beta}{1-\beta} = \frac{\alpha_h}{1-\alpha_h} = \frac{0.9337}{1-0.9337} = 14.08$$

The slip ratio is calculated as follows:

$$S = 1 + E_1\left[\left(\frac{y}{1+yE_2}\right) - yE_2\right]^{1/2}$$

$$\Rightarrow S = 1 + 0.6157 \times \left[\left(\frac{14.08}{1+14.08 \times 0.01289}\right) - 14.08 \times 0.01289\right]^{1/2} = 3.11$$

Finally, the void fraction is given by:

$$\alpha = \frac{x}{x + S(1-x)\rho_v/\rho_l} = \frac{0.5}{0.5 + 3.11 \times (1-0.5) \times 76.5/1077} = 0.8191$$

Yashar et al. (2001)

The Froude rate defined by Yashar et al. (2001) is evaluated as follows:

$$Ft = \left[\frac{G^2x^3}{(1-x)\rho_v^2 gD}\right]^{0.5} = \left[\frac{400^2 \times 0.5^3}{(1-0.5) \times 76.5^2 \times 9.81 \times 1 \times 10^{-3}}\right]^{0.5} = 26.4$$

The Martinelli parameter was calculated above, $X_{tt} = 0.3346$. The void fraction is calculated in terms of these two parameters as follows:

$$\alpha = \left[1 + \frac{1}{F_t} + X_{tt}\right]^{-0.321} = \left[1 + \frac{1}{26.4} + 0.3346\right]^{-0.321} = 0.9034$$

The void fractions predicted by the different models and correlations for all the conditions of interest are shown below.

The results shown above indicate that the void fractions predicted by several of the models are quite close to each other. A graph depicting these values is shown in Section 6.3. Caution must be exercised, however, because, as stated by El Hajal et al. (2003) and Thome et al. (2003), $(1-\alpha)$ is an indicator of the film thickness in annular flow, and small changes in void fraction imply large changes in liquid film thickness, which therefore alters the heat transfer coefficient. It should also be noted that many of the correlations assume annular flow, and although some of the authors recommend that they can be used irrespective of the flow regime, in the flow regimes specific to microchannels, this should be validated further.

Example 6.3 Pressure drop calculation

Compute the frictional pressure gradient during condensation of refrigerant R-134a flowing through a 1-mm diameter tube at a mass flux of 300 kg/m^2 s, a mean quality of 0.5, and a pressure of 1500 kPa. Use the correlations by Lockhart

G kg/m^2 s	*x*	Homogeneous	Baroczy (1965)	Zivi (1964)	Lockhart and Martinelli (1949)	Thom (1964)	Steiner (1993)	El Hajal et al. (2003)	Smith (1969)	Premoli et al. (1971)	Yashar et al. (2001)
150	0.2	0.7788	0.5981	0.5932	0.7647	0.636	0.6274	0.7004	0.6527	0.5964	0.7296
150	0.5	0.9337	0.8059	0.8536	0.8875	0.8748	0.8461	0.8892	0.858	0.7909	0.8904
150	0.8	0.9826	0.9205	0.9589	0.9504	0.9655	0.9489	0.9656	0.9566	0.9017	0.9622
400	0.2	0.7788	0.5981	0.5932	0.7647	0.636	0.6769	0.7267	0.6527	0.6239	0.7597
400	0.5	0.9337	0.8059	0.8536	0.8875	0.8748	0.8675	0.9002	0.858	0.8191	0.9034
400	0.8	0.9826	0.9205	0.9589	0.9504	0.9655	0.9555	0.969	0.9566	0.9264	0.9676
750	0.2	0.7788	0.5981	0.5932	0.7647	0.636	0.6922	0.7347	0.6527	0.6433	0.7691
750	0.5	0.9337	0.8059	0.8536	0.8875	0.8748	0.8737	0.9034	0.858	0.8414	0.9071
750	0.8	0.9826	0.9205	0.9589	0.9504	0.9655	0.9574	0.9699	0.9566	0.9445	0.9692

and Martinelli (1949), Friedel (1979), Chisholm (1973), Mishima and Hibiki (1996b), Lee and Lee (2001), Tran et al. (2000), Wang et al. (1997b), Chen et al. (2001), Wilson et al. (2003), Souza et al. (1993), Cavallini et al. (2001, 2002a), and Garimella et al. (2005) and compare the values predicted by each model.

Refrigerant properties

Saturation temperature $T_{sat} = 55.21°C$
Surface tension $\sigma = 0.00427$ N/m

Liquid Phase	Vapor Phase
$\rho_l = 1077$ kg/m^3	$\rho_v = 76.5$ kg/m^3
$\mu_l = 1.321 \times 10^{-4}$ kg/m s	$\mu_v = 1.357 \times 10^{-5}$ kg/m s

Some common parameters that are used by several correlations are calculated below first.

Liquid Reynolds number,

$$Re_l = \frac{GD(1-x)}{\mu_l} = \frac{300 \times 0.001 \times (1-0.5)}{1.321 \times 10^{-4}} = 1136$$

Vapor Reynolds number,

$$Re_v = \frac{GDx}{\mu_v} = \frac{300 \times 0.001 \times 0.5}{1.357 \times 10^{-5}} = 11054$$

Liquid-only Reynolds number,

$$Re_{lo} = \frac{GD}{\mu_l} = \frac{300 \times 0.001}{1.321 \times 10^{-4}} = 2271$$

Vapor-only Reynolds number,

$$Re_{vo} = \frac{GD}{\mu_v} = \frac{300 \times 0.001}{1.357 \times 10^{-5}} = 22108$$

For this illustration, the single-phase friction factors are calculated using the Churchill (1977) correlation:

$$f = 8\left[\left(\frac{8}{Re}\right)^{12} + \left\{\left[2.457 \times \ln\left(\frac{1}{(7/Re)^{0.9} + 0.27\varepsilon/D}\right)\right]^{16} + \left(\frac{37530}{Re}\right)^{16}\right\}^{-1.5}\right]^{1/12}$$

Assuming a smooth tube, that is, $\varepsilon = 0$, $f_1 = 0.05635$, $f_v = 0.03016$, $f_{lo} = 0.03047$, and $f_{vo} = 0.0252$. The corresponding single-phase pressure gradients are given by:

$$\left(\frac{dP}{dz}\right)_1 = \frac{f_1 G^2 (1-x)^2}{2D\rho_1} = \frac{0.05635 \times 300^2 (1-0.5)^2}{2 \times 0.001 \times 1077} = 588.5 \text{ Pa/m}$$

$$\left(\frac{dP}{dz}\right)_v = \frac{f_v G^2 x^2}{2D\rho_v} = \frac{0.03016 \times 300^2 \times 0.5^2}{2 \times 0.001 \times 76.5} = 4436 \text{ Pa/m}$$

$$\left(\frac{dP}{dz}\right)_{lo} = \frac{f_{lo} G^2}{2D\rho_1} = \frac{0.03047 \times 300^2}{2 \times 0.001 \times 1077} = 1373 \text{ Pa/m}$$

$$\left(\frac{dP}{dz}\right)_{vo} = \frac{f_{vo} G^2}{2D\rho_v} = \frac{0.0252 \times 300^2}{2 \times 0.001 \times 76.5} = 14823 \text{ Pa/m}$$

Lockhart and Martinelli (1949)

The Martinelli parameter is given by:

$$X = \left[\frac{(dP/dz)_l}{(dP/dz)_v}\right]^{1/2} = \left[\frac{588.5}{4436}\right]^{1/2} = 0.3642$$

Two-phase multiplier is given by:

$$\phi_1^2 = 1 + \frac{C}{X} + \frac{1}{X^2}; \quad C = \begin{cases} & \text{Liquid} & \text{Vapor} \\ 20 & \text{Turbulent} & \text{Turbulent} \\ 12 & \text{Laminar} & \text{Turbulent} \\ 10 & \text{Turbulent} & \text{Laminar} \\ 5 & \text{Laminar} & \text{Laminar} \end{cases}$$

In this case:

$$\phi_1^2 = 1 + \frac{12}{X} + \frac{1}{X^2} = 1 + \frac{12}{0.3642} + \frac{1}{0.3642^2} = 41.4865$$

The two-phase pressure gradient is:

$$\frac{\Delta P}{L} = \phi_1^2 \left(\frac{dP}{dz}\right)_1 = 41.4865 \times 588.5 = 24,414 \text{ Pa/m}$$

Friedel (1979)

The parameters E, F, and H are evaluated as follows:

$$E = (1-x)^2 + x^2 \frac{\rho_l f_{vo}}{\rho_v f_{lo}}$$

$$\Rightarrow E = (1-0.5)^2 + 0.5^2 \times \frac{1077 \times 0.0252}{76.5 \times 0.03047} = 3.161$$

$$F = x^{0.78} \times (1-x)^{0.24} = 0.5^{0.78} \times (1-0.5)^{0.24} = 0.4931$$

$$H = \left(\frac{\rho_l}{\rho_v}\right)^{0.91} \left(\frac{\mu_v}{\mu_l}\right)^{0.19} \left(1 - \frac{\mu_v}{\mu_l}\right)^{0.7}$$

$$\Rightarrow H = \left(\frac{1077}{76.5}\right)^{0.91} \left(\frac{1.357 \times 10^{-5}}{1.321 \times 10^{-4}}\right)^{0.19} \left(\frac{1.357 \times 10^{-5}}{1.321 \times 10^{-4}}\right)^{0.7} = 6.677$$

The two-phase mixture density is calculated as follows:

$$\rho_{TP} = \left(\frac{x}{\rho_v} + \frac{1-x}{\rho_l}\right)^{-1} = \left(\frac{0.5}{76.5} + \frac{1-0.5}{1077}\right)^{-1} = 142.8\ \text{kg}/m^3$$

The Froude and Weber numbers are given by:

$$Fr = \frac{G^2}{gD\rho_{TP}^2} = \frac{300^2}{9.81 \times 0.001 \times 142.8^2} = 449.6$$

$$We = \frac{G^2 D}{\rho_{TP\sigma}} = \frac{300^2 \times 0.001}{142.8 \times 4.27 \times 10^{-3}} = 147.5$$

The resulting two-phase multiplier is now calculated as follows:

$$\phi_{lo}^2 = E + \frac{3.24 \times FH}{Fr^{0.045} We^{0.035}} = 3.161 + \frac{3.24 \times 0.4931 \times 6.677}{449.6^{0.045} \times 147.5^{0.035}} = 9.967$$

Finally, the two-phase pressure gradient is given by:

$$\frac{\Delta P}{L} = \phi_{lo}^2 \left(\frac{\Delta P}{L}\right)_{lo} = 9.967 \times 1273 = 12,688\ \text{Pa/m}$$

Chisholm (1973)

The ratio of the vapor- to liquid-only pressure gradients is given by:

$$Y = \sqrt{\frac{(dP/dz)_{vo}}{(dP/dz)_{lo}}} = \sqrt{\frac{14.823}{1273}} = 3.412$$

The two-phase multiplier is given by:

$$\phi_{lo}^2 = 1 + (Y^2 - 1)[Bx^{(2-n)/2}(a-x)^{(2-n)/2} + x^{2-n}]$$

where the parameter B is given by:

$$B = \begin{cases} \dfrac{55}{\sqrt{G}} & \text{for} \quad 0 < Y < 9.5 \\ \dfrac{520}{Y\sqrt{G}} & \text{for} \quad 9.5 < Y < 28 \\ \dfrac{15,000}{Y^2\sqrt{G}} & \text{for} \quad 28 < Y \end{cases}$$

and n is the power to which Reynolds number is raised in the friction factor:

$$n = \begin{cases} 1 & Re_{lo} \le 2100 \\ 0.25 & Re_{lo} \ge 2100 \end{cases}$$

Thus, for the current example,

$$B = \frac{55}{\sqrt{G}} = \frac{55}{\sqrt{300}} = 3.175 \text{ and } n = 0.25$$

The two-phase multiplier can now be calculated as follows:

$$\begin{aligned} \phi_{lo}^2 &= 1 + (Y^2 - 1) \times [3.175x^{(2-0.25)/2}(1-x)^{(2-0.25)/2} + x^{2-0.25}] \\ &= 1 + (3.412^2 - 1) \times [3.175 \times 0.5^{(2-0.25)/2}(1-0.5)^{(2-0.25)/2} + 0.5^{2-0.25}] \\ &= 14.213 \end{aligned}$$

The pressure gradient is evaluated as follows:

$$\frac{\Delta P}{L} = \phi_{lo}^2 \left(\frac{\Delta P}{L}\right)_{lo} = 14.213 \times 1273 = 18,093 \text{ Pa/m}$$

Mishima and Hibiki (1996b)

Martinelli parameter (calculated above) is $X = 0.3642$. The two-phase multiplier is given by:

$$\phi_l^2 = 1 + \frac{C}{X} + \frac{1}{X^2}$$

For small-diameter tubes, Mishima and Hibiki recommend the following expression for C:

$$C = 21(1 - e^{-0.319D})$$

$$\Rightarrow C = 21(1 - e^{-0.319 \times 1}) = 5.736$$

With this value of C, the two-phase multiplier is given by:

$$\Rightarrow \phi_l^2 = 1 + \frac{5.736}{X} + \frac{1}{X^2} = 1 + \frac{5.736}{0.3642} + \frac{1}{0.3642^2} = 24.285$$

The two-phase pressure gradient is then calculated as follows:

$$\frac{\Delta P}{L} = \phi_1^2 \left(\frac{dP}{dz}\right)_1 = 24.285 \times 588.5 = 14,292 \text{ Pa/m}$$

Lee and Lee (2001)

The Martinelli parameter, calculated above, is $X = 0.3642$. For the C parameter in the two-phase multiplier, Lee and Lee suggested $C = A\lambda^q \psi^r Re_{lo}{}^S$, where, $\lambda = (\mu_1^2/\rho_1 \sigma D), \psi = (\mu_1 j/\sigma)$, and j = liquid slug velocity, also given by $j = j_v + j_L$. The value of constants A, q, r, and S are determined based on the following table:

Flow Regime							
Liquid	**Vapor**	***A***	***q***	***r***	**S**	***X*-range**	***Re*$_{lo}$ Range**
Laminar	Laminar	6.833×10^{-8}	−1.317	0.719	0.557	0.776–14.176	175–1480
Laminar	Turbulent	6.185×10^{-2}	0	0	0.726	0.303–1.426	293–1506
Turbulent	Laminar	3.627	0	0	0.174	3.276–79.415	2606–17642
Turbulent	Turbulent	0.408	0	0	0.451	1.309–14.781	2675–17757

It should be noted that surface tension effects are important only in the laminar–laminar flow regimes. For all other flow regimes, the exponents of ψ and λ are almost zero. Thus, for these flow regimes, C is merely a function of Re_{lo}. The above table also provides the range of X and Re_{lo} for the data from which these constants were determined. The liquid Reynolds number for this case, calculated above, is $Re_{lo} = 2271$. For the current example (laminar liquid film and turbulent vapor core), the coefficient C is evaluated as follows:

$$C = 6.185 \times 10^{-2} \times Re_{lo}^{0.726} = 6.185 \times 10^{-2} \times 2271^{0.726} = 16.9$$

The two-phase multiplier is therefore:

$$\phi_1^2 = 1 + \frac{16.9}{X} + \frac{1}{X^2} = 1 + \frac{16.9}{0.3642} + \frac{1}{0.3642^2} = 54.953$$

Finally, the two-phase pressure gradient is calculated as follows:

$$\frac{\Delta P}{L} = \phi_1^2 \left(\frac{dP}{dz}\right)_1 = 54.953 \times 588.5 = 32,340 \text{ Pa/m}$$

Tran et al. (2000)

Tran et al. modified the two-phase multiplier correlation of Chisholm (1973) to the following expression, which includes the confinement number:

$$\phi_{lo}^2 = 1 + (4.3Y^2 - 1) \times N_{\text{conf}} x^{0.875}(1 - x)^{0.875} + x^{1.75}$$

where the confinement number is given by:

$$N_{\text{conf}} = \frac{[\sigma/(g(\rho_l-\rho_v))]^{1/2}}{D} = \frac{[(4.27\times10^{-3})/(9.81(1077-76.5))]^{1/2}}{0.001} = 0.6595$$

for the current example. With $Y = 3.412$, calculated above,

$$\phi_{lo}^2 = 1 + (4.3\times3.412^2 - 1)\times0.6595\times0.5^{0.875}\times(1-0.5)^{0.875} + 0.5^{1.75} = 10.916$$

The two-phase pressure gradient is evaluated in the usual manner:

$$\frac{\Delta P}{L} = \phi_{lo}^2\left(\frac{\Delta P}{L}\right)_{lo} = 10.916\times1373 = 13,897\ \text{Pa/m}$$

Wang et al. (1997b)

According to their correlation, for $G<200\ \text{kg/m}^2\ \text{s}$ $\phi_v{}^2 = 1 + 9.4X^{0.62} + 0.564X^{2.45}$. For $G>200\ \text{kg/m}^2$ s, they modified the parameter C in the Lockhart–Martinelli (1949) correlation as follows:

$$C = 4.566\times10^{-6}X^{0.128}\times Re_{lo}^{0.938}\left(\frac{\rho_l}{\rho_v}\right)^{-2.15}\left(\frac{\mu_l}{\mu_v}\right)^{5.1}$$

In this example, $G = 300\ \text{kg/m}^2$ s; therefore:

$$C = 4.566\times10^{-6}\times0.3642^{0.128}\left(\frac{1077}{76.5}\right)^{-2.15}\left(\frac{1.321\times10^{-4}}{1.357\times10^{-5}}\right)^{5.1} = 2.099$$

The two-phase multiplier is then given by:

$$\phi_l^2 = 1 + \frac{2.099}{X} + \frac{1}{X^2} = 1 + \frac{2.099}{0.3642} + \frac{1}{0.3642^2} = 14.304$$

And the two-phase pressure gradient is:

$$\frac{\Delta P}{L} = \phi_l^2\left(\frac{dP}{dz}\right)_l = 14.304\times588.5 = 8416\ \text{Pa/m}$$

Chen et al. (2001)

Chen et al. modified the homogenous model and the Friedel (1979, 1980) correlation to determine the pressure drop in microchannels. The modification to the Friedel correlation is illustrated here. Thus,

$$\frac{dP}{dz} = \left.\frac{dP}{dz}\right|_{\text{Friedel}}\Omega;\quad \Omega = \begin{cases} \dfrac{0.0333\times Re_{lo}^{0.45}}{Re_v^{0.09}(1+0.4\exp(-Bo))} & Bo<2.5 \\[2ex] \dfrac{We^{0.2}}{(0.5+0.06Bo)} & Bo\geq2.5 \end{cases}$$

where Weber number $We = (G^2D/\sigma\rho_m)$ and Bond number $Bo = g(\rho_l - \rho_v)((D/2)^2/\sigma)$. For this case:

$$Bo = g(\rho_l - \rho_v)\left(\frac{(D/2)^2}{\sigma}\right) = 9.81(1077 - 76.5)\left(\frac{(0.001/2^2)}{4.27 \times 10^{-3}}\right) = 0.5748$$

Because $Bo < 2.5$, the modifier to the Friedel correlation is given by:

$$\Rightarrow \Omega = \frac{0.0333 \times Re_{lo}^{0.45}}{Re_v^{0.09}(1 + 0.4\exp(-Bo))} = \frac{0.0333 \times 2271^{0.45}}{11054^{0.09}(1 + 0.4\exp(-0.5748))} = 0.3808$$

Therefore, the two-phase pressure gradient is:

$$\frac{dP}{dz} = \left.\frac{dP}{dz}\right|_{Friedel} \Omega = 12688 \times 0.3808 = 4831 \text{ Pa/m}$$

Wilson et al. (2003)

Wilson et al. recommended the liquid-only two-phase multiplier of Jung and Radermacher (1989):

$$\phi_{lo}^2 = 12.82 X_{tt}^{-1.47}(1-x)^{1.8}$$

With the Martinelli parameter (calculated above), $X_{tt} = 0.3346$, the two-phase multiplier is

$$\phi_{lo}^2 = 12.82 \times 0.3346^{-1.47}(1-0.5)^{1.8} = 18.413$$

which yields a two-phase pressure gradient of

$$\frac{\Delta P}{L} = \phi_{lo}^2\left(\frac{dP}{dz}\right)_{lo} = 18.413 \times 1273 = 23{,}434 \text{ Pa/m}$$

Souza et al. (1993)

The two-phase liquid multiplier given by them is as follows:

$$\phi_l^2 = 1.376 + C_1 X_{tt}^{-C_2}$$

where, for $Fr_l < 0.7$,

$$C_1 = 4.172 + 5.48 Fr_l - 1.564 Fr_l^2$$

$$C_2 = 1.773 - 0.169 Fr_l$$

and for $Fr_l > 0.7$, $C_1 = 7.242$; $C_2 = 1.655$. In the present case, the Froude number is:

$$Fr_l = \frac{G}{\rho_l\sqrt{gD}} = \frac{300}{1077\sqrt{9.81 \times 0.001}} = 2.812$$

Therefore, $C_1 = 7.242$; $C_2 = 1.655$, which yields a two-phase multiplier of:

$$\phi_1^2 = 1.376 + 7.242 \times 0.3346^{-1.655} = 45.725$$

For the corresponding single-phase friction factor, they recommend the Colebrook (1939) friction factor:

$$\frac{1}{\sqrt{f_1}} = -2\log\left(\frac{\varepsilon/D}{3.7} + \frac{2.51}{Re_1\sqrt{f_1}}\right)$$

Thus for smooth tubes:

$$\frac{1}{\sqrt{f_1}} = -2\log\left(0 + \frac{2.51}{1136\sqrt{f_1}}\right) \Rightarrow f_1 = 0.05983$$

The single-phase liquid pressure drop is calculated as follows:

$$\left(\frac{dP}{dz}\right)_1 = \frac{f_1 G^2 (1-x)^2}{2D\rho_1} = \frac{0.05983 \times 300^2(1-0.5)^2}{2 \times 0.001 \times 1077} = 624.8 \text{ Pa/m}$$

Finally, the two-phase pressure drop, using the two-phase multiplier is given by:

$$\frac{\Delta P}{L} = \phi_1^2 \left(\frac{dP}{dz}\right)_1 = 45.725 \times 624.8 = 28,568 \text{ Pa/m}$$

Cavallini et al. (2001, 2002a)

Cavallini et al. recommended modifications to the Friedel (1979) correlation. The parameter E is the same as in his correlation, $E = 3.161$. The parameter F is modified to $F = x^{0.6978}$, which in this case is $F = 0.5^{0.6978} = 0.6165$. The modified parameter H is given by:

$$H = \left(\frac{\rho_1}{\rho_v}\right)^{0.3278} \left(\frac{\mu_v}{\mu_1}\right)^{-1.181} \left(1 - \frac{\mu_v}{\mu_1}\right)^{3.477}$$

For the present condition,

$$H = \left(\frac{1077}{76.5}\right)^{0.3278} \left(\frac{1.357 \times 10^{-5}}{1.321 \times 10^{-4}}\right)^{-1.181} \left(1 - \frac{1.357 \times 10^{-5}}{1.321 \times 10^{-4}}\right)^{3.477} = 23.99$$

The Weber number is defined in terms of the gas-phase density:

$$We = \frac{G^2 D}{\rho_v \sigma} = \frac{300^2 \times 0.001}{76.5 \times 4.27 \times 10^{-3}} = 275.5$$

The two-phase multiplier is then calculated as follows:

$$\phi_{lo}^2 = E + \frac{1.262 \times FH}{We^{0.1458}} = 3.161 + \frac{1.262 \times 0.6165 \times 23.99}{275.5^{0.1458}} = 11.391$$

The two-phase pressure drop is calculated based on this multiplier as follows:

$$\frac{\Delta P}{L} = \phi_{lo}^2 \left(\frac{\Delta P}{L}\right)_{lo} = 11.391 \times 1273 = 14,501 \text{ Pa/m}$$

Garimella et al. (2005)

The case of $G = 300 \text{ kg/m}^2\text{ s}$ and $x = 0.5$ represents annular flow, according to the transition criteria of Coleman and Garimella (2000a,b, 2003). While Garimella et al. developed a multi-regime (including intermittent/discrete wave and annular/mist/disperse wave) model to span the entire quality range as described in Section 6.4.4, for this condition, the annular flow model applies. For the annular flow model, the individual phase Reynolds numbers are computed in terms of the cross-sectional areas occupied by the phases. Therefore, the void fraction is first calculated using the Baroczy (1965) correlation:

$$\alpha = \left[1 + \left(\frac{1-x}{x}\right)^{0.74} \left(\frac{\rho_v}{\rho_l}\right)^{0.65} \left(\frac{\mu_l}{\mu_v}\right)^{0.13}\right]^{-1}$$

For the present case,

$$\alpha = \left[1 + \left(\frac{1-0.5}{0.5}\right)^{0.74} \left(\frac{76.5}{1077}\right)^{0.65} \left(\frac{1.321 \times 10^{-4}}{1.357 \times 10^{-5}}\right)^{0.13}\right]^{-1} = 0.8059$$

Therefore, the liquid and vapor Re values are given by:

$$Re_l = \frac{GD(1-x)}{(1+\sqrt{\alpha})\mu_l} = \frac{300 \times 0.001 \times (1-0.5)}{(1+\sqrt{0.8059})1.321 \times 10^{-4}} = 598.5$$

$$Re_v = \frac{GDx}{\mu_v \sqrt{\alpha}} = \frac{300 \times 0.001 \times 0.5}{1.357 \times 10^{-5} \times \sqrt{0.8059}} = 12316$$

The friction factor for the laminar film is $f_l = 64/Re_l = 64/598.5 = 0.1069$. The turbulent vapor friction factor is calculated using the $f_v = 0.316 \times Re^{-0.25}{}_v$ $= 0.316 \times 12316^{-0.25} = 0.03$. With these friction factors, the corresponding single-phase pressure gradients are given by:

$$\left(\frac{dP}{dz}\right)_l = \frac{f_l G^2 (1-x)^2}{2D\rho_l} = \frac{0.1069 \times 300^2 (1-0.5)^2}{2 \times 0.001 \times 1077} = 1117 \text{ Pa/m}$$

$$\left(\frac{dP}{dz}\right)_v = \frac{f_v G^2 x^2}{2D\rho_v} = \frac{0.03 \times 300^2 \times 0.5^2}{2 \times 0.001 \times 76.5} = 4412 \text{ Pa/m}$$

The Martinelli parameter is calculated from the definition as follows:

$$X = \left[\frac{(dP/dz)_l}{(dP/dz)_v}\right]^{1/2} = \left[\frac{1117}{4412}\right]^{1/2} = 0.5031$$

The mean liquid velocity is given by:

$$j_l = \frac{G(1-x)}{\rho_l(1-\alpha)} = \frac{300 \times (1-0.5)}{1077 \times (1-0.8059)} = 0.7172 \text{ m/s}$$

This velocity is used to evaluate the surface tension parameter:

$$\psi = \frac{j_l \mu_l}{\sigma} = \frac{0.7172 \times 1.321 \times 10^{-4}}{4.27 \times 10^{-3}} = 0.02218$$

The interface friction factor is then calculated as follows:

$$\frac{f_i}{f_l} = AX^a Re_l^b \psi^c$$

Since $Re_l < 2100$, $A = 1.308 \times 10^{-3}$; $a = 0.4273$; $b = 0.9295$; $c = -0.1211$, which yields:

$$\frac{f_i}{f_l} = 1.308 \times 10^{-3} \times X^{0.4273} \times Re_l^{0.9295} \times \psi^{-0.1211}$$

$$= 1.308 \times 10^{-3} \times 0.5031^{0.4273} \times 598.5^{0.9295} \times 0.02218^{-0.1211}$$

$$= 0.5897$$

The resulting interface friction factor is $f_i = 0.5897 \times f_l = 0.5897 \times 0.1069 = 0.06306$. The interfacial friction factor is used to determine the pressure gradient as follows:

$$\frac{\Delta P}{L} = \frac{1}{2} f_i \frac{G^2 x^2}{\rho_v \alpha^{2.5}} \frac{1}{D}$$

$$= \frac{1}{2} \times 0.06306 \times \frac{300^2 \times 0.5^5}{69.82 \times 0.8509^{2.5}} \times \frac{1}{0.001}$$

$$= 15{,}907 \text{ Pa/m}$$

A comparison of the pressure drops predicted by each of these correlations is shown below. It can be seen that the predicted pressure drops vary considerably, from 4.8 to 32.3 kPa. This large variation is attributed to the considerably different two-phase multipliers developed by the various investigators.

The only recommendation that can be made is to choose a model that is based on the geometry, fluid, and operating conditions similar to those of interest for a given application. This information is available in Section 6.4 as well as in Table 6.3.

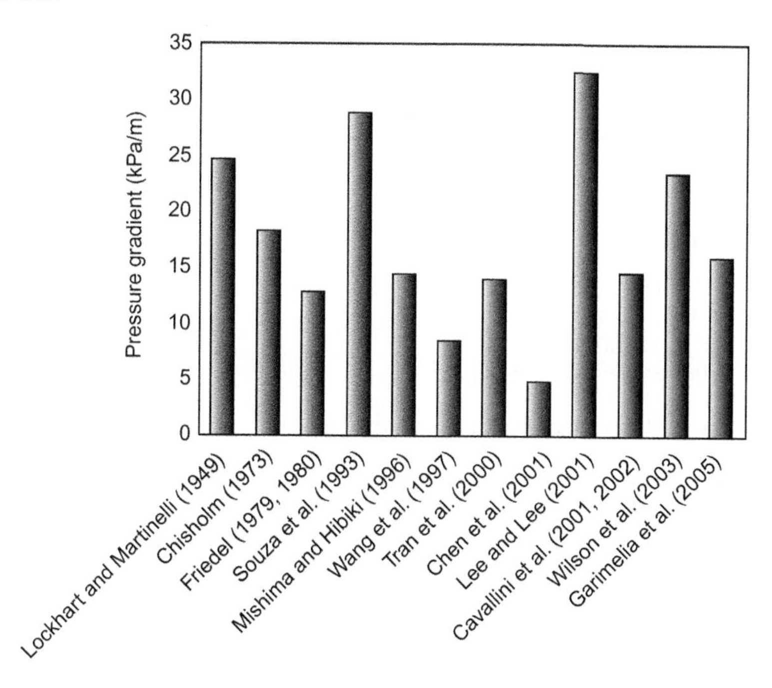

Example 6.4 Calculation of heat transfer coefficients

Compute the heat transfer coefficient for condensation of refrigerant R-134a flowing through a 1-mm tube at a mass flux of 300 kg/m^2 s, a mean quality of 0.5, and a pressure of 1500 kPa. The wall temperature of the tube is 52°C. Use the correlations by Shah (1979), Soliman et al. (1968), Soliman (1986), Traviss et al. (1973), Dobson and Chato (1998), Moser et al. (1998), Boyko and Kruzhilin (1967), Cavallini et al. (2002a), and Bandhauer et al. (2006) and compare the values predicted by each model.

Refrigerant properties

Saturation temperature $T_{sat} = 55.21°C$
Critical pressure $P_{crit} = 4059$ kPa
Reduced pressure $P_{red} = 0.37$
Surface tension $\sigma = 0.00427$ N/m
Heat of vaporization $h_{lv} = 145{,}394$ J/kg

Liquid Phase	Vapor Phase
$\rho_l = 1077$ kg/m^3	$\rho_v = 76.5$ kg/m^3
$\mu_l = 1.321 \times 10^{-4}$ kg/m s	$\mu_v = 1.357 \times 10^{-5}$ kg/m s
$Cp_l = 1611$ J/kg K	$Cp_v = 1312$ J/kg K
$h_l = 131{,}657$ J/kg	$h_v = 277{,}051$ J/kg
$k_l = 0.06749$ W/m K	$k_v = 0.0178$ W/m K
$Pr_l = 3.152$	$Pr_v = 1$

Shah (1979)

Applicability range: $0.002 < P_{red} < 0.44$

$21°C < T_{sat} < 310°C$

$3 < V_v < 300$ m/s

$10.83 < G < 210.56$ kg/m^2 s

The velocity of the vapor phase is given by:

$$V_v = \frac{xG}{\rho_v} = \frac{0.5 \times 300}{76.5} = 1.961 \text{ m/s}$$

In the present case, the vapor velocity and mass flux are beyond the range of applicability of the correlation.

The liquid-only Reynolds number Re_{lo} was calculated in Example 6.3 to be $Re_{lo} = 2271$. The liquid-only heat transfer coefficient is calculated as follows:

$$h_{lo} = 0.023 Re_{lo}^{0.8} Pr_l^{0.4} \times \frac{k_l}{D}$$

$$= 0.023 \times 2271^{0.8} \times 3.152^{0.4} \times (0.06749/0.001)$$

$$= 1190 \text{ W/m}^2 \text{ K}$$

From this single-phase heat transfer coefficient, the condensation heat transfer coefficient is obtained by applying the multiplier as follows:

$$h = h_{lo}\left[(1-x)^{0.8} + \frac{3.8x^{0.76}(1-x)^{0.04}}{P_{red}^{0.38}}\right]$$

$$= 1190\left[(1-0.5)^{0.8} + \frac{3.8 \times 0.5^{0.76}(1-0.5)^{0.04}}{(1500/4059)^{0.38}}\right]$$

$$= 4474 \text{ W/m}^2 \text{ K}$$

Soliman et al. (1968)

For this correlation, a quality change per unit length is also required, and for illustrative purposes it is assumed that $\Delta x = 0.10$ over a length of 0.3048 m. This correlation was based on data for the following range of conditions:

$$6.096 < V_v < 304.8 \text{ m/s}$$

$$0.03 < x < 0.99$$

$$1 < Pr < 10$$

The vapor velocity here is 1.961 m/s, which is outside the range of applicability of the correlation.

The vapor-only Reynolds number Re_{vo} was calculated in Example 6.3 to be 22,114, and the liquid Reynolds number $Re_l = 1136$. F_f represents the effect of two-phase friction:

$$F_f = 0.0225 \times \frac{G^2}{\rho_v} Re_{vo}^{-0.2} \left[x^{1.8} + 5.7 \left(\frac{\mu_l}{\mu_v} \right)^{0.0523} (1-x)^{0.47} x^{1.33} \left(\frac{\rho_v}{\rho_l} \right)^{0.261} + 8.11 \left(\frac{\mu_l}{\mu_v} \right)^{0.105} (1-x)^{0.94} x^{0.86} \left(\frac{\rho_v}{\rho_l} \right)^{0.522} \right]$$

For the present case,

$$F_f = 0.0225 \times \frac{300^2}{76.5} \times 22114^{-0.2} \times \left[0.5^{1.5} + 5.7 \left(\frac{1.321 \times 10^{-4}}{1.357 \times 10^{-5}} \right)^{0.0523} (1-0.5)^{0.47} \times 0.5^{1.33} \left(\frac{76.5}{1077} \right)^{0.261} + 8.11 \left(\frac{1.321 \times 10^{-4}}{1.357 \times 10^{-5}} \right)^{0.105} (1-0.5)^{0.94} \times 0.5^{0.86} \left(\frac{76.5}{1077} \right)^{0.522} \right]$$

$$= 6.999$$

F_m represents the effect of momentum change:

$$F_m = \frac{DG^2}{4\rho_v} \left(-\frac{dx}{L} \right) \left[2(1-x) \left(\frac{\rho_v}{\rho_l} \right)^{2/3} + \left(\frac{1}{x} - 3 + 2x \right) \left(\frac{\rho_v}{\rho_l} \right)^{4/3} + (2x - 1 - \beta x) \times \left(\frac{\rho_v}{\rho_l} \right)^{1/3} + \left(2\beta - \frac{\beta}{x} - \beta x \right) \left(\frac{\rho_v}{\rho_l} \right)^{5/3} + 2(1 - x - \beta + \beta x) \left(\frac{\rho_v}{\rho_l} \right) \right]$$

where

$$\beta = \begin{cases} 2 & \text{for} \quad Re_l \le 2000 \\ 1.25 & \text{for} \quad Re_l > 2000 \end{cases}$$

For the present case,

$$F_m = \frac{0.001 \times 300^2}{4 \times 76.5}\left(-\frac{0.1}{0.3048}\right)$$

$$\times \left[2(1-0.5)\left(\frac{76.5}{1077}\right)^{2/3} + \left(\frac{1}{0.5} - 3 + 2 \times 0.5\right)\left(\frac{76.5}{1077}\right)^{4/3}\right.$$

$$+ (2 + 0.5 - 1 - 2 \times 0.5)\left(\frac{76.5}{1077}\right)^{1/3} + \left(2 \times 2 - \frac{2}{0.5} - 2 \times 0.5\right)\left(\frac{76.5}{1077}\right)^{5/3}$$

$$\left. + 2(1 - 0.5 - 2 + 2 \times 0.5)\left(\frac{76.5}{1077}\right)\right]$$

$$= 0.03144$$

F_a represents the effect of the axial gravitational field on the wall shear stress. In case of horizontal condensation in microchannels, this term can be neglected. The net shear stress is therefore:

$$F_0 = F_f + F_m \pm F_a$$
$$= 6.999 + 0.03144$$
$$= 7.031$$

The heat transfer coefficient is given by:

$$h = 0.036 \times \frac{k_l \times \rho_l^{0.5}}{\mu_l} \times Pr_l^{0.65} \times F_0^{1/2}$$

For the present case:

$$h = 0.036 \times \frac{0.06749 \times 1077^{0.5}}{1.321 \times 10^{-4}} \times 3.152^{0.65} \times 7.031^{1/2}$$

$$= 3377 \text{ W/m}^2 \text{ K}$$

Soliman (1986)

This correlation should only be used for mist flow ($We > 30$), which is not the case here, but the calculations are shown below to illustrate the procedure and investigate the value of the resulting heat transfer coefficient. With $Re_l = 1136$,

$Re_v = 11{,}057$, $X_{tt} = 0.3346$, $\phi_v = 1 + 1.09 \times X^{0.039}{}_{tt} = 2.044$, all illustrated above or in the previous example, the Weber number can be calculated as follows:

$$We = \begin{cases} 2.45 \times Re_v^{0.64} \dfrac{(\mu_v^2/\rho_v \sigma D)^{0.3}}{\phi_v^{0.4}} & \text{for} \quad Re_l \leq 1250 \\ 0.85 \times Re_v^{0.79} \left(\dfrac{\mu_v^2}{\rho_v \sigma D}\right)^{0.3} \left[\left(\dfrac{\mu_v}{\mu_l}\right)^2 \left(\dfrac{\rho_l}{\rho_v}\right)\right]^{0.084} \left(\dfrac{X_{tt}}{\phi_v^{2.55}}\right)^{0.157} & \text{for} \quad Re_l \leq 1250 \end{cases}$$

For $Re_l = 1136$:

$$We = 2.45 \times Re_v^{0.64} \frac{(\mu_v^2/\rho_v \sigma D)^{0.3}}{\phi_v^{0.4}}$$

$$= 2.45 \times 11{,}057^{0.64} \frac{(((1.357 \times 10^{-5})^2)/(76.5 \times 0.00427 \times 0.001))^{0.3}}{2.044^{0.4}} = 9.509$$

Clearly, the flow is not in the mist flow regime. The mixture viscosity required for the heat transfer coefficient is given by:

$$u_m = \left(\frac{x}{\mu_v} + \frac{1-x}{\mu_l}\right)^{-1} = \left(\frac{0.5}{1.357 \times 10^{-5}} + \frac{1-0.5}{1.321 \times 10^{-4}}\right)^{-1} = 2.46 \times 10^{-5}\ \text{kg/m s}$$

With this mixture viscosity, the corresponding Reynolds number is:

$$Re_m = \frac{GD}{\mu_m} = \frac{300 \times 0.001}{2.46 \times 10^{-5}} = 12{,}193$$

With the Reynolds number calculated, the Nusselt number is given by:

$$Nu = 0.00345 \times Re_m^{0.9} \left(\frac{\mu_v h_{lv}}{k_v (T_{sat} - T_{wall})}\right)^{1/3}$$

which for the present case yields:

$$Nu = 0.00345 \times 12193^{0.9} \left(\frac{1.357 \times 10^{-5} \times 145394}{0.0178(55.21-52)}\right)^{1/3}$$

$$= 53.46$$

The two-phase heat transfer coefficient is therefore:

$$h = Nu \frac{k_l}{D} = 53.46 \times \frac{0.06749}{0.001} = 3608\ \text{W/m}^2\ \text{K}$$

Traviss et al. (1973)

This correlation was developed from data for the condensation of R-22 in an 8-mm tube over the following range of conditions:

$$25°C < T_{sat} < 58.3°C$$

$$161.4 < G < 1532 \text{ kg/m}^2 \text{ s}$$

The required liquid Reynolds number and Martinelli parameter are $Re_l = 1136$ and $X_{tt} = 0.3346$, respectively. With these quantities, the correlation for the shear-related term F in the liquid film is calculated as follows:

$$F = \begin{cases} 0.707 \times Pr_l Re_l^{0.5} & \text{for} \quad Re_l < 50 \\ 5Pr_l + 5\ln(1 + Pr_l(0.09636 Re_l^{0.585} - 1)) & \text{for} \quad 50 \le Re_l \le 1125 \\ 5Pr_l + 5\ln(1 + 5Pr_l) + 2.5\ln(0.00313 Re_l^{0.812}) & \text{for} \quad Re_l > 1125 \end{cases}$$

For the present case:

$$\begin{aligned} F &= 5Pr_l + 5\ln(1 + 5Pr_l) + 2.5\ln(0.00313 Re_l^{0.812}) \\ &= 5 \times 3.152 + 5\ln(1 + 5 \times 3.152) + 2.5\ln(0.00313 \times 1136^{0.812}) \\ &= 29.72 \end{aligned}$$

Finally, the heat transfer coefficient is calculated as follows:

$$\begin{aligned} h &= \frac{k_l}{D}\left(\frac{1}{X_{tt}} + \frac{2.85}{X_{tt}^{0.476}}\right)\left(\frac{0.15 \times Pr_l \times Re_l^{0.9}}{F}\right) \\ &= \frac{0.06749}{0.001}\left(\frac{1}{0.3346} + \frac{2.85}{0.3346^{0.476}}\right)\left(\frac{0.15 \times 3.152 \times 1136^{0.9}}{29.72}\right) \\ &= 4700 \text{ W/m}^2 \text{ K} \end{aligned}$$

Dobson and Chato (1998)

The relevant basic parameters (calculated above) are as follows: $Re_l = 1136$, $Re_{vo} = 22{,}114$, and $X_{tt} = 0.3346$. The void fraction is calculated using Zivi's (1964) model:

$$\alpha = \left[1 + \frac{1-x}{x}\left(\frac{\rho_v}{\rho_l}\right)^{2/3}\right]^{-1} = \left[1 + \frac{1-0.5}{0.5}\left(\frac{76.5}{1077}\right)^{2/3}\right]^{-1} = 0.8536$$

The Galileo number was calculated in Example 6.1 for the present case, resulting in $Ga = 478{,}182$. The modified Soliman Froude number is then calculated using the following equation:

$$Fr_{so} = \begin{cases} 0.025 Re_l^{1.59} \left(\dfrac{1 + 1.09 \times X_{tt}^{0.039}}{X_{tt}} \right) \dfrac{1}{Ga^{0.5}} & \text{for} \quad Re_l \le 1250 \\ 1.26 Re_l^{1.04} \left(\dfrac{1 + 1.09 \times X_{tt}^{0.039}}{X_{tt}} \right) \dfrac{1}{Ga^{0.5}} & \text{for} \quad Re_l \le 1250 \end{cases}$$

For the present case, with $Re_l = 1136$,

$$Fr_{so} = 0.025 Re_l^{1.59} \left(\frac{1 + 1.09 \times X_{tt}^{0.039}}{X_{tt}} \right)^{1.5} \frac{1}{Ga^{0.5}}$$

$$= 0.025 \times 1136^{1.59} \left(\frac{1 + 1.09 \times 0.3346^{0.039}}{0.3346} \right)^{1.5} \frac{1}{478,182^{0.5}}$$

$$= 39.37$$

Since $Fr_{so} > 20$, the annular flow correlation proposed by them is used:

$$Nu_{annular} = 0.023 \times Re_l^{0.8} \times Pr_l^{0.4} \left(1 + \frac{2.22}{X_{tt}^{0.89}} \right)$$

Substituting the relevant parameters, the Nusselt number for the present case is:

$$Nu = 0.023 \times 1136^{0.8} \times 3.152^{0.4} \left(1 + \frac{2.22}{0.3346^{0.89}} \right)$$

$$= 69.98$$

The resulting heat transfer coefficient is:

$$h = Nu \times \frac{k_l}{D} = 69.68 \times \frac{0.06749}{0.001} = 4703 \text{ W/m}^2 \text{ K}$$

Moser et al. (1998)

This correlation was based on data from tubes with $3.14 < D < 20$ mm. The various phase Reynolds numbers are $Re_l = 1136$, $Re_{lo} = 2271$, and $Re_{vo} = (GD/\mu_v) = (300 \times 0.001/1.357 \times 10^{-5}) = 22,114$. The two-phase homogeneous density is given by:

$$\rho_{tp} = \left(\frac{x}{\rho_v} + \frac{1-x}{\rho_l} \right)^{-1} = \left(\frac{0.5}{76.5} + \frac{1-0.5}{1077} \right)^{-1} = 142.8 \text{ kg/m}^3$$

The corresponding Froude and Weber numbers are:

$$Fr_{\text{tp}} = \frac{G^2}{gD\rho_{\text{tp}}^2} = \frac{300^2}{9.81 \times 0.001 \times 142.8^2} = 449.6$$

$$We_{\text{tp}} = \frac{G^2 D}{\sigma \rho_{\text{tp}}} = \frac{300^2 \times 0.001}{0.00427 \times 142.8} = 147.5$$

For the calculation of the equivalent Reynolds number, the required friction factors are obtained as follows:

$$f_{\text{lo}} = 0.079 \times Re_{lo}^{-0.25} = 0.079 \times 2271^{-0.25} = 0.01144$$

$$f_{\text{vo}} = 0.079 \times Re_{\text{vo}}^{-0.25} = 0.079 \times 22114^{-0.25} = 0.006478$$

Constants required for the two-phase multiplier evaluation are calculated as follows:

$$A_1 = (1-x)^2 + x^2\left(\frac{\rho_1}{\rho_{\text{v}}}\right)\left(\frac{f_{\text{vo}}}{f_{\text{lo}}}\right) = (1-0.5)^2 + 0.5^2\left(\frac{1077}{76.5}\right)\left(\frac{0.006478}{0.01144}\right) = 2.243$$

$$A_2 = x^{0.78}(1-x)^{0.24}\left(\frac{\rho_1}{\rho_{\text{v}}}\right)^{0.91}\left(\frac{\mu_{\text{v}}}{\mu_1}\right)^{0.19}\left(1-\frac{\mu_{\text{v}}}{\mu_1}\right)^{0.70}$$

$$= 0.5^{0.78}(1-0.5)^{0.24}\left(\frac{1077}{76.5}\right)^{0.91}\left(\frac{1.357\times10^{-5}}{1.321\times10^{-4}}\right)^{0.19}\left(1-\frac{1.357\times10^{-5}}{1.321\times10^{-4}}\right)^{0.70}$$

$$= 3.293$$

The two-phase multiplier is obtained from the parameters calculated above:

$$\phi_{lo}^2 = A_1 + \frac{3.24 \times A_2}{Fr_{\text{tp}}^{0.045} \times We_{\text{tp}}^{0.035}} = 2.243 + \frac{3.24 \times 3.293}{449.6^{0.045} \times 147.5^{0.035}} = 3.008$$

The equivalent Reynolds number is calculated in terms of this multiplier as follows:

$$Re_{\text{eq}} = \phi_{lo}^{8/7} \times Re_{lo} = 3.008^{8/7} \times 2271 = 7996$$

The Nusselt number is calculated in terms of this equivalent Reynolds number as follows:

$$Nu = \frac{0.0994^{0.126 \times Pr_1^{-0.448}} Re_1^{-0.113 \times Pr_1^{-0.563}} \times Re_{\text{eq}}^{1+0.11025 \times Pr_1^{-0.448}} \times Pr_1^{0.815}}{(1.58 \times \ln Re_{\text{eq}} - 3.28)(2.58 \times \ln Re_{\text{eq}} + 13.7 \times Pr_1^{2/3} - 19.1)}$$

For the present case,

$$Nu = \frac{0.0994^{0.126\times3.152^{-0.448}} \times 1136^{-0.113\times3.152^{-0.563}} \times 7996^{1+0.11025\times3.152^{-0.448}} \times 3.152^{0.815}}{(1.58\times\ln 7996 - 3.28)(2.58\times\ln 7996 + 13.7\times3.152^{2/3} - 19.1)}$$

$$= 55.77$$

Finally, the heat transfer coefficient is given by:

$$h = Nu \times \frac{k_1}{D} = 55.77 \times \frac{0.06749}{0.001} = 3764\ \mathrm{W/m^2\,K}$$

Boyko and Kruzhilin (1967)

This correlation relates the heat transfer coefficient in single-phase flow with the entire fluid flowing as a liquid to the two-phase heat transfer coefficient based on the mixture density, $\rho_m = 142.8\ \mathrm{kg/m^3}$, calculated above. The bulk and wall Prandtl numbers are also required. Here the Prandtl number at the refrigerant pressure and wall temperature, $Pr_w = 3.144$. The single-phase heat transfer coefficient is calculated as follows:

$$h_0 = 0.024 \times \frac{k_1}{D} \times \left(G \times \frac{D}{\mu_1}\right)^{0.8} \times Pr_1^{0.43} \times \left(\frac{Pr_1}{Pr_w}\right)^{0.25}$$

For the present case,

$$h_0 = 0.024 \times \frac{0.06749}{0.001} \times \left(300 \times \frac{0.001}{1.321 \times 10^{-4}}\right)^{0.8} \times 3.152^{0.43} \times \left(\frac{3.152}{3.144}\right)^{0.25}$$

$$= 1286\ \mathrm{W/m^2\,K}$$

The two-phase heat transfer coefficient is obtained as follows:

$$h = h_0\sqrt{\frac{\rho_1}{\rho_m}}1268\sqrt{\frac{1077}{142.8}} = 3531\ \mathrm{W/m^2\,K}$$

Cavallini et al. (2002a)

The applicable flow regime is first found from the Martinelli parameter, $X_{tt} = 0.3346$, and the dimensionless gas velocity:

$$j_g^* = \frac{Gx}{\sqrt{Dg\rho_v(\rho_1 - \rho_v)}} = \frac{300 \times 0.5}{\sqrt{0.001 \times 9.81 \times 76.5(1077 - 76.5)}} = 5.473$$

Based on the transition criteria presented in Example 6.1, the flow is in the annular regime. The two-phase multiplier and the pressure gradient required for the calculation of the shear stress were presented in Example 6.3. Thus, $\Delta P/L = 14{,}500$ Pa/m. The shear stress is given by:

$$\tau = \left(\frac{\mathrm{d}p}{\mathrm{d}z}\right)_f \frac{D}{4} = 14,500 \times \frac{0.001}{4} = 3.625 \text{ Pa}$$

The dimensionless film thickness is based on the liquid-phase Reynolds number:

$$\delta^+ = \begin{cases} \left(\dfrac{Re_1}{2}\right)^{0.5} & \text{for } Re_1 \leq 1145 \\ 0.0504 \times Re_1^{7/8} & \text{for } Re_1 > 1145 \end{cases}$$

In the present case, $Re_1 = 1136$, which yields:

$$\delta^+ = \left(\frac{Re_1}{2}\right)^{0.5} = \left(\frac{1136}{2}\right)^{0.5} = 23.83$$

The dimensionless temperature is given by:

$$T^+ = \begin{cases} \delta^+ Pr_1 & \delta^+ \leq 5 \\ 5\left\{Pr_1 + \ln\left[1 + Pr_1\left(\dfrac{\delta^+}{5} - 1\right)\right]\right\} & 5 < \delta^+ < 30 \\ 5\left[Pr_1 + \ln(1 + 5Pr_1) + 0.495 \ln\left(\dfrac{\delta^+}{30}\right)\right] & \delta^+ \geq 30 \end{cases}$$

For the present case,

$$T^+ = 5\left\{Pr_1 + \ln\left[1 + Pr_1\left(\frac{\delta^+}{5} - 1\right)\right]\right\}$$

$$= 5\left\{3.152 + \ln\left[1 + 3.152\left(\frac{23.83}{5} - 1\right)\right]\right\} = 28.54$$

Finally, the heat transfer coefficient is calculated as follows:

$$h = \frac{\rho_1 C_{\mathrm{pl}}(\tau/\rho_1)^{0.5}}{T^+} = \frac{1077 \times 1611 \times (3.625/1077)^{0.5}}{28.54} = 3527 \text{ W/m}^2\text{ K}$$

Bandhauer et al. (2006)

This model was based on heat transfer data for $0.5 < D < 1.5$ mm and $150 < G < 750$ kg/m^2 s, and the underlying pressure drop model was based on data for a much wider range of hydraulic diameters ($0.4 < D < 4.9$ mm). The pressure gradient calculation was discussed in Example 6.3, and resulted in a value of $\Delta P/L = 15{,}907$ Pa/m. The corresponding interfacial shear stress is:

$$\tau_\mathrm{i} = \left(\frac{\Delta P}{L}\right)\frac{D_\mathrm{i}}{4} = \left(\frac{\Delta P}{L}\right)\frac{\sqrt{\alpha}\times D}{4}$$

$$= 15.907\times 10^3 \times \frac{\sqrt{0.8059}\times 0.001}{4} = 3.57\ \mathrm{Pa}$$

The friction velocity is calculated from the interfacial shear stress as follows:

$$u^* = \sqrt{\frac{\tau_\mathrm{i}}{\rho_\mathrm{l}}} = \sqrt{\frac{3.57}{1077}} = 0.05756\ \mathrm{m/s}$$

The film thickness is directly calculated from the definition of the void fraction as follows:

$$\delta = (1-\sqrt{\alpha})\frac{D}{2} = \left(1-\sqrt{0.8059}\right)\frac{0.001}{2} = 5.115\times 10^{-5}\ \mathrm{m}$$

This thickness is used to obtain the dimensionless film thickness as follows:

$$\delta^+ = \frac{\delta\times\rho_\mathrm{l}\times u^*}{\mu_\mathrm{l}} = \frac{5.115\times 10^{-5}\times 1077\times 0.05756}{1.321\times 10^{-4}} = 24.02$$

The turbulent dimensionless temperature is given by:

$$T^+ = \begin{cases} 5Pr_\mathrm{l} + 5\ln\left[Pr_\mathrm{l}\left(\dfrac{\delta^+}{5}-1\right)+1\right] & \text{if } Re_\mathrm{l} < 2100 \\[2ex] 5Pr_\mathrm{l} + 5\ln(5Pr_\mathrm{l}+1) + \displaystyle\int_{30}^{\delta+}\frac{dy^+}{((1/Pr_\mathrm{l})-1)+\dfrac{y^+}{5}(1-(y^+/R^+))} & \text{if } Re_\mathrm{l} \geq 2100 \end{cases}$$

Thus, for the current example, with $Re_\mathrm{l} = 598.5$:

$$T^+ = 5Pr_\mathrm{l} + 5\ \ln\left[Pr_\mathrm{l}\left(\frac{\delta^+}{5}-1\right)+1\right]$$

$$= 5\times 3.152 + 5\ \ln\left[3.152\left(\frac{24.02}{5}-1\right)+1\right] = 28.58$$

Therefore, the heat transfer coefficient is:

$$h = \frac{\rho_l \times C_{P_l} \times u^*}{T^+} = \frac{1077 \times 1611 \times 0.05756}{28.58} = 3495 \text{ W/m}^2 \text{ K}$$

The heat transfer coefficients predicted by the above models are depicted in the following graph.

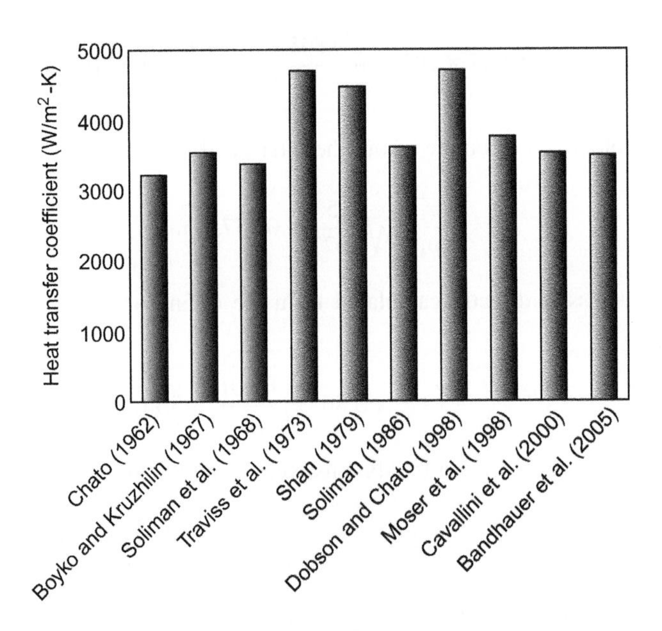

It can be seen from this graph that there are considerable variations between the heat transfer coefficients predicted by these different models and correlations. One of the main reasons for these variations is that most of these correlations have been applied in this case outside their ranges of applicability, either in the mass flux, the phase velocities, or parameters such as the Weber and Froude numbers, as applicable. But the most important reason for the differences is that all except the Bandhauer et al. (2006) model were based primarily on tube diameters in the ~8 mm range, with a few correlations being proposed based on limited data on ~3 mm tubes. The channel under consideration in this example is a 1-mm channel.

6.7 Exercises

1. Plot simulated air–water data points at different equally spaced combinations of gas and liquid superficial velocities in the range 0.1–100 m/s, and 0.01–10.0 m/s, respectively, as is typically done in air–water studies. Now plot on the same graph refrigerant R-134a (at 40°C) points for the mass flux

range 100–800 kg/m^2 s in 100 kg/m^2 s increments and quality increments of 0.10; that is, compute the vapor and liquid superficial velocities for these G, x combinations and plot on the air–water superficial velocity graph. Assume a 3-mm ID channel. Use this graph to comment on extending air–water flow regime maps to refrigerant condensation. Propose appropriate means to transform air–water results to refrigerant predictions based on comparisons of the relevant properties.

2. Flow regime maps are often plotted on the basis of superficial velocities, in which the phases are assumed to occupy the entire channel cross-section. While this is a commonly used practice out of convenience, what are the disadvantages of choosing superficial velocities instead of actual phase velocities? Quantify for a range of flow patterns.

3. Conduct a dimensional analysis to identify the dimensionless parameters that would quantify transitions between the four principal regimes, intermittent, wavy, annular, and mist flow, by considering viscous, surface tension, inertia, and gravity forces. Conduct parametric analyses to choose the most significant parameters across the scales.

4. Estimate liquid film thickness as a function of three different void fraction models in annular flow, and its effect on the heat transfer coefficient. Refer to El Hajal et al. (2003) and Thome et al. (2003) for guidance. How would your conclusions change for intermittent or stratified flows?

5. One of the uncertainties often overlooked in reporting pressure drop results in microchannels is the effect tolerances in tube diameter. Conduct a parametric analysis of the change in pressure drop in a 1-mm channel during annular flow condensation of refrigerant R-22 across the quality range $0.3 < x < 0.8$. Use a reasonable range of tolerances in the tube diameter to illustrate these effects. Use a mass flux of 400 kg/m^2 s, at a saturation temperature of 45°C. Comment on how this variation would change at evaporating conditions (e.g., 5°C).

6. Develop a quantitative indicator of momentum coupling of liquid and vapor phases in condensing flows based on the papers of Lee and Lee (2001), the origin of the C coefficient expression developed by Chisholm (1967) for the Lockhart–Martinelli (1949) two-phase multiplier approach, and the discussions in the papers by Kawahara et al. (2002) and Chung and Kawaji (2004). Estimate numerical values of this measure for 10 μm $< D_H <$ 10 mm.

7. Estimate the difference in saturation temperature decrease due to the flow of refrigerant R-22 in a 0.5-mm channel at a mass flux of 200 kg/m^2 s by computing the condensation in 10% quality increments and accounting for

frictional and deceleration contributions. Assume an appropriate bulk-wall temperature difference. Plot the pressure drop in each increment, and the corresponding decrease in saturation temperature. Use two different pressure drop models, one based on air–water studies, another based on refrigerant condensation.

8. Microchannels are especially attractive due to the high heat transfer coefficients they offer. In measuring these heat transfer coefficients, one has to measure the heat transfer rate using, typically, a rise in coolant temperature, and a temperature difference between the refrigerant and the wall. Assume channel $D_H = 0.5$ mm. Assume further that the heat transfer coefficient needs to be measured in increments of $\Delta x = 0.10$. First estimate the heat transfer rate to achieve this condensation at this mass flux. How accurately can this heat transfer rate be measured in a single microchannel? What changes in coolant temperature would be required to effect this heat transfer across the test section? Is this realistic? Estimate the uncertainties in heat transfer rate measurement using reasonable thermocouple and mass flow rate measurement uncertainties. In an attempt to improve the uncertainties, multiple channels in parallel are proposed. How many channels would be required to achieve accurately measurable heat transfer rates? What are the drawbacks of using multiple channels? Next, estimate the heat transfer coefficient in condensing R-134a at 450 kg/m^2 s and 50°C using one of the available models. What refrigerant-wall temperature difference would this cause in a channel that accomplishes the 10% quality change? (You will have to choose a reasonable test section length.) How accurately can this temperature difference be measured? What would be the resulting heat transfer coefficient uncertainty? How would the uncertainty vary with mass flux? Quantify. How would the uncertainty change if R-134a is substituted with steam?

9. It has been stated in this chapter that boundary-layer analyses in the liquid film and two-phase multiplier approaches are essentially equivalent methods of analyzing annular flow heat transfer. Refer to Dobson and Chato (1998) for background on this matter. Conduct parallel derivations of these two approaches to point out the essential similarities. Also, comment on any significant differences.

10. Most heat transfer models recommend simply extending annular flow models to the intermittent flow. Develop a methodology for conducting this extension on a rational basis by accounting for the slug/bubble frequency.

11. Will intermittent flow heat transfer coefficients be higher or lower than those in annular flow? Why? Will they be higher or lower than those in single-phase gas or single-phase liquid flow? Why? Demonstrate through

approximate models, using R-134a condensing at a 35°C through a 200-μm channel at 100 kg/m^2 s.

12. Assuming that condensing flows would go through a progression of annular-mist-homogeneous flow as the reduced pressure is increased, identify the most important fluid properties and the corresponding dimensionless parameters, and investigate changes in these quantities as the pressure is increased from $0.2 \times p_r$ to $0.9 \times p_r$. How would the pressure drop and heat transfer be affected in this progression?

13. Design an R-22 condenser that transfers 3 kW at a saturation temperature of 55°C, with the refrigerant entering as a saturated vapor and leaving as a saturated liquid. Dry air flowing in crossflow at 4.8E-01 m^3/s, with an inlet temperature of 35°C and a heat transfer coefficient of 50 W/m^2-K, is the cooling medium. Use copper tubing with copper fins on the air side that yield a factor of 10 surface area enhancement. The efficiency of the fins can be assumed to be 90%. Choose the tubing size and wall thickness from standard tables such that the drop in refrigerant saturation temperature (due to the pressure drop) is less than 1.5°C. Account for the flow regime variation over the length of the heat exchanger. Choose quality and/or tube length increments of reasonable size that demonstrate the flow phenomena and the heat transfer coefficient and pressure drop variations. Assume a horizontal orientation.

14. Repeat Problem 13 by changing the design to an optimal parallel-serpentine arrangement of extruded aluminum tubes with twenty 0.7-mm microchannel tubes with the same amount of air-side aluminum fin surface area enhancement as in Problem 13. Thus, decrease the number or tubes per pass as condensation proceeds so that a larger flow area is provided in the initial passes to account for the higher flow velocities, and decrease the number of parallel tubes in subsequent passes. Compare the sizes (total material volume of each heat exchanger including tube and fin volume) resulting from these two approaches (Problems 13 and 14) and comment.

References

Agarwal, A., 2006. Heat Transfer and Pressure Drop During Condensation of Refrigerants in Microchannels (Ph.D. dissertation). GWW School of Mechanical Engineering, Georgia Institute of Technology, Atlanta, GA.

Agarwal, A., Bandhauer, T.M., Garimella, S., 2010. Measurement and modeling of condensation heat transfer in non-circular microchannels. Int. J. Refrig. 33 (6), 1169–1179.

Akers, W.W., Rosson, H.F., 1960. Condensation inside a horizontal tube. Chem. Eng. Prog. S. Ser. 56 (30), 145–150.

Akers, W.W., Deans, H.A., Crosser, O.K., 1959. Condensation heat transfer within horizontal tubes. Chem. Eng. Prog. S. Ser. 55 (29), 171–176.

Ali, M.I., Sadatomi, M., Kawaji, M., 1993. Adiabatic two-phase flow in narrow channels between two flat plates. Can. J. Chem. Eng. 71 (5), 657–666.

Alves, G.E., 1954. Cocurrent liquid–gas flow in a pipe-line contactor. Chem. Eng. Prog. 50 (9), 449–456.

Ananiev, E., Boyko, L., Kruzhilin, G., 1961. Heat transfer in the presence of steam condensation in a horizontal tube. Int. Dev. Heat Transfer 2, 290–295.

Andresen, U., 2007. Supercritical Gas Cooling and Near-Critical-Pressure Condensation of Refrigerant Blends in Microchannels (Ph.D. dissertation). GWW School of Mechanical Engineering, Georgia Institute of Technology, Atlanta, GA.

Armand, A.A., 1946. The resistance during the movement of a two-phase system in horizontal pipes. Izv. Vses. Teplotekh. Inst. 1, 16–23 (AERE-Lib/Trans 828).

Azer, N., Abis, L., Soliman, H., 1972. Local heat transfer coefficients during annular flow condensation. ASHRAE Trans. 78 (2), 135–143.

Bae, S., Maulbetsch, J.S., Rohsenow, W.M., 1968. Refrigerant Forced-Convection Condensation Inside Horizontal Tubes. MIT Heat Transfer Laboratory, Cambridge, MA.

Baker, O., 1954. Simultaneous flow of oil and gas. Oil Gas J. 53 (12), 185–190.

Bandhauer, T.M., Agarwal, A., Garimella, S., 2006. Measurement and modeling of condensation heat transfer coefficients in circular microchannels. J. Heat Transfer 128 (10), 1050–1059.

Barnea, D., Luninski, Y., Taitel, Y., 1983. Flow pattern in horizontal and vertical two phase flow in small diameter pipes. Can. J. Chem. Eng. 61 (5), 617–620.

Baroczy, C., 1966. Systematic correlation for two-phase pressure drop. Chem. Eng. Prog. Symp. Ser 62 (64), 232–249.

Baroczy, C., Sanders, V.D., 1961. Pressure Drop for Flowing Vapors Condensing in a Straight Horizontal Tube. Atomics International, Canoga Park, CA.

Baroczy, C.J., 1965. Correlation of liquid fraction in two-phase flow with application to liquid metals. Chem. Eng. Prog. S. Ser. 61 (57), 179–191.

Beattie, D.R.H., Whalley, P.B., 1982. A simple two-phase frictional pressure drop calculation method. Int. J. Multiphase Flow 8 (1), 83–87.

Blangetti, F., Schlunder, E., 1978. Local heat transfer coefficients on condensation in a vertical tube. Sixth International Heat Transfer Conference, Toronto, Canada, 2, pp. 437–442.

Boyko, L.D., Kruzhilin, G.N., 1967. Heat transfer and hydraulic resistance during condensation of steam in a horizontal tube and in a bundle of tubes. Int. J. Heat Mass Transfer 10 (3), 361–373.

Brauner, N., Maron, D.M., 1992. Identification of the range of 'small diameters' conduits, regarding two-phase flow pattern transitions. Int. Commun. Heat Mass Transfer 19 (1), 29–39.

Breber, G., Palen, J.W., Taborek, J., 1980. Prediction of horizontal tubeside condensation of pure components using flow regime criteria. J. Heat Transfer 102 (3), 471–476.

Butterworth, D., 1975. A comparison of some void–fraction relationships for co-current gas–liquid flow. Int. J. Multiphase Flow 1 (6), 845–850.

Carey, V., 2008. Liquid–Vapor Phase-Change Phenomena. Taylor & Francis Group, New York, NY.

Carnavos, T., 1980. Heat transfer performance of internally finned tubes in turbulent flow. Heat Transfer Eng. 1 (4), 32–37.

Carnavos, T.C., 1979. Cooling air in turbulent-flow with multi-passage internally finned tubes. Mech. Eng. 101 (3), 96–97.

Carpenter, F.G., Colburn, A.P., 1951. The effect of vapor velocity on condensation inside tubes. Proceedings of the General Discussion of Heat Transfer. ASME, pp. 20–26.

Cavallini, A., Zecchin, R., 1974. A dimensionless correlation for heat transfer in forced convection condensation. Fifth International Heat Transfer Conference, Tokyo, Japan, vol. 3, pp. 309–313.

Cavallini, A., Del Col, D., Doretti, L., Longo, G.A., Rossetto, L., 2000. Heat transfer and pressure drop during condensation of refrigerants inside horizontal enhanced tubes. Int. J. Refrig. 23 (1), 4–25.

Cavallini, A., Censi, G., Del Col, D., Doretti, L., Longo, G.A., Rossetto, L., 2001. Experimental investigation on condensation heat transfer and pressure drop of new HFC refrigerants (R134a, R125, R32, R410A, R236ea) in a horizontal smooth tube. Int. J. Refrig. 24 (1), 73–87.

Cavallini, A., Censi, G., Del Col, D., Doretti, L., Longo, G.A., Rossetto, L., 2002a. Condensation of halogenated refrigerants inside smooth tubes. HVAC&R Res. 8 (4), 429–451.

Cavallini, A., Censi, G., Del Col, D., Doretti, L., Longo, G.A., Rossetto, L., 2002b. A tube-in-tube water/zeotropic mixture condenser: design procedure against experimental data. Exp. Therm. Fluid Sci. 25 (7), 495–501.

Cavallini, A., Del Col, D., Doretti, L., Matkovic, M., Rossetto, L., Zilio, C., 2005a. Condensation heat transfer and pressure gradient inside multiport minichannels. Heat Transfer Eng. 26 (3), 45–55.

Cavallini, A., Del Col, D., Doretti, L., Matkovic, M., Rossetto, L., Zilio, C., 2005b. A model for condensation in minichannels. ASME Summer Heat Transfer Conference. ASME, San Francisco, CA, pp. 297–304.

Cavallini, A., Del Col, D., Doretti, L., Matkovic, M., Rossetto, L., Zilio, C., et al., 2006a. Condensation in horizontal smooth tubes: a new heat transfer model for heat exchanger design. Heat Transfer Eng. 27 (8), 31–38.

Cavallini, A., Doretti, L., Matkovic, M., Rossetto, L., 2006b. Update on condensation heat transfer and pressure drop inside minichannels. Heat Transfer Eng. 27 (4), 74–87.

Cavallini, A., Del Col, D., Matkovic, M., Rossetto, L., 2009. Frictional pressure drop during vapour–liquid flow in minichannels: modelling and experimental evaluation. Int. J. Heat Fluid Flow 30 (1), 131–139.

Chato, J.C., 1962. Laminar film condensation inside horizontal and inclined tubes. ASHRAE J. 4, 52–60.

Chen, I., Kocamustafaogullari, G., 1987. Condensation heat transfer studies for stratified, cocurrent two-phase flow in horizontal tubes. Int. J. Heat Mass Transfer 30 (6), 1133–1148.

Chen, I.Y., Yang, K.S., Chang, Y.J., Wang, C.C., 2001. Two-phase pressure drop of air–water and R-410A in small horizontal tubes. Int. J. Multiphase Flow 27 (7), 1293–1299.

Chen, S.J., Reed, J.G., Tien, C.L., 1984. Reflux condensation in a two-phase closed thermosyphon. Int. J. Heat Mass Transfer 27 (9), 1587–1594.

Chen, S.L., Gerner, F.M., Tien, C.L., 1987. General film condensation correlations. Exp. Heat Transfer 1 (2), 93–107.

Chexal, B., Lellouche, G., 1986. Full-range drift–flux correlation for vertical flows. Revision 1, Electric Power Research Inst.. Nuclear Power Div, Palo Alto, CA.

Chexal, B., Lellouche, G., Horowitz, J., Healzer, J., 1992. A void fraction correlation for generalized applications. Prog. Nucl. Energy 27 (4), 255–295.

Chexal, B., Maulbetsch, J., Santucci, J., Harrison, J., Jensen, P., Peterson, C., et al., 1996. Understanding Void Fraction in Steady and Dynamic Environments. Electric Power Research Institute, Palo Alto, CA.

Chexal, B., Merilo, M., Maulbetsch, J., Horowitz, J., Harrison, J., Westacott, J.C., et al., 1997. Void Fraction Technology for Design and Analysis. Electric Power Research Institute, Palo Alto, CA.

Chisholm, D., 1967. A theoretical basis for the Lockhart–Martinelli correlation for two-phase flow. Int. J. Heat Mass Transfer 10 (12), 1767–1778.

Chisholm, D., 1973. Pressure gradients due to friction during the flow of evaporating two-phase mixtures in smooth tubes and channels. Int. J. Heat Mass Transfer 16 (2), 347–358.

Chisholm, D., 1983. Two-Phase Flow in Pipelines and Heat Exchangers. G. Godwin in Association with the Institution of Chemical Engineers, London: New York, NY.

Chisholm, D., Sutherland, L., 1969. Prediction of pressure gradients in pipeline systems during two-phase flow. Proceedings of the Institution of Mechanical Engineers. SAGE Publications, pp. 24–32.

Chitti, M., Anand, N., 1996. Condensation heat transfer inside smooth horizontal tubes for R-22 and R-32/125 mixture. HVAC&R Res. 2 (1), 79–100.

Chitti, M.S., Anand, N.K., 1995. An analytical model for local heat-transfer coefficients for forced convective condensation inside smooth horizontal tubes. Int. J. Heat Mass Transfer 38 (4), 615–627.

Chun, K.R., Seban, R.A., 1971. Heat transfer to evaporating liquid films. J. Heat Transfer 93 (4), 391–396.

Chung, P.M.Y., Kawaji, M., 2004. The effect of channel diameter on adiabatic two-phase flow characteristics in microchannels. Int. J. Multiphase Flow 30 (7–8), 735–761.

Chung, P.M.Y., Kawaji, M., Kawahara, A., Shibata, Y., 2004. Two-phase flow through square and circular microchannels—effects of channel geometry. J. Fluids Eng. 126 (4), 546–552.

Churchill, S.W., 1977. Friction-factor equation spans all fluid-flow regimes. Chem. Eng. 84 (24), 91–92.

Churchill, S.W., Usagi, R., 1974. A standardized procedure for the production of correlations in the form of a common empirical equation. Ind. Eng. Chem. Fund. 13 (1), 39–44.

Cicchitti, A., Lombardi, C., Silvestri, M., Soldaini, G., Zavattarelli, R., 1960. Two-phase cooling experiments: pressure drop, heat transfer and burnout measurements. Energ. Nucl. 7 (6), 407–425.

Cioncolini, A., Thome, J.R., Lombardi, C., 2009. Unified macro-to-microscale method to predict two-phase frictional pressure drops of annular flows. Int. J. Multiphase Flow 35 (12), 1138–1148.

Clarke, N.N., Rezkallah, K.S., 2001. A study of drift velocity in bubbly two-phase flow under microgravity conditions. Int. J. Multiphase Flow 27 (9), 1533–1554.

Colebrook, C.F., 1939. Turbulent flow in pipes, with particular reference to the transition region between the smooth and rough pipe laws. J. ICE 11 (4), 133–156.

Coleman, J.W., Garimella, S., 1999. Characterization of two-phase flow patterns in small diameter round and rectangular tubes. Int. J. Heat Mass Transfer 42 (15), 2869–2881.

Coleman, J.W., Garimella, S., 2000a. Two-phase flow regime transitions in microchannel tubes: the effect of hydraulic diameter. ASME Heat Transfer Division. American Society of Mechanical Engineers, Orlando, FL, pp. 71–83.

Coleman, J.W., Garimella, S., 2000b. Visualization of two-phase refrigerant flow during phase change. Thirty-Fourth National Heat Transfer Conference. ASME, Pittsburgh, PA.
Coleman, J.W., Garimella, S., 2003. Two-phase flow regimes in round, square and rectangular tubes during condensation of refrigerant R134a. Int. J. Refrig. 26 (1), 117–128.
Collier, J.G., Thome, J.R., 1994. Convective Boiling and Condensation. Clarendon Press; Oxford University Press, Oxford; New York, NY.
Cornwell, K., Kew, P.A., 1993. Boiling in small parallel channels. Energy Efficiency in Process Technology. Springer Verlag, pp. 624–638.
Da Riva, E., Del Col, D., 2012. Numerical simulation of laminar liquid film condensation in a horizontal circular minichannel. J. Heat Transfer 134 (5), 051019–8.
Damianides, C., Westwater, J., 1988. Two-phase flow patterns in a compact heat exchanger and in small tubes, Second UK National Conference on Heat Transfer, vol. 11. Glasgow, Scotland, pp. 1257–1268.
Del Col, D., Torresin, D., Cavallini, A., 2010. Heat transfer and pressure drop during condensation of the low GWP refrigerant R1234yf. Int. J. Refrig. 33 (7), 1307–1318.
Dittus, W., Boelter, L.M.K., 1930. Heat transfer in automobile radiators of the tubular Type. Univ. Calif.– Publ. Eng. 2 (13), 443–461.
Dobson, M.K., 1994. Heat Transfer and Flow Regimes during Condensation in Horizontal tubes (Ph.D. dissertation). Mechanical and Industrial Engineering, University of Illinois at Urbana-Champaign, Urbana-Champaign, IL.
Dobson, M.K., Chato, J.C., 1998. Condensation in smooth horizontal tubes. J. Heat Transfer 120 (1), 193–213.
Dobson, M.K., Chato, J.C., Hinde, D.K., Wang, S.P., 1994. Experimental evaluation of internal condensation of refrigerants R-12 and R-134a, Proceedings of the ASHRAE Winter Meeting, January 23–26, New Orleans, USA, pp. 744–754.
Dukler, A., Wicks III, M., Cleveland, R., 1964. Frictional pressure drop in two-phase flow: a comparison of existing correlations for pressure loss and holdup. AIChE J. 10 (1), 38–43.
Dukler, A.E., 1960. Fluid mechanics and heat transfer in vertical falling film systems. Chem. Eng. Prog. S. Ser. 56 (30), 1–10.
Dukler, A.E., Hubbard, M.G., 1975. A model for gas–liquid slug flow in horizontal and near horizontal tubes. Ind. Eng. Chem. Fundam. 14 (4), 337–347.
Eckels, S.J., Doerr, T.M., Pate, M.B., 1994. In-tube heat transfer and pressure drop of R-134a and ester lubricant mixtures in a smooth tube an a micro-fin tube: part I—evaporation. ASHRAE Annual Meeting. ASHRAE, Orlando, FL, pp. 265–282.
Ekberg, N.P., Ghiaasiaan, S.M., Abdel-Khalik, S.I., Yoda, M., Jeter, S.M., 1999. Gas–liquid two-phase flow in narrow horizontal annuli. Nucl. Eng. Des. 192 (1), 59–80.
El Hajal, J., Thome, J.R., Cavallini, A., 2003. Condensation in horizontal tubes. Part 1: two-phase flow pattern map. Int. J. Heat Mass Transfer 46 (18), 3349–3363.
Elkow, K., Rezkallah, K., 1997a. Statistical analysis of void fluctuations in gas–liquid flows under 1 g and μg conditions using a capacitance sensor. Int. J. Multiphase Flow 23 (5), 831–844.
Elkow, K., Rezkallah, K., 1997b. Void fraction measurements in gas-liquid flows under under 1 g and μg conditions using capacitance sensors. Int. J. Multiphase Flow 23 (5), 815–829.
Elkow, K.J., Rezkallah, K.S., 1996. Void fraction measurements in gas–liquid flows using capacitance sensors. Meas. Sci. Technol. 7 (8), 1153–1163.
Feng, Z.P., Serizawa, A., 1999. Visualization of two-phase flow patterns in an ultra-small tube. Eighteenth Multiphase Flow Symposium of Japan. Suita, Osaka, Japan, pp. 33–36.

Feng, Z.P., Serizawa, A., 2000. Measurement of steam-water bubbly flow in ultra-small capillary tube, Thirty-Seventh National Heat Transfer Symposium of Japan, vol. 1. Heat Transfer Society of Japan, pp. 351–352.

Friedel, L., 1979. Improved friction pressure drop correlations for horizontal and vertical two phase pipe flow (Paper E2). European Two Phase Flow Group Meeting, Ispra, Italy.

Fukano, T., Kariyasaki, A., 1993. Characteristics of gas–liquid two-phase flow in a capillary tube. Nucl. Eng. Des. 141 (1–2), 59–68.

Fukano, T., Kariyasaki, A., Kagawa, M., 1989. Flow patterns and pressure drop in isothermal gas–liquid concurrent flow in a horizontal capillary tube, National Heat Transfer Conference, vol. 4. ASME, Philadelphia, PA, pp. 153–161.

Galbiati, L., Andreini, P., 1992. Flow pattern transition for vertical downward two-phase flow in capillary tubes. Inlet mixing effects. Int. Commun. Heat Mass Transfer 19 (6), 791–799.

Garimella, S., 2004. Condensation flow mechanisms in microchannels: basis for pressure drop and heat transfer models. Heat Transfer Eng. 25 (3), 104–116.

Garimella, S., Bandhauer, T.M., 2001. Measurement of condensation heat transfer coefficients in microchannel tubes. ASME International Mechanical Engineering Congress and Exposition. American Society of Mechanical Engineers, New York, NY, November 11–16, 2001, pp. 243–249.

Garimella, S., Fronk, B.M., 2012. Single and multi-constituent condensation of fluids and mixtures with varying properties in microchannels. ECI Eighth International Conference on Boiling and Condensation. Heat Transfer, Lausanne, Switzerland, 3–7 June.

Garimella, S., Wicht, A., 1995. Air-Cooled Condensation of Ammonia in Flat-Tube, Multi-Louver Fin Heat Exchangers, vol. **320**. ASME Publications, pp. 47–58.

Garimella, S., Killion, J.D., Coleman, J.W., 2002. An experimentally validated model for two-phase pressure drop in the intermittent flow regime for circular microchannels. J. Fluids Eng. 124 (1), 205–214.

Garimella, S., Agarwal, A., Coleman, J.W., 2003a. Two-phase pressure drops in the annular flow regime in circular microchannels. Twenty-First IIR International Congress of Refrigeration. International Institute of Refrigeration, Washington, DC.

Garimella, S., Killion, J.D., Coleman, J.W., 2003b. An expermentally validated model for two-phase pressure drop in the intermittent flow regime for noncircular microchannels. J. Fluids Eng. 125 (5), 887–894.

Garimella, S., Agarwal, A., Killion, J.D., 2005. Condensation pressure drop in circular microchannels. Heat Transfer Eng. 26 (3), 28–35.

Gibson, A.H., 1913. On the motion of long air-bubbles in a vertical tube. Philos. Mag. 26 (156), 952–965.

Gnielinski, V., 1976. New equations for heat and mass-transfer in turbulent pipe and channel flow. Int. Chem. Eng. 16 (2), 359–368.

Goto, M., Inoue, N., Ishiwatari, N., 2001. Condensation and evaporation heat transfer of R410A inside internally grooved horizontal tubes. Int. J. Refrig. 24 (7), 628–638.

Goto, M., Inoue, N., Yonemoto, R., 2003. Condensation heat transfer of R410A inside internally grooved horizontal tubes. Int. J. Refrig. 26 (4), 410–416.

Govier, G., Short, W.L., 1958. The upward vertical flow of air-water mixtures: II. Effect of tubing diameter on flow-pattern, holdup and pressure drop. Can. J. Chem. Eng. 36 (5), 195–202.

Govier, G., Radford, B.A., Dunn, J.S.C., 1957. The upward vertical flow of air-water mixtures: I. Effect of air and water rates on flow-pattern, holdup and pressure drop. Can. J. Chem. Eng. 35 (5), 58–70.

Gregorig, R., 1962. Verfahrenstechnisch günstigere führung der mittel der Wärmeübertragung beim verdampfen und kondensieren. Int. J. Heat Mass Transfer 5 (3–4), 175–188.

Griffith, P., Lee, K.S., 1964. The stability of an annulus of liquid in a tube. J. Basic Eng. 86, 666.

Haraguchi, J., Koyama, S., Fujii, T., 1994a. Condensation of refrigerants HCFC22, HFC134a and HCFC123 in a horizontal smooth tube (1st report, proposal of empirical expressions for the local frictional pressure drop). Transac. JSME (B) 60 (574), 239–244.

Haraguchi, J., Koyama, S., Fujii, T., 1994b. Condensation of refrigerants HCFC22, HFC134a and HCFC123 in a horizontal smooth tube (2nd report, proposal of empirical expressions for the local heat transfer coefficient). Transac. JSME (B) 60 (574), 245–252.

Hashizume, K., 1983. Flow pattern, void fraction and pressure drop of refrigerant two-phase flow in a horizontal pipe—I. Experimental data. Int. J. Multiphase Flow 9 (4), 399–410.

Hashizume, K., Ogawa, N., 1987. Flow pattern, void fraction and pressure drop of refrigerant two-phase flow in a horizontal pipe. III: Comparison of the analysis with existing pressure drop data on air/water and steam/water systems. Int. J. Multiphase Flow 13 (2), 261–267.

Hashizume, K., Ogiwara, H., Taniguchi, H., 1985. Flow pattern, void fraction and pressure drop of refrigerant two-phase flow in a horizontal pipe—II: analysis of frictional pressure drop. Int. J. Multiphase Flow 11 (5), 643–658.

Hetsroni, G., 1982. Handbook of Multiphase Systems. Hemisphere Pub. Corp., New York, NY.

Hewitt, G.F., Roberts, D., 1969. Studies of Two-Phase Flow Patterns by Simultaneous X-Ray and Flash Photography. Atomic Energy Research Establishment, Harwell, England.

Hewitt, G.F., Shires, G.L., Bott, T.R., 1994. Process Heat Transfer. CRC Press, New York, NY.

Hibiki, T., Mishima, K., 1996. Approximate method for measurement of phase-distribution in multiphase materials with small neutron-attenuation using a neutron beam as a probe. Nucl. Instrum. Meth. A. 374 (3), 345–351.

Hibiki, T., Mishima, K., 2001. Flow regime transition criteria for upward two-phase flow in vertical narrow rectangular channels. Nucl. Eng. Des. 203 (2–3), 117–131.

Hibiki, T., Mishima, K., Nishihara, H., 1997. Measurement of radial void fraction distribution of two-phase flow in a metallic round tube using neutrons as microscopic probes. Nucl. Instrum. Meth. A. 399 (2–3), 432–438.

Hosler, E.R., 1967. Flow Patterns in High Pressure Two-Phase (Steam-Water) Flow with Heat Addition. Bettis Atomic Power Lab., Pittsburgh, PA.

Hosler, E.R., 1968. Flow patterns in high pressure two-phase (steam-water) flow with heat addition. AIChE Symposium Series, pp. 54–66.

Hurlburt, E.T., Newell, T.A., 1999. Characteristics of refrigerant film thickness, pressure drop, and condensation heat transfer in annular flow. HVAC&R Res. 5 (3), 229–248.

Ide, H., Matsumura, H., Fukano, T., 1995. Velocity characteristics of liquid lumps and its relation to flow patterns in gas–liquid two-phase flow in vertical capillary tubes. ASME/JSME Fluids Engineering and Laser Anemometry Conference and Exhibition. ASME, Hilton Head, SC.

Ide, H., Matsumura, H., Tanaka, Y., Fukano, T., 1997. Flow patterns and frictional pressure drop in gas–liquid two-phase flow in vertical capillary channels with rectangular cross section. Nippon Kikai Gakkai Ronbunshu, B Hen/Trans. Jp. Soc. Mech. Eng. B 63 (606), 452–460.

Infante Ferreira, C., Newell, T., Chato, J., Nan, X., 2003. R404A condensing under forced flow conditions inside smooth, microfin and cross-hatched horizontal tubes. Int. J. Refrig. 26 (4), 433–441.

Ishii, M., 1977. Drift flux model and derivation of kinematic constitutive laws. Two-Phase Flows and Heat Transfer: Proceedings of NATO Advanced Study Institute, August 16–27, Istanbul, Turkey, Kakaç, S., Mayinger, F., North Atlantic Treaty Organization. Scientific Affairs Division, Washington, Hemisphere Pub. Corp, vol. 1, pp. 187–208.

Jaster, H., Kosky, P.G., 1976. Condensation heat-transfer in a mixed flow regime. Int. J. Heat Mass Transfer 19 (1), 95–99.

Jiang, Y., Garimella, S., 2001. Compact air-coupled and hydronically coupled microchannel heat pumps. ASME International Mechanical Engineering Congress and Exposition. ASME, New York, NY, pp. 227–239.

Jones Jr., O.C., Zuber, N., 1975. Interrelation between void fraction fluctuations and flow patterns in two-phase flow. Int. J. Multiphase Flow 2 (3), 273–306.

Jung, D.S., Radermacher, R., 1989. Prediction of pressure-drop during horizontal annular-flow boiling of pure and mixed refrigerants. Int. J. Heat Mass Transfer 32 (12), 2435–2446.

Kariyasaki, A., Fukano, T., Ousaka, A., Kagawa, M., 1991. Characteristics of time-varying void fraction in isothermal air–water cocurrent flow in horizontal capillary tube. Trans. JSME 57 (B) (544), 4036–4043.

Kariyasaki, A., Fukano, T., Ousaka, A., Kagawa, M., 1992. Isothermal air–water two-phase up and downward flows in a vertical capillary tube (1st report, flow pattern and void fraction). Trans. JSME (Ser. B) 58, 2684–2690.

Kattan, N., Thome, J.R., Favrat, D., 1998a. Flow boiling in horizontal tubes. Part 1—development of a diabatic two-phase flow pattern map. J. Heat Transfer 120 (1), 140–147.

Kattan, N., Thome, J.R., Favrat, D., 1998b. Flow boiling in horizontal tubes. Part 2—new heat transfer data for five refrigerants. J. Heat Transfer 120 (1), 148–155.

Kattan, N., Thome, J.R., Favrat, D., 1998c. Flow boiling in horizontal tubes. Part 3—development of a new heat transfer model based on flow pattern. J. Heat Transfer 120 (1), 156–165.

Kawahara, A., Chung, P.M.Y., Kawaji, M., 2002. Investigation of two-phase flow pattern, void fraction and pressure drop in a microchannel. Int. J. Multiphase Flow 28 (9), 1411–1435.

Kawahara, A., Sadatomi, M., Okayama, K., Kawaji, M., Chung, P.M.Y., 2005. Effects of channel diameter and liquid properties on void fraction in adiabatic two-phase flow through microchannels. Heat Transfer Eng. 26 (3), 13–19.

Keinath, B.L., Garimella, S., 2010. Bubble and film dynamics during condensation of refrigerants in minichannels, Fourteenth International Heat Transfer Conference (IHTC14), vol. **2**. ASME, Washington, DC, pp. 177–186.

Keinath, B.L., Garimella, S., 2011. Void fraction and pressure drop during condensation of refrigerants in minichannels. Sixth International Berlin Workshop on Transport Phenomena with Moving Boundaries, Berlin.

Kew, P.A., Cornwell, K., 1997. Correlations for the prediction of boiling heat transfer in small-diameter channels. Appl. Therm. Eng. 17 (8–10), 705–715.

Kim, M.H., Shin, J.S., 2005. Condensation heat transfer of R22 and R410A in horizontal smooth and microfin tubes. Int. J. Refrig. 28 (6), 949–957.

Kim, N.H., Cho, J.P., Kim, J.O., Youn, B., 2003. Condensation heat transfer of R-22 and R-410A in flat aluminum multi-channel tubes with or without micro-fins. Int. J. Refrig. 26 (7), 830–839.

Kim, S.-M., Mudawar, I., 2012. Flow condensation in parallel micro-channels. Part 2: heat transfer results and correlation technique. Int. J. Heat Mass Transfer 55 (4), 984–994.

Kim, S.-M., Kim, J., Mudawar, I., 2012. Flow condensation in parallel micro-channels. Part 1: experimental results and assessment of pressure drop correlations. Int. J. Heat Mass Transfer 55 (4), 971–983.

Kolb, W.B., Cerro, R.L., 1991. Coating the inside of a capillary of square cross-section. Chem. Eng. Sci. 46 (9), 2181–2195.

Kolb, W.B., Cerro, R.L., 1993. The motion of long bubbles in tubes of square cross section. Phys. Fluids A: Fluid Dyn. 5, 1549.

Kosky, P., 1971. Thin liquid films under simultaneous shear and gravity forces. Int. J. Heat Mass Transfer 14, 1220–1224.

Kosky, P.G., Staub, F.W., 1971. Local condensing heat transfer coefficents in the annula flow regime. AIChE J. 17 (5), 1037–1043.

Koyama, S., Yu, J., 1996. Condensation heat transfer of pure refrigerant inside an internally grooved horizontal tube. JAR Annual Conference, pp. 173–176.

Koyama, S., Kuwahara, K., Nakashita, K., 2003a. Condensation of refrigerant in a multiport channel. First International Conference on Microchannels and Minichannels. ASME, Rochester, NY.

Koyama, S., Kuwahara, K., Nakashita, K., Yamamoto, K., 2003b. An experimental study on condensation of refrigerant R134a in a multi-port extruded tube. Int. J. Refrig. 26 (4), 425–432.

Koyama, S., Lee, J., Yonemoto, R., 2004. An investigation on void fraction of vapor–liquid two-phase flow for smooth and microfin tubes with R134a at adiabatic condition. Int. J. Multiphase Flow 30 (3), 291–310.

Kureta, M., Kobayashi, T., Mishima, K., Nishihara, H., 1998. Pressure drop and heat transfer for flow-boiling of water in small-diameter tubes. JSME Int. J. Ser. B 41 (4), 871–879.

Kureta, M., Akimoto, H., Hibiki, T., Mishima, K., 2001. Void fraction measurement in subcooled-boiling flow using high-frame-rate neutron radiography. Nucl. Technol. 136 (2), 241–254.

Kureta, M., Hibiki, T., Mishima, K., Akimoto, H., 2003. Study on point of net vapor generation by neutron radiography in subcooled boiling flow along narrow rectangular channels with short heated length. Int. J. Heat Mass Transfer 46 (7), 1171–1181.

Kutateladze, S.S., 1963. Fundementals of Heat Transfer. Academic Press, New York, NY.

Lee, H.J., Lee, S.Y., 2001. Pressure drop correlations for two-phase flow within horizontal rectangular channels with small heights. Int. J. Multiphase Flow 27 (5), 783–796.

Lin, S., Kwok, C., Li, R.-Y., Chen, Z.-H., Chen, Z.-Y., 1991. Local frictional pressure drop during vaporization of R-12 through capillary tubes. Int. J. Multiphase Flow 17 (1), 95–102.

Lockhart, R., Martinelli, R., 1949. Proposed correlation of data for isothermal two-phase, two-component flow in pipes. Chem. Eng. Prog. 45 (1), 39–48.

Lombardi, C., Carsana, C.G., 1992. A dimensionless pressure drop correlation for two-phase mixtures flowing uphill in vertical ducts covering wide parameter range. Heat Technol. 10, 125–141.

Lowry, B., Kawaji, M., 1988. Adiabatic vertical two-phase flow in narrow flow channels, Twenty-Fifth National Heat Transfer Conference, vol. 84. AIChE, Houston, TX, pp. 133–139.

Mandhane, J.M., Gregory, G.A., Aziz, K., 1974. A flow pattern map for gas–liquid flow in horizontal pipes. Int. J. Multiphase Flow 1 (4), 537–553.

McAdams, W.H., 1954. Heat Transmission. McGraw-Hill, New York, NY.
McAdams, W.H., Woods, W.K., Bryan, R.L., 1942. Vaporization inside horizontal tubes-II-benzene–oil-mixtures. Trans. ASME, 64.
Mikheev, M.A., 1956. Heat Transfer Fundamentals. Gosenergoizdat, Moscow.
Miropolsky, Z.L., 1962. Heat transfer during condensation of high pressure steam inside a tube. Teploenergetika 3, 79–83.
Mishima, K., Hibiki, T., 1996a. Quantitative limits of thermal and fluid phenomena measurements using the neutron attenuation characteristics of materials. Exp. Therm. Fluid Sci. 12 (4), 461–472.
Mishima, K., Hibiki, T., 1996b. Some characteristics of air–water two-phase flow in small diameter vertical tubes. Int. J. Multiphase Flow 22 (4), 703–712.
Mishima, K., Hibiki, T., 1998. Development of high-frame-rate neutron radiography and quantitative measurement method for multiphase flow research. Nucl. Eng. Des. 184 (2–3), 183–201.
Mishima, K., Ishii, M., 1984. Flow regime transition criteria for upward two-phase flow in vertical tubes. Int. J. Heat Mass Transfer 27 (5), 723–737.
Mishima, K., Hibiki, T., Nishihara, H., 1993. Some characteristics of gas–liquid flow in narrow rectangular ducts. Int. J. Multiphase Flow 19 (1), 115–124.
Mishima, K., Hibiki, T., Nishihara, H., 1997. Visualization and measurement of two-phase flow by using neutron radiography. Nucl. Eng. Des. 175 (1–2), 25–35.
Moriyama, K., Inoue, A., Ohira, H., 1992a. Thermohydraulic characteristics of two-phase flow in extremely narrow channels (the frictional pressure drop and heat trasnfer of adiabatic two-component flow, analytical model). Heat Transfer Jpn. Res. 21 (8), 823–837.
Moriyama, K., Inoue, A., Ohira, H., 1992b. Thermohydraulic characteristics of two-phase flow in extremely narrow channels (the frictional pressure drop and heat trasnfer of boiling two-phase flow, analytical model). Heat Transfer Jpn. Res. 21 (8), 838–856.
Moser, K.W., Webb, R.L., Na, B., 1998. A new equivalent Reynolds number model for condensation in smooth tubes. J. Heat Transfer 120 (2), 410–417.
Müller-Steinhagen, H., Heck, K., 1986. A simple friction pressure drop correlation for two-phase flow in pipes. Chem. Eng. Process. 20 (6), 297–308.
Nebuloni, S., Thome, J.R., 2010. Numerical modeling of laminar annular film condensation for different channel shapes. Int. J. Heat Mass Transfer 53 (13–14), 2615–2627.
Nebuloni, S., Thome, J.R., 2012. Numerical modeling of the conjugate heat transfer problem for annular laminar film condensation in microchannels. J. Heat Transfer 134, 5.
Nema, G., Garimella, S., Fronk, B.M., 2013. Flow Regime transitions during condensation in microchannels. Int. J. Refrig. Rev.
Nitheanandan, T., Soliman, H., Chant, R., 1990. A proposed approach for correlating heat transfer during condensation inside tubes. ASHRAE Trans. 96 (Pt 1), 230–248.
Nusselt, W., 1916. Die Oberflächenkondensation des Wasserdampfes. VDI Zeitschrift 60, 541–546 and 569–575.
Oliemans, R., Pots, B., Trompe, N., 1986. Modelling of annular dispersed two-phase flow in vertical pipes. Int. J. Multiphase Flow 12 (5), 711–732.
Owens, W., 1961. Two-phase pressure gradient. ASME Int. Dev. Heat Transfer, 363–368.
Palen, J., Breber, G., Taborek, J., 1979. Prediction of flow regimes in horizontal tube-side condensation. Heat Transfer Eng. 1 (2), 47–57.
Palm, B., 2001. Heat transfer in microchannels. Microscale Thermophys. Eng. 5 (3), 155–175.

Petukhov, B.S., 1970. Heat transfer and friction in turbulent pipe flow with variable physical properties. Adv. Heat Transfer 6, 503–564.

Premoli, A., DiFrancesco, D., Prina, A., 1971. A dimensionless correlation for the determination of the density of two-phase mixtures. Termotecnica (Milan) 25 (1), 17–26.

Qu, W., Mudawar, I., 2003. Flow boiling heat transfer in two-phase micro-channel heat sinks—I. Experimental investigation and assessment of correlation methods. Int. J. Heat Mass Transfer 46 (15), 2755–2771.

Rabas, T., Arman, B., 2000. Effect of the exit condition on the performance of in-tube condensers. Heat Transfer Eng. 21 (1), 4–14.

Revellin, R., Thome, J.R., 2007. Adiabatic two-phase frictional pressure drops in microchannels. Exp. Therm. Fluid Sci. 31 (7), 673–685.

Ribatski, G., Wojtan, L., Thome, J.R., 2006. An analysis of experimental data and prediction methods for two-phase frictional pressure drop and flow boiling heat transfer in micro-scale channels. Exp. Therm. Fluid Sci. 31 (1), 1–19.

Richardson, B.L., 1959. Some problems in horizontal two-phase two-component flow (Ph.D. dissertation). Mechanical Engineering, Purdue University, West Lafayette, IN.

Rohsenow, W.M., Webber, J.H., Ling, A.T., 1957. Effect of vapor velocity on laminar and turbulent-film condensation. Am. Soc. Mech. Eng. Transac. 78 (8), 1637–1642.

Rosson, H., Meyers, J., 1965. Point values of condensing film coefficients inside a horizontal tube. Chem. Eng. Prog. S. Ser. 61 (59), 190–199.

Rouhani, S.Z., Axelsson, E., 1970. Calculation of void volume fraction in the subcooled and quality boiling regions. Int. J. Heat Mass Transfer 13 (2), 383–393.

Rouhani, S.Z., Sohal, M.S., 1983. Two-phase flow patterns: a review of research results. Prog. Nucl. Energ. 11 (3), 219–259.

Sacks, P., 1975. Measured characteristics of adiabatic and condensing single-component two-phase flow of refrigerant in a 0.377-in diameter horizontal tube. ASME Winter Annual Meeting, Houston, TX.

Sadatomi, M., Sato, Y., Saruwatari, S., 1982. Two-phase flow in vertical noncircular channels. Int. J. Multiphase Flow 8 (6), 641–655.

Sardesai, R., Owen, R., Pulling, D., 1981. Flow regimes for condensation of a vapour inside a horizontal tube. Chem. Eng. Sci. 36 (7), 1173–1180.

Schlichting, H., Gersten, K., 2000. Laminar-turbulent transition. Boundary Layer Theo., 415–590.

Serizawa, A., Feng, Z., 2001. Two-phase flow in micro-channels, Proceedings of the Fourth International Conference on Multiphase Flow, New Orleans, LA.

Serizawa, A., Feng, Z., Kawara, Z., 2002. Two-phase flow in microchannels. Exp. Therm. Fluid Sci. 26 (6–7), 703.

Shah, M.M., 1976. New correlation for heat transfer during boiling flow through pipes. ASHRAE Trans. 82 (Pt 2), 66–86.

Shah, M.M., 1979. A general correlation for heat transfer during film condensation inside pipes. Int. J. Heat Mass Transfer 22 (4), 547–556.

Smith, S.L., 1969. Void Fractions in Two-Phase Flow: A Correlation Based upon an Equal Velocity Head Model. Proc. Inst. Mech. Eng. 184 (1), 647–657.

Soliman, H., 1982. On the annular-to-wavy flow pattern transition during condensation inside horizontal tubes. Can. J. Chem. Eng. 60 (4), 475–481.

Soliman, H., 1986. The mist-annular transition during condensation and its influence on the heat transfer mechanism. Int. J. Multiphase Flow 12 (2), 277–288.

Soliman, H.M., Schuster, J.R., Berenson, P.J., 1968. A general heat transfer correlation for annular flow condensation. J. Heat Transfer 90 (2), 267–276.
Souza, A.L., Chato, J.C., Wattelet, J.P., Christoffersen, B.R., 1993. Pressure drop during two-phase flow of pure refrigerants and refrigerant–oil mixtures in horizontal smooth tubes. Twenty-Ninth National Heat Transfer Conference. ASME, Atlanta, GA, New York, NY, pp. 35–41.
Souza, D.A.L., Pimenta, M.D.M., 1995. Prediction of pressure drop during two-phase flow of pure and mixed refrigerants. ASME/JSME Fluids Engineering and Laser Anemometry Conference and Exhibition. ASME, Hilton Head, SC, pp. 161–171.
Steiner, D., 1993. Heat Transfer to boiling saturated liquids (VDI Heat Atlas). Dusseldorf. VDI-Gessellschaft Verfahrenstechnik und Chemieingenieurwesen (GCV), Germany.
Sugawara, S., Katsuta, K., Ishihara, I., Muto, T., 1967. Consideration on the pressure loss of two-phase flow in small-diameter tubes. Fourth National Heat Transfer Symposium of Japan. Heat Transfer Society of Japan, Japan, pp. 169–172.
Suo, M., Griffith, P., 1964. Two-phase flow in capillary tubes. Trans. ASME. J. Basic Eng. 86 (3), 576–582.
Tabatabi, A., Faghri, A., 2001. A new two-phase flow map and transition boundary accounting for surface tension effects in horizontal miniature and micro tubes. J. Heat Transfer 123 (5), 958–968.
Taitel, Y., Dukler, A.E., 1976. A model for predicting flow regime transitions in horizontal and near horizontal gas-liquid flow. AIChE J. 22 (1), 47–55.
Taitel, Y., Bornea, D., Dukler, A.E., 1980. Modelling flow pattern transitions for steady upward gas–liquid flow in vertical tubes. AIChE J. 26 (3), 345–354.
Tandon, T., Varma, H., Gupta, C., 1982. A new flow regimes map for condensation inside horizontal tubes. J. Heat Transfer 104, 763.
Tandon, T., Varma, H., Gupta, C., 1985. Prediction of flow patterns during condensation of binary mixtures in a horizontal tube. J. Heat Transfer 107 (2), 424–430.
Tandon, T., Varma, H., Gupta, C., 1995. Heat transfer during forced convection condensation inside horizontal tube. Int. J. Refrig. 18 (3), 210–214.
Tang, L., 1997. Empirical Study of New Refrigerant Flow Condensation Inside Horizontal Smooth and Micro-fin Tubes (Ph.D. thesis). University of Maryland, College Park, MD.
Thom, J.R.S., 1964. Prediction of pressure drop during forced circulation boiling of water. Int. J. Heat Mass Transfer 7 (7), 709–724.
Thome, J.R., El Hajal, J., Cavallini, A., 2003. Condensation in horizontal tubes. Part 2: new heat transfer model based on flow regimes. Int. J. Heat Mass Transfer 46 (18), 3365–3387.
Tran, T., Chyu, M.-C., Wambsganss, M., France, D., 2000. Two-phase pressure drop of refrigerants during flow boiling in small channels: an experimental investigation and correlation development. Int. J. Multiphase Flow 26 (11), 1739–1754.
Traviss, D.P., Rohsenow, W.M., 1973. Flow regimes in horizontal two-phase flow with condensation. ASHRAE Trans. 79 (2), 31–39.
Traviss, D.P., Rohsenow, W.M., Baron, A.B., 1973. Forced-convection condensation inside tubes: a heat transfer equation for condenser design. ASHRAE Trans. 79 (Pt 1), 157–165.
Triplett, K., Ghiaasiaan, S., Abdel-Khalik, S., LeMouel, A., McCord, B., 1999a. Gas–liquid two-phase flow in microchannels. Part II: void fraction and pressure drop. Int. J. Multiphase Flow 25 (3), 395–410.

Triplett, K., Ghiaasiaan, S., Abdel-Khalik, S., Sadowski, D., 1999b. Gas–liquid two-phase flow in microchannels. Part I: two-phase flow patterns. Int. J. Multiphase Flow 25 (3), 377–394.

Troniewski, L., Ulbrich, R., 1984. Two-phase gas–liquid flow in rectangular channels. Chem. Eng. Sci. 39 (4), 751–765.

Tung, K.W., Parlange, J.-Y., 1976. Note on the motion of long bubbles in closed tubes-influence of surface tension. Acta Mech. 24 (3), 313–317.

Unesaki, H., Hibiki, T., Mishima, K., 1998. Verification of neutron radioscopic measurement of void fraction by Monte Carlo simulation. Nucl. Instrum. Meth. A 405 (1), 98–104.

Ungar, E.K., Cornwell, J.D., 1992. Two-phase pressure drop of ammonia in small diameter horizontal tubes. Sixteenth Aerospace Ground Testing Conference. AIAA, Nashville, TN, pp. 1.

van Driest, E.R., 1956. On turbulent flow near a wall. J. Aeronaut. Sci. 23, 1007–1011.

Wadekar, V., 1990. Flow boiling—a simple correlation for convective heat transfer component. Ninth International Heat Transfer Conference, Jerusalem, Israel, p. 87.

Wallis, G.B., 1969. One-Dimensional Two-Phase Flow. McGraw-Hill, New York, NY.

Wambsganss, M., Jendrzejczyk, J., France, D., 1991. Two-phase flow patterns and transitions in a small, horizontal, rectangular channel. Int. J. Multiphase Flow 17 (3), 327–342.

Wambsganss, M., Jendrzejczyk, J., France, D., 1994. Determination and characteristics of the transition to two-phase slug flow in small horizontal channels. J. Fluids Eng. 116 (1), 140–146.

Wambsganss, M.W., Jendrzejczyk, J.A., France, D.M., Obot, N.T., 1992. Frictional pressure gradients in two-phase flow in a small horizontal rectangular channel. Exp. Therm. Fluid Sci. 5 (1), 40–56.

Wang, C.-C., Chiang, C.-S., Lin, S.-P., Lu, D.-C., 1997a. Two-phase flow pattern for R-134 a inside a 6. 5-mm(0. 25-in.) smooth tube. ASHRAE Trans. 103 (1), 803–812.

Wang, C.-C., Chiang, C.-S., Lu, D.-C., 1997b. Visual observation of two-phase flow pattern of R-22, R-134a, and R-407C in a 6.5-mm smooth tube. Exp. Therm. Fluid Sci. 15 (4), 395–405.

Wang, H.S., Rose, J.W., 2004. Film condensation in horizontal triangular section microchannels: a theoretical model. Proceedings of the Second International Conference on Microchannels and Minichannels (ICMM2004). American Society of Mechanical Engineers, Rochester, NY, Jun 17–19, New York, NY, pp. 661–666.

Wang, H.S., Rose, J.W., 2005. A theory of film condensation in horizontal noncircular section microchannels. J. Heat Transfer 127 (10), 1096–1105.

Wang, H.S., Rose, J.W., 2011. Theory of heat transfer during condensation in microchannels. Int. J. Heat Mass Transfer 54, 2525–2534.

Wang, H.S., Rose, J.W., Honda, H., 2004. A theoretical model of film condensation in square section horizontal microchannels. Chem. Eng. Res. Des. 82 (4), 430–434.

Wang, W.-W., Radcliff, T.D., Christensen, R.N., 2002. A condensation heat transfer correlation for millimeter-scale tubing with flow regime transition. Exp. Therm. Fluid Sci. 26 (5), 473–485.

Wattelet, J.P., Chato, J., Souza, A., Christofferson, B., 1994. Evaporative characteristics of R-12, R-134a, and a mixture at low mass fluxes. ASHRAE Winter Meeting. ASHRAE, New Orleans, LA, pp. 603–615.

Webb, R.L., Ermis, K., 2001. Effect of hydraulic diameter on condensation of R-134A in flat, extruded aluminum tubes. J. Enhanc. Heat Transfer 8 (2), 77–90.

Webb, R.L., Lee, H., 2001. Brazed aluminum condensers for residential air conditioning. J. Enhanc. Heat Transfer 8, 1.

Weisman, J., Duncan, D., Gibson, J., Crawford, T., 1979. Effects of fluid properties and pipe diameter on two-phase flow patterns in horizontal lines. Int. J. Multiphase Flow 5 (6), 437–462.

Wilmarth, T., Ishii, M., 1994. Two-phase flow regimes in narrow rectangular vertical and horizontal channels. Int. J. Heat Mass Transfer 37 (12), 1749–1758.

Wilson, M.J., Newell, T.A., Chato, J.C., Infante Ferreira, C.A., 2003. Refrigerant charge, pressure drop, and condensation heat transfer in flattened tubes. Int. J. Refrig. 26 (4), 442–451.

Winkler, J., Killion, J., Garimella, S., 2012a. Void fractions for condensing refrigerant flow in small channels. Part II: void fraction measurement and modeling. Int. J. Refrig. 35 (2), 246–262.

Winkler, J., Killion, J., Garimella, S., Fronk, B.M., 2012b. Void fractions for condensing refrigerant flow in small channels. Part I: literature review. Int. J. Refrig. 35 (2), 219–245.

Xu, J., 1999. Experimental study on gas–liquid two-phase flow regimes in rectangular channels with mini gaps. Int. J. Heat Fluid Flow 20 (4), 422–428.

Xu, J., Cheng, P., Zhao, T., 1999. Gas–liquid two-phase flow regimes in rectangular channels with mini/micro gaps. Int. J. Multiphase Flow 25 (3), 411–432.

Yan, Y.-Y., Lin, T.-F., 1999. Condensation heat transfer and pressure drop of refrigerant R-134a in a small pipe. Int. J. Heat Mass Transfer 42 (4), 697–708.

Yang, C.-Y., Shieh, C.-C., 2001. Flow pattern of air–water and two-phase R134a in small circular tubes. Int. J. Multiphase Flow 27 (7), 1163–1177.

Yang, C.Y., Webb, R.L., 1996a. Condensation of R-12 in small hydraulic diameter extruded aluminum tubes with and without micro-fins. Int. J. Heat Mass Transfer 39 (4), 791–800.

Yang, C.Y., Webb, R.L., 1996b. Friction pressure drop of R-12 in small hydraulic diameter extruded aluminum tubes with and without micro-fins. Int. J. Heat Mass Transfer 39 (4), 801–809.

Yang, C.Y., Webb, R.L., 1997. A predictive model for condensation in small hydraulic diameter tubes having axial micro-fins. J. Heat Transfer 119 (4), 776–782.

Yashar, D., Wilson, M., Kopke, H., Graham, D., Chato, J., Newell, T., 2001. An investigation of refrigerant void fraction in horizontal, microfin tubes. HVAC&R Res. 7 (1), 67–82.

Zhang, M., Webb, R.L., 2001. Correlation of two-phase friction for refrigerants in small-diameter tubes. Exp. Therm. Fluid Sci. 25 (3–4), 131–139.

Zhao, T., Bi, Q., 2001a. Co-current air–water two-phase flow patterns in vertical triangular microchannels. Int. J. Multiphase Flow 27 (5), 765–782.

Zhao, T., Bi, Q., 2001b. Pressure drop characteristics of gas–liquid two-phase flow in vertical miniature triangular channels. Int. J. Heat Mass Transfer 44 (13), 2523–2534.

Zivi, S., 1964. Estimation of steady-state steam void–fraction by means of the principle of minimum entropy production. J. Heat Transfer 86, 247.

Zuber, N., Findlay, J.A., 1965. Average volumetric concentration in two-phase flow systems. J. Heat Transfer 87, 453–468.

Zukoski, E.E., 1966. Influence of viscosity, surface tension, and inclination angle on motion of long bubbles in closed tubes. J. Fluid. Mech. 25 (Part 4), 821–837.

Zürcher, O., Thome, J., Favrat, D., 1999. Evaporation of ammonia in a smooth horizontal tube: heat transfer measurements and predictions. J. Heat Transfer 121 (1), 89–101.

CHAPTER

7 Biomedical Applications of Microchannel Flows

Michael R. King
Department of Biomedical Engineering, Cornell University, Ithaca, NY, USA

7.1 Introduction

As ever more is learned about the molecular and cellular biology of human and model organism cells in the contexts of normal physiological function and disease pathology, it is the role of bioengineers to apply this knowledge toward the improvement of public health and the alleviation of human suffering. The control and manipulation of living cells outside of the body presents technical challenges in which physical conditions such as temperature and pH must be maintained close to their physiological values of 37°C and 7.4, respectively; if these quantities deviate by as much as 10%, cells cannot survive for long periods of time. With greater deviations from physiological baseline, proteins and other macromolecules themselves begin to degrade. The complete sequencing of the human genome at the start of the twenty-first century has engendered an increased interest in the development of high-throughput microfluidic technologies to assay thousands of genes simultaneously. The interest in transporting, manipulating, and synthesizing biomolecules in smaller and smaller conduits is motivated by a desire to automate biochemical protocols and eliminate some of the most mundane technical tasks, as well as a need to operate on extremely small sample sizes: for instance, to identify individuals based on minute samples in forensic crime scenes or to detect biochemical factors that may be produced in the body's tissues at nanomolar concentrations.

Microfluidics and microchannel transport processes have also led to the development of new research tools. We now have the ability to peer inside of microscopic cells, measure the interior rheology of their cytoplasm, and visualize the co-localization of specific proteins within various organelles. Micron-sized pipettes are able to enter a cell, remove its nuclear material, and inject this DNA "soup" into a different cell to achieve cloning of animals and humans (in the case of humans, therapeutic rather than reproductive). Another active area of research is in the precise measurement of the specific adhesion of cells with other contacting surfaces. Many of the surface receptor proteins presented by cells are themselves exquisitely sensitive detectors of soluble chemical concentration and force

© 2014 Elsevier Ltd. All rights reserved.

transducers that can initiate rapid chain reactions within the cell compartments. In this chapter I discuss some of the different technological applications of microchannels and minichannels to measure the strength and frequency of individual cell adhesion, grow cells to replace vital organs, peel cells from protein-coated surfaces using well-defined forces, and characterize the deformability of different cell types. In each case, we apply fundamentals from transport phenomena to understand and model these biological applications.

7.2 Microchannels to probe transient cell adhesion under flow

Microchannels with functionalized biochemical surfaces have been quite useful in elucidating the molecular mechanisms governing the transient adhesion between circulating white blood cells (leukocytes) and the blood vessel wall during inflammation. The selectins are a family of adhesion molecules that recognize a tetrasaccharide moiety called sialyl Lewis^x presented by many glycoprotein ligands expressed on cell surfaces, and are involved in the trafficking of leukocytes throughout the systemic and lymphatic circulations (Lasky, 1995; Ebnet and Vestweber, 1999). The selectin–carbohydrate bond is of high specificity and mechanical strength, low affinity, and exhibits fast rates of formation and dissociation. The rate of dissociation is a dominant physical parameter, distinct for each of the three selectins P-, E-, and L-selectin, and has been demonstrated to exhibit an exponential dependence on the force loading on the bond (Smith et al., 1999). The same molecular mechanisms that mediate acute inflammatory responses under physiological conditions are also responsible for chronic inflammation, vessel occlusion, and other disease states featuring abnormal accumulations of adherent leukocytes or platelets (Ramos et al., 1999). Many of the biological complexities of these myriad surface interactions between leukocytes, platelets, and endothelial cells have been revealed using idealized microchannel assays and semiquantified using knockout mice. There are opportunities to advance our understanding of selectin biology in the context of inflammatory and cardiovascular disease by applying engineering tools to integrate this knowledge into descriptive and predictive models of blood cell interactions in more realistic settings.

In addition to creating a well-defined flow field to study the balance between fluid forces and chemical adhesion of blood cells, microchannels, minichannels, and corresponding numerical simulations have been successfully developed to more closely mimic flow irregularities found in the diseased human circulatory system. The deposition of platelets on surfaces in complex flow geometries such as stenosis, aneurysm, or flow over a protrusion, has been characterized *in vitro* by several groups (Karino and Goldsmith, 1979; Tippe et al., 1992; Schoephoerster et al., 1993; Slack and Turitto, 1994). These studies support the idea that flow nonidealities are important in modulating thrombotic phenomena, yet lack the spatial and temporal resolution necessary to support detailed cellular-scale modeling. In hypercholesterolemic rabbits and mice, monocytes preferentially deposit around aortic and celiac orifices due to local flow characteristics in these regions (Back et al., 1995; Walpola et al., 1995; Nakashima et al., 1998; Iiyama et al., 1999; Truskey et al., 1999). Endothelial cell layers respond to disturbed shear fields (e.g., flow over a rectangular obstacle) by upregulating expression of adhesion molecules, resulting in a general increase in the number of cells recruited downstream of such an obstacle (DePaola et al., 1992; Pritchard et al., 1995; Barber et al., 1998). Multiple groups have attempted to model the particulate flow of blood through complex vessel geometries with and without particle deposition at the walls, while neglecting hydrodynamic cell–cell interactions (Perktold, 1987; Lei et al., 1997). Despite theoretical limitations, such studies have been applied toward providing recommendations for surgical reconstruction procedures and the design of commercial grafts that minimize particle deposition based on stresses exerted by the fluid on the wall surfaces. Others have computationally studied shear activation and surface deposition of platelets by treating the cells as infinitesimal points (Bluestein et al., 1999; Kuharski and Fogelson, 2001). The Multiparticle Adhesive Dynamics (MAD) simulation (King and Hammer, 2001a,b) joins a rigorous calculation of multiparticle fluid flow with a realistic model for specific cell–cell adhesion. Work with MAD has revealed biophysical mechanisms that control dynamic cell adhesion under flow, which have been verified experimentally both in microchannel experiments (King and Hammer, 2003) and *in vivo* (King et al., 2003). Further understanding of the rate of atherogenesis and risk of acute thrombotic events will occur through careful computer and *in vitro* models that are developed to consider the near-wall cell–cell hydrodynamic interactions that occur during cell deposition in diseased vessel geometries.

7.2.1 Different types of microscale flow chambers

The most common types of microchannels used in cell adhesion and flow studies are the radial flow chamber (Kuo and Lauffenburger, 1993; Goldstein and DiMilla, 1998), and the parallel-plate flow chamber (Sung et al., 1985; Palecek et al., 1997). Such systems have found widespread use for the following applications:

1. measurement of transient adhesion of circulating (i.e., blood) cells to reactive surfaces under flow;
2. use of well-defined shear stresses to measure the detachment force of strongly adherent cells;
3. measurement of the biological response of a cell layer to long-term exposure to physiological levels of steady or pulsatile shear stress; and
4. monitoring the real-time formation of biofilms on flow-exposed surfaces.

Figure 7.1 shows a diagrammatic representation of the radial and parallel-plate flow chamber geometries. The main advantage of the radial flow chamber comes from the fact that as the fluid introduced at the center of the chamber flows radially outward, it passes through a larger and larger cross-section around the perimeter, and thus the linear velocity of the fluid decreases smoothly as it flows from

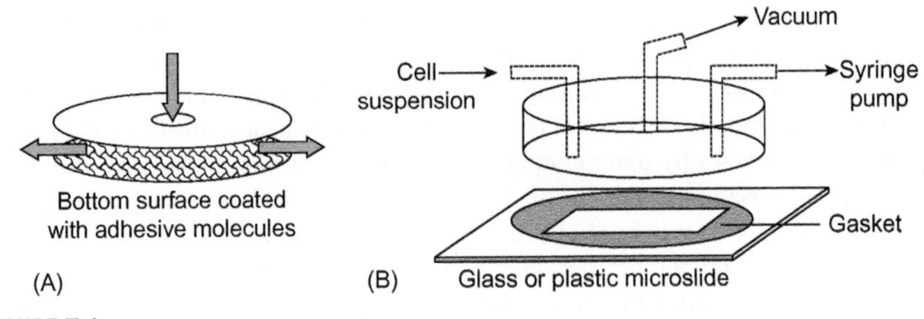

FIGURE 7.1

Diagram of (A) radial and (B) parallel-plate cell perfusion chambers.

inner to outer regions. As a result, the fluid samples a wide range of wall shear stresses, from a high value near the center to a minimum value at the outer edge of the chamber. If the entire surface of interest is imaged from below at low magnification, one can obtain interaction data such as the number of adherent cells or the cell translational velocities for a range of shear stress values at once rather than having to perform repeated experiments at many different flow rates. It is straightforward to show (see Problem 7.1) that the wall shear stress in the radial flow assay will exhibit the following dependence on radial position r:

$$\tau = \frac{3Q\mu}{\pi r H^2} \tag{7.1}$$

where Q is the volumetric flow rate, μ is the fluid viscosity, and H is the spacing between the upper and lower surfaces.

A more detailed diagram of the parallel-plate flow chamber assay is depicted in Figure 7.1B. This shows a common design, where a reusable upper surface is brought into contact with a rubber gasket laid over a disposable glass or plastic microslide. The microslide can be incubated with relevant adhesion or extracellular matrix proteins of interest, or else cultured with cells until a confluent monolayer is achieved. The upper surface of the flow chamber is fitted with luer fittings that attach to inlet and outlet streams to allow for the perfusion of cell media, or suspensions of cells or microsphere particles coated with their own appropriate molecules. A third fitting is commonly attached to a vacuum line, which for typical chamber sizes and physiological fluid flow rates is sufficient to hold together the top and bottom of the chamber without the use of bolts or mechanical clamps. The flow domain is defined by a rectangular, cut-out region in the rubber gasket, which necessarily overlaps with the inlet and outlet ports (but not with the vacuum port). Most cell perfusion chambers are specially designed so that the outer dimensions are compatible with standard sizes of polystyrene tissue culture plates, round glass coverslips, or multiwall plates, which facilitate the exchange of materials and protocols between different laboratories. Many such parallel-plate cell perfusion assays are now commercially available, in smaller and smaller volumes to minimize the amount of rare and valuable materials needed to conduct each experiment. Table 7.1 shows a sampling of the parallel-plate chambers currently available commercially, along with the reported ranges of channel dimension.

7.2.2 Inverted systems: well-defined flow and cell visualization

The flat lower wall of the parallel-plate microchannel provides an excellent surface to study cellular adhesive interactions. As discussed below, the slightly greater density of cells relative to saline buffer or cell media ($\approx$ water) causes sedimentation of cells to the lower surface. Thus, relatively dilute systems of 10^6 cells/ml (volume fraction $\sim 5 \times 10^{-4}$ for 10-μm-diameter cells) can be used while still obtaining a large number of individual interactions for study. Although the geometry of the microcirculation more closely resembles circular glass capillaries, and capillaries can be commercially purchased with very accurate inner diameters ranging from 3 to 1000 μm, round capillaries have some disadvantages compared to planar channels. The circular outer geometry of capillaries creates lensing effects, which interfere with accurate imaging of the inner surface. Submersion of the entire capillary within index of refraction-matched objective oil improves this to some degree. However, due to sedimentation effects, the surface area available for cellular interactions is less in the capillary geometry. For these reasons as well as to simplify theoretical modeling, the parallel-plate geometry is the most popular method for studying cell–surface interactions under flow. High-quality images of cells adhesively attaching to molecular surfaces can be obtained as shown in Figure 7.2, and cells grown under a variety of conditions can be exposed to fluid flow or brought into contact with suspended particles of interest.

A crucial element of bioengineering experiments in parallel-plate flow chambers is that near the lower wall of the microchannel, the parabolic fluid velocity profile can be approximated as linear shear flow. In most cases this assumption is valid, and simplifies the analysis. Figure 7.3 shows a comparison of parabolic

Table 7.1 Some Commercially Available Microchannels and Minichannels for Cell Perfusion

Model	Manufacturer	Channel Height H (μm)
Stovall flow cell	Stovall Life Science, Inc.	400
DH-40i micro-incubator	Warner Instruments, Inc.	800–980
ECIS flow system	Applied BioPhysics	400
Flow chamber system FCS	Provitro	300
Vacu-cell	C & L Instruments, Inc.	250
Circular flow chamber	Glycotech Corporation	125–250
BioFlux Live Cell Imaging System	Fluxion	70
Focht chamber system 2	Bioptechs	50–1000

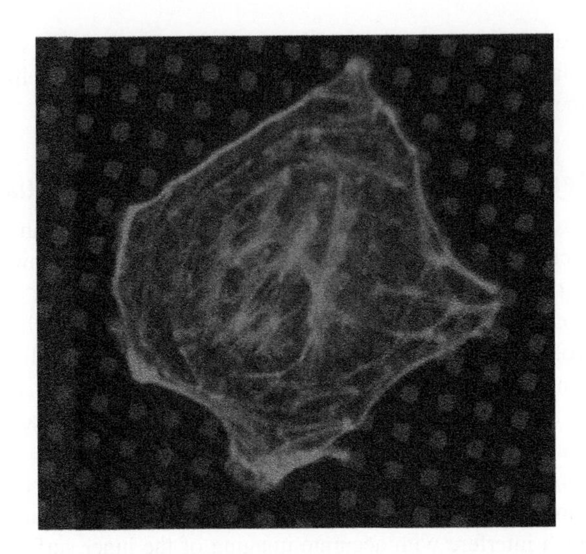

FIGURE 7.2

Bovine aortic endothelial cell fluorescently labeled with phalloidin, which shows the interior cytoskeleton of the cell. The cell is adhering to a 3×3 μm micropattern of fibronectin protein, also fluorescently labeled. Such cell growth experiments can be performed on the lower surface of a parallel-plate microchannel, to later expose cells to fluid shear stress or to easily exchange the fluid environment.

Source: *Image courtesy of M. Mancini, K. Fujiwara, G. Csucs, and M. King.*

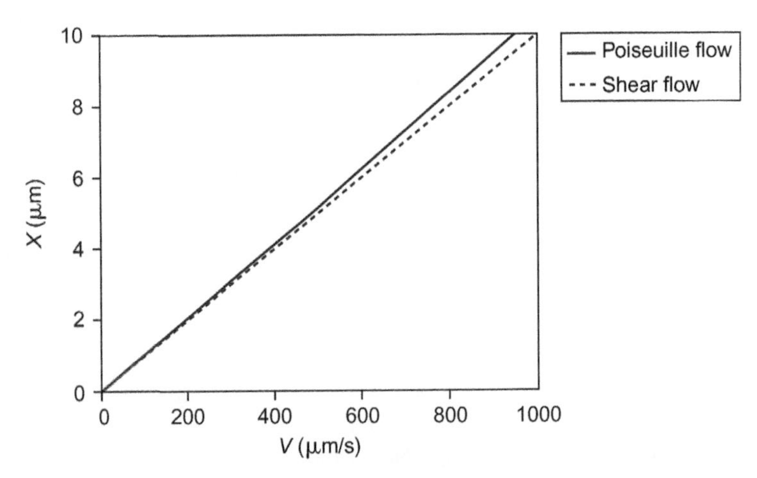

FIGURE 7.3

Comparison of parabolic Poiseuille flow and linear shear flow in the vicinity of the lower wall in a 200-μm microchannel. The wall shear rate has been set at $100\ s^{-1}$.

versus linear flow in a 200-μm channel, within 10 μm of the lower wall. This distance is relevant to most cell adhesion studies, where cells above this distance are readily identified and ignored since they are too far from the molecular lower surface to interact adhesively. In Figure 7.3 the parabolic velocity lies within 5% of the linear approximation for $x \leq 10$ μm.

Different optical techniques are useful for tracking the motion of individual cells near the surface when physiological concentrations of blood cells are used. For instance, it is often difficult to obtain clear images through whole blood in either bright field or phase contrast imaging modes, since visualization of the lower surface on an inverted microscope using a trans-illumination modality requires light transmission from the condenser through the entire sample of the dense suspension of red cells. Epifluorescence mode, on the other hand, is illuminated from below, and thus there is minimal light loss due to the red cells flowing above the surface. Additionally, for large-scale cell deposition studies it is not necessary to resolve individual cells; rather, an intensity average over the entire view can be taken and related to the average concentration of fluorescently labeled cells within the image.

Many fluorescent dyes are available for different blood cells that do not affect their surface adhesiveness, such as fluorochrome carboxyfluorescein diacetate succinimidyl ester for blood platelets, or calcein-AM for leukocytes.

Despite the obvious advantages of the inverted microchannel geometry, one must be aware of anomalous behavior that may be less important within the body. Most previous studies of leukocyte adhesion under flow have been performed with dilute leukocyte suspensions in a parallel-plate flow chamber, with observation of the lower surface on an inverted microscope. There it is recognized that gravitational sedimentation promotes initial contact between the cells and the reactive surface. Indeed, in similar dilute experiments Lawrence et al. (1997) showed that inverting the flow chamber completely abolishes new interactions, but previously rolling cells will continue to roll under flow. *In vivo*, in physiological concentrations of red blood cells (RBCs), the situation is quite different, with red cells migrating toward the center of the vessel and displacing the less deformable leukocytes to the near-wall region (a process called "margination"; see, for instance, Goldsmith and Spain, 1984). An erythrocyte-depleted plasma layer, containing leukocytes and platelets, is formed adjacent to the vessel wall. Cell collisions can promote initial contact between leukocytes and the vessel wall; this may help to explain how roughly equal numbers of rolling leukocytes are visible on the upper and lower walls of intact mouse and hamster microvessels. There is a subtle effect of gravity in real vessels, however, as evidenced by a study by Bishop et al. (2001), who observed the flow of whole blood in horizontally and vertically oriented microvessels in rat spinotrapezius muscle. They observed a symmetrical plasma layer in vertically oriented venules, whereas in horizontally oriented venules the plasma layer formed near the upper wall. Similarly, gravitational effects of whole blood flow in rectangular channels can be minimized by orienting the channel vertically (Abbitt and Nash, 2001), yet when optically imaging leukocytes through regions >40 μm, the same complications of fluorescence labeling occur as described above. A theoretical comparison of gravitational effects and normal forces at the tips of leukocyte microvilli suggest that surface interactions can dominate over gravity under the proper conditions (Zhao et al., 2001). To summarize, gravity plays a more important role in microchannel flow chamber experiments compared to *in vivo*. However, controlled experiments can be performed by density-matching the solution to the suspended cells, via addition of high-molecular-weight dextran (shown to not interfere with selectin adhesion; Chen and Springer, 2001) to separate out any artifactual effects of gravitational sedimentation.

7.2.3 Lubrication approximation for a gradually converging (or diverging) channel

One potential application of microchannels in the study and manipulation of circulating cells is the use of very small-gap channels ($H \leq 10\ \mu m$) to continuously exert high mechanical stresses on 8–10-μm blood cells or blood cell precursors as they pass through such a device. Indeed, such mechanical stresses, which approximate the close packing of cells within the bone marrow, may in fact be necessary for proper maturation of RBCs and explain why production of healthy red cells has proven to be so challenging *in vitro* (Waugh et al., 2001). In such small microchannels created by the orientation of two parallel plates in close proximity to each other, whether the gap is maintained by rubber gaskets or particulate spacers, one must be concerned about the exact alignment of the bounding walls. Specifically, a slight tilt to either surface with respect to the primary flow direction will create either a converging or a diverging flow, which could significantly influence bulk flow characteristics and design issues such as the total pressure drop across the microchannel. If polystyrene slides are used as the bounding surfaces, flexing of the wall at high pressures could also result in a deviation from parallel alignment.

Figure 7.4 shows a diagram of the gradually converging channel geometry, symmetric about the centerline $x = 0$. The corresponding plane Poiseuille flow solution for the velocity profile is

$$v_z = \frac{H^2}{8\mu}\left(-\frac{dp}{dz}\right)\left[1-\left(\frac{2x}{H}\right)^2\right] \tag{7.2}$$

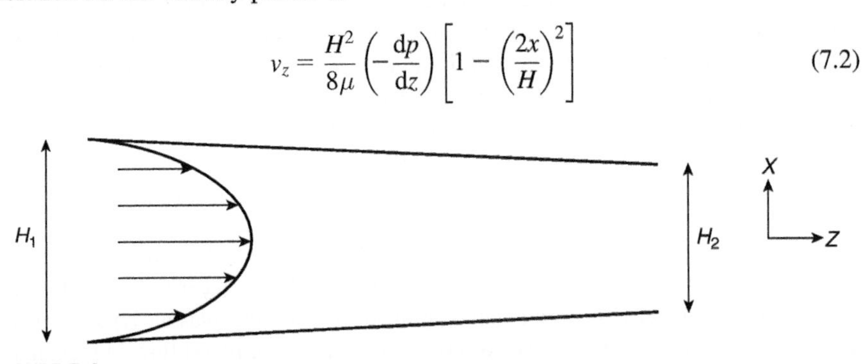

FIGURE 7.4

Pressure-driven flow through a gradually converging channel. In the general case of a curved upper surface, the position of the upper wall is defined by $x = H(z)$.

where the differential version of the pressure drop is used, to allow generalization to a slowly varying channel cross-section. Similarly, for plane Poiseuille flow, the flow rate per unit width q is

$$q = \frac{H^3}{12\mu}\left(-\frac{dp}{dz}\right) \tag{7.3}$$

For a small variation in the channel spacing, $(H_1 - H_2)/H_1 \ll 1$, it can be assumed that the flow will deviate from the plane Poiseuille solution only slightly. Thus, we are justified in assuming that Eqs. (7.2) and (7.3) will remain valid for the case depicted in Figure 7.4, only with H as a slowly varying function of z. Since the flow rate through any cross-section must be a constant, we conclude that the local pressure gradient must have the following spatial dependence:

$$\frac{dp}{dz} \sim \frac{1}{H^3(z)} \tag{7.4}$$

Thus, rearranging Eq. (7.3) for the local pressure drop in the gradually converging channel yields the expression:

$$\frac{dp}{dz} = -\frac{12\mu q}{H^3(z)} \tag{7.5}$$

Equation (7.5) can be integrated over the length of the channel $L = z_2 - z_1$, to give the overall pressure drop across the converging channel,

$$\Delta p = -12\mu q \int_{z_1}^{z_2} H^{-3}(z)\, dz \tag{7.6}$$

For a simple linear variation in channel height of the form

$$H(z) = H_1 + \left(\frac{H_1 - H_2}{z_1 - z_2}\right)(z - z_1) \tag{7.7}$$

the solution to Eq. (7.6) becomes

$$\Delta p = -6\mu q\left(\frac{L}{H_1 - H_2}\right)\left[\frac{1}{H_2^2} - \frac{1}{H_1^2}\right] \tag{7.8}$$

This can be compared to the pressure drop across a uniform planar channel:

$$\Delta p = -12\mu q\left(\frac{L}{H}\right)\frac{1}{H^2} \tag{7.9}$$

Figure 7.5 shows the ratio of the total pressure drop across the converging channel to the total pressure drop in a truly parallel channel evaluated either at the initial height (upper, solid curve) or at the average height (lower, dashed curve). Note that using the average height in Eq. (7.9) yields an estimate for the pressure drop in a converging channel that is accurate to within 90%.

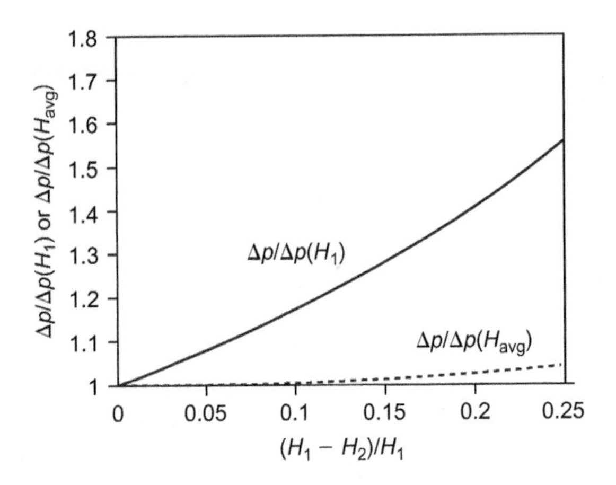

FIGURE 7.5

Plot of the ratio of the pressure drop over a gradually converging channel to the pressure drop in a parallel channel evaluated at either the initial height (upper, solid line) or the average height (lower, dashed), as a function of the dimensionless height difference. Note from the upper curve that a dimensionless height difference of $(H_1 - H_2)/H_1 = 0.25$ results in an increase in total pressure drop of 55% relative to a parallel channel with a spacing of H_1 (upper curve); however, evaluating the pressure drop at the mean channel height (lower curve) captures all but 5% of this increase.

7.3 Blood capillaries and "optimal bumpiness" for minimization of flow resistance

Blood flow in the body proceeds through the vascular tree from the heart to the aorta, then the arteries, smaller arterioles, to the capillary bed, then up again through the venules, larger veins, and back to the heart. Capillaries represent the majority of the surface area of the vascular network, and they play a crucial role in efficient mass and heat transfer. A typical human has millions of arterioles and billions of capillary vessels. Most of the total pressure drop and flow resistance occurs in the arterioles and capillaries. Wang (2006) obtained a closed form solution for the three-dimensional Stokes flow through a tube with periodic bumps. Interestingly, analysis of this model and comparison of its predictions to typical human histology data and mouse blood vessels show that many capillaries, venules, and arterioles display an optimal spacing of bumps to minimize flow resistance, thereby lowering the necessary pumping capacity necessary for the heart.

The model of Wang consists of a cylindrical tube whose wall has periodic bumps described by the dimensionless equation:

$$r = 1 + \varepsilon\, sin(n\theta) sin(\alpha z) \tag{7.10}$$

where the radial (r) and longitudinal (z) coordinates have been scaled with the mean tube radius a, the bump amplitude $\varepsilon = b/a \ll 1$, n is the circumferential wavenumber, $\alpha = 2\pi a/l$, and l is the longitudinal wavelength. Figure 7.6 shows this bumpy tube surface for parameter values typical for capillaries (left) and arterioles (right). Wang performed a regular perturbation in the small parameter ε and solved for the velocity distribution in terms of modified Bessel functions up to the second-order correction to the mean flow rate. The most interesting aspect of this solution is that it is predicted that for a given dimensionless bump area,

$$A = \pi^2 / n\alpha \tag{7.11}$$

there exists an optimal circumferential wavenumber n for which the flow resistance is minimized.

Human histology micrographs available in the literature yield some geometric measurements for comparison with the Stokes flow theory (King, 2007). Capillaries can be approximated as quasi-periodic bumpy tubes with circumferential wavenumber $n = 1$, since in these smallest vessels a single endothelial cell wraps around the tube circumference and the cell nucleus projects into the vessel lumen to create a smooth, sinusoidal bump. A review of histological sections finds inner diameters from 5.86 to 11.6 μm, with average longitudinal wavelengths l of 38.5–48.4 μm. Histological sections of the larger arterioles, however, appear with ruffled inner surfaces comprised of a well-defined circumferential and longitudinal periodicity. Arterioles with a diameter range of 35.3–40.0 μm were determined to have circumferential wavenumbers of 26 down to 25, and a circumferential to longitudinal aspect ratio for the bumps of 2.38. Quite remarkably, when the optimal circumferential wavenumber that minimizes flow resistance is plotted as a function of dimensionless bump area, a power law relationship is exhibited and the measured parameters for capillaries and arterioles fall squarely on this predicted curve (King, 2007).

A more thorough characterization of microvessel geometry was performed by imaging fluorescently labeled arterioles and venules in skeletal muscle of the living mouse (Sumagin et al., 2008). The longitudinal versus circumferential aspect ratio and lengths of hundreds of endothelial cells were measured, which when combined with the local vessel diameter and typical endothelial cell shape assumptions yields

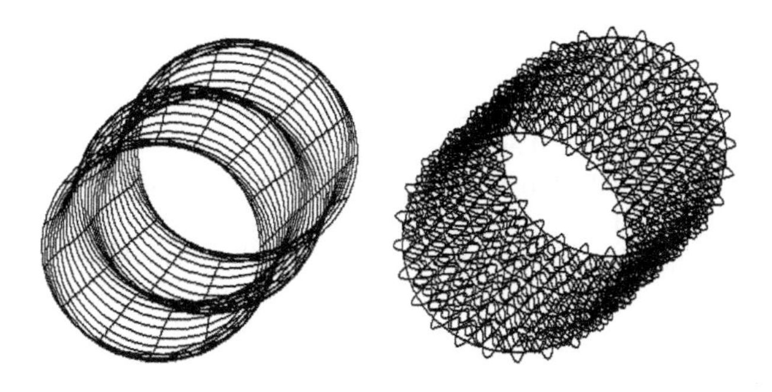

FIGURE 7.6

Computer plot of representative capillary surface on the left ($n = 1$, $\varepsilon = 0.15$, $\alpha = 0.5$), and a representative arteriole surface on the right ($n = 25$, $\varepsilon = 0.12$, $\alpha = 10$).

the wavenumber in both directions along with the bump (representing the cell nucleus) area. When compared against the predicted optimal circumferential wavenumber, venular endothelial cells were found to agree well over two decades of dimensionless bump area. The measured circumferential wavenumber for resting arterioles lay well above the optimum value predicted from the Stokes flow solution, however. It must be noted that the layer of smooth muscle cells surrounding arterioles dynamically regulate their diameter, in response to vasoactive signals such as nitric oxide. When a tissue bed is deoxygenated, these arterioles will rapidly dilate to a diameter approximately double the resting diameter. Interestingly, when the arteriolar endothelial cell geometry is recompared to the theoretical prediction at the dilated diameter, there is very good agreement. Thus, the mouse data indicate that the arteriole vessel geometry is evolved to minimize flow resistance when the vessels are dilated, that is, when the body is trying to provide the largest blood flow rate to the local capillary bed. This is the most advantageous situation, and it agrees with our intuitive expectation for these vessels.

7.4 Circular cross-section microchannels for blood flow research

As readily appreciated in the other chapters of this book, most microchannel research has focused on channels of rectangular cross-section, due to their ease of manufacture through photolithography of polydimethylsiloxane (PDMS) and other methods. However, blood vessels are, of course, circular in cross-section, and much of the important biorheological phenomena that occur in whole blood flowing at the microscale—such as margination and the Fahraeus–Lindqvist effect—are dependent on this cylindrical geometry. Professor Russell Carr of the University of New Hampshire has pioneered a fabrication method of creating circular microchannels, or microchannel branches of various forms, in PDMS (Fenton et al., 1985). The fabrication approach is depicted in the images shown in Figure 7.7 (de Guillebon et al., 2012). The first step is to cure a layer of PDMS onto a glass microscope slide (Figure 7.7, top). Then, a wire structure is created using wire (e.g., diameter = 50 μm) and carnauba wax (Figure 7.7, center). Two more layers of PDMS are then cured over the wire structure. The wires are retracted from the PDMS, leaving circular microchannels in their tracks (Figure 7.7, lower). The channels are then washed free of wax. Finally, the channels can be incubated with materials to achieve the desired coatings for flow experiments: for instance, E-selectin protein to examine cancer cell adhesion to the blood vessel wall, and milk protein to block nonspecific cell adhesion with the wall.

Figure 7.8 shows examples of a finished Y-, V-, and T-shaped branch, respectively. Note that in the Y- and V-branches, a wire remains in place in the PDMS. A current research question of great interest is how aggregates of cancer cells, or heterogeneous aggregates of cancer cells and blood cells (leukocytes and platelets), travel through microvessels and vessel branches under flow. Figure 7.9 is a

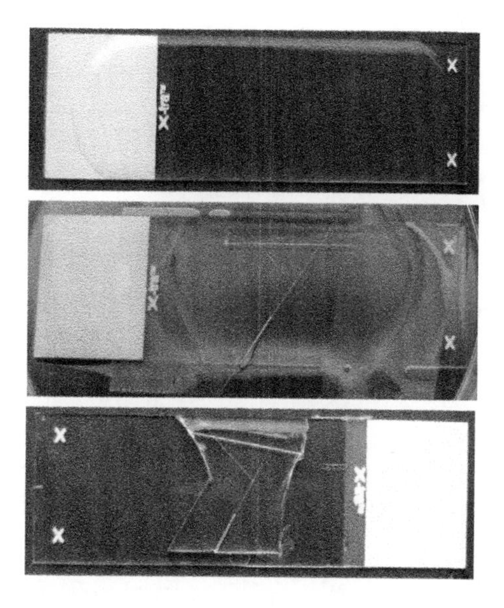

FIGURE 7.7

Steps in the fabrication of branched circular microchannels in PDMS.

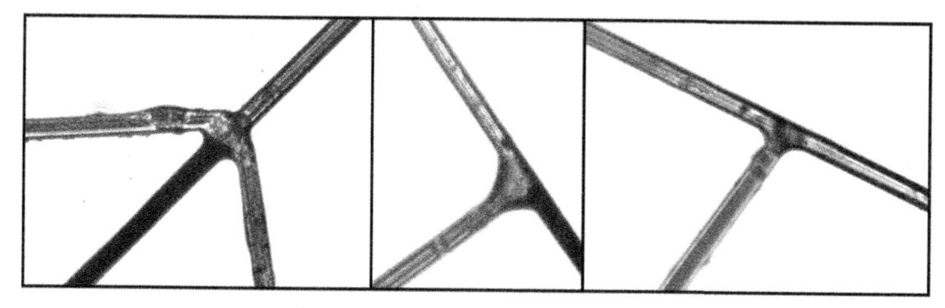

FIGURE 7.8

Representative branched circular microchannels. From the top: Y-, V- and T-shaped, respectively.

sequence of images taken 1 min apart, showing clusters of COLO 205 colon carcinoma cells flowing through a straight circular microchannel of 50 μm diameter (de Guillebon et al., 2012). Flow is from left to right. The cells were first cultured by hanging drop method for 16–20 h at a density of 2.3 million cells per milliliter to encourage aggregate formation (Agastin et al., 2011). When single cells or small aggregates are suspended in saline buffer and perfused through an E-selectin/milk-coated flow network, a number of trends are observed. The average rolling velocity of single COLO 205 cells or small aggregates is found to monotonically increase with aggregate size. This can be understood by noting that for small aggregates, the local curvature of the cell surface and cell contact area with the channel wall

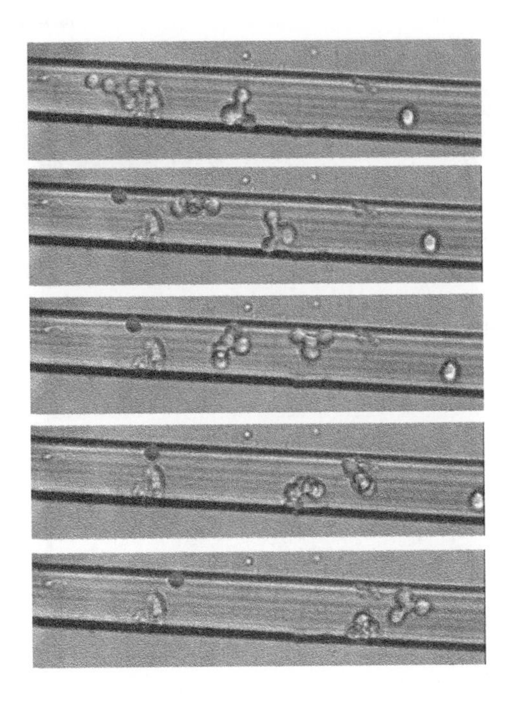

FIGURE 7.9

Sequential frames from an experiment showing cancer cell aggregates perfused through a circular microchannel. Frames taken 1 min apart.

remain constant as the aggregate increases in size. Thus the adhesion energy, or equivalently the number of protein–ligand bonds, is expected to stay constant. Meanwhile, the hydrodynamic drag force (the so-called "Stokes drag") will increase with aggregate size. The net result will then be to increase the rolling velocity for larger aggregates, as higher drag forces start to overcome chemical adhesion. The velocities of aggregates are found to vary as a function of time, with aggregates alternating between periods of high and low velocities. For instance, single cells exhibit a much lower velocity standard deviation than tetramers, even when normalizing with average velocity. Aggregates also show a much greater "deviation" velocity in the circumferential direction while rolling on the channel (vessel) wall, compared to spherical single cells, which exhibit a velocity in the axial direction alone. Finally, aggregates such as these are found to easily change their shape and deform under shear force, as demonstrated in Figure 7.9.

7.5 Nanoscale roughness in microtubes: effects on cell adhesion and biological applications

As discussed in previous chapters of this book, surface roughness in micro- and minichannels affects flow parameters such as pressure drop and the nature of the boundary conditions assumed when modeling a given flow problem. Surface roughness can also have a profound influence on how biological cells interact with microchannel surfaces in flow-based applications. These effects have been explored through the development of thin, stable coatings to immobilize a layer of nanoparticles or nanotubes on the interior surfaces of flow devices (Han et al., 2010). Titanium (IV) butoxide or poly L-lysine can both be used as an adhesive layer on which to attach either colloidal silica nanospheres or halloysite nanotubes. Solved Example 7.1 shows a representative calculation to determine the experimental coating thickness achieved. These coatings have been shown to withstand physiological levels of shear stress and sustained contact with blood, and they are bendable to facilitate implantation strategies.

Nanostructured surfaces with surface height between 10 and 1000 nm are sufficient to cause dramatic differences in cellular adhesion and protein absorption. A valid question to ask is whether such surface features also affect the bulk flow properties of fluid through typical microscale flow devices used for biological cell applications. Flow rate measurements and particle tracking experiments through 300-μm-ID microtubes coated with a monolayer of halloysite nanotubes have been performed to address this question (Hughes and King, 2010). Microtubes of length 50 cm and ID = 300 μm, either uncoated or coated with halloysite nanotubes, were attached to a reservoir of distilled water at four heights ranging from 49 to 84 cm, and the fluid height in the reservoir held constant for 5-min intervals while the total amount of effluent was measured by weight. The flow rates were well predicted by the Hagen–Poiseuille equation, and difference between tubes was 0.18% for the highest reservoir height and 2.1% at the lowest reservoir height.

Fluorescent microspheres of diameter 1.9 μm were perfused through smooth or nanotube-coated microtubes, and their motion tracked through an inverted epifluorescence microscope. Interestingly, microspheres flowing over the nanostructured surface translated at a significantly higher velocity, consistent with maintaining an average position on a streamline farther from the surface compared to microspheres flowing over the smooth microtube surface. Comparison of the Stokes sedimentation velocity of these spheres to the characteristic convective velocity suggests that the largest peak heights of the nanostructured surface, as measured with atomic force microscopy, define the average height at which the particles will travel relative to the underlying surface. If this interpretation is correct, then the following Stokes flow solution of a spherical particle translating near a wall in shear flow should predict the velocity of suspended particles just "skimming" the tops of nanostructured peaks:

$$\frac{U}{hS} \sim \frac{0.7431}{0.6376 - 0.200\, ln(\delta/a)} \tag{7.12}$$

where U is the microsphere velocity, h is the distance between the center of the microsphere and the wall, S is the shear rate, d is the distance between the microsphere surface and the wall (the separation distance), and a is the microsphere radius. Indeed, when this equation was applied to the prediction of microsphere

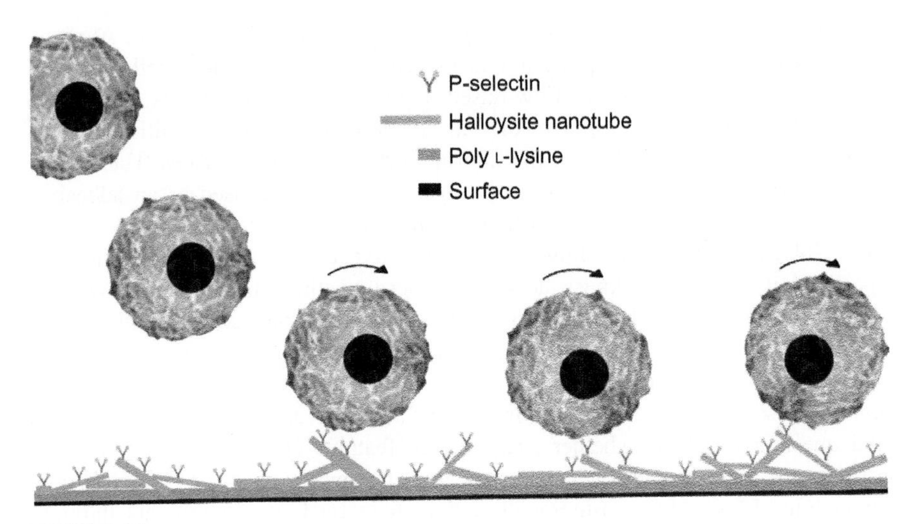

FIGURE 7.10

Schematic of a halloysite nanotube layer functionalized with selectin adhesion protein, for the purpose of adhesively capturing flowing cells.

velocities in halloysite-coated microtubes, excellent agreement was observed with no adjustable parameters.

One important biological application of nanostructured surface coatings in microscale flow devices that has received much recent attention is in the capture and enrichment of rare cell subpopulations from mixtures of cells under flow (Figure 7.10). As an example, there is great current interest in isolating circulating tumor cells (CTCs), rare cancer cells dispersed in peripheral blood that accompany some of the most metastatic cancers such as those originating from breast and prostate. These CTCs can be as rare as one cell per mL of blood, surrounded by millions of white blood cells and billions of RBCs. CTC count in blood has been shown to correlate well with the progression of metastatic cancer, and is a strong predictor of patient survival. The state of the art in CTC applications is to capture intact, living CTCs from blood samples using microscale flow devices and adhesion proteins or specific antibodies, to screen different drug treatments or perform single-cell genetic analysis. Microtubes coated with halloysite nanotubes, then functionalized with a combination of E-selectin adhesion protein and antibodies against epithelial surface markers, have proven to be quite effective in isolating viable CTCs from patient blood samples (Hughes et al., 2012). The advantages conferred by the nanostructured halloysite layer are threefold:

1. Increase the surface area of the flow device, resulting in a two-fold increase in total adhesion protein absorbed.
2. Nanotubes penetrate the lubrication layer and capture gradually sedimenting cells before they have completely settled to the surface.

3. Repel white blood cells (the main contaminant in this application) from the surface and prevent them from activating and spreading.

This last effect is the most surprising outcome of the halloysite experiments, and it moves CTC isolation into a purity range of 60–80%, sufficient for most genetic and live cell assays. On smooth surfaces such as glass or plastic, leukocytes will rapidly spread, extending pseudopod extensions and increasing their adhesive contact area, whereas on halloysite they will retain their spherical resting shape indefinitely. This effect shares similarities with the well-known "lotus leaf" effect of superhydrophobicity, where water drops will sit atop hierarchical nanostructured surfaces at zero contact angle with no wetting. Perhaps in the future some design principles from the wetting/superhydrophobicity field can be used to engineer improved surfaces for controlling cellular response. As discussed in Chapter 3, there are many ways to quantify the roughness of a surface, such as the peak-to-peak distance, the root-mean-squared feature height, and so on. For the suppression of the leukocyte spreading response, it appears that the mean inter-peak distance is the best predictor of the number of captured cells and cell contact area.

7.6 Microchannels and minichannels as bioreactors for long-term cell culture

Bioreactors are liquid vessels in which live populations of suspended or surface-attached cells can be maintained for extended periods of time, either for expansion and differentiation of a desired cell type, or for the mass production of a biomolecule produced by the cells. Bioreactors usually feature gas and/or liquid exchange, to replenish nutrients to the cells as they are consumed and to remove waste and other products made by the cells. While commercial cell-based production of small molecules (e.g., ethanol) or proteins involves the scale-up of bioreactors to >1000-gallon tanks, there is much interest in miniaturizing bioreactors for use in bioartificial organ replacement of the pancreas (Sullivan et al., 1991), liver (Juaregui et al., 1997), and kidney (Cieslinski and Humes, 1994). These bioartificial organs, in which one or more live cell populations are combined with artificial engineered materials, must replace the following functions of the native organs (Fournier, 1999):

1. *Pancreas:* The main purpose of the pancreas is to regulate glucose levels in the body by hosting insulin-producing cells in small clusters of cells called the "islets of Langerhans." There is a great need for the restoration of pancreatic function among those afflicted with insulin-dependent diabetes mellitus (IDDM).
2. *Liver:* The human liver is a large (~1500 g), highly vascularized organ that receives 25% of the cardiac output and performs a variety of crucial life

functions. The liver stores and releases excess glucose, is the primary organ for fat metabolism, and produces virtually all of the plasma proteins other than antibodies. The main engineering challenge in developing a bioartificial liver replacement is that the organ is home to nearly 250 billion hepatocyte cells, each around 25 μm in diameter!

3. *Kidney:* Among the vital responsibilities of the kidneys are filtration of waste and its removal from the blood, regulation of red cell production within the bone marrow, and rapid control of blood pressure through the production of vasoactive molecules.

Microchannels and minichannels provide an excellent opportunity to maximize surface area available for monolayer cell attachment, while enabling good temperature regulation and rapid exchange of liquid media. Successful miniaturization of bioreactors for organ replacement could result in a device that may be comfortably worn outside the body, or even potentially implanted within the abdominal cavity. Another potential advantage of microchannels as bioreactors and the local flow environments that they are able to produce is that many cell types such as osteoblasts (bone cells) and endothelial cells (blood vessel lining) have evolved to prefer sustained exposure to shear stresses >1 dyne/cm and exhibit normal phenotype only in such an environment.

7.6.1 Radial membrane minichannels for hematopoietic blood cell culture

Hematopoietic stem and precursor cells (HSPC) in the bone marrow are able to differentiate and produce all of the different types of blood cells in the body. Bone marrow cells grown *ex vivo* maintain this pluripotent nature, and there is great interest in expanding blood cell populations in bioreactors for later transplantation to treat various blood disorders. The difficulty in growing large quantities of mammalian cells in culture is that oxygen is rapidly consumed and must be continuously replenished somehow. Peng and Palsson (1996) examined the potential of a radial membrane bioreactor to culture primary human mononuclear cells subjected to several growth factors (interleukin-3, granulocyte-macrophage colony-stimulating factor, erythropoietin) known to induce production of white and red blood cells. Figure 7.11 shows a diagram of a radial membrane bioreactor. Cell media saturated with oxygen enters the chamber from the left-hand side, at a well-developed laminar velocity $u(x)$. Oxygen in the media is consumed by the cells near the lower surface. The fluid near the upper, gas-permeable membrane is assumed to be in equilibrium with the gas and at the saturation oxygen concentration C_S. In such a system, the fluid velocity is chosen to be very low—typical transit times equal about 1.33 days. For comparison, the diffusivity of oxygen in protein solutions at 37°C has been measured to be $D = 2.69 \times 10^{-5}$ cm^2/s (Goldstick, 1966). For a channel height of 3 mm, the characteristic time for diffusion from the upper membrane to the cell layer at the bottom of the channel is 1 h. Thus, in this case, convective transport can be

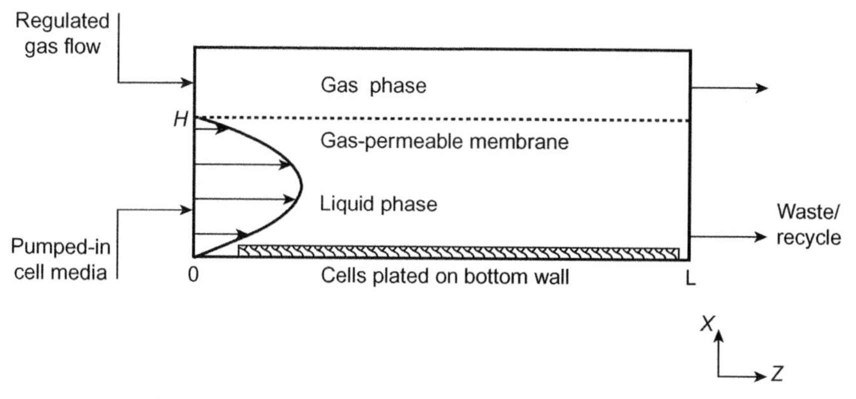

FIGURE 7.11

Minichannel membrane bioreactor.

neglected and the oxygen transport is modeled as a one-dimensional diffusion equation of the form:

$$D\frac{\partial^2 C}{\partial x^2} = 0 \tag{7.13}$$

The boundary condition at the upper membrane is that the fluid is saturated with oxygen:

$$C = C_S \quad @x = H \tag{7.14}$$

and at the lower wall the flux is equal to the cellular uptake rate N_O:

$$N_O = D\frac{\partial C}{\partial x} \quad @x = 0 \tag{7.15}$$

Subject to these two conditions, the solution to Eq. (7.13) is simply a linear concentration profile:

$$C(x) = C_S + \frac{N_O}{D}(x - H) \tag{7.16}$$

The uptake rate N_O obeys Michaelis–Menten kinetics given by:

$$N_O = \frac{qXC_O}{K_m + C_O} \tag{7.17}$$

where C_O is the oxygen concentration at the lower wall (i.e., the oxygen concentration experienced by the cells), q is the oxygen uptake rate on a per-cell basis, and X is the cell density. The Michaelis constant K_m, which can be interpreted as the concentration at which the uptake rate is half of maximum, has been measured to be about 1–5% of C_S (Ozturk, 1990). This implies that oxygen consumption operates at near zeroeth-order kinetics. Substituting Eq. (7.17) into Eq. (7.16) and evaluating the solution at $x = 0$ yields a quadratic expression for C_O:

$$C_O^2 + (K_m - C_S + qXH/D)C_O - C_S K_m = 0 \tag{7.18}$$

In terms of the fractional saturation $\sigma = C/C_S$, Eq. (7.18) becomes,

$$\sigma_0^2 + (\kappa + \phi - 1)\sigma_0 - \kappa = 0 \tag{7.19}$$

where we have introduced two dimensionless parameters, a dimensionless Michaelis constant $k = K_m/C_S$, and

$$\phi = qXH/DC_S = \frac{H^2/D}{HC_S/qX} \tag{7.20}$$

which is the ratio of the characteristic time for diffusion to the characteristic time for cellular oxygen uptake. The parameter ϕ depends on the cell density, which changes with time as the cells on the surface divide and double in number every 24–48 h. There is only one physically reasonable root to Eq. (7.19), since the other will always be negative. This fractional saturation design equation can be used to determine the range of parameter values for which the cells receive sufficient oxygen to support proliferation.

7.6.2 The bioartificial liver: membranes enhance mass transfer in planar microchannels

Many types of cells can be grown *in vitro* more successfully when co-cultured in the presence of a second cell type such as 3T3-J2 fibroblasts, which are robust and can produce vital nutrients that support the growth of the cells of interest. Tilles et al. (2001) studied the co-culture of rat hepatocytes (liver cells) with 3T3-J2 fibroblasts in microchannel bioreactors ($H = 85–500$ μm). Their analysis of the effects that an oxygenating membrane has on hepatocyte growth and protein production (Figure 7.12) provides a useful set of design equations for the development of microchannel membrane-based bioreactors.

For simplicity, we assume uniform, or "plug," flow within the microchannel (see Problem 7.5 for a modification of this approximation). The flow direction is defined with the z-coordinate, and oxygen diffusion occurs in the x-direction

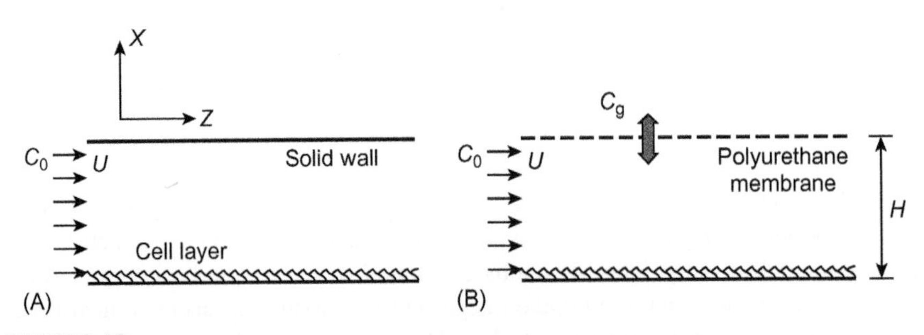

FIGURE 7.12

Microchannel bioreactors (A) without and (B) with an internal membrane oxygenator.

normal to the wall and is consumed by hepatocytes at $x = 0$. Once the oxygen concentration $C(x, z)$ is nondimensionalized with the inlet concentration C_{in}, and the x and z variables are nondimensionalized with the height (H) and length (L) of the bioreactor, respectively, then the dimensionless steady state convection–diffusion equation becomes

$$\frac{\partial C^*}{\partial z} = \frac{\gamma}{Pe}\frac{\partial^2 C^*}{\partial x^2} \tag{7.21}$$

where we have introduced the dimensionless length ratio $\gamma = L/H$, and the Péclet number $Pe = UH/D$. Here, U is the average velocity and D is the oxygen diffusivity. The boundary conditions for Eq. (7.21) are

$$\begin{aligned} &C^* = 1 && @z = 0, 0 \le x \le 1 \\ &\frac{\partial C^*}{\partial x} = Da && @x = 0, 0 \le z \le 1 \\ &\frac{\partial C^*}{\partial x} = \begin{cases} 0 & @x = 1, 0 \le z \le 1 \\ Sh[C_g^* - C^*] & @x = 1, 0 \le z \le 1 \end{cases} \end{aligned} \tag{7.22}$$

where there exist two different flux conditions at the upper surface depending on whether there is no membrane oxygenation, or with membrane oxygenation. We have introduced two additional dimensionless groups, the Damkohler number

$$Da = \frac{V_m HX}{DC_{in}} \tag{7.23}$$

and the Sherwood number

$$Sh = \frac{\sigma H}{D} \tag{7.24}$$

where the zeroeth-order uptake of oxygen by the hepatocytes is defined as V_m, the surface density of cells as X, and σ is the membrane permeability to oxygen. We have nondimensionalized the gas phase oxygen concentration with C_{in} as $C_g^* = C_g/C_{in}$. A similar differential system has been solved analytically be Carslaw and Jaeger (1959), which can be applied to yield the cell surface oxygen concentration along the length of the bioreactor, either without internal membrane oxygenation:

$$C^*(0,z) = 1 - z\frac{\gamma}{Pe}Da - \frac{Da}{3} + \frac{2Da}{\pi^2}\sum_{n=1}^{\infty}\frac{(-1)^n}{n^2}\cos(n\pi)\exp\left[-\frac{\gamma n^2\pi^2 z}{Pe}\right] \tag{7.25}$$

or with internal membrane oxygenation:

$$C^*(0,z) = C_g^* - Da - \frac{Da}{Sh} + \sum_{n=1}^{\infty} B_n \exp\left[-\frac{\gamma\lambda_n^2 z}{Pe}\right] \tag{7.26}$$

In Eq. (7.26), the coefficients B_n are defined by:

$$B_n = \frac{4Da(1-\cos\lambda_n) + 4\lambda_n \sin\lambda_n\left(1 - C_g^* + Da/Sh\right)}{\left[\lambda_n \sin 2\lambda_n + 2\lambda_n^2\right]} \tag{7.27}$$

and λ_n are the roots of the transcendental equation:

$$\tan\lambda_n = \frac{Sh}{\lambda_n} \tag{7.28}$$

Example 7.2 (see page 532) illustrates the use of these design equations (7.25)–(7.28).

7.6.3 Oxygen and lactate transport in micro-grooved minichannels for cell culture

As stated in Section 7.6.1, there is considerable interest in culturing and expanding hematopoietic stem cell populations *ex vivo*, for later autologous (donor = recipient) or allogeneic (donor ≠ recipient) transplantation for the treatment of hematological or immunological disorders and to treat complications of cancer therapies. In their native environment in the bone marrow, hematopoietic cells are not strongly adherent to their surroundings but are retained in the marrow, in part due to adhesion to a second type of cell called stromal cells. It has been recognized that *ex vivo* cell perfusion systems for long-term culture that are able to retain weakly adherent cells would hold many advantages over co-culture systems (Sandstrom et al., 1996). Horner et al. (1998) analyzed the solute transport in a micro-grooved minichannel designed to retain non-adherent cells within microscale cavities oriented perpendicular to the main flow, as depicted in Figure 7.13. Flow through the main minichannel generates a form of lid-driven flow within each cavity known as nested Moffatt eddies (Moffatt, 1964). These eddies provide convection-enhanced mass transfer between the cells resting at the bottom of each cavity and the main chamber, while sheltering the cells from the high wall shear stresses that would be experienced in a comparable flat-bottomed channel.

For a relatively wide channel of several centimeters, the aspect ratio is such that the third dimension (vorticity direction) can be neglected and the problem treated as two-dimensional. As in our previous bioreactor examples, we take the flow direction as z and the velocity gradient direction as x. At steady state, the Navier–Stokes equation in two dimensions simplifies to:

$$\begin{aligned} \rho\left(v_x\frac{\partial v_x}{\partial x} + v_z\frac{\partial v_x}{\partial z}\right) &= -\frac{\partial p}{\partial x} + \mu\left(\frac{\partial^2 v_x}{\partial x^2} + \frac{\partial^2 v_x}{\partial z^2}\right) \\ \rho\left(v_x\frac{\partial v_z}{\partial x} + v_z\frac{\partial v_z}{\partial z}\right) &= -\frac{\partial p}{\partial x} + \mu\left(\frac{\partial^2 v_z}{\partial x^2} + \frac{\partial^2 v_z}{\partial z^2}\right) \end{aligned} \tag{7.29}$$

Note that unlike previous examples where the flow was assumed unidirectional in the z-direction and diffusion in the z-direction was neglected relative to

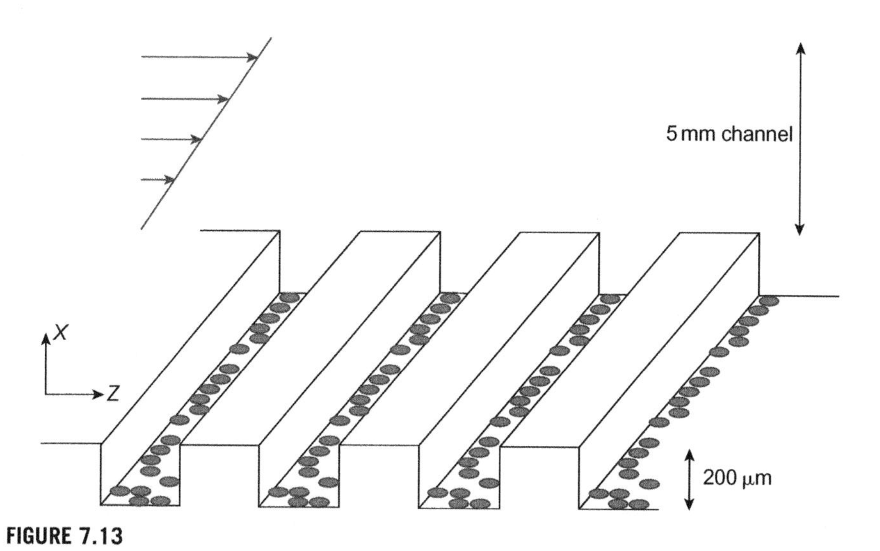

FIGURE 7.13

Micro-grooved, minichannel bioreactor for the culture of weakly adherent cells. Note that the cells are sheltered within the grooves and do not become entrained in the flow. The exterior flow induces Moffett eddies that exchange solutes between the groove cavities and the minichannel region.

convection, leading to simplifications to the governing equations, the lid-driven cavity flow is expected to be fully two-dimensional and these full equations must be integrated. The solute balance for species i, at steady state, takes the form:

$$v_x \frac{\partial C_i}{\partial x} + v_z \frac{\partial C_i}{\partial z} = -D_i \left(\frac{\partial^2 C_i}{\partial x^2} + \frac{\partial^2 C_i}{\partial z^2} \right) \tag{7.30}$$

The solutes of main biological interest (oxygen and lactate) are dilute small molecules that do not appreciably affect the fluid density and viscosity. Thus, the fluid velocity field may be solved independently of the solute balance equation. It is reasonable to impose a well-developed parabolic velocity profile at the channel inlet, provided that the cell-containing region occurs after the entrance length given by:

$$\frac{L_{\text{ent}}}{H} = aRe \tag{7.31}$$

where the Reynolds number is defined for channel flow as:

$$Re = \frac{\rho \langle v \rangle H}{\mu} \tag{7.32}$$

In Eq. (7.31), the coefficient a is equal to 0.034 to ensure that the centerline velocity is within 98% of the theoretical Poiseuille flow limit (Bodoia and Osterle, 1961), and the angled brackets denote an averaged quantity. For typical bioreactor conditions, the entrance length is reached within a fraction of one channel height ($\leq 0.05H$). No-slip conditions are enforced at the upper, lower, and micro-groove walls. The general flux condition at the lower wall of the micro-groove cavities, plated with hematopoietic cells, is obtained by setting the surface solute flux equal to a Michaelis–Menten kinetic rate of solute consumption or production (e.g., oxygen or lactate, respectively). This boundary condition at the cell surface takes the form:

$$N_0 = -D\frac{\partial C}{\partial x} = \frac{qXC_0}{K_m + C_0} \tag{7.33}$$

where a sign change from Eq. (7.15) is introduced for a local coordinate system within the cavity corresponding to an upper cavity opening at $x = 0$ and the positive x-coordinate directed downward. The system of Eqs. (7.29), (7.30), and (7.33), subject to no-slip at the wall and given inlet velocity and solute concentration conditions, can be readily solved for the spatial concentration profiles using a commercial computational fluid dynamics (CFD) code such as *FIDAP* (Fluent, Inc., Lebanon, NH). Horner et al. (1998) have used full numerical analysis of the micro-grooved minichannel system to derive constitutive equations that describe the solute concentration within the groove cavities as a function of various physical parameters such as flow rate, channel dimensions, and zeroeth-order solute consumption (production) rate.

7.7 Microspherical cavities for cell sorting and tumor growth models

As discussed elsewhere in this book, channels and microcavities with rectangular cross-sections are much easier to fabricate using machining, etching, or photolithography techniques. Additionally, if it is desired to create microcavities with openings smaller than the inner dimensions, this often requires multilayer or multistep fabrication processes. Such rectangular channels and cavities are useful for many fluidic applications. Biological cells, however, can sense and respond to surface curvature, and some cell types require surface curvature to exhibit their natural behavior. Additionally, since many types of cells sense fluid shear stress and can become damaged by fluid forces, it is often advantageous to create microcavities for cell culture that are connected to the outside environment by smaller pores, to shelter cells from shear stress while still allowing for some exchange of dissolved nutrients. Giang et al. (2007, 2008) described a simple process for the fabrication of smooth spherical cavities in PDMS, termed gas expansion molding (GEM). The process utilizes a mold of trenches such as those created in a silicon wafer using deep reactive ion etching (DRIE) for use in reverse molding of features. The process is depicted in Figure 7.14; it can be contrasted with the standard process of reverse molding in PDMS. In the conventional process, the PDMS polymer is

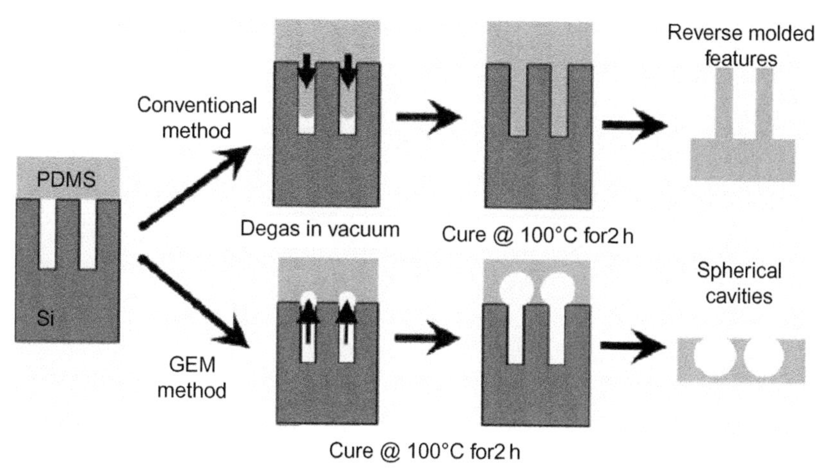

FIGURE 7.14

Illustration of the GEM fabrication method to produce spherical cavities in PDMS.

degassed in vacuum prior to pouring over the silicon mold. After curing, the PDMS layer can then be peeled off to reveal the reverse of the etched pattern on the master, such as an array of posts. If the degassing step is omitted, however, then the gas contained in the trench nucleates with gas dissolved in the PDMS and expands to form a spherical cavity above the trench opening. The spherical cavities are reproducible, and the diameter can be controlled by the trench cross-section, spacing between trenches, and PDMS thickness. A trench width of $\sim 100\,\mu m$ will typically produce a cavity of diameter $\sim 300\,\mu m$, and the process seems to be insensitive to the opening shape (trench cross-section) as square, circular, and triangular trenches all produce spherical cavities. Theoretical modeling of the GEM process is an interesting heat and mass transfer problem involving moving boundaries, which has not yet been addressed.

As discussed in the previous section for square grooves, when the PDMS layer with spherical cavities is assembled as the lower surface of a parallel-plate flow chamber, a lid-driven cavity flow is generated within the cavity. Figure 7.15 shows the velocity field generated within one of these cavities, as calculated in COMSOL (Agastin et al., 2011). It is evident that spherical cavities lack the corners and dead spaces present in square cavities, thus promoting more uniform exchange of solutes between main channel and cavity. Figure 7.16 shows the streamlines from a simulation of a 2×3 array of cavities. The fluid mechanics in each cavity is unaffected by this spacing of arrays in the PDMS; however, in large arrays the suspended cells and proteins will tend to become depleted as one moves in the axial direction of the main channel flow. These spherical microcavity arrays have been explored as a platform for cell separations, as depicted in Figure 7.17. One application is to coat the upper flat surface with an adhesion

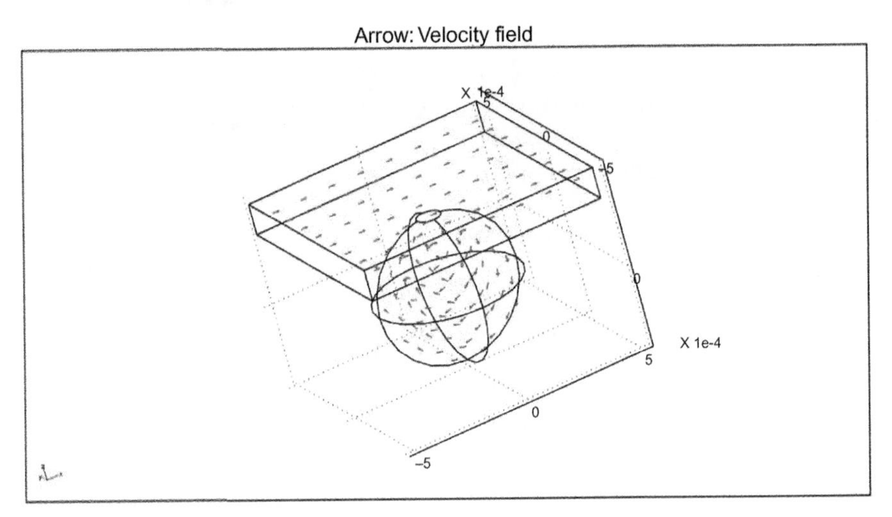

FIGURE 7.15

Quiver plot of the velocity field produced by flow over the circular opening in a spherical microcavity.

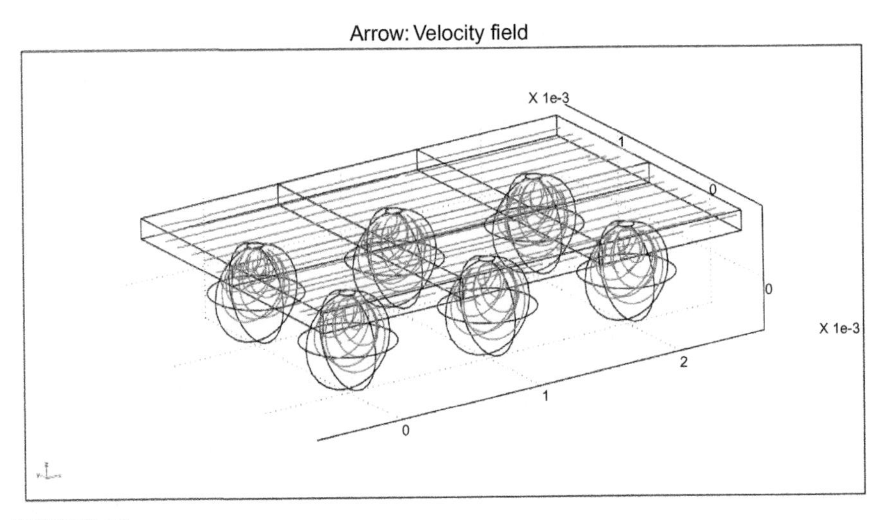

FIGURE 7.16

Streamline plot of flow through a 2 × 3 array of spherical cavities with parallel-plate channel flow above the microcavity openings.

molecule that will induce target cells (e.g., tumor cells dispersed in blood) to slowly roll along the upper surface. Once encountering a microcavity opening, the target cell will drop into the cavity for later harvesting or cell culture. Using a "vacuum-assisted coating" process, in which a protein solution will not enter into the cavities until placed in a vacuum chamber, the upper surface and inner cavity surfaces can be effectively coated with different molecules (Giang et al., 2008). This approach has been shown to enrich target cells under flow from 20% up to 59% purity, or 75% up to 99%, when E-selectin binding COLO 205 cells were mixed with nonbinding cells (Agastin et al., 2011).

Once biological cells are introduced into the spherical microcavities, they can be cultured and expanded there. Figure 7.18 shows live COLO 205 cells in a microbubble at 0, 24, and 48 h after capture. COMSOL simulations show operating flow rates that produce adequate nutrient exchange (Figure 7.19), and chemical gradients comparable to tumor spheroid models (Chandrasekaran and King, 2012). Figure 7.20 shows a 3-D confocal image stack of a spheroid of COLO 205 cells labeled with a fluorescent live cell dye CellTracker, at (left to right) 0, 3, and 5 days. A small number of dead cells were found using propidium iodide dye at

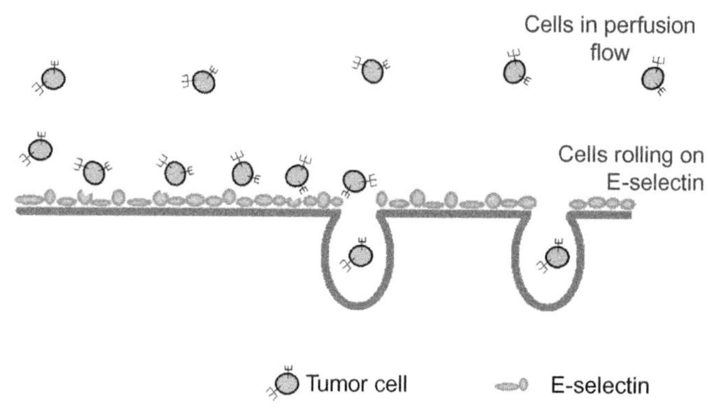

FIGURE 7.17

Schematic showing the cell sorting approach wherein target cells bind and roll on the upper selectin-coated surface, enter into the spherical microcavities, and are retained there for growth and expansion in culture.

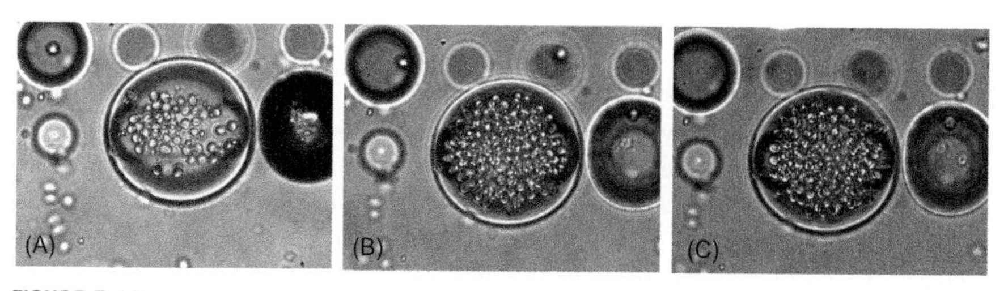

FIGURE 7.18

Micrograph of COLO 205 cancer cells captured from flow and sequestered in a spherical microcavity, at 0 h (A), 24 h (B), and 48 h (C).

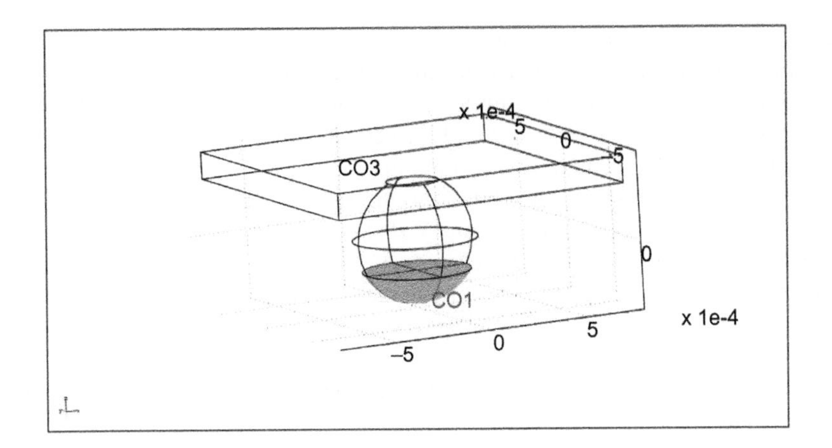

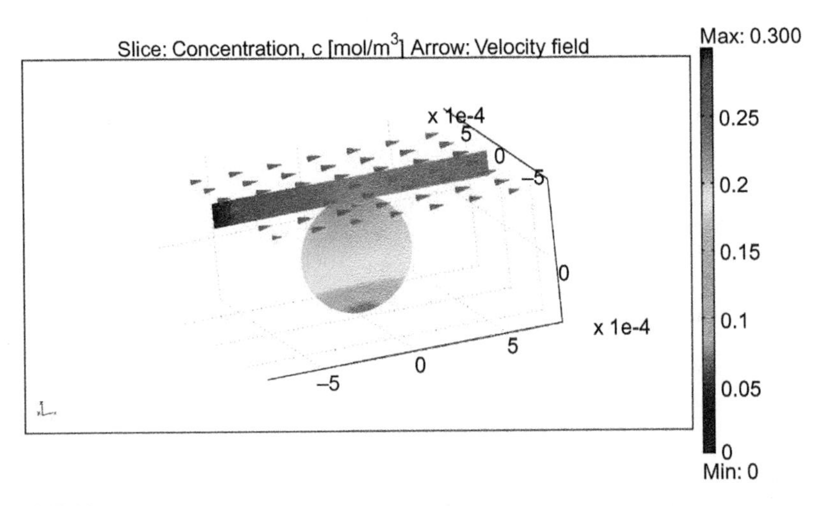

FIGURE 7.19

Mass transport model of cells in the bottom region of a microcavity consuming and producing solute, solved in COMSOL. The upper image shows the cell-containing region, and the lower image shows a steady state concentration profile.

the bottom of the spheroid, <2% of cells at day 0, up to 5% of cells on day 5. The value of creating large arrays of microcavities containing live tumor spheroids is for enabling personalized medicine—the ability to screen different drug concentrations on cell lines (or, preferably, primary patient cells) in a high-throughput, automated manner. Figure 7.21 shows a demonstration of this concept, where the response of COLO 205 spheroids to the chemotherapeutic drug doxorubicin has been measured. Note the different response measured for traditional monolayer cell culture, compared to the more realistic 3-D tumor spheroid culture, illustrating the importance of these more realistic microscale bioreactor systems.

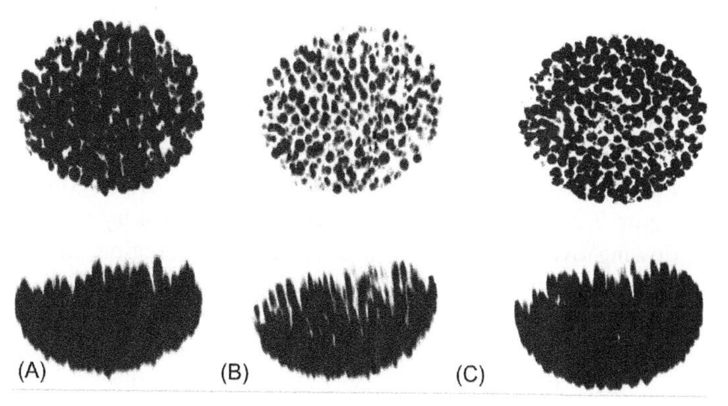

FIGURE 7.20

Tumor spheroid viability inside a spherical microcavity. Confocal image stacks taken at 0 days (A), 3 days (B), and 5 days (C) after seeding.

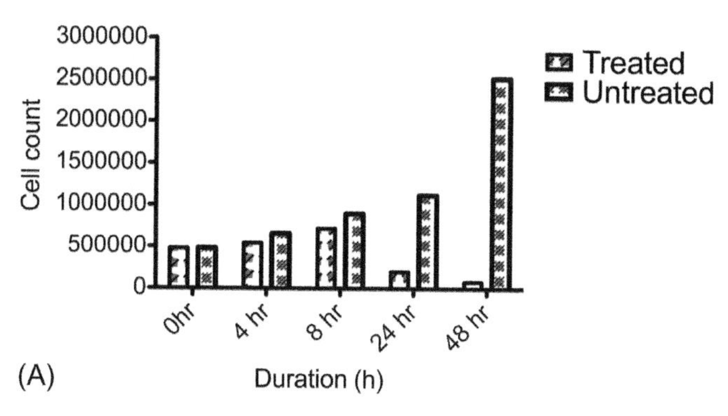

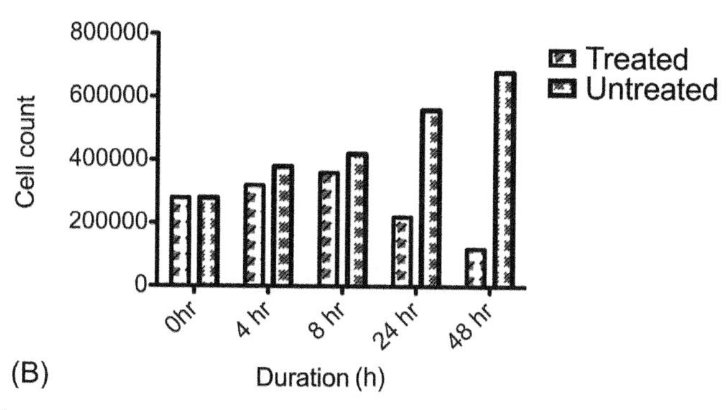

FIGURE 7.21

Response of monolayer (A) or spheroid (B) cultured COLO 205 cells to doxorubicin chemotherapy treatment.

7.8 Generation of normal forces in cell detachment assays

Hydrodynamic flow in parallel-plate microchannels is often used to detach firmly adhered cells from substrates, as a measure of the strength of integrin:matrix protein bonds (Figure 7.22). However, the high flow rates necessary to detach spread cells, when flowing over a flat cell with a raised nuclear region, may produce a negative pressure over the nucleus (the so-called "Bernoulli effect") that can potentially equal or exceed the shear forces that one is trying to generate (Figure 7.23). Thus, here we model this problem as inviscid, irrotational flow over a wavy wall, as developed by Michelson (1970), keeping in mind that this will only be valid for Reynolds numbers much greater than unity (often achieved in previous experimental studies, as discussed below).

Figure 7.22 depicts the dimensions of a typical cell detachment flow chamber. From the literature, the force required for cell detachment from an adhesive substrate can be estimated as between 10^{-9} and 10^{-7} N. In saline buffer ($\mu = 1$ cP), this requires flow rates of $U_{avg} = 1-100$ cm/s. In such an experiment, the highest Reynolds number studied would be

$$Re = \frac{Q\omega}{\nu} = 256 \tag{7.34}$$

Thus, not only is inertia important in this microscale system; it may be expected to dominate over viscous effects.

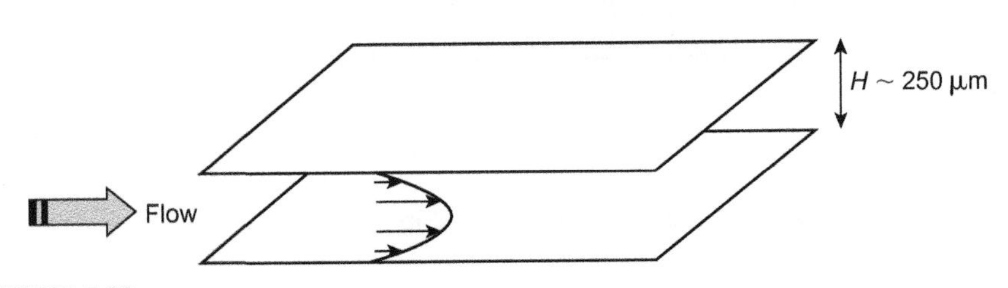

FIGURE 7.22

Microchannel geometry for cell detachment.

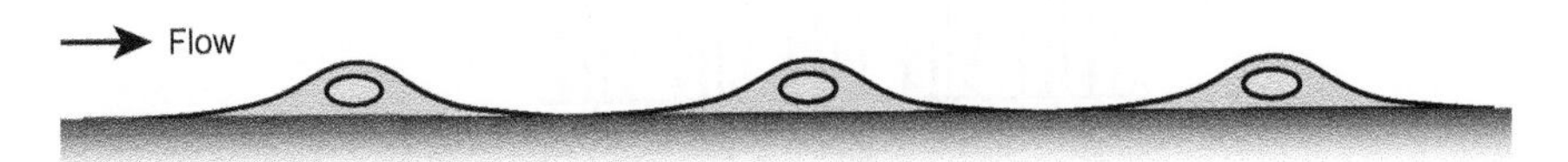

FIGURE 7.23

The nuclei of adherent cells project into the flow field.

7.8.1 Potential flow near an infinite wall

Our starting point for modeling the flow over a cell monolayer is the set of nonlinear Euler's equations:

$$
\begin{aligned}
u\frac{\partial u}{\partial x} + v\frac{\partial u}{\partial y} + \frac{1}{\rho}\frac{\partial p}{\partial x} &= -\frac{\partial U}{\partial x} \\
u\frac{\partial v}{\partial x} + v\frac{\partial v}{\partial y} + \frac{1}{\rho}\frac{\partial p}{\partial y} &= -\frac{\partial U}{\partial y}
\end{aligned}
\tag{7.35}
$$

which describe steady, two-dimensional *frictionless* flows subject to conservative forces (i.e., derived from a potential U) such as gravity.

When a rigid body rotates at angular speed ω, the velocity components of a point located at (x, y) are

$$u = -\omega y, \quad v = \omega x \tag{7.36}$$

Eliminating x and y from Eq. (7.36) yields a definition for the vorticity ω,

$$2\omega = \left\{ \frac{\partial v}{\partial x} - \frac{\partial u}{\partial y} \right\} \tag{7.37}$$

a measure of the local fluid rotation. Kelvin's theorem states that vorticity is a constant in incompressible, frictionless flows, when the only forces are derived from potentials such as gravity. Therefore, Kelvin's theorem gives us an additional relation:

$$\frac{\partial v}{\partial x} - \frac{\partial u}{\partial y} = 0 \tag{7.38}$$

along with the continuity equation for incompressible flow,

$$\frac{\partial u}{\partial x} + \frac{\partial v}{\partial y} = 0 \tag{7.39}$$

We define the velocity potential $\phi(x, y)$ as

$$u = -\frac{\partial \phi}{\partial x}, \; v = -\frac{\partial \phi}{\partial y} \tag{7.40}$$

which identically satisfies the irrotational condition. Substituting this definition into the continuity equation yields

$$\frac{\partial^2 \phi}{\partial x^2} + \frac{\partial^2 \phi}{\partial y^2} = 0 \tag{7.41}$$

Thus, the velocity potential for plane irrotational flow of an incompressible fluid is a solution to Laplace's equation. This equation is solved for the uniform flow past a flat wall in Example 7.3 (see page 534).

7.8.2 Linearized analysis of uniform flow past a wavy wall

To model the disturbance flow generated by a uniform flow over a periodic layer of cells, one of the streamlines of uniform flow is modified to be everywhere only nearly parallel to the x-axis. We expect that the velocities elsewhere will differ slightly from U, and the associated pressure variations will be small. Figure 7.24 shows a slightly curved boundary wall of sinusoidal form

$$y = h \sin \lambda x \tag{7.42}$$

The length of the whole cycle of the sine wave is $L = 2\pi/\lambda$, and here h is the wave height. We require that h be small, as $h \ll L$, which corresponds to an assumption of "slight waviness." The flow velocity components are assumed to differ only slightly from $(U, 0)$ and can be written as

$$u = U + u', \quad v = v' \tag{7.43}$$

where the primes indicate velocity perturbations that vanish with h, and

$$u' \ll U, \quad v' \ll U \tag{7.44}$$

Because of the irrotationality condition,

$$\frac{\partial v'}{\partial x} - \frac{\partial u'}{\partial y} = 0 \tag{7.45}$$

the velocity potential $\phi(x, y)$ may be taken to refer only to the perturbations

$$u' = -\frac{\partial \phi}{\partial x}, \quad v' = -\frac{\partial \phi}{\partial y} \tag{7.46}$$

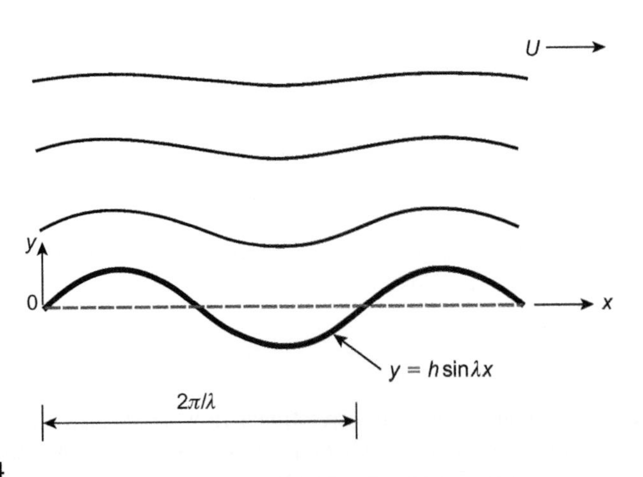

FIGURE 7.24

Geometry of the linearized analysis near a wavy wall.

This yields the following expression for the total flow:

$$u = U - \frac{\partial \phi}{\partial x}, \quad v = -\frac{\partial \phi}{\partial y} \tag{7.47}$$

Bernoulli's theorem can now be applied:

$$\frac{u^2}{2} + \frac{p}{\rho} + \text{potential} = \text{constant} \tag{7.48}$$

which is valid at different points along a single streamline. Thus, the linearized version of the governing equation becomes

$$p(x,y) - p_0 = \rho U \frac{\partial \phi}{\partial x} \tag{7.49}$$

where we have subtracted the quantity $p_0 + \rho U^2 = \text{cst}$ evaluated far from the wall. In Eq. (7.49), p_0 denotes the pressure in those parts of the flow where the velocity reaches its uniform value U.

The velocity potential is a harmonic function given by the equation

$$\frac{\partial^2 \phi}{\partial x^2} + \frac{\partial^2 \phi}{\partial y^2} = 0 \tag{7.50}$$

We must now find the particular solution to Eq. (7.50) out of the many general solutions to Laplace's equation. First, we note the far field condition $\phi(x, \propto) = 0$. At the boundary, we enforce that the slope of the velocity vector equals the boundary slope:

$$\frac{v}{u} = h\lambda \cos \lambda x \tag{7.51}$$

From our earlier development, the left-hand side of Eq. (7.51) is given by

$$\text{LHS} = -\frac{\partial \phi / \partial y}{U - \partial \phi / \partial x} = -\frac{1}{U}\frac{\partial \phi}{\partial y} \tag{7.52}$$

where we have neglected higher-order terms. We may write an approximate boundary condition evaluated at $y = 0$:

$$-\frac{\partial \phi}{\partial y}(x,0) = Uh\lambda \cos \lambda x \tag{7.53}$$

Note that the differential equation, boundary conditions, and pressure relationship are all linear in ϕ-derivatives. This problem admits a separable solution of the form:

$$\begin{aligned} X(x) &= A \cos kx + B \sin kx \\ Y(y) &= C \exp(-ky) + D \exp(ky) \end{aligned} \tag{7.54}$$

which combine to give the following potential function:

$$\phi(x, y) = \{A \cos kx + B \sin kx\}\{C \exp(-ky) + D \exp(ky)\} \tag{7.55}$$

where the term $D \exp(ky)$ vanishes in order to satisfy the far field condition. We are thus left with two remaining constants AC and BC, and can set $C = 1$ without loss of generality. The boundary condition at $y = 0$ takes the form

$$Uh\lambda \cos \lambda x = k\{A \cos kx + B \sin kx\} \tag{7.56}$$

which, upon manipulation, yields the result

$$Uh\lambda = kA, \quad B = 0, \quad k = \lambda \tag{7.57}$$

These constants can be inserted into our general solution to give

$$\phi(x, y) = Uh \exp(-\lambda y)\cos \lambda x \tag{7.58}$$

for the velocity potential, and the following expression for the pressure at any point in the flow:

$$p(x, y) = p_0 - \rho U^2 h\lambda \sin \lambda x \exp(-\lambda y) \tag{7.59}$$

Thus, the linear approximation to the pressure on the boundary wall $y = 0$ is

$$p(x, 0) = p_0 - \rho U^2 h\lambda \sin \lambda x \tag{7.60}$$

Equation (7.60) implies that the pressure is a minimum at the peaks of the wavy wall ($\lambda x = \pi/2$) and reaches a maximum at the wave troughs, as depicted in Figure 7.25.

Typical parameter values for cell detachment experiments yield the following estimate for the peak (negative) pressure above the cell nucleus:

$$\rho U^2 h\lambda = (1000\ \text{kg/m}^3)(1\ \text{m/s})^2(2 \times 10^{-6}\ \text{m})(2\pi/20 \times 10^{-6}\ \text{m}) \tag{7.61}$$

Thus, the pressure above the cell nucleus is predicted to have magnitude $p = 628\ \text{N/m}^2$, or, assuming a characteristic cell area of $100\ \mu\text{m}^2$, a negative pressure of magnitude $F = |63\ \text{nN}|$. This value for the normal force induced by the Bernoulli effect is of the same order as the shear forces generated by the imposed flow! Reanalysis of many previous experiments using cell detachment assays shows the generation of large normal forces in both the microchannel geometry

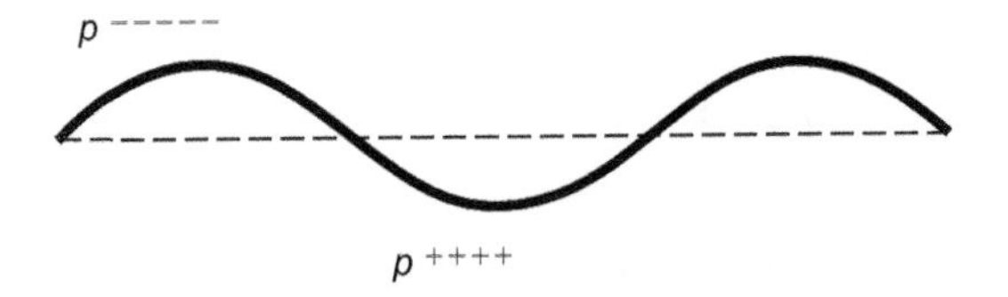

FIGURE 7.25

Pressure distribution at the surface of the wall.

(Van Kooten et al., 1992; Schnittler et al., 1993; Sank et al., 1994; Thompson et al., 1994; Wechezak et al., 1994; Kapur and Rudolph, 1998; Sirois et al., 1998; Chan et al., 1999; Malek et al., 1999; Grandas et al., 2001) and the spinning disk assay (Garcia et al., 1997; Boettiger et al., 2001; Deligianni et al., 2001; Miller and Boettiger, 2003). To reduce the Reynolds number, and thus reduce the normal forces relative to shear forces, one should simply increase the fluid viscosity with a high-molecular-weight molecule such as dextran. For instance, a 7.5 wt% dextran solution, of molecular weight $MW = 2 \times 10^6$ g/mol, would result in a relative viscosity of

$$\frac{\mu}{\mu_0} = 20.5$$

7.9 Small-bore microcapillaries to measure cell mechanics and adhesion

The human circulatory system represents a challenging system for engineering modeling, due to the wide variation in vessel geometry, linear flow velocity, and Reynolds numbers, starting at the major aortic vessels down to the smallest microscopic capillaries. Table 7.2 shows the range of these values in a typical human, assuming a constant blood viscosity of 0.035 P and mean peak velocities reported for the pulsatile flow of arterial flow.

It is evident that the human circulation covers the entire range of flow behavior from creeping (non-inertial) flow up to fully developed turbulence. We see in the arterioles, capillaries, and venules that the length scales span the range of fluid physics encountered in microchannel and minichannel flow. In the largest vessels of the body, blood behaves as a non-Newtonian fluid, and constitutive

Table 7.2 Typical Flow Parameters in the Human Systemic Circulation

Structure	Diameter (cm)	Blood Velocity (cm/s)	Tube Reynolds Number
Ascending aorta	2.0–3.2	63	3600–5800
Descending aorta	1.6–2.0	27	1200–1500
Large arteries	0.2–0.6	20–50	110–850
Arterioles	0.001–0.015	0.5–1.0	0.014–0.43
Capillaries	0.0005–0.001	0.05–0.1	0.0007–0.003
Venules	0.001–0.02	0.1–0.2	0.0029–0.11
Large veins	0.5–1.0	15–20	210–570
Vena cavae	2.0	11–16	630–900

Source: *Data from Cooney, 1976; Whitmore, 1968; Schneck, 2000.*

equations are usually successful in describing behavior (e.g., blunted velocity profiles) at the continuum level. At the scale of the smallest vessels, however, the particulate nature of blood and its constituent cells must be taken into account (King et al., 2004). In the capillaries of the lungs, for instance, the mechanical properties of individual white blood cells significantly influence the transit time through these vessels (Bathe et al., 2002).

Suspensions of erythrocytes (RBCs) flowing through microcapillaries ($d \sim 40\ \mu m$) are well known to concentrate at the tube center, resulting in velocity profiles blunted from parabolic. Trajectories of individual RBCs have been visualized both *in vitro* (Goldsmith and Marlow, 1979) and *in vivo* (Lominadze and Mchedlishvili, 1999), showing radial drift velocities and random walks that attenuate as the cell approaches the centerline. The single-file motion of RBCs through small capillaries ($d \sim 4-20\ \mu m$) has been modeled theoretically subject to various simplifying assumptions: $x-y$ two-dimensionality (Sugihara-Seki et al., 1990), symmetry about the center axis (Secomb and Hsu, 1996), or approximation via multipole expansion (Olla, 1999). The radial drift of RBCs is known to produce a cell-free plasma layer in microvessels (Yamaguchi et al., 1992), suggesting that RBCs have only a secondary influence on the dynamics of leukocytes once they are displaced to the vessel wall (Goldsmith and Spain, 1984). Furthermore, leukocytes have a dramatic influence on total blood viscosity in the microcirculation despite their relatively low numbers in blood as compared to RBCs (Helmke et al., 1998), highlighting the need to better understand the motion of leukocytes through microcapillaries.

Example 7.4 (see page 534) provides an estimate of the blood transit time through a capillary, where the majority of oxygen exchange takes place.

7.9.1 Flow cytometry

Flow cytometry is a powerful analysis tool in biomedical research, utilizing microscale flow for automated single cell characterization and sorting. The concentration of any fluorescently labeled molecule on the surface (or interior, following permeabilization) of the cell can be measured relative to some standard, and these cells can furthermore be sorted according to the concentration of this marker or according to size. Antibodies against many of the known receptors on the surface of human or animal cells can be obtained commercially. Figure 7.26 shows a schematic of a typical system. The cell suspension (possibly containing a mixture of different cell types) is introduced through a circular conduit. This conduit opens into a larger flow domain containing a flowing annulus of "sheath fluid" which acts to focus the stream of cells into a single-file configuration within a narrow jet of fluid ($<100\ \mu m$ diameter). This stream of cells then passes through a quartz chamber where they are characterized for their scattering and fluorescent properties. Most commonly, a beam of light from an argon laser is used to excite the fluorophore of interest. The emitted light from this fluorophore is then collected at one of two photomultiplier tubes. Downstream of these optics, the outlet stream is ruptured into a series of droplets, each containing a single cell,

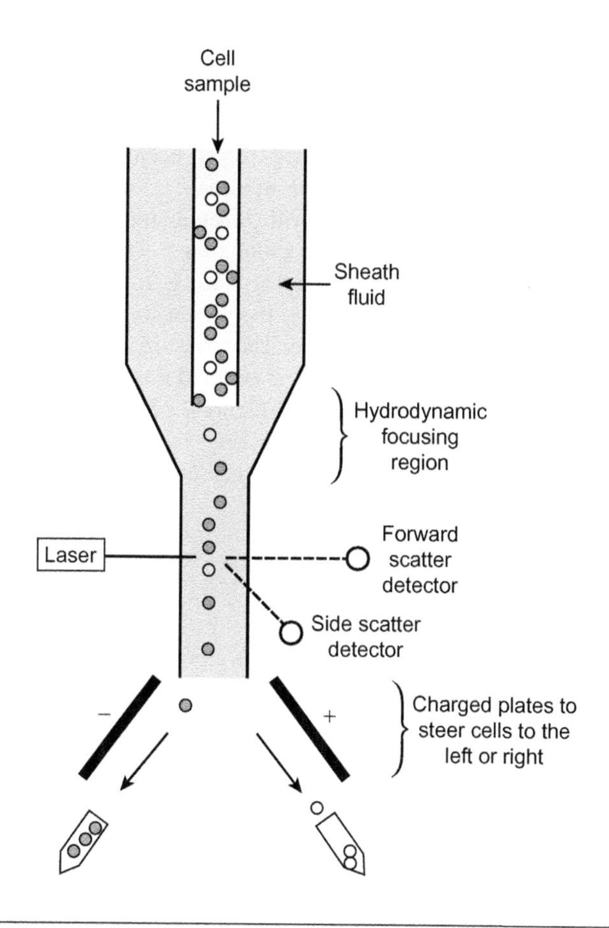

FIGURE 7.26

Schematic of a flow cytometer, and fluorescence-activated cell sorting (FACS).

via piezoelectric vibration. An electrical field is then used to steer positively or negatively charged individual droplets (a charge imposed depending on the previous fluorescence intensity of the corresponding cell) into multiple collection vessels. The whole process is automated so that commercially available flow cytometers can sort cells at a rate of over 10^4 cells/s. In addition to characterizing individual cells for clinical applications, flow cytometry has been used to measure the size of small blood cell aggregates, where up to pentuplets of neutrophils can be successfully identified (Simon et al., 1990; Neelamegham et al., 2000).

7.9.2 Micropipette aspiration

Small-bore glass microcapillaries are often used either to characterize the mechanics and deformability of individual cells or to test the surface adhesion of cells contacting other cells or artificial bead surfaces (Lomakina et al., 2004; Spillmann et al., 2004). Capillaries with inner diameter less than the major diameter of cells such as blood cells are used to partially aspirate the cell, with the projection length of the cell within the capillary interior being a measure of the cortical tension of the cell exterior (Herant et al., 2003). Once the cell is ejected from the micropipette, its shape recovery can also yield information on the viscoelastic parameters of the cell (Dong et al., 1998). Capillaries with inner diameter larger than the cell diameter can be used to translate the cell back and forth within the pipette to bring a cell into repeated contact with another nearby surface. In this manner, the probability for cell adhesion can be determined in the presence of various biological stimuli. Shao and Hochmuth (1996) calculated the pressure drop caused by the gap flow in the annular region between a spherical cell and a larger cylindrical micropipette. The geometry considered is depicted in Figure 7.27.

Assuming axisymmetric flow, and neglecting inertia, yields the following form of the Navier–Stokes equations in cylindrical coordinates:

$$\frac{1}{r}\frac{\partial}{\partial r}(ru) + \frac{\partial w}{\partial z} = 0$$

$$\frac{\partial p}{\partial r} = \mu\left(\frac{\partial^2 u}{\partial r^2} + \frac{1}{r}\frac{\partial u}{\partial r} - \frac{u}{r^2} + \frac{\partial^2 u}{\partial z^2}\right) \tag{7.62}$$

$$\frac{\partial p}{\partial z} = \mu\left(\frac{\partial^2 w}{\partial r^2} + \frac{1}{r}\frac{\partial w}{\partial r} + \frac{\partial^2 w}{\partial z^2}\right)$$

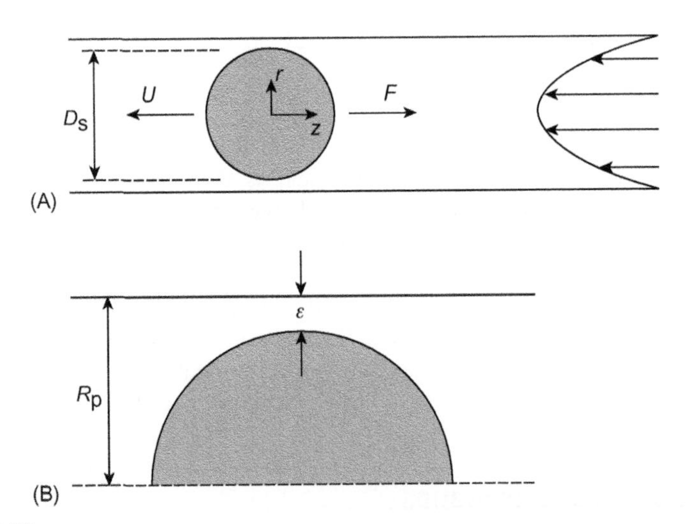

FIGURE 7.27

(A) Spherical cell positioned at the centerline within a larger micropipette. (B) Close-up of the upper half of the cylindrical region.

where p is the pressure, u and w are the local fluid velocities in the r- and z-directions, respectively, and μ is the fluid viscosity. No-slip conditions are enforced at all boundaries, and the total volumetric flow rate through any circular cross-section of the pipette is a constant Q. After considerable manipulations, useful analytical expressions can be obtained relating the total pressure drop Δp, sphere velocity U, and external force acting on the sphere F for two simplified cases: (i) stationary sphere, and (ii) force-free sphere. First, the force necessary to hold a sphere stationary within a pressure-driven flow in a micropipette is:

$$F = \frac{\pi R_p^2 \Delta p}{(1 + (4/3)\bar{\varepsilon}) + ((8(L_{eq} - D_s))/R_p)((2\sqrt{2}/9\pi)\bar{\varepsilon}^{5/2} - (8/9\pi^2)\bar{\varepsilon}^3)} \tag{7.63}$$

where L_{eq} is the total pipette length, and the gap spacing has been nondimensionalized with the pipette radius. For a neutrally buoyant or force-free sphere, on the other hand, the relationship between the total pressure drop and the sphere velocity is given by:

$$\Delta p = \frac{\mu U}{R_p}\left[\left(\frac{4\sqrt{2}\pi}{\bar{\varepsilon}^{1/2}} - \frac{71}{5}\right) + \frac{8(L_{eq} - D_s)}{R_p}(1 - (4/3)\bar{\varepsilon})\right] \tag{7.64}$$

7.9.3 Particle transport in rectangular microchannels

The transport of suspended cells in rectangular microchannels is of interest in the analysis of red cell dimensions (Gifford et al., 2003), the study of blood coagulation and thrombotic vessel occlusion (Kamada et al., 2004), measurement of intracellular signaling in Jurkat cells (Li et al., 2004), and viral-based transfection (Walker et al., 2004). When the channel height approaches the same order of magnitude as the diameter of the transported cells, the particles are influenced by the presence of the walls and the assumption that the particles travel with the local fluid velocity no longer remains valid. Using a boundary-integral algorithm, Staben et al. (2003) numerically calculated the motion of spherical and spheroidal particles within a rectangular channel. Both the translational and rotational velocities of individual particles are affected by the presence of the walls of the channel. As can be seen in Figure 7.28, this retardation effect can be quite pronounced when the cell diameter is 20–90% of the entire gap width. Interestingly, because the particle centers are excluded from a region one radius thick at the upper and lower walls (where the fluid velocity is smallest), the average particle velocity for a collection of spheres can actually be greater than the average velocity of the suspending fluid itself. Specifically, Staben et al. showed that randomly distributed particles with diameter $\leq 82\%$ of the gap height will have an average velocity greater than the average fluid velocity, and that this effect is most pronounced for particles with diameter $d = 0.42\ H$, which will travel at a velocity 18% greater than the average fluid velocity.

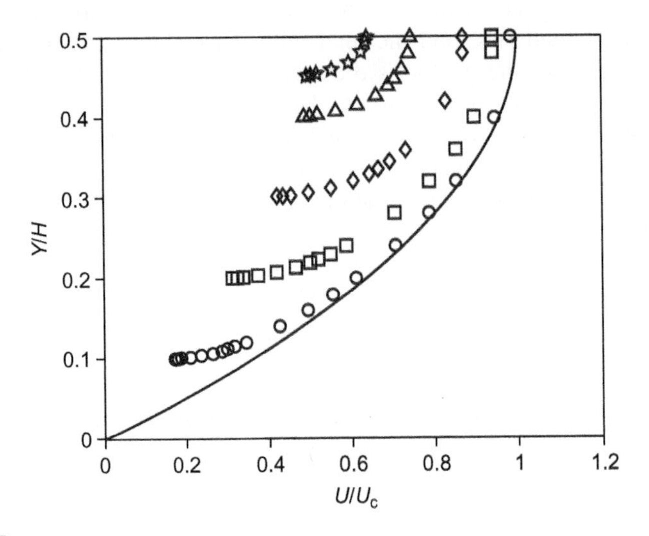

FIGURE 7.28

The dimensionless translational velocity of a sphere in a two-dimensional Poiseuille flow in a rectangular channel. Solid curve represents the fluid velocity, and the symbols represent particles, which, when nondimensionalized with the height of the channel, have the following diameters: 0.2 (circles), 0.4 (squares), 0.6 (diamonds), 0.8 (triangles), and 0.9 (stars). All velocity curves are symmetric about $Y/H = 0.5$.

Source: *Data from Staben et al. (2003).*

7.10 Solved examples

Example 7.1

Can differences in mass measured on an analytical balance be used to determine the average coating thickness inside a polymeric microtube designed for cell adhesion applications? A thin layer of poly L-lysine (PLL) was coated on the inside of microrenethane microtubes of inner diameter 300 μm, outer diameter 600 μm, and length 50 cm. In 10 separate trials, the average mass difference between coated and uncoated microtubes was measured and is presented in Table 7.3. From the table, the mean difference is calculated to be 0.80 ± 0.03 ($\pm$ SE). Figure 7.29 shows the cylindrical coating geometry. The coating volume V is calculated as:

$$V = \Delta \text{Mass}/\rho_{\text{PLL}} \tag{7.65}$$

and the PLL density has been experimentally measured to be $\rho_{\text{PLL}} = 1.1012\ \text{g/cm}^3$. Based on Figure 7.29, it is clear that this calculated coating volume of

Table 7.3 Mass Measurements to Determine Poly L-Lysine Coating Thickness

Empty Microtube Mass (mg)	PLL-Coated Microtube Mass (mg)	ΔMass (mg)
117.4	118.2	0.80
117.4	118.0	0.60
117.4	118.2	0.80
117.3	118.2	0.90
116.7	117.5	0.80
116.8	117.6	0.80
116.7	117.5	0.80
116.8	117.6	0.80

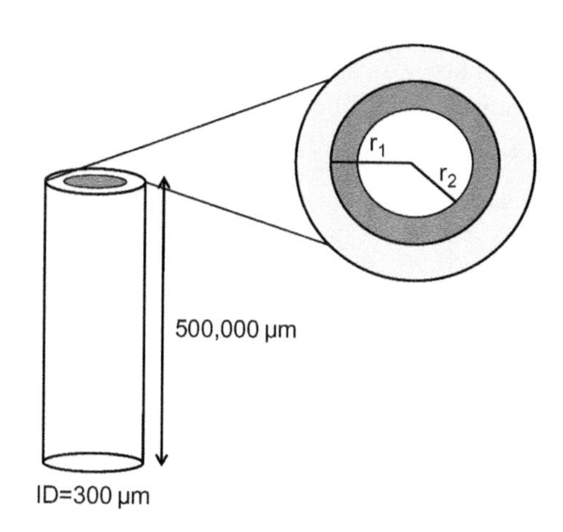

FIGURE 7.29

Geometry for calculating the coating thickness inside a microtube flow device.

$V = 7.27 \times 10^8\ \mu m^3$ must be set equal to the following expression in terms of the inner (r_2) and outer (r_1) radius of the coating:

$$V = \pi h(r_1^2 - r_2^2) \tag{7.66}$$

where r_1 is the known microtube ID, $r_1 = 150\ \mu m$, and $h = 50\ cm = 5 \times 10^5\ \mu m$. This yields the unknown inner radius to be $r_2 = 148.4\ \mu m$, for an average coating thickness of $r_1 - r_2 = 1.55\ \mu m$. Note that the reported random coil PLL cylinder dimensions at neutral pH are ~150 nm after absorption (Hawkins et al., 2006; Idiris et al., 2000), this suggests that the measured PLL coating is about 10 molecules in thickness.

Example 7.2

Compare graphically the two solutions describing the spatial dependence of the cell surface oxygen concentrations without and with an internal membrane oxygenator, for parameter values of $\gamma = 100$, $Da = 0.25$, $Pe = 25$, $Sh = C_g = 1$.

We will use the Matlab programming language to calculate and display our two solutions.

First, we plot the left- and right-hand sides of our transcendental Eq. (7.28) to approximately locate the value of the first few roots λ_n, using the commands:

```
lambda = linspace (1e-4, 12, 1000);
semilogy (lambda, tan (lambda), lambda, 1./lambda, '-')
```

where minor figure formatting commands have been omitted. Note that the lambda vector was started at 10^{-4} to avoid the singularity at $\lambda = 0$. Figure 7.30 shows this plot, where it is evident that the two sides of the transcendental equation cross each other at around 1, 3, 6, 9, and it is clear that the higher-order roots will be separated by a distance of about 3.14. Now that we know the approximate location of the first few roots, we can use a built-in nonlinear root-finding routine to find these more accurately. The following commands accomplish this:

```
guess = [1,3,6,9,12];
for i = 1: length (guess)
        lambda_n(i) = fzero ('tan(x) -1/x', guess (i));
end
```

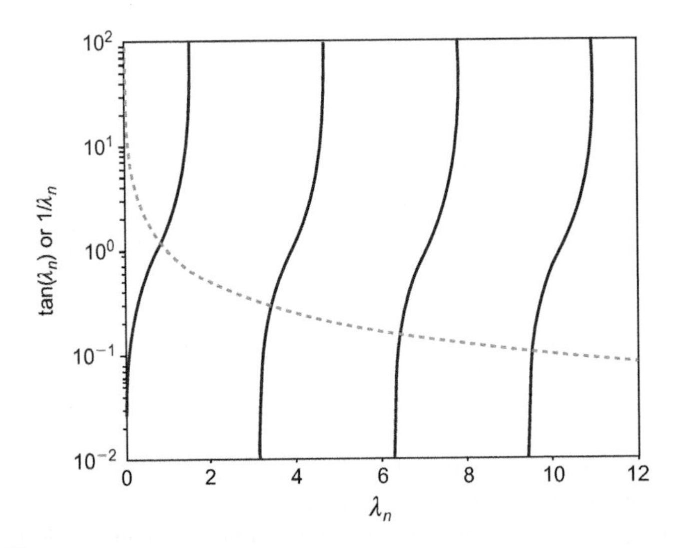

FIGURE 7.30

Matlab plot to determine the approximate location of the first few roots to the transcendental equation for λ_n.

Executing the above commands produces the following roots for the first five λ_n's: 0.8603, 3.4256, 6.4373, 9.5293, 12.6453. Now we are ready to calculate the two surface concentration distributions and compare them. We will store the first five eigenfunctions in the second through sixth rows of two matrices, called `C_no_mem` and `C_mem`, and sum them up before plotting:

```
z = linspace (0, 1, 100); Pe = 25; Da = 0.25;
C_no_mem (1, :) = 1 − z*100/Pe*Da − Da/3;
C_mem (1, :) = (1 − Da − Da)*ones (size (x));
for n = 1:5
    C_no_mem (n + 1, :) = 2*Da/pi^2* (−1) ^n/n^2* cos (n*pi) * ...
    exp( − 100 *n ^2*pi^2*z/Pe);
Bn = (4*Da* (1 − cos (lambda_n (n))) + ...
    4* lambda_n (n) *sin (lambda_n (n)) * (1 − 1 + Da))/...
    (lambda_n (n) *sin (2*lambda_n (n)) + 2*lambda_n (n)^2);
  C_mem (n + 1,:) = Bn*exp (−100*lambda_n(n) ^ 2*z/Pe);
end
plot (z, sum (C_no_mem), z, sum (C_mem))
```

Figure 7.31 shows these results, where it is evident that the microchannel without membrane oxygenation will be fully depleted of oxygen at the surface at a dimensionless position of about $z = 0.9$.

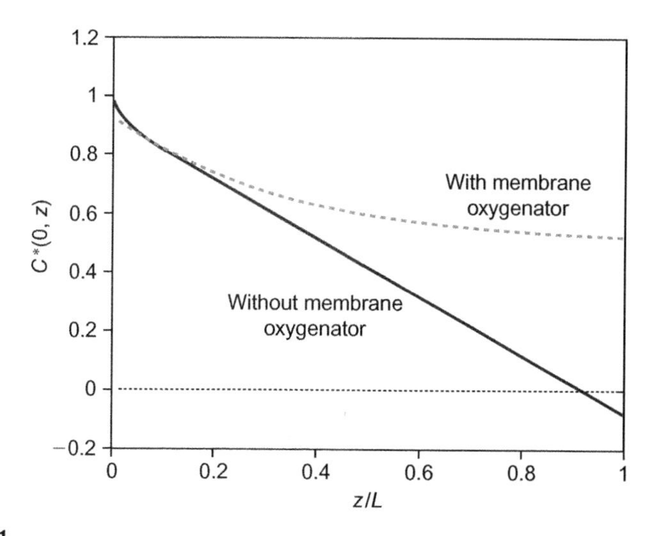

FIGURE 7.31

Results from the analytical model of oxygen transport in a microchannel bioreactor. Note that without a membrane oxygenator, the cell surface will be completely devoid of oxygen before the end of the channel is reached.

Example 7.3

As an example we consider the case of uniform potential flow. This serves as the base state for our analysis of the flow over a wavy cell surface in Section 7.4.2. In uniform flow, the potential is given by:

$$\phi_1 = -Ux \tag{7.67}$$

where U is a constant. The corresponding velocity components are then given by:

$$\begin{aligned} u_1 &= -\frac{\partial \phi_1}{\partial x} = U = \text{constant} \\ v_1 &= -\frac{\partial \phi_1}{\partial y} = 0 \end{aligned} \tag{7.68}$$

as expected.

Example 7.4

Consider the demand for effective mass transport in the microcirculation. Cooney (1976) showed through simple scaling arguments that the rate of mass transport across blood capillaries must be quite effective owing to the relatively short residence time of blood in these vessels. The typical capillary diameter is about 8 μm, which allows deformable red and white blood cells ($d \sim 8-10$ μm) to squeeze through these smallest vessels. The number of systemic capillaries in the body can be estimated to be about $n = 10^9$, and the cardiac output about $Q = 5$ l/min. Thus, the linear velocity of blood flow in the capillaries can be readily calculated as:

$$v = \frac{Q}{n\pi r^2} = \frac{5000\ \text{cm}^3/\text{min}}{10^9(3.142)(4 \times 10^{-4}\ \text{cm})^2} = 9.95\ \text{cm/min} = 0.166\ \text{cm/s} \tag{7.69}$$

If we assume that each capillary is 1 mm long, the characteristic blood residence time in the capillary is thus:

$$t = \frac{1\ \text{mm}}{1.66\ \text{mm/s}} = 0.6\ \text{s} \tag{7.70}$$

Thus, mass transfer of metabolites and waste products through the capillary wall cannot encounter much barrier to mass transport since so little time for exchange is provided before the blood begins flowing back through the venous tree to the heart.

7.11 Practice problems

Problem 7.1

Calculate the velocity profile for a radial flow assay as a function of r and z, in terms of the volumetric flow rate and the channel dimensions. Neglect the central region and assume that the inlet radius is small. Then, use your solution for the velocity field to derive the wall shear stress expression given by Eq. (7.1).

Problem 7.2

Consider a slowly converging channel. Assuming a desired gap height of $H = 8\ \mu m$, a channel width and length of $W = 25$ mm and $L = 75$ mm, respectively, but an actual gap height achieved of $H_1 = 8$ and $H_2 = 7\ \mu m$, calculate the required pressure head by neglecting the additional pressure drop across fittings and tubing. Based on your value, will gravity-driven flow or a commercially available syringe pump be more appropriate for this application? Noting that the wall shear stress varies as the inverse of H squared, will the average wall shear stress within the channel be simply that for a straight channel using the average channel height? Calculate the average wall shear stress in such a channel, for arbitrary height change, by computing the area-averaged integral of the wall shear stress down the length of the flow domain.

Problem 7.3

Calculate the total pressure drop in a microchannel that experiences a symmetric buckling of the upper and lower walls in a parabolic shape, in terms of the fluid properties, flow rate, the initial/final height, and the height at the center of the channel. How much error is introduced by only measuring the gap height in the center of the channel and assuming that the gap is constant?

Problem 7.4

Perform a literature search to find micrograph images of neovasculature in solid tumors to determine the geometric parameters such as vessel diameter and spacing between endothelial cells. Compare these values to the plot of optimal vessel dimensions to minimize flow resistance presented in Sumagin et al. (2008). How close to the optimum geometry are these new cancer-associated capillaries? Explain in biological terms why this might be so.

Problem 7.5

Reformulate the model of Section 7.6.2 for a parabolic velocity profile, instead of assuming a uniform velocity. For the parameter values of Table 7.4, obtain a numerical solution to this modified system of equations, and calculate how much error is introduced by the simplified velocity profile. Show that without a membrane oxygenator, a fraction of the hepatocyte culture will not receive sufficient oxygen, and report this fraction.

Problem 7.6

Calculate the force exerted on a stationary spherical cell with diameter $9\ \mu m$, within a cylindrical micropipette of inner diameter $10\ \mu m$, in terms of the total

Table 7.4 Parameter Values for Hepatocyte/3T3 Fibroblast Co-culture in a Microchannel Bioreactor

Parameter	Value
H	100 μm
α	10^5 hepatocytes/cm^2
Q	0.1 ml/min
U	7×10^{-2} cm/s
V_m	0.4 nmol/s per 10^6 cells
σ	5 μm/s
D	2×10^{-5} cm^2/s
Pe	30
Da	0.1
Sh	0.25

pressure drop. Compare this to the solution if curvature is neglected and the flow within the gap is approximated as planar channel flow.

Problem 7.7

If particles enter a microchannel at the centerline height, and slowly sediment toward slower-moving streamlines close to the lower wall as they flow through the chamber, will this result in a net accumulation of particles within the flow chamber? Use the results of Section 7.9.3 to estimate the extent of this effect for 12-μm-diameter particles with density 1.05 g/cm^3, suspended in water. Assume that the channel has height 30 μm and length 1.0 cm, and neglect the effect of the side walls (i.e., two-dimensional channel). How does this differ from the case when the transported particles are infinitesimal points?

Problem 7.8

Utilizing the Stokes flow solution in Eq. (7.12), determine the height of nanostructured surface to cause 10-μm spherical cells 10% denser than water to flow at a 30% increased velocity relative to cells flowing over a smooth surface. Can you specify the maximum separation distance between nanoscale surface features to achieve this velocity enhancement, in terms of the density difference between cell and fluid, and the cell velocity?

Problem 7.9

A thin coating of the adhesive titanium (IV) butoxide was created on the interior surface of microrenethane tubing of ID = 300 μm and length = 50 cm. Ten microtubes were coated and the mass measured before and after coating, for an average mass difference of 1.75 ± 043 mg (± SE). Search for the density of titanium butoxide and then perform a calculation similar to Solved Example 7.1 to determine the average coating thickness.

References

Abbitt, K.B., Nash, G.B., 2001. Characteristics of leukocyte adhesion directly observed in flowing whole blood *in vitro*. Br. J. Haematol. 112, 55–63.

Agastin, S., Giang, U.-B.T., Geng, Y., DeLouise, L.A., King, M.R., 2011. Continuously perfused microbubble array for 3D tumor spheroid model. Biomicrofluidics 5, 024110.

Back, M.R., Carew, T.E., Schmid-Schoenbein, G.E., 1995. Deposition pattern of monocytes and fatty streak development in hypercholesterolemic rabbits. Atherosclerosis 116, 103–115.

Barber, K.M., Pinero, A., Truskey, G.A., 1998. Effects of recirculating flow on U-937 cell adhesion to human umbilical vein endothelial cells. Am. J. Physiol. 275, H591–H599.

Bathe, M., Shirai, A., Doerschek, C.M., Kamm, R.D., 2002. Neutrophil transit times through pulmonary capillaries: the effects of capillary geometry and fMLP-stimulation. Biophys. J. 83, 1917–1933.

Bishop, J.J., Nance, P.R., Popel, A.S., Intaglietta, M., Johnson, P.C., 2001. Erythrocyte margination and sedimentation in skeletal muscle venules. Am. J. Physiol. Heart Circ. Physiol. 281, H951–H958.

Bluestein, D., Gutierrez, C., Londono, M., Schoephoerster, R.T., 1999. Vortex shedding in steady flow through a model of an arterial stenosis and its relevance to mural platelet deposition. Ann. Biomed. Eng. 27, 763–773.

Bodoia, J.R., Osterle, J.F., 1961. Finite difference analysis of plane Poiseuille and Couette flow developments. Appl. Sci. Res. A10, 265–276.

Boettiger, D., Lynch, L., Blystone, S., Huber, F., 2001. Distinct ligand-binding modes for integrin alpha(v)beta(3)-mediated adhesion to fibronectin versus vitronectin. J. Biol. Chem. 276, 31684–31690.

Carslaw, H.S., Jaeger, J.C., 1959. Conduction of Heat in Solids. Oxford University Press, London.

Chan, B.P., Bhat, V.D., Yegnasubramanian, S., Reichert, W.M., Truskey, G.A, 1999. An equilibrium model of endothelial cell adhesion via integrin-dependent and integrin-independent ligands. Biomaterials 20, 2395–2403.

Chandrasekaran, S., King, M.R., 2012. Gather round: in vitro spheroids as improved models of *in vivo* tumors. J. Bioeng. Biomed. Sci. 2, e109.

Chen, S., Springer, T.A., 2001. Selectin receptor-ligand bonds: formation limited by shear rate and dissociation governed by the Bell model. Proc. Natl. Acad. Sci. USA 98, 950–955.

Cieslinski, D.A., Humes, H.D., 1994. Tissue engineering of a bioartificial kidney. Biotechnol. Bioeng. 43, 678–681.

Cooney, D.O., 1976. Biomedical Engineering Principles: An Introduction to Fluid, Heat, and Mass Transport Processes. Marcel Dekker, Inc., New York, NY.

de Guillebon, A., Geng, Y., Carr, R., King, M.R., 2012. Adhesion of circulating tumor microemboli to E-selectin in circular branched microvessels. In: Proceedings of the Tenth International Conference on Nanochannels, Microchannels, and Minichannels, 8–12 July, Puerto Rico, U.S.A.

DePaola, N., Gimbrone Jr., M.A., Davies, P.F., Dewey Jr., C.F., 1992. Vascular endothelium responds to fluid shear stress gradients. Arterioscler. Thromb. 12, 1254–1257.

Deligianni, D.D., Katsala, N.D., Koutsoukos, P.G., Missirlis, Y.F., 2001. Effect of surface roughness of hydroxyapatite on human bone marrow cell adhesion, proliferation, differentiation and detachment strength. Biomaterials 22, 87–96.

Dong, C., Skalak, R., Sung, K.L., Schmid-Schonbein, G.W., Chien, S., 1998. Passive deformation analysis of human leukocytes. J. Biomech. Eng. 110, 27–36.

Ebnet, K., Vestweber, D., 1999. Molecular mechanisms that control leukocyte extravasation: the selectins and the chemokines. Histochem. Cell Biol. 112, 1–23.

Fenton, B.M., Carr, R.T., Cokelet, G.R., 1985. Nonuniform red cell distribution in 20 to 100 mm bifurcations. Microvasc. Res. 29, 103–126.

Fournier, R.L., 1999. Basic Transport Phenomena in Biomedical Engineering. Taylor & Francis, Philadelphia.

Garcia, A.J., Ducheyne, P., Boettiger, D., 1997. Quantification of cell adhesion using a spinning disk device and application to surface-reactive materials. Biomaterials 18, 1091–1098.

Giang, T.-B.T., King, M.R., DeLouise, L.A., 2008. Microfabrication of bubbular cavities in PDMS for cell sorting and microcell culture applications. J. Bionic Eng. 5, 308–316.

Giang, U.-B.T., Lee, D., King, M.R., DeLouise, L.A., 2007. Microfabrication of cavities in polydimethylsiloxane using DRIE silicon molds. Lab Chip 7, 1660–1662.

Gifford, S.C., Frank, M.G., Derganc, J., Gabel, C., Austin, R.H., Yoshida, T., et al., 2003. Parallel microchannel-based measurements of individual erythrocyte areas and volumes. Biophys. J. 84, 623–633.

Goldsmith, H.L., Marlow, J.C., 1979. Flow behavior of erythrocytes. II. Particle motions in concentrated suspension of ghost cells. J. Colloid Interface Sci. 71, 383–407.

Goldsmith, H.L., Spain, S., 1984. Margination of leukocytes in blood flow through small tubes. Microvasc. Res. 27, 204–222.

Goldstein, A.S., DiMilla, P.A., 1998. Comparison of converging and diverging radial flow for measuring cell adhesion. AIChE J. 44, 465–473.

Goldstick, T.K., 1966. Diffusion of Oxygen in Protein Solution (Ph.D. Thesis). University of California, Berkeley.

Grandas, O.H., Costanza, M.J., Donnell, R.L., Reddick, T.T., Carroll, R.C., Stevens, S.L., et al., 2001. Effect of retroviral transduction of canine microvascular endothelial cells on beta(1) integrin subunit expression and cell retention to PTFE grafts. Cardiovasc. Surg. 9, 595–599.

Han, W., Allio, B.A., Foster, D.G., King, M.R., 2010. Nanoparticle coatings for enhanced capture of flowing cells in microtubes. ACS Nano 4, 174–180.

Hawkins, K.M., Wang, S.S.S., Ford, D.M., Shantz, D.F., 2006. Poly-L-lysine template silicas: using polypeptide secondary structure to control oxide pore architectures. J. Am. Chem. Soc. 126, 9112–9119.

Helmke, B.P., Sugihara-Seki, M., Skalak, R., Schmid-Schönbein, G.W., 1998. A mechanism for erythrocyte-mediated elevation of apparent viscosity by leukocytes *in vivo* without adhesion to the endothelium. Biorheology 35, 437–448.

Herant, M., Marganski, W.A., Dembo, M., 2003. The mechanics of neutrophils: synthetic modeling of three experiments. Biophys. J. 84, 3389–3413.

Horner, M., Miller, W.M., Ottino, J.M., Paoutsakis, E.T., 1998. Transport in a grooved perfusion flat-bed bioreactor for cell therapy applications. Biotechnol. Prog. 14, 689–698.

Hughes, A.D., King, M.R., 2010. Use of naturally occurring halloysite nanotubes for enhanced capture of flowing cells. Langmuir 26, 12155–12164.

Hughes, A.D., Mattison, J., Western, L.T., Powderly, J.D., Greene, B.T., King, M.R., 2012. Microtube device for selectin-mediated capture of viable circulating tumor cells from blood. Clin. Chem. 58, 846–853.

Idiris, A., Alam, M.T., Idai, A., 2000. Spring mechanics of alpha-helical polypeptide. Protein Eng. 13, 763–770.

Iiyama, K., Hajra, L., Iiyama, M., Li, H., DiChiara, M., Medoff, B.D., et al., 1999. Patterns of vascular cell adhesion molecule-1 and intercellular adhesion molecule-1 expression in rabbit and mouse atherosclerotic lesions and at sites predisposed to lesion formation. Circ. Res. 85, 199–207.

Juaregui, H.O., Mullon, C.J.P., Solomon, B.A., 1997. Extracorporeal artificial liver support. In: Lanza, R.P., Langer, R., Chick, W.L. (Eds.), Principles of Tissue Engineering. Landes R.G., Boulder, CO, pp. 463–479.

Kamada, H., Hattori, K., Hayashi, T., Suzuki, K., 2004. *In vitro* evaluation of blood coagulation and microthrombus formation by a microchannel array flow analyzer. Thromb. Res. 114, 195–203.

Kapur, R., Rudolph, A.S., 1998. Cellular and cytoskeleton morphology and strength of adhesion of cells on self-assembled monolayers of organosilanes. Exp. Cell Res. 244, 275–285.

Karino, T., Goldsmith, H.L., 1979. Aggregation of human platelets in an annular vortex distal to a tubular expansion. Microvasc. Res. 17, 217–237.

King, M.R., 2007. Do blood capillaries exhibit optimal bumpiness? J. Theor. Biol. 249, 178–180.

King, M.R., Hammer, D.A., 2001a. Multiparticle adhesive dynamics: interactions between stably rolling cells. Biophys. J. 81, 799–813.

King, M.R., Hammer, D.A., 2001b. Multiparticle adhesive dynamics: hydrodynamic recruitment of rolling cells. Proc. Natl. Acad. Sci. USA. 98, 14919–14924.

King, M.R., Hammer, D.A., 2003. Hydrodynamic recruitment of rolling cells *in vitro*. Biophys. J. 84, 4182.

King, M.R., Kim, M.B., Sarelius, I.H., Hammer, D.A., 2003. Hydrodynamic interactions between rolling leukocytes *in vivo*. Microcirculation 10, 401–409.

King, M.R., Bansal, D., Kim, M.B., Sarelius, I.H., Hammer, D.A., 2004. The effect of hematocrit and leukocyte adherence on flow direction in the microcirculation. Ann. Biomed. Eng. 32, 803–814.

Kuharski, A.L., Fogelson, A.L., 2001. Surface-mediated control of blood coagulation: the role of binding site densities and platelet deposition. Biophys. J. 80, 1050–1074.

Kuo, S.C., Lauffenburger, D.A., 1993. Relationship between receptor/ligand binding affinity and adhesion strength. Biophys. J. 65, 2191–2200.

Lasky, L.A., 1995. Selectin–carbohydrate interactions and the initiation of the inflammatory response. Annu. Rev. Biochem. 64, 113–139.

Lawrence, M.B., Kansas, G.S., Kunkel, E.J., Ley, K., 1997. Threshold levels of fluid shear promote leukocyte adhesion through selectins (CD62L, P, E). J. Cell Biol. 136, 717–727.

Lei, M., Kleinstreuer, C., Archie, J.P., 1997. Hemodynamic simulations and computer-aided designs of graft–artery junctions. J. Biomech. Eng. 119, 343–348.

Li, P.C.H., de Camprieu, L., Cai, J., Sangar, M., 2004. Transport, retention and fluorescent measurement of single biological cells studied in microfluidic chips. Lab Chip 4, 174–180.

Lomakina, E.B., Spillmann, C.M., King, M.R., Waugh, R.E., 2004. Rheological analysis and measurement of neutrophil indentation. Biophys. J. 87, 4246–4258.

Lominadze, D., Mchedlishvili, G., 1999. Red blood cell behavior at low flow rate in microvessels. Microvasc. Res. 58, 187–189.

Malek, A.M., Zhang, J., Jiang, J.W., Alper, S.L., Izumo, S., 1999. Endothelin-1 gene suppression by shear stress: pharmacological evaluation of the role of tyrosine kinase, intracellular calcium, cytoskeleton, and mechanosensitive channels. J. Mol. Cell Cardiol. 31, 387–399.

Michelson, I., 1970. The Science of Fluids. Van Nostrand Reinhold Co., New York, NY.

Miller, T., Boettiger, D., 2003. Control of intracellular signaling by modulation of fibronectin conformation at the cell–materials interface. Langmuir 19, 1723–1729.

Moffatt, H.K., 1964. Viscous and resistive eddies near a sharp corner. J. Fluid Mech. 18, 1–18.

Nakashima, Y., Raines, E.W., Plump, A.S., Breslow, J.L., Ross, R., 1998. Upregulation of VCAM-1 and ICAM-1 at atherosclerosis-prone sites on the endothelium in the ApoE-deficient mouse. Arterioscler. Thromb. Vasc. Biol. 18, 842–851.

Neelamegham, S., Taylor, A.D., Shankaran, H., Smith, C.W., Simon, S.I., 2000. Shear and time-dependent changes in Mac-1, LFA-1, and ICAM-3 binding regulate neutrophil homotypic adhesion. J. Immunol. 164, 3798–3805.

Olla, P., 1999. Simplified model for red cell dynamics in small blood vessels. Phys. Rev. Lett. 82, 453–456.

Ozturk, S., 1990. Characterization of Hybridoma Growth, Metabolism, and Monoclonal Antibody Production (Ph.D. Thesis). University of Michigan, Ann Arbor, MI.

Palecek, S.P., Loftus, J.C., Ginsberg, M.H., Lauffenburger, D.A., Horwitz, A.F., 1997. Integrin ligand binding properties govern cell migration speed through cell–substratum adhesiveness. Nature 385, 537–540.

Peng, C.-A., Palsson, B.O., 1996. Determination of specific oxygen uptake rates in human hematopoietic cultures and implications for bioreactor design. Ann. Biomed. Eng. 24, 373–381.

Perktold, K., 1987. On the paths of fluid particles in an axisymmetrical aneurysm. J. Biomech. 20, 311–317.

Pritchard, W.F., Davies, P.F., Derafshi, Z., Polacek, D.C., Tsao, R., Dull, R.O., et al., 1995. Effects of wall shear stress and fluid recirculation on the localization of circulating monocytes in a three-dimensional flow model. J. Biomech. 28, 1459–1469.

Ramos, C.L., Huo, Y.Q., Jung, U.S., Ghosh, S., Manka, D.R., Sarembock, I.R., et al., 1999. Direct demonstration of P-selectin- and VCAM-1-dependent mononuclear cell rolling in early atherosclerotic lesions of apolipoprotein E-deficient mice. Circ. Res. 84, 1237–1244.

Sandstrom, C.E., Bender, J.G., Miller, W.M., Papoutsakis, E.T., 1996. Development of novel perfusion chamber to retain nonadherent cells and its use for comparison of human "mobilized" peripheral blood mononuclear cell cultures with and without irradiated stroma. Biotechnol. Bioeng. 50, 493–504.

Sank, A., Wei, D., Reid, J., Ertl, D., Nimni, M., Weaver, F., Yellin, A., Tuan, T.L., 1994. Human endothelial-cells are defective in diabetic vascular-disease. J. Surg. Res. 57, 647–653.

Schneck, D.J., 2000. An outline of cardiovascular structure and function. In: Bronzino, J.D. (Ed.), Biomedical Engineering Handbook, Vol. 1. CRC Press, USA.

Schnittler, H.J., Franke, R.P., Akbay, U., Mrowietz, C., Drenckhahn, D., 1993. Improved *in vitro* rheological system for studying the effect of fluid shear-stress on cultured-cells. Am. J. Physiol. 265, C289–C298.

Schoephoerster, R.T., Oynes, F., Nunez, G., Kapadvanjwala, M., Dewanjee, M.K., 1993. Effects of local geometry and fluid dynamics on regional platelet deposition on artificial surfaces. Arterioscler. Thromb. 13, 1806–1813.

Secomb, T.W., Hsu, R., 1996. Motion of red blood cells in capillaries with variable cross-sections. J. Biomech. Eng. 118, 538–544.

Shao, J.-Y., Hochmuth, R.M., 1996. Micropipette suction for measuring pico newton forces of adhesion and tether formation from neutrophil membranes. Biophys. J. 71, 2892–2901.

Simon, S.I., Chambers, J.D., Sklar, L.A., 1990. Flow cytometric analysis and modeling of cell–cell adhesive interactions: the neutrophil as a model. J. Cell Biol. 111, 2747–2756.

Sirois, E., Charara, J., Ruel, J., Dussault, J.C., Gagnon, P., Doillon, C.J., 1998. Endothelial cells exposed to erythrocytes under shear stress: an *in vitro* study. Biomaterials 19, 1925–1934.

Slack, S.M., Turitto, V.T., 1994. Flow chambers and their standardization for use in studies of thrombosis. Thromb. Haemost. 72, 777–781.

Smith, M.J., Berg, E.L., Lawrence, M.B., 1999. A direct comparison of selectin-mediated transient, adhesive events using high temporal resolution. Biophys. J. 77, 3371–3383.

Spillmann, C.M., Lomakina, E., Waugh, R.E., 2004. Neutrophil adhesive contact dependence on impingement force. Biophys. J. 87, 4237–4245.

Staben, M.E., Zinchenko, A.Z., Davis, R.H., 2003. Motion of a particle between two parallel plane walls in low-Reynolds-number Poiseuille flow. Phys. Fluids 15, 1711–1733.

Sugihara-Seki, M., Secomb, T.W., Skalak, R., 1990. Two-dimensional analysis of two-file flow of red cells along capillaries. Microvasc. Res. 40, 379–393.

Sullivan, S.J., Maki, T., Borland, K.M., Mahoney, M.D., Solomon, B.A., Muller, T.E., et al., 1991. Biohybrid artificial pancreas: long term implantation studies in diabetic, pancrea tectomized dogs. Science 252, 718–721.

Sumagin, R., Brown, C.W., Sarelius, I.H., King, M.R., 2008. Microvascular endothelial cells exhibit optimal aspect ratio for minimizing flow resistance. Ann. Biomed. Eng. 36, 580–585.

Sung, L.A., Kabat, E.A., Chien, S., 1985. Interaction energies in lectin-mediated erythrocyte aggregation. J. Cell Biol. 101, 652–659.

Thompson, M.M., Budd, J.S., Eady, S.L., James, R.F., Bell, P.R., 1994. Effect of pulsatile shear-stress on endothelial attachment to native vascular surfaces. Br. J. Surg. 81, 1121–1127.

Tilles, A.W., Baskaran, H., Roy, P., Yarmush, M.L., Toner, M., 2001. Effects of oxygenation and flow on the viability and function of rat hepatocytes cocultured in a microchannel flat-plate bioreactor. Biotechnol. Bioeng. 73, 379–389.

Tippe, A., Reininger, A., Reininger, C., Riess, R., 1992. A method for quantitative determination of flow induced human platelet adhesion and aggregation. Thromb. Res. 67, 407–418.

Truskey, G.A., Herrmann, R.A., Kait, J., Barber, K.M., 1999. Focal increases in vascular cell adhesion molecule-1 and intimal macrophages at atherosclerosis-susceptible sites in the rabbit aorta after short-term cholesterol feeding. Arterioscler. Thromb. Vasc. Biol. 19, 393–401.

van Kooten, T.G., Schakenraad, J.M., van der Mei, H.C., Busscher, H.J., 1992. Influence of substratum wettability on the strength of adhesion of human fibroblasts. Biomaterials 13, 897–904.

Walker, G.M., Ozers, M.S., Beebe, D.J., 2004. Cell infection within a microfluidic device using virus gradients. Sens. Actuators B Chem. 98, 347–355.

Walpola, P.L., Gotlieb, A.I., Cybulsky, M.I., Langille, B.L., 1995. Expression of ICAM-1 and VCAM-1 and monocyte adherence in arteries exposed to altered shear stress. Arterioscler. Thromb. Vasc. Biol. 15, 2–10.

Wang, C.Y., 2006. Stokes flow through a tube with bumpy wall. Phys. Fluids 18, 078101.

Waugh, R.E., Mantalaris, A., Bauserman, R.G., Hwang, W.C., Wu, J.H.D., 2001. Membrane instability in late-stage erythropoiesis. Blood 97, 1869–1875.

Wechezak, A.R., Viggers, R.F., Coan, D.E., Sauvage, L.R., 1994. Mitosis and cytokinesis in subconfluent endothelial-cells exposed to increasing levels of shear-stress. J. Cell Physiol. 159, 83–91.

Whitmore, R.L., 1968. Rheology of the Circulation. Pergammon, Oxford.

Yamaguchi, S., Yamakawa, Y., Niimi, H., 1992. Cell-free plasma layer in cerebral microvessels. Biorheology 29, 251–260.

Zhao, Y.H., Chien, S., Weinbaum, S., 2001. Dynamic contact forces on leukocyte microvilli and their penetration of the endothelial glycocalyx. Biophys. J. 80, 1124–1140.

Index

Note: Page numbers followed by "*f*" and "*t*" refer to figures and tables, respectively.

D

E

F

L

M

T

U

V

W

Y

Z

Printed and bound by CPI Group (UK) Ltd, Croydon, CR0 4YY
08/06/2025
01896868-0016